The Mechanosensory Lateral Line

S. Coombs P. Görner H. Münz
Editors

The Mechanosensory Lateral Line

Neurobiology and Evolution

With 262 Illustrations

Springer-Verlag
New York Berlin Heidelberg
London Paris Tokyo

Sheryl Coombs
Parmly Hearing Institute
Loyola University of Chicago
Chicago, Illinois 60626
USA

Peter Görner
Faculty of Biology
Department of Neurophysiology
University of Bielefeld
Bielefeld, Federal Republic of Germany

Heinrich Münz
Faculty of Biology
Department of Neurophysiology
University of Bielefeld
Bielefeld, Federal Republic of Germany

Library of Congress Cataloging-in-Publication Data
The mechanosensory lateral line : neurobiology and evolution / Sheryl Coombs, Peter Görner, and Heinrich Münz, editors.
p. cm.
Includes bibliographies.
ISBN-13:978-1-4612-8157-3 e-ISBN-13:978-1-4612-3560-6
DOI: 10.1007/978-1-4612-3560-6

1. Lateral line organs—Congresses. 2. Lateral line organs--Evolution—Congresses. 3. Mechanoreceptors—Congresses. 4. Fishes—Physiology—Congresses. 5. Fishes—Evolution--Congresses. 6. Amphibians—Physiology—Congresses. 7. Amphibians--Evolution—Congresses. I. Coombs, Sheryl. II. Görner, Peter. III. Münz, Heinrich, 1949–
QL639.1.M43 1989
597'.01'82—dc19 88-24978

Printed on acid-free paper.

Cover design depicts elaborately branched lateral line canals found in the marine sculpin, *Myoxycephalis niger*, and has been simplified and redrawn from the work of A. Neyelov (published by Nauka, U.S.S.R.).

Typeset by Publishers Service, Bozeman, Montana.

9 8 7 6 5 4 3 2 1

ISBN-13:978-1-4612-8157-3

This volume is dedicated to Sven Dijkgraaf

Dear frog and fish, or newt and shark,
You needn't worry when it's dark;
You'll escape or dine just fine,
Svenning with your lateral line.

From Chapter 32

Preface

This volume represents the published proceedings of an international conference on the Neurobiology and Evolution of the Mechanosensory Lateral Line System held August 31 to September 4, 1987, at the Center for Interdisciplinary Research at the University of Bielefeld, West Germany. The goal of this conference was to bring together researchers from all over the world to share information about a major aquatic sensory system, the evolution and function of which have largely remained an enigma since the 18th century.

The "lateral line" or "lateralis" system has been used as an umbrella term to describe what originally (without the aid of modern anatomical techniques) looked like a series of pits, grooves, and lines on the head and trunk of fishes and some amphibians. For at least the past 30 years, however, it has been recognized that the lateralis system comprises not one, but at least two functional classes of receptors: mechanoreceptors and electroreceptors. The relative ease with which the appropriate stimulus could be defined and measured for the electroreceptive class has resulted in an explosion of information on this submodality during the past 20 years. As a result, there is little ambiguity about the overall function of the electrosensory system, now generally regarded as an independent system in its own right.

A similarly clear definition for the function of the mechanosensory lateralis system has not been as forthcoming. Questions about the function or functions of the mechanoreceptive lateral line have been difficult to address on several counts, including the lack of a comparable system in human beings and difficulties in determining and measuring the precise stimuli to which the system responds. These difficulties are even further exacerbated by the fact that the mechanosensory lateral line system and the octaval (auditory and vestibular) system have in common the same type of receptor cell, the hair cell, and show, under certain circumstances, physiological responses to the same stimulus. As a result, the mechanoreceptive lateral line system has been regarded by many as part of yet another umbrella system, the octavolateralis system. Thus, not only has there been no clear understanding of the overall function of the mechanosensory lateralis system (relative to the auditory and electrosensory systems), but there has not even been a dedicated vocabulary with which to discuss it.

At the time of the 1985 conference, Sensory Biology of Aquatic Animals (organized by J. Atema, R.R. Fay, A.N. Popper, and W.N. Tavolga), for which a small enclave of lateral line researchers gathered, it became apparent that interest in and enthusiasm for mechanosensory lateral line research had accelerated rapidly over the preceding 5 years and that the number of new investigators attracted to the field had significantly expanded. It was also evident that both German and American investigators had independently begun plans for organizing a conference devoted exclusively to mechanosensory lateral line research, demonstrating a clear bilateral interest in such a conference. A collaboration between German and American organizers ensued, and this book is a direct outgrowth of that collaboration.

Since the last major lateral line conference and volume in 1966 (*Lateral Line Detectors*, organized and edited by P. Cahn), work on the mechanosensory lateral line system has progressed on several fronts. Major technological advances, primarily in neuroanatomical and stimulus-measuring techniques, have led to significant strides in our understanding of (1) the nature of the stimuli to which the mechanoreceptive lateral line responds, (2) the organization of the system in the peripheral and central nervous system, and (3) the functional and behavioral significance of the system.

In organizing the conference and the book, we wanted to summarize these accomplishments by choosing contributors whose research focused primarily on the function and evolution of the mechanosensory lateral line system. At the same time, we wanted to highlight research areas of global neurobiological interest, such as those dealing with mechanisms of sensory processing or development. Moreover, we wanted to include results from very recent experiments and descriptions of ongoing research programs as well as reviews of research progress since the last volume on lateral line research. As a result, each contribution revolves around a common theme, but is also a unique blend of literature review and current research, with the proportions of each differing according to the areas of research. Our hope is that this volume will not only serve as a reference volume for anyone interested in the mechanosensory lateral line and related sensory systems, but that it will also stimulate interest in an easily accessible sensory system as a valuable model system for addressing basic yet diverse neurobiological questions.

Chicago, Illinois — Sheryl Coombs
Bielefeld, F.R.G — Peter Görner
Bielefeld, F.R.G. — Heinrich Münz

Acknowledgments

As conference organizers, we would like to thank the Deutsche Forschungsgemeinschaft (4850/35/87) and the National Science Foundation (INT-8610374) under the U.S. Federal Republic of Germany Cooperative Science Program, the Office of Naval Research (contract N00014-87-J-0180), the Center for Interdisciplinary Research (ZIF), and, finally, the Max Goldenberg Foundation for funds to support the conference. Without financial support from these agencies, our conference would clearly have been impossible to convene.

In addition, we would like to thank the staff at ZIF which hosted the conference. Particular thanks go to the director of ZIF, Dr. W. Prinz, for handling all the on-site organizational aspects of the conference. Special thanks also to Ms. Beth Langer and others at Parmly Hearing Institute for their help in photocopying and handling the tremendous volume of mail through various phases of the conference and subsequent book preparation.

As editors, we thank Ms. Helen Johnson for her excellent editorial assistance and Dr. Jacqueline Webb for her meticulous preparation of the taxonomic index for the volume.

Finally, as scientists, we would like to extend our extreme gratitude to the conference participants and contributors to this book. Although often badgered by us with a stream of memorandums, rules and regulations, and deadlines, they came through like troopers, with good advice, good humor, dedication, and, most importantly, timely and significant contributions to our understanding of the lateral line system.

S.C.
P.G.
H.M.

Contents

Contributors

Kenneth J. Balak, Békésy Laboratory of Neurobiology, University of Hawaii, Honolulu, HI 96822, USA

Michael A. Barry, Department of BioStructure and Function, University Connecticut Health Center, Farmington, CT 06032, USA

Michael V.L. Bennett, Department of Neuroscience, Albert Einstein College of Medicine, Yeshiva University, Bronx, NY 10461, USA

John H.S. Blaxter, Scottish Marine Biological Association, Dunstaffnage Marine Research Laboratory, Oban, Argyll, PA 34 4AD, Scotland

Horst Bleckmann, Faculty of Biology, Department of Neurophysiology, University of Bielefeld, D-4800 Bielefeld 1, Federal Republic of Germany

Elke Blübaum-Gronau, Institut für Tierphysiologie der Universität, D-6300 Giessen, Federal Republic of Germany

David Bodznick, Department of Biology, Wesleyan University, Middletown, CT 06457, USA

Robert L. Boord, School of Life and Health Sciences, University of Delaware, Newark, DE 19716, USA

Paula C. Borden, Békésy Laboratory of Neurobiology, University of Hawaii, Honolulu, HI 96822, USA

Bernd-Ulrich Budelmann, Marine Biomedical Institute and Department of Otolaryngology, University of Texas Medical Branch, Galveston, TX 77550-2772, USA

Theodore H. Bullock, Department of Neurosciences, School of Medicine, University of California-San Diego, La Jolla, CA 92093, USA

Barbara Claas, Faculty of Biology, Department of Neurophysiology, University of Bielefeld, D-4800 Bielefeld 1, Federal Republic of Germany

Sheryl Coombs, Parmly Hearing Institute, Loyola University of Chicago, Chicago, IL 60626, USA

Jeffrey T. Corwin, Department of Otolaryngology, University of Virginia School of Medicine, Charlottesville, VA 22904, USA

Paul D. Danielson, Department of Biology, Williams College, Williamstown, MA 02167, USA

Eric J. Denton, Marine Biological Association of the United Kingdom, Plymouth, PL1 2PB, England

Sven Dijkgraaf, Maliebaan 63, Utrecht 3581CE, The Netherlands, Laboratorium UOOR, Vergelukende Fysiologie, Utrecht, Holland

Andreas Elepfandt, Department of Biology, Universität Konstanz, D-7750 Konstanz, Federal Republic of Germany

Per S. Enger, Institute of Biology, University of Oslo, 0316 Oslo 3, Norway

Richard R. Fay, Parmly Hearing Institute, Loyola University of Chicago, Chicago, IL 60626, USA

Bernd Fritzsch, Faculty of Biology, Department of Neurophysiology, University of Bielefeld, D-4800 Bielefeld 1, Federal Republic of Germany

Lee A. Fuiman, Marine Science Institute, University of Texas at Austin, Port Aransas, TX 78373, USA

Peter Görner, Fakultät für Biologie, University of Bielefeld, D-4800 Bielefeld 1, Federal Republic of Germany

John A.B. Gray, Marine Biological Association of the United Kingdom, Plymouth PL1 2PB, England

El-sayed Hassan, Institut für Zoologie, Johannes Gutenberg Universität, D-6500 Mainz, Federal Republic of Germany

John Janssen, Biology Department, Loyola University of Chicago, Chicago, IL 60626, USA

Jørgen Mørup Jørgensen, University of Aarhus, Institute of Zoology and Zoophysiology, DK-8000 Aarhus C, Denmark

Ad. J. Kalmijn, Scripps Institute of Oceanography, University of California-San Diego, La Jolla, CA 92093, USA

Alfons B.A. Kroese, Laboratory of Medical Physics, Academic Medical Center, University of Amsterdam, Amsterdam 1105 AZ, The Netherlands

Catherine A. McCormick, Department of Biology, Oberlin College, Oberlin, OH 44074, USA

Gloria E. Meredith, Department of Anatomy, Free University Medical School, 1007 MC Amsterdam, The Netherlands

Walter K. Metcalfe, Institute of Neuroscience, University of Oregon, Eugene, OR 97403, USA

Claudia Mohr, Faculty of Biology, Department of Neurophysiology, University of Bielefeld, D-4800 Bielefeld 1, Federal Republic of Germany

John C. Montgomery, Department of Zoology, University of Auckland, Auckland, New Zealand

Heinrich Münz, Faculty of Biology, Department of Neurophysiology, University of Bielefeld, D-4800 Bielefeld 1, Federal Republic of Germany

Sietse M. van Netten, Department of Otolaryngology, College of Physicians and Surgeons, Columbia University, New York, NY 10032, USA

R. Glenn Northcutt, SIO Neurobiology Unit, Department of Neurosciences, A-001, University of California-San Diego, La Jolla, CA 92093, USA

Christopher Platt, Department of Zoology, University of Maryland, College Park, MD 20742, USA

Arthur N. Popper, Department of Zoology, University of Maryland, College Park, MD 20742, USA

Barry L. Roberts, Department of Zoology, Biological Center, University of Amsterdam, 1098 SM Amsterdam, The Netherlands

Olav Sand, Institute of Biology, University of Oslo, 0316 Oslo 3, Norway

Nico A.M. Schellart, Academic Medical Centre, University of Amsterdam, Amsterdam 1105 AZ, The Netherlands

Gottfried Tittel, Institute für Tierphysiologie der Universität, Wartweg 95, D-6300 Giessen, Federal Republic of Germany

Jacqueline F. Webb, SIO Neurobiology Unit, Department of Neurosciences, A-001, University of California-San Diego, La Jolla, CA 92093, USA

Udo Will, Faculty of Biology, Department of Neurophysiology, University of Bielefeld, D-4800 Bielefeld 1, Federal Republic of Germany

Klaus E. Zittlau, Faculty of Biology, Department of Neurophysiology, University of Bielefeld, D-4800 Bielefeld 1, Federal Republic of Germany

Steven J. Zottoli, Department of Biology, Williams College, Williamstown, MA 02167, USA

Part I Introduction

1
A Brief Overview of the Mechanosensory Lateral Line System and the Contributions to This Volume

SHERYL COOMBS, PETER GÖRNER, and HEINRICH MÜNZ

The mechanosensory lateral line system can be most easily distinguished from its electrosensory counterpart by the morphology of its end organ, the neuromast, which consists of hair cells and associated support cells (see Jørgensen, Chapter 6, for further details). Historically, at least four classes of end organs, distinguished primarily by surrounding support structures, have been classified as mechanosensory lateral line organs: (1) *superficial neuromasts* found on the skin surface in cartilaginous and bony fishes and some aquatic and semiterrestrial amphibians; (2) *canal neuromasts* enclosed in tubes formed by cartilage in elasmobranch fishes and by bone or scale in most bony fishes; (3) *spiracular organs* housed in diverticula of the hyoid pouch and found in most nonteleost bony fishes; and (4) *vesicles of Savi*, enclosed pouches containing a cluster of neuromasts and found in some elasmobranchs. The majority of the work contained in this book focuses on what is known about the first two classes of neuromasts (see Dijkgraaf 1963 and Coombs et al. 1988 for further reviews on morphological variation within these classes), but the chapter by Barry and Bennett is devoted exclusively to spiracular organs and the vesicles of Savi.

Despite long-standing questions about functional overlap between the inner ear and lateral line (see Sand 1981 and Kalmijn, Chapter 9, for review), most researchers now recognize that the mechanosensory lateral line subserves several different behavioral functions independent of audition, including surface feeding (Bleckmann et al., Chapter 25; Görner and Mohr, Chapter 27; Elepfandt, Chapter 26), schooling behavior (Partridge and Pitcher 1979; Partridge 1981) obstacle avoidance (Von Campenhausen et al., 1981; Hassan, Chapter 10), and subsurface detection of prey (Coombs and Janssen, Chapter 15; Enger et al., Chapter 29; Montgomery, Chapter 28). Moreover, it is now becoming apparent that the mechanoreceptive lateral line responds to relative movement between the fish and surrounding water at very close distances (less than a few centimeters) rather than to propagated sound pressure waves over longer distances (Enger et al., Chapter 29; Kalmijn, Chapter 9; Platt et al., Chapter 32). Finally, recent neuroanatomical studies have indicated that mechanosensitive lateral line pathways to and in the central nervous system are for the most part separated from those of the auditory and electrosensory systems (Boord and Montgomery, Chapter 16;

McCormick, Chapter 17; Will, Chapter 18), although these pathways also feed into central integration areas along with other sensory systems (Schellart and Kroese, Chapter 21).

Although there is a growing consensus that the mechanosensory lateral line system is an important sensory system in its own right with unique contributions to the animal's behavior (Dijkgraaf, Chapter 2), questions about its evolutionary history and relationship to the electrosensory and auditory systems remain largely unanswered. In light of these questions, research into the development of the lateralis system in various fish groups has been resumed after nearly 50 years of inactivity (Northcutt, Chapter 3; Webb, Chapter 4), and comparative neuroanatomical studies to look at the extent of morphological overlap between these systems have been accelerating at a rapid rate (Boord and Montgomery, Chapter 16; Fritzsch, Chapter 5; Jorgensen, Chapter 6; McCormick, Chapter 17; Will, Chapter 18). Questions about the role of developmental constraints and functional adaptation in the evolution of diverse lateral line morphologies and distributions are being raised (Northcutt, Chapter 3; Webb, Chapter 4). Similarly, there is research interest in the functional significance of morphological variation (Coombs and Janssen, Chapter 15; Denton and Gray, Chapter 11; Münz, Chapter 14).

Techniques for measuring and quantifying surface and subsurface wave stimuli have greatly facilitated research on the sensory capabilities of lateral line systems, and for the first time, systematic and quantitative descriptions of sensory capabilities have been made (Bleckmann et al., Chapter 25; Coombs and Janssen, Chapter 15; Elepfandt, Chapter 26; Görner and Mohr, Chapter 27; Hassan, Chapter 10). These studies have indicated that although the mechanosensory lateralis system responds to different stimuli than the auditory system, its capacity for detecting and discriminating signals can be quite similar to that of other hair cell systems (Platt et al., Chapter 32). The functional distinction between the mechanosensory lateral line and auditory system is due, in large part, to a greater understanding of the hydrodynamics of the stimulus and the importance of flow patterns in the design and function of lateral line systems (Denton and Gray, Chapter 11; Kalmijn, Chapter 9; Hassan, Chapter 10; Van Netten and Kroese, Chapter 12).

Last but not least, the superficial location of lateral line organs that has made this system so easily accessible and attractive as a model system continues to attract research in a wide range of neurobiological disciplines. The mechanoreceptive lateral line system is being used to ask questions about developmental (Metcalfe, Chapter 7), evolutionary (Northcutt, Chapter 3; Webb, Chapter 4), and regenerative (Corwin et al., Chapter 8) mechanisms; about sensory transduction (Kroese and Van Netten, Chapter 13; Van Netten and Kroese, Chapter 12); about sensory processing by the peripheral and central nervous system (Bleckmann and Bullock, Chapter 19; Boord and Montgomery, Chapter 16; Claas et al., Chapter 20; Coombs and Janssen, Chapter 15; McCormick, Chapter 17; Münz, Chapter 14; Roberts and Meredith, Chapter 22; Schellart and Kroese, Chapter 21; Will, Chapter 18; Zotolli and Danielson, Chapter 23); and about neural mechanisms of behavior (Blaxter and Fuiman, Chapter 24; Bleckmann et al., Chapter 25; Elepfandt, Chapter 26; Görner and Mohr, Chapter 27; Montgomery, Chapter 28).

Beginning with this general overview, Part I of the book introduces past and present areas of lateral line research, with an added historical perspective coming from a personal review by Dr. Sven Dijkgraaf. In Chapter 2 he has summarized some of the very earliest lateral line experiments, starting in the mid-1800s, and has recalled some of his own, very elegant experiments which have led to a significant understanding of the function and behavioral significance of the mechanosensory lateral line system.

Part II of this book contains a selection of chapters summarizing what is known about the functional morphology, ontogeny, and evolution of the lateral line system in the major groups where it is found. This section in spearheaded by a uniquely comprehensive and meticulously detailed treatment of the phylogenetic distribution of the mechanosensory lateral line system in all living and fossil vertebrate groups for which it has been described.

The next several sections of the book deal primarily with the encoding of sensory information in the nervous system, starting at the periphery with the functional anatomy and physiology of the end organs and their peripheral innervation (Part III), progressing to the physiological and neuroanatomical organization of the central nervous system (Part IV), and ending with behavioral outcome, signal analysis, and overall function of the lateral line system (Part V).

Part VI of the book examines the function and anatomy of two highly specialized groups of lateral line receptors, the spiracular organs and the vesicles of Savi, and other systems thought to be either functionally or evolutionarily related to the mechanosensory lateral line system. Finally, in Part VII, Dr. Theodore Bullock ends the book with his personal perspective on lateral line research—where it's been and where it ought to go.

References

Coombs S, Janssen J, Webb J (1988) Diversity of lateral line systems: Evolutionary and functional considerations. In: Atema Y, Fay RR, Popper AN, Tavolga WN (eds) Sensory Biology of Aquatic Animals. New York: Springer-Verlag, pp. 553–595.

Dijkgraaf S (1963) The functioning and significance of the lateral line organs. Biol Rev 38:51–105.

Partridge B (1981) Lateral line function and the internal dynamics of fish schools. In: Tavolga WN, Popper AN, Fay RR (eds) Hearing and Sound Communication in Fishes. New York: Springer-Verlag, pp. 515–522.

Partridge B, Pitcher T (1979) The sensory basis of fish schools: Relative roles of lateral line and vision. J Comp Physiol 135:315–325.

Sand O (1981) The lateral line and sound reception. In: Tavolga WN, Popper AN, Fay RR (eds) Hearing and Sound Communication in Fishes. New York: Springer-Verlag, pp. 459–480.

Von Campenhausen C, Riess I, Weissert R (1981) Detection of stationary objects by the blind cave fish *Anoptichthys jordani*. J Comp Physiol 143:369–374.

2
A Short Personal Review of the History of Lateral Line Research

Sven Dijkgraaf

When the organizers of this conference asked me to give a short personal historical review of the development of research work on the lateral line system, I felt of course honored and gladly accepted, be it not without a certain timidity, because among a group of mostly young or middle-aged conference attendants still engaged in active research, I felt somewhat like a living fossil. Therefore I have confined my review to the origin and first results of physiological research.

Until the middle of the 19th century, the lateral line system, or rather the lateral canals, was generally considered a mucus-producing organ, causing the slimy surface of fishes. That the canals contain sense organs was discovered by the German anatomist Franz Leydig. In his paper "Ueber die Schleimkanäle der Knochenfische," which appeared in 1850, he described the big neuromasts, or *Nervenknöpfe*, as he called them, in the very wide canals on the head of the ruff (*Acerina cernua*) (Fig. 2.1). These neuromasts are easily detectable with the naked eye.

The next important discovery was made by another German, Franz Eilhard Schulze, in 1861. He studied fish and amphibian larvae under the microscope and saw and recognized for the first time free neuromasts with their cupulae. In a subsequent paper of 1870, Schulze also noticed the movement of the cupulae caused by water impact, and he presumed that these *Seitenorgane*, as he coined them, might be stimulated by gross water movements striking against the fish body or vice versa, as well as by low-frequency vibrations propagated through the water such as sound waves.

During the second half of the 19th century, a period in which Darwinism made morphology and embryology dominate the scientific interest of biologists, the lateral line system was generally thought to be a receptor for low-frequency sound, and it was correspondingly called *ein akzessorisches Gehörorgan*—an accessory organ of hearing. This opinion was, however, apart from Schulze's presumptions, based merely on the anatomical relationship of the lateral line system and the labyrinth taken as the principal organ of hearing.

About 1900, animal physiology began to attract interest once again. One of the pioneers in this domain was the American zoologist George Howard Parker. In 1904, he published a paper in which he concluded—on the basis

FIGURE 2.1. Head of the ruff *Acerina cernua* showing innervated canal neuromasts after removal of the skin. From Leydig (1850).

of experiments—that water vibrations of about six per second represented an adequate stimulus for the lateral line. Although Parker was a deservedly famous research worker, it is amusing to see how this six-per-second figure came into life: it was simply the frequency with which the wooden table bearing Parker's heavy aquarium swung to and fro when given a push. The fish, which were startled by the push, were less (or no more) startled after elimination of the lateral lines. However, just as Parker's method of stimulation was open to serious criticism, so was his technique of lateral line elimination, and his conclusions were therefore not convincing.

The man who discovered the true function of the lateral line system was the German ichthyologist Bruno Hofer in 1908. He found, among other things, that a pike with damaged eyesight reacted clearly when a weak water current struck its body locally. Unilateral removal of the lateralis nerve of the trunk abolished this sensitivity, and it did so only at the operated side. Yet Hofer's clear experimental results remained largely unnoticed, presumably for three reasons.

First, his work was published in the local journal on fishery biology entitled "Berichte der königlich-bayerischen biologischen Versuchsstation München." This journal was founded by Hofer himself, but it disappeared after a few issues had been published.

Second, Hofer considered the rheotactic orientation of fishes against currents in rivers as the main behavioral task of the lateral lines. He wrote, "Without this sense organ, all the fishes would in the end be swept out of the rivers." And this was certainly not right.

But the third and most important reason may have been the rather generally accepted conviction—dating from long before Hofer—about the use of the lateral line system as an organ for the detection of low-frequency sound waves—a conviction that as I have said, was up to this time based mainly on anatomical and embryological arguments.

The first physiological test did not support this generally accepted opinion—on the contrary. In 1923, Karl von Frisch discovered that catfish (*Ictalurus*) were sensitive to sound by training them to respond to sound, and he published these

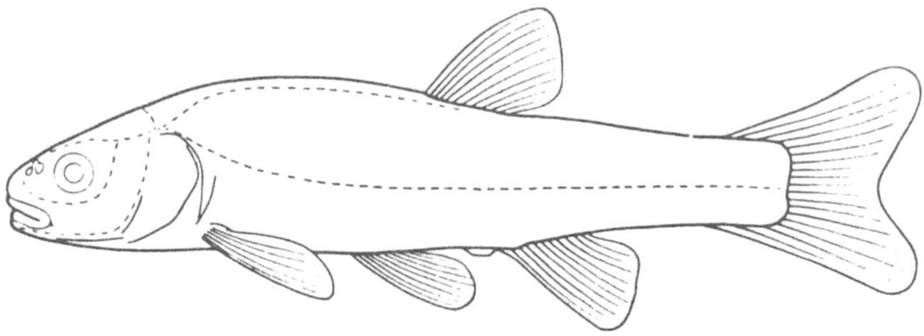

FIGURE 2.2. Topography of lateral lines in the minnow, *Phoxinus phoxinus*. From Dijkgraaf (1952).

first results under the title "Ein Zwergwels, der kommt, wenn man ihm pfeift." In the following years, von Frisch extensively studied the sense of hearing in the minnow (*Phoxinus phoxinus*). By neat surgical elimination of the sense organs involved, he found that sound reception resides in the pars inferior of the labyrinth. Only tones of very low frequency, for instance 16 cycles per second, were still perceived after removal of this sense organ. However, additional elimination of the lateral line system did not at all abolish the sensitivity of the fish for this low-frequency tone. Von Frisch concluded, therefore, that the lateral lines are not used for sound reception, including sound of low frequency (von Frisch and Stetter 1932). And when I repeated these experiments later with marine fish species, like *Gobius*, I got quite similar results (Dijkgraaf 1950).

Because I was asked to give a *personal* historical review, I would now like to indicate briefly how my research with the lateral line system began. As a student of the natural sciences at the University of Vienna, in 1931, I earned a doctoral thesis under the direction of a well-known professor there. He was a paleontologist and was particularly interested in the behavior of extinct animals. He called this special scientific domain "paleobiology." The subject he gave me as a thesis was the following: In many teleost species, the trunk lateral line shows a dorsally directed bend in its foremost part. According to my professor, this bend should be regarded as a relic from bilaterally compressed ancestors with a high body when they (for some reason or another) regained their normal body shape in the course of evolution. I had to find arguments in favor of this explanation.

I was not very happy about this task. And when I studied the morphological literature, I soon found evidence that the displacement of the anterior part of the trunk line was connected with the position of the pectoral fin. When the fin was inserted ventrally, as, for instance, in most cyprinids, there was no bend at all in the trunk line (Fig. 2.2). But when the pectoral fin was placed higher up at the side of the body so that it beat against the body wall, the lateral line in this region was diverted (Fig. 2.3). For me, this was a satisfactory physiological explanation, but my professor was anything but a physiologist. So I looked for more morphological evidence against his idea, and I was happy when I found the pictures of

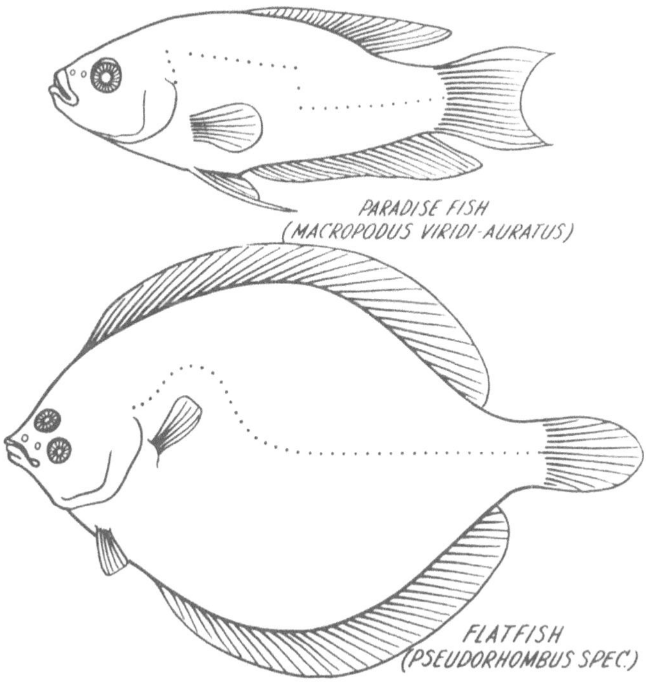

FIGURE 2.3. Dorsal displacement of anterior part of trunk lateral line in several fishes to avoid region of water disturbance caused by movements of the pectoral fin. From Dijkgraaf (1963). Reprinted with permission of Cambridge University Press.

a flatfish that had a normal pectoral fin at its left side (functionally, its upper side) but showed a complete lack of any pectoral fin at the right (or lower) side. The trunk lateral line performed a neat bend around the fin at the left side but ran completely straight at the finless lower side (Goode and Bean 1896, Fig. 357). This was exactly the kind of proof I needed to convince my professor that his evolutionary explanation was wrong—at least so I thought. But when, with youthful enthusiasm I showed him the pictures, he only said, "Well, if you don't like this thesis, I can give you another one." After this experience, however, I was no longer interested in this professor as my "Doktorvater."

In the meantime, I had done some experiments with blinded minnows in the bathtub of my host in Vienna. These experiments showed clearly that the lateral line plays only a minor part in the rheotactic orientation of fish in streaming water. Now, my host was the surgeon Otto von Frisch, a brother of Karl von Frisch. And when the latter heard about my bathtub experiments, he asked me to come to his laboratory in Munich. Since he had found that the lateral line system was not used in sound reception, he was looking for a student who would investigate its biological function.

As you know, in my view, which I expressed first in 1933 in my dissertation in German, and which I repeated 30 years later in a review paper in English, the lateral lines are mainly used as a *Ferntastsinn*—a sense of touch at some distance. The system enables fishes and amphibians to perceive other moving animals as prey, enemies, or sexual partners, without really touching them. Of course, this ability is of particular interest when vision is hampered—for instance, in turbid water or at night. Of lesser importance is the ability to detect motionless objects or obstacles at some distance, and in rheotactic orientation the lateral lines play only a minor role, as I have mentioned. One should expect the lateral lines to be extensively used by schooling fish, but most earlier authors concluded that fish schools are predominantly kept together by eyesight and, perhaps, also by some scent during darkness. More recent work, however, shows that the lateral line system plays an important part in schooling behavior (Partridge and Pitcher 1980).

Here I would like to discuss a point that has been neglected in the past and that has repeatedly caused some misunderstanding. When people speak about the function of a sense organ, they can mean two different things: either the function of a sense organ in the life of the animal as a whole, in other words, the perceptive task for which it is used by the animal; or the function of the isolated sense organ, including the receptive capabilities of its sense cells and so on. To study the first kind of function, only behavioral responses are appropriate. And until the time of my dissertation, about 1933, only such reactions had been studied. The conclusions mentioned so far, therefore, merely concern the use of lateral line systems by fishes and amphibians.

To study the second type of function, the recording of receptor potentials or spikes in the afferent nerve fibers is required, and on the basis of such studies—beginning with the electrophysiological work of Hudson Hoagland in the years 1932–35—a much larger spectrum of stimuli, including low-frequency vibrations, turned out to be effective. Hoagland also discovered the so-called spontaneous activity of the lateral line organs, and then other discoveries, such as the microstructure of the hair cell and the polarity of its response, greatly increased our insight into the physiological mechanism of these sense organs. However, electrophysiological results do not give conclusive evidence as to whether the stimulus when applied and yielding a response will ever be used by the animal in an overt behavioral response.

Some confusion has also arisen about the physical nature of the behaviorally adequate stimulus. In front of an object moving (or being moved) underwater, two phenomena may occur (Fig. 2.4). First is a local displacement of water particles, which will have to flow aside and around the object to make room for it. This phenomenon, caused by the pressure of an object, includes a strictly local rise of water pressure, as well as a certain amount of particle displacement in the direction in which an object moves. In 1933, I called these phenomena *Stauungserscheinungen*, or damming phenomena. They occur under all circumstances including in front of an object moving with a constant velocity. There is no doubt that stimuli of this kind are behaviorally adequate to the lateral lines.

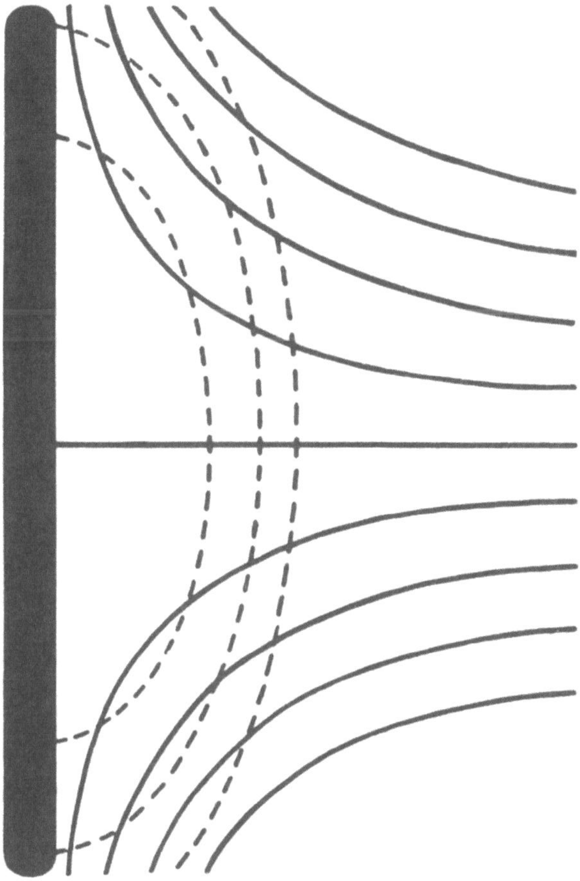

FIGURE 2.4. Damming pressure (interrupted lines) and lateral water displacement (solid lines) in front of a disk (black, shown in section), moving to the right. After Prandtl, from Dijkgraaf (1963). Reprinted with permission of Cambridge University Press.

Second, quite another phenomenon may occur when there is a marked velocity *change* of the moving object: this may cause a compression wave, propagating itself in all directions with sound velocity—that is, about 1,450 m-s in water. In contrast to the first phenomenon, this second depends on the compressibility of water, and it does not occur at all when an object moves with constant velocity.

A sound source underwater may cause both phenomena: local damming and self-propagating pressure waves. Harris and van Bergeijk (1962) called the damming phenomena in this particular case "near-field effects of acoustical sources in water." Within certain limits, water motions of this kind may well represent behaviorally effective stimuli to the lateral lines. Nevertheless, sound in the usual meaning of compression waves seems behaviorally inadequate, and quick stimulus repetition, as from a usual sound source, seems not particularly characteristic for most natural lateral line stimuli. Therefore, I am still reluctant to describe the lateral line system as an organ of hearing as, for example, Pumphrey's "short-range auditory system" of 1950 implies. Other chapters in this volume (e.g., Coombs and Janssen Chapter 15, Enger et al. Chapter 29) address this question in the context of more recent findings.

To conclude my review, I'd like to clear up one more misunderstanding about adequate stimuli for the lateral line system. Since we know that currentlike water motions acting locally on the fish body stimulate the lateral lines and cause behavioral responses, it is quite clear that water motions created by the fish itself as it swims should be well within the perceptive range of the system. However, I found experimentally that stimuli of this kind are not used by the animal to regulate its locomotion (Dijkgraaf 1934, 1963). As a reaction to this statement, two of my American friends and colleagues wrote, "Dijkgraaf eliminated the functioning of the lateral line in the detection of locomotor currents. We fail to see the difference between water displacements produced by 'moving obstacles' and those produced by the hydrodynamics of fish locomotion" (Tavolga and Wodinsky 1963). To this I replied, "Though the authors may fail to see the difference, the integrative centers of the fish brain evidently see it" (Dijkgraaf 1964). This was, of course, not meant to suggest that the authors' brains were anyhow inferior to those of a fish, but only to indicate the kind of difference the authors did not see: water displacements produced by fish locomotion are strictly dependent on the fish's own muscle actions. The animal can "expect" them to occur when it moves actively, whereas the water displacements produced by the moving bodies are caused by forces unknown to the fish and come, therefore, quite "unexpectedly." Reactions to self-generated stimuli of the first kind – in the German literature known as *Reafferenz* – are usually suppressed, either peripherally, through efferent nerve impulses, or centrally, by some process within the brain. Clearly, evidence for such a mechanism is accumulating, as presented by Roberts and Meredith in Chapter 22.

References

Dijkgraaf S (1934) Untersuchungen über die Funktion der Seitenorgane an Fischen. Z Vergl Physiol 20:162–214.

Dijkgraaf S (1950) Untersuchungen über die Funktionen des Ohrlabyrinths bei Meeresfischen. Physiol Comp Oecol 2:81–106.

Dijkgraaf S (1952) Bau und Funktionen der Seitenorgane und des Ohrlabyrinths bei Fischen. Experientia 8:205–216.

Dijkgraaf S (1963) The functioning and significance of the lateral-line organs. Biol Rev 38:51–105.

Dijkgraaf S (1964) The supposed use of the lateral line as an organ of hearing in fish. Experientia 20:586–587.

Frisch K von (1923) Ein Zwergwels, der kommt, wenn man ihm pfeift. Biol Zbl 43:439–446.

Frisch K von, Stetter H (1932) Untersuchungen über den Sitz des Gehörsinnes bei der Elritze. Z Vergl Physiol 17:686–801.

Goode GB, Bean TH (1896) Oceanic ichthyology. Mem Mus Comp Zool Harvard Coll 22, Fig. 357.
Harris GG, Bergeijk WA van (1962) Evidence that the lateral line organ responds to near-field displacements of sound sources in water. J Acoust Soc Am 34:1831-1841.
Hoagland H (1932) Impulses from sensory nerves of catfish. Proc Natl Acad Sci USA 18:701-705.
Hoagland H (1933) Electrical responses from lateral-line nerves of catfish. I. J Gen Physiol 16:695-714.
Hoagland H (1933) Quantitative analysis of responses from lateral-line nerves of fishes. II. J Gen Physiol 16:715-732.
Hoagland H (1934) Electrical responses from the lateral-line nerves of catfish. III. J Gen Physiol 17:77-82.
Hoagland H (1935) Pacemakers in Relation to Aspects of Behavior. New York: Macmillan, 138 pp.
Hofer B (1908) Studien über die Hautsinnesorgane der Fische. I. Die Funktion der Seitenorgane bei den Fischen. Ber Kgl Bayer Biol Versuchsstation Munchen 1:115-164.
Leydig F (1850) Ueber die Schleimkanäle der Knochenfische. Mull Arch Anat Physiol 170-181.
Parker GH (1904) The function of the lateral line organs in fishes. Bull US Bur Fish 24:185-207.
Partridge BL (1982) The structure and function of fish schools. Sci Am 246:114-123.
Partridge BL, Pitcher TJ (1980) The sensory basis of fish schools: Relative roles of lateral line and vision. J Comp Physiol 135:315-325.
Pumphrey RJ (1950) Hearing. Symp Soc Exp Biol 4:3-18.
Schulze FE (1861) Ueber die Nervenendigung in den sogenannten Schleimkanälen der Fische und ueber entsprechende Organe der durch Kiemen athmenden Amphibien. Arch Anat Physiol Lpz 759-769.
Schulze FE (1870) Ueber die Sinnesorgane der Seitenlinie bei Fischen und Amphibien. Arch Mikrosk Anat 6:62-68.
Tavolga WN, Wodinsky J (1963) Auditory capacities in fishes. Pure tone thresholds in nine species of marine teleosts. Bull Mus Nat Hist 126:177-240.

Part II Morphology, Phylogeny, and Ontogeny

3
The Phylogenetic Distribution and Innervation of Craniate Mechanoreceptive Lateral Lines

R. Glenn Northcutt

1. Introduction

Phylogenetic studies can be undertaken with a number of aims: (1) to describe the history of traits; (2) to discern form-function relationships that may not be evident from examining the traits in a single species; (3) to generate hypotheses regarding the evolutionary mechanisms responsible for the history of the traits; and (4) to detect gaps in our knowledge of the traits (Gans 1985; Northcutt 1985a, 1986a). The following analysis involves each of these aims but is confined to the distribution and innervation of mechanoreceptive neuromasts. It does not attempt to deal with related receptors and organs or the phylogeny of the octavolateralis system as a whole, although I have discussed electroreceptors and the inner ear in previous publications (Northcutt 1980, 1986a,b).

There are a number of assumptions that underlie this analysis. The first is that it is impossible to recognize the polarity of a trait–i.e., whether it is primitive (plesiomorphic) or derived (apomorphic)–unless one can establish the ancestral condition (morphotype), and a cladistic approach (Eldredge and Cracraft 1980; Wiley 1981; Northcutt 1984, 1985a) is the only valid method for doing so.

Second, a reasonably well-corroborated hypothesis of craniate genealogy (Fig. 3.1) is assumed to be valid. The bony fishes (osteichthyans) are the sister group of the extinct acanthodians, and cartilaginous fishes and the extinct placoderms are, in turn, sister groups of the acanthodians and bony fishes, respectively. Following current cladistic interpretations (Rosen et al. 1981; Panchen and Smithson 1987), the osteichthyans are believed to be a monophyletic taxon comprising two sister groups, the ray-finned fishes (actinopterygians) and the lobe-finned fishes and their descendants (sarcopterygians). Further, the living sarcopterygians (the actinistian *Latimeria*, the lungfishes, and the tetrapods) are regarded phylogenetically as an unresolved trichotomy (Panchen and Smithson 1987). Uncertainty also exists regarding the phyletic relationships of the placoderms. It is very possible that these taxa do not constitute a monophyletic group, but they will be treated as such in this analysis because of our lack of information regarding their phylogenetic relationships (Moy-Thomas and Miles 1971).

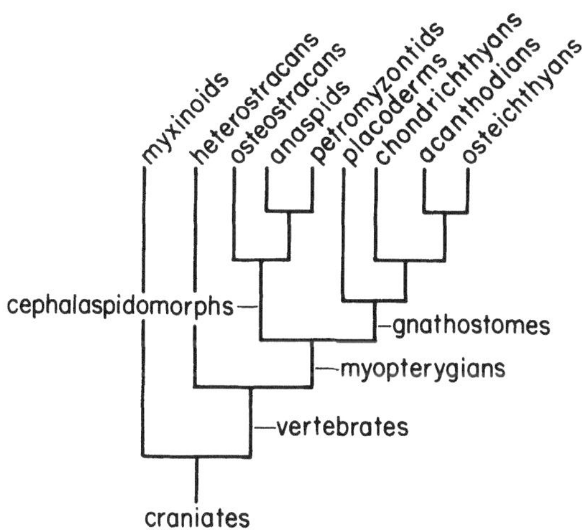

FIGURE 3.1. Cladogram of the suspected interrelationships of extinct and living groups of craniates.

Finally, the fossil and living agnathans do not constitute a monophyletic group, but the cephalaspidomorphs, consisting of the extinct osteostracans and anaspids, as well as the extant lampreys, are the sister group of jawed vertebrates (Janvier 1981), and the extinct heterostracans are, in turn, the sister group of all other vertebrates (myopterygians). The hagfishes (myxinoids) are not considered vertebrates, as they do not exhibit any shared derived traits (synapomorphies) with cephalaspidomorphs or myopterygians and are thus considered the sister group of vertebrates (Løvtrup 1977; Janvier 1978; Hardisty 1979).

The third assumption is that at least some dorsolateral ectodermal placodes give rise embryonically to both mechanoreceptive neuromasts and the first-order ganglionic neurons that innervate these receptors (Platt 1896; Stone 1922; Winklbauer and Hausen 1983). Although this developmental relationship has been demonstrated only for some dorsolateral placodes in amphibians, it is assumed to be the plesiomorphic pattern of development for craniates.

Assumption number four is that the sequence of ontogenetic development of head neuromasts and their accessory structures is critical in considering the possible role of heterochrony in the phylogeny of lateral lines. The dorsolateral placodes are believed to arise and elongate within the deepest cellular layer of the ectoderm and subsequently differentiate into neuromast primordia (Platt 1896; Johnson 1917; Stone 1933; Winklbauer and Hausen 1983), at which time the nonincorporated placodal cells may remain as a continuous band or be broken up into cellular islands, depending on the mitotic activity of the placodal cells relative to that of the cells of the general ectoderm (Winklbauer and Hausen 1983). Developing neuromasts erupt to the surface of the ectoderm, at which time they are similar in appearance to the lines of superficial neuromasts observed in many adult anamniotes (Ruud 1920; Pehrson 1922; Janssen et al. 1987; Coombs et al. 1988). If development proceeds beyond this stage, ectodermal ridges of uncertain embryonic origin arise parallel to the neuromasts, constituting a stage of development (Johnson 1917; Ruud 1920; Pehrson 1922; Janssen et al. 1987; Webb 1988) comparable to the neuromast-lined open grooves of the lateral lines in many adult vertebrates, or the ectodermal ridges may fuse middorsally, forming closed canals except at periodic points along the length of the groove where failure of the ridges to fuse dorsally results in open pores (Johnson 1917; Ruud 1920; Pehrson 1922; Lekander 1949; Janssen et al. 1987; Webb 1988).

The fifth assumption is that an out-group analysis and geological precedence have priority over ontogenetic precedence as criteria in establishing the polarity of lateral line traits, as there is a substantial fossil record for many traits, and comparable developmental studies do not exist. In this context, it is assumed that accessory line structures, such as pits, grooves, or canals, housed neuromasts in fossil taxa as comparable structures do in living taxa.

With these assumptions, the variation in the number and distribution of lateral lines is compared among taxa of all major craniate radiations in order to pose hypotheses regarding lateral line homologies and to reconstruct the initial pattern of organization that characterized the first gnathostomes and craniates. The innervation of the lateral lines is reviewed, and the data are used to examine the phylogeny of the lateral line nerves and their relationship to the dorsal cranial nerves. Finally, reductive or proliferative trends in lateral lines are noted, and possible evolutionary mechanisms are discussed.

2. Phyletic Distribution of the Lateral Lines

The head and trunk of most anamniotic vertebrates are characterized by lines of mechanoreceptive organs (neuromasts) that arise embryonically from a series of ectodermal placodes (Fig. 3.2). These placodes are commonly termed dorsolateral placodes to distinguish them from a more ventrolateral series of placodes, termed the epibranchial placodes, that are involved with the formation of the gustatory system (see Northcutt and Gans 1983, for a recent review). Neuromast lines usually consist of singly placed neuromasts with the major axes of their sensory maculae aligned with the major axis of the line (Platt 1896; Johnson 1917; Janssen et al. 1987; Coombs et al. 1988; Webb 1988). However, neuromasts may occur in clusters (stitches), and their sensory maculae may sometimes be oriented at right angles to the major axis of the line , as in salamanders (Platt 1896; Lannoo 1985, 1987a,b). A survey of the number of neuromasts within the lines and the orientation of their maculae (Platt 1896; Johnson 1917; Janssen et al. 1987; Lannoo 1987a,b; Coombs et al. 1988) suggests that the pattern of single neuromasts with maculae oriented with the long axis of the line is the plesiomorphic condition for vertebrates.

Neuromast lines may consist of single neuromasts located at the bottom of shallow pits (pit lines), in grooves of the epidermis (groove lines), or enclosed in canals (canal lines). An extensive nomenclature (see Coombs et al. 1988, for the most inclusive review) exists for various types of neuromasts housed in pits (pit

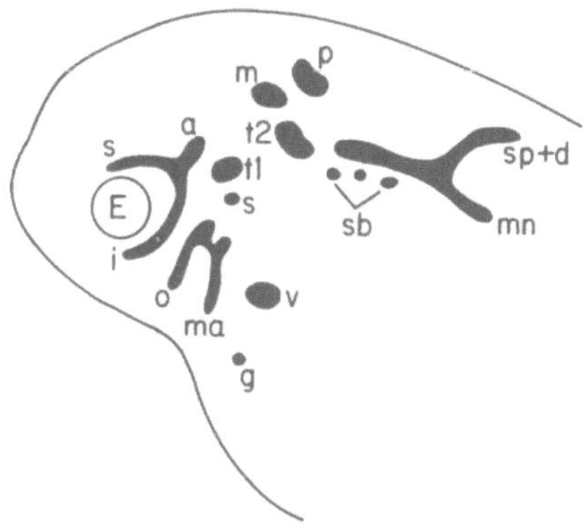

FIGURE 3.2. Schematic drawing of a lateral view of the head of an embryonic *Protopterus*, illustrating the positions of the epidermal placodes that give rise to various portions of the lateral line system. Key: a, anterior pit line; E, lens placode; g, gular line; i, infraorbital line; m, middle pit line; ma, mandibular line; mn, main trunk line; o, oral line; p, posterior pit line; s, spiracular organ or supraorbital line; sb, transitory suprabranchial placodes; sp + d, supratemporal and dorsal trunk lines; t1, anterior portion of temporal (otic) line; t2, posterior portion of temporal line; v, ventral trunk line. Drawing after Northcutt (1986a), based on data from Pehrson (1949).

organs), but it is unclear how many types of pit organs occur in nonteleost vertebrates, and much of the variation in pit organs may be restricted to teleosts.

Neuromast lines in vertebrates, particularly on the head, are extremely complex and form an interconnecting network (Fig. 3.3) that varies from group to group (Norris 1925; Allis 1934; Holmgren 1942; Stensiö 1947; Holmgren and Pehrson 1949). The nomenclature for various segments of the lateral line network is based in part on topography and in part on differences in placodal origin and innervation. The first descriptions of the lateral lines (Steno, 1664; Leydig 1850; Garman 1888) were based on the topographical position of these lines; for example, lines above and below the eye were termed the supraorbital and infraorbital lines, respectively. Later, as researchers became concerned with comparing lines among taxa and determining their phylogenetic history, the placodal origin and innervation of various lines became important (Platt 1896; Allis 1934; Landacre 1910, 1916; Stensiö 1947). If individual lateral lines could be defined as having a separate placodal origin and innervation, it would be possible to recognize the precise extent of individual lines, and it would be relatively easy to reconstruct their phylogenetic history. Unfortunately, this is not possible because multiple lines—such as the oral, mandibular, and other cheek lines—arise from a single placode and are innervated by a single cranial nerve ramus, whereas other lines, such as the supra- and infraorbital lines, appear to arise from a single placode (Platt 1896; Stone 1922) and are innervated by different rami. The nomenclature currently used is primarily descriptive, and the embryonic origin

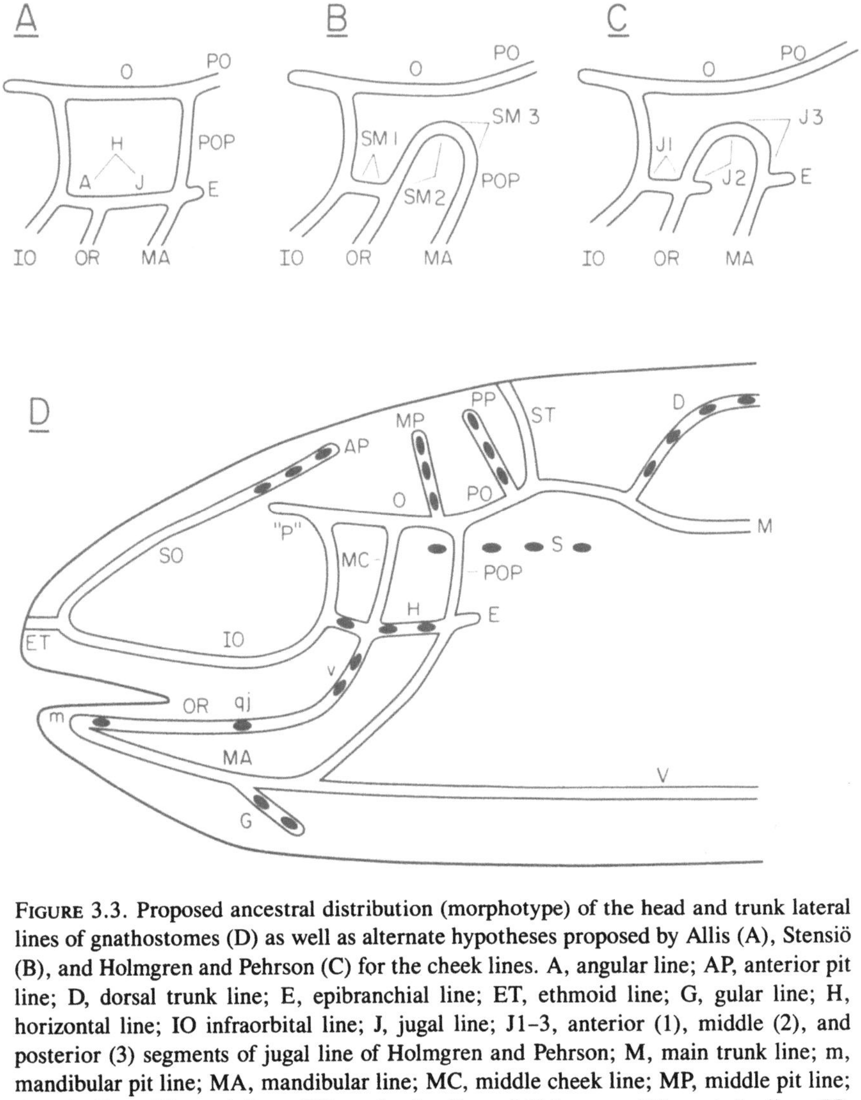

FIGURE 3.3. Proposed ancestral distribution (morphotype) of the head and trunk lateral lines of gnathostomes (D) as well as alternate hypotheses proposed by Allis (A), Stensiö (B), and Holmgren and Pehrson (C) for the cheek lines. A, angular line; AP, anterior pit line; D, dorsal trunk line; E, epibranchial line; ET, ethmoid line; G, gular line; H, horizontal line; IO infraorbital line; J, jugal line; J1–3, anterior (1), middle (2), and posterior (3) segments of jugal line of Holmgren and Pehrson; M, main trunk line; m, mandibular pit line; MA, mandibular line; MC, middle cheek line; MP, middle pit line; O, otic line; OR, oral line; "P", profundus line of Holmgren; PO postotic line; PP, posterior pit line; POP, preopercular line; qj, quadratojugal pit line; S, spiricular line; SM1–3, anterior (1), middle (2), and posterior (3) segments of supramaxillary line of Stensiö; SO, supraorbital line; ST, supratemporal line; V, ventral trunk line; v, vertical pit line. Black ovals indicate lines or portions of lines that may occur as superficial (pit) neuromasts in nonteleost ray-finned fishes.

and innervation of lines or segments of lines are used as additional criteria in recognizing possible homologues in different taxa. In this review, these criteria (topography, embryology, and innervation), as well as the cladistic criterion of out-group analysis, are used to recognize possible lateral line homologues and infer their phylogenetic history.

The number and distribution of lateral lines in gnathostomes are examined first, then variation in these lines is analyzed to test several models of lateral line morphotypy. A similar analysis is then applied to agnathans, and the two morphotypes are compared in an attempt to construct a morphotype for the earliest craniates.

2.1 The Gnathostomes

The lateral lines on the head and trunk of gnathostomes can be arbitrarily grouped into the following categories: (1) supra- and infraorbital lines surrounding the eyes; (2) pit lines located dorsally on the head; (3) lateral lines of the trunk, of which the temporal segment of the main line continues rostrally onto the head; and (4) the lateral lines of the postorbital head (cheek region) and lower jaw.

2.1.1 The Supraorbital and Infraorbital Lines

Supraorbital and infraorbital lines occur in all gnathostome radiations (Table 3.1) and must have been present in the earliest gnathostomes (Fig. 3.3D). The supraorbital line is a single continuous line in all living gnathostomes and usually connects with the infraorbital line rostrally on the snout and caudally above the orbit, from where these lines continue as a main trunk line. However, placoderms (Fig. 3.4A), acanthodians (Fig. 3.5), palaeoniscoid actinopterygians, and even embryos of some living actinopterygians, *Amia* (Allis 1889) and *Lepisosteus* (Hammarberg 1937), possess a disjunct supraorbital line divided into rostral and caudal segments, with the posterior end of the rostral segment (AP, Fig. 3.3D), passing dorsal and caudal to the more ventrally lying caudal segment ("P", Fig. 3.3D). Given the distribution of a disjunct supraorbital line, it is clear that this is the plesiomorphic condition for gnathostomes and that a continuous line in chondrichthyans and modern bony fishes is an apomorphic condition that has probably evolved independently in the two radiations.

A major part of the anterior segment of the supraorbital line is present as a canal in all gnathostome clades except living amphibians, but the extreme caudal portion of the anterior segment always occurs as a pit line (Fig. 3.3D) and may never have existed as a canal in any gnathostome radiation, with the possible exception of some placoderms. This pit line, usually termed the anterior pit line (dorsal supraorbital line of Coombs et al. 1988), occurs in all gnathostome groups except chondrichthyans and, perhaps, actinistians.

In placoderms the caudal segment of the supraorbital line coursed medially (Fig. 3.4A) and has been termed a pineal or central line. Holmgren (1942) sug-

TABLE 3.1. Distribution of gnathostome lateral lines.

Lateral line	Placoderms	Chondrichthyans	Acanthodians	Actinopterygians	Actinistians	Dipnoans	Amphibians
ET	?	g/c	+	c	c	–	–
SO	g/c	c	g/c	c	c	g/c	p
IO	g/c	c	c	c	c	g/c	p
AP	g/c	–	c	p	?	p	p
MP	g	?	g/c	p	?	p	p
PP	g	p	g/c	p	?	p	?
ST	g/p	c	+	c	c	g/c	?
O	g/c	c	c	c	c	g/c	?
PO	g/c	c	c	c	c	g/c	?
MC	g	p	g	–	?	p	p
S	?	s	?	s	?	s	–
H	g/c	g/c	g/c	p	?	p	p
OR	?	g/c	g	p	?	g/c	p
MA	?	g/p	g	c	c	g/c	p
G	?	–	+	p	?	p	p
POP	g/c	g/p	c	c	?	g/c	p
E	–	g	–	–	–	–	–
D	g/p	p	–	p	?	p	p
M	g/c	c	g/c	c	c	g/c	p
V	?	p	g	–	?	p	p

Key: c, canal; g, groove; g/c, groove and/or canal in a single species or group; g/p, groove and/or pit in a single species or group; p, pit line; s, spiracular organ; +, line present, but condition uncertain; –, line absent; ?, presence of line uncertain. See Figure 3.3 for abbreviations for lateral lines.

gested that the pineal line was a separate lateral line, innervated by a lateral line component of the trigeminal nerve, based on the presence of a supposedly similar line in lampreys whose innervation was uncertain at that time (see Section 2.2) and his observation of a transitory trigeminal nerve twig in embryos of *Squalus* (Holmgren 1940). Given the wide occurrence of a disjunct supraorbital line in fossil gnathostomes and the development of the supraorbital line from at least two placodes in elasmobranchs (Holmgren 1940) and ray-finned fishes (Allis 1889; Lekander 1949), it seems more probable that the pineal line in placoderms is homologous to the caudal segment of the supraorbital line in other gnathostomes, and the existence of an additional trigeminal-related lateral line should therefore be viewed with skepticism.

Although the supraorbital and infraorbital lines generally interconnect caudal and dorsal to the orbit in gnathostomes, the caudal course of the supraorbital line in ratfishes is quite different (Garman 1888; Cole 1896). In these animals the supraorbital line connects caudally to the supratemporal commissure, and the infraorbital line connects caudally with the otic segment of the main trunk line. This autapomorphic (uniquely derived) connection of the supraorbital line in ratfishes may be a consequence of the hypertrophy of the eyes with concomitant distortion of the postorbital head region.

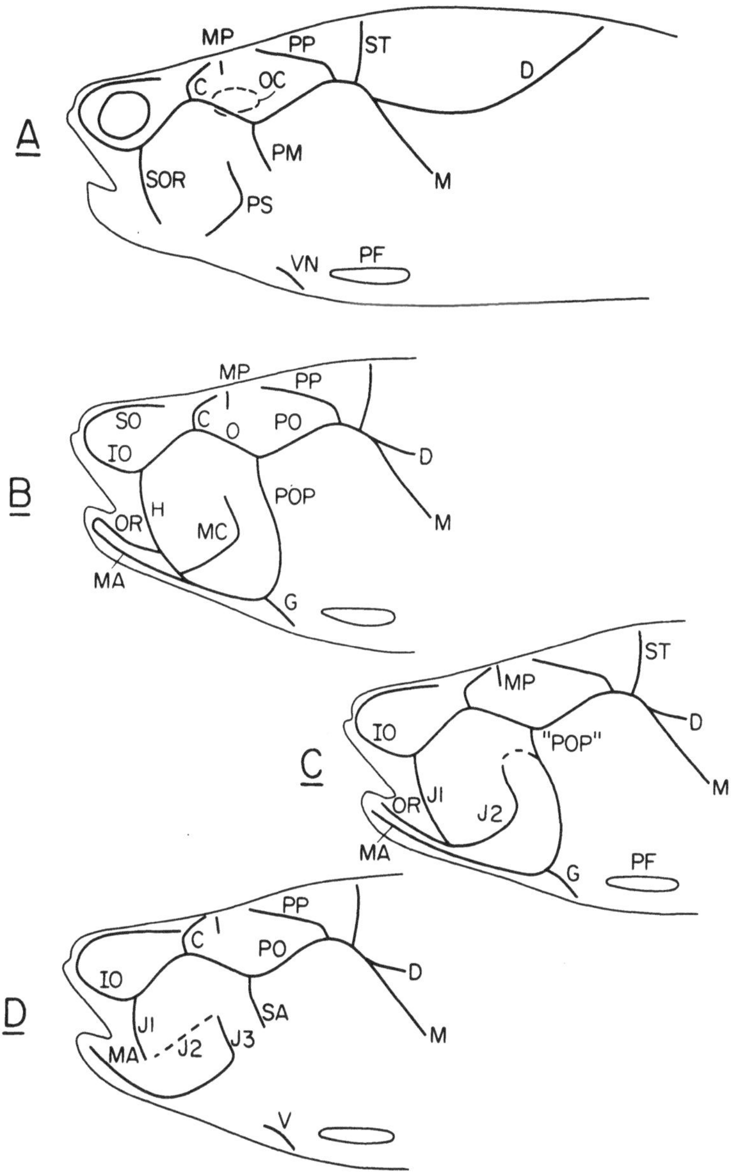

FIGURE 3.4. Distribution of lateral lines in an arthrodire placoderm, *Coccosteus* (A), after Miles and Westoll (1968), and several hypotheses (B–D) indicating possible lateral line homologies with other gnathostomes. Abbreviations for lines same as in Figure 3.3 with the following additions: C, central line, OC, otic capsule; PF, pectoral fin; PM, postmarginal line; "POP", autapomorphic preopercular line; PS postsuborbital line; SA, scapular line; SOR, supraoral line; VN, ventral neuromast line.

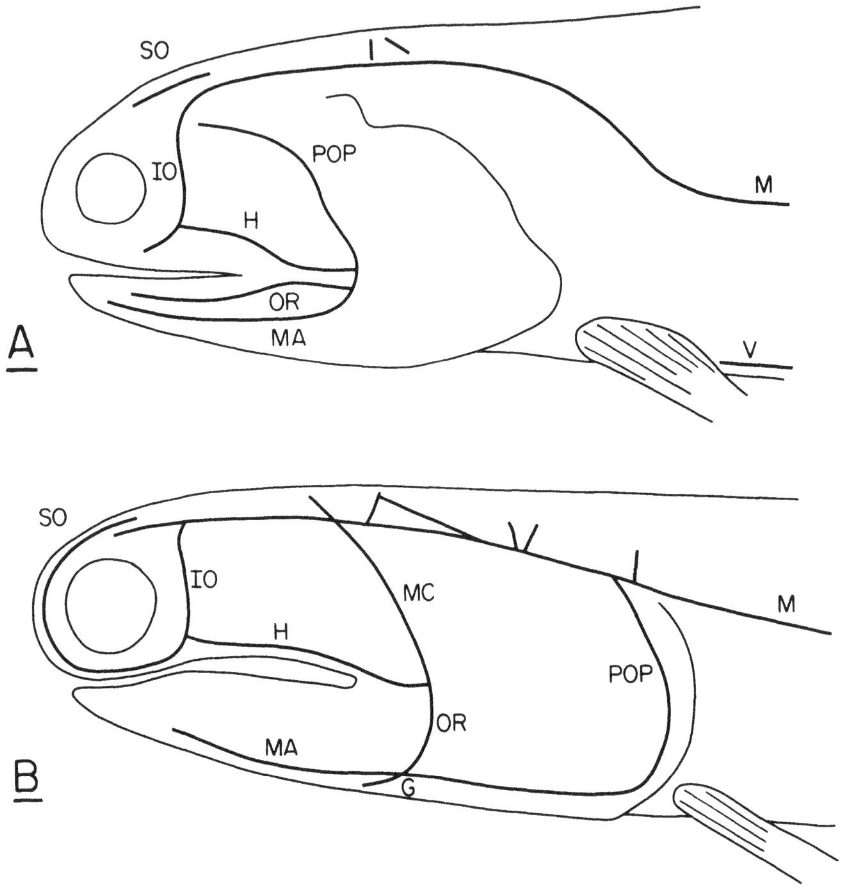

FIGURE 3.5. Distribution of lateral lines, after Watson (1937), in two acanthodian genera, *Euthacanthus* (A) and *Acanthodes* (B), indicating possible lateral line homologies with other gnathostomes. Abbreviations same as in Figure 3.3.

The rostral course of the supraorbital and infraorbital lines is highly variable among gnathostomes, but the wide occurrence of a commissural connection (Table 3.1) suggests that an ethmoid commissure (Fig. 3.3D) probably occurred in the earliest gnathostomes. Given the shape of the rostral tip of the skull in placoderms, it is likely that the snout was composed of soft tissues, as in sharks, and therefore no details of the lateral lines on the snout of placoderms are preserved in the fossils. However, ethmoid commissures occur in some members of all other gnathostome radiations, except for lungfishes (Fig. 3.6) and amphibians (Fig. 3.7). Among gnathostomes, however, there is considerable variation regarding the position on the snout where the supraorbital and infraorbital lines

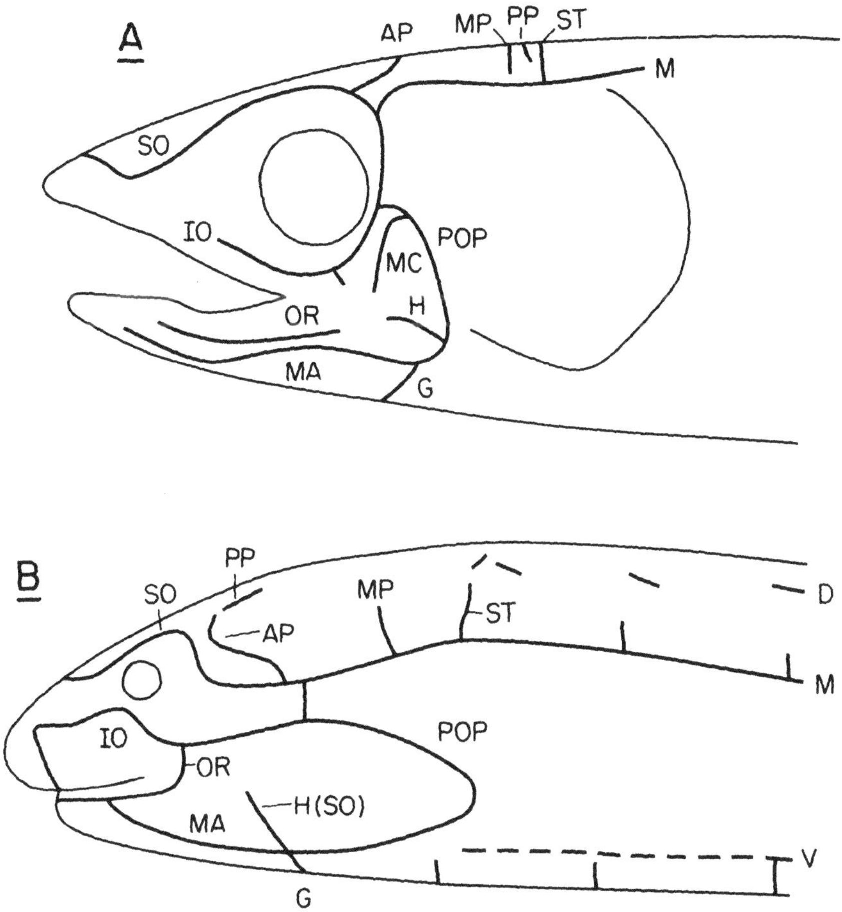

FIGURE 3.6. Distribution of lateral lines in a Devonian lungfish, *Dipterus* (A), after Westoll (1963), and the living African lungfish *Protopterus* (B) indicating possible lateral line homologies with other gnathostomes. Abbreviations are the same as those in Figure 3.3 with the addition of a secondary oral line (SO).

meet (Gardiner 1963; Jollie 1969). In elasmobranchs, the supraorbital and infraorbital lines meet caudal to the posterior nasal opening, whereas in primitive actinopterygians and crossopterygian bony fishes, the supraorbital and infraorbital lines meet between the nasal openings and rostral to the anterior opening, respectively.

There is considerable variation in the course of these lines on the snout in chondrichthyans and actinopterygians. In chondrichthyans secondary loops may occur, and it is not clear whether both lines are involved in these autapomorphic proliferations. In actinopterygians secondary connections between the lines may occur caudal to the olfactory organ, as may reduction or loss of segments of the

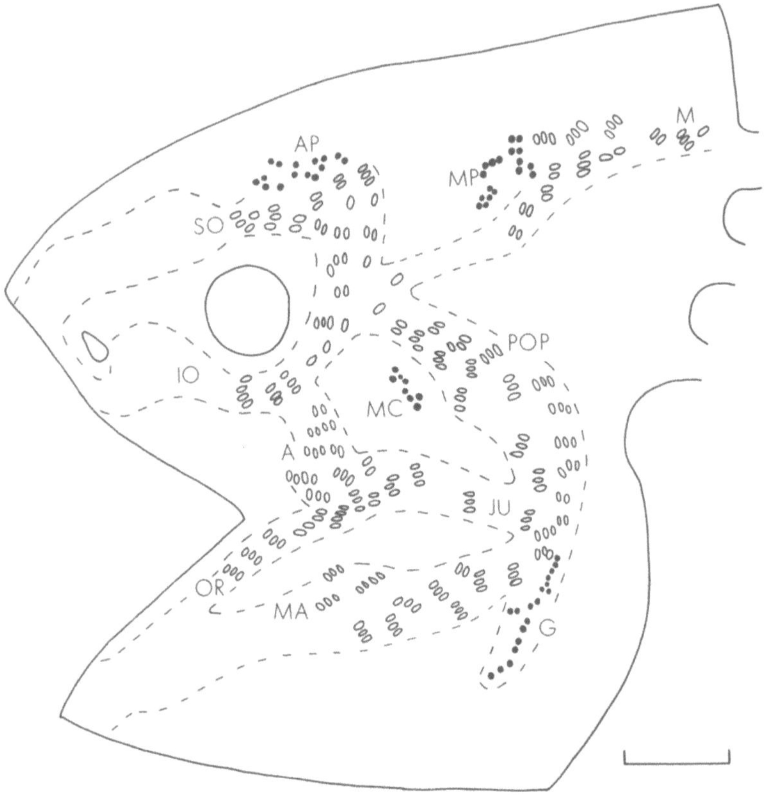

FIGURE 3.7. Distribution of neuromasts (open ovals) and pit organs (solid circles) in a juvenile axolotl. All mechanoreceptive lateral line organs are superficial in axolotls, and the borders of individual lines are indicated by dashed lines. Abbreviations are the same as those in Figure 3.3. Bar scale equals 2 mm.

canals. Reductions of the canals may result in pit lines that have been given various names (tectal lines—Jarvik 1947; nasal lines—Pehrson 1947; supraorbital rostral replacement lines—Coombs et al. 1988; rostral lines—Pehrson 1945; ethmoidal lines—Nelson 1972; rostral fork replacement lines—Coombs et al. 1988), but it is unclear how many times such canal reductions have occurred, which canal segments have been involved, and, therefore, the homologous relationships of these pit lines.

In the earliest gnathostomes the infraorbital line consisted of a single groove or canal that connected with the supraorbital line rostrally on the snout, approximately ventral to the orbit, and again caudally above the orbit ("P", Figs. 3.3D, 3.4, 3.5A). However, ratfishes (holocephalans) and many teleost actinopterygians appear to have evolved autapomorphies involving the infraorbital lines. In ratfishes two lateral line grooves run ventral to the orbit (Garman 1888). The more dorsal groove occupies the usual position of the infraorbital line in other gnathostomes and continues caudally and dorsally around the orbit to connect with the main trunk groove; the more ventral groove courses caudally to connect with an obliquely oriented groove that terminates on the lower jaw and proximally at the caudal end of the more dorsally located "infraorbital" groove. According to Cole (1896), the neuromasts of both the dorsal and ventral grooves are innervated by the ramus buccalis and thus should be interpreted as arising from the same placode (Fig. 3.2) that gives rise to a single infraorbital line in other gnathostomes. Given the peculiar presence of these lines, as well as the problematical process of homologizing lines of the lower jaw (Allis 1934; Holmgren 1942; Stensiö 1947), it would be useful to investigate the development of these lines in ratfishes and to reexamine their innervation.

In some teleosts the lateral line system of the infraorbital region may consist of a dorsally located infraorbital canal as well as one or more accessory infraorbital lines of free neuromasts (Lekander 1949; Coombs et al. 1988; Puzdrowski 1988); in some others the infraorbital line itself may be present only as a pit line (Pehrson 1944; replacement lines of Coombs et al. 1988). Given the distribution of the infraorbital line as a canal and the absence of accessory pit lines in other gnathostomes, the accessory lines must be viewed as teleost autapomorphies.

2.1.2 Dorsal Pit Lines of the Head

Neuromasts that do not occur within canals or grooves are commonly termed superficial neuromasts. Although such neuromasts have been classified in a number of ways (Lekander 1949; Srivastava and Srivastava 1968; Coombs et al. 1988), it is clear that superficial neuromasts occurring in depressions of the skin (pit lines) are restricted to five regions of the head and are termed the anterior, middle, and posterior pit lines (Fig. 3.3D). The remaining lines occur on the cheek—middle pit lines—and lower jaw—gular pit lines—and are discussed in Section 2.1.4.

Although present as a distinct trait in many actinoperygians, lungfishes (Fig. 3.6), and salamanders (Figs. 3.7, 3.8B,C), the anterior pit line appears to have been primitively the caudalmost portion of the rostral canal or groove segment of the supraorbital line in placoderms and acanthodians (see Section 2.1.1). Similarly, there is evidence that the middle and posterior pit lines of modern gnathostomes were represented in placoderms (Miles 1966; Miles and Westoll 1968) and acanthodians (Watson 1937) by canals and/or grooves.

In living gnathostomes, pit lines are easily recognized, as the anterior pit line arises as a distinct dorsal diverticulum of the supra- and infraorbital placode after the supraorbital portion of the placode has elongated above the eye (Pehrson 1949); the anterior pit line is always innervated by the first dorsally arising branch of the superficial ophthalmic ramus (Platt 1896; Norris 1925; Song 1986). The middle pit line appears to rise from a separate placode (Fig. 3.2) (Pehrson 1949; Hammarberg 1937) and is the only pit line in nonteleost gnathostomes that is innervated by the middle lateral line nerve (supratemporal glosso-

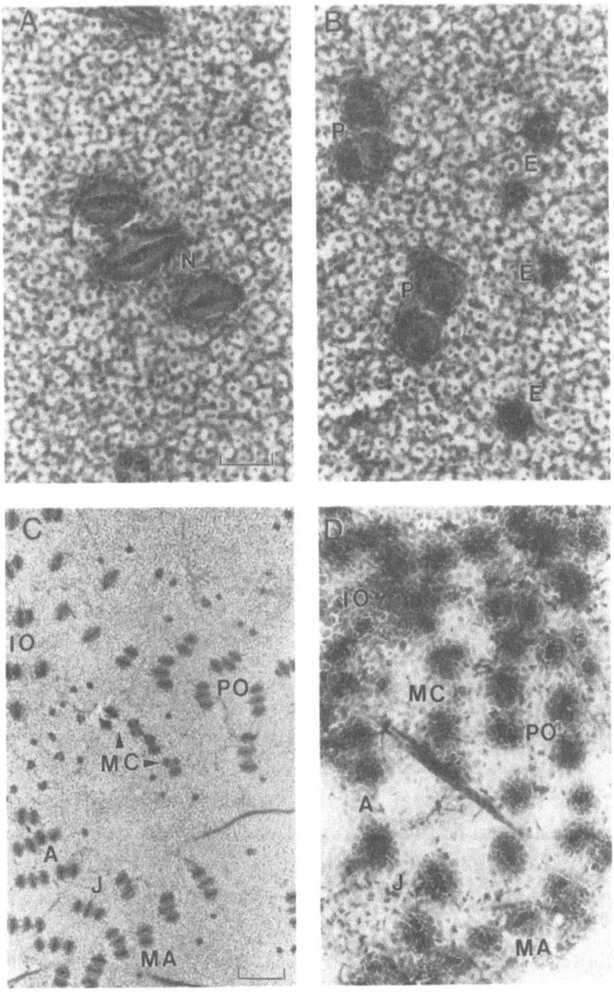

FIGURE 3.8. Photomicrographs of neuromasts of the supraorbital (A) and anterior pit (B) lines in a juvenile axolotl. Note that pit line neuromasts (P) are smaller than the neuromasts (N) of the other superficial lines but can be distinguished from ampullary electroreceptors (E) that usually occur adjacent to neuromast lines in axolotls. The cheek region (C) of axolotls contains an additional pit line (MC), not previously reported, which arises as the most rostral vertical row (D) of neuromast primordia formed from the hyoid placode as seen in a stage 43 embryo. Dorsal and rostral are to the top and left side, respectively, of D and C. Bar scale equals 50 μm (A,B,D) and 500 μm (C). A, angular line; IO, infraorbital line; J, jugal line; MA, mandibular line; MC, middle cheek line; PO, preopercular line.

pharyngeal ramus of Norris 1925). It is not clear whether the posterior pit line arises from a separate placode primitively: Pehrson (1949) indicates that the bulk of the line arises from a separate placode in *Protopterus* (Fig. 3.2), but Hammarberg (1937) did not detect a separate placode in *Lepisosteus*. In any case, the posterior pit line and supratemporal line are innervated by branches of the supratemporal ramus of the posterior lateral line nerve (supratemporal vagal ramus of Norris 1925). Thus it is difficult to discriminate between the posterior pit line and the supratemporal line if the latter is reduced to a pit line or if either line is lost.

An anterior pit line has not been observed in any living chondrichthyan. Although Allis (1923) reported a middle pit line in *Etmopterus* (= *Spinax*) and such a pit line may exist in embryonic *Squalus* (Landacre 1916), it has been reported rarely. A posterior pit line appears to occur commonly in sharks (Johnson 1917; Allis 1923; Holmgren 1942), but pit lines have not been reported in batoids or holocephalans (Garman 1888; Allis 1934).

Anterior, middle, and posterior pit lines occur widely in actinopterygians (Norris 1925) and in dipnoans, where the posterior pit line (Fig. 3.6B) has migrated to a position rostral to that of the middle pit line (Pehrson 1949); it is unclear whether or not they exist in *Latimeria* (Millot and Anthony 1965). They have been reported in salamanders (Allis 1934; Stensiö 1947), but their exact number is uncertain. Although all neuromasts are superficial in salamanders, it is possible to identify large elliptical shaped neuromasts (Fig. 3.8A), comparable to canal neuromasts in other gnathostomes (Coombs et al. 1988), that form lines (Fig. 3.7) comparable to the canal lines in other gnathostomes (Fig. 3.3D). There are also smaller oval neuromasts (Fig. 3.8B,C) that are homologous to pit line neuromasts in other gnathostomes.

Although these pit line neuromasts are superficially similar to ampullary electroreceptors (Fig. 3.8B) in being oval, they are approximately twice the diameter of electroreceptors. They always remain on the surface, they are innervated in a manner identical to that of pit line organs in other gnathostomes, and physiologically they are mechanoreceptors (R.G. Northcutt and H. Bleckmann, unpublished observations).

An anterior pit line occurs in *Ambystoma* (Fig. 3.7) and in *Necturus* (Platt 1896) as the dorsalmost organs of the caudal supraorbital line; the anterior pit line is innervated by the first dorsally arising branch of the superficial ophthalmic ramus (Platt 1896; Coghill 1902) as in actinopterygians (Norris 1925). More caudally a second cluster of pit organs (Fig. 3.7) is innervated by the supratemporal ramus of the middle lateral line nerve (Platt 1896) and thus appears to be homologous to the middle pit line in other gnathostomes. Immediately caudal and adjacent to the middle pit line in *Ambystoma* (Fig. 3.7), there is an additional cluster of pit organs that, in turn, abuts a short row of obliquely oriented larger neuromasts. Stensiö (1947) did not recognize a posterior pit line in *Necturus*, but Platt (1896) noted that lateral line organs immediately caudal to those of the middle pit line are innervated by the supratemporal branch of the posterior lateral line nerve. It is therefore possible that the obliquely oriented neuromasts in *Ambystoma* are homologous to the neuromasts of the supratemporal line and that the pit organs immediately rostral in *Ambystoma* are homologous to the organs of the posterior pit line in other gnathostomes.

2.1.3 Trunk Lines

A cladistic analysis of the number and position of trunk lines in gnathostomes (Table 3.1) suggests that a dorsal, main, and ventral trunk line (Fig. 3.3D) existed in the earliest gnathostomes. In living gnathostomes the main trunk line begins immediately caudal to the orbit, courses around the otic capsule to continue onto the trunk, and usually reaches the tip of the caudal fin. Embryological evidence (Fig. 3.2) indicates that the main trunk line comprises three divisions: the otic and postotic segments as well as the trunk proper. In sharks (Ruud 1920; Holmgren 1940), lungfishes (Pehrson 1949), and salamanders (Platt 1896; Stone 1922), at least three different placodes contribute neuromasts that form the main trunk line. A placode (t1) immediately caudal to the supra-infraorbital placode forms the otic segment of the main trunk line and consists of one or more neuromasts that are innervated by an otic ramus of the anterior lateral line nerve. More caudally a second placode (t2) forms the postotic segment of the main trunk line and consists of one or more neuromasts that are innervated by the middle lateral line nerve. The otic and postotic segments are collectively referred to as the temporal portion of the main trunk lateral line (Pehrson 1949; Northcutt 1986b). The supratemporal line or commissure, the dorsal trunk line, and the main trunk line (Fig. 3.3D) arise from a third placode, whose neuromasts are innervated by the posterior lateral line nerve. A ventral trunk line (Fig. 3.3D) arises from a separate placode located caudal to the hyoid placode (Fig. 3.2), and its neuromasts are also innervated by the posterior lateral line nerve.

The main trunk line probably formed a groove and/or canal in the earliest gnathostomes as it does in most living gnathostomes, with the exception of some chondrichthyans such as the holocephalans (Garman 1888) and *Chlamydoselachus* (Allis 1923), lepidosirenid lungfishes (Holmgren 1942), and amphibians (Kingsbury 1895; Lannoo 1987a,b). Rostral segments of both the dorsal and ventral trunk lines may have existed as grooves in the earliest gnathostomes, as some placoderms (Miles and Westoll 1968) exhibit this pattern, but these lines were probably represented as pit lines along most of their length as they are in living gnathostomes (Table 3.1).

A dorsal pit line has not been observed in acanthodians (Watson 1937), and it is not known whether *Latimeria* possesses any pit lines (Millot and Anthony 1965). The distribution of a ventral trunk line among gnathostomes is more variable. Placoderms possessed a transversely oriented groove or pit line (Fig. 3.4A) immediately rostral to the pectoral fins. This line may have been homologous to the short ventral trunk line that exists in some sharks and batoids (Disler 1977), or it may have been homologous to the gular line (see Section 2.1.4) in other gnathostomes. The ventral trunk line of elasmobranchs (the isolated pit lines adjacent to the yolk sac of Johnson 1917; the ventral field of Disler 1977) consists of a short transverse or obliquely running pit line located approximately at the level of the last gill slit and innervated by a ventral ramus of the posterior lateral line nerve. If the similarly located pit line in placoderms is homologous to this short, transversely oriented line in elasmobranchs, the elongation of this line in acanthodians (Fig. 3.5), dipnoans (Fig. 3.6), and salamanders, as well as the loss of a ventral line in actinopterygians, is an apomorphic condition.

Some teleosts do possess a ventrally located trunk line, as well as extensions and disruptions of the dorsal and main trunk lines (Coombs et al. 1988); however, as these characters do not occur in other actinopterygians (Allis 1889; Norris 1925; Jarvik 1947; Song 1984), they must be interpreted as teleost autapomorphies.

2.1.4 Cheek and Lower Jaw Lines

The postorbital (cheek) region of the head exhibits considerable variation in the number and distribution of lateral lines, and much of the literature (Allis 1934; Holmgren 1942; Stensiö 1947; Holmgren and Pehrson 1949; Coombs et al. 1988) comprises attempts to recognize the homologues of the various cheek lines among vertebrates. An analysis of this literature reveals that two alternate morphotypes have been proposed for gnathostomes: a morphotype based on the distribution and variation of the cheek lines in adults of various taxa (Allis 1934; Holmgren 1942; Coombs et al. 1988), and a morphotype based on the ontogeny of the cheek lines (Stensiö 1947; Holmgren and Pehrson 1949), primarily in a salamander, *Necturus* (Platt 1896), the spiny dogfish, *Squalus* (Holmgren 1940), and the lungfishes *Neoceratodus* and *Protopterus* (Pehrson 1949; Holmgren and Pehrson 1949).

Allis (1934) was the first to review the earlier literature on the diversity of cheek lines in gnathostomes. Based on his own earlier work (Allis 1923) on the frilled shark, *Chlamydoselachus*, which he viewed as a primitive species little changed from its Devonian ancestors, he proposed that the cheek region (Fig. 3.3A) consisted of a horizontal line (hyomandibular line of Norris and Hughes 1920) divided into an anterior portion (the angular line of Garman 1888) and a posterior portion (the jugular or jugal line of Garman 1888) by a ventrally coursing oral line. Allis (1934) termed the entire horizontal line the jugal line and the ventrally directed oral line the oralomandibular line. In *Chlamydoselachus* both lines usually exist as deep grooves, but in other adult sharks that have been described, the horizontal line forms a canal, whereas frequently only a rostral remnant of the oral line occurs, termed the "mandibular" canal (Norris and Hughes 1920). In most sharks the horizontal cheek line appears to end blindly rather than connecting to the preoperculomandibular line as in *Chlaymdoselachus* (Merritt Hawkes 1906). However, Allis (1934) concluded that ancestrally the horizontal line probably connected to the preoperculomandibular line and that the absence of this connection in most living sharks is a derived condition.

In *Chlamydoselachus* the preoperculomandibular line forms a deep groove that is divided into dorsal and ventral segments by a short, caudally directed groove (E in Fig. 3.3A) that houses a single neuromast (Allis 1923). In many sharks this line frequently extends to the edge of the first gill slit or even more caudally onto the branchial region (Garman 1888). Allis did not name this line, but it was subsequently termed a secondary posterior branch of the supramaxillary line by Stensiö (1947) and the opercular line by Holmgren and Pehrson (1949). This line will be termed the epibranchial line (E) in this review to distinguish it from a more dorsally occurring line in dipnoans, termed the spiracular or suprabranchial line, and from cheek lines in teleosts, termed the opercular lines (Coombs et al. 1988).

Unlike in *Chlamydoselachus*, in most sharks the preoperculomandibular line does not exist as a groove but as a pit line, or it may even be absent (Garman 1888; Allis 1923, 1934; Norris and Hughes 1920). Furthermore, the preoperculomandibular line does not appear to connect dorsally with the postotic segment of the temporal line but rather ends ventral to the spiracular cleft. However, Allis (1934) assumed that the preoperculomandibular line formed such a connection dorsally in primitive gnathostomes and only secondarily formed a connection with the infraorbital line in relation to changes in the articulation of the hyomandibular apparatus.

In 1942 Holmgren discovered a new pit line on the cheek in late embryonic stages of the frilled shark, *Chlamydoselachus*; he termed this pit line the middle cheek line. This new line could be traced rostrally from the dorsal end of the preoperculomandibular groove, then ventrally down the cheek to connect with the horizontal line immediately rostral to the point where the oral line issues from the horizontal line. Holmgren reviewed White's (1935) work on the fossil acanthodians (Fig. 3.5) and concluded that these taxa also possessed a middle cheek line that connected dorsally to the temporal line in the primitive condition (Figs. 3.3D, 3.5B).

Holmgren generated a new morphotype for the cheek line of gnathostomes, based on these observations, and argued that the earliest gnathostomes possessed two vertical cheek lines: an anterior line that arose dorsally from the otic line and coursed across the cheek as a middle cheek line, and whose ventral continuation below the horizontal cheek line was the oral line; and a posterior line that arose dorsally from the post-otic line and coursed across the cheek as the preoperculomandibular line (Fig. 3.3D). Subsequently, a cheek line termed the postsuborbital groove (Fig. 3.4A) in some placoderms (Ørvig 1960; Miles and Westoll 1968) and an unnamed pit line (Westoll 1963) in the Devonian lungfish *Dipterus* have been described and represent potential homologues of a middle cheek line (Figs. 3.4B, 3.6A).

In 1947 Stensiö, in a monographic work, reviewed the embryology and variation of the cheek lines in gnathostomes and concluded that the cheek region in the earliest gnathostomes did not possess vertical and horizontal cheek lines, as Holmgren (1942) claimed, but consisted of a *single* obliquely running line (supramaxillary line) that began rostrally at the infraorbital line and ended caudally at the preoperculomandibular line (Fig. 3.3B).

Stensiö realized that data regarding the topography and innervation of the cheek lines were of limited value in evaluating homologies among cheek lines because of the extensive number and variable position of these lines among taxa and because all cheek lines are innervated by a single branch (ramus mandibularis externus) of the hyomandibular trunk (Platt 1896; Allis 1897; Norris and Hughes 1920; Norris 1925). In reviewing the developmental studies of the cheek lines in *Ambystoma* (Stone 1922), *Necturus* (Platt 1896), and *Squalus* (Holmgren 1940), Stensiö concluded that all cheek lines in these taxa arise from a *single* hyoid placode (Fig. 3.9). Thus it was impossible to homologize cheek lines among different taxa based on differences in their placodal origins. However, he believed that it was possible to discern a common pattern of development in the hyoid placode of the dorsolateral series in *Necturus* and *Squalus*. In both taxa the hyoid placode (Fig. 3.9) first appears as an elliptical thickening in the ectoderm and subsequently alters its configuration to form an inverted-V-shaped structure. The ventral half of the anterior limb of the inverted V bifurcates to form a dorsally directed angular line (anterior division of the supramaxillary line of Stensiö) and a ventrally directed oral line. The posterior limb of the inverted V appears to elongate to form the preoperculomandibular line.

In *Necturus*, Stensiö believed that the dorsal half of the anterior limb rapidly elongates to form a horizontal segment (middle division of the supramaxillary line of Stensiö) and a vertical segment (posterior division of the supramaxillary line of Stensiö). He failed to identify a posterior segment in the development of the hyoid placode in *Squalus* and concluded that the developmental pattern in *Necturus* must represent the primitive developmental pattern and that in *Squalus* the dorsal portion of the posterior limb of the hyoid placode includes the anlagen of both the posterior segment of the supramaxillary line and the dorsal segment of the preoperculomandibular line. Thus Stensiö generated a new morphotype of the lateral lines of the cheek in gnathostomes by assuming that the developmental pattern he discerned in *Necturus* represented a primitive developmental pattern for living gnathostomes and must also have represented the *adult* condition in ancestral gnathostomes. He concluded that ancestral gnathostomes possessed a single obliquely directed line (Fig. 3.3B) whose anterior segment is recognizable as the angular line in living taxa and whose middle and/or posterior segments represent the "middle cheek line" of Holmgren (1942). This interpretation required Stensiö to argue that the jugal line observed in sharks and amphibians either represented a secondary extension of the anterior segment of the supramaxillary line (sharks) or was homologous to the middle segment of the supramaxillary line (anurans). In accounting for the variation of cheek lines in other living vertebrates, as well as in extinct radiations, it was necessary for Stensiö to propose numerous changes in the position of portions of his supramaxillary line, changes for which there is little or no fossil evidence.

Holmgren and Pehrson (1949), in reexamining Holmgren's (1940) reconstructions of the development of *Squalus*, on which Stensiö (1947) based part of his interpretation of a supramaxillary line, concluded that a caudally directed protuberance on the posterior limb of the hyoid placode in *Squalus* (Fig. 3.9) represented a vestige of the suprabranchial line in sharks rather than the anlage

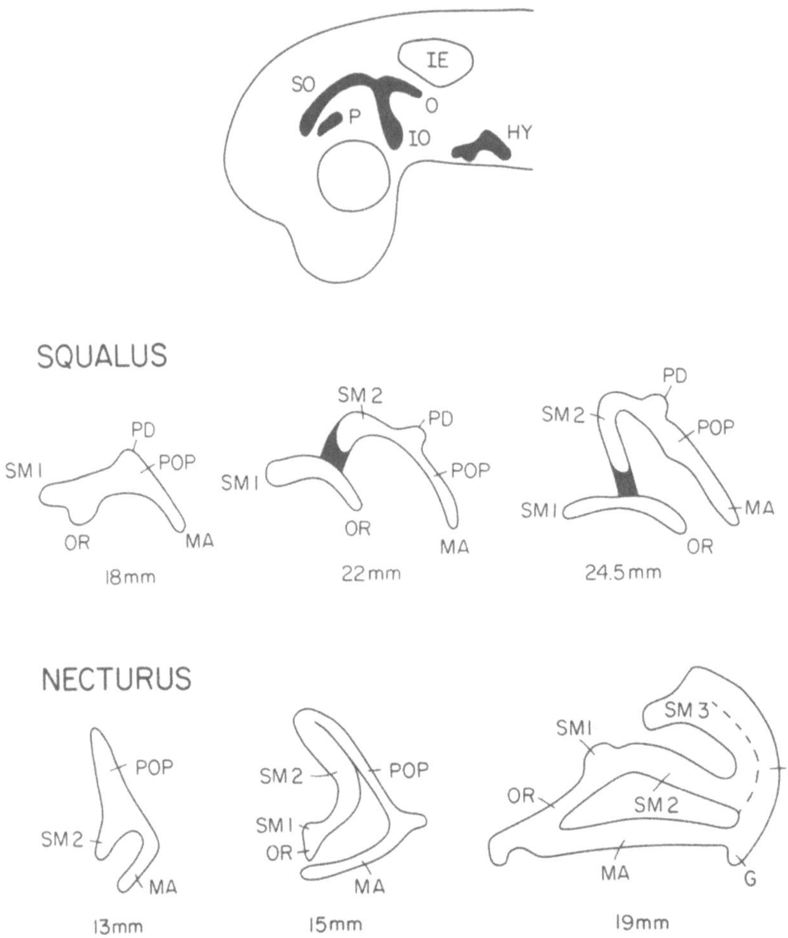

FIGURE 3.9. Comparable stages (expressed as embryonic body length) in the differentiation of the dorsolateral placode in the hyoid region of *Squalus* (Holmgren 1940) and *Necturus* (Platt, 1896, as interpreted by Stensiö 1947) that gives rise to cheek lateral lines. Solid black segments of hyoid placode indicated in *Squalus* are segments reinterpreted by Holmgren and Pehrson (1949) as regions that do not give rise to lateral lines. Outline at top of figure is a lateral view of the head of an 18-mm embryo of *Squalus* illustrating the position of the hyoid placode relative to other major features of the head. Dashed line in hyoid placode of 19-mm stage *Necturus* indicates a fold in the "supramaxillary" line as interpreted by Stensiö (1947). HY, hyoid placode; IE, inner ear; IO, anlage of infraorbital line; MA, anlage of mandibular line; O, anlage of otic line; OR, anlage of oral line; P, anlage of profundus line of Holmgren; PD, dorsal process of hyoid placode; POP, anlage of preopercular line; SM1, anlage of anterior segment of supramaxillary line of Stensiö; SM2, middle segment of supramaxillary line; SM3, posterior segment of supramaxillary line; SO anlage of supraorbital line.

of the dorsal segment of the preoperculomandibular line as suggested by Stensiö (1947). Therefore, they concluded that only ray-finned fishes possess a preopercular line and that the dorsal segment of Stensiö's preopercular line is, in fact, the dorsal portion of the anterior limb (posterior division of Stensiö's supramaxillary line) of the hyoid placode. Based on their reassessment of Holmgren's earlier developmental study of *Squalus* and Pehrson's developmental studies of *Amia* (1922) and lungfishes (1949), they concluded that the earliest gnathostomes possessed a single, obliquely directed cheek line (Fig. 3.3C), which they termed the jugal line, divided into anterior, middle, and posterior segments. They claimed that the anterior segment corresponded to Stensiö's anterior supramaxillary line and thus to the angular line in living vertebrates, that the middle segment corresponded to Stensiö's middle supramaxillary line or to the "middle cheek line" of Holmgren (1942), and that the posterior segment corresponded to Stensiö's "preopercular" line in anamniotes other than ray-finned fishes. Thus they concluded that all living anamniotes exhibit a jugomandibular line, whereas only ray-finned fishes possess a new proliferation center giving rise to a preoperculomandibular line. Like Stensiö (1947), Holmgren and Pehrson interpreted the jugal line (Allis 1934) of sharks as a secondary extension of the anterior division of their jugal (supramaxillary) line.

Equally important, they had to argue that the caudal part of the middle segment (SM2) of Stensiö's supramaxillary line had fused with Stensiö's preopercular line (posterior, J3, segment of the jugal line of Holmgren and Pehrson, 1949) in salamanders and that no true preopercular line exists in placoderms or acanthodians. The postmarginal line in placoderms (Fig. 3.4A) that was reported to be homologous to the preopercular line in living anamniotes (Stensiö 1925; Holmgren 1942) was reinterpreted as a uniquely derived (autapomorphic) line, termed the scapular line, and the middle cheek and preopercular lines in the acanthodians (Fig. 3.5B) were reinterpreted as the middle (J2) and posterior (J3) segments of a jugal line whose dorsal portions had secondarily fused with the temporal segment of the main trunk line.

The morphotypes of cheek lateral lines constructed by Stensiö (1947) and Holmgren and Pehrson (1949) are based ultimately on these authors' interpretations of the embryology of cheek lines in salamanders. They must assume that the embryology and distribution of these lines in adult salamanders represents the plesiomorphic condition for gnathostomes or that a derived embryology, ultimately responsible for a derived pattern of lateral line distribution in adult salamanders, can be used to infer an ancestral adult condition.

A reexamination of the distribution (Fig. 3.7) and development (Fig. 3.8) of the cheek lines in the axolotl, *Ambystoma mexicanum*, reveals several points. A horizontal line of neuromasts runs across the cheek, interconnecting the infraorbital and mandibular lines and resulting in a pattern that could be interpreted as angular and jugal segments divided by an oral line (Fig. 3.7). A caudally located vertical line, which has been interpreted as a preopercular line, consists of anterior and posterior neuromast series with different orientations, consistent with the interpretation that the jugal line continues caudally as the anterior series of the more caudal vertical line. However, earlier studies failed to note an additional vertical line of pit organs (Figs. 3.7, 3.8C), here termed the middle cheek pit line. In axolotls, all neuromast lines consist of clusters or stitches (Lannoo 1985) that arise embryologically with the formation of a single line of neuromast primordia. These first neuromasts (primary neuromasts) give rise subsequently to the other neuromasts (secondary neuromasts) of the stitch. Immediately following the formation of the primary neuromasts (Fig. 3.8D), the cheek region in axolotls exhibits one horizontal row of neuromasts and three vertical rows of neuromasts. At this and earlier stages there is no evidence of an elongated dorsal segment of an anterior limb of the hyoid placode as drawn by Stensiö (1947). The hyoid placode has the shape of an inverted Y, rather than an inverted V, and the body of the inverted Y is much thicker than the anterior and posterior limbs. It is the body of the Y that gives rise to the three rows of vertically oriented neuromasts; the anterior limb gives rise to the horizontal and oral lines, and the posterior limb gives rise to the mandibular line. If the middle cheek pit line in axolotls is a plesiomorphic trait of salamanders, the development of the hyoid placode as reported by Stensiö (1947) and Holmgren and Pehrson (1949) could be reinterpreted as evidence that yet another cheek line existed in ancestral gnathostomes or that the supramaxillary line of Stensiö, or the jugal line of Holmgren and Pehrson, consisted of an additional segment with two or more secondary connections with adjacent lines. This type of speculation appears to be limited only by the imagination of an observer and his or her ability to juggle complex geometries. A conservative interpretation of the development of the hyoid placode would suggest that the dorsal half of the placode may give rise to more than one line, with the anteroventral and posteroventral segments giving rise to horizontal-oral and mandibular lines, respectively. Given the wildly varying interpretations that can be partly supported by the embryology and distribution of cheek lines, the only valid method for reconstructing a morphotype for the earliest gnathostomes is an out-group analysis of the distribution of cheek lines among living and extinct gnathostomes.

An out-group analysis of the number and position of cheek lines in gnathostome radiations (Table 3.1) suggests that the morphotype reconstructed in Figure 3.3D most accurately reflects the cheek lateral lines in ancestral gnathostomes. There is evidence for a horizontal cheek line in some members of all living and extinct gnathostome radiations.

The position of the supraoral line (Fig. 3.4A) in placoderms is consistent with the position of an angular segment of the horizontal cheek line (Fig. 3.4B). However, there is no trace of lines preserved on the lower cheek region in placoderms, and it is also possible to interpret the supraoral line as a homologue of the anterior segment of a jugal or supramaxillary line (Fig. 3.4C,D). If the supraoral line is interpreted as the anterior segment of a supramaxillary line, one must assume that the posterior segment of the supramaxillary line was present as a pit line in placoderms, for which there is no evidence, or that it was lost (autapomorphy) in placoderms (Fig. 3.4C). Interpreting the supraoral line as a jugal line also requires that the middle segment of this line was present as a pit line, or was lost and that the postmarginal line is not homologous to the preopercular line of other gnathostomes but is an autapomorphy (scapular line) in placoderms.

Sharks possess a distinct horizontal cheek line divided into angular and jugal segments (Garman 1888; Allis 1923). However, the condition in batoids and ratfishes is uncertain, owing to the complexity of the poorly studied lateral lines in these taxa.

The course of a presumed horizontal cheek line in acanthodians is consistent with the interpretation of a complete horizontal line in *Euthacanthus* (Fig. 3.5A) but requires that a middle cheek line was present as a pit line or was lost. The position of the horizontal line in *Acanthodes* (Fig. 3.5B) is consistent with the interpretation that a complete horizontal line was present and the oral line lost or that the oral line secondarily connected to the mandibular line with reduction or loss of the jugal portion of the horizontal line. Interpretation of the horizontal line in *Euthacanthus* as an anterior segment of a supramaxillary line (Stensiö 1947) requires one to interpret the posterior part of the horizontal line as part of the oral line, to postulate an oblique middle segment (SM2) interconnecting the horizontal and preopercular lines, and to assume the loss of the posterior segment (SM3). Interpreting the horizontal line of *Acanthodes* as an anterior segment of a jugal line (Holmgren and Pehrson 1949) requires one to interpret the middle cheek line and preopercular lines as homologues of the middle and posterior segments of a jugal line, segments that have formed secondary dorsal connections with the main trunk line with loss or fusion of parts of the middle and posterior segments of the jugal line.

Actinopterygians commonly possess a horizontal pit line (Allis 1889; Norris 1925; Jarvik 1947; Song 1984) that runs across the cheek between the infraorbital and preopercular lines, consistent with its being homologous with the horizontal cheek line in other gnathostomes. However, Stensiö (1947) argued that this pit line in *Amia* is homologous to his anterior segment of the supramaxillary line, the middle and posterior segments being severely reduced or lost, whereas Holmgren and Pehrson (1949) argued that the horizontal pit line in actinopterygians is a neomorph, the anterior and posterior segments of the jugal line are lost, and the middle segment of the jugal line is homologous with the vertical pit line.

The cheek region in the living actinistian, *Latimeria*, exhibits a preoperculomandibular canal that does not connect dorsally with the temporal segment of the main trunk canal but ends dorsally as a short, rostrally directed canal close to the ventral border of the postorbital segment of the infraorbital canal. This rostrally directed segment of the preopercular canal could be interpreted as a horizontal cheek line, but this segment, as well as the vertical extension of the "preopercular" canal, is reportedly innervated by the buccal ramus (Millot and Anthony 1965) rather than the hyomandibular ramus as in other bony fishes (Norris 1925). Either the innervation of these canals has been incorrectly reported in *Latimeria* or they represent actinistian autapomorphies.

The cheek lines of the living dipnoans are particularly complex, and most analyses have suggested that secondary oral and mandibular lines occur as dipnoan autapomorphies (Holmgren 1942; Stensiö 1947; Holmgren and Pehrson 1949). Lepidosirenid lungfishes (Fig. 3.6B) possess a short, dorsally directed groove that arises from the mandibular groove, and *Neoceratodus* has a pit line in the comparable position as well as an additional dorsally and obliquely directed pit line termed the secondary mandibular line. These secondary lines have been interpreted as autapomorphic subdivisions of the middle jugal line (Holmgren and Pehrson 1949) or as an autapomorphic division (secondary mandibular line) of the original oral line and the homologue of the middle division of the supramaxillary line (Stensiö 1947). However, the Devonian lungfish *Dipterus* exhibits a pattern of cheek lines that suggests a simpler explanation. Reconstructions of the cheek region in *Dipterus* (Fig. 3.6A) indicate that a well-developed preoperculomandibular canal existed in this taxon and that vertical and horizontal pit lines were closely associated with the anterior margin of the preoperculomandibular canal. A short canal connected with the infraorbital canal was interpreted by Westoll (1963) as the "oralomandibular line." The simplest interpretation of these data is that *Dipterus* possessed both horizontal and middle cheek lines, that the secondary oral pit line or groove in living lungfishes is homologous to the jugal segment of the horizontal line in other gnathostomes, and that the secondary mandibular pit line in *Neoceratodus* is homologous to the middle cheek line in other gnathostomes and has been lost in lepidosirenid lungfishes.

Salamanders (Figs. 3.7, 3.8D) and larval anurans (Holmgren and Pehrson 1949), as discussed above, exhibit a horizontal pit line with angular and jugal segments that is most obviously interpreted as the homologue of the horizontal line in other gnathostomes.

There is evidence for a middle cheek line in all gnathostome radiations except actinopterygians and actinistians (Table 3.1). In terms of their position and interconnections with adjacent lines, the postsuborbital groove (Fig. 3.4A) in placoderms, the middle cheek pit line in sharks (Holmgren 1942), the middle cheek groove in *Acanthodes*, the unnamed vertical cheek pit line in *Dipterus* (Westoll 1963), the "secondary mandibular" pit line in *Neoceratodus* (Holmgren and Pehrson 1949), and the middle cheek pit line in *Ambystoma* (Fig. 3.7) are consistent with the hypothesis that ancestral gnathostomes possessed such a line. Given the position and/or termination of this line in placoderms and acanthodians, it is probable that this line existed as a canal or open groove in the earliest gnathostomes, that it connected dorsally to the otic segment of the main trunk line (Fig. 3.3D) or far dorsally onto the infraorbital line (Figs. 3.4B, 3.5B), that the middle cheek canal or groove was secondarily reduced to a pit line or lost in the remaining vertebrate radiations, and that the dorsal terminus of the middle cheek canal secondarily abutted the preopercular line in sharks and dipnoans. Given the presence of this line in lungfishes and amphibians, it is possible that a middle cheek line is also present in *Latimeria* and that it was not observed and reported (Millot and Anthony 1965) because of the poorly preserved state of the epidermis in the specimen. Alternate interpretations of the homologues (segments of a jugal or supramaxillary line) of the gnathostome lines, interpreted here as a middle cheek line, and their attendant problems have been discussed earlier in this section.

Although no trace of lateral line canals or grooves has been detected on the lower jaw in placoderms (Stensiö 1925; Ørvig 1960; Miles and Westoll 1968), the lower jaw of the earliest gnathostomes was probably traversed by oral and mandibular lines (Fig. 3.3D) that arose dorsally from the horizontal cheek line and may have connected rostrally at the tip of the jaw, as such lines occur in *Chlamydoselachus* (Allis 1923), lungfishes (Fig. 3.6), and salamanders (Fig. 3.7), and vestiges of an oral line occur in all gnathostome radiations with the possible exception of actinistians. An extensive oral groove occurs in ratfishes and some sharks (Garman 1888; Cole 1896; Allis 1923) but is reduced rostrally to a short canal in *Mustelus* and *Squalus* (Allis 1934), may have been reduced or lost in some acanthodians (Fig. 3.5B), and may never have occurred as a continuous line in actinopterygians.

In many ray-finned fishes, a discontinuous series of pit lines, termed the vertical, quadratojugal, and mandibular lines (Fig. 3.3D), occur in a position comparable to that of the oral lines in other gnathostomes and have been interpreted as the homologues of segments of the oral lines (Allis 1934; Holmgren 1942; Stensiö 1947). However, as noted above, Holmgren and Pehrson (1949) considered the vertical pit line in lungfishes, as well as that in ray-finned fishes, to be an autapomorphy derived from the ventral portion of an ancestral middle segment of their jugal line, and the quadratojugal pit line as also derived in part from the middle segment and in part from a second autapomorphy. Coombs et al. (1988) list the vertical (their median mandibular line) and quadratojugal (their ventral mandibular line) pit lines of ray-finned fishes as replacement lines for canals lost in evolution and quote the opinions of Holmgren and Pehrson (1949) but do not comment on whether they believe this interpretation is correct.

Holmgren and Stensiö's interpretation of these pit lines in ray-finned fishes requires that only amphibians, among all bony fishes, are considered to retain a plesiomorphic condition of the middle cheek lines. Viewed in the context of a cladistic analysis, this interpretation of homologies leads to certain conclusions: the horizontal and oral lines in amphibians are patristically homologous to the horizontal and oral lines in ancestral gnathostomes, but they represent a case of parallel homoplasy as regards the horizontal and oral lines in all other living gnathostomes, as the condition in bony fishes would be considered a derived pattern, with the amphibian pattern arising owing to reversal. If this were the case, there would be an additional complication in using the embryology of the hyoid placode in amphibians to establish a morphotype (Stensiö 1947; Holmgren and Pehrson 1949). The simplest explanation at present, based on topography and innervation, as well as the configuration of the hyoid placode at an early stage of development in *Lepisosteus* (Hammarberg 1937), is that the vertical, quadratojugal, and mandibular pit lines of ray-finned fishes are homologous to segments of the oral line in other gnathostome radiations.

The phylogeny of the mandibular line in gnathostomes is particularly conservative (Table 3.1). Although its presence has not been established in placoderms (Miles and Westoll 1968), it occurs as a distinct groove or canal in other gnathostome radiations, except for elasmobranchs, where it occurs as a pit line in some species (Johnson 1917; Norris and Hughes 1920) and has been lost in many others (Garman 1888). However, Garman's claim that many sharks possess a well-developed oral canal, with loss of the mandibular line, should be reexamined owing to his failure to detect the preoperculomandibular pit line in *Squalus* and *Mustelus* (Johnson 1917; Allis 1934).

A medially directed gular line may have issued from the caudal third of the mandibular line in the earliest gnathostomes (Fig. 3.3D), as a gular pit line occurs in all living gnathostome radiations except cartilaginous fishes (Table 3.1). A groove or trace of a pit line occurs in a comparable position in placoderms (Fig. 3.4A) and acanthodians (Fig. 3.5B) but has been termed a ventral line in placoderms and could be interpreted as a ventral trunk line. In *Coccosteus* this ventral line is located rostral to the pectoral fin and, more importantly, fuses across the midline with its counterpart to form a commissure at right angles to the mandibular lines (Miles and Westoll 1968), rather than passing caudally as a continuation of the mandibular line (Figs. 3.4A, 3.6B), as does a ventral trunk line in cartilaginous fishes (Disler 1977). In this context it is not clear whether the short, caudally directed canal issuing from the mandibular canal in *Latimeria* is homologous to the gular or ventral trunk line in other gnathostomes. Finally, it is not clear whether the gular line ancestrally arose from a separate dorsolateral placode, as the gular line in amphibians appears to arise from the hyoid placode (Platt 1896), whereas the gular line in lungfishes (Fig. 3.2) arises from a separate placode (Pehrson 1949).

The dorsal continuation of the horizontal line is termed a preopercular line, and an out-group analysis suggests that ancestrally it connected dorsally with the temporal segment of the main trunk canal (Fig. 3.3D), as a groove or canal (postmarginal line) exists in this position in placoderms (Fig. 3.4A), in some acanthodians (Fig. 3.5B), and in actinopyterygians. However, the preopercular line ends caudally at the level of the spiracular cleft in some sharks (Johnson 1917; Allis 1934), curves rostrally to end immediately caudal to the postorbital segment of the infraorbital line in some acanthodians (Fig. 3.5A), or connects to the infraorbital line in sarcopterygians (Fig. 3.7). The highly variable dorsal position of the preopercular line in different gnathostomes, and the relatively late development of the dorsal segment of this line in ray-finned fishes, led Holmgren and Pehrson (1949) to conclude that the preopercular line in gnathostomes other than ray-finned fishes was actually a posterior segment of the jugal line and that the preopercular line in ray-finned fishes was a newly developed line (autapomorphy) for ray-finned fishes. At present there is no embryological evidence that a separate placode, or a separate proliferation zone of the hyoid placode, gives rise to the preopercular line in ray-finned fishes. Furthermore, a ventrodorsal maturation gradient is consistent with the migration and differentiation of a single, ventrally located hyoid placode. This is particularly true as similar caudorostral gradients are observed in the oral and mandibular lines in amphibians.

Teleosts, however, do possess a variable number of pit lines caudal to the preoperculomandibular canal; these pit lines are termed opercular lines (Lekander 1949; Coombs et al. 1988), and they appear to be autapomorphic for teleosts, as there is no trace of them in other gnathostome radiations, with the possible exception of the pleural or scapular lines in ratfishes and batoids (Garman 1888; Holmgren 1942). In this context, it should be noted that the scapular lines in the ratfish *Callorhynchus* was described by Holmgren (1942) as arising from the postorbital segment of the infraorbital line and homologized to a preopercular line, then subsequently used as evidence (Holmgren and Pehrson 1949) that the postmarginal line in placoderms—which does not arise from the infraorbital line but far caudally from the temporal segment of the main trunk line—was homologous to the scapular line in *Callorhynchus* and thus could not be a preopercular line!

Finally, two additional longitudinal (epibranchial and spiracular) lines may have existed in ancestral gnathostomes (Fig. 3.3D). The only evidence for the existence of an epibranchial line is the caudally directed epibranchial line in sharks. Because of the uncertain existence of a preoperculomandibular pit line in most sharks, it is unclear whether this line is an expanded jugal line or an extension of the preopercular line. In either case, this unique condition is probably an autapomorphy for elasmobranchs, rather than a sympleisomorphy for gnathostomes.

A spiracular neuromastlike organ occurs in the spiracular pouch of chondrichthyans, nonteleost actinopterygians, and dipnoans and is thus probably a symplesiomorphy for gnathostomes. Pehrson (1949) noted that the spiracular organ in lungfishes arises from a distinct placode (Fig. 3.2) and that in embryos of *Neoceratodus* there is a spiracular line of four neuromasts, of which only the most rostral persists to form the adult spiracular organ. The widespread occurrence of a spiracular organ in gnathostomes (Table 3.1), the occurrence of a transitory spiracular line in some lungfishes, and the occurrence of a comparable longitudinal pit line in lampreys (Fig. 3.10A) suggest that ancestral gnathostomes may have possessed a spiracular line that arose immediately rostral or caudal to the preopercular line, depending on the initial dorsal terminus of the preopercular line.

2.2 The Agnathans

The living (Fig. 3.10) and extinct (Figs. 3.11, 3.12) agnathans (jawless craniates) exhibit considerable variation in the lateral lines, which is not surprising, considering recent comparative studies (Løvtrup 1977; Hardisty 1979; Janvier 1981; Jarvik 1980), most of which suggest that jawless craniates (Fig. 3.1) are not a monophyletic clade but several clades whose relationships are as complex as those of gnathostomes. The living agnathans (Fig. 3.10) appear to exhibit the entire range of lateral line organization seen among craniates. Lampreys (Fig. 3.10A) possess only neuromasts distributed as pit lines (Allis 1934; Holmgren 1942), whereas eptatretid hagfishes (Fig. 3.10B) possess short, transversely and vertically oriented grooves that do not contain neuromasts (Fernholm 1985), and myxinid hagfishes appear to have no lateral lines. These data could be used to support the hypothesis that eptatretid hagfishes exhibit a plesiomorphic condi-

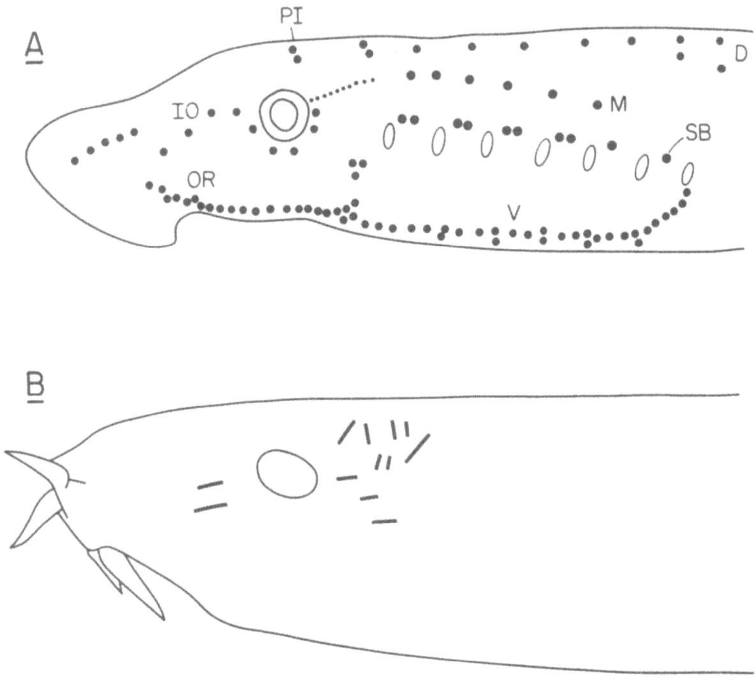

FIGURE 3.10. Distribution of neuromasts (solid circles) in the European freshwater lamprey, *Lampetra fluviatilis* (A), modified from Holmgren (1942), and suspected lateral lines in a Pacific hagfish, *Eptatretus stouti* (B), after Fernholm (1985). D, dorsal trunk line; IO, infraorbital line; M, main trunk line; OR, oral line; PI, pineal line; SB, suprabranchial line; V, ventral head line.

tion in which elongated placodes form only single-celled sensory receptors and that lampreys retain the plesiomorphic pattern of compound sensory organs (neuromasts) as pit lines. However, these data could also be used to support a different hypothesis: that the lateral line system in living agnathans is highly regressive (paedomorphic). Unfortunately, we have no details on the development of the lateral line system in hagfishes and lampreys, and our information regarding the innervation of this system in both taxa is too ambiguous to facilitate testing these hypotheses or hypotheses of lateral line homologies with gnathostomes. For these reasons, topographical comparison of the lateral line system in living and extinct agnathans is the primary approach, and this too is confounded by the fact that the relevant out-groups to hagfishes (noncraniate chordates) do not possess lateral lines.

2.2.1 CEPHALASPIDOMORPHS

The lateral lines on the head and trunk of lampreys occur only as pit lines (Allis 1934; Holmgren 1942). A well-developed pit line (Fig. 3.10A), usually

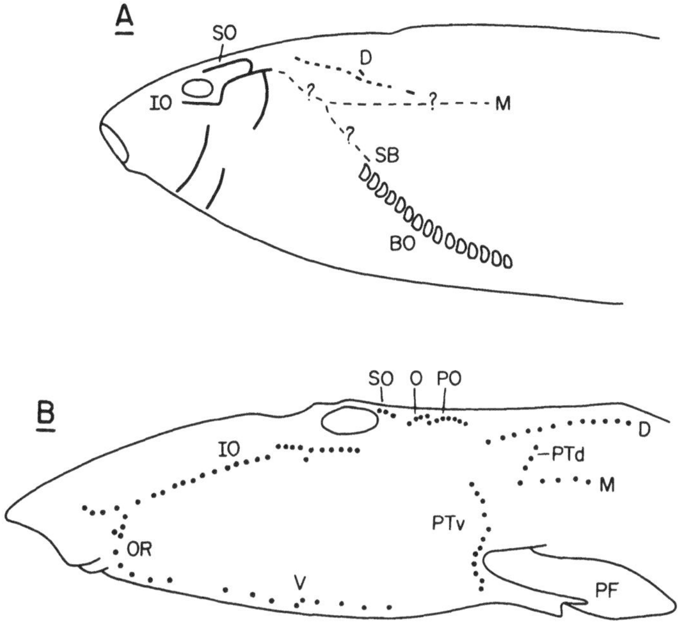

FIGURE 3.11. Distribution of lateral lines in an extinct anaspid (A), modified from Smith (1957), and osteostracan (B), modified from Janvier (1974). BO, branchial opening; D, dorsal trunk line; IO, infraorbital line; O, otic line; OR, oral line; PF, pectoral fin; PO, postotic line; PTd, dorsal segment of posterior transverse zonal line; PTv, ventral segment of posterior transverse zonal line; SO, supraorbital line; V, ventral head line; ?, sulci or deep sutures that may have housed neuromasts.

interpreted as an infraorbital line, begins just caudal to the eye, continues rostrally onto the oral hood, and is believed to be innervated by a buccal ramus (Allis 1934), as in gnathostomes. A second pit line (oral line) arises from the middle segment of this line and courses ventrally and caudally to terminate at approximately the last gill cleft. This ventral line has been divided into two segments on the basis of innervation (Johnston 1905): an oral line, believed to be innervated by the hyomandibular ramus, and a ventral line proper, whose rostral and caudal halves are believed to be innervated by ventral rami of the middle and posterior lateral line nerves, respectively. The position and innervation of the oral line in lampreys suggest that it is a homologue of either the oral or mandibular line in gnathostomes. Although the middle lateral line nerve in gnathostomes is claimed to innervate the gular line (Norris 1925), experimental studies of *Ambystoma* (R.G. Northcutt, unpublished observations) and *Lepisosteus* (Song 1986) demonstrate that the gular line is innervated by the hyomandibular ramus and that the middle lateral line nerve does not possess a ventral ramus. If the rostral segment of the ventral pit line of lampreys is indeed innervated by a ventral ramus of the middle lateral line nerve, there appears to be no homologue of this ventral

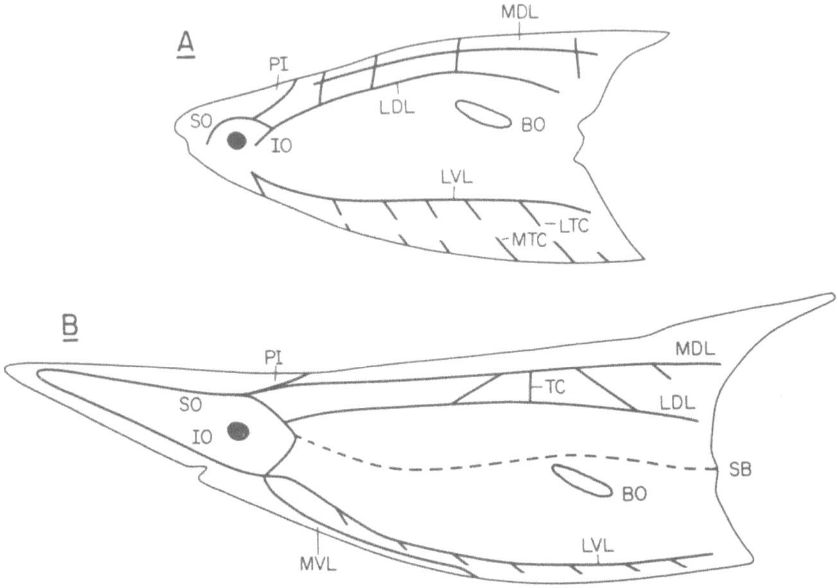

FIGURE 3.12. Distribution of lateral lines in an extinct cyathaspidid (A) and pteraspidid (B) heterostracan, modified from Denison (1964) and White (1935), respectively. BO, branchial opening; IO, infraorbital line; LDL, laterodorsal longitudinal line; LTC, lateral transverse commissure or line; LVL, lateroventral longitudinal line; MDL, mediodorsal longitudinal line; MTC, medial tranverse commissure or line; MVL, medioventral longitudinal line; PI, pineal line; SB, suprabranchial line; SO, supraorbital line; TC, transverse commissure.

segment in gnathostomes. Given the position and innervation of the caudal segment of the ventral line in lampreys, it is probably homologous to the ventral trunk line in gnathostomes, particularly as a comparable pit line extends onto the trunk in some fossil osteostracans (Janvier 1974).

There is considerable confusion regarding the existence of a supraorbital line in lampreys. Johnston (1905) interpreted the anterior half of what has usually been termed the infraorbital line (Allis 1934; Holmgren 1942) as a supraorbital line and claimed that it was innervated by the profundus nerve. A more caudally and dorsally located line, later termed the supraorbital line by Allis (1934) and the pineal line by Holmgren (1942), was believed by Johnston (1905) to be innervated by the superficial ophthalmic ramus of the trigeminal nerve. Mainly on theoretical grounds, Holmgren (1942) concluded that primitively each branchiomeric nerve possessed a lateralis ramus that innervated a distinct lateral line. Thus the profundus nerve possessed a superficial ophthalamic ramus that innervated a profundus line immediately above the eye, a pattern that Holmgren observed in only one of several specimens of *Lampetra* (= *Petromyzon*) *fluviatilis*, a pineal line (Fig. 3.10A), which he believed was innervated by a superficial ophthalmic ramus of the trigeminal nerve, and a supraorbital line, lost in lampreys but, when present, innervated by a superficial ophthalmic ramus of the facial nerve in other vertebrates.

Holmgren's hypothesis that each branchiomeric nerve possesses a lateralis component has been refuted in numerous gnathostomes (McCready and Boord 1976; McCormick 1981; Fritzsch 1981; Puzdrowski 1988), and Ronan and Northcutt (1987) have recently demonstrated experimentally that an anterior lateral line nerve exists in lampreys and consists of superficial ophthalmic, buccal, and hyomandibular rami. There is thus no evidence for lateralis rami existing in the profundus, trigeminal, or facial cranial nerve. Unfortunately, Ronan and Northcutt (1987) did not follow the peripheral rami of the anterior lateral line nerve. If Johnston (1905) is correct regarding the innervation of the anterior segment of the "infraorbital" line in *Petromyzon*, this segment of the line, as well as the pineal line, should be considered homologous to segments of the supraorbital line in gnathostomes, and only the caudal half of the "infraorbital" line in lampreys should be considered the homologue of the infraorbital line in gnathostomes.

Otic and postotic segments of a main trunk line appear to be represented in lampreys by a short pit line running obliquely and caudally from the dorsocaudal margin of the eye; this pit line is believed to be innervated by an otic branch of the anterior lateral line nerve and, possibly, by a branch of the middle lateral line nerve. The ganglia of the lateral line nerves in lampreys are interconnected by a recurrent ramus (Johnston 1905), and recent experiments (Ronan and Northcutt 1987) have demonstrated that the majority of the fibers labeled in the recurrent ramus possess cell bodies located in the ganglion of the anterior lateral line nerve, cell bodies whose peripheral processes innervate trunk electroreceptors. However, there are a small number of frequently labeled recurrent fibers that are believed to innervate neuromasts dorsal to the recurrent ramus. This is consistent with Johnston's claim that the neuromasts of the temporal region are innervated by otic and post-otic rami of the anterior and middle lateral line nerves.

The temporal segment of the main trunk continues caudally in lampreys (Fig. 3.10A) and is innervated by a lateral ramus of the posterior lateral line nerve, but this pit line extends only the length of the branchial region. A more dorsally located line begins slightly caudal to the pineal line (Fig. 3.10A) and continues along the trunk to reach the caudal fin. This more dorsal line is topographically comparable to the dorsal trunk line in gnathostomes, but its innervation is unclear. Johnston (1905) recognized a dorsal ramus arising from the ganglion of the posterior lateral line nerve that appeared to him to innervate the first neuromast of the dorsal line. However, he then indicated that the lateral ramus of the posterior lateral line nerve courses dorsally to run adjacent to the remaining neuromasts of the dorsal pit line rather than running adjacent to and innervating the neuromasts of the pit line interpreted as the main trunk pit line by Holmgren (1942). In fact, Johnston (1905) did not identify a pit line in *Petromyzon* in the position of the pit line identified as the main trunk pit line in *Lampetra* by Holmgren (1942). This line of neuromasts clearly exists in *Ichthyomyzon* (unpublished observations), and its innervation, as well as the distribution and innervation of comparable lines in other genera of lampreys, should be examined to establish possible gnathostome homologues of the dorsal and main trunk lines in lampreys.

Lampreys also exhibit one additional longitudinal head line represented by a series of pit organs located above the gill slits (Fig. 3.10A) and termed the suprabranchial line (Holmgren 1942). The neuromasts of this line appear to be innervated by multiple branches of the ventral ramus of the posterior lateral line nerve that course with the branchiomeric nerves innervating the gill pouches (Alcock 1899). The suprabranchial line in lampreys appears to be homologous to a comparable line observed in some heterostracans (White 1946) and, possibly, to the epibranchial line in sharks (Fig. 3.3D). However, if the epibranchial line in sharks is innervated by the hyomandibular ramus, and the presumed innervation of the suprabranchial line in lampreys is correct, these lines must have arisen independently from different placodes and are thus examples of convergent homoplasy.

The sister group of lampreys appears to be the extinct anaspids (Fig. 3.1). The exoskeleton of anaspids consisted of small plates and scales of reduced thickness, compared to those of other extinct agnathans, with only the inner layer of the corium ossified. Consequently, only traces of the lateral lines are preserved in the form of a few grooves and traces of pit lines (Smith 1957). Lateral lines (Fig. 3.11A) apparently existed as grooves above and below the eye, but if they continued more rostrally on the head, they must have existed as pit lines that left no trace on the plates of the skull. The rostral segment of the groove below the orbit most likely corresponded to the infraorbital line in other vertebrates, but interpreting the caudal segment of the "infraorbital" groove, as well as the groove caudal and dorsal to the orbit, is more problematical.

Smith (1957) homologized the vertical postorbital groove of anaspids to the posterior pit line of gnathostomes on the basis of the presumed position of the otic capsule. In one specimen of the anaspid *Pharyngolepis*, he observed lateral depressions in the cranial roof, between the orbits and the vertical lateral line grooves, which he interpreted as resulting from dermal elements of the skull crushing the otic capsules. Unfortunately, no trace of the otic capsules or other portions of the neurocranium are preserved in known specimens of *Pharyngolepis*. If Smith's interpretation of the postorbital vertical groove in *Pharyngolepis* is correct, then the posterior segment of the groove below and caudal to the orbit would necessarily be homologous to the otic and postotic segments of the main trunk line. The rostrally directed groove above the orbit could then be interpreted as an apomorphic line in anapsids, as a middle cheek line, or as a supraorbital line that connected to the posterior pit line. If Smith is incorrect regarding the position of the otic capsule, and if it was located caudal to the postorbital vertical groove, this groove must be interpreted as a supraorbital line, with the otic and postotic segments located more caudally.

Similar problems plague interpretations of the other cheek lines in *Pharyngolepis*. A ventrally directed cheek groove (Fig. 3.11A) issuing from the horizontally oriented postorbital groove can be homologized to the preopercular line in gnathostomes if the otic capsule was located more caudally, or to the ventral trunk line in gnathostomes if the otic capsule was located more rostrally. One or two additional obliquely oriented cheek lines parallel the most caudal vertical "preopercular" cheek line in different specimens of *Pharyngolepis*. The most rostral line is comparable to the oral pit line in lampreys and, possibly, the medial ventral longitudinal line in other fossil agnathans (Fig. 3.11B, 3.12). The variable, more caudal cheek line in some specimens of *Pharyngolepis* may represent a ventral segment of the "preopercular" line that connects more dorsally with the main trunk line. If this interpretation is correct, then the "preopercular" line in *Pharyngolepis* is homologous to the lateral ventral longitudinal line in heterostracans (Fig. 3.12), and the line interpreted by Smith (1957) as a posterior pit line in anaspids should be interpreted as a supraorbital line.

Smith (1957) noted a series of short grooves running transversely across the scales in *Pharyngolepis*, and he homologized these with the main trunk line in other vertebrates; however, he also noted that the groove formed by the union of the supra- and infraorbital lines may have continued more caudally as a suture or groove between rows of plates, dividing caudally into a dorsal longitudinal suture or groove and a ventral, obliquely running suture or groove that terminates at the level of the first gill slit (Fig. 3.11A). If these disruptions in the plate rows housed lateral lines, the short series of grooves that Smith (1957) identified as a main trunk line should be reinterpreted as a dorsal trunk line, and the more ventral horizontal and obliquely running grooves as main and suprabranchial lines, respectively. Clearly, resolving these various interpretations of the lateral line system of anaspids will require additional information on the position of the otic capsule and the nature of the disruptions in the plate rows of the caudal head region.

The sister group of lampreys and anaspids appears to be the extinct osteostracans (Fig. 3.1). The exoskeleton of osteostracans typically consisted of a dorsally undivided bony head shield, extending a variable distance down the trunk, and a large number of bony plates that covered the ventral surface of the head and extended onto the trunk as a series of scale rows (Stensiö 1927, 1932, 1958; Moy-Thomas and Miles 1971). Not only was the dermis heavily ossified, so was much of the perichondrial connective tissue of the neurocranium, and it is therefore possible to trace the course of the cranial nerves as well as that of the lateral line grooves and pits (Stensiö 1927, 1932; Janvier 1974).

The distribution of the lateral lines in lampreys (Fig. 3.10A) and osteostracans (Fig. 3.11B) is remarkably similar. In both radiations a short pit line occurs dorsal and caudal to the orbit, and a long, slightly disjunct pit line occurs below the orbit. The dorsal pit line (postpineal or pineal line) of osteostracans has been homologized to the supraorbital line of lampreys and interpreted by Janvier (1974) to have been innervated by a superficial ophthalmic ramus of the trigeminal nerve. The entire line ventral to the orbit in ostracoderms has been homologized to the infraorbital line in lampreys and gnathostomes and interpreted by Janvier (1974) to have been innervated by a buccal ramus of the facial nerve. However, given Johnston's claim (1905) that the rostral and caudal segments of the disjunct infraorbital line in *Petromyzon* are innervated by different rami, and the recent experimental observations of *Ichthyomyzon* and *Petromyzon* by Ronan and Northcutt (1987), it is possible that the postpineal line and rostral segment of the "infraorbital" line in osteostracans are homologous to the pineal and rostral "infraorbital" lines in lampreys and to portions of the supraorbital line in gnathostomes.

Many osteostracans, like lampreys, possessed a short, laterally coursing oral pit line (Fig. 3.11B) that arose from the rostral segment of the "infraorbital" line. The osteostracan oral line may have ended on the lateral margin of the shield, or it may have continued ventroposteriorly on the plates of the oralobranchial chamber; in either case, it appears to have been innervated by a hypomandibular ramus (Janvier, 1974). The position of the oral line in osteostracans is comparable to that of portions of the medioventral longitudinal line in heterostracans (Fig. 3.12B).

Two short pit lines occur on the shield roof, adjacent to the inner ear, in osteostracans – a rostral transverse postorbital line and a caudal longitudinal line, the latter interpreted by Janvier (1974) as a cephalic segment of the main trunk line. The transverse postorbital line was interpreted by Janvier (1974) as a possible homologue of the short line of three pit organs observed by Holmgren (1942) in a single specimen of *Lampetra*, or as an osteostracan autapomorphy. It is also possible that this line is homologous to the otic segment of the main trunk line in other vertebrates. If this line is so interpreted, then the more caudal line, interpreted by Janvier (1974) as a cephalic segment of the main trunk line, should be homologized to the postotic segment of the main trunk line in other vertebrates. Alternatively, if the transverse postorbital line is viewed as an osteostracan autapomorphy, the more caudal line should be interpreted as homologous to both the otic and postotic segments of the main trunk line in other vertebrates.

Immediately caudal to the postotic pit line (cephalic trunk segment of Janvier), dorsal and lateral longitudinal pit lines continue onto the trunk. The dorsal longitudinal pit line occurs rarely in known osteostracans, and its caudal extent is uncertain, but its position suggests that it is the osteostracan homologue of the dorsal trunk line in other vertebrates. The more lateral longitudinal pit line has been interpreted as a main trunk line by Janvier (1974), who noted that this line usually lies caudal to a transverse pit line, termed by him the posterior transverse zonal line and frequently divided into dorsal and ventral segments (Fig. 3.11B). Thus the main trunk line of osteostracans apparently consisted of four segments: postotic segment, dorsal division of a posterior transverse zonal segment, trunk line proper, and ventral division of a posterior transverse zonal segment. In some osteostracans, the ventral division of the posterior transverse zonal segment continues caudally to a series of posteriorly directed processes or cornu and is probably an autapomorphy for osteostracans.

In at least one osteostracan, *Tremataspis*, there are traces of two additional longitudinal lines on the ventral surface of the head and trunk: rostrally, a short longitudinal pit line, located medial to the gill slits, parallels the more medially located oral line; more caudally and laterally, a longer longitudinal pit line occurs along most of the length of the abdominal portion of the shield (Robertson 1938; Janvier 1974). These more ventrolateral lines may be homologous to segments of the lateroventral longitudinal lines in heterostracans (Fig. 3.12B), or the rostral line may be apomorphic for osteostracans and the caudal segment comparable to the ventral trunk line in other vertebrates.

A suprabranchial pit line comparable to that in lampreys has not been described in any osteostracan. However, as discussed above, there is some evidence for such a line in anaspids, and a distinct suprabranchial line has been described in one heterostracan (White 1946). If such a line existed in osteostracans, it would be difficult to detect, as the head shields of osteostracans are extremely broad and flat, with the gill pouches opening near the ventrolateral margin of the shield.

2.2.2 Heterostracans

The sister group of cephalaspidomorphs appears to be the extinct heterostracans (Fig. 3.1), whose remains are the earliest undisputed vertebrate fossils. The exoskeleton typically consisted of bony shields, composed of at least dorsal, branchial, and ventral plates, and the trunk was covered by bony scales (Moy-Thomas and Miles 1971). Although the dermal skeleton was heavily ossified, there is no trace of the neurocranium in the fossil heterostracans thus far discovered, and we therefore have no details regarding the nature of the cranial nerves. The lateral line system (Fig. 3.12) was well developed and consisted of canals lying in the middle layer of the dermal exoskeleton and opening to the exterior by pores, as in most gnathostomes (White 1935; Denison, 1964).

A second group of extinct agnathans, the thelodonts, are frequently allied with the heterostracans. Unfortunately, these fishes are poorly understood: the entire body was covered with small denticles, and only traces of the lateral line system have been observed (Gross 1968). The phyletic relationship of heterostracans and thelodonts is uncertain. The two groups are frequently treated as a single clade, termed the pteraspidomorphs, but there are few if any synapomorphic traits to support such a contention.

The basic pattern of the lateral line system on the head of heterostracans is a series of longitudinal canals connected by transverse commissures (Fig. 3.12). Mediodorsal and laterodorsal longitudinal lines (Fig. 3.12) usually occur on the dorsal plate of heterostracans (White 1935; Stensiö 1958; Denison 1964). Commonly one or both canals continue rostrally to connect with the supra- and infraorbital canals, which meet rostrally in pteraspidids (Fig. 3.12B) but not in cyathaspidids (Fig. 3.12A). The cyathaspidids (Fig. 3.12A) are the earliest known heterostracans with more than fragmentary remains, and they appear to retain more plesiomorphic heterostracan traits than do the more derived pteraspidids (Fig. 3.12B); however, at least one cyathaspidid, *Irregulareaspis*, exhibits one of the most derived lateral line systems known in any vertebrate. All the longitudinal and transverse canals in *Irregulareaspis* appear to have formed multiple secondary branches, so that the entire dermal skeleton was honeycombed with lateral line canals (Denison 1964). One of these secondary canals, a short, longitudinal canal, paralleled the supraorbital canal, connecting with it caudally, and has been cited (Denison 1964; Smith 1957) as evidence of a profundus canal (Holmgren 1942); however, this canal does not occur in other cyathaspidids (White 1935), and it is probably an autapomorphic trait of *Irregulareaspis*.

The caudal extent of the mediodorsal and laterodorsal canals in heterostracans is uncertain. White (1935) suspected that both canals continued onto the trunk in *Pteraspis* but was able to trace only the mediodorsal longitudinal canal onto the trunk in a single specimen. Denison (1964) noted that lateral line pores occur on the trunk scales of cyathaspidids but are rare, and that at least one of the dorsal lines continued onto the trunk. Unfortunately, he did not specify which of the lines did so. Given the position of these lines, it is probable that the mediodorsal and laterodorsal lines in heterostracans are homologous to the dorsal and main trunk lines, respectively, in gnathostomes. If White correctly observed that only the mediodorsal longitudinal line continued onto the trunk, heterostracans and lampreys may be identical regarding the caudal extent of these lines, and this pattern may be plesiomorphic for vertebrates.

Heterostracans, like many other agnathans (Figs. 3.10–3.12), possessed a distinct, obliquely directed pineal canal arising from the supraorbital canal and connecting in some specimens with the mediodorsal longitudinal canal. The pineal line in heterostracans is similar to the caudal portion of the rostral segment of the supraorbital line in placoderms, acanthodians, and actinopterygians and may thus be homologous to the anterior pit line in gnathostomes (Fig. 3.3D).

At least four considerably variable transverse commissural canals characterized the early cyathaspidids (Fig. 3.12A). In the fossils, these transverse lines may occur as a series of short, disrupted canals, not connected with the dorsal longitudinal lines, or as continuous canals that meet at the dorsal midline to form complete commissures as well as extend ventrally to end below the laterodorsal longitudinal line. In pteraspidids these lines are reduced to three short canals interconnecting the mediodorsal and laterodorsal longitudinal canals. In pteraspidids, these lines are reduced to three short canals interconnecting the mediodorsal and laterodorsal canals (Fig. 3.12B) caudal to the otic capsules (White 1935). Similarities between the dorsal transverse canals of heterostracans and the dorsal pit lines and supratemporal lines of gnathostomes have been noted by most workers (Allis 1934; Holmgren 1942; Simth 1957; Stensiö 1958; Denison 1964). The position and connections of the heterostracan lines are consistent with their being homologous to the dorsal pit lines of gnathostomes, but the absence of any information on their innervation, as well as the probable absence of these lines in lampreys, suggests that they may be heterostracan apomorphies.

A longitudinal canal on the branchial plate of heterostracans (Fig. 3.12B) has been reported in only one genus. In *Phialaspis* (White 1946), a suprabranchial canal appears to have run the length of the branchial plate, connecting rostrally with a transverse commissure that linked it to the laterodorsal and mediodorsal longitudinal lines and then connecting with the infraorbital canal. This is the only fossil evidence that early vertebrates may have possessed a suprabranchial line comparable to that in lampreys (Fig. 3.10A) and, perhaps, an epibranchial line comparable to that in sharks.

The ventral plate in most heterostracans exhibits lateroventral and medioventral longitudinal lines (Fig. 3.12). The lateroventral line is the most conservative, and there is evidence that it continued onto the trunk in some taxa (White 1935; Denison 1964) and may thus have been homologous to the ventral trunk line in other vertebrates. The medioventral longitudinal line in heterostracans is more variable and may consist of a short groove or pit line, fused at the midline, approximately halfway back on the ventral plate (Fig. 3.12B), or it may consist of short, obliquely directed grooves as in many cyathaspidids (Fig. 3.12A). Equally short transverse commissural grooves are associated with both ventral longitudinal lines in heterostracans but appear to have interconnected the longitudinal lines only in *Irregulareaspis*, where they formed an extensive network covering the entire ventral plate (Dension 1964).

2.2.3 Myxinoids

In 1907 Ayers and Worthington reported that *Eptatretus stouti* possessed fully formed and functional lateral line grooves. Close examination of the skin immediately rostral and caudal to the cornea of *Eptatretus* (Fig. 3.10B) reveals a series of short grooves: rostral to the eye these form a horizontal series; caudal to the eye they form a ventral horizontal series as well as a more dorsal vertical series. The number of these grooves varies greatly within a species and even between the left and right sides of an individual. In *Eptatretus stouti* there are two to five grooves in each series (Fernholm 1985).

Ayers and Worthington (1907) described the lateral lines of *Eptatretus* as deep, narrow grooves, within the outer layer of the epidermis, whose floor comprised a continuous series of spindle-shaped cells that they interpreted as neuromasts innervated by lateralis rami of the trigeminal and vagal nerves. Recently, Kishida et al. (1987) reported that the lateral line grooves in *Eptatretus burgeri* are innervated by anterior and posterior lateral line nerves. These workers labeled both nerves with horseradish peroxidase and traced their projections centrally where they end in a nucleus in the rostral medulla that Kishida et al. termed the medial nucleus of the area acousticolateralis. However, although these workers claim that the lateral line nerves in *Eptatretus* innervate the lateral line grooves peripherally, they do not present any evidence to support this contention and appear to have relied on the earlier description of Ayers and Worthington (1907). In addition, they apparently were not aware that scanning and transmission electron microscopic examination of the lateral line grooves in *Eptatretus* has revealed that these grooves comprise only small mucous cells (Fernholm 1985). Neither neuromasts nor single sensory cells with any morphological resemblance to the hair cells of other lateral line organs occur within the grooves; equally important, there is no evidence that the grooves are innervated.

Examination of whole head skin mounts of *Eptatretus stouti* (unpublished observations) has confirmed Fernholm's observations that no specialized receptor cells or compound receptors occur within the grooves of these hagfish. Instead, the grooves are lined with rather generalized cuboidal epithelial cells that are similar to elongated dorsolateral placodes in amphibians prior to the differentiation of neuromast primordia; it is possible that the grooves are, in fact, arrested elongated lateral line placodes. If this is the case, it is also possible that lateral line ganglion cells and their neurites have not differentiated, as lateral line nerves appear to arise from the same placodes as the lateral line organs they innervate (Stone 1922; Winklbauer and Hausen 1983, Metcalfe, Chapter 7). The cranial nerve that Kishida et al. (1987) recognize as an anterior lateral line nerve may, in fact, be the sensory root of the facial nerve. Although a small nerve in myxinoids has been recognized as a facial nerve (Worthington 1906; Jansen (1930), this nerve occupies the usual position of the motor root of the facial nerve and is much smaller than a typical vertebrate facial nerve. It is possible that myxinoids have a dorsally displaced and hypertrophied facial sensory root associated with increased gustatory sensation and that the medial octavolateralis nucleus (Bullock et al. 1983; Kishida et al. 1987) is a nucleus similar to the facial lobe in actinopterygians. Similarly, the nerve identified as a posterior lateral line nerve by Kishida et al. (1987) may be the glossopharyngeal nerve (Worthington 1906; Jansen 1930).

3. The Lateral Lines of the Earliest Craniates

The plesimorphic traits of the lateral lines shared by all four craniate clades (myxinoids, heterostracans, cephalaspidomorphs, and gnathostomes) would constitute the morphotype of this system in the earliest craniates. Unfortunately, it is not clear that myxinoids possess lateral lines. Atlantic myxinoids exhibit no trace of lateral lines (Fernholm 1985), and Pacific eptatretids have only a series of short epidermal grooves that do not contain neuromasts and may or may not be innervated by "lateral line" nerves.

It is unfortunate that myxinoids have no fossil record and that we have no details regarding the development of myxinoid placodes embryonically. As there are no lateral lines in the chordates that comprise sister groups to the craniates, conclusions regarding the earliest morphotype of the lateral lines are strongly affected by our interpretation of the myxinoid pattern. If we conclude that myxinoids have no lateral lines, a cladistic analysis based on out-group comparisons would require that we further conclude that the lateral lines arose with the origin of vertebrates rather than craniates (Fig. 3.1). However, if one concludes that eptratretid hagfishes retain more plesiomorphic traits than myxinoid hagfishes (Hardisty 1979; Northcutt 1985b), one could further argue that ancestral myxinoids possessed a well-developed lateral line system, comparable to that in vertebrates, and that modern eptatretids retain vestiges of the system, whereas myxinoids, with complete loss of the lateral lines, represent the extreme in a paedomorphic trend. A third interpretation must also be considered: that eptatretids do possess lateral lines with single-celled mechanoreceptors innervated by lateral line nerves, and this condition is plesiomorphic for craniates. Clearly, additional studies are needed to resolve questions regarding the "lateral line" grooves of eptatretids. Given the ambiguities in myxinoids, the remainder of this analysis will be restricted to vertebrate clades, and their features will be used to construct a morphotype for vertebrates, a morphotype that may also apply to all craniates.

Both gnathostomes (Fig. 3.3D) and heterostracans (Fig. 3.12) possess extensive lateral lines that form canals and grooves, whereas cephalaspidomorphs (Figs. 3.10A, 3.11) possess lateral lines that are fewer in number and usually represented as pit lines. Parsimony requires us to conclude that extensive lines, represented as grooves and/or canals, characterized the ancestral pattern for vertebrates, and that cephalaspidomorphs exhibit a derived pattern of reductions, probably due to paedomorphy—as opposed to concluding that cephalaspidomorphs retain a plesiomorphic pattern and that heterostracans and gnathostomes have independently evolved canals and a larger number of lines.

The earliest vertebrates probably possessed supra- and infraorbital canals, as all three clades exhibit such lines. The supraorbital canal may or may not have been disjunct in the earliest vertebrates, as a disjunct line occurs in the placodermal and acanthodian gnathostomes (Fig. 3.3D) and, possibly, cephalaspidomorphs, whereas a continuous canal occurs in heterostracans, chondrichthyans, and osteichthyans.

A pineal canal probably existed in the earliest vertebrates, as heterostracans and cephalaspidomorphs possess this line. The pineal line in agnathans may be homologous to the anterior pit line in gnathostomes, but further embryological and experimental neuroanatomical studies are needed to test this hypothesis.

The infraorbital canal of the earliest vertebrates probably connected both rostrally and caudally with the supraorbital canal, as this pattern occurs in some heterostracans (Fig. 3.12B) and most gnathostomes (Fig. 3.3D). If this was the plesiomorphic condition, then the poorly developed circumorbital pit lines of cephalaspidomorphs should be interpreted as apomorphic. Otherwise, one would have to argue that well-developed and connected supra- and infraorbital lines evolved independently in heterostracans and gnathostomes.

Lateroventral and medioventral longitudinal lines, in the form of canals or grooves, probably occurred on the oralobranchial surface of the head in the earliest vertebrates and probably connected rostrally with the infraorbital canal, as such lines occur in heterostracans (Fig. 3.12) and some osteostracans (Janvier 1974). It is impossible to determine the phylogenetic history of the oralobranchial lines without further information on the placodal origin and innervation of the oral and ventral lines in lampreys (Fig. 3.10A). If the oral line of lampreys is innervated by the hyomandibular ramus of the anterior lateral line nerve and arises from the hyoid placode, as does the oral line in gnathostomes (Fig. 3.9), and if the ventral line in lampreys is innervated by a ventral ramus of the posterior lateral line nerve and arises from a placode caudal to the hyoid placode, as does the ventral trunk line in gnathostomes (Fig. 3.2), these lines could be interpreted as homologues and, in turn, possibly homologous to the rostral and caudal segments of the lateroventral longitudinal line in heterostracans (Fig. 3.12B). In this context, the similar position of the medioventral longitudinal line in heterostracans and the mandibular line in gnathostomes would be consistent with such homology. Alternatively, it is possible that the oralobranchial lines in heterostracans arose from secondary proliferations of a placode that gave rise to the supra- and infraorbital lines (Fig. 3.2) and are thus homoplasous to the oral and mandibular lines in gnathostomes.

Given the course of the lateroventral longitudinal line in heterostracans, its rostral segment probably arose from a hyoid placode, whereas its caudal segment could have arisen from a second placode, as does the ventral trunk line in gnathostomes (Fig. 3.2). If the hyoid placode in heterostracans and cephalaspidomorphs developed as in gnathostomes (Fig. 3.9), but the body of the Y was oriented rostrally, rather than dorsocaudally as in gnathostomes, the hyoid placode would have exhibited dorsal and ventral limbs directed caudally. Thus the body of the placode could have connected to the infraorbital line, and the dorsal and ventral limbs could have formed the rostral segment of the lateroventral longitudinal line and the medioventral longitudinal line, respectively. If the long axis of the hyoid placode was rotated caudally during the modifications of the mandibular and hyoid arches involved with the origin of gnathostomes, the dorsal and ventral limbs of the hyoid placode in agnathans would be homologous to the posterior and anterior limbs, respectively, in gnathostomes. In this context, the mandibular and oral lines in gnathostomes (Fig. 3.3D) would be interpreted as homologues of the medioventral and rostral segments, respectively, of the lateroventral longitudinal lines in heterostracans, and the remaining cheek lines in gnathostomes would have to be interpreted as apomorphic lines derived from the body of the hyoid placode.

A morphotype of the original oralobranchial surface was reconstructed by Holmgren (1942) and Holmgren and Pehrson (1949): purportedly, there were two longitudinal lines on this surface in the earliest vertebrates, as in heterostracans. However, Holmgren originally believed that the rostral segment of the lateroventral line and the entire medioventral line were lost with the origin of gnathostomes and that new oral and preoperculomandibular lines evolved by the ventral extension of transverse commissural lines associated with the laterodorsal longitudinal line. Later Holmgren and Pehrson (1949) argued that the oral and mandibular lines of gnathostomes were homologous to the rostral and caudal segments, respectively, of the medioventral line in heterostracans, whereas the jugal line in gnathostomes was homologous to the rostral segment of the lateroventral line in heterostracans. Holmgren's earlier hypothesis requires that the rostral oralobranchial lines of the earliest vertebrates arose from the same placode as the supra-and infraorbital lines, that these lines were lost in gnathostomes, and that the hyoid placode rotated in gnathostomes. His later hypothesis in collaboration with Pehrson (1949) requires that the rostral oralobranchial lines of the earliest vertebrates arose from a hyoid placode that did not rotate with the origin of gnathostomes but did develop differently to produce new cheek lines. Thus data regarding the number and orientation of the placodes that give rise to the oralobranchial pit lines of lampreys are critical in evaluating all hypotheses regarding the initial pattern of these lines and their phylogeny in vertebrates.

The dorsolateral surface of the head in the earliest vertebrates must have contained two if not three longitudinal lines: dorsal, main, and suprabranchial lines. The dorsal trunk line may have originally extended as far rostrally as the pineal line, as in heterostracans (Fig. 3.12) and, possibly, lampreys (Fig. 3.10A). Initially this may have been the only dorsal line to extend onto the trunk, as appears to be the case in lampreys (Fig. 3.10A). If so, a rostral terminus of the dorsal line, caudal to the otic capsule, as well as the caudal continuation of the main line onto the trunk in gnathostomes, should be viewed as apomorphic traits.

The main trunk line in the earliest vertebrates probably arose from three separate placodes, as in gnathostomes (Fig. 3.2), as indicated by the suspected innervation of the otic and postotic segments of the corresponding pit lines in lampreys (Johnston 1905). The innervation of the more caudal neuromasts of the main line in lampreys should be determined, as Johnston did not recognize these neuromasts in *Petromyzon*, and their innervation in *Ichthyomyzon* and *Lampetra* is not known. If these more caudal neuromasts are innervated by a lateral ramus of the posterior lateral line nerve and arise from a separate placode, as in gnathostomes, the apomorphic features of the main trunk line in gnathostomes must be due to heterochronic changes in the development of this placode.

The earliest vertebrates may have possessed a suprabranchial line, but the only real evidence for such a line is its occurrence in one genus of heterostracans (White 1946) and the presence of neuromasts in a comparable position in lampreys (Fig. 3.10A). Again, the placodal origin and innervation of these neuromasts should be closely examined. If they arise from any placode other than the hyoid placode, the suprabranchial pit line in lampreys cannot be homologous to the epibranchial line in elasmobranchs.

4. Innervation of the Lateral Lines

The lateral lines of the head and trunk are innervated by cranial nerves. The phylogeny of these nerves is problematical, because interpretations are deeply embedded in theoretical considerations regarding vertebrate head segmentation (Balfour 1878; Goodrich 1918; Jarvik 1980; Bjerring 1984). Two hypotheses have been proposed concerning cranial nerve fibers that innervate lateral lines: (1) these fibers are one of several functional components that constitute a dorsal series of cranial nerves that also innervate mesodermal segments (branchiomeres) of the head and (2) collections of these fibers constitute a separate series of cranial nerves that innervate lateral lines and are only superficially and peripherally associated with the dorsal cranial nerves.

In 1878, Balfour noted that the spinal nerves of a small shark, *Scyliorhinus* (= *Scyllium*), initially develop as separate dorsal and ventral elements and that these elements innervate the hypomeric and epimeric divisions, respectively, of a single trunk segment. He believed that the somites and branchiomeres of the head were formed by rostral continuations of the trunk epimeres and hypomeres, respectively, and concluded that all cranial nerves, except the special sensory nerves (olfactory, optic, and otic), were serially homologous to the dorsal and ventral spinal nerves of embryonic gnathostomes and adult lampreys. He thus concluded that the cranial nerves of vertebrates consist of a dorsal series of nerves (profundus, trigeminal, facial, glossopharyngeal, and vagal) and a ventral series of nerves (oculomotor, trochlear, abducens, and hypoglossal). In 1878 almost nothing was known about the sensory and motor nature of these nerves. These details were established over the next 40 years in a series of elegant studies of the peripheral distribution and suspected innervation of the cranial nerves in a broad range of vertebrate species (Pinkus 1895; Strong 1895; Cole 1896; Allis 1897; Herrick 1899, 1900, 1901; Coghill 1902; Johnston 1905; Merritt Hawkes 1906; Landacre 1916; Norris and Hughes 1920; Norris 1925). Many of these workers noted that the dorsal cranial nerves contain the same somatic and visceral sensory fiber components as the dorsal spinal nerves, in addition to sensory fibers that innervate lateral line organs and taste buds. These last two components were interpreted as specialized somatic (lateral line organs) and visceral (taste buds) components that were retained in the dorsal head nerves but subsequently lost in dorsal spinal nerves (Strong 1895). Accepting Strong's conclusions, many subsequent workers hypothesized that ancestrally, all dorsal cranial nerves must have had lateralis and gustatory components that were retained in all dorsal nerves (Holmgren 1940; Holmgren and Pehrson 1949), modified in some dorsal nerves (Jarvik 1980; Bjerring 1984),or lost in some dorsal nerves (Norris 1925; Romer 1970).

It was clear early in this century that spinal nerves possess a single sensory ganglion, which arises embryonically from neural crest, but that many dorsal cranial nerves (facial, glossopharyngeal, and vagal nerves) possess multiple sensory ganglia that arise from distinctly different embryonic sources. All dorsal cranial nerves (the profundus may be an exception) possess a neural crest-derived sensory ganglion whose cells innervate head tissues comparable to trunk tissues that are innervated by neural crest-derived sensory ganglia of the dorsal spinal nerves. However, the neural crest-derived ganglia of the facial, glossopharyngeal, and vagal nerves are closely associated with *additional* sensory ganglia that are derived from dorsolateral and epibranchial placodes that innervate lateral line organs and taste buds, respectively (Platt 1896; Landacre 1910, 1916; Stone 1922). Not only do the ganglion cells that innervate lateral line organs and taste buds arise from embryonic tissues different from those of the neural crest-derived ganglia of the dorsal cranial nerves, but the fibers of the placodally derived ganglia project to and terminate in the medulla separately from the neural crest-derived ganglion cells of the dorsal cranial nerves (Maler et al. 1973a,b; Luiten 1975; Boord and Campbell 1977; Northcutt 1979a,b; McCormick 1981; Finger and Tong 1984; Meredith 1984; New and Northcutt 1984a,b: Morita and Finger 1985; Puzdrowski 1987, 1988; Ronan and Northcutt 1987).

The embryological and hodological data are consistent with the hypothesis that the head of craniates consists of four series of cranial nerves: (1) a dorsal and

(2) a ventral series of nerves that represent serial homologues of the dorsal and ventral spinal nerves of adult lampreys and embryonic gnathostomes; (3) a series of chemoreceptor nerves (terminal, olfactory, and gustatory nerves); and (4) a series of octavolateralis nerves (otic and lateral line nerves). In this context, the optic nerve is not considered a member of any cranial nerve series, as it is really a central neural tract that, along with the retina, has evaginated from the brain.

The hypothesis that lateral line fibers are components of dorsal cranial nerves and the hypothesis that these fibers constitute a separate series of cranial nerves are both plausible. The dorsal nerve hypothesis, however, is less plausible, as it requires that gustatory and lateral line components were lost in several dorsal cranial nerves and all dorsal spinal nerves, or that new components were added to all dorsal cranial nerves, with subsequent loss of these components in some dorsal nerves, or that new components were added to only some dorsal cranial nerves. In any case, there is no evidence that dorsal spinal nerves ancestrally contained lateral line and gustatory fibers or that all dorsal cranial nerves ever contained both lateral line and gustatory components, as neither the profundus nor the trigeminal nerve in any living craniate possesses lateral line and gustatory fibers or lateral line and gustatory ganglia. Furthermore, the dorsal cranial nerve hypothesis requires that the terminal nerve be interpreted as a dorsal nerve that has lost visceral sensory and lateral line components, and the olfactory nerve be interpreted as a second dorsal nerve that has lost all components except a visceral sensory component. Since this nerve does not possess a ganglion, the olfactory epithelium must be interpreted as a modified ganglion (Jarvik 1980). There are additional problems in explaining the occurrence of an otic nerve. If this nerve is interpreted as a separate dorsal cranial nerve, one must argue for the loss of all other components, as well as all embryological traces of a branchiomeric segment that was present ancestrally; alternatively, one must argue that the otic nerve is a specialized subdivision of the lateralis component of the facial or glossopharyngeal nerve and has become secondarily separated from these nerves.

In contrast, the hypothesis most compatible with the condition observed in craniates also requires us to postulate the fewest transformations: the octavolateralis and gustatory organs and their nerves are apomorphic characters of the head that arose phylogenetically as the result of new inductive interactions, and the octavolateralis and gustatory ganglia and fibers are only superficially associated with dorsal cranial nerves. Although the dorsal nerve hypothesis of Strong (1895) is the one most frequently cited, particularly in secondary sources, it should be noted that the hypothesis that lateral line nerves are a separate series of cranial nerves was proposed equally early (Pollard 1892; Cole 1896). In an elegant analysis Cole (1896) listed six reasons why lateral line nerves could not be considered components of the dorsal cranial nerves and noted that even Strong (1895) implied the same by denying the segmental nature of the lateral line system. Cole's observations have been confirmed by all subsequent experimental studies, and the main unanswered question concerns the number of lateral line nerves, not their existence as a separate series.

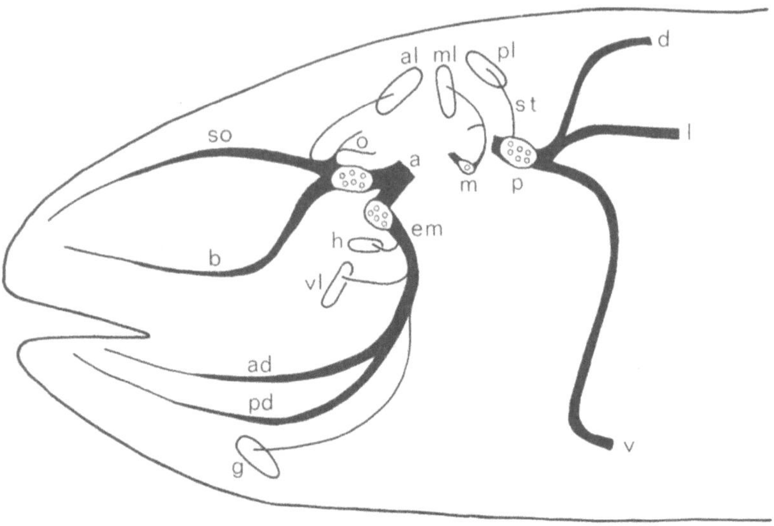

FIGURE 3.13. Possible number and innervation pattern of lateral line nerves in ancestral gnathostomes. Small open circles within nerve rami indicate position of ganglion cells. a, Anterior lateral line nerve complex; ad, anterior division of external mandibular ramus of anteroventral lateral line nerve; al, anterior pit line; b, buccal ramus of anterodorsal lateral line nerve; d, dorsal ramus of posterior lateral line nerve; em, external mandibular ramus of anteroventral lateral line nerve; g, gular line; h, horizontal pit line; l, lateral ramus of posterior lateral line nerve; m, middle lateral line nerve; ml, middle pit line; o, otic ramus of anterodorsal lateral line nerve; p, posterior lateral line nerve; pd, posterior division of external mandibular ramus of anteroventral lateral line nerve; pl, posterior pit line; so, superficial ophthalmic ramus of anterodorsal lateral line nerve; st, supratemporal ramus of posterior lateral line nerve; v, ventral ramus of posterior lateral line nerve; vl, vertical pit line.

Although experimental studies emphasizing the central projections of the lateral line sensory fibers (e.g., Maler et al. 1973a,b; Boord and Campbell 1977; McCormick 1981; New and Northcutt 1984a; Boord and Montgomery, Chapter 16. McCormick, Chapter 17; Will, Chapter 18) have recognized only anterior and posterior lateral line cranial nerves, studies of *Carassius* recently completed (Puzdrowski 1988) and studies of *Ambystoma*, *Lepisosteus*, and *Polypterus* under way (R.G. Northcutt, J. Song, and J.F. Webb) in my laboratory indicate that a larger number of lateral line nerves exist in living vertebrates and must have occurred in the earliest craniates.

The "anterior lateral line nerve" (Fig. 3.13) usually consists of two ganglia (Cole 1896; Coghill 1902; Johnston 1905; Norris and Hughes 1920; Norris 1925; Puzdrowski 1988) and sometimes three (Norris and Hughes 1920; McCready and Boord 1976) that appear to arise from separate placodes (Stone 1922; Pehrson 1949). The dorsal ganglion gives rise to dorsal and ventral rami termed the superficial ophthalmic and buccal rami, respectively (Fig. 3.13). A small, unnamed branch of the superficial ophthalmic ramus innervates the anterior pit line (Allis 1897; Norris 1925; Song 1986), and the bulk of the fibers innervate the neuromasts of the supraorbital line. A small otic ramus that innervates one or more neuromasts of the otic line and the spiracular organ may arise from the superficial ophthalmic ramus or dorsal ganglion, or it may be present as a separate ganglion and ramus (McCready and Boord 1976). The buccal ramus innervates the neuromasts of the infraorbital line.

The lateral lines of the cheek and lower jaw are innervated by fibers of a more ventral ganglion (Fig. 3.13) that may arise from the hyoid placode (Platt 1896; Pehrson 1949). The peripherally coursing fibers of this ganglion are traditionally identified as the external mandibular ramus of the hyomandibular trunk of the facial nerve (Norris 1925; Norris and Hughes 1920), and anterior and posterior divisions of the external mandibular ramus innervate the oral and preoperculo-mandibular lines, respectively (Fig. 3.13). The horizontal, middle cheek, and gular lines are innervated by unnamed smaller rami arising from the hyomandibular trunk (Allis 1897; Norris and Hughes 1920; Norris 1925; Song 1986), as is the autapomorphic opercular line in teleosts (Puzdrowski 1988).

Given the embryonic origin and the innervation of the lateral lines on the rostral half of the head, it appears likely that the earliest vertebrates possessed at least two separate lateral line nerves that might be termed the anterodorsal and anteroventral lateral line nerves (Puzdrowski 1988). The anterdorsal nerve would have innervated the neruomasts above and below the orbit, and the anteroventral nerve would have innervated the neuromasts of the cheek and lower jaw. It is possible that two additional nerves (otic and spiracular) also occurred if each placode that gives rise to lateral line organs (Fig. 3.2) also gives rise to ganglion cells that innervate these organs. Unfortunately, there is not sufficient information to determine the plesiomorphic number of dorsolateral placodes in craniates or to establish whether or not every placode gives rise to both sensory organs and nerves.

The classical descriptions of the innervation of the lateral lines of gnathostomes (Allis 1897; Coghill 1902; Landacre 1916; Norris and Hughes 1920; Norris 1925) claimed that a bundle of fibers closely associated with the glossopharyngeal nerve innervated one or more neuromasts of the postotic line, possibly the gular line, and, when present, the middle pit line. However, recent experimental studies (McCormick 1981; Meredith 1984) utilizing degeneration and horseradish peroxidase methods failed to demonstrate these projections. Allis (1889) reported that this lateral line branch in *Amia* is only superficially associated with the glossopharyngeal nerve and that the lateralis fibers possess a separate ganglion, pass into the neurocranium through a separate foramen, and enter the brain as a root distinct from the glossopharyngeal nerve. Allis's observations have been corroborated by recent studies of *Carassius* (Puzdrowski 1988) and *Lepisosteus* (J. Song and R.G. Northcutt, unpublished observations) utilizing histological serial sections of entire heads and Sihler or Sudan black preparations that render the head transparent while staining the cranial nerves. These studies confirm the earlier observations, except that the gular line is innervated by a branch of the hyomandibular trunk rather than by the "glossopharyngeal" lateralis fibers. Equally important, application of horseradish peroxidase to this "glossopharyngeal" ramus in gars labels fibers in the medial octavolateralis nucleus (J. Song and R.G. Northcutt, unpublished observations) in a manner identical to that resulting from similar applications to other lateral line nerves. These fibers were probably not detected in previous experimental studies, because the lateralis fibers are associated with the glossopharyngeal nerve for only a short distance. McCormick (1981) cut the glossopharyngeal root immediately prior to its entrance into the brain, a point at which the lateralis fibers have already moved away from the glossopharyngeal nerve to enter the brain as a separate root. Meredith (1984) applied horseradish peroxidase to the glossopharyngeal nerve distal to its ganglion, a point where the lateralis fibers have separated from the glossopharyngeal nerve to innervate the postotic line.

Given the evidence that a distinct bundle of fibers with a separate ganglion innervates neuromasts of the middle pit and postotic lines, and that these organs arise from one or more placodes separate from those that give rise to the otic and trunk lines, it is probable that a separate middle lateral line nerve (Fig. 3.13) existed in the earliest vertebrates and is retained in living vertebrates that possess a middle pit line and/or a postotic line.

The "posterior lateral line nerve" (Fig. 3.13) of gnathostomes usually consists of a single ganglion from which arise three major rami: a supratemporal ramus that innervates the neuromasts of the posterior pit and supratemporal lines, a dorsal ramus that innervates the dorsal trunk line, and a lateral ramus that innervates the main trunk line (Strong 1895; Allis 1897; Herrick 1899; Landacre 1916; Norris and Hughes 1920; Norris 1925; Puzdrowski 1988). Elasmobranchs (Norris and Hughes 1920; Disler 1977), salamanders (Coghill 1902), and lungfishes (Pinkus 1895; Northcutt 1986b) possess an additional ramus, the ventral ramus that innervates the ventral trunk line (Fig. 3.13). Although the lateral lines innervated by the rami of the posterior lateral line nerve arise from three separate placodes (Fig. 3.2), only sharks exhibit any trace of multiple ganglia that would indicate that the posterior lateral line nerve is actually a compound nerve. The supratemporal ramus bears a distinct ganglion in *Mustelus* (Norris and Hughes 1920), and the remaining rami arise from a more caudal ganglionic mass that is constricted into rostral and caudal subdivisions in both *Mustelus* and *Squalus* (Landacre 1916; Norris and Hughes 1920). The dorsal ramus arises from the more rostral subdivision, whereas the lateral ramus appears to arise from both ganglionic subdivisions. The origin of the ventral ramus has not been indicated (Norris and Hughes 1920).

Examination of the number of ganglia associated with the rami of the nerves that innervate the postotic head and trunk lines in lungfishes and salamanders does not resolve the question of how many lateral line nerves originally innervated these regions of the body. Both lungfishes and salamanders possess a single lateral line ganglion that gives rise not only to the supratemporal, dorsal, lateral, and ventral rami of the "posterior lateral line nerve" but also to a ramus that bears a distinct ganglion (middle lateral line nerve) in other anamniotes and innervates the posterior pit and postotic lines (Pinkus 1895; Coghill 1902; Northcutt 1986b). Fusion of ganglia that are separate in other anamniotes is even more extreme in amphibians than in lungfishes. Tadpoles (Strong 1895) and salamanders (Coghill 1902) exhibit a ganglionic complex that includes the ganglionic cells of the middle and posterior lateral line nerves, as well as all ganglionic cells of the glossopharyngeal and vagal nerves.

Given evidence that the lateral lines of the postotic head and trunk arise from at least three separate placodes and that two distinctly separate lateral line ganglia (supratemporal and trunk ganglia) exist in *Mustelus*, it is possible that at least two, perhaps three, separate lateral line nerves occurred caudal to the middle lateral line nerve in the earliest vertebrates. This hypothesis can be tested only by further study of the innervation of the lateral lines in other vertebrates that possess three trunk lines, such as lampreys, and, equally important, by more detailed examination of the development of the lateral lines of the postotic head and trunk and the origin of the ganglion cells that innervate these lines.

5. Discussion

As noted in the Introduction, phylogenetic studies have multiple aims beyond describing the history of traits. These aims frequently include discerning form–function relationships and proposing and testing hypotheses regarding the mechanisms responsible for the traits examined. Often the process of formulating hypotheses, let alone testing them, reveals gaps in our knowledge and thus alters the direction of future research. In the following sections each of these points will be considered: the history of lateral lines and their innervation, as inferred from the morphotypes generated; possible mechanisms that might account for the observed variation in these traits in descendants of the earliest vertebrates; and the kinds of information still needed to further our understanding of the phylogeny of lateral lines.

5.1 Phylogeny of Lateral Lines and Their Innervation

No extinct or living craniate radiation exhibits all lateral line traits purported to exist in ancestral craniates or vertebrates (Sections 2 and 3 and Table 3.1). Heterostracans, among agnathans, and *Chlamydoselachus* and sarcopterygians, among gnathostomes, appear to retain the largest number of ancestral lateral lines. However, in each radiation there are reductions in the number of lines as well as loss of accessory traits (grooves and canals), and frequently the course of lines is changed or, more rarely, new lines are added.

Myxinid hagfishes do not appear to possess lateral lines, whereas the more plesiomorphic eptatretid hagfishes appear to retain only traces of these lines. This suggests that hagfishes, among living anamniotic craniates, exhibit the most severely reduced lateral line pattern. Similarly, cephalaspidomorphs appear to exhibit reductions in the supraorbital and transverse lines and in most oralobranchial lines. Many placoderms (Fig. 3.4) may have retained a large number of plesiomorphic lines for gnathostomes, but the lack of information regarding the presence and course of lines on the lower cheek and jaw (Ørvig 1960; Miles and Westoll 1968) limits interpretations. Unfortunately, acanthodians (Fig. 3.5) are too poorly known for us to accurately assess whether or not dorsal pit lines, as well as middle cheek, dorsal, and ventral trunk lines, were reduced or lost in many taxa. However, chondrichthyans exhibit loss of the anterior pit, gular, and, probably, middle pit lines, and most chondrichthyans have reduced and/or lost the middle cheek, oral, and ventral trunk lines. Similarly, some sarcopterygians exhibit reduction and/or loss of the ethmoid, middle cheek, otic, and postotic lines, and most actinopterygians exhibit loss and/or reduction of the anterior and middle pit, middle cheek, and horizontal, oral, and ventral trunk lines (Table 3.1).

In the earliest vertebrates, lateral lines occurred as grooves or canals (see Sections 2 and 3), but these lines exist only as superficial neuromasts in lampreys (Holmgren 1942), salamanders (Lannoo 1987a), and some teleosts (Lekander 1947). Furthermore, the dorsal pit lines of the head, the middle cheek line, and the dorsal and ventral trunk lines of living gnathostomes occur only as superficial lines. There is thus reduction or loss of accessory traits in a large number of lateral lines in members of all living vertebrate radiations.

Changes in the orientation of existing lines, or the origin of new lateral lines, occurred with the origin of gnathostomes. The lateral lines on the lateroventral surface of the head of agnathans (Figs. 3.10–3.12) arise from the infraorbital line and course caudally, whereas these lines in gnathostomes arise primarily from the cheek region and course rostrally on the lower jaw (Fig. 3.3D). Not only has the orientation of these lines changed, but gnathostomes exhibit more cheek lines than do agnathans. These changes may have occurred by a rotation of the long axis of the hyoid placode and redifferentiation of the main body of the placode, or they may have resulted from alterations in both the supra-infraorbital and hyoid placodes (Section 3). Similar but less dramatic changes must have occurred in the development of the supraorbital line of gnathostomes to account for disjunct and continuous supraorbital lines (Section 2.1.1). In each case, there is insufficient information on the number and subsequent development of the placodes that give rise to lateral lines to account for these changes.

The occurrence of accessory lateral lines on the head and trunk of teleosts (Lekander 1947; Coombs et al. 1988) is another striking example of the origin of new lines. The presence of these lines indicates that changes have occurred in teleost development, but unfortunately the nature of these changes is uncertain. The new lines may be generated by changes in the differentiation of preexisting placodes or by further division of primary neuromasts giving rise to secondary neuromasts that migrate to form the new lines (Lekander 1947). These new lines may even form by the peripheral processes of lateral line ganglion cells inducing neuromasts from general ectoderm directly, rather than by neuromasts arising from placodes, which appears to be the conservative developmental sequence for vertebrates. Clearly the development of these new lateral lines in teleosts must be examined, as their presence may indicate that major changes have occurred in the basic developmental program for lateral line formation.

Some reduction in the number of distinct cranial nerves that innervate lateral lines also appears to characterize the phylogenetic history of this system. There is evidence that the lateral lines on the pro-otic surface of the head of ancestral vertebrates were innervated by anterodorsal, anteroventral, and, perhaps, otic and spiracular lateral line nerves (Section 4). Whereas most vertebrates have at least two distinctly separate lateral line ganglia, apparently arising from different placodes, the centrally projecting fibers of these cells enter the brain at a single point, as loosely associated rootlets or even a single root. Similarly, a middle lateral line nerve with a distinct ganglion occurs in most vertebrates, but its root also enters the brain immediately adjacent to, or even as a rostral part of, the posterior lateral line nerve.

Separate lateral line ganglia associated with the rami of the "posterior lateral line nerve" are known only in some sharks (Landacre 1916; Norris and Hughes 1920). There is thus some evidence that lateral lines on the postotic surface of the head and trunk were originally innervated by at least three (middle, supratemporal, and dorsolateral) and perhaps four (plus ventral) lateral line nerves. If this hypothesis is considered valid, the number of lateral line nerves is more conservative than the number of lateral lines in vertebrate phylogeny, and the main trends are the secondary fusion of the ganglia and entering roots of these nerves and the frequent loss of the ventral lateral line nerve. Alternatively, lateral line nerve rami can be viewed as consisting ancestrally of two nerves (anterior and posterior nerves) whose rami became more differentiated; or all lateral line nerve fibers are components of dorsal cranial nerves that have increasingly segregated from their original nerves. All available evidence (Section 4) supports the hypothesis that lateral line nerves are a distinct series, separate from dorsal cranial nerves, and there are sufficient data to argue that at least three, if not more, lateral line nerves existed in the earliest vertebrates.

5.2 Form–Function Relationships

A cladistic analysis of the variation in the number of lateral lines among vertebrates and the condition of these lines—whether neuromasts are located superficially or in grooves and/or canals—reveals a number of recurring patterns: (1) neuromasts in lampreys, lepidosirenid lungfishes (all neuromasts in *Lepidosiren* and most in *Protopterus*), anuran larvae, salamanders, and some teleosts have independently evolved from one condition to another—i.e., from being housed in grooves and/or canals to occurring as lines of superficial neuromasts; (2) reductions and changes in the course of lateral lines, particularly on the cheek, have occurred independently in most vertebrate radiations; (3) increases in the width of lateral lines have occurred by the addition of neuromasts, as in some salamanders (Lannoo 1987a), or increases in the width of canals have occurred without an increase in the number of neuromasts, as in some teleosts (Coombs et al. 1988); (4) new lines of neuromasts occur as accessory lines, adjacent to phylogenetically older lines, or neomorphic lines (opercular lines) occur in regions without a history of line formation.

Canal loss and replacement with superficial neuromasts have been correlated with still-water habitats and/or sedentary behavior (Dijkgraaf 1962; Marshall 1971). Coombs et al. (1981) have suggested that low-noise environments may favor the evolution of widened canals and an increase in superficial neuromasts. Blaxter (1987) has suggested that superficial neuromasts may function as proprioreceptors of swimming velocity; there may thus be correlations with locomotion. Finally, Lannoo (1987a) has suggested that lotic (flowing water) environments may favor an increase in the number of primary neuromasts, as well as longitudinal stitches, whereas lentic (still water) environments may favor transverse stitch formation in salamanders.

Such correlations are pointed out frequently in comparative studies and represent a form of reductionism that often obscures our recognition of higher-order interactions by suggesting that the biological significance of the attributes of a character is known. In examining a system such as the neuromasts of the lateral lines, we discover that some species exhibit only superficial neuromasts, whereas others possess only neuromasts in canals, and we assume that natural selection has favored each condition; i.e., we would look for correlations between superficial (or canal) neuromasts and some environmental factor.

For example, if we discovered that superficial neuromasts had lower thresholds and followed higher frequencies than canal neuromasts, we might argue that superficial neuromasts were selected for because these attributes conferred some biological advantage in the animal's environment; in fact, such a correlation might exist. This panadaptationist view of evolution forces us to reduce an organism to multiple characters and to argue that each character is optimally adaptive for its particular function. Unfortunately, if we discovered that superficial neuromasts had higher thresholds and only followed lower frequencies than canal neuromasts, other correlations would be found, and equally plausible explanations would be presented. Any given character may arise, or any character may have certain attributes, owing to constraints other than selection (see Section 5.3), or selection may have favored other characters that are genetically linked to a character being examined, without that character itself being selected for. Many life history characters of an organism (e.g., metamorphosis, rate of maturity, parental care, longevity) are under strong directional selection, and it is possible that many existing traits are the result of selection acting on other traits. For example, it is possible that development was independently truncated (paedomorphosis) in the ancestors of lampreys, salamanders, and some teleosts, resulting in the presence of only superficial neuromast lines in these taxa. In this case, selection did not act directly on the neuromasts to produce their present position; rather, their position is the consequence of selection acting on other traits. Therefore, the relevant correlation may not be superficial neuromasts and some environmental factor, the relevant correlation may be the occurrence of only superficial neuromasts in taxa with a large number of other paedomorphic characters. Rather than conferring some advantage in a particular environment, superficial neuromasts, as opposed to groove or canal neuromasts, may be the consequence of selection favoring a shorter life history, so that lateral line development, as well as the development of many other characters, is truncated. In these circumstances, neuromasts located superficially, rather than in grooves or canals, might be neutral or even mildly deleterious characters.

Similarly, it is difficult to imagine the selective advantage of such limited canal formation in *Protopterus*, where only the posterior segments of the supraorbital and infraorbital lines form canals, as opposed to the condition in *Lepidosiren*, where all neuromasts occur as superficial lines or in grooves. The lateral lines in these taxa form a morphocline, with canal formation in *Neoceratodus* representing the plesiomorphic condition and the condition in *Protopterus* and *Lepidosiren* representing successive stages in a paedomorphic trend. Superficial neuromasts may not be directly selected for in lungfishes or amphibians; rather, selection may be acting to truncate development, as a shorter developmental period appears to be highly advantageous in an environment of rapidly desiccating ponds and marshes. In such a case, development of the lateral lines, as well as that of many other characters, would be truncated relative to that in the ancestral adult condition.

Finally, it is unlikely that ecological factors can be causally correlated with certain taxonomic features such as the presence of an oral and a mandibular line as pit lines, or the reduction or loss of one or the other in different vertebrate radiations. These patterns may be examples of different morphological "themes" (bounded morphological domains of Alberch 1980) in which characters occur in specific patterns, but the patterns themselves are discontinuous among living vertebrate groups, and transitional forms are not seen even among extinct taxa. In fact, transitional patterns may never have existed because of architectural and/or developmental constraints (see Section 5.3). These patterns may not be the result of competition among groups, in which a particular pattern of lateral line distribution is favored over another; instead, they may be the result of developmental and/or historical constraints and may represent different but equally satisfactory solutions to perceiving low-frequency stimuli.

Not all patterns of change need be, nor are they likely, the result of selection acting on other traits or the result of chance. The transformation from narrow to widened canals, increases in the number of neuromasts within lines, and occurrences of new lines may all be due to direct selective pressures, but some aspects of these transformations could also be due to peramorphic trends in related traits such as dermal bone. Future considerations of form-function relationships of the lateral lines should go beyond the search for correlations at the immediate level of the lateral lines. Researchers should also consider these organs within a broader context: as elements in suits of morphological characters that have figured into more encompassing life history strategies. Equally important, correlations should be examined for evidence of causality (Gould and Lewontin 1979; Gould and Vrba 1982; Northcutt 1988).

5.3 Evolutionary Constraints and Lateral Line Variation

Biological evolution is generally considered to comprise processes that generate morphological variation due to natural selection and chance, acting differentially on phenotypic differences that arise by mutations and recombinations of genotypes. However, other processes that limit and channel variation appear to exist, as discrete sets of morphological transformations occur time and again, whereas certain potential morphological patterns never occur (Wright 1931; Gould and Lewontin 1979; Alberch 1980). Architectural, historical, and developmental constraints, as well as selection, appear to limit the direction and nature of morphological transformations (Gould and Lewontin 1979; Alberch 1980; Dumont and Robertson 1986; Northcutt 1988).

Numerous architectural constraints are imposed on lateral line variation by the physics of water flow across surfaces and in grooves and canals (Denton and Gray 1988; Kalmijn, Chapter 9). Furthermore, hair cells appear to represent a historical constraint, as ciliated ectodermal cells constitute the only available class of receptors among deuterostomes. In this context, similarities in the mechanoreceptive organs of protochordates (Jørgensen, Chapter 6) and the acceleration-sensitive mechanoreceptive systems of invertebrates (Budelmann 1988) offer additional opportunities to examine the relative roles of selectional, architectural, and historical constraints on the evolution of low-frequency, acceleration-sensitive receptors.

Historical (phylogenetic) constraint may account for the pattern of embryonic development of multiple trunk lines in many teleosts. In nonactinopterygian anamniotes, the ventral trunk line appears to arise from a separate, ventrally located placode, whereas the dorsal and main trunk lines arise from a single, more dorsally located placode (Fig. 3.2). In those teleosts that possess multiple trunk lines, even ventrally located trunk lines arise from a dorsally situated placode (Lekander 1949). Multiple trunk lines may be adaptive, but natural selection does not explain the differences in their developmental origin. Nonteleost actinopterygians have neither ventral trunk lines nor ventral placodes (Hammarberg 1937). It thus appears that historical constraint (loss of ventral placodes in ancestral actinopterygians) limits the embryonic origin of ventral trunk lines to a more dorsal placode when selection acts to increase the number of trunk lines in some teleosts; that is, historical constraint, not selective pressure, accounts for the embryonic origin of ventral trunk lines in teleosts.

Complex inductive processes among numerous ectodermal, including neural crest and plate, and mesodermal tissues are probably involved in several developmental sequences: the differentiation of dorsolateral placodes from ectoderm at specific sites on the head, their subsequent migration and differentiation into neuromast primordia, the development of grooves and/or canals, and the development of the cranial nerves that innervate the lateral lines organs. Unfortunately, almost nothing is known about the induction of dorsolateral placodes; however, if this induction sequence is as complicated as that of many other ectodermally derived organs—such as teeth, feathers, or hair—the phylogenetic origin of new lateral lines by the addition of new placodes may be highly unlikely, as it would require major changes in developmental programs involving several tissues. Similarly, the number of lateral line nerves and their peripheral innervation patterns may be highly conservative, as they arise from dorsolateral placodes and are thus similarly constrained.

There is little information regarding the clues that are used by migrating placodes and that determine the final course and length of lateral lines. However, these variables are clearly influenced by growth rates of the surrounding ectoderm (Winklbauer and Hausen 1983), centers and configuration of dermal ossification (Pehrson 1922; Lekander 1949; Graham-Smith 1978), and position of connective tissue septa such as the horizontal skeletal septa and myosepta (see Webb, Chapter 4, for an extensive review). Thus, many aspects regarding the course and extent of lateral lines may not be under direct selection but may be the result of selection acting to alter closely associated characters mediating feeding, respiratory, and locomotive behavior. In this context, it is possible that those changes in orientation of the oralbranchial lines that occurred with the origin of gnathostomes, as well as much of the variation in gnathostome cheek lines, are not adaptive but are the consequence of changes underlying head structure associated with feeding and respiration.

Groove and canal formation is another aspect of the development of lateral lines that clearly involves multiple tissue interactions. Lateral line canals consist of a membranous canal composed of cuboidal epithelial cells, surrounded by cartilaginous rings or housed in bony canals. At present, it is unclear whether the membranous canal arises from placodal tissue, as do the neuromasts within the canal, or from surrounding ectoderm. The surrounding cartilaginous rings or bony elements could arise from neural crest or mesenchymal cells of mesodermal origin. At any rate, canal formation appears to involve the interaction of at least two and perhaps more tissues. Neuromasts in grooves and/or canals appear to be more sensitive than superficial neuromasts (Münz and Claas 1983; Denton and Gray 1988), so it is difficult to believe that selection would favor superficial neuromasts over canal neuromasts, notwithstanding the various environmental correlations purported to account for them in various vertebrates. In these cases, selection may have favored truncation of development, with concomitant loss of dermal bone, so that canal formation is constrained developmentally, rather than being less preferential than superficial neuromasts.

In some teleosts, such as *Carassius*, however, the number of superficial neuromasts far exceeds that of canal neuromasts (Puzdrowski 1988), and selection thus again appears to have favored superficial neuromasts. However, selection may be acting only to increase neuromast number, and the fact that the increase occurs in superficial neuromasts may again be due to developmental constraints. If canal and dermal bone formation are closely linked, after formation of canals, a substantial increase in neuromast number could occur only by division of superficial neuromasts. Even here, scale formation may impose a constraint on the development of superficial neuromasts.

5.4 Future Directions

Every aspect of this review could benefit from additional information, particularly the kinds of information needed to understand the relative role of different constraints on the phylogeny of lateral lines. Continued research on the cranial nerves of hagfishes, particularly on the peripheral structures innervated by the "anterior and posterior lateral line nerves," and on the embryology of these structures is needed to determine whether the lateral lines and nerves of hagfishes exhibit severe paedomorphosis or retain many plesiomorphic characters of the earliest craniates. Present knowledge of the distribution and innervation of the lines of superficial neuromasts in lampreys is also severely limited: data on their distribution are based on *Lampetra* (Holmgren 1942); data on their innervation are based on *Petromyzon* (Johnston 1905). Comparative studies of the lateral lines in larvae and adults of several genera, as well as more detailed experimental studies on the innervation of the lateral lines, are needed to determine the plesiomorphic pattern for petromyzontids, as this is the only living clade of jawless fishes that constitutes an out-group for comparisons with gnathostomes. An understanding of the number and development of dorsolateral placodes in petromyzontids is critical in determining the placodal morphotype for vertebrates and in testing the hypothesis that oralobranchial lines of heterostracans, cephalaspidomorphs, and gnathostomes are homologous.

Continued research on the possible occurrence of lateral lines on the lower cheek and jaws of placoderms, as well as research on the number and condition of lateral lines in acanthodians, is critical for testing hypotheses of lateral line morphotypy in gnathostomes. Particular attention should be directed toward describing the cheek lines in these taxa as well as those in early sarcopterygians for evidence of a middle cheek line. In this context, the distribution and innervation of pit lines in *Latimeria* are important, as are details regarding anuran larvae, to establish the variation in sarcopterygian lateral lines.

Further examination of lateral line variation and innervation is needed in sharks and, particularly, ratfishes and batoids. Most current information is based on studies done in the 19th century, and it is not possible to homologize many of the lateral lines in ratfishes and batoids to those in sharks, much less to those in other gnathostomes. Again, detailed embryological studies of the number and subsequent development of dorsolateral placodes in the cartilaginous fishes are needed; we do not know the total number of placodes for even a single species, even though chondrichthyans constitute the only living clade of gnathostomes that form an out-group for comparisons with osteichthyans.

An understanding of the tissue interactions responsible for the formation and subsequent differentiation of placodes is critical to assessing the relative role of selection and other constraints on the phylogeny of lateral lines. Appropriate studies might involve amphibian larvae, as the general features of lateral line development are well established for some amphibian species (Stone 1922), and amphibian larvae possess large placodes and tolerate extensive experimental manipulation. Details of canal formation must also be examined in actinopterygians, however, as most sarcopterygians exhibit limited canal and dermal bone development. The development of extensive accessory superficial neuromast lines and neomorphic lines in some teleosts should be reexamined, as these taxa represent one of the few possible cases of a peramorphic trend in lateral line phylogeny, as opposed to more common paedomorphic trends. Such examination should reveal whether lateral line hypertrophy involves de novo placode development or only heterochronic alterations in preexisting developmental patterns.

Finally, continued behavioral, morphological, and physiological studies are needed to sample the staggering array of diversity in teleost lateral lines. We need to recognize and test correlations at all levels of biological complexity, from subcellular to populational, in order to understand the evolutionary processes responsible for lateral line diversity.

6. Conclusions

The lateral lines of the earliest craniates probably consisted of an extensive series of canals that have been reduced to placodal vestiges in eptatretid hagfishes and completely lost in myxinid hagfishes. Heterostracans, among jawless vertebrates, and *Chlamydoselachus* and sarcopterygians, among jawed vertebrates, appear to retain the largest number of plesiomorphic lateral line characters. The earliest vertebrates probably possessed supra- and infraorbital canals, but it is uncertain whether the supraorbital canal was complete or disjunct in these forms. A pineal canal probably issued from the caudal segment of the supraorbital canal and may have been homologous to the anterior pit line of gnathostomes. Alternatively, the pineal line was lost with the origin of gnathostomes, and the anterior pit line is the caudal portion of the rostral segment of the disjunct supraorbital canal of early gnathostomes.

The infraorbital canal of the earliest vertebrates was probably connected, both rostrally and caudally, with the supraorbital canal. The circumorbital pit lines of osteostracans and petromyzontids may represent apomorphic reductions in the supraorbital and, possibly, infraorbital lines. Autapomorphic accessory lines have evolved in teleosts in association with both the supra- and infraorbital lines.

Lateroventral and medioventral longitudinal lines appear to have occurred on the oralobranchial surface of the head of the earliest vertebrates and probably connected rostrally with the infraorbital canal. The oralobranchial lines of jawless fishes may be homologous to the oral and mandibular lines of gnathostomes, or the gnathostome lines may represent neomorphic lines. Information on the number and differentiation of placodes of the oralobranchial region in petromyzontids is needed for us to discriminate between these two hypotheses. The oral and mandibular lines of gnathostomes are subject to repeated reduction and/or loss. The oral line is reduced to three short pit lines (mandibular, quadratojugal, and vertical) in actinopterygians and is a rostral remnant in some sharks. The mandibular line may be present as a pit line or even lost in some sharks.

The earliest craniates apparently possessed no cheek lines except a possible longitudinally directed suprabranchial line. This line is known in a single genus of heterostracans and, in extant forms, occurs only in petromyzontids. It may be homologous to the transitory spiracular line in embryonic lungfishes. The earliest gnathostomes apparently possessed two vertical cheek lines (middle cheek and preopercular) and one horizontal cheek line (Fig. 3.3D). These lines appear to be apomorphic for gnathostomes and probably arose by changes in the differentiation of the body of the hyoid placode. The middle line is reduced and/or lost in most chondrichthyans and all actinopterygians. The horizontal line is present as a canal in sharks but is reduced to a pit line in all living osteichthyans. The preopercular line is thus the most conservative cheek line in gnathostomes, and even it is reduced to a pit line in most sharks; its dorsal connections are also highly variable in gnathostomes. It may connect dorsally with the middle cheek line (some sharks), with the dorsal segment of the infraorbital line (most sarcopterygians), or with the postotic segment of the main trunk line (placoderms, acanthodians, actinopterygians). This variation in dorsal connections might suggest that the preopercular lines in gnathostomes are homoplasous, but their occurrence in all gnathostome radiations and their apparent origin from a single placode suggest that the preopercular lines are homologous in gnathostomes despite their highly variable dorsal course.

The dorsolateral surface of the head of the earliest craniates probably exhibited dorsal and main trunk lines. Originally, the dorsal line may have extended as far rostrally as the pineal line and may have been the only line to extend onto the trunk. A number of short, dorsally directed, transverse lines arose from the otic region of the main trunk line and are probably homologous to the middle pit, posterior pit, and supratemporal lines of gnathostomes.

It appears that a ventral trunk line existed in the earliest craniates but has been reduced in living chondrichthyans and lost in actinopterygians. A medially directed gular line arises from the ventral trunk line in living osteichthyans. It appears to have been present in the extinct acanthodians but apparently does not exist in chondrichthyans. Its possible occurrence in placoderms is problematical, and the gular line may thus be a synapomorphic character of acanthodians and osteichthyans.

The cranial nerves that innervate lateral lines appear to constitute a separate series of cranial nerves rather than components of a dorsal cranial nerve series: (1) the embryonic origin of their ganglia is different from that of the dorsal series; (2) the lateral line nerves project to medullary centers different from those of the dorsal cranial nerves. There is evidence that the earliest craniates possessed at least four lateral line nerves (anterodorsal, anteroventral, middle, and posterior) and perhaps as many as seven (plus otic, spiracular, and ventral).

Lateral line nerves innervate the lateral lines as follows. (1) The anterodorsal nerve innervates the supraorbital, infraorbital, and anterior pit line. (2) The anteroventral nerve innervates the oral, mandibular, horizontal, middle cheek, gular, and preopercular lines. (3) An otic nerve innervates the spiracular organ and the otic segment of the main trunk line and may have consisted, originally, of two separate lateral line nerves. (4) A middle lateral line nerve innervates the middle pit and post-otic segment of the main trunk lines. (5) A posterior lateral line nerve complex innervates the posterior pit, supratemporal, and all trunk lines and may originally have consisted of three separate lateral line nerves. Most variation in the lateral lines and their innervation among extinct and living craniates can be attributed to changes in the timing of developmental events—heterochronic changes, with paedomorphic (regressive) changes being most frequent. Peramorphic changes do occur, however, such as the accessory lines in skates and some teleosts.

Although numerous behavioral, ecological, and physiological factors correlate with various patterns of lateral line distribution and their accessory structures (grooves and canals), the relative roles of different evolutionary constraints (architectural, developmental, historical, and selectional) as possible causal factors are unclear. Thus many, if not most, aspects of the distribution, function, history, innervation, and organization of the lateral lines of craniates remain to be examined.

Acknowledgments. Much of the research reported herein was supported by NIH grants NS24669 and NS24869. Various members of my laboratory were kind enough to read the manuscript and offer their comments: Richard Puzdrowski, Jiakun Song, Georg Striedter, Heinrich Vischer, and Jacqueline Webb. My colleague, Theodore H. Bullock, as usual, offered perceptive suggestions, and Marcia Farr showed great and much appreciated tenacity in retrieving many of the older references. I was also fortunate to have available the excellent illustration service that exists at Scripps Institution of Oceanography. Finally, Mary Sue Northcutt was associated in many phases of the research and the compilation of the manuscript.

References

Alberch P (1980) Ontogenesis and morphological diversification. Am Zool 20:653–667.

Alcock R (1899) The peripheral distribution of the cranial nerves of ammocoetes. 1. The branchial nerves, and the innervation of the lateral line system. J Anat Physiol 33:131–153.

Allis EP (1889) The anatomy and development of the lateral line system in *Amia calva*. J Morphol 2:463–567.

Allis EP (1897) The cranial muscles and cranial and first spinal nerves in *Amia calva*. J Morphol 12:487–808.

Allis EP (1923) The cranial anatomy of *Chlamydoselachus anguineus*. Acta Zool 4:123–221.

Allis EP (1934) Concerning the course of the laterosensory canals in recent fishes, pre-fishes and *Necturus*. J Anat 68:361–415.

Ayers H, Worthington J (1907) The skin end-organs of the trigeminus and lateralis nerves of *Bdellostoma dombeyi*. Am J Anat 7:327–336.

Bailey SW (1937) An experimental study of the origin of lateral-line structures in embryonic and adult teleosts. J Exp Zool 76:187–233.

Balfour FM (1878) A Monograph on the Development of Elasmorbranch Fishes. London: Macmillan.

Bjerring HC (1984) Major anatomical steps toward craniotedness: A heterodox view based largely on embryological data. J Vert Palenotol 4:17–29.

Blaxter JHS (1987) Structure and development of the lateral line. Biol Rev 62:471–514.

Boord RL, Campbell CBG (1977) Structural and functional organization of the lateral line system of sharks. Am Zool 17:431–441.

Budelmann B-U (1988) Morphological diversity of equilibrium receptor systems in aquatic invertebrates. In: Atema J, Fay RR, Popper AN, Tavolga WN (eds) Sensory Biology of Aquatic Animals. New York: Springer-Verlag, pp 757–782.

Bullock TH, Bodznick D, Northcutt RG (1983) The phylogenetic distribution of electroreception: Evidence for convergent evolution of a primitive vertebrate sense modality. Brain Res Rev 6:25–46.

Coghill GE (1902) The cranial nerves of *Amblystoma tigrinum*. J Comp Neurol 12:205–289.

Cole FJ (1896) On the cranial nerves of *Chimaera monstrosa* (Linn); with a discussion of the lateral line system and of the morphology of the chorda tympani. Trans R Soc Edinb 38:631–680.

Coombs S, Janssen J, Webb JF (1988) Diversity of lateral line systems: Evolutionary and functional considerations, In: Atema J, Fay RR, Popper AN, Tavolga WN (eds) Sensory Biology of Aquatic Animals. New York: Springer-Verlag, pp 553–593.

Dension RH (1964) The Cyathaspididae: A family of Silurian and Devonian jawless vertebrates. Fieldiana Geol 13:309–473.

Dijkgraaf S (1962) The functioning and significance of the lateral line organs. Biol Rev 38:51–105.

Disler NN (1977) The Lateral Line System Sense Organs of Sharks (Elasmobranchii). Moscow: Science Publications.

Dumont JPC, Robertson RM (1986) Neuronal circuits: An evolutionary perspective. Science 233:849–853.

Eldredge N, Cracraft J (1980) Phylogenetic Patterns and the Evolutionary Process. New York: Columbia University Press.

Fernholm B (1985) The lateral line system of cyclostomes. In: Forman RE, Gorbman A, Dodd JM, Olsson R (eds) Evolutionary Biology of Primitive Fishes. New York: Plenum Press, pp 113–122.

Finger TE, Tong SL (1984) Central organization of eighth nerve and mechanosensory lateral line systems in the brainstem of ictalurid catfish. J Comp Neurol 229:129–151.

Fritzsch B (1981) The pattern of lateral line afferents in urodeles. A horseradish peroxidase study. Cell Tissue Res 218:581–594.

Gans C (1985) Scenarios: Why? In: Foreman RE, Gorbman A, Dodd JM, Olsson R (eds) Evolutionary Biology of Primitive Fishes. New York: Plenum Press, pp 1–10.

Gardiner BG (1963) Certain palaeoniscoid fishes and the evolution of the snout in actinopterygians. Bull Br Mus (Nat Hist) 8:257–325.

Garman S (1888) On the lateral canal system of Selachia and Holocephala. Bull Mus Comp Zool 17:57–119.

Goodrich ES (1918) On the development of the segments of the head in *Scyllium*. Q J Micro Sci 63:1–30.

Gould SJ, Lewontin RC (1979) The spandrels of San Marco and the Panglossian paradigm: A critique of the adaptationist programme. Proc R Soc Lond B 205:581-598.

Gould SJ, Vrba ES (1982) Exaptation – a missing term in the science of form. Paleobiology 8:4-15.

Graham-Smith W (1978) On the lateral lines and dermal bones in the parietal region of some crossopterygian and dipnoan fishes. R Soc Lond Philos, Trans 282:41-105.

Gross W (1968) Porenschuppen und Sinneslinien des Thelodontiers *Phlebolepis elegans* Pander. Palaontol Z 42:131-146.

Hammarberg F (1937) Zur Kenntnis der ontogenetischen Entwicklung des Schädels von *Lepidosteus platystomus*. Acta Zool 18:209-337.

Hardisty MV (1979) Biology of the Cyclostomes. London: Chapman and Hall.

Herrick CJ (1899) The cranial and first spinal nerves of *Menidia*; a contribution upon the nerve components of the bony fishes. J Comp Neurol 9:153-455.

Herrick CJ (1900) A contribution upon the cranial nerves of the codfish. J Comp Neurol 10:265-318.

Herrick CJ (1901) The cranial nerves and cutaneous sense organs of the North American siluroid fishes. J Comp Neurol 11:177-249.

Holmgren N (1940) Studies on the head in fishes; embryological, morphological, and phylogenetical researches. Acta Zool 21:51-267.

Holmgren N (1942) General morphology of the lateral sensory line system of the head in fish. K Svenska Vet-Ak Hdl Oefv 20:51-46.

Holmgren N, Pehrson T (1949) Some remarks on the ontogenetical development of the sensory lines on the cheek in fishes and amphibians. Acta Zool 30:249-314.

Jansen J (1930) The brain of *Myxine glutinosa* J Comp Neurol 49:359-507.

Janssen J, Coombs S, Hoekstra D. Platt C (1987) Anatomy and differential growth of the lateral line system of the mottled sculpin, *Cottus bairdi* (Scorpaeniformes: Cottidae). Brain Behav Evol 30:210-229.

Janvier P (1974) The sensory line system and its innervation in the Osteostraci (Agnatha, Cephalaspidomorphi). Zool Scripta 3:91-99.

Janvier P (1978) Les negeorires paires des Ostéostracés et la position systématique des Céphalaspidomorphes. Ann Paleontol 64:113-142.

Janvier P (1981) The phylogeny of the craniata, with particular reference to the significance of fossil "agnathans." J Vert Paleontol 1:121-159.

Jarvik E (1947) Notes on the pit-lines and dermal bones of the head in *Polypterus*. Zool Bidr Upps 25:60-78.

Jarvik E (1980) Basic Structure and Evolution of Vertebrates, 2 Vols. London: Academic Press.

Johnson SE (1917) Structure and development of the sense organs of the lateral canal system of selachians (*Mustelus canis* and *Squalus acanthias*). J Comp Neurol 28: 1-74.

Johnston JB (1905) The cranial nerve components of *Petromyzon*. Morphol Jahrb 34:149-203.

Jollie M (1969) Sensory canals of the snout of actinopterygian fishes. Trans Ill St. Acad Sci 62:61-69.

Kingsbury BF (1895) The lateral line system of sense organs in some American Amphibia, and comparison with the dipnoans. Trans Am Microsc Soc 17:115-145.

Kishida R, Goris RC, Nishizawa H, Koyama H, Kadota T, Amemiya F (1987) Primary neurons of the lateral line nerves and their central projections in hagfishes. J Comp Neurol 264:303-310.

Landacre FL (1910) The origin of the cranial ganglia in *Ameiurus*. J Comp Neruol 20:309–411.
Landacre FL (1916) The cerebral ganglia and early nerves of *Squalus acanthias*. J Comp Neurol 27:19–67.
Lannoo MJ (1985) Neuromast topography in *Ambystoma* larvae. Copeia 1985:535–539.
Lannoo MJ (1987a) Neuromast topography in urodele amphibians. J Morphol 191:247–263.
Lannoo MJ (1987b) Neuromast topography in anuran amphibians. J Morphol 191:115–129.
Lekander B (1949) The sensory line system and the canal bones in the head of some *Ostariophysi*. Acta Zool 30:1–131.
Leydig F (1850) Ueber die Schleimkanäle der Knochenfische. Archiv Anat Physiol 1850:170–181.
Løvtrup S (1977) The Phylogeny of Vertebrata. London: Wiley.
Luiten PGM (1975) The central projections of the trigeminal, facial and anterior lateral line nerves in the carp (*Cyprinus carpio* L.). J Comp Neurol 160:399–418.
Maler L, Karten HJ, Bennett MVL (1973a) The central connections of the posterior lateral line nerve of *Gnathonemus petersii*. J Comp Neurol 151:57–66.
Maler L, Karten HJ, Bennett MVL (1973b) The central connections of the anterior lateral line nerve of *Gnathonemus petersii*. J Comp Neurol 151:67–84.
Marshall NB (1971) Explorations in the Life of Fishes. Cambridge, MA: Harvard University Press.
McCormick CA (1981) Central projections of the lateral line and eighth nerves in the bowfin, *Amia calva*. J Comp Neurol 197:1–15.
McCready PJ, Boord RL (1976) The topography of the superficial roots and ganglia of the anterior lateral line nerve of the smooth dogfish, *Mustelus canis*. J Morphol 150:527–538.
Meredith GE (1984) Peripheral configuration and central projections of the lateral line system in *Astronotus ocellatus* (Cichlidae): A nonelectroreceptive teleost. J Comp Neurol 228:342–358.
Merritt Hawkes OA (1906) The cranial and spinal nerves of *Chlamydoselachus anguineus* (Gar.). Proc Zool Soc Lond 1906:959–991.
Miles RS (1966) *Protitanichthys* and some other coccosteomorph arthrodires from the Devonian of North America. Kungl Svenska Vetens Handl 10:1–49.
Miles RS, Westoll TS (1968) The placoderm fish *Coccosteus cuspidatus* Miller ex Agassiz from the Middle Old Red Sandstone of Scotland I. Descriptive morphology. Trans R Soc Edinb 67:373–476.
Millot J, Anthony J (1965) Anatomie de *Latimeria chalumnae*, Vol II; Systéme Nerveaux et Organes des Sens. Paris: CNRS.
Morita Y, Finger TE (1985) Topographic and laminar organization of the vagal gustatory system in the goldfish, *Carassius auratus*. J Comp Neurol 218:187–201.
Moy-Thomas JA, Miles RS (1971) Palaeozoic Fishes. Philadelphia: Saunders, 259 pp
Münz H, Claas B (1983) The functional organization of neuromasts in the lateral-line system of a cichlid fish. In: Ewert J-P, Capranica RR, Ingle DJ (eds) Advances in Vertebrate Neuroethology. New York: Plenum, pp 301–307.
Nelson GJ (1972) Cephalic sensory canals, pitlines, and the classification of escocoid fishes, with notes on galaxiid and other teleosts. Am Mus Novitates 2392:1–49.
New JG, Northcutt RG (1984a) Central projections of the lateral line nerves in the shovel nose sturgeon. J Comp Neurol 225:129–140.

New JG, Northcutt RG (1984b) Primary projections of the trigeminal nerve in two species of sturgeon: *Acipenser oxyrhynchus* and *Scaphirhynchus platorynchus*. J Morhpol 182:125–136.
Norris HW (1925) Observations upon the peripheral distribution of the cranial nerves of certain ganoid fishes (*Amia*, *Lepidosteus*, *Polyodon*, *Scaphirhynchus* and *Acipenser*). J Comp Neurol 39:345–432.
Norris HW, Hughes SP (1920) The cranial, occipital, and anterior spinal nerves of the dogfish, *Squalus acanthias*. J Comp Neurol 31:293–401.
Northcutt RG (1979a) Experimental determination of the primary trigeminal projections in lampreys. Brain Res 163:323–327.
Northcutt RG (1979b) Central projections of the eighth nerve in lampreys. Brain Res 167:163–167.
Northcutt RG (1980) Central auditory pathways in anamniotic vertebrates. In: Popper AN, Fay RR (eds) Comparative Studies of Hearing in Vertebrates. New York: Springer-Verlag, pp 79–118.
Northcutt RG (1984) Evolution of the vertebrate nervous system: Patterns and processes. Am Zool 24:701–716.
Northcutt RG (1985a) Central nervous system phylogeny: Evaluation of hypotheses. In: Duncker H-R, Fleischer G (eds) Functional Morphology in Vertebrates. Stuttgart: Gustav Fischer, pp 497–505.
Northcutt RG (1985b) The brain and sense organs of the earliest vertebrates: reconstruction of a morphotype. In: Foreman RE, Gorbman A, Dodd JM, Olsson R (eds) Evolutionary Biology of Primitive Fishes. New York: Plenum, pp 81–112.
Northcutt RG (1986a) Evolution of the octavolateralis system: evaluation and heuristic value of phylogenetic hypotheses. In: Ruben RW, Van De Water TR, Rubel EW (eds) The Biology of Change in Otolaryngology. New York: Elsevier, pp 3–14.
Northcutt RG (1986b) Electroreception in non-teleost bony fishes. In: Bullock TH, Heiligenberg W (eds) Electroreception. New York: Wiley, pp 257–285.
Northcutt RG (1988) Sensory and other neural traits and the adaptationist program: mackerels of San Marco? In: Atema J, Fay RR, Popper AN, Tavolga WN (eds) Sensory Biology of Aquatic Animals. New York: Springer-Verlag, pp 869–883.
Northcutt RG, Gans C (1983) The genesis of neural crest and epidermal placodes: a reinterpretation of vertebrate origins. Q Rev Biol 58:1–28.
Ørvig T (1960) New finds of acanthodians, arthrodires, crossopterygians, ganoids and dipnoans in the upper middle Devonian calcereous flags (Oberer Plattenkalk) of the Bergisch Gladbach-Paffrath trough. Palanotol Z 34:295–335.
Panchen AL, Smithson TR (1987) Character diagnosis, fossils and the origin of tetrapods. Biol Rev 62:341–438.
Pehrson T (1922) Some points in the cranial development of teleostomian fishes. Acta Zool 3:1–63.
Pehrson T (1944) The development of latero-sensory canal bones in the skull of *Esox lucius*. Acta Zool 25:135–157.
Pehrson T (1945) The system of pit organ lines in *Gymnarchus niloticus*. Acta Zool 26:1–8.
Pehrson T (1947) Some new interpretations of the skull in *Polypterus*. Acta Zool 28:399–345.
Pehrson T (1949) The ontogeny of the lateral line system in the head of dipnoans. Acta Zool 30:153–182.
Pinkus F (1895) Die Hirnnerven des *Protopterus annectens*. Morphol Arbeit 4:275–346.

Platt JB (1896) Ontogenetic differentiations of the ectoderm in *Necturus*. J Microsc Sci 38:485–547.
Pollard HB (1892) The lateral line system in siluroids. Zool Jahrbuch Anat 5:525–551.
Puzdrowski RL (1987) The peripheral distribution and central projections of the sensory rami of the facial sensory nerve in goldfish, *Carassius auratus*. J Comp Neurol 259:382–392.
Puzdrowski RL (1988) The peripheral distribution and central projections of the trigeminal, facial and lateral line nerves in goldfish, *Carassius auratus*. Ph.D. Dissertation, University of Michigan, Ann Arbor.
Robertson GM (1938) The Tremataspidae. Am J Sci 35:172–206.
Romer AS (1970) The Vertebrate Body, 4th Ed. Philadelphia: Saunders.
Ronan M, Northcutt RG (1987) Primary projections of the lateral line nerves in adult lampreys. Brain Behav Evol 30:62–81.
Rosen DE, Forey PL, Gardiner BG, Patterson C (1981) Lungfishes, tetrapods, paleontology, and plesiomorphy. Bull Am Mus Nat Hist 167:163–275.
Ruud G (1920) Uber Hautsinnesorgane bei *Spinax niger* Bon. II. Die embryologische Entwicklung. Zool Jahrb Anat Ontog 41:459–546.
Smith IC (1957) New restorations of the heads of *Pharyngolepis oblongus* Kiaer and *Pharyngolepis kiaeri* Sp. nov., with a note on their lateral-line systems. Norsk Geol Tidsskrift 37:373–417.
Song J (1984) The distribution of lateral line receptors in garfish (Holostei). Am Zool 24:134A.
Song J (1986) Peripheral distribution of the lateral line nerves of garfish (Ginglymoidi). Am Zool 26:85A.
Srivastava CBL, Srivastava MDL (1968) Lateral line organs in some teleosts: *Cirrhina mrigala* Ham. Buch. (Cyprinidae), *Ophicephalus* (Channa) *punctatus* Bloch (Characidae) and *Gobius striatus* Day (Gobiidae) J Comp Neurol 134:339–352.
Steno (Stennonis) N (1664) De muscalis et glandulis observationum specimen cum duabus epistolis quarum una ad. Guil. Pisonum de anatome Rajae, etc. Halniae (cited by Johnson 1917).
Stensiö E (1925) On the head of the macropetalichthyids, with certain remarks on the head of the other arthrodires. Pub Field Mus Nat Hist (Geol) 4:89–197.
Stensiö E (1927) The Downtonian and Devonian vertebrates of Spitsbergen. I. Cephalaspidae. Skr Svalb Ish No 12:1–391.
Stensiö E (1932) The Cephalaspids of Great Britain. London: British Museum (Natural History).
Stensiö E (1947) The sensory lines and dermal bones of the cheek in fishes and amphibians. K Svenska Vet-AT Hdl Oefv 24:1–195.
Stensiö E (1958) Les cyclostomes fossiles ou ostracodermes. In: Grassé P-P (ed) Traité de Zoologie, Vol 13. Paris: Masson, pp 173–425.
Stone LS (1922) Experiments on the development of the cranial ganglia and the lateral line sense organs in *Amblystoma punctatum*. J Exp Zool 35:421–496.
Stone LS (1933) The development of lateral-line sense organs in amphibians observed in living and vital-stained preparations. J Comp Neurol 57:507–540.
Strong OS (1895) The cranial nerves of amphibia. J Morphol 10:101–231.
Watson DMS (1937) The ancantodian fishes. R soc Lond Phil Trans B 228:49–146.
Webb JF (1988) Comparative morphology and evolution of the lateral line system in the labroid fishes (Pisces: Perciformes). Ph.D. Dissertation, Boston University, Boston.

Westoll TS (1963) On the evolution of the dipnoi. In: Jepsen GL, Mayr E, Simpson GG (eds) Genetics, Paleontology, and Evolution. New York: Atheneum, pp 121–184.
White EI (1935) The Ostracoderm *Pteraspis* Kner and the relationships of the agnathous vertebrates. R Soc Lond Phil Trans B 225:381–457.
White EI (1946) The genus *Phialaspis* and the "*Psammosteus* limestones." Q J Geol Soc Lond 101:207–242.
Wiley EO (1981) Phylogenetics. New York: Wiley.
Winklbauer R. Hausen P (1983) Development of the lateral line system in *Xenopus laevis*. I. Normal development and cell movement in the supraorbital system. J Embryol Exp Morphol 76:265–281.
Worthington J (1906) The descriptive anatomy of the brain and cranial nerves of *Bdellostoma dombeyi*. Q J Microsc Sci 49:137–181.
Wright S (1931) Evolution in Mendelian populations. Genetics 16:97–159.

4
Developmental Constraints and Evolution of the Lateral Line System in Teleost Fishes

JACQUELINE F. WEBB

1. Introduction

1.1 Constraints and Adaptation in Evolutionary Theory

Evolution has been described as the result of two independent and sequential processes: the origin of variation and natural selection. The phenotypic variation upon which natural selection acts is, in turn, a result of processes that generate as well as those that limit variation (Alberch 1980, 1982b). The processes that generate variation (mutation, recombination) have been the mainstay of evolutionary thought for the past century (Mayr 1982), but many processes limit and channel variation to produce "morphological gaps" (Alberch 1980,1982b). These gaps or "potential morphospaces" remain unoccupied, not because of the absence of appropriate selection pressures, but because historical (phylogenetic) and ontogenetic constraints on morphological change limit morphological variation (Liem and Wake 1985). Consideration of the contributions of these constraints in morphological evolution (e.g., Gould and Lewontin 1979; Alberch 1980) challenges the adaptationist program that has prevailed in the field of evolutionary biology for the past century.

Developmental constraints determine and limit sources of morphological variation by channeling the course of development which, in turn, imposes discontinuities on that variation (Alberch 1980; Gould 1982). Developmental constraints, one category of "internal determinants of design," determine and limit variation on which natural selection (one "external determinant of design"– Liem and Wake 1985) can subsequently act. In doing so, constraints play just as important a role in evolution as those processes that generate variation (Alberch 1980; Gould and Lewontin 1979). In addition, pleiotropic linkages probably play an important role in maintaining phenotypic characteristics that are not the direct result of natural selection and may be nonadaptive (Clutton-Brock and Harvey 1979).

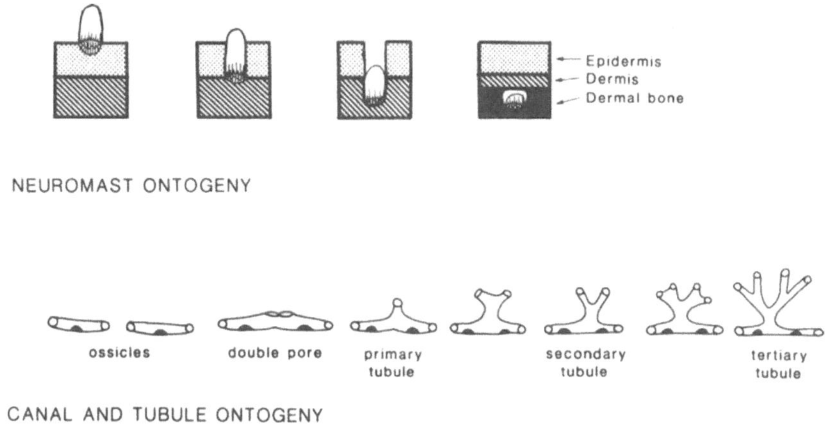

FIGURE 4.1. Schematic representation of neuromast and canal and tubule ontogeny (following Allis, 1889). See text for details.

1.2 Determinants of Lateral Line Morphology

Because of the obvious functional nature of sensory systems, the morphology of the lateral line system has often been viewed as the product of natural selection and functional adaptation. The faults in this view, however, have been pointed out by Northcutt (1988). Adaptive arguments based on correlation between particular ecological or behavioral characteristics and morphological characteristics in the lateral line system have been made for the relationship between pectoral fin placement and the course of the trunk canal, the presence of extensive superficial neuromasts as opposed to canal neuromasts, and the position of the gap characteristic of the disjunct trunk canal (see Webb 1988). These hypotheses, based on correlations, have been constructed largely in the absence of comparative functional data. Such correlations are valuable, but they represent only the first step in establishing the presence of biological adaptation (Bock 1980; Northcutt 1988).

The lateral line system is the product of complex developmental interactions involving (1) the dorsolateral placodes, which give rise to neuromasts and their nervous innervation (Stone 1922; Northcutt and Gans 1983; Northcutt 1986, Chapter 3); (2) dermal bones, which support the lateral line canals; and (3) axial musculature [geometry] which may serve as a guide for placodal migration. Further, the patterns and processes of placodal migration, neuromast ontogeny and proliferation, scale formation, and lateral line canal formation interact during ontogeny to determine adult lateral line morphology (Fig. 4.1). Superficial neuromasts (presumptive canal neuromasts) are transformed into canal neuromasts through a process of invagination, ossicle formation, and ossicle fusion. It is therefore essential to carefully assess these and other developmental processes, before the evolution of the lateral line system (including functional adaptation) can be understood.

2. Developmental Constraints in Lateral Line Evolution

The location, shape, orientation, and topography* of superficial neuromasts determine the potential range of their responses. In addition, differences in size and orientation of neuromasts within canals and variation in canal structure are thought to be responsible for the functional properties of the lateral line system (Denton and Gray 1983, 1988, Chapter 11; Kalmijn 1988). It is therefore tempting to speculate on adaptive explanations for these morphological features, but it is important to keep in mind that they are limited by specific patterns of ontogenetic changes that occur during neuromast and canal development.

2.1 Neuromast and Canal Location

2.1.1 Neuromast and Canals on the Head

The location of canal and superficial neuromasts is the result of the migration of dorsolateral placodes from the head, where they arise (Stone 1922; Pehrson 1949; Northcutt 1986). Canal neuromast location on the head is rather conservative, and is developmentally tied to the configuration of dermal bones (Graham-Smith 1978; Hanken and Hall 1985). Although the causal relationship between neuromast location and the location of dermal bone ossifications remains unclear and requires further experimentation, it appears that the lateral line canals and dermal bones in which they are contained are formed as two separate components (Lekander 1949; Reno 1961). The position of presumptive head canal neuromasts in larval and juvenile fishes predicts their positions within the bony canals of the adult animal (see Fig. 4.2).

Many superficial neuromasts in teleosts are not presumptive canal neuromasts and remain superficial throughout life; they usually occur in rows or lines on the head (Fig. 4.2; Peters 1973; Disler 1960; Lekander 1949). The homologues of these superficial neuromasts have not yet been determined in out-groups, although they may be homologous to neuromasts of canals that have been secondarily reduced or lost in teleosts (see Northcutt, Chapter 3 for discussion of this issue). If this is the case, the position of these neuromasts will be determined by canal position with the possibility of subsequent evolutionary modification. Variation in and limitations on their location cannot be assessed until such homologies are established.

2.1.2 Neuromasts and Canals on the Trunk

Like the neuromasts of the head canals, the location of trunk canal neuromast is also established by placodal migration. Trunk canal neuromasts are not

---

*Topography is defined here as the three-dimensional topography or surface landscape of neuromasts. This differs from Lannoo (1987), where topography is defined as the morphology and distribution of neuromasts.

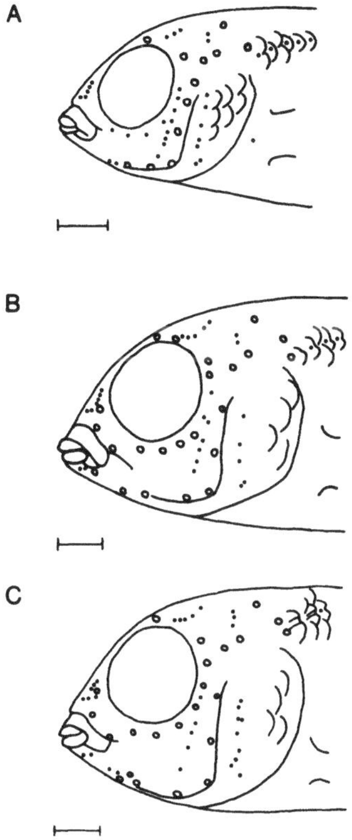

FIGURE 4.2. Head canal pores and superficial neuromasts in an ontogenetic series of formation in *Cichlasoma nigrofasciatum*: (A) 11.5 mm SL; (B) 16.5 mm SL; (C) 22 mm SL. Head canal pores (open circles) and superficial neuromasts (solid dots). Traced from scanning electron micrographs. Scale bar equals 2 mm.

necessarily formed in a continuous rostrocaudal linear series. In the zebrafish (*Brachydanio rerio*; Metcalfe et al., 1985), in *Eigenmannia* (H. Vischer, personal communication), and in *Clupea* (Blaxter 1987), a small number of neuromasts is initially formed rostrocaudally along the entire length of the trunk, and additional neuromasts are then added between these primary neuromasts to form the complete lateral (presumptive trunk canal) neuromast series.

Accessory superficial neuromasts often accompany presumptive canal neuromasts as the dorsoventrally oriented component of orthogonal neuromast pairs (see Coombs et al. 1988; Webb 1988). In some fishes, extensive superficial neuromast proliferation is evident and may occur on all scales of the body (e.g., herring–Blaxter et al. 1983; *Carassius*–Puzdrowski 1988). From their innervation it appears that these superficial neuromasts are organized in lines or clusters on individual scales (Münz 1979; Marshall and Staiger 1975; Puzdrowski, personal communication). The developmental relationships between canal and superficial neuromasts are still unclear.

The course and placement of the trunk canal are apparently correlated with the configuration of the myomeres (axial musculature). The distribution of trunk canal neuromasts along landmarks of myomere configuration indicates that myomere configuration may provide a matrix of migration cues for placodal tissue (although this is apparently not the case in amphibians; Lannoo, personal communication). Metcalfe (personal communication) has shown that in mutants of zebrafish lacking a horizontal septum, the migrating placodal tissue does not follow a horizontal course along the body but runs onto the yolk, suggesting that the cue for the course of placodal migration (e.g., horizontal septum) has been lost. The systematic distribution of a complete straight trunk canal overlying the horizontal septum and the presence of only one presumptive canal neuromast associated with each myomere indicate that these features are probably primitive teleost characters (personal observation).

In the few species examined, one presumptive canal neuromast is found on each myomere at the level of the horizontal septum or the dorsal myomeric flexure (Allis 1889; Metcalfe et al. 1985; Webb 1988). The trunk canal neuromasts are largely confined to the level of the horizontal septum (between the epaxial and hypaxial musculature) or to the dorsal and ventral myomeric flexures. The anterior segment of the trunk canal in cichlids, pomacentrids, and wrasses is located at the level of the dorsal posterior flexure of the myomeres and the posterior section of the canal in the cichlids and wrasses overlies the horizontal septum. The series of pitted scales (presumably homologous to the posterior trunk canal segment) in the pomacentrids are located along the horizontal septum.

In the cichlids *Cichlasoma nigrofasciatum* and *Oreochromis aureus*, each lateral line scale forms beneath a single presumptive trunk canal neuromast (Webb 1988). There is one canal neuromast in the canal segment in each lateral line scale on the trunk of the other teleosts in which it has been specifically examined (Eigenmannia–H. Vischer, personal communication; *Oreochromis aureus*, *Cichlasoma nigrofasciatum*–Webb 1988). Therefore, interspecific differences in lateral line scale number (which are often associated with changes in environmental parameters; Hubbs 1922; Täning 1952) must be tightly correlated with changes in the number of neuromasts along the trunk canal series in order to maintain a continuous canal.

The position of myomeric flexures along which the trunk canal lies is, in part, determined by body shape. Body shape may therefore impose constraints on trunk canal morphology and position. This is clearly illustrated in the pomacentrids and cichlids, where interspecific body shape differences and changes in trunk canal pattern are highly correlated (Fig. 4.3A,B). In both groups, the anterior trunk canal segment typically ends at or before the level of the posterior base of the dorsal fin (Fig. 4.3A,C), where the scale row in which the canal is contained also stops. This is also the level at which the body narrows down into the caudal peduncle. The pomacentrids and cichlids typically have rather deep bodies with relatively narrow caudal peduncles. The pomacentrid *Azurina* (Fig. 4.3B) and the cichlids *Bathybates* (Fig. 4.3D) and *Hemibates* are unusual in that they have torpedo-shaped bodies in which the depth of the caudal peduncle is not much less than that of the trunk (Fig. 4.3). Instead of ending at the posterior end of the trunk, the scale row in which the anterior trunk canal segment is contained

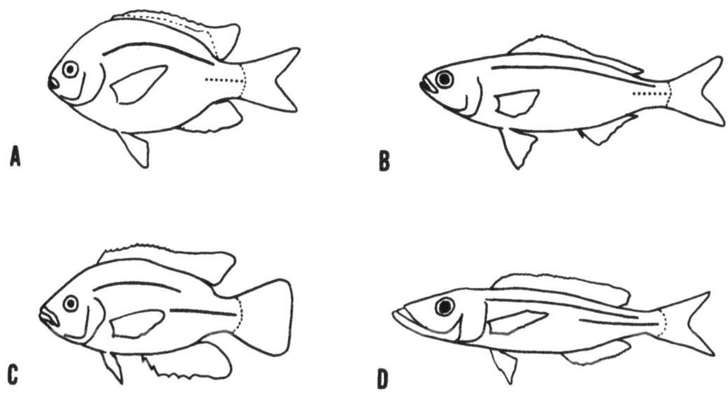

FIGURE 4.3. Variation in trunk canal pattern and body shape in (A) *Abudefduf saxatilis* (Pomacentridae), (B) *Azurina eupalama* (Pomacentridae), (C) generalized cichlid, and (D) *Bathybates leo* (Cichlidae). Solid lines represent canals, dots represent superficial neuromasts.

continues onto the caudal peduncle, carrying with it an extension of the trunk canal segment (Webb 1988).

It therefore appears that trunk canal pattern, placement, and length are largely a product of the combined limitations imposed by myomere geometry and body shape. There are, however, aspects of trunk canal morphology for which the relationship to myomere geometry is yet unknown. The relationship between the position of the trunk canal in teleosts with arched, multiple, dorsally displaced, ventrally displaced, zigzag (in some tetraodontiforms), and decurved (in some cyprinids and characoids) trunk canal patterns (Coombs et al. 1988; Webb 1988) and their respective myomere configurations needs to be examined. In addition, it has been suggested that a functional adaptation in response to a change in the hydrodynamic environment on the fish surface due to the shift of the pectoral fin onto the flank, of diversification of fin use patterns in swimming, or respiratory flow emanating from the operculum (see Webb 1988) may be responsible for the course of the trunk canal. The extent to which these potential functional demands have modified the placement of trunk canals is not known; none of these hypotheses have yet been tested directly.

2.2 Patterns of Neuromast Ontogeny

Although hair cell orientation, which reflects physiological polarization (Flock 1967), appears to be fixed early in ontogeny, variation in sensory strip, cupular shape, and neuromast topography may be the basis for ontogenetic and phylogenetic variation in neuromast function.

The axis of best physiological sensitivity for the hair cells in a given neuromast is presumably established during hair cell differentiation from placodal tissue.

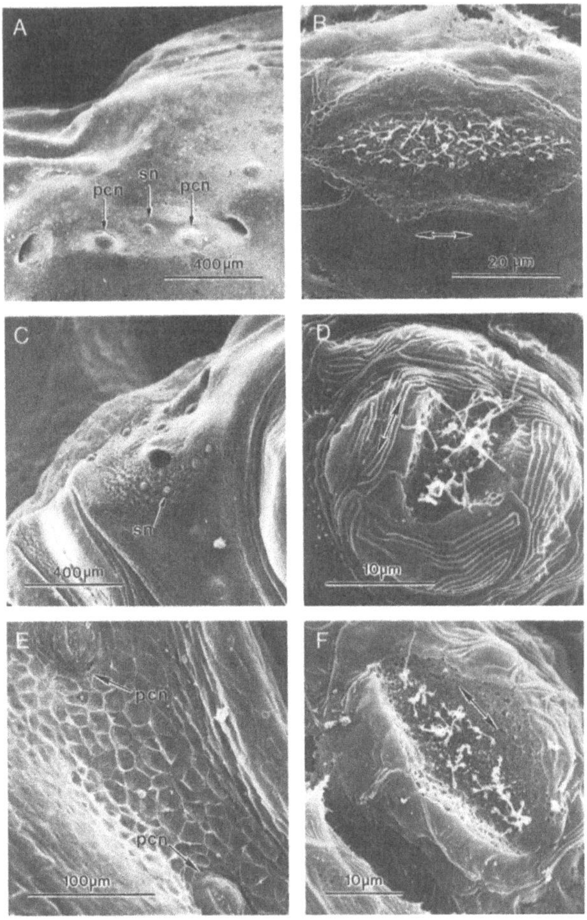

FIGURE 4.4. Scanning electron micrographs of neuromasts in *Oreochromis aureus*. (A) Presumptive temporal canal neuromasts with (B) closeup of left presumptive canal neuromast. (C) Superficial neuromast of circumnasal series with (D) closeup of superficial neuromast noted in C. (E) Presumptive infraorbital canal neuromast rostral to orbit with (F) closeup of ventral presumptive canal neuromast. (Note: hair cell polarization indicated by arrow.) pcn, presumptive canal neuromast; sn, superficial neuromast.

According to available data, hair cell orientation is fixed and does not change during neuromast ontogeny in fishes (Janssen et al. 1987; Webb 1988; Vischer communication). This implies that when neuromasts form during embryonic development, their physiological axes are fixed and follow the axis of the canal in which they will be contained in the adult (see Fig. 4.4).

Canal neuromasts have physiological polarizations parallel to the axis of the canal in which they lie (Coombs et al. 1988). Superficial neuromasts accessory to canals tend to have physiological polarizations either parallel or perpendicular

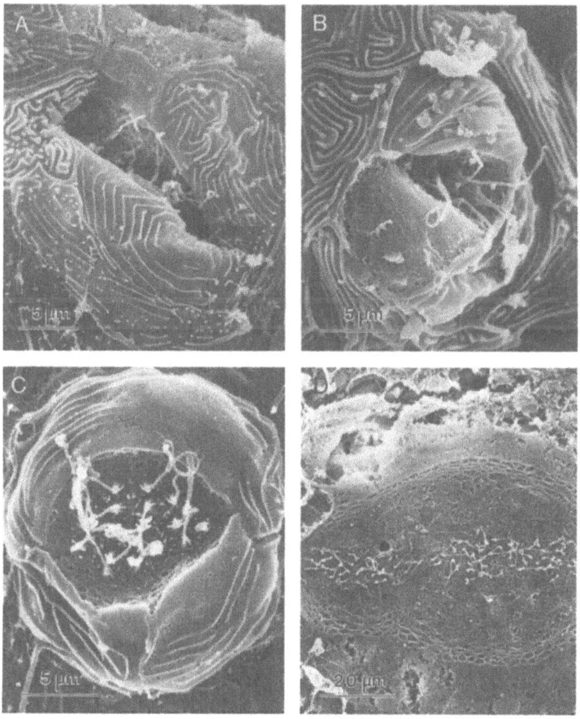

FIGURE 4.5. Ontogenetic changes in neuromast topography in *Sarotherodon aureus*. (A) Presumptive infraorbital canal neuromast (9 mm SL). (B) Presumptive trunk canal neuromast (6 mm SL). (C) Presumptive trunk canal neuromast (8 mm SL). (D) Presumptive trunk canal neuromast (17 mm SL).

to the long axis of the animal. In contrast, superficial neuromasts, like the ones that encircle the nares in the cichlid, *Oreochromis aureus* (Fig. 4.4), and other fishes (damselfishes; Webb, unpublished) have variable orientations; their homologues in other taxa have not been identified. It would be inappropriate to suggest adaptive explanations for the physiological orientation of presumptive canal neuromasts in larval of juvenile fishes. They are canal neuromast precursors, and their orientation is related to their function as canal neuromasts in the adult.

Distinct changes in neuromast topography take place during neuromast ontogeny and lateral line canal formation. In the cichlids *Oreochromis aureus* and *Cichlasoma nigrofasciatum*, for example, presumptive head and trunk canal neuromasts and superficial neuromasts on the head, trunk, and caudal fin erupt through the epithelium as papillate neuromasts (Fig. 4.5A). Their sensory strip is sunken between bolsters of tissue that resemble the guard cells of plant stomata (Fig. 4.5B). As presumptive canal neuromasts mature, they flatten out until they are flush with the epithelial surface surrounding them (Fig. 4.5D).

Interspecific and intraspecific variation in neuromasts in adult fishes may represent various stages in neuromast development (Coombs et al. 1988). For example, presumptive canal neuromasts that are papillate and round may not be the result of selection for round, papillate neuromasts; they may resemble neuromasts in young and larval fishes and may therefore represent early stages in neuromast ontogeny. Although there may be a functional adaptive component to larval lateral line morphology, the sequence of lateral line ontogeny must be kept in mind. The appearance of superficial neuromasts always precedes canal formation and so superficial neuromasts will therefore be present in young fishes. Whether the system is functional in young fishes and whether superficial neuromasts are specifically adaptive and the result of selection in these young fishes are two separate issues.

The shape of the sensory strip (distribution of hair cells) and cupula vary ontogenetically and interspecifically depending on the type of lateral line canal pattern possessed by the fish. In teleosts with narrow canals and in those with branched tubule patterns (see Webb 1988), the major axis of the canal neuromast run parallel with the axis of the canal in which it is contained. In addition, the shape of the cupular base (easily visualized with vital stains, Blaxter et al. 1983; Marshall 1986) is similar to that of the sensory strip. In *Oreochromis aureus* and *Cichlasoma nigrofasciatum*, the sensory strip of presumptive canal neuromasts elongates before canal formation via the addition of hair cells along the long axis of the future canal (Fig. 4.5D).

In teleosts with widened head canals, the outline of the cupular base does not necessarily reflect the distribution of hair cells. The long axis of the cupulae runs across the major axis of the canal, but hair cell orientation remain parallel with the length of the canal (Coombs et al. 1988; Janssen et al. 1987; Vischer, personal communication). The ontogeny of neuromasts in widened canal systems has only been examined in *Notopterus* (Kapoor 1970) and in *Eigenmannia* (Vischer, personal communication).

Superficial neuromasts form lines or stitches on the head and trunk, but their ontogeny has not been studied in detail in teleosts with the exception of cichlids. Three rows of superficial neuromasts accompany the trunk canal segment in each lateral line scale in various species of cichlids (Peters 1973; Münz 1979). One row of superficial neuromasts is oriented perpendicular to the canal neuromasts; it arises as a single neuromast that is part of an orthogonal pair present before canal formation. Two other rows of neuromasts arise later, during or soon after canal formation, and are oriented parallel to the canal neuromast. In each neuromast within these rows, the long axis of the sensory strip is parallel to the physiological polarity of the hair cells, and the neuromasts are oriented end to end like a string of sausages (Webb 1988). The transverse stitches found in some amphibians, in which neuromasts are lined up side by side as a result of neuromast proliferation in an axis transverse to hair cell orientation (Lannoo 1987), are unknown in fishes to date.

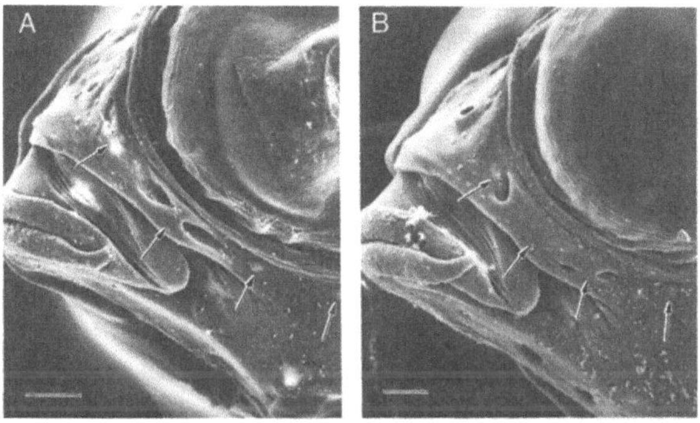

FIGURE 4.6. Progressive enclosure of infraorbital canal in the lacrimal bone in *Cichlasoma nigrofasciatum*. (A) 9 mm specimen, (B) 11 mm specimen. Arrows indicate position of consecutive presumptive infraorbital canal neuromasts. (Scale bar = 200 μm).

2.3 Order of Canal Formation

Canal formation is initiated around individual neuromasts on both the head and trunk. Ridges of epithelial tissue form lateral to the neuromast and fuse over the neuromast to form a canal segment, which subsequently ossifies in most cases (Fig. 4.6). This has been observed in ostariophysans (Lekander 1949), a cottid (Janssen et al. 1987), two species of cichlids (Webb 1988), and a sternopygid (Vischer, personal communication) as well as a number of other actinopterygian fishes (Jollie 1969). Canal formation continues as individual canal segments grow along the canal axis and fuse leaving a pore to the surface at the point of fusion. This process establishes the pore-neuromast-pore pattern characteristic of the lateral line canals of teleosts (Fig. 4.1). In fishes with branched head canal tubules, the location of pores at the ends of tubules is independent of neuromast position, which is limited to the main canals and does not change during canal ontogeny (e.g., clupeoids – Hoss and Blaxter 1982; Blaxter et al. 1983; Best and Gray 1982; Cottidae – Vischer, personal communication).

Because canal formation starts around individual neuromasts, the order of formation of a canal may have as many variations as there are neuromasts. In teleosts, however, only certain patterns are observed. Neuromast enclosure occurs sequentially within the presumptive canal neuromast series of a given canal, although the starting point of enclosure and the direction of this sequence may vary interspecifically (e.g., Disler 1960; Lekander 1949; Figs. 4.2, 4.6). Given the hypothesized developmental relationship between lateral line canals on the head and dermal bone (Hanken and Hall 1985), the order of formation of the various head canals and the order of the ossification of the dermal bones bearing those canals are probably correlated, but this needs to be specifically investigated.

The order of formation of the trunk canal appears to be more conservative. In teleosts in which it has been considered, complete trunk canal pattern forms

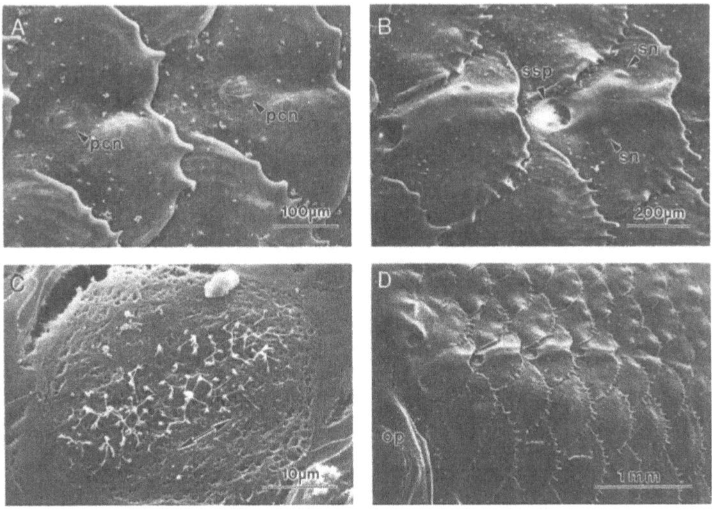

FIGURE 4.7. Trunk canal enclosure in *Cichlasoma nigrofasciatum*. (A) Before canal formation–note shape of neuromast (14 mm SL specimen). (B) After canal formation (17 mm SL specimen). (C) Closeup of presumptive trunk canal neuromast. Hair cell polarization (arrow) and long axis of sensory strip are parallel to the long axis of canal. (D) Overview of anterior trunk canal segment showing rostrocaudal direction of canal formation (17 mm SL specimen). op, operculum; pcn, presumptive canal neuromast; sn, superficial neuromast; ssp, suprascalar pore.

rostrocaudally through the sequential formation of canal segments in adjacent scales of a single horizontal scale row after neuromast distribution is established by placodal migration (e.g., *Cottus bairdi*--Janssen et al. 1987; *Eigenmannia*–Vischer, personal communication). The disjunct pattern, found exclusively in the perciform fishes (Coombs et al. 1988; Webb 1988), is a result of a major change in the order of canal formation. This process has apparently decoupled from the direction of placodal migration, which is presumed to occur rostrocaudally from the head where the placodes arise. A disjunct pattern occurs with only a few exceptions throughout the Cichlidae (only *Teleogramma*, *Gobiocichla* have a complete canal). It has been shown that in cichlids the disjunct trunk canal, unlike a complete canal, forms not in one, but in two directions. The anterior trunk canal segment (Fig. 4.7), located on the dorsal myomeric segment, forms rostrocaudally, but formation of the posterior trunk canal segment, located over the horizontal septum, starts in the last lateral line scale on the caudal peduncle and proceeds rostrally (Peters 1973; Webb 1988; Fig. 4.8). The existence of a second origin of canal formation in the last lateral line scale on the caudal peduncle and the subsequent bidirectional formation of the two canal segments provide the opportunity for the diversification of trunk canal patterns and as such should be considered a key innovation in the Cichlidae (sensu Liem 1973; Webb 1988). The order of canal segment formation in other taxa with a disjunct pattern (e.g.,

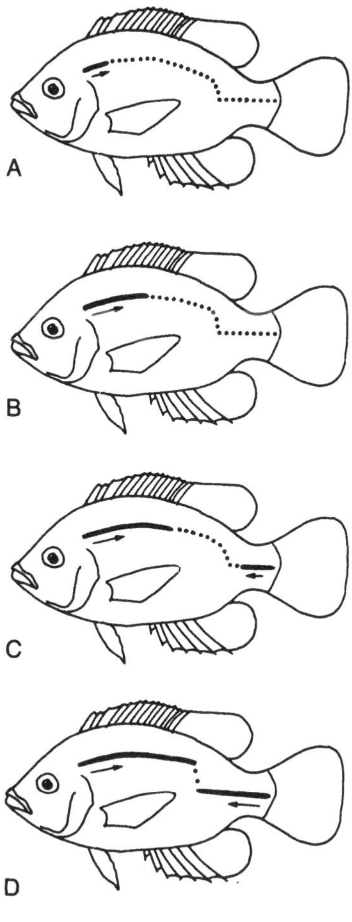

FIGURE 4.8. Schematic representation of bidirectional trunk canal formation in a cichlid with a simple disjunct canal pattern. (A–B) Enclosure of trunk canal starts rostrally and continues caudally. (C) While the enclosure of the anterior half of the canal continues, the posterior half of the canal begins to enclose in a caudo-rostral direction. (D) Canal enclosure occurs in two separate scale rows and is complete when the two canal halves just meet, leaving a vertical gap of two scale rows. Solid lines represent canals, dots represent superficial neuromasts.

labrids, serranoids, anabantoids) has not been investigated, but it seems likely that a similar mechanism may be involved in at least some of these groups.

3. Heterochrony as a Mechanism of Lateral Line Evolution

Heterochrony, the alteration of the order and timing of developmental events (Alberch et al. 1979) is thought to be a common evolutionary mechanism in vertebrates (Gould 1977) and has been documented in many groups of fishes (Hubbs 1926; Fink 1981; Weitzman, in preparation). Heterochrony has been implicated as a mechanism in the evolution of tooth attachment modes in teleosts (Fink 1981), osteology, pigmentation patterns and various aspects of internal anatomy in centrarchids (Mabee 1987), and the lateral plates of sticklebacks (Gasterosteidae – Bell 1981). It has also been suggested as a mechanism in the ontogeny of polymorphism in the cichlid *Cichlasoma managuense* (Meyer 1987).

The lateral line system is ideally suited for a quantitative test of heterochrony according to the guidelines set forth by Alberch et al. (1979). The morphology of the head and trunk lateral line system varies along an ontogenetic spectrum. Neuromasts are discrete, repeated units which appear to be metamerically arranged on the trunk. Events in the development of the system (neuromast formation, neuromast proliferation, canal enclosure, and tubule formation) are causal, temporally ordered, and discrete. Additionally, the transformation of superficial neuromasts into canal neuromasts as a consequence of canal formation is easily visualized at the gross morphological level. The order of canal formation, the time of onset and offset, and the rate of canal formation can be quantified and concisely expressed in terms of the time of enclosure of individual neuromasts within canal series on the head and trunk. Phylogenetic hypotheses produced as a result of cladistic revisions will allow heterochrony to be specifically tested (Fink 1982). By superimposing morphological characteristics on an established cladogram, morphological changes due to shifts in developmental timing should become apparent. A quantitative test of heterochrony will require an assessment of the time of onset, time of offset, and rate of development for each defined developmental event in each of the species on the cladogram (Alberch et al. 1979).

3.1 Timing of Canal Formation

Although a quantitative analysis of the timing of lateral line canal formation has not been done, there is qualitative evidence that the relative timing of canal formation varies considerably (Lekander 1949). In addition, it appears that variations in lateral line system morphology may be the result of changes in the timing of canal formation. Given the process canal formation, which involves the gradual transformation of superficial neuromasts to canal neuromasts, reduced head canals or an incomplete trunk canal pattern in which presumptive canal neuromasts (superficial neuromasts) are present can be described as the result of the deceleration or truncation of canal formation (paedomorphosis as a result of neoteny or progenesis—Alberch et al. 1979). This would explain the absence of canals in cobitids (Lekander 1949), for example, and the presence of an incomplete trunk canal pattern in miniature cichlids and in *pomacentrids* (Fig. 4.9; Webb 1988). Similarly, a system of branched head canal tubules, a disjunct trunk canal pattern with extensive overlap (Cichlidae—Webb 1988), and even multiple canals may be the result of the acceleration or prolongation of head and trunk canal formation (peramorphosis— Alberch et al. 1979).

The ontogeny of the widened head canals found in euteleosts has not been studied, although a similar pattern has been examined in the osteoglossomorph *Notopterus* (Kapoor 1970). In osteoglossomorphs, the widened head canals appear to be the result of reduced ossification of the canal roof (a potential paedomorphic trend) associated with a simple or a branched canal system. The ballooning out of canal segments or tubules to form sinuslike sacs is probably a secondary specialization. These large cavities have important functional

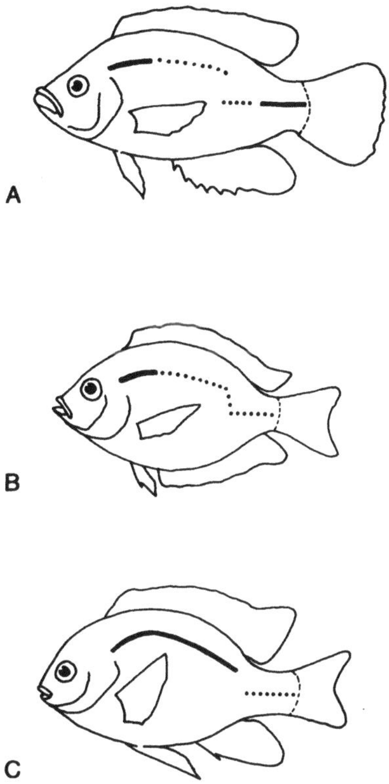

FIGURE 4.9. Schematic representation of trunk canal patterns in (A) a juvenile cichlid, (B) an adult miniature cichlid (*Etroplus*), and (C) an adult pomacentrid (damselfish). Solid lines represent canals, dots represent superficial neuromasts.

implications (Denton and Gray 1988; Chapter 11). In contrast, these features are not present in the ostoglossomorph *Hiodon*, which is considered to have a large number of primitive characteristics (Greenwood 1973; Webb 1988).

3.2 Order of Canal Formation

The order of formation of the major head canals (infraorbital, supraorbital, preopercular, mandibular, temporal, otic, postotic and the various commissures) varies among the taxa in which canal formation has been noted (*Engraulis mordax*–O'Connell 1981; *Clupea harengus*–Blaxter 1987; *Notopterus*–Kapoor 1964; ostariophysans–Lekander 1949; *Ophicephalus punctatus*–Kapoor 1961; centrarchids–Mabee 1987; cichlids–Peters 1973; Webb 1988).

A model of heterochronic variation in head canal system morphology would predict that in fishes with reduced head canal patterns, those head canal segments that do not form are the last to form in sister taxa that have complete head canal patterns and are lost though terminal deletion (truncation, paedomorphosis). This has been demonstrated in the head canal system of centrarchid fishes (Mabee 1987). A quantitative analysis of the order of canal formation defined by the time of onset or offset still needs to be done in a phylogenetic context in order to test this hypothesis. Similarly, a change in the order of canal formation is the basis for the disjunct triuk canal pattern found in the cichlids (see above).

3.3 Changes in Lateral Line Morphology as a Consequence of Heterochronic Alteration of Other Morphological Features

Apparent heterochronic alterations in lateral line morphology may not be the result of direct selection but the result of pleiotropic linkages with other features such as body size. One might suggest that a reduced head canal pattern in miniature fishes is the result of selection against complete canals in which a low Reynolds number (Re) would effectively prevent cupular movement inside small canals in response to stimuli of normal amplitude. In addition, the correlation between a reduced head canal pattern and a low-noise habitat in many teleost groups has reinforced the hypothesis that superficial neuromasts are selectively advantageous over canal neuromasts in fishes living in a low-noise habitat (Dijkgraaf 1962). Alternatively, a nonadaptive explanation is suggested. Miniature fishes possess many progenic characters (Alberch et al. 1979), one of which, by definition, is small size (Weitzman and Vari, in preparation), which is thought to be the product of selection in these fishes. Münz (1986) has demonstrated that the rate of lateral line development is not linked to age but to body size in cichlids. It therefore follows that a paedomorphic lateral line morphology would be expected in miniature fishes; as such it would not be the result of direct selection at all but the result of a linkage between the object of selection (body size) and the development of the lateral line system which is directly linked to it.

In South American freshwater fishes, miniaturization has occurred in 84 species, has evolved independently as many as 33 times in 5 orders (clupeiforms, characiforms, siluriforms, cyprinodontiforms, and preciforms) and is accompanied by progenic reduction in the canal system of the head and trunk in each of these lineages (Weitzman, personal communication). Other examples of pleiotropic linkages with heterochronic changes in unrelated morphological features have been studied in detail in amphibians (Hanken 1983; Alberch and Gale, 1985; Wake 1986). This is not to say that adaptive morphologies in the lateral line system cannot or do not exist. For instance, superficial neuromasts are proliferated in blind cave fishes and the underside of the head of some flatfishes. These correlations suggest an adaptive explanation, but until the function of superficial neuromasts in these fishes is understood, a hypothesis of adaptation cannot be formally tested.

4. Summary and Conclusions

The importance of developmental constraints in morphological evolution is a major thrust in evolutionary theory today. To date, the study of lateral line systems has not generally considered these factors but has centered on adaptive and functional explanations of morphology. A study of the ontogeny, diversity, and phylogeny of the lateral line system indicates, however, that developmental constraints play a significant role in the determination of lateral line morphology.

To fully explain the evolution of the lateral line system or any other organ system, adaptive and functional explanations must be considered, but only after constraints that limit morphological possibilities are assessed. Morphological patterns in the lateral line system may not be the direct result of natural selection. Functional adaptation via natural selection can only operate within the context provided by, or the morphological range limited by, developmental processes.

The ability to place the observed morphological diversity of the lateral line system on an ontogenetic spectrum suggests that evolutionary changes in lateral line morphology occur as a result of heterochrony, the alteration of developmental order, and timing. Finally, it is suggested that the lateral line system is well suited for a quantitative test of heterochrony as a mechanism of evolution.

Acknowledgments. I am grateful to Jørgen Jørgensen, Michael Lannoo, Paula Mabee, Margaret McFall-Ngai, Glenn Northcutt, Rick Puzdrowski, Richard Rosenblatt, Georg Streidter, and Mario Wullimann for helpful discussions and for critically reading and greatly improving earlier drafts of the manuscript. Scanning electron microscopy was partially supported by NINCDS grant NS15090 and ONR contract N0001487K0604 to Dr. Arthur Popper, Georgetown University, and a Sigma Xi Grant-in-Aid for Research to the author. Trisha Rice (Museum of Comparative Zoology) provided assistance with scanning electron microscopy. Robert Zeller was of invaluable help in the preparation of photographic plates. This manuscript was prepared during tenure as an NIH Post-Doctoral Fellow (PHS F2NS08283A).

References

Alberch P (1980) Ontogenesis and morphological diversification. Am Zool 20:653–657.

Alberch P (1982a) Developmental constraints in evolutionary processes. In: Bonner JT (ed) Evolution and Development. Berlin: Springer-Verlag, pp. 313–332.

Alberch P (1982b) The generative and regulatory role of development in evolution. In: Mossakowski D, Roth G (eds) Environmental Adaptation and Evolution. Stuttgart; Gustav Fischer, pp. 19–23.

Alberch P, Gale EA (1985) A development analysis of an evolutionary trend: Digital reduction in amphibians. Evolution 39:8–23.

Alberch P, Gould SJ, Oster GF, Wake DB (1979) Size and shape in ontogeny and phylogeny. Paleobiology 5:296–317.

Allis EP (1889) The anatomy and development of the lateral line system in *Amia calva*. J Morphol 2:463–567.

Bell MA (1981) Lateral plate polymorphism and ontogeny of the complete plate morphology of the threespine stickleback (*Gasterosteus aculeatus*). Evolution 35(1):67–74.

Best ACG, Gray JAB (1982) Nerve fibre and receptor counts in the sprat utriculus and lateral line. J Mar Biol Assoc UK 62:201–213.

Blaxter JHS (1987) Structure and development of the lateral line. Biol Rev 62:471–514.

Blaxter JHS, Gray JAB, Best ACG (1983) Structure and development of the free neuromasts and lateral line system of the herring. J Mar Biol Assoc UK 63:247–260.
Bock W (1980) The recognition of biological adaptation. Am Zool 20:217–227.
Clutton-Brock TH, Harvey PH (1979) Comparison and adaptation. Proc R Soc Lond B 205:547–565.
Coombs S, Janssen J, Webb JF (1988) Diversity of lateral line systems: Evolutionary and functional considerations, In: Atema J, Fay RR, Popper AN, Tavolga WN (eds) Sensory Biology of Aquatic Animals. New York: Springer-Verlag, pp. 553–593.
Denton EJ, Gray JAB (1983) Mechanical factors in the excitation of clupeid lateral lines. Proc R Soc Lond B 218:1–26.
Denton EJ, Gray JAB (1988) Mechanical factors in the excitation of the lateral lines of fishes. In: Atema J, Fay RR, Popper AN, Tavolga WN (eds) Sensory Biology of Aquatic Animals. New York: Springer-Verlag, pp. 595–617.
Disler NN (1960) Lateral Line Sense Organs and Their Importance in Fish Behavior. Academy of Sciences of the USSR, Severtsov Institute of Animal Morphology (translated from Russian). Jerusalem: Israel Program for Scientific Translations, 1971.
Fink WL (1981) Ontogeny and phylogeny of tooth attachment modes in actinopterygian fishes. J Morphol 167:167–184.
Fink WL (1982) The conceptual relationship between ontogeny and phylogeny. Palaeobiology 8:254–264.
Flock A (1967) Ultrastructure and function in the lateral line organs. In: Cahn P (ed) Lateral Line Detectors. Bloomington: Indiana University Press, pp. 163–197.
Gould SJ (1977) Ontogeny and Phylogeny. Cambridge, MA: Belknap–Harvard University Press.
Gould SJ (1982) Change in developmental timing as a mechanism of macroevolution. In: Bonner JT (ed) Evolution and Development. Berlin: Springer-Verlag, pp. 333–346.
Gould SJ, Lewontin RC (1979) The spandrels of San Marco and the Panglossian paradigm: A critique of the adaptationist programme. Proc R Soc Lond B . 205:581–598.
Graham-Smith W (1978) On the lateral lines and dermal bones in the parietal region of some crossopterygian and dipnoan fishes. Phil Trans R Soc Lond 282:41–105.
Greenwood PH (1973) Interrelationships of Osteoglossomorphs. In: Greenwood PH, Miles RS, Patterson C (eds) Interrelationships of Fishes. New York: Academic Press, pp. 307–332.
Hall BK, Hanken J (1985) Foreword to DeBeer, GR: The Development of the Vertebrate Skull. Chicago: University of Chicago Press, pp. vii–xxviii.
Hanken J (1983) Miniaturization and its effects on cranial morphology in plethodontid salamanders, genus *Thorius* (Amphibia, Plethodontidae). II. The fate of the brain and sense organs and their role in skull morphogenesis and evolution. J Morphol 177:255–268.
Hoss DE, Blaxter JHS (1982) Development and function of the swimbladder–inner ear–lateral line system in the Atlantic menhaden, *Brevoortia tyrannus* (Latrobe). J Fish Biol 20:131–142.
Hubbs CL (1922) Variations in the number of vertebrae and other meristic characters of fishes correlated with the temperature of water during development. Am Nat 56:360–372.
Hubbs CL (1926) The structural consequences of modifications of the developmental rate in fishes, considered in reference to certain problems of evolution. Am Nat 60:57–81.
Janssen J, Coombs S, Hoekstra D, Platt C (1987) Anatomy and differential growth of the lateral line system of the mottled sculpin, *Cottus bairdi* (Scorpaeniformes: Cottidae). Brain Behav Evol 30:210–229.

Jollie M (1969) Sensory canals of the snout of actinopterygian fishes. Trans Ill St Acad Sci 62:61–69.

Kalmijn A (1988) Hydrodynamic and acoustic field detection. In: Atema J, Fay RR, Popper AN, Tavolga WN (eds) Sensory Biology of Aquatic Animals. New York: Springer-Verlag, pp. 83–130.

Kapoor AS (1961) The time and order of formation of sensory canals in the fishes *Ophicephalus punctatus* (Ophicephalidae) and *Wallago attu* (Siluridae). Copeia 1961: 176–181.

Kapoor AS (1964) Functional morphology of latero-sensory canals in the Notopteridae (Pisces). Acta Zool 65:77–91.

Kapoor AS (1970) Development of dermal bones related to sensory canals of the head in fishes *Ophicephalus punctatus* Bloch (Ophicephalidae) and *Wallago attu* B. & Schn. (Siluridae). Zool J Linn Soc 49:69–97.

Lannoo MJ (1987) Neuromast topography in urodele amphibians. J Morphol 191:247–263.

Lekander B (1949) The sensory line system and the canal bones in the .head of some Ostariophysi. Acta Zool 30:1–131.

Liem KF, Wake DB (1985) Morphology: Current approaches and concepts. In: Hildebrand M, Bramble DM, Liem KF, Wake DB (eds) Functional Vertebrate Morphology. Cambridge, MA: Belknap–Harvard University Press, pp. 366–377.

Mabee PM (1987) Phylogenetic change and ontogenetic interpretation in the family Centrarchidae (Perciformes: Centrarchidae). PhD Dissertation, Duke University, Durham, NC.

Marshall NB, Staiger JC (1975) Aspects of the structure, relationships, and biology of the deep-sea fish *Ipnops murrayi* (Family Bathypteroidae). Bull Mar Sci 25:101–111.

Marshall NJ (1986) Structure and general distribution of free neuromasts in the black goby, *Gobius niger*. J. Mar Biol Assoc UK 66:323–333.

Mayr E (1982) Growth of Biological Thought. Cambridge, MA: Belknap–Harvard University Press.

Metcalfe WK, Kimmel CB, Schabtach E (1985) Anatomy of the posterior lateral line system in young larvae of the zebrafish. J Comp Neurol 233:377–389.

Meyer A (1987) Phenotypic plasticity and heterochrony in *Cichlasoma managuense* (Pisces, Cichlidae) and their implications for speciation in cichlid fishes. Evolution 41:1357–1369.

Münz H (1979) Morphology and innervation of the lateral line system in *Sarotherodon niloticus* (L.) (Cichlidae, Teleostei). Zoomorphology 93:73–86.

Münz H (1986) What influences the development of canal and superficial neuromasts? Ann Mus R Afr Centr Sci Zool 251:85–89.

Northcutt RG (1986) The brain and sense organs of the earliest vertebrates: Reconstruction of a morphotype. In: Foreman RE, Gorbman A, Olsson R (eds) Evolutionary Biology of Primitive Fishes. New York: Plenum Press, pp. 81–112.

Northcutt RG (1988) Sensory and other neural traits and the adaptationist program: Mackerels of San Marco? In: Atema J, Fay RR, Popper AN, Tavolga WN (eds) Sensory Biology of Aquatic Animals. New York: Springer-Verlag, pp. 869–883.

Northcutt RG, Gans C (1983) The genesis of neural crest and epidermal placodes: A reinterpretation of vertebrate origins. Rev Biol 58:1–28.

O'Connell CP (1981) Development of organ systems in the northern anchovy, *Engraulis mordax*, and other teleosts. Am Zool 21:429–446.

Pehrson T (1949) The ontogeny of the lateral line system in the head of dipnoans. Acta Zool 30:153-182.

Peters HM (1973) Anatomie und Entwicklungsgeschichte des Lateralissystems von *Tilapia*. Z Morphol Tiere 74:89-161.

Puzdrowski RL (1988) The peripheral distribution and central projections of the trigeminal, facial and lateral line nerves in goldfish, *Carassius auratus*. PhD Dissertation, University of Michigan, Ann Arbor, MI.

Reno HW (1961) The infraorbital canal, its lateral-line ossicles and neuromasts in the minnows *Notropis volucellus* and *N. buchanani*. Copeia 1961:403-413.

Stone LS (1922) Experiments on the development of the cranial ganglia and the lateral-line sense organs in *Amblystoma punctatum*. J Exp Zool 35:421-496.

Täning AV (1952) Experimental study of meristic characters in fishes. Biol Rev 27: 169-193.

Wake MH (1986) The morphology of *Idiocranium russeli* (Amphibia: Gymnophiona) with comments on miniaturization through heterochrony. J Morphol 189:1-16.

Webb JF (1988) Comparative morphology and evolution of the lateral line system in the labroid fishes (Pisces: Perciformes). PhD Dissertation, Boston University, Boston, MA.

5
Diversity and Regression in the Amphibian Lateral Line and Electrosensory System

BERND FRITZSCH

Introduction

Amphibians possess, at least as larvae, a system of placodally derived secondary sensory cells aggregated into organs on the head and the trunk. Similar to many fishes (Northcutt 1986), two amphibian orders, urodeles and gymnophionans, possess two types of organs: mechanoreceptive neuromasts (the lateral line) and electroreceptive ampullary organs (the electrosensory system). The third order, the anurans, possess only neuromasts (Fritzsch and Münz 1986; Fig. 5.1a). Present in all aquatic larvae and some in utero embryos, these organs are lost in many, but not all, amphibians during metamorphosis (Wahnschaffe et al. 1987). Moreover, in some amphibians, without free-living larvae, none of these organs seem to be formed, and they are absent in all amniotic vertebrates (Escher 1925).

Because the apical parts of the secondary sensory cells of these organs are directly exposed to the environment, it is conceivable that prolonged unprotected exposure to air may be detrimental for the sensory cells. In addition, available data indicate that lateral line organs function in water and are likely nonfunctional in air. They may be present in postmetamorphic, terrestrial salamanders (Escher 1925; Fritzsch and Wahnschaffe 1983), however, covered by epidermis and most likely nonfunctional. Consequently, the reduction and loss of the lateral line and electroreceptive system in tetrapods represent a case of regressive evolution (i.e., evolutionary reduction) of organs that cannot adapt to airborne stimuli. The aim of this chapter is to provide a survey of our knowledge concerning the diversity in structure and regression of the lateral line and electrosensory system and their innervation in amphibians. A comparison with the lateral line and electrosensory system of extant Sarcoptergii is provided to evaluate the amphibian data.

1.1 The Neuromast

In extant amphibians, the mechanoreceptive lateral line organs—*Nervenhügel*, or neuromasts (Kingsbury 1895)—occur singly or in groups forming lines on the head and the trunk. The longer, sensitive axis of a neuromast consists of a

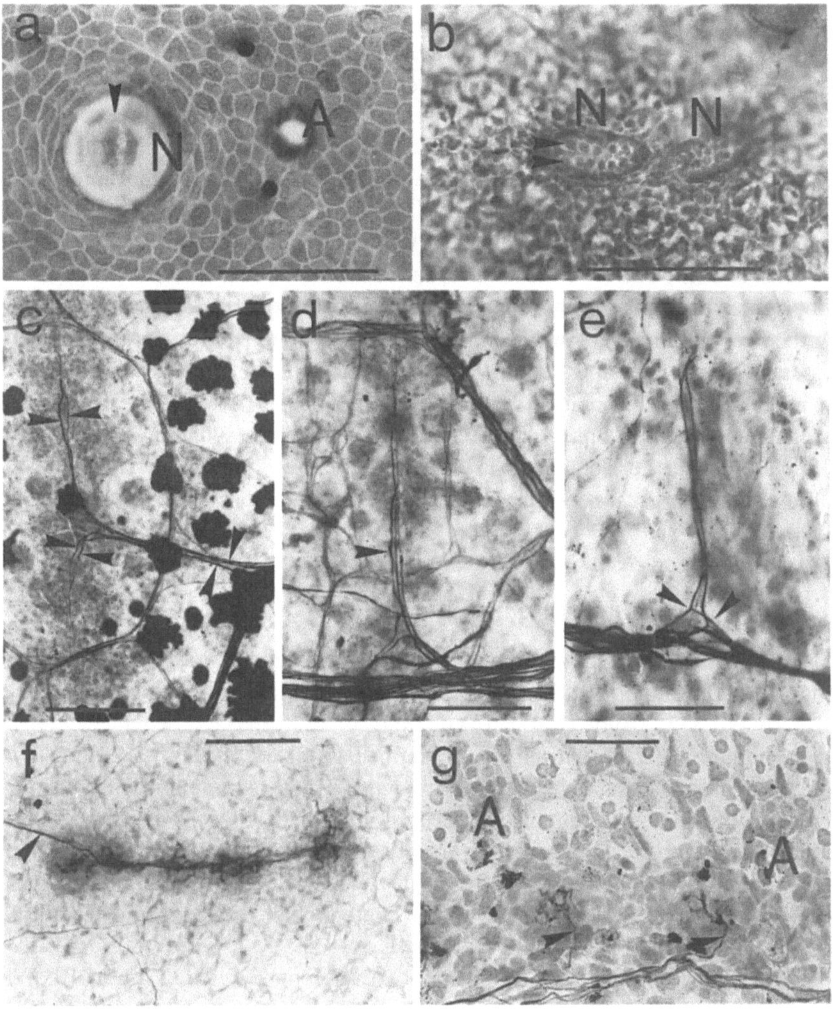

FIGURE 5.1. The flat-mounted skin of *Ichthyophis kohtaoensis* (a), *Ascaphus truei* (b), *Xenopus laevis* (c,d,f), and *Ambystoma mexicanum* (e,g) is shown. Note the neuromast (N) with its longitudinal arrangement of hair cell processes (arrowhead) besides an ampullary organ (A) in *Ichthyophis*. In *Ascaphus* (b), two neuromasts, longitudinally arranged in the direction of hair cell rows (arrowheads), are visible, in *Xenopus*, a stitch may be innervated by more than two affrents, each of which bifurcates so that an individual neuromast seems to be innervated by two fibers (c,d, arrowheads). A stitch is also supplied by an efferent fiber (f, arrowhead) that branches to reach each neuromast. In *Ambystoma*, fibers to a neuromast may be collaterals of fibers heading for other neuromasts or stitches (e, arrowheads). Efferent fibers innervate neuromasts (g, arrowheads) but never ampullary organs (A). Cresyl violet stain (a,b,g), silver stain (c,d,e), and reaction for acetylcholine esterase (f,g). Bars indicate 100 μm.

longitudinal group of pear-shaped sensory hair cells (eight to 33 cells; Lannoo 1987a) polarized into two opposite directions. Supporting cells extend between the sensory cells from the basement membrane to the apical surface. They are surrounded by mantle cells which separate the organ from the epidermis. Depending on the thickness of the epidermis, the neuromasts lie either at the epidermal surface (young larvae) or in a groove (adults, premetamorhpic larvae) but never in canals, as in many fishes. The kinocilium and the stereovilli of each hair cell protrudes into a long cupula which overlies all hair cells in a given organ. At least two thick afferent fibers and, in most cases, one additional thin efferent fiber enter each neuromast (Fig. 5.1c-e). The afferent fibers can be visualized with silver staining (Zakon 1984); the efferents, with a reaction for acetylcholine esterase (Karnovsky and Roots 1964; Fig. 5.1f,g). Each afferent fiber branches to reach all hair cells polarized in one direction (Russell 1976). Afferent terminals are characterized by a presynaptic sphere surrounded by synaptic vesicles in the hair cell. For further details on the ultrastructure, development, and regeneration of lateral line organs, see Jörgensen (Chapter 6) and Corwin et al. (Chapter 8), this volume.

1.2 Stitch Formation and Functional Units

During growth, the total number of hair cells on an animal may increase to more than 20,000 (Lannoo 1987a). Moreover, neuromasts may divide to form stitches (Harris and Milne 1966; Fig. 5.1b) or fields of neuromasts (Escher 1925; Reno and Middleton 1973). A stitch may be defined as all neuromasts descending from one primary neuromast. When there are two afferents per stitch, each neuromast receives the same afferents and efferents. It is likely that all hair cells of one polarity are innervated by one afferent fiber (Görner and Mohr Chapter 27) and constitute a functional unit. The term functional unit is used here in analogy to the electrosensory system (Zakon 1984) where all hair cells of single or multiple organs are supplied by a single afferent fiber.

However, when there are more than two afferents per stitch (up to 10; Will et al. 1985), different subgroups of neuromasts are innervated by different afferents (Fig. 5.1c,d). Because each stitch develops from a single primary neuromast (Winklbauer and Hausen 1983), this could mean that some primary neuromasts receive more than two afferents that are later largely withdrawn. Alternatively, all neuromasts may initially receive two afferents, and some afferents may develop later and innervate preferentially larger primary organs. In *Ambystoma* and in the tail organs of *Xenopus*, the pattern of innervation is such that one fiber may branch to innervate more than one stitch (Fig. 5.1e). These data show that in amphibians the general pattern of innervation of stitches (i.e., two afferent fibers with their set of oppositely polarized hair cells) is highly variable.

Likewise, the presence or absence of stitches and their orientation vary among amphibians. No stitches are found in gymnophionans (Hetherington and Wake 1979), some anurans, and some urodeles (Lannoo 1987a,b). Longitudinal stitches occur in some urodeles (Fig. 5.2) and in premetamorphic *Ascaphus* (Fig. 5.1b). Both longitudinal and transverse stitches are present in Ambystomatidae. Almost all anurans have transverse stitches. Longitudinal groups of neuromasts, presumably representing stitches, are known in lungfish (Kingsbury 1895; Fahrenholz 1929) and may occur in sharks and teleosts (Münz, Chapter 14). This distribution indicates that longitudinal stitches may be primitive for sarcopterygians. Absence of stitch formation may be interpreted as the retention of a larval character, whereas formation of transverse stitches may be regarded as derived. Only two independent changes—from primitive longitudinal to derived transverse stitches in Ambystomatidae and in nonleiopelmid anurans—are necessary to accommodate the known variability of stitch formation in amphibians (Fig. 5.2). This hypothesis is at variance with Lannoo (1987b), who regarded the transverse stitches of Ambystomatidae and anurans as primitive for amphibians. The arguments of Lannoo (1987b) are partly based on ecological and partly on fossil evidence. He argues that within the canals of ancestral amphibians, there might have been no stitch formation. However, formation of longitudinal stitches may have been possible in canals. Clearly, the innervation pattern of canals of *Latimeria* and the Australian lungfish *Neoceratodus* needs to be known to settle this question. In addition, we need ontogenetic data that demonstrate what causes the polarization of stitches and of neuromasts.

1.3 Ampullary Organs

In addition to neuromasts, many urodeles and some gynophionans have ampullary electroreceptors on the head, whereas anurans lack these organs (Fritzsch and Münz 1986). In contrast to neuromasts, the sensory cells in ampullary organs are not directionally polarized and are grouped around the ampullary lumen. They may possess either microvilli alone (urodeles) or a kinocilium surrounded by microvilli (gymnophionans; Wahnschaffe et al. 1985). In urodeles, ampullary organs divide to form circular groups of organs (Fritzsch and Münz 1986). In contrast to the polarized organization of neuromasts at the level of the apical part of a single hair cell, the orientation of hair cells within a neuromast, and the orientation of neuromasts within a stitch, ampullary organs show a circular arrangement at all of these levels. In addition, ampullary organs lack at least a cholinergic efferent innervation (Fig. 5.1g) and are supplied by a single afferent fiber. Ampullary organs are also found in lungfish, and the rostral organ of *Latimeria* may consist of modified ampullary organs (Northcutt 1986). The structural differences between neuromasts and ampullary organs are presented in more detail by Jörgensen (Chapter 6).

Most likely, ampullary organs derive from the same placodal material as neuromasts (Northcutt, Chapter 3), but the necessary events to induce the formation of these organs from anterior placodal material are completely unknown. It is also not known why ampullary organs are absent in anurans and most teleosts. The potential to develop these organs may either be lost entirely or only be suppressed in anurans. That some gymnophionans have only ampullary organs

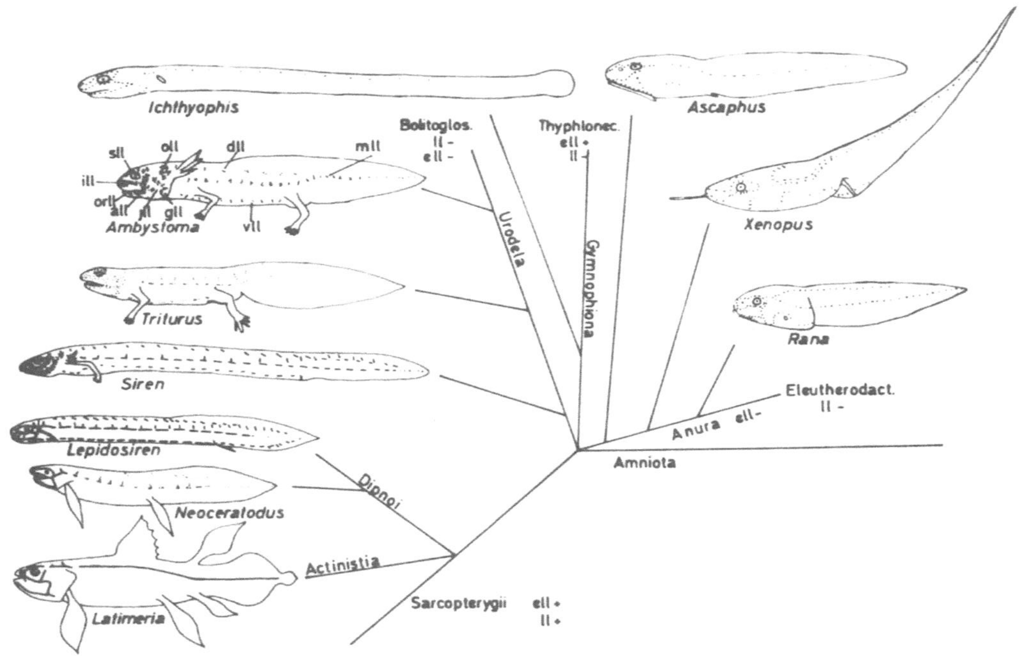

FIGURE 5.2. The distribution of lateral line neuromasts (ll+) and ampullary organs (ell+) is shown for Sarcopterygii. There are differences in lateral line numbers on the trunk and the head between Actinistia and Dipnoi. Some amphibians (*Siren*, *Xenopus*) have four trunk lines, but three trunk lines prevail. Rudimentation leads to two lines (*Ascaphus*) or one line (*Ichthyophis*). Note the longitudinal stitches in *Lepidosiren* and *Siren* and the changes in stitch orientation in *Ambystoma* and *Xenopus*. Direct developing urodels (*Bolitoglossa*) do not develop neuromasts (ll−) and ampullary organs (ell−); some gymnophionans (*Typhlonectes*) develop no neuromasts (ll−) but ampullary organs (ell+), and all anurans lack ampullary organs (ell−). The lateral line organs of anurans may either disappear during metamorphosis or never develop (*Eleutherodactylus*; ll−). Neither neuromasts nor ampullary organs seem to develop in amniotes. Continuous lines indicate canals; short lines, stitches of more than three neuromasts; and dots, individual neuromasts. The distribution of ampullary organs is not shown. Abbreviations: all, angular line; dll, dorsal line; gll, gular line; ill, infraorbital line; jll, jugular line; mll, medial line; oll, occipital line; orll, oral line; sll, supraorbital line; vll, ventral line. Drawings are modified according to Escher (1925) and our own data.

(Fritzsch and Wake 1986; McCormick and Braford 1987), that anurans have only neuromasts, and that ampullary organs do not develop from posterior placodes indicate some ontogenetic and phylogenetic independence for both types of organs. This independence may justify a reconsideration of the term lateral line, which commonly refers to both the mechanosensory and the electrosensory system. In this chapter, the term lateral line will be used exclusively for the system of canal and free mechanoreceptive neuromasts, and the term electrosensory system will be used for electroreceptive organs.

2. Neuromast Organization in Amphibians

2.1 The Trunk

Neuromasts or stitches on the trunk form three lines in most amphibians—the dorsal, medial, and ventral lines. On the head, they form an occipital line and a few other short lines (Shelton 1970). All the trunk lines are supplied with nerve fibers from the posterior lateral line nerve originating from the glossopharyngeal and vagal complex. These nerve fibers form the superior, medial, ventral, and auricular rami (Escher 1925). The urodele *Siren* has a fourth line (Fig. 5.2), which apparently is a caudal elongation of the occipital group (Fritzsch 1988a) as it is in pipids (Escher 1925). The nerves run subcutaneously except in *Siren*, where they run deeper in myosepta. On the body of both urodeles and anurans, all primary neuromasts are oriented similarly, neuromasts in the middle and ventral body lines are oriented rostrocaudally, and neuromasts in the dorsal line are oriented dorsoventrally. Because the formation of stitches is either longitudinal or transverse (Lannoo 1987b), this results in a different overall orientation of stitches in urodeles (except Ambystomatidae) and anurans (Fig. 5.2).

Examples of reductions are the shortened dorsal line in *Bombina* and the loss of one line in *Ascaphus* (Fig. 5.2). A further reduction occurs in larval gymnophionans which have lost two lines so that a single line of neuromasts exists on the body (Sarasin and Sarasin 1887; Hetherington and Wake 1979) supplied by one nerve (Norris and Hughes 1918), a finding confirmed by this author for *Ichthyophis kohtaoensis* (Fig. 5.2). Another type of reduction leads to trunk lines that do not extend beyond the hind legs, as in the in utero larvae of the live-bearing *Salamandra atra* (Escher 1925).

In agreement with Goeldi (1898), this author recognizes at least four lines on the trunk of the lungfishes *Protopterus* and *Lepidosiren*, with neuromasts and longitudinal stitches oriented in a way similar to *Siren* (Fig. 5.2). In contrast to amphibians, only the ventral nerve runs subcutaneously, whereas the main trunk nerve runs alongside the notochord as in sharks (Fritzsch, unpublished). In the Australian lungfish *Neoceratodus*, at least three lines may be present (Fahrenholz 1929; Jarvick 1980). Only one trunk canal was described in the coelacanth fish *Latimeria* (Millot and Anthony 1965), and only a single lateral line nerve in the horizontal myoseptum is known.

The data in lungfish indicate that the presence of four trunk lines may be primitive for sarcopterygians, retained in some lungfish, one urodele, and some anurans. Out-group comparison indicates that the position of the trunk lateral line nerve close to the notochord (lungfish) is primitive. The position in myosepta (*Latimeria*, *Siren*) may be regarded as intermediate, and the subcutaneous position found in all other amphibians may be regarded as further derived. The variability in trunk lines may be achieved by a reduction in numbers from this primitive pattern. The absence of canals in derived lungfish and all amphibians, which have a comparable semiterrestric life-style (Escher 1925), may be caused by pedomorphic changes, because all canal neuromasts originate as superficial neuromasts during development (Münz, chapter 14). It is likely that this pedomorphic loss of canals may have been reached independently in the lungfish and the amphibian lineage.

2.2 The Head

Escher (1925) described four major lines on the head of amphibians (Fig. 5.2), supplied by distinct rami of the dorsal and ventral lateral line ganglia associated with the trigeminal gasserian ganglion and the facial geniculate ganglion, respectively: (1) the supraorbital line supplied by the superficial ophthalmic ramus; (2) the infraorbital line supplied by the buccal ramus; (3) the jugular, angular, and oral line supplied by the external mental ramus ("first and second jugular and primary oral line"—Holmgren and Pehrson 1949; "mandibular ramus"—Northcutt 1986); and (4) the postorbital and gular line supplied by the internal mental ramus ("third jugular and primary mandibular line"—Holmgren and Pehrson, 1949; "opercular ramus"—Northcutt 1986).

Although this pattern is generally accepted for urodeles and is sustained by embryological data, there is little agreement with respect to the nomenclature (Holmgren and Pehrson 1949). More recently a further subdivision of the lateral lines has been proposed, particularly rostral to the eyes (Lannoo 1987a). Among anurans, there appears to be more variability in the pattern of lines, and there is even more disagreement on the nomenclature, especially of the smaller lines. As is clear from the work of Escher (1925) and Lannoo (1987b), the lines on the cheek vary with respect to the size and position of the mouth. Furthermore, some lines become reduced, in particular the jugular and the gular lines (Escher 1925; Lannoo 1987b). Although named differently, the lines of gymnophionans appear to be homologous to the lines of other amphibians (Hetherington and Wake 1979), and they are innervated by four nerve branches, as in urodeles and anurans (Norris and Hughes 1918).

Four main lines, partly consisting of canals and partly of free neuromasts, are described in lungfish (Holmgren and Pehrson 1949). The innervation is much like that of amphibians (Northcutt 1986). Only three main lateral line canals are described in *Latimeria* (Millot and Anthony 1965), and the present pit lines contain no neuromasts (Hensel 1986). As with the trunk lateral line system, the head lateral line systems of lungfish and amphibians show similarities, whereas that of *Latimeria* is different. The interpretation of the latter as derived is supported by out-group comparison (Webb, Chapter 4) and the absence of ampullary organs or a rostral organ on the lower jaw (Northcutt 1986).

In conclusion, the entire pattern of neuromast arrangement on the head is in anurans the most derived among amphibians, as compared to the pattern that exists in lungfish, urodeles, and gymnophionans. This is also true with respect to the loss of ampullary organs in the anterior lateral line system. Both changes may be related to the new trophical niche that tadpoles explore as omnivorous, nonpredatory larvae (Duellman and Trueb 1985).

3. Central Pathways

As in lungfish and *Latimeria* (Northcutt 1986), the anterior lateral line nerve is composed of a dorsal and a ventral root in larval and most adult urodeles and larval gymnophionans (Fritzsch 1981; Fritzsch et al. 1985). The posterior lateral line nerve in all amphibians and the anterior lateral line nerve in larval and adult anurans have only a single root (Fritzsch et al. 1984a, Will et al. 1985). The dorsal root of urodeles, gymnophionans, and lungfish is formed by afferents of the ampullary organs which end in the dorsal nucleus (Fritzsch 1981; Northcutt 1983). In urodeles and gymnophionans, the afferents of the ventral roots, of the anterior and the posterior lateral lines each form two longitudinal fascicles which run alongside the intermediate nucleus (Fig. 5.3). These bundles, two from the trigeminal gasserian ganglion, two from the facial geniculate ganglion, and two from the vagal jugular ganglion, run adjacent but clearly separate from each other. The two afferents of a neuromast enter a different fascicle each, and it has been suggested that each fascicle is formed by fibers innervating hair cells of only on polarity (Fritzsch 1981). Electrophysiological data to support this suggestion are not available, however, and we do not know whether the segregated innervation of hair cells of one polarity has functional meaning and is retained beyond the second-order neurons of the alar plate.

A similar fasciculation and ordering exist in gymnophionans (Fritzsch et al. 1985), in lungfish (Northcutt 1983) and in many subteleosts (Bodznick, Chapter 33). In contrast, the organization of fibers into two fascicles is rare among anurans (Fritzsch et al. 1984a) and teleosts (McCormick, Chapter 17). In *Xenopus*, the lateral line afferents have numerous collaterals which form a complicated pattern that excludes a simple topographic relationship between peripheral neuromasts and their terminal fields (Will et al. 1985). Clearly, the topography of neuromasts is not reflected in any obvious way in the pattern of afferents, and little is known about the longitudinal organization of the second-order neurons, except that the rostral and caudal parts of the intermediate nucleus project to slightly different parts of the torus semicircularis (Will, Chapter 18). A better knowledge of this organization is necessary for an understanding of the information processing that leads to directional responses (Claas et al., Chapter 20).

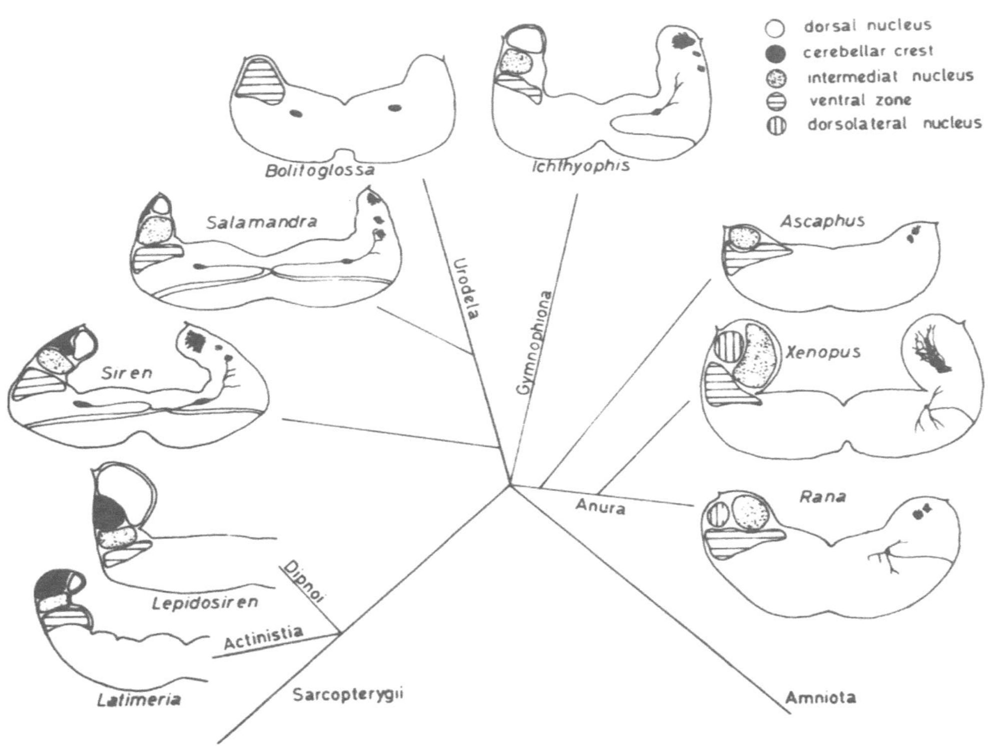

FIGURE 5.3. The nuclei of the alar plate, the pattern of lateral line afferents, and the efferent cells are shown in this drawing. Note the similar alar plate pattern in Dipnoi, Actinistia, and *Siren*. *Salamandra* has a lateral line only as larvae, whereas *Bolitoglossa* never develops a lateral line and lacks identifiable dorsal and intermediate nuclei. A projection of ampullary organs to the dorsal nucleus is also present in larval gymnophionans but is absent in all anurans. The octavolateralis efferent cells are bilateral with bilateral collaterals only in urodeles but are unilateral in gymnophionans and anurans, with characteristic differences in the course of the axon (arrowhead). *Ascaphus* seems to lack neuromast efferents. *Bolitglossa* may have retained the lateral line efferents because of their connection to the inner ear, whereas *Rana* may lose efferents innervating exclusively the lateral line during metamorphosis. Compiled after Fritzsch (1988a,b) and Will and Fritzsch (1988).

In all amphibians and the lungfish, the afferents are confined to the ipsilateral intermediate nucleus and do not extend beyond the rhombencephalic alar plate (Fritzsch 1988a). Besides lack of ampullary organs, anurans are the only amphibians that deviate from the lungfish/urodelan pattern, because their neuromast afferents may lack a fasciculated termination and form numerous long collaterals.

4. Lateral Line Efferents

In amphibians, most neuromasts receive at least one efferent fiber (Russell 1976). This fiber branches to reach all the sensory cells of a neuromast and all the neuromasts of a stitch, as can be shown using acetylocholine esterase stain (Fig. 5.1f,g). Ultrastructurally, the efferent terminal is characterized by electron-lucent vesicles and a subsynaptic cistern in the sensory cell (Russell 1976). Few such terminals are found in neuromasts on the head of gymnophionans (Wahnschaffe et al. 1985).

The cells of origin of lateral line efferents lie in the rhombencephalic reticular formation near the facial motor nucleus (Will 1982; Fritzsch and Crapon de Caprona 1984). They occur either bilaterally (urodeles) or unilaterally (anurans, gymnophionans). In gymnophionans and urodeles, some cells lie near the nucleus raphe, adjacent to the medial longitudinal fascicle. In particular, the paedomorphic plethodontid salamanders have more medially located cells than other salamanders. The axons of efferent cells form a medial loop in urodeles and gymnophionans, but they exit straight laterally in anurans (Fig. 5.3). In urodeles, the collaterals of efferents from the supraorbital line reach the contralateral lateral line and both labyrinths (Claas et al. 1981), where they establish efferent contacts on hair cells (Fritzsch and Wahnschaffe 1987). Common lateral line and inner-ear efferents have been suggested for anurans (Will 1982), whereas the available data on gymnophionans present no evidence either way (Fritzsch et al. 1984b).

Out-group comparison with chondrichthyes (Meredith and Roberts 1986) suggests that the presence of bilateral and common inner-ear/lateral line efferents may be primitive for gnathostome vertebrates. Interestingly, no efferents have been found with horseradish peroxidase for the lateral line of lampreys (Fritzsch et al. 1984b; Ronan and Northcutt 1987) or the inner ear of hagfishes (Amemiya et al. 1985). This absence may reflect a selective loss of efferents or, alternatively, efferents may be primitively absent in the lateral line and the inner ear of agnathans.

The efferent innervation of the inner ear only in lampreys and of inner ear and lateral line in gnathostome vertebrates may be regarded as a stepwise outgrowth of the efferent axon to the peripheral target. The presence of diencephalic lateral line efferents only in teleosts (Roberts and Meredith, Chapter 22) indicates that fiber outgrowth of central neurons to a peripheral target may happen, and supports the notion that this may have happened primitively for the medullary efferents, too. Among amphibians, only urodeles display the primitive bilateral distribution, whereas, for unknown reasons, gymnophionans and anurans show different degrees of reduction, including unilateralization of pericarya.

Furthermore, in *Ascaphus*, lateral line efferent cells may be absent, because none could be filled with HRP, whereas inner-ear efferents were readily labeled with HRP (Fritzsch et al. 1987). A further reduction may occur during metamorphosis in amphibians that lose their lateral line system. For example, it is conceivable that efferents cells that lack collaterals to the inner ear may degenerate in anurans after metamorphic loss of their target, the neuromasts. The function of efferents is discussed by Roberts and Meredith (Chapter 22).

5. Phylogenetic Loss

As revealed by comparison with gnathostome vertebrates, lungfish may represent the primitive sarcopterygiian pattern of the lateral line system, in addition found in *Siren*. Although *Latimeria* may be regarded as derived by loss of some components of the lateral line, it has a unique course of the posterior lateral line nerve in myosepta shared only by *Siren* among sarcopterygiians. This indicates that *Siren* most probably represents the primitive lateral line pattern of extant amphibians.

Among urodeles, there is reduction of the dorsalmost line, loss of stitch formation, and change of the mode of stitch formation. In utero embryos of the entirely terrestrial *Salamandra atra* have no stitches and shorter trunk lines before they lose their organs during metamorphosis (Escher 1925). Plethodontid salamanders, with a development in the egg on land (direct developers), apparently never have a lateral line system (Dent 1942) but are certainly insufficiently surveyed. More recent data (Wake, Roth, and Nishikawa, personal communication) indicate the presence of some lateral line fibers in some direct developing plethodontids. In gymnophionans, larvae show a reduced lateral line system (no stitches, a single trunk line) and seem to lose it during metamorphosis. Some in utero embryos and direct-developing embryos have ampullary organs (Fritzsch and Wake 1986) but no neuromasts. This lack contrasts with the presence of neuromasts and, presumably, ampullary organs in embryos of *Salamandra atra* (Escher 1925).

As stated above, the system of placodally derived organs of anurans is derived from the primitive pattern, as evidenced by the absence of the ampullary organs. Like gymnophionans, most anurans lose their lateral line during metamorphosis, but there are exceptions to this rule in many anuran families (Fritzsch et al. 1987). As in urodeles and gymnophionans, direct developers may completely suppress the formation of neuromasts (Lynn 1942), but knowledge is insufficient in this respect. Whether or not live-bearers develop neuromasts, as is known for *Pipa* embryos, is also largely unknown. Because the life-styles of anurans are very different, a vast source of as yet unexplored differences in the pattern of lateral line reduction may exist.

6. Metamorphic Loss

Metamorphosis leads to the loss of ampullary organs and neuromasts in gymnophionans (Norris and Hughes 1918) and some urodeles (Escher 1925; Fritzsch 1988a) and of the neuromasts in anurans (Wright 1951). Most urodeles that undergo metamorphosis retain the larval lateral line system and cover the partially dedifferentiated organs with epidermis during postlarval terrestrial life (Fritzsch and Wahnschaffe 1983), similar to lungfish and *Siren* during aestivation (Reno and Middleton 1973). This indicates that terrestrial life precludes function but does not necessarily lead to degeneration of neuromasts. When these animals enter the water during breeding or after flooding, they shed their skins, including the covers of the neuromasts. The retained neuromasts may play some as yet unknown role in the courtship of these animals or may be used in foraging. In anurans, the retention of a substantial part of the lateral line system is related to an aquatic life-style or at least to aquatic breeding (Fritzsch et al. 1987). All anurans lose at least the tail neuromasts. How they regulate the differential retention of some neuromasts is unknown.

The pattern of metamorphic lateral line loss is different among amphibians (Escher 1925). The lateral line nerves degenerate long after the neuromasts have disappeared in *Salamandra* (Göppert 1929) but degenerate much faster in anurans (Wahnschaffe et al. 1987). However, some lateral line afferents are retained in adult *Salamandra* (Fritzsch 1988a), a urodele that seems to lose all lateral line organs. Interestingly, the degeneration of the central pathways includes many second-order neurons in anurans (Wahnschaffe et al. 1987; Fritzsch et al. 1988), whereas in urodeles and gymnophionans, which lose all or some neuromasts, the second-order neurons apparently all survive metamorphosis (Will and Fritzsch 1988).

7. Conclusion

In extant amphibians, the lateral line system shows regressive tendencies that lead to reduction and loss of parts of the primitive amphibian lateral line and electrosensory system (rudimentation; Sewertzoff 1931). Part of this rudimentation is due to paedomorphosis: the canal formation is lost, individual lines are shortened or lost, stitches do not form, and bilateral distribution of efferent cells and their collaterals is lost. The reasons for this rudimentation are not clear, particularly because the functional significance of the fully developed structures are not yet clear. Likewise, neither the function of the two polarities of hair cells and afferents nor the reason for the maintenance of this organization in central nerve fascicles in all amphibians is understood. It is also unclear what advantage is offered by the segregated terminations of these fascicles of lateral line afferents in urodeles and gymnophionans and the changes of this organization in most anurans. It can be reasonably hypothesized that loss of ampullary organs in anurans may be related to changes in feeding behavior (Fritzsch and Münz 1986).

The metamorphic loss of the lateral line system may be interpreted as a loss of an embryonic organ like the tadpole tail (aphanisia; Sewertzoff 1931). This event shows interesting differences in the three amphibian orders. Primitively, the lateral line organs, when exposed to air, may have been covered with epidermal cells as in aestivating lungfish and *Siren*. Most modern urodels, which later return to water for mating, cover their lateral line organs and may thereby display the presumably primitive type of amphibian metamorphosis. Metamorphic events that lead to the complete loss of organs and most or all afferents, as in some urodeles and in gymnophionans, may be interpreted as an advanced type of metamorphosis independently evolved in these two orders. The loss of neuromasts, afferents, second-order cells, and perhaps efferent cells in many anurans during metamorphosis may be considered as a further derived type of amphibian metamorphosis present only in this amphibian order. The retention of neuromasts, afferents, and second-order neurons in some anurans either is primitive for anurans or may be interpreted as being even further derived (Fritzsch et at. 1987).

A complete suppression of lateral line formation has been developed independently in some direct-developing animals of all three amphibian orders and in amniotes but not among in utero embryos of amphibians. It would be interesting to know which constraints are imposed on the formation of neuromasts by an in ovo development to understand the way in which the development of an old sensory system was suppressed and presumably lost completely during evolution of amniotes.

Finally, it should be pointed out that the loss of the ability of ancestral amniotes to form lateral line organs may have been beneficial for the formation of auditory nuclei of modern amniotes. It has been proposed that the second-order lateral line neurons of the alar plate may have been rearranged after the loss of lateral line afferents and may have contributed to the formation of the auditory nuclei (Larsell 1934). However, the details of this supposed rearrangement are not yet clear (Fritzsch 1988b).

Acknowledgments. This work was supported by the Deutsche Forschungsgemeinschaft (Fr 572). I thank Dr. M.-D. Crapon de Caprona for linguistic advice, Dr. P. Görner for his continuous encouragement, and Drs. M. Lannoo, C.A. McCormick, and H. Münz for insightful discussions and comments on an earlier draft.

References

Amemiya F, Kishida R, Goris RC, Onishi H, Kusunoki T (1985) Primary vestibular projections in the hagfish, *Eptatretus burgeri*. Brain Res. 337: 73–79.

Claas B, Fritzsch B, Münz H (1981) Common efferents to the lateral line and labyrinthine systems in aquatic vertebrates. Neurosci Lett 27:231–235.

Dent JN (1942) The embryonic development of *Plethodon cinereus* as correlated with the differentiation and functioning of the thyroid gland. J Morphol 71:577–601.

Duellman WE, Trueb L (1985) Biology of Amphibians. New York: McGraw-Hill, p. 670.
Escher K (1925) Das Verhalten der Seitenlinienorgane der Wirbeltiere und ihrer Nerven beim Übergang zum Landleben. Acta Zool (Stockh) 6:307–414.
Fahrenholz C (1929) Uber die "Drüsen" und die Sinnesorgane in der Haut der Lungenfische. Z Mikr-Anat Forsch 16:55–74.
Fritzsch B (1981) The pattern of lateral line afferents in urodeles. A horseradish peroxidase study. Cell Tis Res 218:581–594.
Fritzsch B (1988a) The inner ear projections of larval and adult urodeles. Brain Behav Evol 37:325–348.
Fritzsch B (1988b) Phylogenetic and ontogenetic origin of the dorsolateral auditory nucleus of anurans. In: Fritzsch B, Ryan M, Wilczynski W, Hetherington T, Walkowiak W (eds) The Evolution of the Amphibian Auditory System. New York: Wiley, pp. 561–585.
Fritzsch B, Crapon de Carpona M-D (1984) The origin of centrifugal inner ear fibers of gymnophions. A horseradish peroxidase study. Neurosci Lett 46:131–136.
Fritzsch B, Münz H (1986) Electroreception in amphibians. In: Bullock TH, Heiligenberg W (eds) Electroreception. New York: Wiley, pp. 483–496.
Fritzsch B, Wahnschaffe U (1983) The electroreceptive ampullary organs of urodeles. Cell Tis Res 229:483-503.
Fritzsch B, Wahnschaffe U (1987) Electron microscopical evidence for bilateral and common lateral line and inner ear efferents in urodeles. Neurosci Lett 81:48–52.
Fritzsch B, Wake MH (1986) A note on the distribution of ampullary organs in gymnophions. J Herpetol 20:90–93.
Fritzsch B, Nikundiwe AM, Will U (1984a) Projection patterns of lateral line afferents in anurans. A comparative study using transganglionic transport of HRP. J Comp Neurol 229:451–469.
Fritzsch B, Crapon de Caprona M-D, Wächtler K, Körtje KH (1984b) Neuroanatomical evidence for electroreception in lampreys. Z Naturforsch 39c:856–858.
Fritzsch B, Wahnschaffe U, Crapon de Caprona M-D, Himstedt W (1985) Anatomical evidence for electroreception in larval *Ichthyophis kohtaoenisis.* Naturwissenschaften 72:102–104.
Fritzsch B, Drewes RC, Ruibal R (1987) On the retention of the lateral line nucleus in adult anurans. Copeia 1987 (1):127–135.
Fritzsch B, Wahnschaffe U, Bartsch U (1988) Metamorphic changes in the octavo-lateralis system of amphibians. In Fritzsch B, Ryan M, Wilczynski W, Hetherington T, Walkowiak W (eds) The Evolution of the Amphibian Auditory System. New York: Wiley, pp. 359–376.
Goeldi EA (1898) On the *Lepidosiren* of the Amazonas, being notes of five specimen obtained 1895–97, and remarks upon an example living in the Pra Museum. Trans Zool Soc Lond 14:413–420.
Göppert E (1929) Untersuchungen am Lateralissystem der Amphibien. Die Kopfganglien der Urodelen vor und nach der Metamorphose. Gegenbaurs Morhp Jahrb 62:507–542.
Harris GG, Milne DC (1966) Input-output characteristics of the lateral line sense organs. of *Xenopus*. J Acoust Soc Am 40:32–42.
Hensel K (1986) Morphologie et interprétation des canaux et canalicules sensoriels céphaliques de *Latimeria chalumnae* Smith, 1939 (Osteichthyes, Crossopterygii, Coelacanthiformes). Bull Mus natn Hist nat, Paris 8:379–407.
Hetherington TE, Wake MH (1979) The lateral line system in larval *Ichthyophis* (Amphibia: Gymnophiona). Zoomorphology 93:209–225.

Holmgren N, Pehrson T (1949) Some remarks on the ontogenetical development of the sensory lines of the cheek in fishes and amphibians. Acta Zool 30:249–314.

Jarvick E (1980) Basic Structure and Evolution of Vertebrates. London: Academic Press, 575 pp.

Karnovsky MJ, Roots L (1964) A "direct-coloring" thiocholine method for cholinesterases. J Histochem Cytochem 12:219–222.

Kingsbury BF (1895) The lateral line system of sense organs in some American Amphibia, and comparison with the dipnoans. Trans Am Microsc Soc 17:115–145.

Lannoo MJ (1987a) Neuromast topography in urodele amphibians. J. Morphol 191:247–263.

Lannoo MJ (1987b) Neuromast topography in anuran amphibians. J. Morphol 191:115–129.

Larsell O (1934) The differentiation of the peripheral and central acoustic apparatus in the frog. J Comp Neurol 60:473–527.

Lynn WG (1942) The embryology of *Eleutherodactylus nubicola*, an anuran which has no tadpole stage. Carnegie Inst Wash Pub 541:27–62.

McCormick CA, Braford MR Jr. (1987) Central connections of the octavolateralis system: Evolutionary considerations. In: Atema J, Fay RR, Popper AN, Tavolga WN (eds) Sensory Biology of Aquatic Animals, New York: Springer-Verlag pp. 773–756.

Meredith GE, Roberts BL (1986) Central organization of the efferent supply to the labyrinthine and lateral line receptors of the dogfish. Neuroscience 17:225–233.

Millot J, Anthony J (1965) Anatomie de *Latimeria chalumnae*, Vol II: Système Nerveux et Organes des Sens. Paris: Centre National de la Recherche Scientifique, 120 pp.

Norris HW, Hughes SP (1918) The cranial and anterior spinal nerves of the caecilian amphibians. J Morphol 31:489–557.

Northcutt RG (1983) The primary lateral line afferents in lepidosirenoid lungfishes. Soc Neurosci Abstr 9:1167.

Northcutt RG (1986) Electroreception in nonteleost bony fishes. In: Bullock TH, Heiligenberg W. (eds) Electroreception. New York: Wiley, pp. 257–285.

Reno HW, Middleton HH (1973) Lateral-line system of *Siren intermedia* Le Conte (Amphibia: Sirenidae), during aquatic activity and aestivation. Acta Zool (Stockh) 54:21–29.

Ronan M, Northcutt RG (1987) Primary projections of the lateral line nerves in adult lampreys. Brain Behav Evol 30:62–81.

Russell IJ (1976) Amphibian lateral line receptors. In: Llinas R, Precht W (eds) Frog Neurobiology. Berlin: Springer-Verlag, pp. 513–550.

Sarasin P, Sarsin F (1887) Ergebnisse naturwissenschaftlicher Forschungen auf Ceylon in den Jahren 1884–1886. Zur Entwicklungsgeschichte und über die Anatomie der cylonesischen Blindwühle *Ichthyophis glutinosus*. Wiesbaden: Kreidels Verlag, p. 320.

Shelton PMJ (1970) The lateral line system at metamorphosis in *Xenopus laevis* (Daudin). J. Embryol Exp Morphol 24:511–524.

Swertzoff AN (1931) Morphologische Gesetzmässigkeiten der Evolution. Jena, DRG: Fischer Verlag, p. 238.

Wahnschaffe U, Fritzsch B, Himstedt W (1985) The fine structure of the lateral line organs of larval *Ichthyophis* (Amphibia: Gymnophiona). J. Morphol 186:369–377.

Wahnschaffe U, Bartsch U, Fritzsch B (1987) Metamorphic changes within the lateral line system of Anura. Anat Embryol 175:431–442.

Will U (1982) Efferent neurons of the lateral line system and the VIIIth cranial nerve in the brainstem of anurans. Cell Tis Res 225:673–685.

Will U, Fritzsch B (1987) The octavus nerve of amphibians: Patterns of afferents and efferents. In :Fritzsch B, Ryan M, Wilczynski W, Hetherington T, Walkowiak W (eds) The Evolution of the Amphibian Auditory System. Chichester, U.K.: Wiley, pp. 159–184.
Will U, Luhede G, Görner P (1985) The area octavo-lateralis in *Xenopus laevis*. I. The primary afferent projections. Cell Tis Res 239:147–161.
Winklbauer R, Hausen P (1983) Development of the lateral line system in *Xenopus laevis*. II. Cell multiplication and organ formation in the supraorbital system. J Embryol Exp Morphol 76:283–296.
Wright MR (1951) The lateral line system of sense organs. Rev Biol 26:264–280.
Zakon HH (1984) Postembryonic changes in the peripheral electrosensory system of weakly electric fish: Addition of receptor organs with age. J. Comp Neurol 228: 557–570.

6 Evolution of Octavolateralis Sensory Cells

JØRGEN MØRUP JØRGENSEN

1. Introduction

According to the acousticolateralis hypothesis (Ayers 1892; Kingsbury 1895; van Bergeijk 1967), the vertebrate inner ear has its evolutionary origin in the lateral line system. The arguments for the hypothesis were based on (1) the origin of the lateralis nerves from the immediate vicinity of the eighth nerve to the inner ear, (2) the general resemblance of the sensory cells of the two systems, (3) the tendency of the canal neuromasts and inner ear to sink below the surface during ontogenetic development, and (4) the ontogenetic development of both systems from a common epibranchial thickening, or placode.

The hypothesis was generally accepted for many years but has recently been challenged by Wever (1976) and Northcutt (1980a). They argue that (1) the neuromasts and inner ear arise from different placodes rather than from a single placode (other types of sense organs are also placode derived), (2) the sensory nerves from the two systems are now known to project to different nuclei, and (3) hair cells are considered to have many features in common with receptor cells in general.

It is clearly important, at intervals, to reexamine hypotheses and see whether they stand up to recent data. The following is a comparison of the structure of the receptor cells belonging to the organs so far united in the acousticolateralis system, now often termed the octavolateralis system, with other vertebrate sensory organs and some invertebrate receptor cells. The hope is, together with other evidence, to give arguments relevant to the acousticolateralis hypothesis and to arrive at some picture of the evolution of hair cell receptors.

2. The Lateral Line Organs

2.1 Distribution of the Organs

Most cartilaginous and bony fishes have free neuromasts in the epidermis and a system of canals in the underlying connective tissue (Dijkgraff 1963; Coombs et al. 1988). Johnson (1917) found that the canal neuromasts in the elasmobranchs

Mustelus canis and *Squalus acanthias* form an almost continuous sensory epithelium. In bony fishes, the canal neuromasts are seen as distinct sensory epithelia positioned with regular intervals.

In myxinoid cyclostomes, lateral line organs have been described from the head of the pacific hagfish *Eptatretus stouti* by Ayers and Worthington (1907) under its old name *Bdellostoma dombei*. Recent examination of these structures, however, has revealed that they do not contain receptor cells or any innervation, so they probably represent a degenerate system (Fernholm 1985; Northcutt, Chapter 3). The other cyclostome class, the lampreys, have free neuromasts distributed on the head and trunk (Johnston 1905; Holmgren 1942).

Water-living amphibian larvae also have free neuromasts but no canal system. In caecilians and anurans, the neuromasts are lost during metamorphosis for species that have a terrestrial phase (Fritzsch, Chapter 5). Most urodeles have the neuromasts covered by epidermis cells during their terrestrial phase but redevelop the neuromasts each spring when they enter the water (Russell 1976).

These sense organs can be visualized by various techniques. In the living animals, methylene blue has been used to disclose the sensory organs and their innervation. A more selective stain is Daspei (dimethylaminostyrylethylpyridylethyl iodide, ICN/K&K Labs, New York) which in fluorescent light gives the receptor cells a bright yellow color on a dark background. Daspei and a related chemical, Dasmei, were originally introduced by Bereiter-Hahn (1976) for staining mitochondria and were subsequently applied to chloride cells in preparations of isolated skin (Marshall and Mishioka 1980).

In the laboratory, I found Daspei useful for staining neuromast sensory cells. Figure 6.1 shows some free neuromasts in two teleosts and a urodele amphibian stained with Daspei. The animals had been kept in a 2-μM solution in normal aquarium water for at least 5 h. As noted by Bereiter-Hahn (1976), the stain is of no or low toxicity. Experiments in the laboratory show that brook lampreys (*Lampetra planeri*), the fishes guppy (*Lebistes reticulatus*) and European eel (*Anguilla anguilla*), and the amphibian (*Triturus vulgaris*) appear unaffected after several days in a solution of the indicated strength, but even much stronger solutions seem harmless to these animals.

2.2 Morphology of the Lateral Line Sensory Organs

Neuromasts consist of sensory, supporting, and basal cells, surrounded by mantle cells (Figs. 6.2a, 6.3). The neuromasts are covered by a gelatinous cupula (probably not in lampreys), which reaches out into the surrounding water or into the canal fluid. The receptor cells occupy the apical part of the sensory epithelium. The supporting and mantle cells reach from the basal lamina to the apical surface, but the basal cells are found only in the lower part of the epithelium.

The receptor cells are secondary sensory cells, which basally form synaptic contacts with nerve endings. As shown in the diagram in Fig. 6.3a,b, there are differences in the shape and contents of the nerve endings. Some have free vesicles and mitochondria; they are the afferent nerve endings. The shape of these

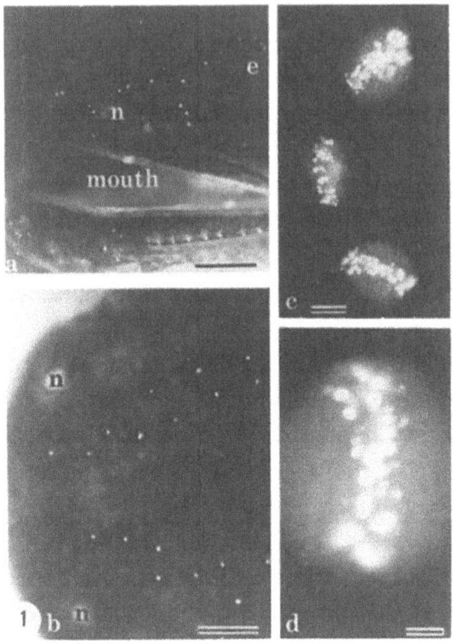

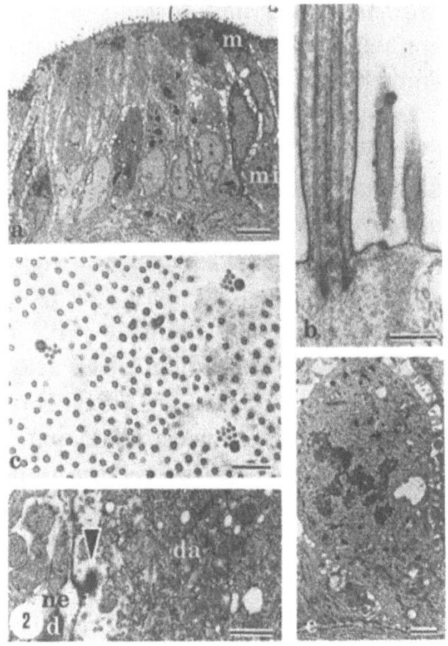

FIGURE 6.1. *(above left)* Distribution of some free neuromasts. The receptor cells appear as light spots after staining with the fluorescent dye Daspei. (a) Front of the head of a 9-cm larva of European eel (*Anguilla anguilla*). Positions of the eye (e) and nostril (n) are indicated. Scale, 0.5 mm. (b) Dorsal side of the anterior part of the head of a 6-cm adult common newt (*Triturus vulgaris*). Position of the nostrils (n) is shown. Scale, 0.5 mm. (c) 3 postorbital neuromasts of a 2 cm guppy (*Lebistes reticulatus*). Scale, 50 μm. (c) Higher magnification of a guppy neuromast. The intensity of the fluorescence varies among different hair cells. Scale, 20 μm.

FIGURE 6.2. *(above right)* Fine structure of the neuromast. (a) Cross section of a neuromast of a 9.5-cm ammocoete larva of the brook lamprey (*Lampetra planeri*). Mantle cells (m) surround the organ. A mitosis (mi) is seen in the lower right corner. Scale, 5 μm. (b) Section of the kinocilium and a few stereovilli. From a canal neuromast of the herring (*Clupea harengus*). Scale, 0.2 μm. (c) Section parallel with the surface of a free neuromast of a brook lamprey (9.5-cm ammocoete larva). Hair bundles of three hair cells are seen. Note the different orientation of the kinocilium relative to the stereovilli. Scale, 1 μm. (d) Basal region of a dark cell (da) of the herring canal neuromast. Note the synaptic body (arrowhead) opposite an afferent nerve ending (ne). Scale, 0.5 μm. (e) Higher magnification of the cell in mitosis from a section adjacent to that shown in Fig. 6.2a. Scale, 1 μm.

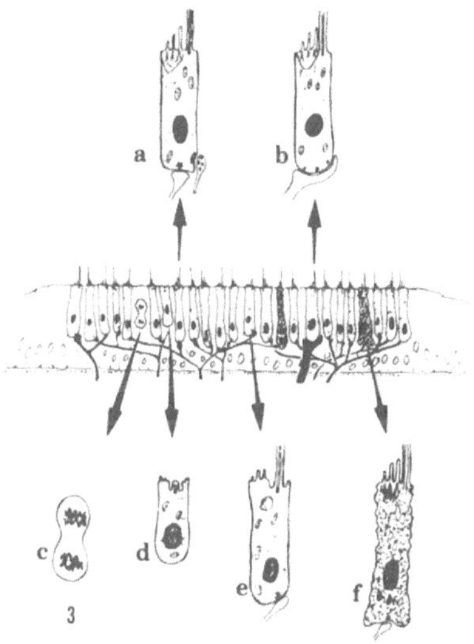

FIGURE 6.3. Components of a teleost canal neuromast. Bouton-innervated hair cells (a) are the most frequent hair cell, but cells innervated by large fibers forming contact with the basal part of the hair cells (b) are occasionally seen. Below are shown some steps in the life cycle of the hair cells with a mitosis (c) and a ciliary bud stage (d) followed by a stage with a longer kinocilium and some higher microvilli at the side of the kinocilium. In (f) is shown a dark cell with an irregular outline and many vacuoles.

endings varies considerably. Katsuki et al. (1952) found in the Japanese eel (*Anguilla japonica*) that the afferent fibers with large diameter innervate the central parts of the neuromast with little branching, whereas fibers of smaller diameter show extensive branching in both peripheral and central areas. Some of the endings of the thicker fibers may cover the basal part of the receptor cell. Such large afferent endings have been found by Trujillo-Cenoz (1961) in the neuromasts of two teleost fishes, by Murray (1955) in *Xenopus* neuromasts, and by Sato and Kawakami (1976) in the neuromasts of *Triturus pyrrhogaster*. In the receptor cell, opposite the afferent nerve ending, there is often a synaptic body surrounded by vesicles. Nerve endings filled with vesicles adjacent to a flat subsynaptic sac in the sensory cell are efferent endings.

The receptor cells are hair cells with an apical tuft of sensory hairs. One of these is the kinocilium with the usual ciliary axoneme of nine peripheral double and two central microtubules. It arises from a basal body with or without a basal foot. The kinocilium stands at the side of a bundle of specialized microvilli, usually termed stereocilia. Because stereocilia are not true cilia (they lack the microtubular axoneme), the term stereovilli, suggested by Neugebauer and Thurm (1985), will be used here.

The stereovilli are graduated in length in an organ-pipe-like arrangement, with the longest stereovilli closest to the kinocilium. Depending on the fixation used, connecting strands between the individual stereovilli and between the tallest stereovilli and the kinocilium may be seen. The basalmost part, the neck, of the stereovillus is narrower than the main part of the stereovillus. Stereovilli always appear as stiff rods, believed to pivot basally. They are filled with microfilaments, which basally condense and continue down into the cuticular plate as rootlets.

The so-called cuticular plate occupies most of the apical part of the receptor cell. Only an area along the margin and around the basal body of the kinocilium is free. The cuticular structure is composed of a finely filamentous material and has no resemblance to the cuticle known from various invertebrates and human nails; it is therefore an unfortunate designation. Below this plate many microtubules and mitochondria are found. Other components of the hair cell cytoplasm are microfilaments, ribosomes, some endoplasmatic reticulum, a few Golgi bodies, and numerous small vesicles.

The structure of the neuromast sensory epithelia in the different water-living anamniotic vertebrates is rather similar. Lamprey neuromasts were described recently by Yamada (1973) and Lane and Whitear (1982). Recent reports on the detailed structure of the neuromasts in the elasmobranchs *Scylliorhinus* and *Mustelus* have been published by Roberts and Ryan (1971) and Hama and Yamada (1977), respectively. Ratfish (*Chimaera monstrosa*) neuromasts have been examined by Ekström von Lubitz (1981), and various teleost neuromasts by Flock (1965), Hama (1965, 1978), Münz (1979), Jørgensen (1985), Marshall (1986), and others. Recent works dealing with the fine structure of various amphibian neuromasts were published by Jande (1966), Jørgensen and Flock (1973), Flock and Jørgensen (1974), Delfino et al. (1984), and Wahnschaffe et al. (1985).

2.3 Light and Dark Hair Cell

Although most hair cells appear rather pale in light and electron microscopic sections, others have a darker-staining cytoplasm (Figs. 6.2d, 6.3). There seems to be a continuous transition from light through gray to very dark cells. The pale cells have a rather smooth outline; the dark cells are irregular. The cytoplasm of the dark hair cells contains many vacuoles. Such dark hair cells have been described in the shark (*Carcharhinus*) canal neuromast by Tester and Kendall (1969) and in the free neuromasts of *Xenopus* by Shelton (1971) and Campantico et al. (1983). Denervation of *Xenopus* neuromasts causes darkening and finally disappearance of the hair cells (Delaveuve, 1974). It is possible that the dark cells represent degenerating receptors, but conclusive evidence is lacking.

The intensity of the fluorescence, after staining in Daspei, among different hair cells varies, as shown in Fig. 6.1d. Whether this can be related to differences in cytologic electron density is currently under examination. Such differences in fluorescence intensity of the mitochondria was observed by Bereiter-Hahn (1976), who suggested that they reflect differences in metabolism.

2.4 Development of Lateral Line Hair Cells

Mitotic figures are occasionally seen, both in the hair cell region and in basal parts of the neuromast (Figs. 6.2a,e; 6.4a). Autoradiographic experiments with ^{3}H-thymidine injected into European eels (*Anguilla anguilla*), 40–60 cm total length, for 14 days consecutively (total dose 1 mCi/kg body weight) demonstrate a rich mitotic activity (Fig. 6.4a,b) in their canal lateral line organs. Labeled

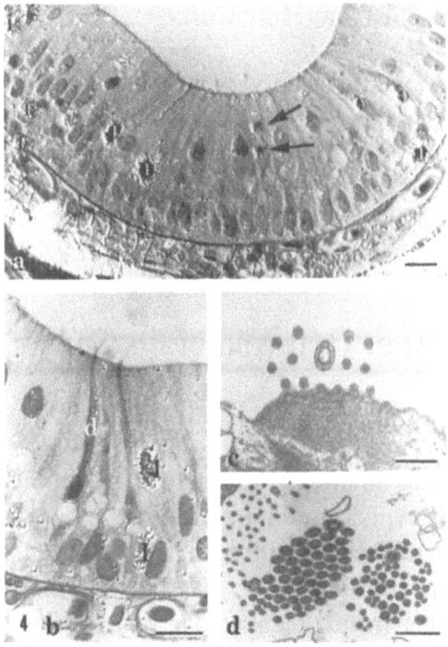

FIGURE 6.4. Developing octavolateralis hair cells. (a) Section of a canal neuromast of a 60-cm eel (*Anguilla anguilla*) that had been injected with ^{3}H-thymidine each day for 14 days. A mitosis (arrows) as well as several labeled nuclei (l) are seen. Interference contrast. Scale, 10 μm. (b) Section of a canal neuromast of an eel treated as in (a). Two labeled nuclei (l) and a dark cell (d) are seen. Scale, 10 μm. (c) Cross section of the sensory hairs of a hair cell of the inner ear of an Atlantic hagfish (*Myxine glutinosa*. The kinocilium is almost surrounded by the stereovilli. Scale, 0.5 μm. (d) Cross section of two hair bundles from the utricular macula in an adult adder (*Vipera berus*). The left bundle is typical. The right has thin stereovilli, which surround the kinocilium. Scale, 1 μm.

nuclei are found in all parts of the neuromast, with a majority in the basal part of the epithelium. Mitosis has previously been observed in the canal neuromasts of the shark *Carcharinus menisorrah* by Tester and Kendall (1969).

Cells with a short central cilium, surrounded by microvilli, as well as cells with thin stereovilli beside a kinocilium, have been found to represent different developmental stages of receptor cells in neuromasts on the regenerating tail of the urodele *Ambystoma mexicanum* (Flock et al., submitted).

In the dogfish *Scyliorhinus canicula*, Roberts and Ryan (1971) found two types of receptor cells. In light of the findings, reported above, there is reason to believe that the second type of receptor cell of Roberts and Ryan (1971), with thin stereovilli around a short cilium, is a developmental stage. It remains to be determined which cells in the neuromast divide; they may be undifferentiated resting cells, but so far we have no certain information (see Corwin et al., Chapter 8).

2.5 Turnover of Lateral Line Hair Cells

The finding of mitosis and different developmental stages as well as a series of cells showing different grades of necrotization may indicate a turnover of the sensory epithelium. Of course, we cannot rule out at present that only some of the cells are dying—e.g., cells that have been unable to obtain proper synaptic connections, or damaged cells.

3. Structure of Other Octavolateralis Sensory Cells

3.1 Inner-Ear Hair Cells

The receptor cells of the inner ear have hair tufts similar to those of the lateral line. In addition to the typical-looking hair bundles, with the kinocilium beside the stereovilli, other arrangements are seen. Figure 6.4c is from the apical region of the macula communis of the Atlantic hagfish (*Myxine glutinosa*). Figure 6.4d shows two hair bundles from the utricular macula of an adult adder (*Vipera berus*): the one is typical; the other has thin stereovilli surrounding the kinocilium and may represent a developmental stage. Similar hair bundles have previously been found in the inner ear of the Atlantic hagfish (Lowenstein and Thornhill 1970) and in the squirrel monkey (Spoendlin 1965).

The formation of the cilium and stereovilli, described in the lamprey inner ear by Thornhill (1972), shows a sequence similar to that described above for the receptor cells of the developing urodele neuromast. During the development of the inner ear of the mallard, mouse, and human, many of the sensory cells have initially a small cilium in the center of the luminal surface, surrounded by microvilli (Chandler 1984; Mbiene et al. 1984; Dechesne and Sans 1985; Mbiene and Sans 1986).

Formation of new hair cells in the inner-ear sensory epithelia of postlarval anamniote vertebrates has been found after injections with ^{3}H-thymidine in sharks (Corwin 1981), the European eel (*Anguilla anguilla*) (Mathiesen, in preparation), and the toad *Bufo marinus* (Corwin 1985, 1986). Injections with ^{3}H-thymidine in adult budgerigars (*Melopsittacus undulatus*) result in labeled hair and supporting cell nuclei in all vestibular sensory epithelia, but none has been found in the basilar papilla, the hearing organ (Jørgensen and Mathiesen 1988). In the developing inner ear of the mouse, terminal mitosis for vestibular sensory organs was found to occur between the 13th gestation day and birth (Ruben 1967; Desmadryl et al. 1985). Corwin (1981, 1985) found most labeled nuclei in a border zone. This contrasts with the findings of Mathiesen (in preparation) and Jørgensen and Mathiesen (1988), where the great majority of labeled nuclei are more central in position.

A hair cell turnover was suggested from an examination of the utricular macula of the caecilian *Ichthyophis glutinosa* by Jørgensen (1981). Dark hair cells, believed to be degenerating, are occasionally seen in inner ear vestibular epithelia of many different vertebrates (Jørgensen, unpublished observations).

3.2 Vesicles of Savi in Rays

These vesicular sense organs are found in *Torpedo* and a few other rays. The sensory epithelium is similar to that of neuromasts with typical hair cells contacted by afferent and efferent nerves (Nickel and Fuchs 1974). They are sensitive to local pressure and low-frequency vibration (Fuchs 1895, Szabo 1968; see also Barry and Bennett, Chapter 30 for further information).

3.3 Spiracular Organ in Cartilaginous and Nonteleost Bony Fishes

This type of sense organ is found in a blind-ending pouch, which opens to the spiraculum in elasmobranchs. In holosteans it is situated in a diverticulum in the hyomandibular cleft. In the dipnoans *Protopterus* and *Lepidosiren*, a corresponding sensory organ (Pinkus's organ) is found in a closed vesicle (Norris and Hughes 1920). In these lungfishes, Agar (1906) found the sense organs to be formed from the epiblastic part of the rudiment of the spiracle.

Recent examinations by Barry and Boord (1984) have shown that the sensory cells of the elasmobranch spiracular organ are hair cells with typical lateral line hair tufts (Barry and Bennett, Chapter 30).

3.4 Paratympanic Organ in Birds

This sense organ is found in a long tube close to the tympanic membrane in birds. It has a thin-walled window toward the middle ear. The sensory epithelium is very similar to that found in lateral line neuromasts and in the inner ear, with typical hair cells (Vitali 1912; Jørgensen 1984b; Giannessi and Pera 1986). Both afferent and efferent nerve endings are present. Degenerating cells have been seen. Calyceal innervation, characteristic for amniote vestibular sensory epithelia, has not been observed. Vitali (1912) claimed the paratympanic organ to be innervated by the facialis nerve. Yntema (1944) postulated from ablation experiments that the paratympanic organ arises from the first epibranchial placode, along with the geniculate ganglion, which innervates the organ. As the middle-ear cavity is considered homologous to the spiracle of elasmobranchs, the paratympanic organ is probably homologous to the spiracular organ. However, no comparable structure has been described in any amphibian or reptile. Nothing is known about the adequate stimulus.

3.5 Multivillous Cells and End Buds in Lampreys

In ammocoetes of *Lampetra planeri*, which have been stained in Daspei, a number of fluorescent cells appear. Figure 6.5a shows such cells along the tail tip of

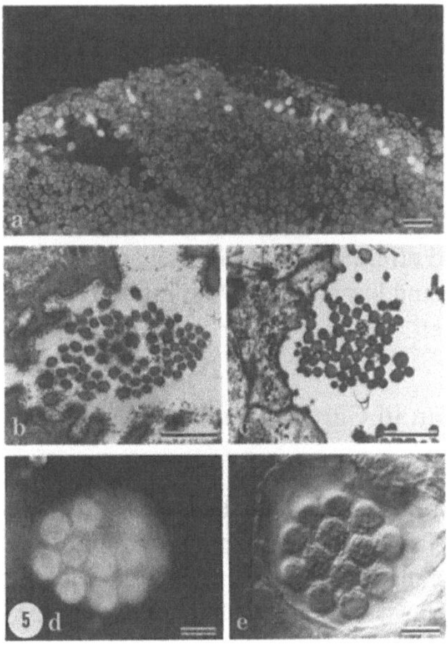

FIGURE 6.5. Structure of other octavolateralis organs. (a) Isolated, presumed multivillous cells from the tail tip of an ammocoete of the brook lamprey (*Lampetra planeri*) treated with Daspei. Scale, 50 μm. (b) Cross section of the hair bundle of a multivillous cell from the ammocoete of a brook lamprey. Only microvilli are present. Scale, 1 μm. (c) Section of a hair bundle of an ampullary receptor of the teleost *Xenomystus*. A cilium is found between the microvilli. Scale, 1 μm. (d) Ampullary organ of the transparent catfish *Kryptopterus* after exposure to Daspei. Scale, 20 μm. (e) Interference-contrast micrograph of an ampullary organ of *Kryptopterus*. Only very short microvilli are seen on the electroreceptors. Scale, 20 μm.

a 9-cm ammocoete. Electron microscopy of this region reveals cells with an apical bundle of microvilli (Fig. 6.5b) but no signs of a kinocilium. The cells, termed multivillous cells by Whitear and Lane (1983), have basal synaptic bodies resembling those found in neuromast and inner-ear hair cells. Only afferent nerve endings have been found.

Cells of similar position were found by Steven (1951) to be innervated by lateral line nerves. He found, confirming Young (1935), that the lateralis nerves innervate light receptors in the tail region. Whitear and Lane (1983) assume that the photoreceptive cells are identical to the multivillous cells.

End buds, sense organs consisting of several cells with many cytological characters in common with the multivillous cells, are found in lampreys after metamorphosis (Whitear and Lane 1983). The end buds have recently been shown to be electroreceptive (Ronan and Bodznick 1986). It is still a question whether the larval fluorescent cells or the multivillous cells are the forerunners of the adult end buds and are electroreceptive.

3.6 Ampullary Organs in Cartilaginous and Nonteleost Bony Fishes

The receptor cells of the ampullae of Lorenzini found in all cartilaginous fishes so far examined are characterized by an apical cilium (Waltman 1966; Szabo et al. 1972). Similar cells are seen in ampullary organs of the chondrostean genera *Scaphirhynchus* (Teeter et al. 1980), *Acipenser*, and *Polyodon* (Jørgensen et al. 1972; Jørgensen 1980). Sensory cells with microvilli around the cilium have been described from the brachiopterygian *Polypterus* (Jørgensen 1982). The dipnoan genera *Protopterus* and *Neoceratodus* also have this type of receptor cell. The microvilli of *Protopterus* receptor cells contain microfilaments which basally condense to form rootlets which continue into the apical part of the sensory cell (Jørgensen 1984a). In all examined organs, a synaptic body or sheet is found opposite afferent nerve endings. No efferent innervation has been observed. Dark receptor cells are occasionally found among the normal-looking cells, and mitosis has also been observed (Jørgensen 1982, 1984a). The extant coelacanth *Latimeria chalumnae* is believed to be electroreceptive (Hetherington and Bemis 1979; Northcutt 1980b). Recent electron microscopic examination has shown ciliated cells in the rostral organ, but the sensory nature of these cells is still uncertain (Jørgensen, unpublished observations). The central projection is to a dorsal nucleus (Northcutt 1980a).

In all the examined ampullary organs, and in bony fishes and amphibians, the receptor cells are separated by supporting cells that contrary to the sensory cells, reach from the basal lamina to the luminal surface. The supporting cells are believed to be involved in secretion of the gelatinous content of the ampulla.

The adequate stimuli for the ampullary organs are low-frequency electric fields (Kalmijn 1974). The receptors are excited by cathodal stimuli, which depolarize the luminal membrane (Bennett and Clusin 1979). They also respond to mechanical deformations and many other stimuli, but perhaps not under natural circumstances (Bullock 1974).

3.7 Ampullary Organs in Amphibia

Ampullary organs are distributed on the head of water-living stages of caecilian and urodele amphibians (Hetherington and Wake 1979; Fritzsch and Wahnschaffe 1983). The caecilian sensory cells have an apical cilium surrounded by a few microvilli (Wahnschaffe et al. 1985). The cave salamander *Proteus* has ampullary receptor cells with numerous microvilli, occasionally a cilium, and always a basal body (Istenic and Bulog 1984). Other urodele species examined have only microvilli (Fritzsch and Wahnschaffe 1983). All sensory cells described have basally a large synaptic body. The urodele ampullary organs are innervated by the anterior lateral line nerve. The fibers end in a dorsal nucleus. The sensory cells respond to low-frequency cathodal stimuli (Fritzsch and Münz 1986).

3.8 Ampullary Organs in Teleost Fishes

Ampullary electroreceptors are found in two very distantly related groups of teleosts. The one group includes the African forms *Mormyridae* and *Xenomystidae*; the other includes the species of the worldwide order *Siluriformes* and the South American *Gymnotiformes* (Bullock 1982). The variation in receptor cell structure is great; a few examples are given in Figures 6.5c,d,e, and 6.6. The luminal membrane may have a few short or numerous long microvilli (Szabo 1974). Only in *Xenomystus* has a short cilium been observed between the microvilli (Jørgensen and Bullock 1987). Basally synaptic bodies of various shapes are found opposite afferent nerve endings. Degenerating receptor cells were occasionally found in ampullary organs of mormyrids by Szamier and Bennett (1974).

Only afferent nerves have been seen. They project to a distinct nucleus that is believed not to be homologous to the electroreceptive nucleus in nonteleosts, and electroreception may have evolved independently in teleosts (Northcutt 1980a; Bullock et al. 1983). The ampullary receptors are sensitive to low-frequency electric fields (Kalmijn 1974). Anodal stimuli depolarize the basal membranes of the electroreceptor (Bennett and Clusin 1979).

3.9 Tuberous Organs in Teleost Fishes

These are found exclusively in electric fishes belonging to the orders *Mormyriformes* and *Gymnotiformes*. Characteristically they are situated in a closed cavity. The structural variation is great, as appears from the two examples in Figures 6.4 to 6.6. Only microvilli have been found luminally. Note the similarity between the tuberous and ampullary electroreceptor in *Gymnarchus* (Mullinger 1969). This suggests that the tuberous organs may be regarded as evolved from ampullary organs, which are "arrested" in their development. No efferent innervation has been observed. The tuberous organs are sensitive to high-frequency electric fields, with many electrophysiologically established types (Bullock 1982).

Anodal stimuli cause graded responses or all-or-none impulses in the basal membranes of the electroreceptor (Bennett and Clusin 1979). For recent reviews of electroreception, see Bullock and Heiligenberg (1986).

3.10 Electroreceptors in Monotreme Mammals

Unexpectedly, electroreception was recently claimed in platypus, *Ornithorhynchus anatinus*. The receptors are believed to be naked endings of trigeminal nerve fibers extending into ampullae of skin glands (Scheich et al. 1986). Such innervated glands have been described from the platypus bill by Andres and von Düring (1984). There is no evidence of any homology with octavolateralis organs.

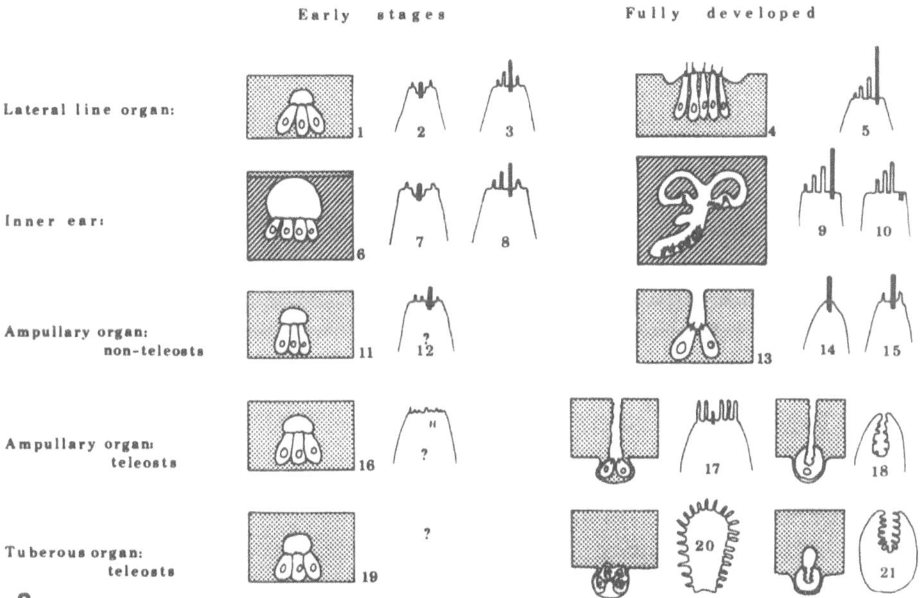

6

FIGURE 6.6. Comparison of early and fully developed stages of some octavolateralis organs. Early during its development, the lateral line organ forms an intraepidermal cavity above the primorial sensory organ **(1)**. The receptor cells form a ciliary bud in the center of the apical surface **(2)**. Later the cilium grows, and some of the microvilli elongate to form thin stereovilli **(3)**. The cavity later opens to the skin surface. The fully developed organ **(4)** has hair cells with the eccentric kinocilium at the side of the stereovilli **(5)**. The sense organs of the inner ear are formed in an otic vesicle **(6)**. At least some of the sensory cells form the stages **(7)** and **(8)**, resembling those of the neuromasts. The hair cells of the fully developed inner ear are similar to those of the neuromast **(9)**. In the mammalian hearing organ, the cilia disappear, leaving only the basal body **(10)**. The early stages in ampullary organ development, **(11, 12)**, are poorly known in nonteleosts. The fully developed organ has connection with the surrounding water **(13)**. The sensory cells may have an apical cilium, as in the ampullae of Lorenzini **(14)** of elasmobranchs, or the cilium may be surrounded by some microvilli, as in *Polypterus* and *Protopterus* **(15)**. The ampullary organs of teleosts are in the few examined species known to form an intraepidermal cavity **(16)** (Sato 1956; Srivastava and Seal 1981). The fully developed organ has in the African knife fish *Xenomystus* receptors with a cilium surrounded by microvilli **(17)**, whereas other teleosts, so far examined, lack the cilium. The apical part may form an invagination with some microvilli **(18)**, as in *Gymnarchus*. Early stages of tuberous organs **(19)** are not known. Later the cells may develop numerous microvilli, as in *Gnathonemus* **(20)** or *Gymnarchus* **(21)**.

4. Protochordate and Hemichordate Epidermal Receptor Cells

4.1 Cephalochordata

The buccal cirrhi of amphioxus (*Branchiostoma lanceolatum*) are equipped with groups of sensory cells, which fluoresce after staining in Daspei (Fig. 6.7a). Similar collared receptor cells are distributed over the entire body (Bone and Best 1978). They are pear-shaped with an apical cilium surrounded by microvilli (Figs. 6.7b,c,d; 6.8). Bone and Best (1978) described basal synapses and thus classified the cells as secondary sensory cells, as did Schulte and Riehl (1977). Baatrup (1981) found basal axonal processes, but no synapses, and considered the cells as primary sensory cells, as previously claimed by Langerhans (1876) and Dogiel (1903). Future examinations will show if more than one cell type exists, or whether the different findings are due to developmental changes of the same cell type.

Some of the collared sensory cells are suggested to be rapidly adapting mechanoreceptors (Bone and Best 1978), since Guthrie (1975) found a response from similar receptors to rapid increase of tension applied to the receptor area.

4.2 Urochordata

In tails of a number of ascidian larvae, intraepithelial nerves are found (Mackie and Bone 1976). In the tail of *Diplosoma macdonaldi*, Torrence and Cloney (1982) described a dorsal and ventral bundle consisting of 50 to 70 naked axons. These originate from receptor cells distributed at irregular intervals along the tail. The receptor cells are collared with a cilium surrounded by basal microvilli. The cilium arises from a deep pocket. It penetrates the inner cuticle and reaches out into the test (Fig. 6.8). Similar collared cells are also found in larval adhesive papillae (Torrence and Cloney 1983).

Cupular sense organs (Fig. 6.8) have been found in the salp *Salpa mucronata* by Lee (1891) and in the siphons of *Ciona* by Millar (1953) and Bone and Ryan (1978). The last authors made a detailed examination, which shows that collared receptor cells extend their cilia 100 μm or more into the cupula. The receptor cells, which lie among supporting cells, have an apical zone containing a granular or fibrous substance resembling the cuticular structure seen in vertebrate mechanoreceptive hair cells. The supporting cells are believed to secrete the cupula. Preliminary experiments indicated that the cupular organs are mechanoreceptive (Bone and Ryan 1978).

The similarity of the collared sensory cells in *Ciona* to the vertebrate hair cells was pointed out by Bone and Ryan (1978), as Reisinger (1969) had done previously in an examination of the arrowworm *Spadella*.

In *Salpa Fusiformis*, Bone et al. (1980) described a modified collar cell, which seems able to transmit electric activity from one individual to other individuals

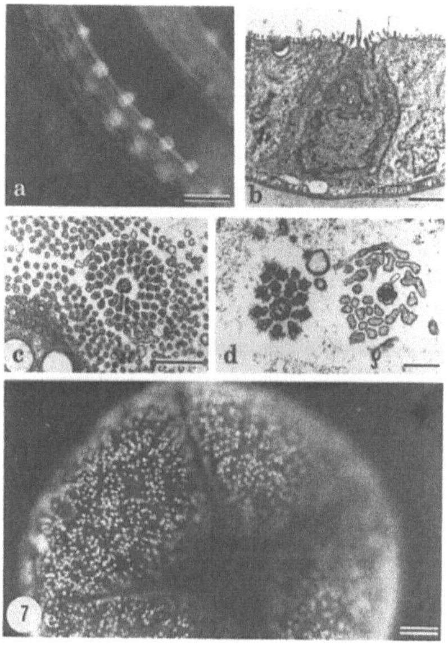

FIGURE 6.7. Protochordate and echinoderm receptor cells. (a) Part of the buccal cirrhi of amphioxus *Branchiostoma* after exposure to Daspei. The sense organs appear as light dots. Scale, 100 μm. (b) Isolated epidermal receptor cell from the rostrum of amphioxus. Scale, 2 μm. (c) Cross section of the hair bundle of a rostral collar cell of amphioxus. A cilium is surrounded by numerous microvilli. Scale, 1 μm. (d) Cross section of two adjacent hair bundles from the rostrum of amphioxus. The cilia are seen in the middle of the bundles. The edged outline of the microvilli is probably caused by the fixative. Scale, 1 μm. (e) End of a tube foot, from the starfish *Asterias rubens*, after exposure to Daspei. Light dots are believed to represent primary collared receptor cells. Scale, 20 μm.

of a salp chain. The cilium, widely branched, forms a synapselike junction with special cells of a neighbor individual (Fig. 6.8).

4.3 Enteropneusta

In nerve stained sections of the epithelium of *Saccoglossus* numerous single primary sensory cells are found (Bullock 1965). The outer endings may be of the collared type, described by Nørrevang (1964) from the skin of *Harrimania* and Saita et al. (1978) from *Glossobalanus*. The basal part of the collared cells has not been described.

4.4 Pterobranchia

In the tentacles of *Rhabdopleura compacta*, Dilly (1972) described collar cells arranged in rows. Nerve fibers were found to have vesicles close to the membrane

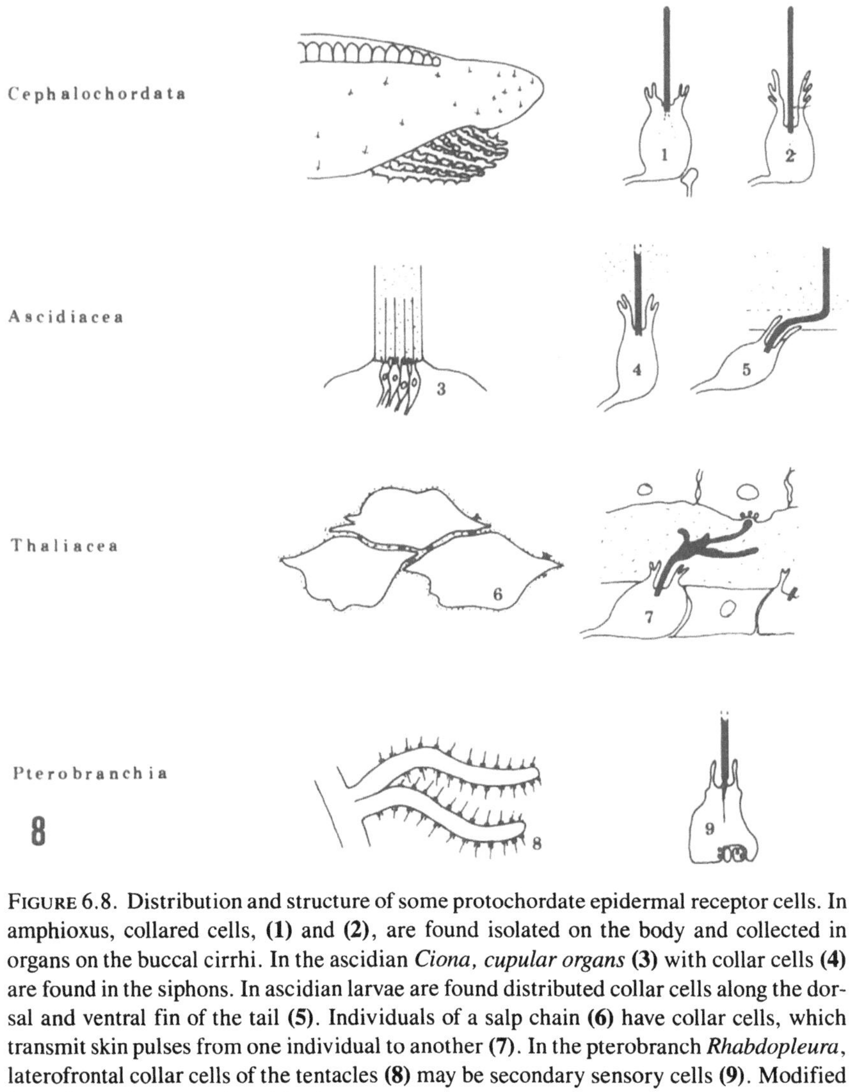

FIGURE 6.8. Distribution and structure of some protochordate epidermal receptor cells. In amphioxus, collared cells, **(1)** and **(2)**, are found isolated on the body and collected in organs on the buccal cirrhi. In the ascidian *Ciona, cupular organs* **(3)** with collar cells **(4)** are found in the siphons. In ascidian larvae are found distributed collar cells along the dorsal and ventral fin of the tail **(5)**. Individuals of a salp chain **(6)** have collar cells, which transmit skin pulses from one individual to another **(7)**. In the pterobranch *Rhabdopleura*, laterofrontal collar cells of the tentacles **(8)** may be secondary sensory cells **(9)**. Modified from the authors mentioned in the text.

of the ciliated cells. Rarely were vesicles found in the collar cell piled against the membrane close to the fibers (Figs. 6.4 to 6.8). Gilmour (1979) suggested that the collared laterofrontal cells may serve as particle detectors. Further investigations are needed to establish whether pterobranchia have secondary mechanoreceptors of this type.

5. Invertebrate Epidermal and Statocyst Sensory Cells

The detailed structure and function of invertebrate receptor cells are not very well investigated, with the exception of some arthropods, cephalopods, and a few other groups (Budelmann, Chapter 31). It is also important to remember that "ordinary" ciliated cells may have a distinct mechanosensitivity, as demonstrated by Thurm (1968) and Murakami and Eckert (1972). This applies to vertebrates as well, but organisms of simpler organization may rely more on less specialized cells.

Figure 6.9 shows selected epidermal and statocyst receptor cells (when possible from larvae), arranged as leaves on a conventional phylogenetic tree. The letters in the following list refer to the letters on the drawing. Encircled cells are statocyst sensory cells.

- **a** Collared receptor cell from a planula larva of the hydrozoan *Pennaria*, described by Martin and Thomas (1980). Similar cells are common on tentacles of the solitary coral *Balanophyllia* (Lyons, 1973).
- **b** Nematocyte from the hydromedusa *Gonionemus* described by Westfall (1970). Mechanical and/or chemical stimuli are necessary to elicit nematocyst discharge.
- **c** Medusae have statocysts described by Horridge (1969). The receptor cells are all collar cells with different numbers of surrounding microvilli.
- **d** Modified collar cell from the lip of *Beroë* (Aronova and Alekseeva 1984). The cell has a cilium and one to three stereovillus-like processes.
- **e** A long cilium about 100 μm in length is the only apical appendage to this receptor cell of *Leucothea* (Horridge 1965).
- **f** Several short cilia only a few micrometers long, *Leucothea* (Horridge 1965).
- **g** The receptors of the nematode *Heterodera* have from one to seven apical cilia (Endo 1980).
- **h** Proceriate and neohabdocoele turbellarians have a collar of eight microvilli (stereovilli) around the cilium (Ehlers and Ehlers 1977).
- **i** Modified collar cells from a ciliated pit in the miracidium of *Fasciola* (Wilson 1970).
- **j** Ciliated receptor cell of the trematode cercaria of *Neophasis* (Køie, 1973).
- **k,l** Mono- and multiciliary collar cells in the pilidium larva of *Lineus* and other species (Storch and Moritz 1971).
- **m** Monociliated collar cells in *Hirudo* (Phillips and Friesen 1982).
- **n** Multiciliated collar cells in *Sabella* and *Hirudo* (Santer and Laverack 1971; Phillips and Friesen 1982).
- **o,p** Monociliated receptor cells of crustaceans and insects (Schmidt and Gnatzy 1984).
- **q,r** Gastropods have receptor cells with exclusively microvilli, and others with one to 30 cilia (Crisp 1971; Zylstra 1972). Similar cells are found in other mollusks.

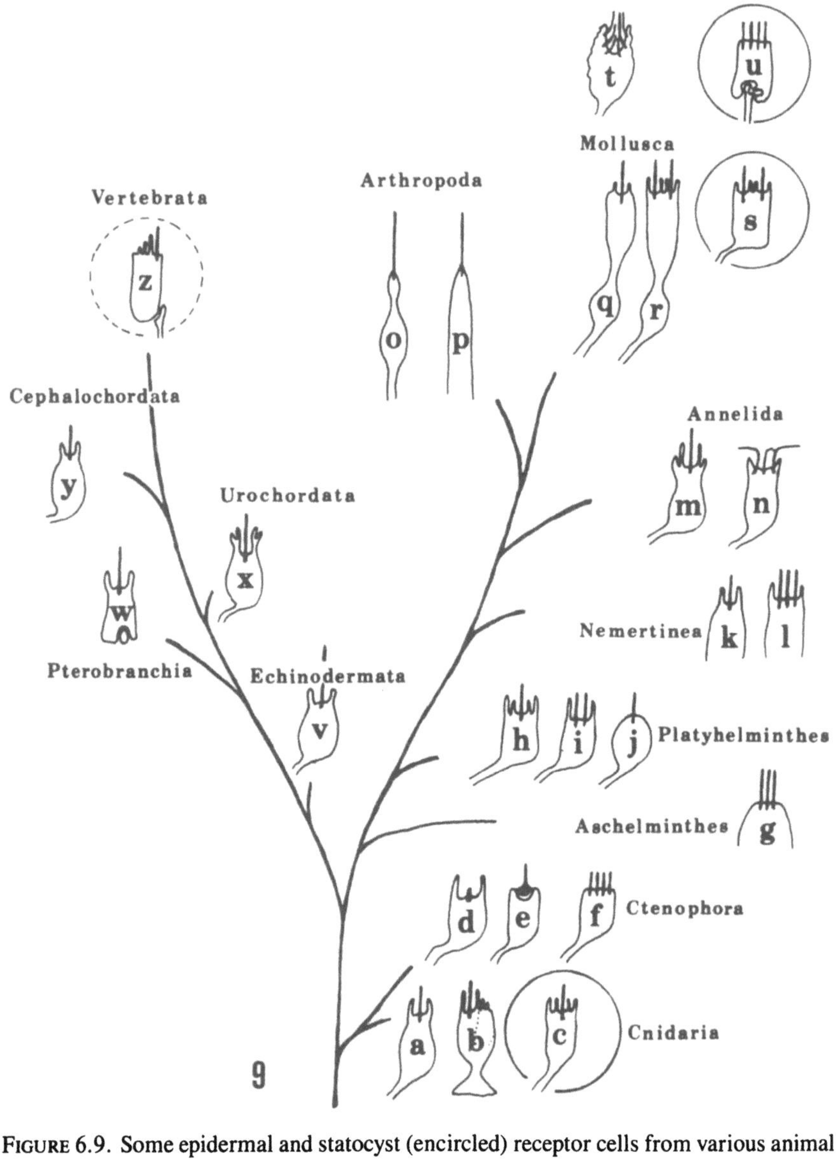

FIGURE 6.9. Some epidermal and statocyst (encircled) receptor cells from various animal groups. See text for explanation.

s Collared multiciliate statocyst receptor cells of the lamellibranch *Pecten* (Barber and Dilly 1969).
t In the cephalopod *Lolliguncula*, Emery (1975) found receptor cells with 20 to 30 cilia and numerous microvilli arising in an invagination.
u Multiciliated secondary receptor cell from the statocyst of the cephalopod *Octopus* (Colmers 1977).
v Collared receptor cell of *Echinus* and *Amphipholis* (Cobb 1967; Whitfield and Emson 1983).
w Collar cell (sensory function uncertain) in *Rhabdopleura* (Dilly 1972).
x Collar cell in *Ciona* (Bone and Ryan 1978).
y Collar cell in *Branchiostoma* (Baatrup 1981).
z Secondary receptor cell from lateral line and inner-ear sensory epithelia of vertebrates.

The receptor cells in protostomes, which form the right side of the diagram, are generally equipped with from one to many apical cilia. Secondary receptor cells have been claimed for many groups, but are best known from cephalopod statocysts. Deuterostomes, to the left in the drawing, are only known to have monociliated epidermal mechanosensory cells. It is interesting to see that some of the best-developed protostome and deuterostome groups, the cephalopods and vertebrates, have vestibular (statocyst or inner-ear) receptor cells built on two different principles. In mollusks the cilia must be involved in the transduction process, whereas in vertebrates, it seems to be the microvilli (stereovilli) (Hudspeth 1983). There is evidence for functional polarity of mollusk statocyst receptor cells, whether primary or secondary (Wolff 1973), as well as vertebrate hair cells (Flock 1965). The scheme, although very incomplete, shows that nothing like the vertebrate mechanoreceptive hair cell has been found in any other animal phylum, and it therefore must be considered unique.

6. Characteristics of the Octavolateralis Sensory Cells

The hair cells of the *lateral line organs*, including the vesicles of Savi, spiracular organs, and paratympanic organs, and the *inner ear* share the following characteristics:

1. The sense organs are believed to develop from placodes lying close to each other. In some lampreys, fishes, and amphibians, the lateralis ganglia are claimed to be formed from parts of the otic placode (Landacre 1910; Yntema 1937; Fisk 1957); in other species they may not do so.
2. The hair cells are secondary sensory cells situated in the apical half or two-thirds of the epithelium, never reaching the basal lamina. They are separated by supporting cells, which reach from the basal lamina to the apical surface.
3. The hair cells are very specialized mechanoreceptors with an organ-pipe-like arrangement of rodlike stereovilli aside a true cilium.

4. The first stage of hair cell development, in at least some animals, may be characterized by a central cilium surrounded by microvilli. This stage is reminiscent of the collared receptor cell seen abundantly among protochordates, hemichordates, and other invertebrate groups.
5. Many of the hair cells are innervated by nerves forming similar types of afferent and efferent nerve endings.
6. The central projection of the different octavolateralis organs are to areas positioned close to each other. They may involve separate nuclei, but cases with overlap are described (Bell 1981; Barry 1987; Will, Chapter 18). Overlap may be plesiomorphic to complete separation.
7. The same efferent nuclei may innervate both the inner ear and lateral line organs, as shown for two teleosts and an amphibian (Claas et al. 1981; Highstein and Baker 1985; Meredith and Roberts 1987).
8. Lateral line canals are formed from superficial neuromasts or primordia, which sink down into the connective tissue (Allis 1889; Omarkhan 1948). Closed vesicles may result, as in the vesicles of Savi. The simplest type of inner ear, as found in *Myxine*, has the shape of a closed ring.
9. Similar function. A bending of the hair tuft toward the kinocilium depolarizes the cell; a bending to the opposite side gives a hyperpolarization (Flock 1965; Hudspeth 1983).
10. Mechanoreceptive hair cells show some sensitivity to electric stimulation, although they are not as sensitive as ampullary and tuberous electroreceptors.

The early development of the ampullary and tuberous organs and the lamprey multivillous cells and end buds is only poorly known. The following characters are shared between the *mechanoreceptive* and *electroreceptive* epidermal organs:

1. The sense organ is covered by a gelatinous substance. The lamprey neuromasts and end buds are probable exceptions.
2. The sensory epithelium is composed of secondary sensory cells separated from each other and the basal lamina by supporting cells.
3. During their ontogenetic development, both types of organs form an intraepidermal cavity above the sensory cells.
4. The structure of the ampullary sensory cell in *Polypterus* and *Protopterus* is believed to be similar to an early stage of the mechanoreceptive hair cell.
5. Synaptic bodies surrounded by synaptic vesicles are found opposite afferent nerve endings.
6. Both types of organs are innervated by lateralis nerves. Electroreceptors have no efferent innervation, a condition shared with the free neuromasts in lampreys and many fishes as well as some inner-ear sensory epithelia—e.g., hagfish inner ear, frog basilar papilla, and inner hair cells of the mammalian organ of Corti. Lack of efferent innervation may be regarded as primitive.
7. Both organ systems project to neighboring parts of nuclei, or to nuclei close to each other.

8. Ampullae of Lorenzini and probably other electroreceptive organs respond to mechanical stimuli, although there is no evidence that they are used as mechanoreceptors.

7. Other Vertebrate Sense Organs

The sense organs described above differ in structure in a number of ways from the other main types of vertebrate sense organs. The cells in *taste buds* may often reach from the basal lamina to the apical surface. A clear distinction between receptor cells and other cell types is not possible. The cells typically have a number of apical microvilli, but cilia have rarely been found. They lack any sort of a synaptic body (Graziadei 1969; Murray 1969).

The *olfactory* epithelia contain receptor cells separated by supporting cells. The receptor cells are primary sensory cells (bipolar neurons) with or without apical cilia (typically six to 150) and with more or less unspecialized microvilli. If cilia are present, they have distal segments characterized by a reduction in diameter and number of microtubules. The mucous layer covering the organs is produced by distinct goblet cells (fish) or the Bowman's glands below the epithelium (amniotes) (Steinbrecht 1969; Reese and Brightman1979).

The retina in the *eye* is formed by evagination of the brain and thus is not placode-derived. It is multilayered with a number of distinct cell types. The photoreceptors have in addition to a modified cilium, forming the outer segment, a distinct synaptic ribbon. Bipolars as well as pinealocytes have similar synaptic structures (Sjöstrand 1953; Vollrath 1981).

When the different sense organs are compared, it is important to consider not only a single cell type, but whole organs or at least parts of organs. The cells termed supporting cells may have a number of important, unknown functions.

8. Evolution of the Octavolateralis Organs

Given the resemblance of mechanoreceptive lateral line and inner-ear organs and electroreceptive organs, compared with other types of sense organs, I find that there is still a basis for collecting these organs under a common heading, the octavolateralis (accousticolateralis) organs, as originally suggested by Ayers (1892), at a time when the electroreceptive organs were unknown. Whether the inner ear arose from the lateral line is uncertain. As mentioned by Northcutt (1986a), geological evidence is inconclusive. The detailed similarity with regard to ontogeny, morphology, and physiology, however, makes it difficult to imagine that the two systems arose independently of each other. Also, it seems most logical to assume that the epidermal lateral line structures, which sense events in the immediate vicinity of the animal, evolved prior to the deeper-situated otic organs, concerned with equilibrium and hearing at larger distances (Kalmijn 1988, Chapter 9).

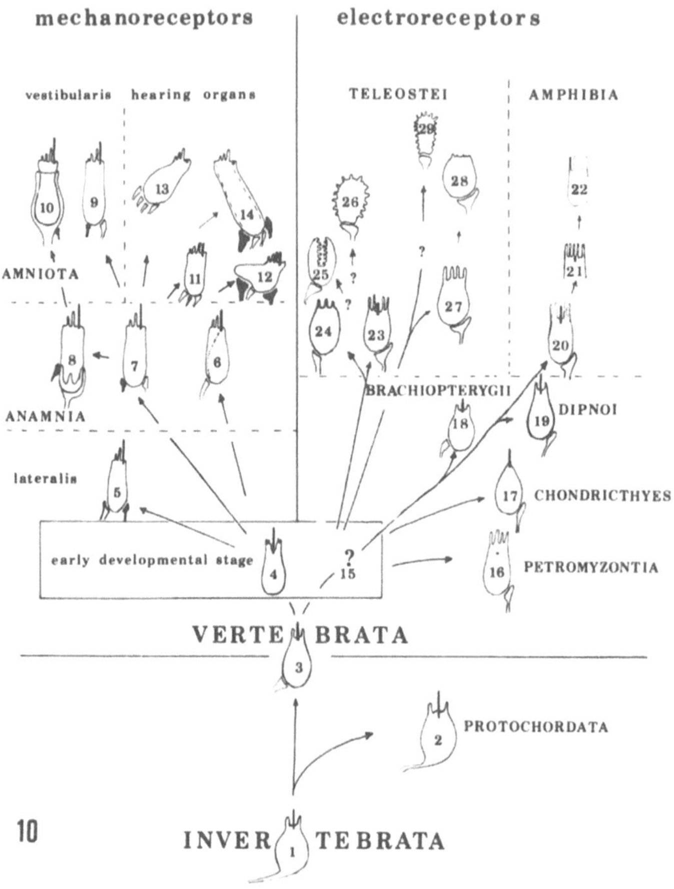

FIGURE 6.10. The suggested evolution of the octavolateralis receptor cells. See text for explanation.

Lissman and Mullinger (1968) suggested that the specialized lateral organs (electroreceptors) have developed from ordinary lateral line organs. Northcutt (1986b) found it reasonable to suspect that teleost electroreceptors arose from free neuromasts.

Figure 6.10 summarizes the ideas of phylogenetic development of the different types of receptor cells. Similar ideas were proposed by Jørgensen (1982). The numbers below refer to the figure. Efferent endings are black.

An invertebrate collared primary sensory cell **(1)** similar to the collar cell found in *Ciona* and *Branchiostoma* **(2)** may form the evolutionary basis for a collared secondary sensory cell without its own axon **(3).** A similar change in innervation is known from muscle cells; in echinoderms and amphioxus, some muscle cells send out processes, which in amphioxus are known to form synapses in the central nervous system (Flood 1966; Cobb 1967a). In vertebrates, the muscle cells are innervated peripherally.

An early stage of hair cell development may be interpreted as a modified collar cell **(4)**. It may develop to a neuromast hair cell **(5)**, or it may form a hair cell like that found in lamprey inner ear **(6)** (Lowenstein et al. 1968). It may also grow to the bouton-innervated hair cell, **(7, 9)**, the most widespread type of receptor cell in the vertebrate inner ear. Hair cells innervated by large candelabra-shaped nerve endings **(8)** are formed during ontogenetic development from bouton-innervated cells, before they reach the calyceal innervation **(10)** (Vasquez-Nin and Sotelo 1968; Favre and Sans 1969; Fink 1974), typical for amniote inner-ear vestibular organs. The cells of the amniote hearing organs may also be derived from **(7)**. In birds, long **(11)** and short **(12)** cells are found in the cochlea. The short hair cells are characterized by large efferent nerve endings (Smith 1985). Similar cells are present in the mammalian organ of Corti, with the modification that the kinocilia disappear during ontogenetic development. The inner hair cells **(13)** have no efferent innervation on the cell body, whereas the outer hair cells **(14)** have large efferent endings. Recent investigations indicate motor activity of the outer hair cells (Flock 1986).

The early stages of ontogenetic development of the different electroreceptors **(15)** are not described, so it is not known whether the lamprey electroreceptor **(16)** ever had an apical cilium. The cells of the ampullaie of Lorenzini in cartilaginous fishes **(17)** have apically only a cilium. Cells similar to the collared receptor cell **(3)** are found in brachiopterygian **(18)** and dipnoan **(19)** fishes. In the ampullary organs of the caecilian *Ichthyophis* and the urodele *Proteus* are found collar cells **(20)**; the latter has in most receptors only a basal body **(21)**. In other urodeles, even the apical basal body has disappeared **(22)**.

The different central projection pattern and polarity of the adequate stimulus of teleost electroreceptors, compared with those of nonteleosts, are some of the main arguments for assuming that teleost electroreceptors have evolved independently. They possibly evolved twice, a belief based on the distribution in two distantly related groups, and on the lack of electroreception in holosteans (Bullock et al. 1983). A collar cell is found in the ampullae of *Xenomystus*, but in ampullary organs of all other teleosts, no cilium has been found. The receptor from a mormyrid fish has only apical microvilli **(24)**; *Gymnarchus* has electroreceptors with microvilli in an invagination, and mormyrid tuberous organs have electroreceptors with large surface areas with microvilli **(26)**. Another line goes to ampullary organs of the catfishes *Ameiurus* **(27)** and *Kryptopterus* **(28)**. A gymnotid tuberous receptor **(29)** may have a large part of the cell membrane, with numerous microvilli, exposed to the intraepidermal cavity.

9. Summary and Conclusions

The lateral line and electroreceptive organs as well as many of the different sense organs of the inner ear are characterized by an overlying gelatinous substance secreted by supporting cells. Support cells reach as the only elements from the basal lamina to the apical surface. The receptors are secondary sensory cells with or without apical cilia, stereovilli, and/or microvilli. They all have synaptic bodies opposite afferent nerve endings. It is concluded that the organs naturally belong to a common system.

The identity of lateral line and inner ear organs with regard to ontogenetical development, structure, and function makes it unlikely that the organs arose independently of each other during evolution. The most logical sequence is that the inner ear arose from a lateral line-like type of organ, in accordance with the acousticolateralis hypothesis.

It is shown that, although the data still are very limited, the morphology of the developing lateral line and inner-ear hair cell and the electroreceptors in some nonteleost fishes and amphibians resemble receptor cells found in protochordate and urochordate animals, and it is suggested that the octavolateralis receptor cells may have evolved from epidermal collar cells found in deuterostome phyla.

Acknowledgments. The excellent technical assistance of Ilse Duun is highly acknowledged. I thank Drs. T.H. Bullock, A. Kalmijn, G.F. Striedter, R.G. Northcutt, J.T. Corwin, and S. Coombs for valuable comments, but the views expressed in this chapter do not necessarily reflect their opinions. The author was in most of his works supported by the Danish Natural Science Research Council.

References

Agar WE (1906) The spiracular gill cleft in *Lepidosiren* and *Protopterus*. Anat Anz 28:298–304.

Allis EP (1988) The anatomy and development of the lateral line system in *Amia calva*. J Morphol 2:463–567.

Andres KH, von Düring M (1984) The platypus bill. A structural and functional model of a pattern-like arrangement of different cutaneous sensory receptors. In: Hamann W, Iggo A (eds) Sensory Receptor Mechanisms. Singapore: World Science, pp. 81–89.

Aronova MZ, Alekseeva TM (1984) Ultrastructural organization of putative chemoreceptor cells in the lips of the ctenophore *Beroë cucumis*. J Evol Biochem Physiol 20:143–149.

Ayers H (1892) Vertebrate cephalogenesis. II. A contribution to the morphology of the vertebrate ear, with a reconsideration of its functions. J Morphol 6:1–360.

Ayers H, Worthington J (1907) The skin end-organs of the trigeminus and lateralis nerves of *Bdellostoma dombeyi*. Am J Anat 7:327–336.

Baatrup E (1981) Primary sensory cells in the skin of amphioxus (*Branchiostoma lanceolatum* (P.). Acta Zool (Stockh) 62:147–157.

Barber VC, Dilly PN (1969) Some aspects of the fine structure of the statocysts of the molluscs *Pecten* and *Pterotrachea*. Z Zellforsch 94:462–478.

Barry, MA (1987) Afferent and efferent connections of the primary octaval nuclei in the clearnose skate, *Raja eglanteria*. J. Comp Neurol 266:457–477.

Barry MA, Boord RL (1984) The spiracular organ of sharks and skates: Anatomical evidence indicating a mechanoreceptive role. Science 226:990–992.

Bell CC (1981) Central distribution of octavolateral afferents and efferents in a teleost (Mormyridae). J Comp Neurol 195:391–414.
Bennett MVL, Clusin WT (1979) Transduction at electroreceptors: Origins of sensitivity. In: Cone R, Dowling JE (eds) Membrane Transduction Mechanisms. New York: Raven Press, pp. 91–116.
Bereiter-Hahn J (1976) Dimethylaminostyrylmethylpyridiniumiodine (Dasmei) as a fluorescent probe for mitochondria in situ. Biochim Biophys Acta 423:1–14.
Bergeijk WA van (1967) The evolution of vertebrate hearing. In: Neff WD (ed) Contributions to Sensory Physiology. New York: Academic Press, pp. 1–49.
Bone Q, Best ACG (1978) Ciliated sensory cells in amphioxus (Branchiostoma). J. Mar Biol Assoc UK 58:479–486.
Bone Q, Ryan KP (1978) Cupular sense organs in *Ciona* (Tunicata: Ascidiacea). J Zool (Lond) 186:417–429.
Bone Q, Anderson PAV, Pulsford A (1980) The communication between individuals in salp chains. I. Morphology of the system. Proc R Soc Lond B 210:549–558.
Bullock TH (1965) Chaetognatha, Pogonophora, Hemichordata, and Chordata Tunicata. In: Bullock TH, Horridge GA (eds) Structure and Function in the Nervous Systems of Invertebrates, Vol. II. San Francisco: Freeman, pp. 1559–1592.
Bullock TH (1974) An essay on the discovery of sensory receptors and the assignment of their functions together with an introduction to electroreceptors. In: Fessard A (ed) Handbook of Sensory Physiology, Vol. III/3: Electroreceptors and Other Specialized Receptors in Lower Vertebrates. New York: Springer-Verlag, pp. 1–12.
Bullock TH (1982) Electroreception. Annu Rev Neurosci 5:121–170.
Bullock TH, Heiligenberg W (eds) Electroreception. New York: Wiley, 1986, 722 pp.
Bullock TH, Bodznick DA, Northcutt RG (1983) The phylogenetic distribution of electroreception: Evidence from convergent evolution of a primitive vertebrate sense modality. Brain Res Rev 6:25–46.
Campantico E, Guastalla A, Pirovano R. (1983) Ultrastructural aspects of the lateral-line organs of the normal and hypophysectomized clawed toad, *Xenopus laevis* (Daudin). Mon Zool Ital (NS) 17:53–163.
Chandler JP (1984) Light and electron microscopic studies of the basilar papilla in the duck, *Anas platyrhynchos*. II. Embryonic development. J Comp Neurol 222:523–542.
Claas B, Fritzsch B, Münz H (1981) Common efferents to the lateral-line and labyrinthine systems in aquatic vertebrates. Neurosci Lett 27:231–235.
Cobb JLS (1967a) The innervation of the ampulla of the tube foot in the starfish (*Astropecten irregularis*). Proc Soc Lond B 168:91–99.
Cobb JLS (1967b) The fine structure of the pedicellariae of *Echinus esculentus* (L.). II. The sensory system. J R Microsc Soc 88:223–233.
Colmers WF (1977) Neuronal and synaptic organization in the gravity receptor system of the statocyst of *Octopus vulgaris*. Cell Tissue Res 185:491–503.
Coombs S, Janssen J, Webb JF (1988) Diversity of lateral line systems: Evolutionary and functional considerations. In: Atema J. Fay RR, Popper AN, Tavolga WN (eds) Sensory Biology of Aquatic Animals. New York: Springer-Verlag, pp. 553–593.
Corwin JT (1981) Postembryonic production and aging of inner ear hair cells in sharks. J Comp Neurol 201:541–553.
Corwin JT (1985) Perpetual production of hair cells and maturational changes in hair cell ultrastructure accompany postembryonic growth in an amphibian ear. Proc Natl Acad Sci USA 82:3911–3915.

Corwin JT (1986) Regeneration and self-repair in hair cell epithelia: Experimental evaluation of capacities and limitations. In: Ruben RW, Van de Water TR, Rubel EW (eds) The Biology of Change in Otolaryngology. New York: Elsevier, pp. 291–304.

Crisp M (1971) Structure and abundance of receptors of the unspecialized external epithelium of *Nassarius reticulatus* (Gastropoda, Prosobranchia). J Mar Biol Assoc UK 51:865–890.

Dechesne CJ, Sans A (1985) Development of vestibular receptor surfaces in human fetuses. Am J Otolaryngol 6:378–387.

Delaveuve B (1974) Mise en évidence d'altérations dans les cellules sensorielles des neuromastes du Xénope aprés section du nerf latéral. Etude ultrastructurale. C R Acad Sci Paris 278:1063–1066.

Delfino G, Brizzi R, Calloni C (1984) Lateral line organs in *Salamandrina terdigitata* (Lacépède, 1788) (Amphibia: Urodela). Z Mikrosk-Anat Forsch Leipzig 98:161–183.

Desmadryl G, Raymond J, Sans A (1985) Histogenesis of the vestibular sensory epithelium in organotypic culture of mouse embryo otocysts: A tritiated thymidine autoradiographic study. Int J Dev Neurosci 3:549–557.

Dijkgraaf S (1963) The functioning and significance of the lateral line organs. Biol Rev 38:51–105.

Dilly PN (1972) The structure of the tentacles of *Rhabdopleura compacta* (Hemichordata) with special reference to neurociliary control. Z Zellforsch 129:20–39.

Dogiel AS (1903) Periphere Nervensystem des Amphioxus (*Branchiostoma lanceolatum*). Anat Hefte 21:145–213.

Ehlers U, Ehlers B (1977) Monociliary receptors in interstitial Proseriata and Neorhabdocoela (Turbellaria Neoophora). Zoomorphology 86:197–222.

Ekström von Lubitz DKJ (1981) Ultrastructure of the lateral-line sense organs of the ratfish, *Chimaera monstrosa*. Cell Tissue Res 215:651–665.

Emery DG (1975) Ciliated sensory neurons in the lip of the squid *Lolliguncula brevis* Blainville. Cell Tissue Res 157:323–329.

Endo BY (1980) Ultrastructure of the anterior neurosensory organs of the larvae of the soybean cyst nematode, *Heterodera glycines*. J Ultrastruct Res 72:349–366..

Favre D, Sans A (1979) Embryonic and postnatal development of afferent innervation in cat vestibular receptros. Acta Otolaryngol 87:97–107.

Fernholm B (1985) The lateral line system of cyclostomes. In: Foreman RE, Gorbman A, Dodd JM, Olsson R (eds) Evolutionary Biology of Primitive Fishes. New York: Plenum, pp. 113–122.

Fink DJ (1974) Morphogenesis of synaptic endings of colossal fibers in the chick vestibular epithelium. Anat Rec 178:354–355.

Fisk A (1957) Experiments on the development of the acoustico-facialis complex and associated structures in the lamprey *Lampetra planeri* (Bloch). Proc Zool Soc Lond 128:267–279.

Flock Å (1965) Electron microscopic and electrophysiological studies on the lateral line canal organ. Acta Otolaryngol (Suppl) 199:1–90.

Flock Å (1971) Sensory transduction in hair cells. In: Lowenstein WR (ed) Handbook of Sensory Physiology, Vol. I: Principles of Receptor Physiology. New York: Springer-Verlag, pp. 396–441.

Flock Å (1986) Mechanical properties of hair cells. In: Flock Å, Wersäll J (eds) Cellular Mechanisms in Hearing. New York: Elsevier, p. 80.

Flock Å, Jørgensen JM (1974) The ultrastructure of lateral line sense organs in the juvenile salamander *Ambystoma mexicanum*. Cell Tissue Res 152:283–292.

Flock Å, Flock B, Jørgensen JM (1988) The development of lateral line organs in the regenerating tail of the urodele *Ambystoma mexicanum*. (Submitted.)
Flood PR (1966) A peculiar mode of muscular innervation in amphioxus. Light and electron microscopic studies of the so-called ventral roots. J Comp Neurol 126:181–217.
Fritzsch B, Münz H (1986) Electroreception in amphibians. In: Bullock TH, Heiligenberg W (eds) Electroreception. New York: Wiley, pp. 483–496.
Fritzsch B, Wahnschaffe U (1983) The electroreceptive ampullary organs of urodeles. Cell Tissue Res 229:483–503.
Fuchs S (1895) Uber die Funktion der unter der Haut liegenden Kanalsysteme bei den Selachiern. Pflugers Arch Ges Phys 59:454–478.
Giannessi F, Pera L (1986) The ultrastructure of the paratympanic organ in the fowl (*Gallus gallus domesticus*). J Anat 147:191–199.
Gilmour THJ (1979) Feeding in pterobranch hemichordates and the evolution of gill slits. Can J Zool 57:1136–1142.
Graziadei PPC (1969) The ultrastructure of vertebrate taste buds. In: Pfaffmann C (ed) Olfaction and Taste. New York: Rockefeller University Press, pp. 315–330.
Guthrie DM (1975) The physiology and structure of the nervous system of amphioxus (the lancelet) *Branchiostoma lanceolatum* Pallas. Symp Zool Soc Lond 36:43–80.
Hama K (1965) Some observations on the fine structure of the lateral line organ of the Japanese sea eel *Lyncozymba nystromi*. J Cell Biol 24:193–210.
Hama K (1978) A study of the fine structure of the pit organ of the common Japanese sea eel *Conger myriaster*. Cell Tissue Res 189:375–388.
Hama K, Yamada Y (1977) Fine structure of the ordinary lateral line organ. II. The lateral line canal organ of spotted shark, *Mustelus manzo*. Cell Tissue Res 176:23–36.
Hetherington TE, Bemis WE (1979) Morphological evidence of an electroreceptive function of the rostral organ of *Latimeria chalumnae*. Am Zool 19:986.
Hetherington TE, Wake MH (1979) The lateral line system in larval *Ichthyophis* (Amphibia: Gymnophiona). Zoomorphology 93:209–225.
Highstein SM, Baker R (1985) Action of the efferent vestibular system on primary afferents in the toadfish, *Opsanus tau*. J Neurophysiol 54:370–384.
Holmgren N (1942) General morphology of the lateral sensory line system of the head in fish. K Svenska Vet-Ak Hdl 20:3–46.
Horridge GA (1965) Non-motile sensory cilia and neuromuscular junctions in a ctenophore independent effector organ. Proc R Soc Lond B 162:333–350.
Horridge GA (1969) Statocysts of medusae and evolution of stereocilia. Tissue Cell 1:341–353.
Hudspeth AD (1983) Mechanoelectrical transduction by hair cells in the acousticolateralis sensory system. Annu Rev Neurosci 6:187–215.
Istenic L, Bulog B (1984) Some evidence for the ampullary organs in the European cave salamander *Proteus anguinus* (Urodela, Amphibia). Cell Tissue Res 235:393–402.
Jande SS (1966) Fine structure of lateral-line organs of frog tadpoles. J Ultrastruct Res 15:496–509.
Johnson SE (1917) Structure and development of the sense organs of the lateral canal system of selachians (*Mustelus canis* and *Squalus acanthias*). J Comp Neurol 28:1–74.
Johnston JB (1905) The cranial nerve components of *Petromyzon*. Morph Jahrb 34:149–203.
Jørgensen JM (1980) The morphology of the Lorenzinian ampullae of the sturgeon *Acipenser ruthenus* (Pisces: Chondrostei). Acta Zool (Stockh) 61:87–92.

Jørgensen JM (1981) On a possible hair cell turn-over in the inner ear of the caecilian *Ichthyophis glutinosus* (Amphibia: Gymnophiona). Acta Zool (Stockh) 62:171–186.

Jørgensen JM (1982) Fine structure of the ampullary organs of the bichir *Polypterus senegalus* Cuvier, 1829 (Pisces: Brachiopterygii) with some notes on the phylogenetic development of electroreceptors. Acta Zool (Stockh) 63:211–217.

Jørgensen JM (1984a) On the morphology of the electroreceptors of the Australian and one African lungfish species. Vidensk Meddr Dansk Naturh Foren 145:77–85.

Jørgensen JM (1984b) Fine structure of the paratympanic organ in the avian middle ear. Acta Zool (Stockh) 65:89–94.

Jørgensen JM (1985) On the fine structure of lateral-line canal organs of the herring (*Clupea harengus*). J Mar Biol Assoc UK 65:751–758.

Jørgensen JM, Bullock TH (1987) Organization of the ampullary organs of the African knife fish *Xenomystus nigri* (Teleostei: Notopteridae). J Neurocytol 16:311–315.

Jørgensen JM, Flock Å (1973) The ultrastructure of lateral line sense organs in the adult salamander *Ambystoma mexicanum*. J Neurocytol 2:133–142.

Jørgensen JM, Mathiesen C (1988) The avian inner ear: Continuous production of hair cells in vestibular sensory organs, but not the auditory papilla. Naturwissenschaften 75:319–320.

Jørgensen JM, Flock Å, Wersäll J (1972) The Lorenzinian ampullae of *Polyodon spathula*. Z Zellforsch 130:363–377.

Kalmijn A (1974) The detection of electric fields from inanimate and animate sources other than electric organs. In: Fessard A (ed) Handbook of Sensory Physiology, Vol. III/3: Electroreceptors and Other Specialized Receptors in Lower Vertebrates. New York: Springer-Verlag, pp. 147–200.

Kalmijn A (1988) Hydrodynamic and acoustic field detection. In: Atema J, Fay RR, Popper AN, Tavolga WN (eds) Sensory Biology of Aquatic Animals. New York: Springer-Verlag, pp. 83–130.

Katsuki Y, Mizuhira V, Yoshino S (1952) On the endorgan of the acousticolateralis system of fish. Jpn J Physiol 2:93–102.

Kingsbury BF (1895) The lateral line system of sense organs in some American Amphibia, and comparison with the dipnoans. Trans Am Microsc Soc 17:115–154.

Køie M (1973) The host-parasite interface and associated structure of the cercaria and adult *Neophasia lageniformis* (Lebour, 1910). Ophelia 12:205–219.

Lane EB, Whitear M (1982) Sensory structures at the surface of fish skin. II. Lateralis system. Zool J Linn Soc 76:19–28.

Landacre FL (1910) The cranial ganglia of *Ameiurus*. J Comp Neurol Psychol 20:309–411.

Langerhans P (1876) Zur Anatomie des *Amphioxus lanceolatus*. Arch Mikrosk Anat 12:290–348.

Lee AB (1891) On a little-known sense-organ in *Salpa*. J Microsc Sci 32:89–97.

Lissmann HW, Mullinger AM (1968) Organization of ampullary electric receptors in Gymnotidae (Pisces). Proc R Soc Lond B 169:345–378.

Lowenstein O, Thornhill RA (1970) The labyrinth of *Myxine*: Anatomy, ultrastructure and electrophysiology. Proc Soc Lond B 176:21–42.

Lowenstein O, Obsorne MP, Thornhill RA (1968) The anatomy and ultrastructure of the labyrinth of the lamprey (*Lampetra fluviatilis*). Proc R Soc Lond B 170:113–134.

Lyons KM (1973) Collar cells in planula and adult tentacle ectoderm of the solitary coral *Balanophyllia regia* (Anthozoa Eupsammiidae). Z. Zellforsch 145:57–74.

Mackie GO, Bone Q (1976) Skin impulses and locomotion in an ascidian tadpole. J Mar Biol Assoc UK 56:751–768.
Marshall NJ (1986) Structure and general distribution of free neuromasts in the black goby, *Gobius niger*. J Mar Biol Assoc UK 66:323–333.
Marshall WS, Nishioka RS (1980) Relation of mitochondria-rich chloride cells to active chloride transport in the skin of a marine teleost. J Exp Zool 214:147–156.
Martin VJ, Thomas MB (1980) Nerve elements in the planula of the hydrozoan *Pennaria tiarella*. J Morhopol 166:27–36.
Mbiene J-P, Sans A (1986) Differentiation and maturation of the sensory hair bundles in the fetal and postnatal vestibular receptors of the mouse: A scanning electron microscope study. J Comp Neurol 254:271–278.
Mbiene J-P, Favre D, Sans A (1984) The pattern of ciliary development in fetal mouse vestibular receptors. Anat Ambryol 170:229–238.
Meredith GE, Roberts BL (1987) Distribution and morphological characteristics of efferent neurons innervating end organs in the ear and lateral line of the European eel. J Comp Neurol 265:494–506.
Millar RH (1953) *Ciona*. LMBC Mem Typ Br Mar Pl Anim 35:1–123.
Mullinger AM (1964) The fine structure of ampullary electric receptors in *Amiurus*. Proc R Soc Lond B 160:345–359.
Mullinger AM (1969) The organization of ampullary sense organs in the electric fish, *Gymnarchus niloticus*. Tissue Cell 1:31–52.
Murakami A, Eckert R (1972) Cilia: Activation coupled to mechanical stimulation by calcium influx. Science 175:1375–1377.
Murray RG (1969) Cell types in rabbit taste buds. In: Pfaffmann C (ed) Olfaction and Taste. New York: Rockefeller University Press, pp. 331–344.
Murray RW (1955) The lateralis organs and their innervation in *Xenopus laevis*. Q J Microsc Sci 96:351–361.
Münz H (1979) Morphology and innervation of the lateral line system in *Sarotherodon niloticus* (L.) (Cichlidae, Teleostei). Zoomorphology 93:73–86.
Neugebauer DC, Thurm U (1985) Interconnections between the stereovilli of the fish inner ear. Cell Tissue Res 240:449–453.
Nickel E, Fuchs S (1974) Organization and ultrastructure of mechanoreceptors (Savi vesicles) in the elasmobranch *Torpedo*. J Neurocytol 3:161–177.
Nørrevang A (1964) Choanocytes in the skin of *Harrimannia kupfferi* (Enteropneusta). Nature 204:398–399.
Norris HW, Hughes SP (1920) The spiracular sense-organ in elasmobranchs, ganoids and dipnoans. Anat Rec 18:205–209.
Northcutt RG (1980a) Central auditory pathways in anamniote vertebrates. In: Popper AN, Ray RR (eds) Comparative Studies of Hearing in vertebrates. New York: Springer-Verlag, pp. 79–118.
Northcutt RG (1980b) Anatomical evidence of electroreception in the coelacanth (*Latimeria chalumnae*). Zbl Vet Med C Anat Histol Embryol 9:289–295.
Northcutt RG (1986a) Evolution of the octavolateralis system: Evaluation and heuristic value of phylogenetic hypotheses. In: Ruben RW, Van de Vater TR, Rubel EW (eds) The Biology of Change in Otolaryngology. New York: Elsevier, pp. 3–14.
Northcutt RG (1986b) Electroreception in non-teleost bony fishes. In: Bullock TH, Heiligenberg W (eds) Electroreception. New York: Wiley, pp. 257–285.
Omarkhan M (1948) The lateral sensory canals of larval *Notopterus*. Proc Zool Soc Lond 118:938–972.

Phillips CE, Friesen WO (1982) Ultrastructure of the water-movement-sensitive sensilla in the medicinal leech. J Neurobiol 13:473–486.
Reese TS, Brightman MW (1970) Olfactory surface and central olfactory connexions in some vertebrates. In: Wolstenholme GEW, Knight J (eds) Taste and Smell in Vertebrates. London: Churchill, pp. 115–143.
Reisinger E (1969) Ultrastrukturforschung und Evolution. Ber Phys Med Ges Würzburg NF 77:5–47.
Roberts BL, Ryan KP (1971) The fine structure of the lateral-line sense organs of dogfish. Proc R Soc Lond B 179:157–169.
Ronan MC, Bodznick D (1986) End buds: Non-ampullary electroreceptors in adult lampreys. J Comp Physiol A 158:9–15.
Ruben RJ (1967) Development of the inner ear of the mouse: A radioautographic study of terminal mitoses. Acta Otolaryngol (Suppl) 220:1–44.
Russell IJ (1976) Amphibian lateral line receptors. In: Llinas R, Precht W (eds) Frog Neurobiology. New York: Springer, pp. 513–550.
Saita A, Castellani LC, Tripepi S (1978) The integument of *Glossobalanus minutus* Kowalevsky (Enteropneusta Ptycoderidae). Ultrastructural analysis. Mon Zool Ital (NS) 12:155–179.
Santer RM, Laverack MS (1971) Sensory innervation of the tentacles of the polychaete *Sabella pavonia*. Z Zellforsch 122:160–171.
Sato A, Kawakami I (1976) The fine structure of the lateral-line organ of larvae of the newt, *Triturus pyrrhogaster*. Annotness Zool Jap 49:131–141.
Satô M (1956) Studies on the pit-organs of fishes. IV. The distribution, histological structure and development of the small pit-organs. Ann Zool J 29:207–212.
Scheich H, Langner G, Tidemann C, Coles RB, Guppy A (1986) Electroreception and electrolocation in platypus. Nature 319:401–402.
Schmidt M, Gnatzy W (1984) Are the funnel-canal organs the "campaniform sensilla" of the shore crab, *Carcinus maenas* (Decapoda, Crustacea)? II. Ultrastructure. Cell Tissue Res 327:81–93.
Schulte E, Riehl R (1977) Elektronenmikroskopische Untersuchungen an den Oralcirren und der Haut von *Branchiostoma lanceolatum*. Helgolander Wiss Meeresunters 29:337–357.
Shelton PMJ (1971) The structure and function of the lateral line system in larval *Xenopus laevis*. J Exp Zool 178:211–232.
Sjöstrand FS (1953) The ultrastructure of the inner segments of the retinal rods of the guinea pig eye as revealed by electron microscopy. J Cell Comp Physiol 42:45–70.
Smith CA (1985) Inner ear. In: King AS, McLelland J (eds) Form and Function in Birds III. London: Academic Press, pp.273–310.
Spoendlin HH (1965) Ultrastructural studies of the labyrinth in squirrel monkeys. In: The Role of the Vestibular Organs in the Exploration of Space. Washington: NASA, SP-77, pp. 7–22.
Srivastava CB, Seal M (1981) Electroreceptors in Indian catfish teleosts. Adv Physiol Sci 31:1–11.
Steinbrecht, RA (1969) Comparative morphology of olfactory receptors. In: Pfaffmann C (ed) Olfaction and Taste. New York: Rockefeller University Press, pp. 3–21.
Steven DM (1951) Sensory cells and pigment distribution in the tail of the ammocoete. J Microsc Sci 92:233–247.
Storch V, Moritz K (1971) Zur Feinstrucktur der Sinnesorgane von *Lineus ruber* O.F. Müller (Nemertini, Heteronemertini). Z Zellforsch 117:212–225.

Szabo T (1968) Analyse morphologique et fonctionelle de l'épithélium d'un m'ecano-récepteur. Actual Neurol 8:131–147.
Szabo T (1974) Anatomy of the specialized lateral line organs of electroreception. In: Fessard A (ed) Handbook of Sensory Physiology, Vol. III/3: Electroreceptors and Other Specialized Receptors in Lower Vertebrates. New York: Springer-Verlag, pp. 13–58.
Szabo T, Kalmijn AJ, Enger PS, Bullock TH (1972) Microampullary organs and a sub-mandibular sense organ in the freshwater ray, *Potamotrygon*. J Comp Physiol 79:15–27.
Szamier RB, Bennett MVL (1974) Special cutaneous receptor organs of fish. VII. Ampullary organ of mormyrids. J Morphol 143:365–384.
Teeter JH, Szamier RB, Bennett MVL (1980) Ampullary electroreceptors in the sturgeon *Scaphirhynchus platorynchus* (Rafinesque). J Comp Physiol 138:213–223.
Tester AL, Kendall JI (1969) Morphology of the lateralis canal system in the shark genus *Carcharhinus*. Pac Sci 23:1–16.
Thornhill RA (1972) The development of the labyrinth of the lamprey (*Lampetra fluviatilis* Linn. 1758). Proc R Soc Lond B 181:175–198.
Thurm U (1968) Steps in the transducer process of mechanoreceptors. Symp Zool Soc Lond 23:199–216.
Torrence SA, Cloney RA (1982) Nervous system of ascidian larvae: Caudal primary sensory neurons. Zoomorphology 99:103–115.
Torrence SA, Cloney RA (1983) Ascidian larval nervous system: Primary sensory neurons in adhesive papillae. Zoomorphology 102:111–123.
Trujillo-Cenóz O (1961) Electron microscope observations on chemo- and mechano-receptor cells of fishes. Z Zellforsch 54:654–676.
Vasquez-Nin GH, Sotelo JR (1968) Electron microscope study of the developing nerve terminals in the acoustic organs of the chick embryo. Z Zellforsch 92:325–338.
Vitali G (1912) Di un interessante derivato dell'ectoderma dells prima fessura branchiale nel passero. Un organo nervoso de senso nell'orecchio medio degli uccelli. Anat Anz 40:631–639.
Vollrath L (1981) The pineal organ. In: Oksche A, Vollrath L (eds) Handbuch der mikroskopischen Anatomie des Menshcen VI/7. Berlin: Springer-Verlag, 665 pp.
Wachtel AW, Szamier RB (1969) Special cutaneous receptor organs of fish. IV. Ampullary organs of the nonelectric catfish, *Kryptopterus*. J Morphol 128:291–308.
Wahnschaffe U, Fritzsch B, Himstedt W (1985) The fine structure of the lateral-line organs of larval *Ichthyophis* (Amphibia: Gymnophiona). J Morhpol 186:369–377.
Waltman B (1966) Electrical properties and fine structure of the ampullary canals of Lorenzini. Acta Physiol Scand 66 (Suppl 264):1–60.
Westfall JA (1970) The nematocyte complex in a hydromedusan, *Gonionemus vertens*. Z Zellforsch 110:457–470.
Wever EG (1976) Origin and evolution of the ear of vertebrates. In: Masterton RB, Bitterman ME, Campbell CBG, Hotton N (eds) Evolution of Brain and Behavior in Vertebrates. Hillsdale, NJ: Lawrence Erlbaum, pp. 89–106.
Whitear M, Lane EB (1983) Multivillous cells: Epidermal sensory cells of unknown function in lamprey skin. J Zool (Lond) 210:259–272.
Whitfield PJ, Emson RH (1983) Presumptive ciliated receptor associated with the fibrillar glands of the spines of the echinoderm *Amphipholis squamata*. Cell Tissue Res 232: 609–624.
Wilson RA (1970) Fine structure of the nervous system and specialized nerve endings in the miracidium of *Fasciola hepatica*. Parasitology 60:399–410.

Wolff HF (1973) Multi-directional sensitivity of statocyst receptor cells of the opisthobranch gastropod *Aplysia limacina*. Mar Behav Physiol 1:361–373.
Yamada Y (1973) Fine structure of the ordinary lateral line organ. I. The neuromast of lamprey, *Entosphenus japonicus* J Ultrastruct Res 43:1–17.
Yamada Y, Hama K (1972) Fine structure of the lateral-line organ of the common eel, *Anguilla japonica*. Z Zellforsch 124:454–464.
Yntema CL (1937) An experimental study of the origin of the cells which constitute the VIIth and VIIIth cranial ganglia and nerves in the embryo of *Amblystoma punctatum*. J Exp Zool 75:75–101.
Yntema CL (1944) Experiments on the origin of the sensory ganglia of the facial nerve in the chick. J Comp Neurol 81:147–167.
Young JZ (1935) The photoreceptors of lampreys. I. Light-sensitive fibers in the lateral line nerves. J Exp Biol 612:229–238.
Zylstra U (1972) Distribution and ultrastructure of epidermal sensory cells in the freshwater snails *Lymnaea stagnalis* and *Biomphalaria pfeifferi*. Neth J Zool 22:283–298.

7
Organization and Development of the Zebrafish Posterior Lateral Line

WALTER K. METCALFE

1. Introduction

Early studies of the developing lateral lines of amphibians demonstrated that a variety of well-orchestrated cellular behaviors give rise to a rather simple morphological pattern in the embryo. In 1904, Harrison first demonstrated that neuromasts of the posterior lateral lines of frogs (*Rana palustris*, *R. Sylvatica*) are derived from cells that migrate from the head region, and that the posterior lateral line nerve connects the posterior lateral line ganglion with the migrating primordium. Later, in studies of the salamander (*Ambystoma punctatum*), Stone (1922) demonstrated that the posterior lateral line ganglion cells and the primordial cells that form neuromasts both arise from a common postotic placode. Cells in the rostral part of the placode become the sensory neurons, and cells in the caudal part form the migratory primordium. As the primordium migrates caudally along the midbody line, clusters of cells are deposited at intervals from the trailing end of the primordium. These cell clusters subsequently differentiate to become the first, or "primary," neuromasts of the midbody line. Thus, the development of the posterior lateral line involves a variety of fundamentally important but poorly understood cellular activities, such as the formation and differentiation of a placode, initiation and guidance of cell migration, and axon outgrowth and guidance. The posterior lateral line system may prove to be a good model system in which to study these phenomena because of its relative simplicity (being a simple linear array of receptors in the embryo) and accessibility (the receptors are located superficially within the epidermis) and because of the considerable background of information that is already available regarding the development of this system.

Although much is known about the development of lateral lines of amphibians, the development of fish lateral lines has been poorly understood until recently, particularly with regard to the origin of neuromasts (for review, see Wright 1951). This chapter will briefly review published reports that show that the posterior lateral line of the zebrafish develops in much the same way as in amphibians (Metcalfe et al. 1985; Metcalfe 1985). Furthermore, new results about the development of the central axons of posterior lateral line ganglion cells will be provided, and where appropriate, speculation about mechanisms that may be involved in the development of this system are included.

The experimental animal used in these studies is the zebrafish, or common zebra danio (*Brachydanio rerio*), a teleost of the family Cyprinidae, which also includes goldfish (*Carassius auratus*). Some of the advantages of studying zebrafish include:

1. Zebrafish are easily maintained in the laboratory using standard aquaculture techniques (Eaton and Farley 1974).
2. A single tank of about 15 adults may lay 30 to 100 eggs daily.
3. The fish are stimulated to spawn by light onset, which can be controlled by the experimenter.
4. The eggs are easily staged during early cleavage stages to provide a time of fertilization within about 15 min.
5. The eggs and embryos are optically clear, which allows direct visualization of embryogenesis by using suitable optical methods such as differential interference contrast optics.
6. The embryos develop rapidly, allowing one to observe significant developmental changes in a relatively short period of time.
7. Cloning (Streisinger et al. 1981) and mutagenesis (Chakrabarti et al. 1983; Walker and Streisinger 1983) procedures have been developed using zebrafish, which allow analysis of the roles of particular genes in development.

2. Developing Neuromasts and Their Pattern of Distribution

There are three phases in the development of the pattern of distribution of neuromasts. First, a row of six neuromasts is laid down by the migrating primordium at approximately five-somite intervals in the embryo; second, additional neuromasts appear so that a single neuromast is present over each of the myotomes of the larva; and third, a vertical row of neuromasts forms on each scale of the midbody line of the adult.

2.1 Origin of Neuromasts

In the embryonic zebrafish, the placode that will give rise to the midbody line of the posterior lateral line can be observed in living embryos by 18 h (h = hours postfertilization at 28.5°C) by using Nomarski (DIC) optics. Close inspection of the postotic region reveals a mass of approximately 100 tightly packed cells lying immediately under the periderm (the outer layer of the skin). Within the next hour of development, this postotic placode splits into two distinct cell masses. The smaller and more rostral cell mass, which remains stationary, is the ganglion of the posterior lateral line. The caudal cell mass, which is now termed the primordium of the midbody line, begins a caudal migration at a rate of about 100 µm/h that will carry some of these cells to the base of the caudal fin by 40 h (Fig. 7.1). The large size of the migrating primordium results in a bump on the surface

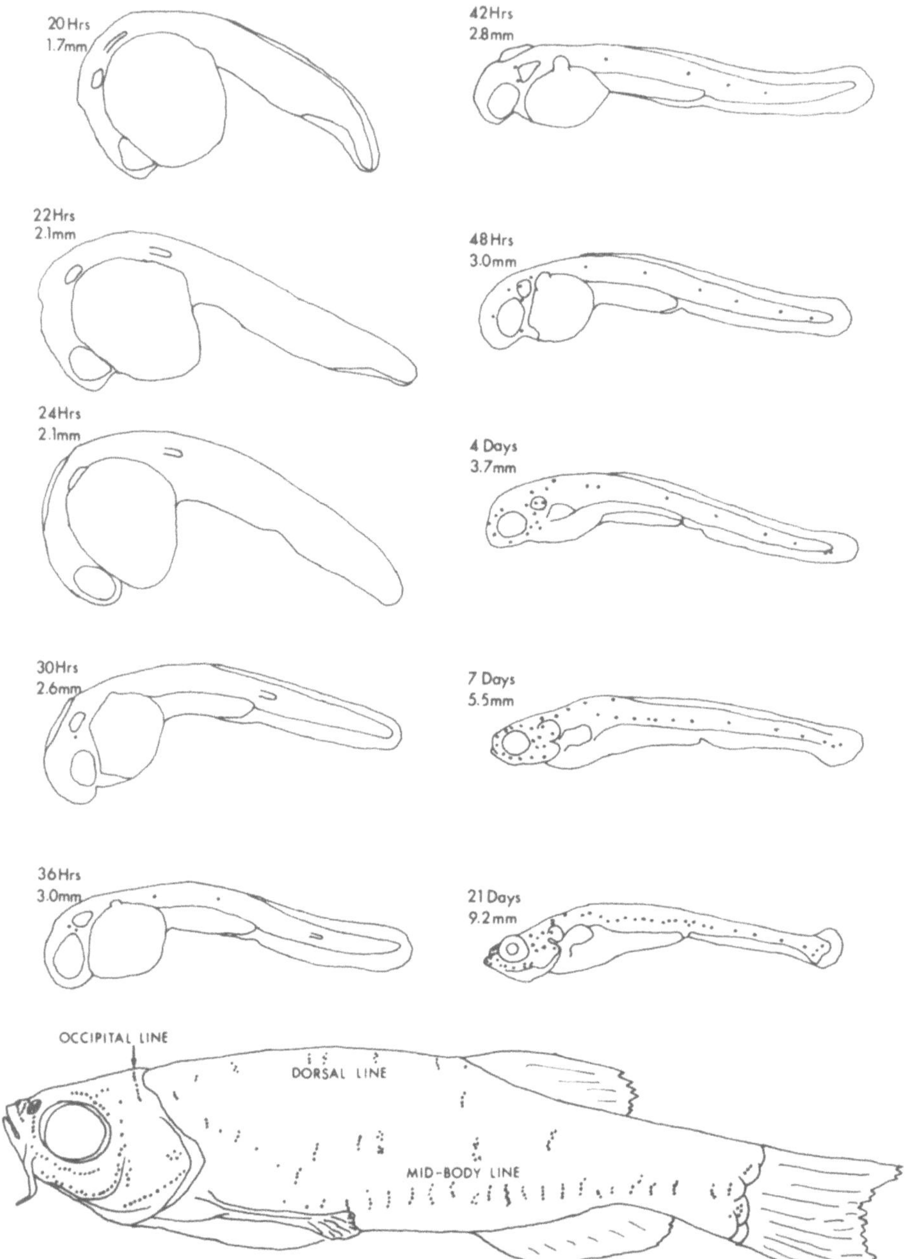

FIGURE 7.1. Drawings showing the positions of the primordium and the formation of the pattern of distribution of neuromasts from 20 h to 50 days. The outline of the migratory primordium is indicated on the embryos from 20 through 36 H. The locations of neuromasts are indicated by dots. The oldest fish (bottom) was 50 days old and 2.1 cm long. (Modified from Metcalfe 1985.)

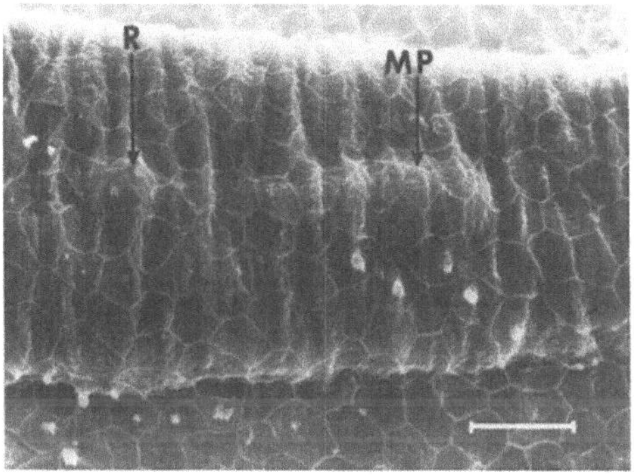

FIGURE 7.2. Scanning electron micrograph of the migratory primordium of a 36-h embryo. The migrating primordium (MP) is teardrop-shaped. A forming neuromast appears as a small round mass (R) behind the trailing end of the primordium. Rostral is to the left; dorsal to the top (the primordium was migrating to the right). Scale bar, 50 μm. (From Metcalfe 1985.)

of the embryo that can be conveniently visualized using the scanning electron microscope (Fig. 7.2). In zebrafish embryos, the skin is a two-layered structure consisting of the outer periderm and an inner epidermis whose cells are more rounded and are underlaid by a basal lamina. The primordium migrates caudally within the epidermis, pushing the epidermal cells aside. During its caudal migration, the primordium leaves behind six clusters of cells that differentiate to become the first, or "primary," neuromasts of the midbody line. Scanning electron microscopy revealed that the apical kinocilia of these six primary neuromasts project outward between cells of the periderm in a rostrocaudal wave that lags behind the migration of the primordium by about 8 h (Metcalfe, 1985).

2.2 Formation of the Adult Pattern of Neuromasts.

After the initial rapid development of the primary neuromasts, new neuromasts appear in a second rostrocaudal wave that follows a much slower time course. Whereas the six primary neuromasts were laid down at approximately five-somite intervals, these new, or "secondary," neuromasts appear between the preexisting primary neuromasts. By 21 days, a linear array of neuromasts has been established; neuromasts are located over 22 of the 30 myotomes of the larvae (Fig. 7.1).

Between 21 and 50 days of development, scales develop on the fish, and two major changes take place in the pattern of distribution of neuromasts. By 50 days, a vertical row of neuromasts appears on most of the 30 scales of the midbody line, and the midbody line is shifted ventrally from the horizontal myoseptum (Fig. 7.1).

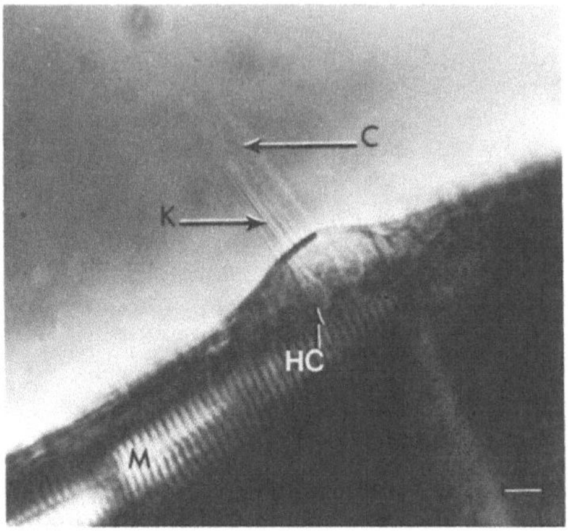

FIGURE 7.3. Neuromast of a live 4-day-old zebrafish larva viewed with Nomarski optics. Kinocilia (K) from hair cells (HC) project into a gelatinous cupula (C). The neuromast appears as a raised mound in the epidermis adjacent to the underlying muscle (M). Scale bar, 10 μm. (From Metcalfe et al. 1985.)

2.3 Neuromast Morphology

Almost all of the neuromasts of the midbody line are free neuromasts in the adult zebrafish. Canal organs are found only on the two or three most rostral scales of the midbody line. Within each neuromast, there are about seven hair cells surrounded by supporting cells. A single kinocilium projects from the apical surface of each hair cell into a gelatinous cupula (Fig. 7.3). Electron microscopy reveals short stereocilia at the apical surface of the hair cells and an accumulation of synaptic vesicles in the basal cytoplasm (Metcalfe et al. 1985). Afferent terminals, arising from the peripheral axons of lateral line sensory neurons, contact the bases of the hair cells and are themselves devoid of synaptic vesicles. Efferent terminals, which arise from lateral line efferent neurons, contain synaptic vesicles and also make direct synaptic contact at the base of the hair cells. Thus, zebrafish neuromasts are typical neuromasts as described by others (Benshalom and Flock 1980; Ekstrom von Lubitz 1981; Jørgenson and Flock 1973; Yamada 1973), except that synaptic ribbons have not been observed in the basal cytoplasm of the hair cells in these young larvae.

3. Sensory Neuron Development

The sensory neurons of the posterior lateral line ganglion (PLLg) are derived from the same postotic placode as that which gives rise to the migratory primordium (Stone 1922). In 5-day-old zebrafish larvae, the PLLg consists of about 15

FIGURE 7.4. Growth cones of the lateral line sensory neurons extend into the migrating primordium. This is a tilted horizontal section from a 22-h embryo, about 4 h after the initiation of migration of the primordium. The lateral line sensory neurons were labeled with horseradish peroxidase (HRP) from a lesion located just rostral to the posterior lateral line ganglion (PLLg). Three growth cones were identified (arrows), two of which extend into the migrating primordium (MP). The rostral boundary of the primordium is indicated (X), although this is not apparent at this focal plane. This is a photomontage of a single 50-μm section, photographed using Nomarski optics. Scale bar, 25 μm. (From Metcalfe 1985.)

to 20 bipolar sensory neurons located in a peripheral ganglion just caudal to the ear (Metcalfe et al. 1985). PLLg sensory neurons appear very early in development; labeling studies using tritiated thymidine have shown that most (about 93%) of these neurons have exited the mitotic cycle (i.e., they fail to incorporate tritiated thymidine) sometime between 8 and 13 h (Metcalfe 1983), which is near the end of gastrulation.

3.1 Growth of the Peripheral Axons of Sensory Neurons

At 18 h, the primordium begins to migrate away from the ganglion. At this time, the PLLg sensory neurons are bipolar cells that have short central axons that extend rostrally in the direction of their point of entry into the hindbrain, and at the same time, they have short peripheral axons that extend caudally into the migratory primordium. During the period of time that the primordium migrates caudally from its origin at the PLLg to the base of the caudal fin (from 18 to 40 h), the most advanced sensory neuron growth cones are found within the migrating primordium (Fig. 7.4). The growth cones advance at a rate of about 100 μm/h, the same as the rate of migration of the primordium. Behind the leading growth cones, growth cones of other PLLg sensory neurons are found at irregular intervals (on the order of 40 μm apart) along the developing posterior lateral line nerve, closely associated with the axons of the more advanced cells. The leading growth cones are located deep within the primordium, apparently close to, or in contact with, the basal lamina of the epidermis.

3.2 Growth of the Central Axons of Sensory Neurons

The central axons of the PLLg sensory neurons project from the posterior lateral line ganglion into the hindbrain. The central axons begin to grow at about the same time as the peripheral axons (about 18 h). The axons grow in a rostral direction outside of the central nervous system until they reach the caudal margin of the otic capsule. The fibers enter the hindbrain at the caudal margin of the otic capsule and grow in a rostral direction along a very narrow longitudinal pathway. By 25 h, the central axons have grown rostrally to the level of the Mauthner cell, where they contact the lateral dendrite (Fig. 7.5) (see also Zottoli and Danielson, Chapter 23). Thus, there is a distinct difference in the rate of growth of the two axons that emerge from these bipolar sensory neurons; the peripheral axons grow at a rate of about 100 μm/h whereas the central axons grow at a rate of about 20 μm/h during the same period in the embryo.

At the light microscope level, there appears to be an accumulation of growth cones at the leading edge of the ingrowing lateral line nerve. This is supported by electron microscopy, since several growth cones appear to be present at the same longitudinal level (Fig. 7.6). This accumulation of growth cones is specific for the developing central projection, since the growth cones of the peripheral axons are distributed along the developing lateral line nerve. Although the reason for this difference is unknown, one possibility is that the slower growth of the leading central fibers allows the following fibers to catch up with them.

Electron microscopy also demonstrates that the growth cones of at least some posterior lateral line central axons contact the basal lamina of the pial surface between the end feet of ependymal cells (Fig. 7.6). Similar contacts between the growth cones of other types of neurons and the basal lamina of the pial surface have been reported in the developing spinal cord of *Xenopus* as well (Nordlander and Singer 1982; Roberts and Patton 1985). Thus, contact with the basal lamina might be an important feature of pathfinding for these axons.

The central axons of the early PLLg neurons grow along a very narrow (about 2 μm wide) longitudinal pathway in the hindbrain (Fig. 7.7). The posterior lateral line pathway is the most dorsal of three parallel longitudinal sensory pathways in the hindbrain of the embryo. The three pathways are, from ventral to dorsal, (1) the lateral longitudinal fascicle, containing descending fibers from trigeminal sensory neurons, descending fibers of several types of reticulospinal neurons (for details, see Metcalfe et al. 1986) and ascending Rohon-Beard fibers; (2) the central projection of the eighth nerve; and (3) the common lateral line pathway, containing fibers of both the anterior and posterior lateral line nerves. Thus, the central projection of the lateral line system is discrete from the central projection of the auditory system even at the earliest times that the axons enter the brainstem. This finding provides additional support for the hypothesis that the

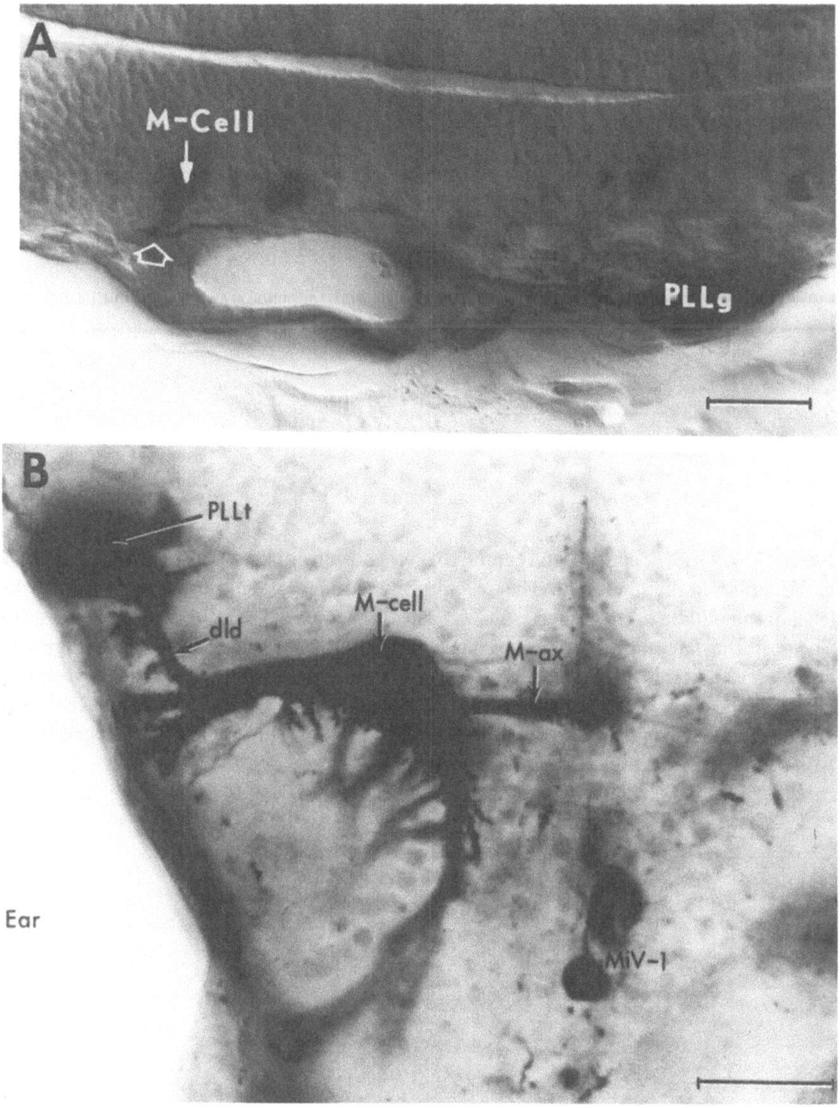

FIGURE 7.5. Lateral line fibers contact the lateral dendrite of the Mauthner cell. (A) Horizontal section of a 25-h embryo. The posterior lateral line fibers and the Mauthner cell (M-Cell) were labeled from a lesion to the trunk and spinal cord. The central axons of the posterior lateral line ganglion (PLLg) approach the otic capsule, the pass out of the plane of section for a short distance. The fibers terminate just rostral to their point of contact with the lateral dendrite of the Mauthner cell (arrowhead). (From previously unpublished work of Bruce Mendelson.) (B) Transverse section from a 5-day-old larva. Posterior lateral line terminals (PLLt) contact the Mauthner cell at the tip of the dorsolateral dendrite (dld). The nerve to the left midbody lateral line nerve and the right medial longitudinal fascicle were labeled with HRP. Scale bars, 25 μm. (From Metcalfe et al. 1985.)

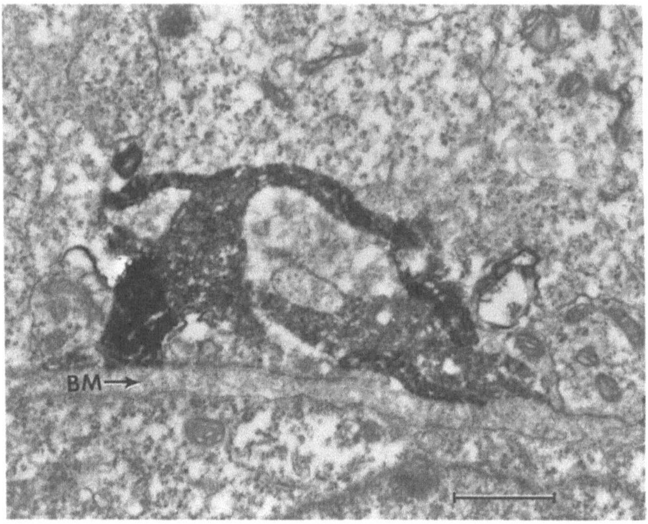

FIGURE 7.6. Growth cones of the central axons of posterior lateral line neurons contact the basal lamina in the hindbrain. This electron micrograph is of a transverse section at the level of the middle of the otic capsule, very near the ends of the ingrowing central axons. The posterior lateral line nerve was labeled with HRP. There appear to be several growth cones present at this level, since there is considerable variation in the density of the HRP label. The growth cones are surrounded by ependymal processes (above) and the basal lamina (BM). Dorsal is to the upper left. Scale bar 1 μm. (From previously unpublished work of Metcalfe, Kimmel, and Schabtach.)

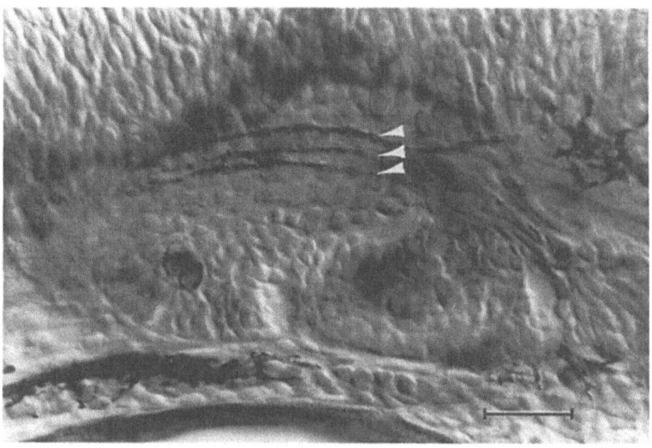

FIGURE 7.7. The monoclonal antibody zn−1 labels three parallel longitudinal pathways in the hindbrain (arrowheads). Sagittal section at the level of the otic capsule in a 29-embryo. Rostral is to the right. Scale bar, 20 μm. (For details on methods, see Trevarrow, 1989.)

lateralis and acoustic systems arose independently in evolution (for discussion, see Northcutt 1980).

4. Speculation Concerning Mechanisms of Development

The finding that growth cones of sensory neurons are located within the migratory primordium is of particular interest. The close relationship of the leading growth cones and the migrating primordium has been consistently observed using a variety of labeling procedures and appears to be an invariant feature of the development of the lateral line system (Metcalfe 1985 and unpublished observations).

Why do the sensory neuron growth cones remain with the migrating primordium? Although the answer to this question is not yet known, the mechanism could be a general one (see Trinkaus 1985) and of fundamental importance. Speculations on a few possible mechanisms follow.

Guidance mechanisms may be general ones such that both the axons and the primordial cells are capable of independently migrating along the same pathway. Alternatively, the migrating primordium could direct the axons, or the axons could direct the migration of the primordium. That growth cones of sensory neurons are never observed caudal to (i.e., leading) the primordium suggests that the growth cones do not direct the migration of the primordium. This suggestion is strengthened by the finding that the primordium can migrate normally in the absence of the lateral line nerve in amphibians (Stone 1931; Tweedle 1977). Thus, the growth cones may be either guided in their peripheral migration by the primordial cells or independently capable of following the same guidance cues as the migrating primordium. These possibilities cannot be distinguished at present, since we do not know whether the growth cones could migrate normally in the absence of the primordium. However, the close and persistent association of the growth cones with the primordium suggests that the most likely possibility is that the growth cones are guided by the migrating primordial cells. This is a particularly interesting possibility, since the hair cells, which are the peripheral targets of the growth cones, are derived from cells of the primordium. That is, the growth cones might be guided by their migrating target cells (or their precursors; see Winklbauer and Hausen 1983).

It is conceivable that the close association of the sensory neuron growth cones with the primordium could be mediated by a trophic factor. Despres and Romand (1987) have reported that nerve growth factor-like immunoreactivity is transiently present in the hair cells of the organ of Corti in newborn rats. Thus, lateral line hair cells (or their precursors) could secrete a trophic factor that maintains the close association of the nerve with the migrating primordium.

Another possibility is that Schwann cells (or their precursors) play a role in guiding the migration of the primordium and/or in maintaining the association of the nerve with the primordium. Schwann cells are generally thought to arise from neural crest, which is a population of cells that are notorious for their ability to migrate over long distances to precise locations in embryos (Weston 1970, 1982).

Stone (1933) observed that Schwann cells are present along the entire length of the nerve connecting the PLLg with the migrating primordium in salamanders, so Schwann cells may be within the primordium, or even leading it. In this regard it is interesting to note that Schwann cells are apparently necessary to guide motor axons over long distances to limb muscles (Harrison 1924; Yntema 1943; Carpenter and Hollyday 1986; Noakes and Bennett 1987; reviewed by Keynes 1987). Furthermore, Schwann cells have been shown to secrete plasminogen activator (Krystosek and Seeds 1984), which is a proteolytic enzyme that may facilitate cell migration by modifying the extracellular environment. Localized protease secretion might be necessary for the lateral line primordium to migrate within the epidermis. Some Schwann cells also secrete nerve growth factor (NGF) (Assouline et al. 1985), which could play a role in maintaining the association of the growth cones with the primordium. More experiments are clearly needed to distinguish between these intriguing possibilities.

5. Concluding Remarks

Major features of the development of posterior lateral lines appear to be the same in zebrafish and amphibians. Primary neuromasts arise from a migratory primordium that originates as a postotic placode, and the sensory ganglion arises from the same placode. The developing posterior lateral line nerve connects the ganglion with the migrating primordium, and in zebrafish, at least, the peripheral growth cones of the sensory neurons comigrate with the primordium. Whereas the studies of Stone (1922, 1931, 1933, 1937) emphasized the role of the primordium in neuromast formation and its autonomous migratory capabilities, the present work has emphasized the development of the sensory neurons (Metcalfe 1985; Metcalfe et al. 1985). However, the remarkable similarities in the patterns of development of the lateral line systems in these diverse organisms suggest that the mechanisms that control their development may be the same in divergent taxa. Further, similar mechanisms may be involved in the development of other neural systems as well.

Acknowledgments. I thank Bill Trevarrow for the monoclonal antibody zn–1, Bruce Mendelson for allowing me to study HRP material he had prepared, Eric Schabtach for his help with the transmission electron microscopy, and Charles Kimmel for his helpful comments on the manuscript. This work was supported by NIH grant NS17963 to Charles Kimmel.

References

Assouline J, Bosch EP, Lim R, Jensen R, Pantazis NJ (1985) Detection of a neurite promoting factor with some similarity to nerve growth factor in conditioned media from rat primary Schwann cells and astroglia. Soc Neruosci Abstr 11:933.

Benshalom G, Flock Å (1980) Ultrastructure of synapses in the lateral line organ. Acta Otolaryngol 90:161–174.

Carpenter EM, Hollyday M (1986) Defective innervation of chick limbs in the absence of presumptive Schwann cells. Soc Neurosci Abstr 12:1210.

Chakrabarti S, Streisinger G, Singer F, Walker C (1983) Frequency of gamma-ray induced specific locus and recessive lethal mutations in mature germ cells of the zebrafish, *Brachydanio rerio*. Genetics 103:109–123.

Despres G, Romand R (1987) Nerve growth factor-like immunoreactivity in the developing rat cochlea. Soc Neurosci Abstr 13:520.

Eaton R, Farley RD (1974) Spawning cycle and egg production of zebrafish, *Brachydanio rerio*, in the laboratory. Copeia 1:195–204.

Ekstrom von Lubitz DKJ (1981) The ultrastructure of lateral-line sense organs of the ratfish, *Chimaera monstrosa*. Cell Tissue Res 215:651-665.

Harrison RG (1904) Experimentelle Untersuchungen uber die Entwicklung der Sinnesorgane der Seitenlinie bei den Amphibian. Arch Mikrosk Anat, Bd 63:35–49.

Harrison RG (1924) Neuroblast versus sheath cell in the development of peripheral nerves. J Comp Neurol 37:123-205.

Jørgensen JM, Flock A (1973) The ultrastructure of lateral line sense organs in the adult salamander *Ambystoma mexicanum*. J Neurocytol 2:133–142.

Keynes RJ (1987) Schwann cells during neural development and regeneration: Leaders or followers? TINS 10(4):137–139.

Krystosek A, Seeds NW (1984) Peripheral neurons and Schwann cells secrete plasminogen activator. J Cell Biol 98:773–776.

Metcalfe WK (1983) Anatomy and development of the zebrafish lateral line system. Doctoral Dissertation, University of Oregon, Eugene.

Metcalfe WK (1985) Sensory neuron growth cones comigrate with posterior lateral line primordial cells in zebrafish. J Comp Neurol 238:218–224.

Metcalfe WK, Kimmel C, Schabtach E (1985) Anatomy of the posterior lateral line system in young larvae of the zebrafish. J Comp Neurol 233:377-389.

Metcalfe WK, Mendelson B, Kimmel CB (1986) Segmental homologies among reticulospinal neurons in the hindbrain of the zebrafish larva. J Comp Neurol 251: 147–159.

Noakes PG, Bennett MR (1987) Growth of axons into developing muscles of the chick forelimb is preceded by cells that stain with Schwann cell antibodies. J Comp Neurol 259:330–347.

Nordlander RH, Singer M (1982) Spaces precede axons in *Xenopus* embryonic spinal cord. Exp Neurol 75:221–228.

Northcutt RG (1980) Central auditory pathways in anamniotic vertebrates. In: Popper AN, Fay RR (eds) Comparative Studies of Hearing in Vertebrates. New York: Springer-Verlag, pp. 79–118.

Roberts A, Patton D (1985) Growth cones and the formation of central and peripheral neurites by sensory neurones in amphibian embryos. J Neurosci Res 13:23–38.

Stone LS (1922) Experiments on the development of the cranial ganglia and the lateral line sense organs in *Ablystoma punctatum*. J Exp Zool 35:421–496.

Stone LS (1931) Studies on the migratory lateral-line primordia of *Amblystoma*. Anat Rec 48:564.

Stone LS (1933) The development of lateral-line sense organs in amphibians observed in living and vital-stained preparations, J Comp Neurol 57:507–540.

Stone LS (1937) Further experimental studies of the development of lateral-line sense organs in amphibians observed in living preparations. J Comp Neurol 68:83–115.
Streisinger G, Walker C, Dower N, Knauber D, Singer F (1981) Production of clones of homozygous diploid zebra fish (*Brachydanio rerio*). Nature 291:293–296.
Trevarrow B (1989) Distribution of differentiating neuronal cells in zebrafish embryos visualized with a monoclonal antibody. (In preparation.)
Trinkaus JP (1985) Further thoughts on directional cell movement during morphogenesis. J Neurosci Res 13:1–19.
Tweedle CD (1977) Ultrastructure of lateral line organs in aneurogenic amphibian larvae (*Ambystoma*). Cell Tissue Res 185(2):191–198.
Walker C. Streisinger G (1983) Induction of mutations by gamma-rays in pregonial germ cells of zebrafish embryos. Genetics 103:125–136.
Weston JA (1970) The migration and differentiation of neural crest cells. Adv Morphogen 8:41–114.
Weston JA (1982) Motile and social behavior of neural crest cells. In: Bellairs R, Curtis A, Dunn G (eds) Cell Behavior. London: Cambridge University Press, pp. 429–469.
Winklbauer R, Hansen P (1983) Development of the lateral line system in *Xenopus laevis*. II. Cell multiplication and organ formation in the supraorbital system. J Embryol Exp Morphol 76:283–296.
Wright MR (1951) The lateral line system of sense organs. Q Rev Biol 26:264–280.
Yamada Y (1973) Fine structure of the ordinary lateral line organ. I. The neuromast of lamprey, *Entosphenus japonicus*. J Ultrastruct Res 43:1–17.
Yntema CL (1943) Deficient efferent innervation of the extremities following removal of neural crest in *Amblystoma*. J Exp Zool 94:319–349.

8 Cellular Events Underlying the Regenerative Replacement of Lateral Line Sensory Epithelia in Amphibians

JEFFREY T. CORWIN, KENNETH J. BALAK,
and PAULA C. BORDEN

In aquatic amphibians, removal of a patch of skin that contains lateral line neuromast organs evokes the production of new neuromasts that differentiate in the approximate location of the organs that were lost. The whole process can be completed in less than 2 weeks, with the number of replacement neuromasts roughly matching the number lost. This chapter attempts to explain some of the cellular mechanisms that lead to the regenerative replacement of lost neuromasts. It also discusses the potential generality of those mechanisms in regard to regeneration of cochlear hair cells, a recently discovered phenomenon that appears to result in recovery of function following profound sensorineural hearing loss in birds.

1. Background

Ross Harrison (1903) first applied microsurgical transplantation techniques to investigate the development of the lateral line system experimentally. By grafting the heads of darkly pigmented embryos of the frog, *Rana sylvatica*, to the embryonic bodies of the lightly pigmented species, *R. palustris*, he determined that the body's lateral line organs originate from a cephalic ectodermal thickening that he termed a placode. That placode produces a migratory primordium of cells that move posteriorly along the embryo's flank, laying down the precursors to the cells of the lateral line neuromasts.

Leon Stone (1933), a colleague of Harrison's, transplanted placodal tissue from embryos that had been treated with the vital dye Nile blue sulfate in place of placodal tissue in unstained host embryos and observed the microscopic development of the dye-containing placodal cells and their progeny in living embryos. He found that the lateral line primordium migrates beneath the two strata in the epidermis of the embryo, causing the epidermis to bulge outward where the primordium is passing. The primordium deposits beadlike groups of cells along its path of migration, and those cells give rise to the sensory and supporting cells of the neuromast organs. After they fragment from the primordium, these small masses of cells become radially arranged, and the apical surfaces of the cells within each mass come together to form a vesicle of extracellular space beneath

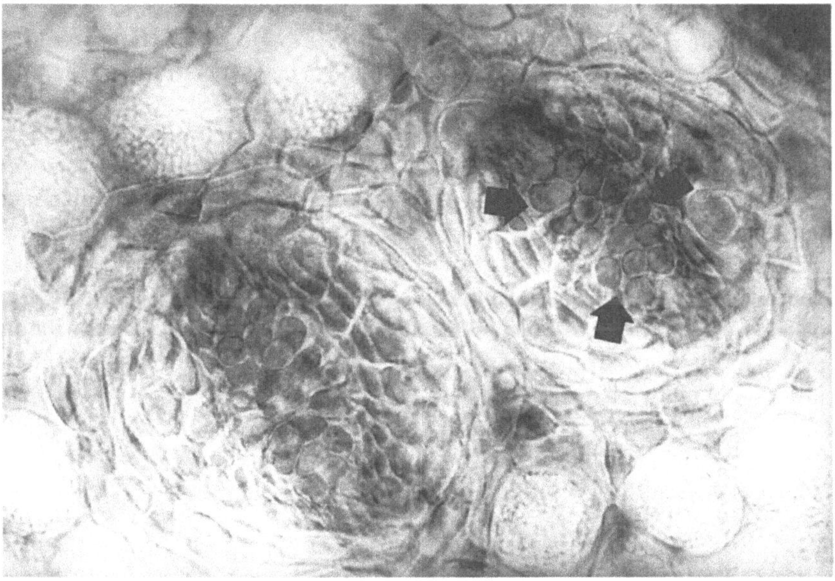

FIGURE 8.1. Two neuromasts from a single stitch in the epidermis of the tail of an axolotl (*Ambystoma mexicanum*, a salamander). In ambystomatid salamanders, the migratory primordium moves down the body, leaving behind small masses of cells that aggregate to form a single neuromast at the site of each future stitch. The single primary neuromast grows and becomes a functional sensory organ before giving off cells that migrate a short distance away to points where they form the additional (secondary) neuromasts that make up the linear group of neuromasts referred to as a stitch. Hair cells, with large spherical nuclei present in the center of each neuromast (arrows), can be distinguished from supporting cells that surround and extend beneath those cells.

the skin surface. As receptor cells form, the epidermal cells over the neuromast flatten and eventually separate, allowing the vesicle to communicate with the exterior, forming the sensory pore of the developing neuromast.

Stone's observations demonstrated that neuromasts are formed by cells that are direct descendants of the preauditory and postauditory cephalic ectodermal placodes, without any contribution of cells from the epidermis that surrounds the differentiated neuromasts (Stone 1933). At that point, he extended his investigations to the formation of accessory neuromasts. He determined that the accessory or secondary neuromasts of each linear group or "stitch" (Fig. 8.1) arose via budding from the first neuromasts that developed from the migratory primordium at the site of each future stitch. According to Stone (1937), the cells of the secondary buds arose from peripheral supporting cells by mitosis and migration.

By removing a patch of skin or amputating a segment of the tail that contained neuromasts, Stone (1937) investigated regenerative events that led to their replacement. He concluded that regenerated neuromasts originated through a budding process similar to that which produced the accessory neuromasts in growing stitches. However, in the regeneration experiments, budding occurred only in the neuromast that was closest to the site of the amputation. Regeneration also involved more extensive cell proliferation and resulted in the formation of a large, placodelike mass of migratory cells. After the formation of the regenerative placode, the subsequent stages closely resembled the late stages in the embryonic formation of primary neuromasts. By grafting together the amputated tail stumps of two salamanders, Stone produced parabiotic pairs which he used to demonstrate that the cells of both the anterior and the posterior poles of a neuromast possess the potential to contribute to the formation of a regenerative placode (Stone 1937). By removing the anterior, neurogenic portion of the postauditory placode in early embryos, Stone (1931) demonstrated that the process of embryonic neuromast development is independent of innervation.

2. Initial Stages of Regeneration

The early events in the regeneration of replacement neuromasts have been examined in detail in several studies subsequent to the investigations by Stone. Speidel (1947, 1948) used normal larvae of anurans and urodeles, and Wright (1947) used hypophysectomized anuran larvae that were unpigmented and incapable of metamorphosis to examine those early reactions. Both reported an initial phase of dedifferentiation, which included the temporary disappearance of the "sensory hairs" that are now recognized as cilia bundles. This preceded the proliferative phase and the migration of the regenerative placode. A recent reinvestigation that employed scanning electron microscopy (SEM) and differential interference contrast (DIC) light microscopy has demonstrated that such a cilia disappearance may be the result of active extrusion of the apical cytoplasm and the cilia arrays of hair cells in the affected neuromasts (Corwin 1985c, 1986). When this occurs, it is followed by a period of cell death that affects the hair cells, but not the supporting cells (Fig. 8.2). However, in contrast to the reports by Speidel and Wright, it has been found that this phase of the regenerative reaction varies among individual cases. When the bath solution is continually changed during observation, and when the site of the amputation is not directly adjacent to a neuromast on the tail stump, the cilia bundles of the neuromast that gives rise to the regenerative placode remain unchanged (Fig. 8.3). No hair cell death occurs in these cases, but regeneration proceeds in spite of this. Thus, it is evident that this form of dedifferentiation is not required for the production of the regenerative placode.

The dedifferentiation reported by Speidel and Wright appears to be a reaction to physiological stress in the affected cells. Since hair cells are tonically active and are particularly sensitive to oxygen concentrations (Brown et al. 1983, Russell and Cowley 1983), interruptions of gas exchange or circulatory flow might be expected to affect these cells adversely. Our observations of living salamanders have shown that circulation is interrupted in the immediate vicinity

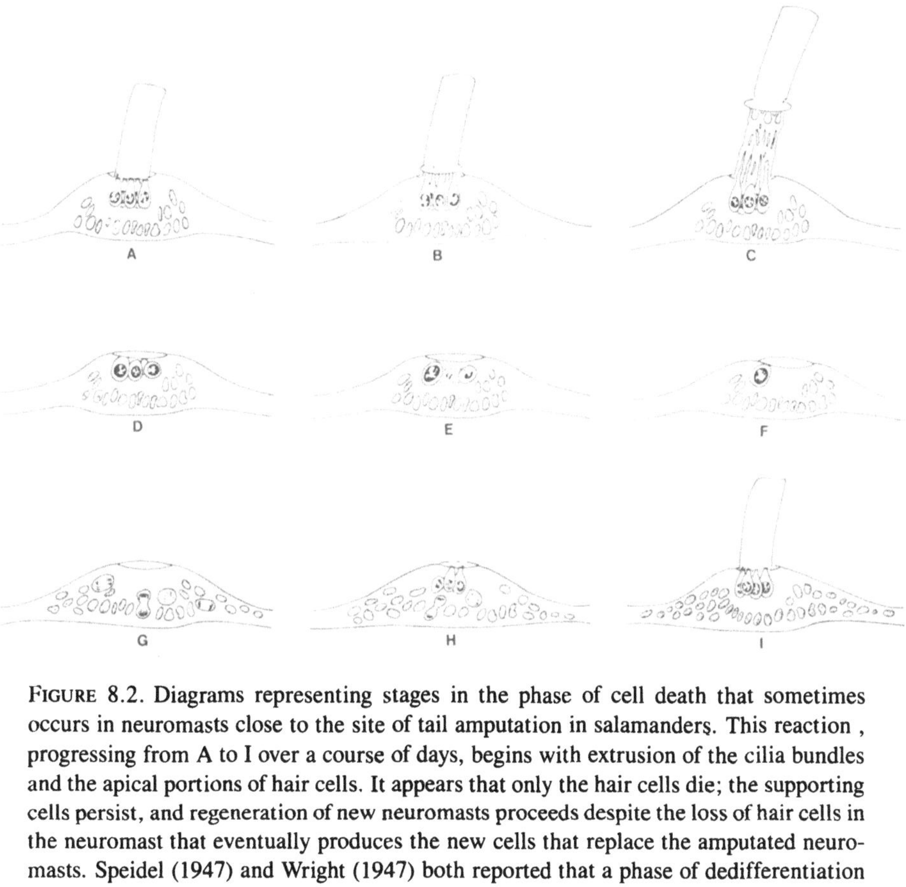

FIGURE 8.2. Diagrams representing stages in the phase of cell death that sometimes occurs in neuromasts close to the site of tail amputation in salamanders. This reaction , progressing from A to I over a course of days, begins with extrusion of the cilia bundles and the apical portions of hair cells. It appears that only the hair cells die; the supporting cells persist, and regeneration of new neuromasts proceeds despite the loss of hair cells in the neuromast that eventually produces the new cells that replace the amputated neuromasts. Speidel (1947) and Wright (1947) both reported that a phase of dedifferentiation and hair cell cilia loss occurred prior to the production of the regenerative placodal cells. It seems likely that the extrusion reaction and the death of hair cells are part of the dedifferentiation that they observed. However, these authors have found that these phenomena are variable in their occurrence and appear to result from physiological stress (see Fig. 8.3).

of an amputation, as coagulation occurs, but collateral vessels allow continued blood flow a few hundred microns away. It is interesting that a similar apical extrusion has recently been reported to occur in hair cells damaged by acoustic overstimulation in the avian cochlea, where it also is suspected that metabolic compromise of the cells may be the cause (Cotanche et al. 1987).

3. Formation of the Regenerative Placode

In many organ systems the key to regeneration appears to be the ability to form a mass of undifferentiated, pluripotent cells capable of differentiating into several different cell types. The fates of these cells, once formed, can generally

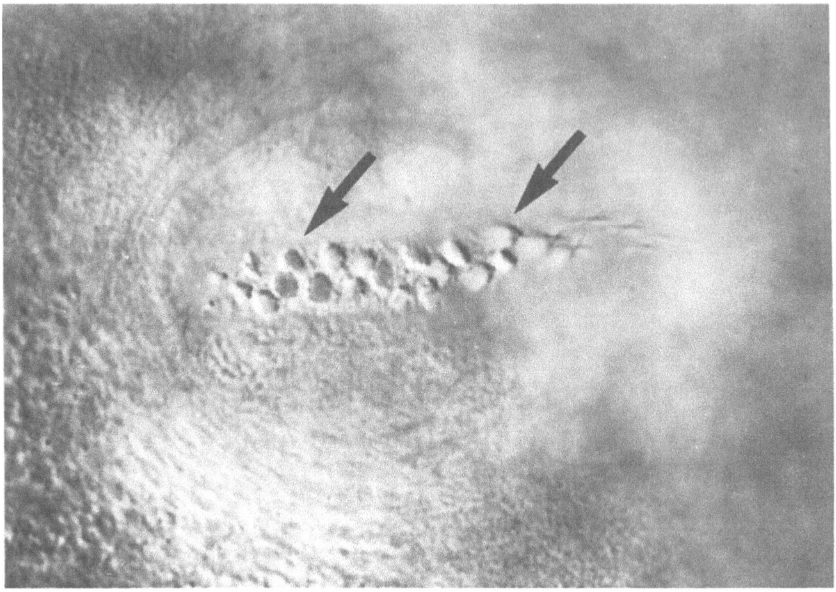

FIGURE 8.3. When the solution that bathes a salamander is replaced with fresh solution regularly, and when amputation of the tip of the tail is not directly adjacent to a neuromast, then hair cells do not die or extrude their cilia. Differential interference contrast (DIC) microscopy has been used to observe the continuing presence of cilia bundles (arrows) of normal hair cells that persist throughout the regeneration of the neuromasts lost through amputation. This has demonstrated that the dedifferentiation phase described by earlier investigators and illustrated in Fig. 8.2 is not essential for the regenerative replacement of neuromasts.

be shaped by regenerative programs of tissue differentiation that resemble the embryonic programs that led to the formation of the original organ. Differentiation programs can exist even in organisms that are not normally capable of regeneration, and they can control regeneration if the production of pluripotent cells can be achieved. This was elegantly demonstrated by Singer (1954) in experiments that used translocation of the cut distal end of the sciatic nerve to produce supranormal innervation of the forelimb in frogs. When followed by amputation of the forelimb, the supranormal innervation resulted in the formation of a mass of proliferative blastemal cells and the subsequent regeneration of an amputated limb in adult frogs that normally do not regenerate limbs.

The normal source of the pluripotent cells that underlie regeneration appears to be the interstitial cell population in *Hydra*, the neoblast cell population in planaria, the cells of the regenerative blastema in the amphibian limb (Goss 1969), and in the lateral line, the cells of the regenerative placode. The undifferentiated cells of that placode proliferate and migrate to the future sites of the replacement neuromasts. The formation of this mass of placodal cells is at present the first recognizable sign of a regenerative response to damage of the

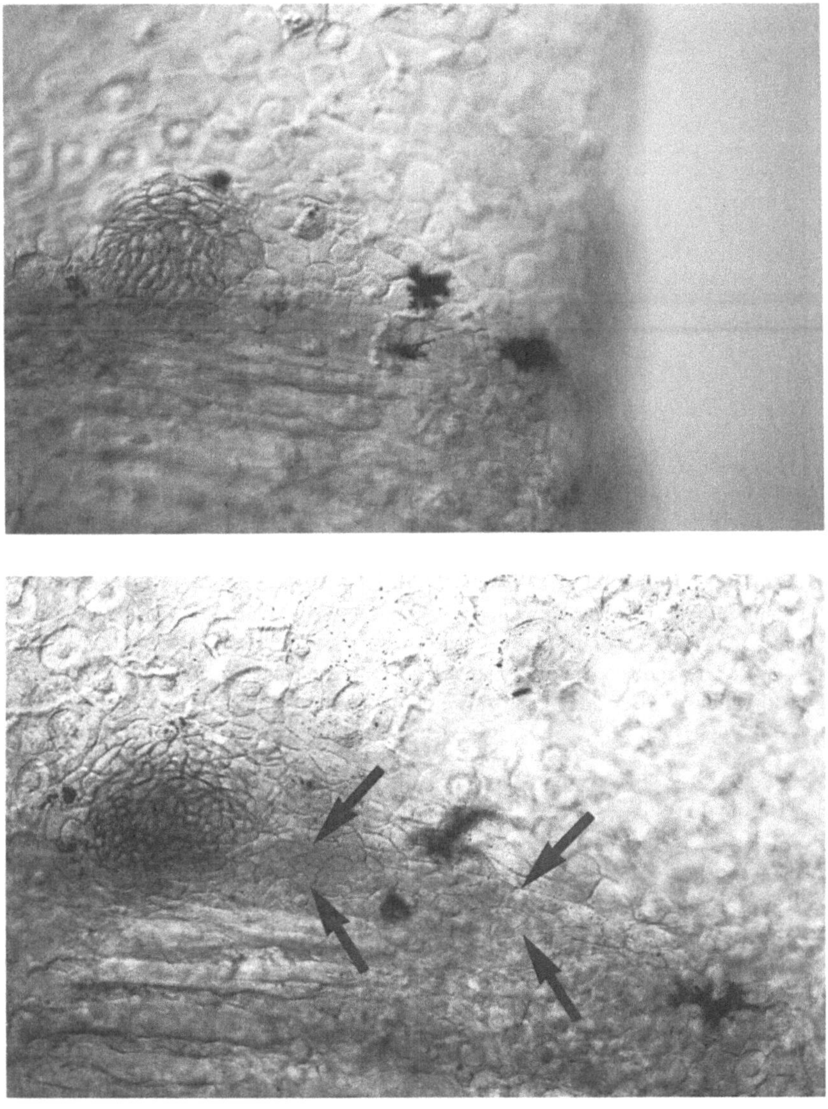

FIGURE 8.4. Two micrographs showing the tail stump of a salamander viewed with differential interference contrast optics one day after amputation of the tip of the tail (A) and 4 days later (B). In (A), the cut end of the tail is visible to the right of the last neuromast. That neuromast is surrounded by normal large epidermal cells that rest above dermal connective tissue. In (B), the last neuromast on the tail stump is again visible, but the edge of the tail has now grown outward through regeneration. The last neuromast on the tail stump has given rise to cells that have begun to form a noticeable regenerative lateral line placode (arrows). The placode is normally observed by the third day after amputation of the tip of the tail in the small (20 mm total length) axolotl salamanders used in these studies.

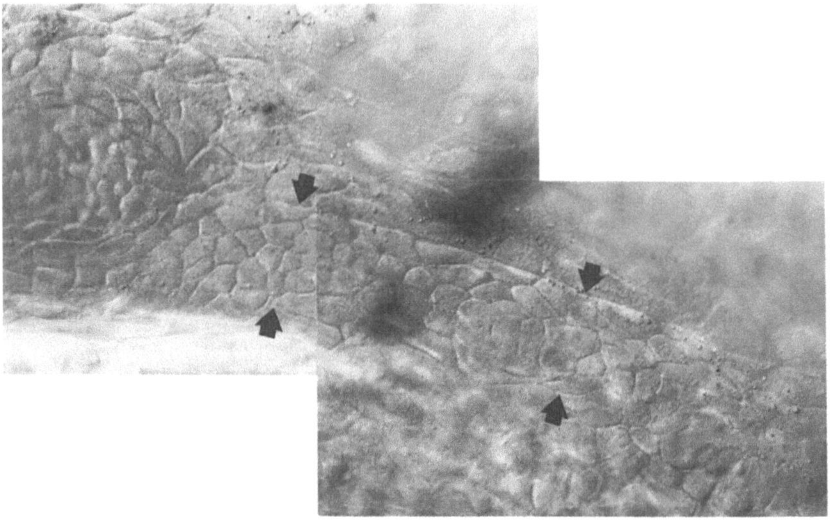

FIGURE 8.5. High-magnification view of a regenerative placode extending from the posteroventral edge of the last neuromast remaining on the stump of the amputated tail in the same living anesthetized axolotl salamander viewed with DIC optics 5 days after amputation. The cells of the placode (arrows) are approximately the same size as the supporting cells of the neuromast but are significantly smaller than the large flat epidermal cells that lie on either side of and above the placode.

lateral line system. Tritiated thymidine injections coupled with autoradiography and DIC microscopy have recently been used in our laboratory to investigate the origin and the growth of this placode in axolotls, *Ambystoma mexicanum*. The cells of the placode are normally first noticeable on the third day after amputation of the tip of the tail, and they appear just above the basal lamina along the posteroventral edge of the neuromast (Fig. 8.4). The cells in the placode divide rapidly, growing in number and moving outward from the neuromast's edge beneath the epidermis and above the basal lamina. As the placode grows, it pushes the epidermis up so that a longitudinal ridge is formed. The cells of the placode are distinguishable because they are much smaller and more spherical than the cells of the surrounding epidermis and approximately the same size as supporting cells in neuromasts (Fig. 8.5). At a gross level, major events in the migration of the cells in the regenerative placode appear similar to those in the migration of cells in the embryonic primordium of the forming lateral line.

Thymidine is a nucleoside precursor to DNA, so injection of radioactive thymidine can be used to identify proliferating cells, since they will take up the label as they replicate their DNA. In our laboratory, tritiated thymidine was injected intraperitoneally in axolotls on the third and the fourth days after tail amputation, and their tails were fixed 12 h later. Autoradiography demonstrated that the majority of the cells throughout the forming regenerative placode had

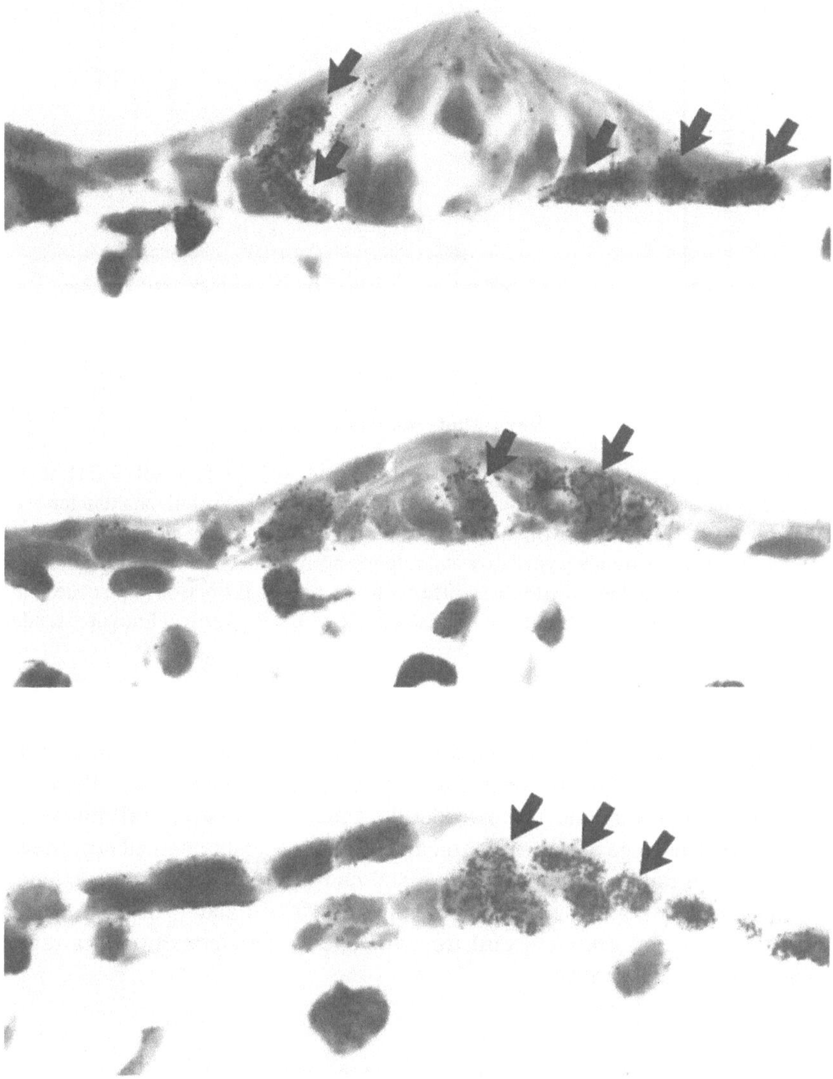

FIGURE 8.6. Microautoradiographs of selected transverse sections along a regenerative placode on the tail of an axolotl salamander. The three representative sections, arranged in a top-to-bottom sequence, are from anterior to posterior levels along the tail. The top section passes through the terminal neuromast, and each successive section passes through the placode extending from it. Silver grains above the nuclei (arrows) of cells show where tritiated thymidine, a radioactively labeled precursor to DNA, has been incorporated. The tritiated thymidine was injected on the third day after amputation of the tip of the tail. Twelve hours later, the salamander was overanesthetized, and the regenerating tail was fixed and prepared for autoradiographic localization of mitotically active cells that had taken up the label during replication of chromosomes.

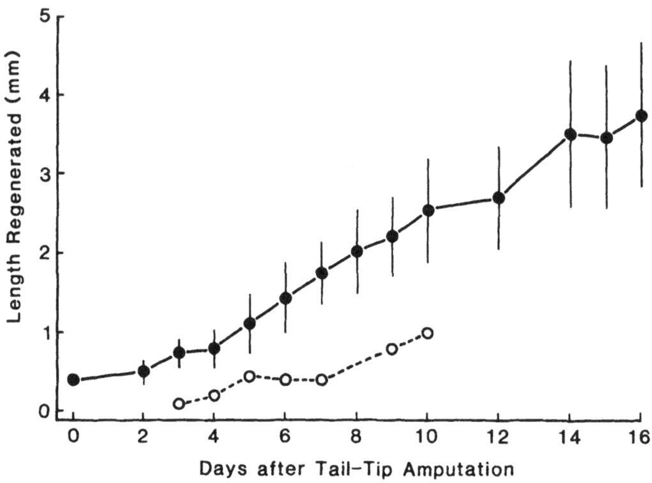

FIGURE 8.7. A graph of the rates of growth of the regenerated tip of the tail (solid line) and of the regenerative lateral line placode (dashed line) from five axolotl salamanders. The tail tip grows at a rate of approximately 250 μm/day. The regenerative placode lags in onset of growth but then extends posteriorly at a rate of approximately 100 μm/day for several days. A period of rapid new neuromast differentiation follows the period of placode extension. The data points are means from five salamanders, and the vertical lines are standard deviations.

taken up the label, so the cells appear to be dividing in all areas of the placode, not just in particular growth zones (Fig. 8.6). Over the next 4 to 7 days, these cells increased in number, and the placode elongated. Its dorsoventral dimensions remained in the range of 80 to 100 μm, and its anterior end remained near the last neuromast on the tail stump, but the distal end of the placodal mass extended posteriorly, growing at a rate of approximately 100 μm/day (Fig. 8.7). At that same time, the tail grew outward from the site of the amputation at a rate of approximately 250 μm/day.

4. Differentiation of Replacement Neuromasts

After the placode has grown for 4 to 7 days, the cells within it begin to aggregate into separate masses that will develop into individual neuromasts in the final phase of the regenerative response. These segregated masses of cells soon form rosettes that are radially symmetrical about a central point. When viewed from the skin surface, the cells of these rosettes appear to be arranged like the sections of an orange. Sato (1976) has reported that the growing processes of the lateral line neurons penetrate the basal lamina in the vicinity of forming neuromasts at the time when rosettes first appear during the embryonic development of this system in salamanders, but the significance of this observation remains unclear.

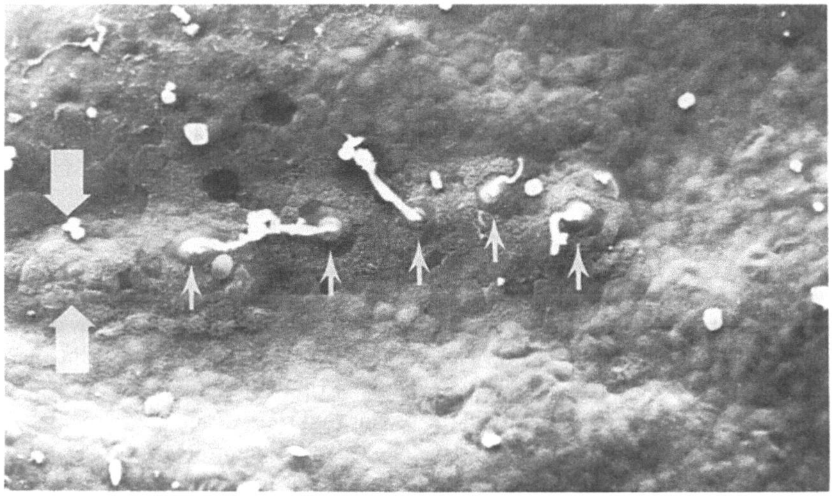

FIGURE 8.8. A scanning electron micrograph of six replacement neuromasts, differentiating on the regenerated tip of the tail of a salamander fixed 9 days after tail amputation. The process of replacement neuromast differentiation progresses faster at the anterior end of the tail stump (to the right in the figure) than at the posterior end (to the left). In this micrograph, the anteriormost neuromasts (five up-pointing arrows) are well differentiated, but the neuromasts at the posterior end of the regenerated group (one is shown and marked by two large arrows) are not well differentiated. The least differentiated of those are the posteriormost ones, and this has been observed in a number of specimens.

If neuronal processes play a role in inducing or organizing the formation of the rosettes that give rise to neuromasts, it is not an essential role, since these organs develop normally in embryos that have been deprived of lateral line innervation (Stone 1931) and in larvae when regeneration occurs after experimental denervation (see the following section).

The rosettes are composed of small cells that resemble the supporting cells of mature neuromasts, but larger cells with spherical nuclei soon form in the centers of the rosettes (Stone 1933, 1937; Sato 1976). These cells are destined to become hair cells, and they begin to differentiate by developing synaptic bodies and a kinocilium even before they reach the surface of the skin (Flock et al. 1973). The kinocilium projects into a small vesicle of space within the forming neuromast (Stone 1933; Flock et al. 1973; Sato 1976). The vesicle eventually penetrates to the surface of the skin between epidermal cells, and the neuromast grows to its final form through the production of more hair cells and supporting cells. The process of neuromast differentiation progresses faster at the anterior end of the regenerative placode than it does at the posterior end, so the anterior neuromasts are more developed than the posterior ones when observed at intermediate stages in regeneration (Fig. 8.8).

The segregation of cell aggregates, the formation of rosettes, the production of hair cells, and their penetration to the surface all occur both in embryonic development and in regeneration. In embryonic development, the cells that participate in these processes originate from cephalic ectodermal placodes, and they arrive at the sites of future neuromasts after proliferating in an aggregate of cells that has been called the migratory primordium. In regeneration, the cells originate from the neuromast just anterior to the site of the amputation, and they proliferate and migrate in an analogous cell aggregate, the regenerative placode.

5. Role of Innervation

The role of innervation in lateral line regeneration was investigated by Speidel (1947, 1948) and Wright (1947). They sectioned the posterior lateral line nerve at a thoracic level prior to amputation of the tip of the tail and observed that regeneration occurred normally in the organs that were denervated. Speidel also found that neuromasts that had developed by regeneration in the absence of innervation could themselves give rise to normal neuromasts when a second phase of regeneration was evoked by another amputation. Those findings have been reinvestigated by electron microscopy, which allowed direct confirmation that the neuromasts that were posterior to the cut in the nerve contained no nerve processes (Jørgensen and Flock 1976).

Denervation did not produce significant effects until 30 or more days after the lateral line nerve was sectioned, and the effects of denervation varied in their time of onset and in their magnitude between individual animals. Speidel and Wright both questioned the existence of neurotrophic phenomena reported from studies of taste bud regeneration (Olmsted 1920a, 1921), and it appears that they both ascribed great significance to the evidence from the lateral line system, interpreting this evidence as arguing against any existence of strong neurotrophic interactions. Now, however, it is known that neurotrophic interactions vary among different elements of the nervous system. It appears that the regeneration of lateral line neuromasts is independent of innervation for the most part, but that innervation is required for long-term maintenance of these organs. A recent reinvestigation has quantified this effect (Fig. 8.9) and has shown that a form of dedifferentiation appears to precede the disappearance of the denervated neuromasts (Borden and Corwin 1985). Neuromasts that have been denervated lose their tissue integrity after some weeks, so that they appear as disaggregating masses with their cells dispersing among the large cells in the surrounding epidermis (Fig. 8.10). Hair cells are absent from neuromasts that have begun to disaggregate. It appears that innervation maintains the organization of neuromasts and may be required either for direct or for indirect long-term trophic maintenance of hair cells. However, innervation is not required for short-term maintenance of hair cells or for their production and differentiation, as studies have demonstrated for both the lateral line system and the ear (Jørgensen and Flock 1976, Van De Water 1976; Corwin and Cotanche 1985).

Regeneration experiments that have examined the effects of unilateral denervation have also provided evidence that neuromasts can actively attract growing nerve processes so that innervation is reestablished. In some cases, spinal nerves

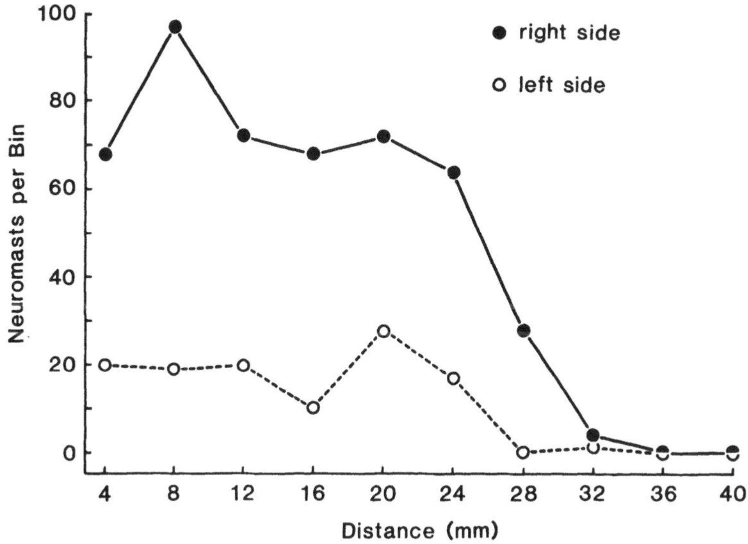

FIGURE 8.9. This graph of the number of neuromasts in 4-mm bins along the length of the tail (from anterior to posterior) illustrates the loss of lateral line neuromasts that can result from long-term denervation caused by transection of the lateral line nerve at a thoracic level in axolotl salamanders. The graph presents pooled data from five axolotls that were 47 to 68 days old at the time that the lateral line nerve on the left side was transected. The right nerve remained intact, so the right-side neuromasts served as internal controls. Following the unilateral denervation, the neuromasts on the left side of the tail began to disaggregate, and in some cases, the hair cells and the neuromasts completely disappeared over the 3-month course of the experiment.

have been attracted to denervated neuromasts; in other cases, the lateral line nerve from the opposite side of the body sends axons across the midline to innervate regenerated neuromasts on the denervated side (Fig. 8.11) (Speidel 1948; Borden and Corwin 1985). In the latter cases, it is suspected that the nerves cross over near the regenerating tail tip where dimensions are small. In cases of long-term denervation, only those neuromasts that have received such an ectopic nerve supply persist on the denervated side of the tail. Thus, it appears that these fibers, while not always appropriate for the organs innervated, do provide a trophic support for them. The mechanism of this nerve attraction is not known, but it has been suggested that the presynaptic apparatus of tonically active hair cells might release neurotransmitter and associated chemicals that would build up in the absence of postsynaptic terminals and could provide gradients for the attraction of growing axons (Corwin 1985a). Extracellular guidance pathways, in the form of adhesive substrates, have been shown to direct growing neurites when innervation develops in other systems (Thiery et al. 1982), and this mechanism may provide an alternative or complementary explanation for some aspects of the development of ectopic innervation. Adhesive guidance channels could result

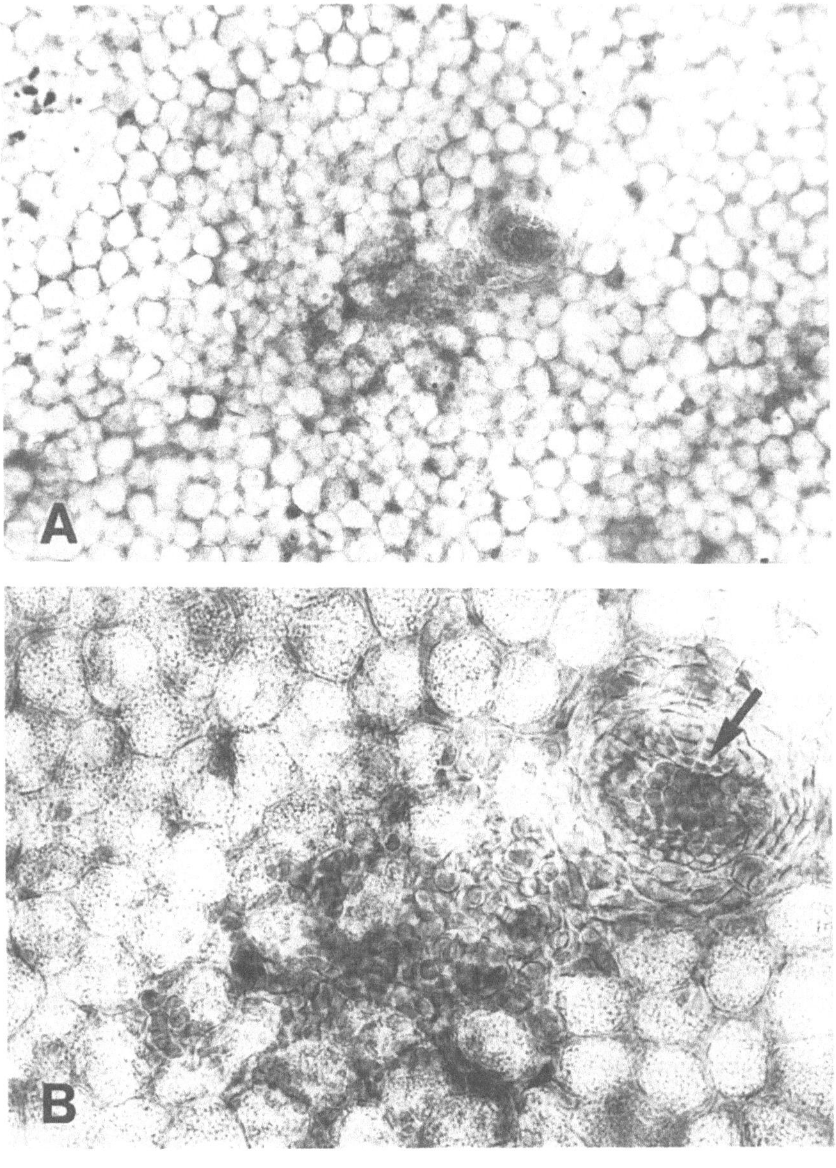

FIGURE 8.10. Two micrographs of a whole mount of the skin from a salamander that was denervated by surgical removal of a 3-mm length of the lateral line nerve at a thoracic level 117 days prior to fixation in glutaraldehyde and staining with OsO_4. (A) A low-magnification view illustrating the lack of innervation and the apparent dedifferentiation of one neuromast in a denervated stitch where an adjacent neuromast and its hair cells remain intact. (B) A higher-magnification view of the intact neuromast, where hair cells can be seen (arrow at the upper right), and where the adjacent small, darkly stained cells of the disaggregated neuromast can be seen scattered among the much larger cells of the epidermis. In salamanders, short-term denervation does not cause loss of neuromasts, but denervation lasting from 1 to 4 months does. The long latency of this effect suggests that the mechanism responsible for the loss of neuromasts may differ from the mechanisms underlying familiar neurotrophic interactions, where the effects of denervation appear much faster.

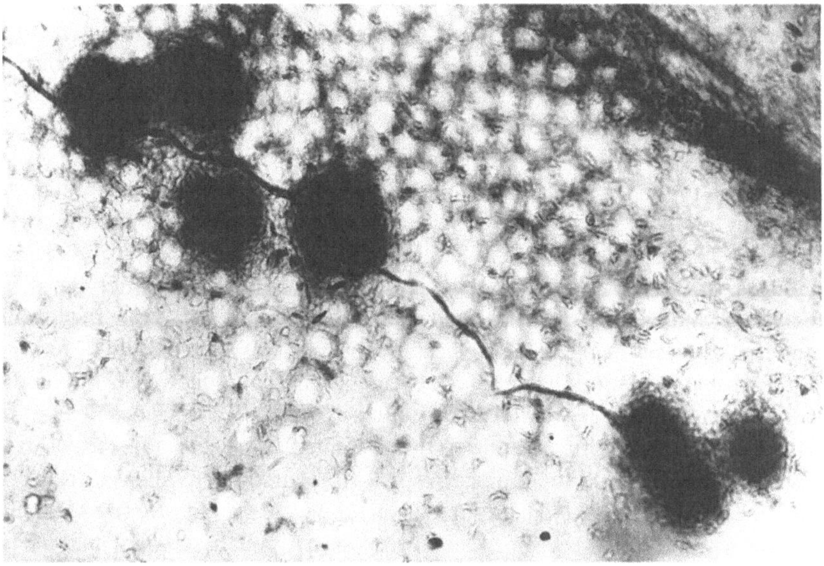

FIGURE 8.11. A micrograph of seven neuromasts that formed through regeneration following amputation and regrowth of the tip of the tail in an axolotl salamander. These neuromasts developed on the side of denervation in a salamander where the nerve on that side was transected at a thoracic level 111 days before fixation. The ectopic nerve that is associated with them may be a twig from the right (intact) lateral line nerve that branched to innervate the distalmost part of the left side of the tail. At the time of fixation, the left lateral line nerve had not regrown to reach the tail, and all of the preexisting neuromasts on that side had disappeared. Only the seven neuromasts that received this ectopic innervation persisted on that side. These manipulations have produced similar results in other specimens, suggesting that ectopic innervation is capable of providing a nonspecific neurotrophic support needed to prevent the disaggregation of neuromasts after their normal innervation has been removed.

in innervation of lateral line neuromasts, if adhesive pathways were formed during regeneration and were intercepted and followed by the growing processes of contralateral lateral line nerves or spinal nerves.

6. Relation to Regeneration in the Ear

In birds and mammals, the production of hair cells in the cochlea normally ceases before birth (Ruben 1967; Katayama and Corwin 1986), so any loss of those hair cells has been presumed to result in a hearing deficit that is irreversible. However, studies of chickens exposed to loud sounds that damage hair cells have suggested that birds may be able to recover from profound hearing loss and hair cell damage through regeneration (Saunders and Tilney 1982; Cotanche 1987), and studies of chickens exposed to an ototoxic aminoglycoside antibiotic have also supported this suggestion (Cruz et al. 1987). Recently, the potential regenerative production of hair cells in young chickens has been investigated through SEM and by administering tritiated thymidine to chickens during a 10-day recovery period following stimulation by a loud tone (Corwin and Cotanche 1987, 1988).

Damage to the hair cells in the cochlear sensory epithelium was produced by exposing pairs or small groups of 9- to 13-day-old white Leghorn chickens to pure tone acoustic stimulation at 1.5 kHz and 115 to 120 dB SPL for 48 h. One chicken from each group was overanesthetized immediately after the sound exposure, and its cochleae were fixed and examined by SEM, so that the location and the extent of the damage produced by the sound exposure could be assessed for each experiment. The other chicken in each pair received ^{3}H-thymidine administered through several daily injections or through continuous delivery for 7 days from two osmotic minipumps that were implanted subcutaneously. All the chickens that received the radioactive DNA precursor were allowed to survive for 10 days after the sound exposure. Then they were overanesthetized and their cochleae were fixed, sectioned, and processed through standard methods of autoradiography (Corwin 1985b). Controls received ^{3}H-thymidine but were not exposed to acoustic overstimulation.

Scanning electron microscopy confirmed that the acoustic stimulus resulted in a limited and reproducible lesion in the cochlear sensory epithelium (Fig. 8.12A). At the site of the lesion, the sensory cilia arrays of hair cells were absent, and hair cells that had been extruded from the epithelium were scattered on its surface, as in earlier studies (Cotanche et al. 1987, Cotanche 1987). The cell surfaces in the region of the lesion were covered by microvilli and resembled the surfaces of supporting cells. Those cells formed a continuous sheet that occupied the sites that normally would have been occupied by hair cells. In cochleae fixed 6 days after the exposure to acoustic overstimulation, small arrays of sensory cilia were observed in the lesioned area (Fig. 8.12B–D). Autoradiographic localization of labeled DNA demonstrated that these small cilia arrays originated from newly produced hair cells. In the region of damage, the nuclei of hair cells and supporting cells were both labeled by the ^{3}H-thymidine that was administered after the onset of acoustic damage (Fig. 8.13). In control cochleae and in the undamaged regions of the experimental cochlea, neither cell type was labeled, so these cells normally would have been considered postmitotic. Yet the damage caused by the loud sound had stimulated their mitosis.

The question remained whether the new, radioactively labeled hair cells were the result of divisions of previously existing hair cells or had differentiated from precursor cells produced by divisions in some other population of cells. The labeling protocol employed could not distinguish between these alternatives because of the long duration of ^{3}H-thymidine administration. For this reason, a new series of experiments was undertaken in the lateral line system, where in vivo observation allowed a direct approach to the question. In these experiments, axolotl salamanders were anesthetized, and the tips of their tails were ampu-

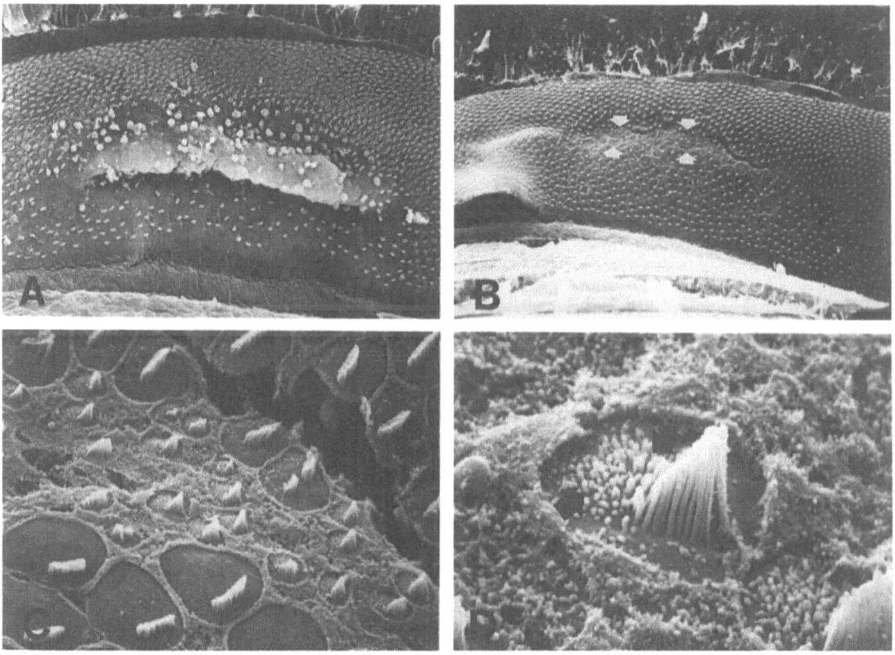

FIGURE 8.12. Scanning electron micrographs of regeneration in the proximal part of the basilar papilla sensory epithelium from the cochleae of chickens that were 10 to 13 days old at the time of acoustic overstimulation. (A) Low-magnification view of the site of an acoustically produced lesion in a cochlea fixed immediately after 48 h of acoustic overstimulation at 1.5 kHz and 115 dB SPL. Hair cell cilia bundles are missing in the area of the lesion, and the cell bodies of extruded dead hair cells are scattered over the epithelial surface at that site. Normal cilia bundles, which appear as small white dots, surround the lesion. (B) Corresponding region in an epithelium fixed 6 days after the same acoustic treatment. Small white dots are now present at the site of the lesion. (C) Higher-magnification view of cell surfaces in the lesion area shown in (B). Small cilia bundles are present on hair cells in the lesion area (arrows). Normal hair cell surfaces surround the area of the lesion. (D) High-magnification view of one of the small hair cell cilia bundles in the lesion area shown in (B) and (C). Small cilia bundles with numerous microvilli (arrow) such as this are normally seen only in embryonically developing ears in chickens. This suggested to Cotanche (1987) that these cilia bundles might be developing on newly regenerated hair cells.

tated. Then under microscopic observation, individual cells in the posteriormost neuromast on the tail stump were penetrated with a glass microelectrode and ionophoretically filled with a nontoxic fluorescent tracer, rhodamine-conjugated dextran (Fig. 8.14). Through DIC microscopy, these living cells were identified as either hair cells or supporting cells. Some authors have proposed names for up to four cell types within the class of cells that we and others recognize as supporting cells. These proposed subcategories may or may not be valid, but for our

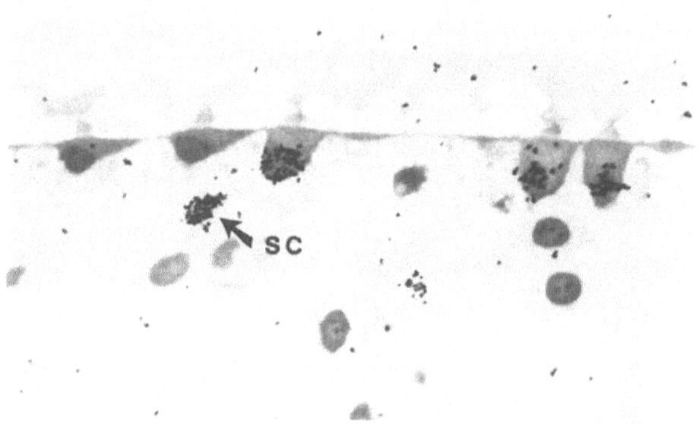

FIGURE 8.13. A micrograph of a transverse section through the site of an acoustically produced lesion in the cochlea of a young chicken that was administered the radioactive DNA precursor tritiated thymidine while recovering for 10 days following the acoustic treatment. The tissue sections where processed autoradiographically, and silver grains marking the presence of radioactive DNA are present over the nuclei of three hair cells at the apical surface and one supporting cell (SC) in the epithelium in this micrograph. This indicates that DNA replication occurred at the site of the lesion and that mitotic production of new hair cells underlies the regeneration of cilia arrays illustrated in Fig. 8.12. Cells in other areas of the lesioned cochleae and in control cochleae that received the isotope, but were not subjected to acoustic damage, were not labeled. It is concluded that the mitotic activity demonstrated at the lesion site is trauma-induced in cells that would otherwise remain mitotically quiescent.

purposes it seemed best to refrain from premature distinction when so little is actually known. The fates of the cells that were labeled by injection and the fates of any of the clones of progeny that developed from their subsequent mitoses could be followed for up to 50 days by observing the intense fluorescence that filled the cytoplasm of only those cells. In no cases were injected hair cells observed to undergo mitosis. Many of the fluorescently labeled supporting cells also remained mitotically quiescent for weeks, some were extruded from the neuromast or were engulfed by macrophages, and a number of supporting cells divided giving rise to fluorescently labeled progeny (Fig. 8.15). The progeny of supporting cells in some cases remained in the original neuromast. In other cases, labeled supporting cells exited from the original neuromast, divided, migrated within the regenerative placode, and became part of the regenerated neuromasts. In one case a supporting cell was observed to divide, giving rise to progeny, one of which differentiated into a hair cell, the other of which differentiated into a supporting cell (Fig. 8.16). This observation, while preliminary and not yet replicated, provides evidence that supporting cell progeny are capable of differentiating as hair cells.

The fate of the progeny of some proliferative cells, such as myoblasts, is strictly determined, and divisions of those cells give rise to new cells which can

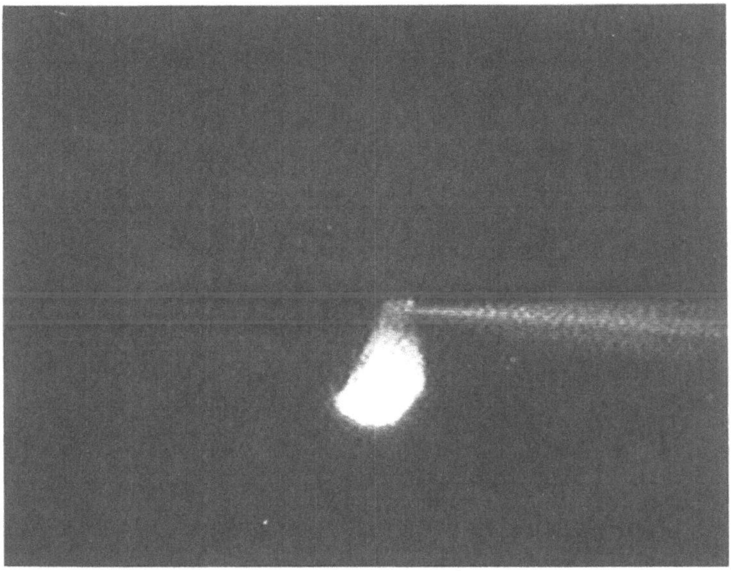

FIGURE 8.14. The lineal relationships of the cell types in neuromasts were investigated by penetrating individual cells with a glass micropipette and filling one cell in an animal with a nontoxic, fluorescently tagged dextran. A supporting cell in a neuromast is shown in this micrograph taken under epifluorescent illumination while the cell was impaled by a microelectrode and iontophoretically filled with the fluorescent tracer. Once labeled, the cell and all of its progeny could be observed for up to 5 weeks by reanesthetizing the salamander and examining it under epifluorescent illumination in a microscope equipped with a silicon intensified target video camera for low-light image enhancement.

only differentiate into the single, fated cell type. It is believed that transcription of the genome in these cells is strictly controlled by local *cis* regulation, so environmental factors do not influence expression of the cell's definitive characteristics (Blau et al. 1985). In contrast, cells such as fibroblasts are able to give rise to progeny that can differentiate in either of several ways – in this case as fibroblasts, chondrocytes, or osteocytes – depending on influences from the environment where the cell develops. In this type of cell, differentiation is apparently influenced by the environment through modification of genome transcription.

The supporting cells of hair cell epithelia appear to have a capacity for multiple lines of differentiation that is similar to that of fibroblasts. Their progeny can differentiate either as supporting cells or as hair cells, depending on environmental conditions. We now believe that the proliferation of supporting cells and the environmentally influenced differentiation of their progeny underlie the replacement of hair cells during regeneration, both in the amphibian lateral line and in the chicken cochlea.

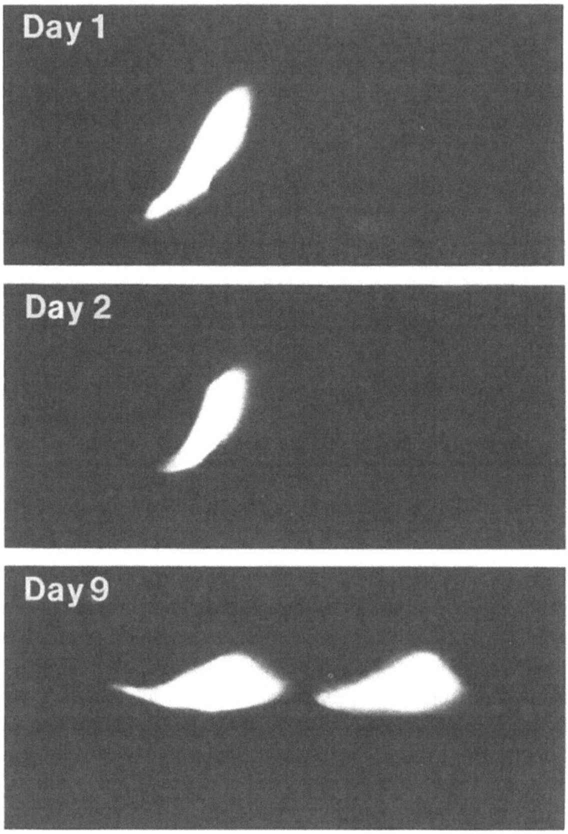

FIGURE 8.15. Many fluorescently labeled cells remained mitotically quiescent during the observation period, but some supporting cells divided, giving rise to progeny which inherited the fluorescent label. The panels in this figure show views from the day of injection (day 1), the next day (day 2), and 1 week after the initial injection of label (day 9) in a salamander tail where a single cell at the posterior edge of a neuromast was labeled by microinjection. One day after the injection, the cell began to move posteriorly, where the final photograph was taken. Then the cell divided, and its two progeny moved further posteriorly. Labeled hair cells have never been observed to divide in these experiments.

It is particularly interesting that in the case of the chicken cochlea, the cells that proliferate in response to trauma are cells that would normally be considered postmitotic. Cumulative DNA labeling experiments in chick embryos have demonstrated that cell production ceases by the eighth day of incubation at the site of the acoustic lesions produced in the study of regeneration (Katayama and Corwin 1986). Hatching occurs 13 days later, and the chickens in the regeneration experiments were not exposed to the acoustic trauma until 10 days after hatching. Yet supporting cells in that region were able to undergo mitosis. This

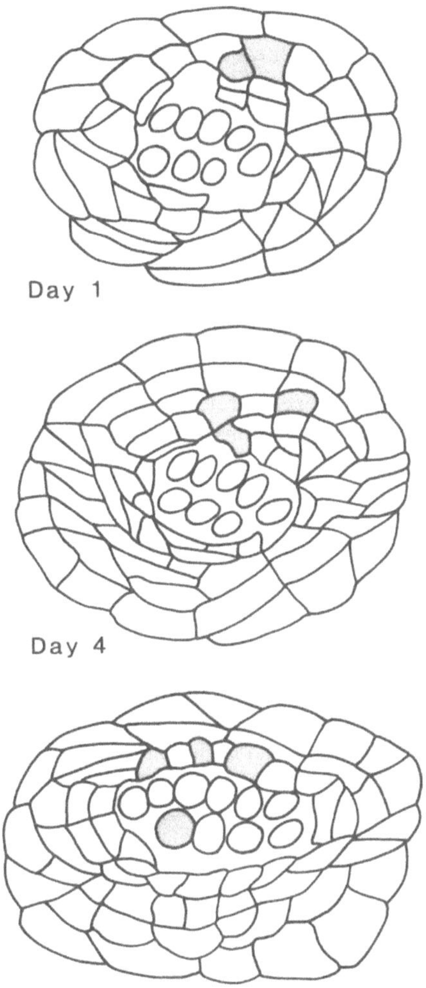

FIGURE 8.16. Drawings illustrating one cell lineage experiment which indicated that supporting cell progeny can give rise to hair cells. On day 1, two supporting cells were labeled by microinjection of fluorescent dextran. Three days later one of those cells divided, producing two new cells. Over the course of the next 3 days, one of the labeled cells moved to the other side of the neuromast and differentiated as a sensory hair cell. On the basis of this and other evidence, it is the conclusion of these authors that the progeny of supporting cell mitoses can differentiate either as supporting cells or as hair cells, depending on their environments. The progeny of fibroblasts have a similar capacity for multiple lines of differentiation. They can differentiate either as fibroblasts, as chondroblasts, or as osteoblasts.

trauma-induced mitosis in a normal, mitotically quiescent population of cells suggests that the absence of proliferation in normal tissue, such as the untraumatized control cochleae that showed no ^{3}H-thymidine labeling, cannot be assumed to reflect the absence of proliferative or regenerative capacities. In the above case, cells that did not normally divide were still capable of division. Comparable results have also been reported on the basis of hair cell regeneration experiments in the cochleae of reproductively mature quail (Ryals and Rubel 1988). In light of this, the absence of supporting cell proliferation in postembryonic mammals should not be judged as firm evidence for the absence of a potential regenerative capacity. It may be that supporting cells in the mammalian cochlea, like supporting cells in the chicken cochlea and the supporting cells in the lateral line neuromasts of axolotls, retain proliferative capacities that normally remain latent during postembryonic life but may be released from their latency under appropriate conditions. The question is whether such conditions exist. If they do, then at some point in the future, their identification and control may allow the induction of hair cell regeneration in human ears and so provide recovery from forms of sensorineural deafness that have generally been considered irreversible.

References

Blau HM, Pavlath GK, Handeman EC, Chiu C-P, Silberstein L, Webster SG, Miller SC, Webster C (1985) Plasticity of the differentiated state. Science 230:768–766.

Borden PC, Corwin JT (1985) Hair cells in the lateral line are trophically dependent on innervation both for maintenance and for regeneration. Soc Neurosci Abstr 11:270.

Brown MC, Nuttall AL, Masta RI, Lawrence M (1983) Cochlear inner hair cells: Effects of transient asphyxia on intracellular potentials. Hear Res 9:131–144.

Corwin JT (1985a) Auditory neurons expand their terminal arbors throughout life and orient toward the site of postembryonic hair cell production in the macula neglecta in elasmobranchs. J Comp Neurol 239:445–452.

Corwin JT (1985b) Perpetual production of hair cells and maturational changes in hair cell ultrastructure accompany postembryonic growth in an amphibian ear. Proc Natl Acad Sci USA 82:3911-3915.

Corwin JT (1985c) Early events in the regeneration of lateral line hair cell epithelia. Biol Bull 169:549–550.

Corwin JT (1986) Regeneration and self-repair in hair cell epithelia: Experimental evaluation of capacities and limitations. In: Ruben RJ, Van de Water TR, Rubel EW (eds) Biology of Change in Otolaryngology. New York: Elsevier, pp. 291–304.

Corwin JT, Cotanche DA (1985) Location-specific patterns of cilia number, cilia length, and bundle orientation differentiate after denervation and transplantation of embryonic cochlea. Soc Neurosci Abstr 11:245.

Corwin JT, Cotanche DA (1987) Regeneration of sensory hair cells in the cochlea of the chicken. Soc Neurosci Abstr 13:539.

Corwin JT, Cotanche DA (1988) Regeneration of sensory hair cells after acoustic trauma. Science 240:1772–1774.

Cotanche DA (1987) Regeneration of hair cell stereocilliary bundles in the chick cochlea following severe acoustic trauma. Hear Res 30:181–196.

Cotanche DA, Corwin JT (1986) Changes in hair cell orientation during the development of the chick basilar papilla. Assoc Res Otolaryngol Abstr 9:81.
Cotanche DA, Saunders JC, Tilney LG (1987) Hair cell damage produced by acoustic trauma in the chick cochlea. Hear Res 25:267–286.
Cruz RM, Lambert PR, Rubel EW (1987) Light microscopic evidence of hair cell regeneration after gentamicin toxicity in chick cochlea. Arch Otolaryngol Head Neck Surg 113:1058–1062.
Flock Å, Jørgensen M, Russell I (1973) The physiology of individual hair cells and their synapses. In: Moller AR (ed) Basic Mechanisms in Hearing. New York: Academic Press, pp. 273–306.
Goss RJ (1969) Principles of Regeneration. New York: Academic Press.
Harrison RG (1903) Experimentelle Untersuchungen über die Entwicklung der Sinnesorgane der Seitenlinie bei den Amphibien. Arch Mickrosk Anat Ent 63:35–149.
Jørgensen JM, Flock Å (1976) Non-innervated sense organs of the lateral line: Development in the regenerating tail of the salamander *Ambystoma mexicanum*. J Neurocytol 5:33–41.
Katayama A, Corwin JT (1986) An experimental determination of the temporal and spatial pattern of hair cell production in the embryonic cochlea of the chicken. Soc Neurosci Abstr 12:1048.
Olmsted JMD (1920a) The nerve as a formative influence in the development of taste buds. J Comp Neurol 31:465–468.
Olmsted JMD (1920b) The results of cutting the seventh cranial nerve in *Ameiurus nebulosus* (LeSucur). J Exp Zool 31:369–401.
Olmsted JMD (1921) Effect of cutting the lingual nerve of the dog. J Comp Neurol 33:149–154.
Ruben RJ (1967) Development of the inner ear of the mouse: A radioautographic study of terminal mitoses. Acta Otolaryngol (Stockh) (Suppl) 220:1–44.
Russell IJ, Cowley EM (1983) The influence of transient asphyxia on receptor potentials in inner hair cells of the guinea pig cochlea. Hear Res 11:373–384.
Ryals BM, Rubel EW (1988) Hair cell regeneration after acoustic trauma in adult *Coturnix* quail. Science 240:1771–1776.
Sato A (1976) Electron microscopic study of the developing lateral-line organ in the embryo of the newt, *Triturus pyrrhogaster*. Anat Rec 186:565–584.
Saunders JC, Tilney LG (1982) Species differences in susceptibility to noise exposure. In: Hamernik RP, Henderson D, Salvi R (eds) New Perspectives on Noise-Induced Hearing Loss. New York: Raven Press, pp. 229–248.
Singer M (1954) Induction of regeneration of the forelimb of the postmetamorphic frog by augmentation of the nerve supply. J Exp Zool 126:419–472.
Speidel CC (1947) Correlated studies of sense organs and nerves of the lateral-line in living frog tadpoles. I. Regeneration of denervated organs. J Comp Neurol 87:29–55.
Speidel CC (1948) Correlated studies of sense organs and nerves of the lateral-line in living frog tadpoles. II. The trophic influence of specific nerve supply as revealed by prolonged observations of denervated and reinnervated organs. Am J Anat 82:277–320.
Stone LS (1931) Studies on the migratory lateral-line primordia in *Amblystoma*. Anat Rec 48 (Suppl) p. 64.
Stone LS (1933) The development of lateral-line sense organs in amphibians observed in living and vital-stained preparations. J Comp Neurol 57:507–540.
Stone LS (1937) Further experimental studies of the development of lateral-line sense organs in amphibians observed in living preparations. J Comp Neurol 68:83–115.

Thiery JP, Duband JL, Delouvée A (1982) Pathways and mechanism of avian trunk neural crest cell migration and localization. Dev Biol 93:324–343.
Van De Water TR (1976) Effects of removal of the statoacoustic ganglion complex upon the growing otocyst. Ann Otol Rhinol Laryngol (Suppl) 33:1–32.
Wright MR (1947) Regeneration and degeneration experiments on lateral line nerves and sense organs in anurans. J Exp Zool 105:221–257.

Part III
Peripheral Processing: Hydrodynamics, Biomechanics and Neural Encoding

9
Functional Evolution of Lateral Line and Inner Ear Sensory Systems

AD. J. KALMIJN

1. Introduction

In detecting moving objects, the fish's lateral line and inner ear support different, but closely related, hydrodynamic functions, suggesting early developments in the vertebrate sense of hearing. In an effort to elucidate the functional evolution of the two sensory systems, this chapter examines the hydrodynamic and acoustic fields in nature, the physics and physiology of the detection process, the evaluation of the sensory data, and the ensuing behavioral responses.

2. Physics and Physiology of Detection Process

To appreciate the functioning of the lateral line and inner ear sensory systems, a basic understanding of the peripheral detection process is essential. This first section begins with a discussion of the physical and physiological events preceding nervous conduction—first in the time domain, then in the frequency domain—followed by a justification of the approach and proper references.

2.1 Time-Domain Description

At the receptor level, the hair cells of the lateral line and inner ear sense organs are inherently displacement detectors (Fig. 9.1a). This follows from the observation that, in response to a wide range of motions, the receptor potentials are proportional to the displacements of the ciliary bundles relative to the cell bodies, independent of the wave form. The hair cells are most sensitive along an imaginary axis marked by the asymmetric position of the single kinocilium relative to the several rows of stereocilia.

At the organ level, most lateral line neuromasts are velocity detectors (Fig, 9.1b). The gelatinous cupulae accommodating the stiff ciliary bundles are driven by the viscous drag forces proportional to the velocity of the water flowing along their sides. The cupulae of other neuromasts may also respond, either partially or solely, to the pressure differences between their fore and aft surfaces,

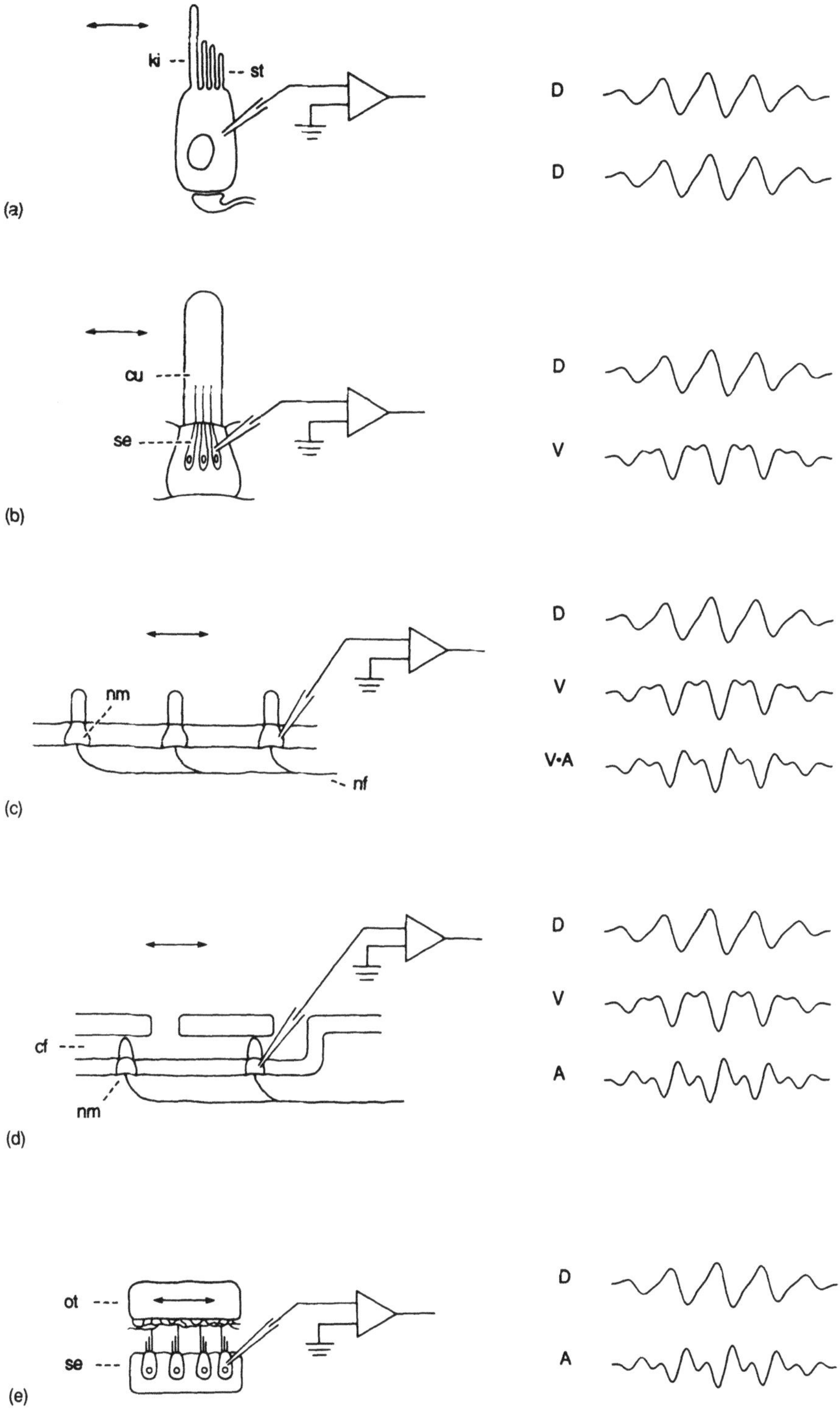
ki
st
D
D
(a)
cu
se
D
V
(b)
nm
nf
D
V
V•A
(c)
cf
nm
D
V
A
(d)
ot
se
D
A
(e)

as will be detailed later. By sliding over the sensory epithelia, the cupulae displace the ciliary bundles, thus stimulating the hair cells.

At the system level, the arrays of free neuromasts rank between velocity and acceleration detectors, since, in parallel flows, the velocity of the water at the surface of the animal is proportional to a fractional derivative of the velocity of the water volume outside the boundary layer (Fig 9.1c). The lateral line canal system usually functions as an acceleration detector, because the velocity of the water in the subcutaneous canals is more nearly proportional to the first full derivative of the velocity of the water volume (Fig. 9.1d).

The paired otolith organs of the inner ear, the utriculus, sacculus, and lagena, serve as linear acceleration detectors (Figs. 9.1e). The displacements of the stiff ciliary bundles result from the inertial forces on the calcareous otoliths or otoconia during acceleration of the animal. The three paired semicircular canals and their cristae ampullares are essentially angular acceleration detectors. By their inertia, the endolymphatic canal fluids develop the pressure differences that displace the large cupulae blocking the ampullary lumina.

Hence, the lateral canal system and the inner ear otolith organs are both acceleration detectors. Nevertheless, they differ in two important aspects. First, the lateral line canal system detects the acceleration of the water relative to the fish, whereas the otolith organs detect the acceleration of the fish per se. Second, with regard to imposed flows, the lateral line system responds to the spatial differences in the acceleration of the water along the length of the sensory arrays, whereas the otolith organs detect the acceleration of the water averaged over the volume occupied by the free-moving fish.

Since the particle accelerations of the imposed, intertially controlled flows are directly proportional to the negative gradients of the pressure distribution, the results can also be expressed in terms of scalar pressure fields. Thus, the lateral line system of unrestrained fish responds to the spatial differences in the steepness of the pressure drop along the sensory arrays, whereas the otolith organs respond to the total pressure drop between opposite surfaces of the fish. Naturally, the animal's own accelerations also give rise to nonuniform pressure gradients stimulating the lateral line.

◄

FIGURE 9.1. Time-domain representations of lateral line and inner ear responses to displacement stimuli applied at cell, organ, and system levels. To distinguish the displacement (D); its first derivative, the velocity (V); its first-and-a-half derivative, the velocity•acceleration (V• A); and its second derivative, the acceleration (A), the stimuli are composed of three sine waves in and 1:2:3 frequency ratio (see also Fig. 9.3f for actual recordings). (a) Lateral line or inner ear hair cell; ki) kinocilium; st) stereocilia; with hair bundle displacement (D) and receptor-potential (D) traces. (b) Lateral line organ or neuromast; cu) cupula; se) sensory epithelium; with cupula displacement (D) and velocity (V) traces; (c) Array of free lateral line neuromasts; nm) neuromast; nf) nerve fibers; with volume displacement (D), velocity (V), and velocity·acceleration (V·A) traces; (d) Lateral line canal with canal organs; nm) neuromasts; cf) canal fluid; with volume displacement (D), velocity (V), and acceleration (A) traces. (e) Inner ear otolith organ; ot) otolith; se) sensory epithelium; with animal displacement (D) and acceleration (A) traces.

The present treatise focuses on the responses of free-moving fish to sources of hydrodynamic and acoustic fields in the water column. The concept of functional significance is founded in the work of Dijkgraaf (Chapter 2); the systems approach is shared with Denton and Gray (Chapter 11) and with van Netten and Kroese (Chapters 12 and 13). Surface waves and substrate vibrations are not discussed here, even though their detection is ruled by many of the same physical principles (Bleckmann et al., Chapter 25; Elepfandt, Chapter 26; Görner and Mohr, Chapter 27).

2.2 Frequency-Domain Description

When expressed in terms of the stimulus parameter that the cell, organ, or system detects, the amplitude response curves are nearly frequency independent, and the phase lag is close to zero for the lower frequencies of interest, for only then will the wave shape be faithfully reproduced in the time domain. Conversely, if the amplitude responses are nearly frequency independent and the phase is close to zero at the lower frequencies, it follows that the curves were expressed in the stimulus parameter that the cell, organ, or system detects (Fig. 9.2a,b).

In short, flat, low-frequency response curves are expected for the lateral line and inner ear hair cells when the receptor potentials are referred to the displacement of the ciliary bundles; for the individual neuromasts, to the velocity of the water at the cupulae; for the system of free neuromasts, to a combination of the velocity and acceleration of the water volume; for the system of canal organs, to the acceleration of the water volume, all relative to the fish; and for the otolith organs, to the acceleration of the fish.

At low frequencies, the motions of the lateral line cupulae and inner ear otoliths are stiffness controlled, because the restoring forces resulting from the rigidity of the ciliary bundles prevail over the frictional and inertial forces associated with the cupular or otolithic motions. However, with increasing frequency, the frictional forces gain in importance and, ultimately, the inertial forces overtake both the stiffness and frictional forces, rendering the amplitude and phase responses highly frequency dependent (Fig. 9.2a,b).

In the intermediate-frequency, friction-controlled region, the amplitude of the response diminishes, with or without initial overshoot, depending on the amount of friction (or, equivalently, the degree of damping), whereas the phase increases, passing through 90°. Thus, about the natural frequency, the lateral line canal system and the inner ear otolith organs act as nearly critically damped integrating accelerometers, indicating the velocity of the stimulus (velocity mode). The inner ear semicircular canals normally operate in the velocity mode.

In the high-frequency, inertia-controlled region, the amplitude of the response gradually approaches zero, whereas the phase tends to 180°. Thus, at high frequencies, the lateral line canal system and the inner ear otolith organs act as double-integrating accelerometers, indicating the displacement of the stimulus (displacement mode). Even though the lateral line system is not expected to be commonly engaged in the detection of sufficiently high frequencies to operate

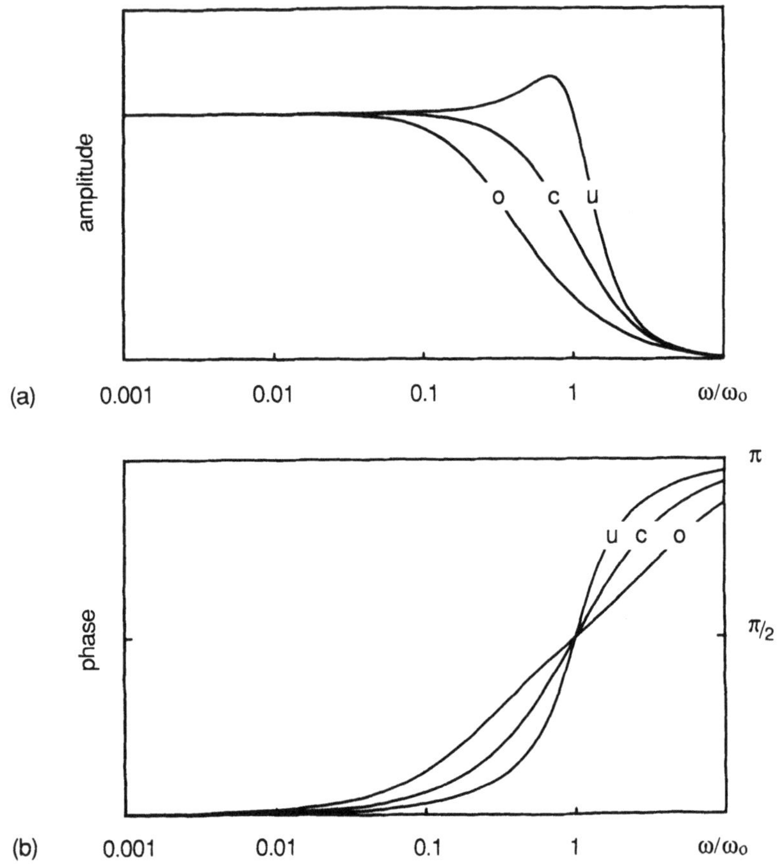

FIGURE 9.2. Frequency-domain representations of lateral line canal and inner ear otolith responses to accelerations of the water volume and of the fish respectively, illustrating the low-frequency acceleration nature of the sensory systems and their higher-frequency velocity and displacement modes of operation. The responses are expressed in terms of cupular displacement per unit of acceleration (on a linear scale) versus ω/ω_0, the frequency ω over the natural frequency ω_0 (on a logarithmic scale). For the arrays of free neuromasts, the responses would have to be expressed per unit of a fractional derivative intermediate between velocity and acceleration (see text). (a) Amplitude response curves. (b) Phase response curves; u) underdamped; c) critically damped; o) overdamped. 1. Acceleration detection: at $\omega/\omega_0 \ll 1$, responses to accelerations well below the natural frequency are stiffness controlled. 2. Single-integrated acceleration detection: at $\omega/\omega_0 \approx 1$, responses to accelerations around the natural frequency are friction controlled; the system operates in the velocity mode. 3. Double-integrated acceleration detection: at $\omega/\omega_0 \gg 1$, responses to accelerations well above the natural frequency are inertia controlled; the system operates in the displacement mode. The natural frequencies of the semicircular canals, the free neuromasts, the canal organs, and the otolith organs are on the order of 1, 40, 80, and 160 Hz, respectively. The semicircular canals are heavily overdamped; all others are nearly critically damped. (After Kalmijn 1988a.)

in the displacement mode, the high-frequency responses of the otolith organs are well documented (Platt et al., Chapter 32).

In the above discussion, it was assumed that the individual sense organs can be meaningfully treated as simple mechanical systems, driven by simple forces. Whereas the inertial forces driving the otolith organs fit this description, the forces driving the lateral line organs are usually complex functions of the water motions in the surrounding medium (Jielof et al. 1952). Nevertheless, comparisons with simple systems, preceded by simple filters, can be highly instructive, even if valid only within limited frequency bands.

In short, at low frequencies, many neuromasts behave approximately as simple, nearly critically damped mechanical systems driven by velocity-dependent forces, which in turn relate to the acceleration or a lesser fractional derivative of the water volume (Denton and Gray 1983; Kalmijn 1988a). For significant deviations from this scheme, see van Netten and Kroese (1987).

2.3 Evidence and Justification

The displacement nature of the hair cell response was demonstrated neurophysiologically for the lateral line canal organs of the fish *Lota vulgaris* by Flock (1965), who displaced the whole cupula, and for the inner ear sacculus of the frog *Rana pipiens* by Hudspeth and Corey (1977), who displaced the individual ciliary bundles. Other authors have invoked a velocity, or even an acceleration, sensitivity at the hair cell level, but their conclusions were often unfounded, as they neglected the complex, differentiating properties of the boundary layer on the surface of the animals (for specifics, see Kalmijn 1988a).

With regard to the mechanical coupling, the ciliary bundles of the receptor cells apparently connect to the cupular stroma or otolithic membrane only at their tips and, when displaced, hinge at their bases (Flock 1971; Flock et al. 1977; Dale 1980). By virtue of their stiffness, the hair bundles tend to restore the resting positions of the cupulae or otoliths, perhaps aided by other structural elements. How precisely the ciliary deflections give rise to the receptor potentials is uncertain (Hudspeth 1982). For a further discussion of the transduction process, see Kroese and van Netten (Chapter 13).

In describing the neuromasts as velocity detectors, the ciliary displacements were regarded as resulting only from the viscous forces acting along the sides of the cupulae. Nevertheless, the forces due to the pressure differences between their leading and trailing surfaces may render the neuromasts also sensitive to the acceleration of the water (van Netten and Kroese 1987, Chapter 12; Kalmijn 1988a). The relative contributions of the pressure-related forces depend on the size and shape of the cupulae. If significant, their effect will become apparent as an enhanced high-frequency response and reduced phase lag.

Where the cupulae completely block the lateral line canals, especially where they form tight partitions between pouchlike canal sections, the pressure differences between adjacent compartments must force the cupulae to follow the displacements of the canal fluid (Coombs et al. 1988). At low frequencies, the system will be stiffness controlled and respond to the pressure differences between neighboring skin pores or, equivalently, to the acceleration of the water volume outside the fish (Denton and Gray 1988; Kalmijn 1988a). For a detailed discussion of the pressure distribution over the skin, see Kuiper (1967) and Gray (1984).

Whereas in the water volume the inertial forces prevail, in the thin hydrodynamic boundary layer on the surface of the skin and in the lateral line canals, viscous forces play a significant role as well (Jielof et al. 1952). In canals 0.28 mm in diameter, the full differentiating tendency was found to be nearly constant up to 80 Hz (Denton and Gray 1983, 1988). At the surface of the skin, however, the fractional differentiating tendency is theoretically predicted to diminish from half to zero with increasing frequency and cupular length (see Fig. 4.13 in Kalmijn 1988a). The effect may be recognized as a reduced high-frequency response and an increased phase lag in the free neuromasts.

The graded (half-to-zero) fractional derivative of the velocity in the time domain is equivalent to a dependence on both the velocity and acceleration amplitudes and to a phase advance of 45° or less in the frequency domain. In this respect, the frequency-domain description is mathematically simpler than the time-domain description. Nevertheless, when midwater fish react to sinusoidal accelerations within the first period of a stimulus train, they evidently operate in the time domain (Blaxter et al. 1981). Surface-feeding fish, however, often await the passage of several wavelengths before responding (Bleckmann et al., Chapter 25).

Even though the discussion of the lateral line and inner ear responses in the frequency domain may be illuminating, if the animals actually operate in the time domain, the conclusions reached in the frequency domain must be transformed back to the time domain as well (Kalmijn 1988a). In particular, in dealing with transient signals, the time-domain representation is of utmost importance. In applying the concepts of frequency analysis, the system is considered linear, which seems reasonable because of the small excursions at the hair cell level.

The detection process is not only expected, but actually known, to be more complicated than depicted here. However, a plausible detection scheme, founded on first physical and physiological principles, is imperative in order to organize the present data and to interpret the results of new experiments. Future data will, in turn, lead to correction and improvement of the current scheme. Part of this interactive process is the search for the proper stimulus parameter, which is the subject of the next section.

3. Results of Neurophysiological Studies

In response to mechanical stimulation, hair cells produce graded receptor potentials, leading to discrete nerve potentials. Hence, electrical recording techniques seem particularly suited for a detailed study of the detection process. After exposing a historical error that has led to considerable confusion, this section presents an effort to clarify the neurophysiological data of earlier years.

3.1 Correction of Erroneous Data

Subsequent to the pioneering studies of Jielof et al. (1952), Harris and van Bergeijk (1962) published the first complete frequency response curves for the lateral line canal organs in the killifish *Fundulus heteroclitus*. They stimulated the

---

FIGURE 9.3. Electrophysiological identification of the proper stimulus parameter for the lateral line canal organs of the killifish *Fundulus heteroclitus*. The original experiments of Harris and van Bergeijk (1962) were repeated to ascertain that the lateral line canal system of the killifish responds to the acceleration, not the displacement, of a vibrating sphere (Kalmijn 1988a). (a) Local-flow field of sphere impinges on head of restrained fish; microphonic potentials are recorded with a pipette electrode inserted into nasal lateral line canal. Arrows indicate motion of sphere and of local flow; vb) vibrating sphere; kf) killifish; nc) nasal canal; el) electrode. (b) Longitudinal section through nasal lateral line canal and single neuromast with electrode lodged in canal mucous. Cupula and water volume move in opposite directions owing to complex properties of boundary layer and viscous nature of driving force, or to stiffness of ciliary bundles if cupula blocks canal lumen; sk) skin; mu) mucous; cu) cupula; se) sensory epithelium; ne) nerve; el) electrode. (c) Top: Plotted in terms of response per unit of displacement (-dB), microphonic potentials yield broad tuning curves; dot marks best frequency. Bottom: Plotted in terms of response per unit of acceleration (-dB), microphonic potentials produce low-pass curves; dot marks rolloff frequency. Flattened, low-frequency part of bottom curve suggests that acceleration is the proper parameter. (d) Summation of nonlinear receptor potentials of oppositely oriented hair cells. 1. Cupula unbiased: summated receptor potentials yield double-frequency microphonics. 2. Cupula biased rostrally, receptor potentials of hair cells having kinocilia in rostral position prevail. 3. Cupula biased caudally: receptor potentials of hair cells having kinocilia in caudal position prevail. Top: Hair-cell bundles with stereocilia and single kinocilia. Middle: Transfer characteristics of oppositely oriented hair cells. Bottom: Responses to sinusoidal displacements indicated by double-headed arrows in transfer characteristics. (e) Microphonic potentials in response to sine wave stimuli. 1. Displacement of vibrating sphere. 2. Microphonic potentials, cupula biased rostrally by slight push on electrode. 3. Microphonic potentials, cupula unbiased. 4. Microphonic potentials, cupula biased caudally by slight pull on electrode. Note that the phases reflect the acceleration, not the displacement, of the sphere. *Down* in 1 corresponds to downward stroke in (a). *Up* in microphonic traces denotes positive (+). (f) Microphonic potentials in response to a ramplike stimulus composed of the first three sine wave terms of the sawtooth Fourier series (the same wave form used for the calculated responses of Fig. 9.1). 1. Ramplike displacement of vibrating sphere. 2. Microphonic potentials for rostrally biased cupula. (g) Microphonic potentials in response to a rounded-off square-wave stimulus. 1. Square-wave displacement of vibrating sphere. 2. Microphonic potentials for rostrally biased cupula. Note that the responses in (f) and (g) reflect the acceleration of the sphere. (Parts a,e,f,g, from recent experiments by Ad. J. and Jelger A. Kalmijn; b,c (top), after Harris and van Bergeijk 1962; c (bottom), after Kalmijn 1988a; d, after Flock 1965.)

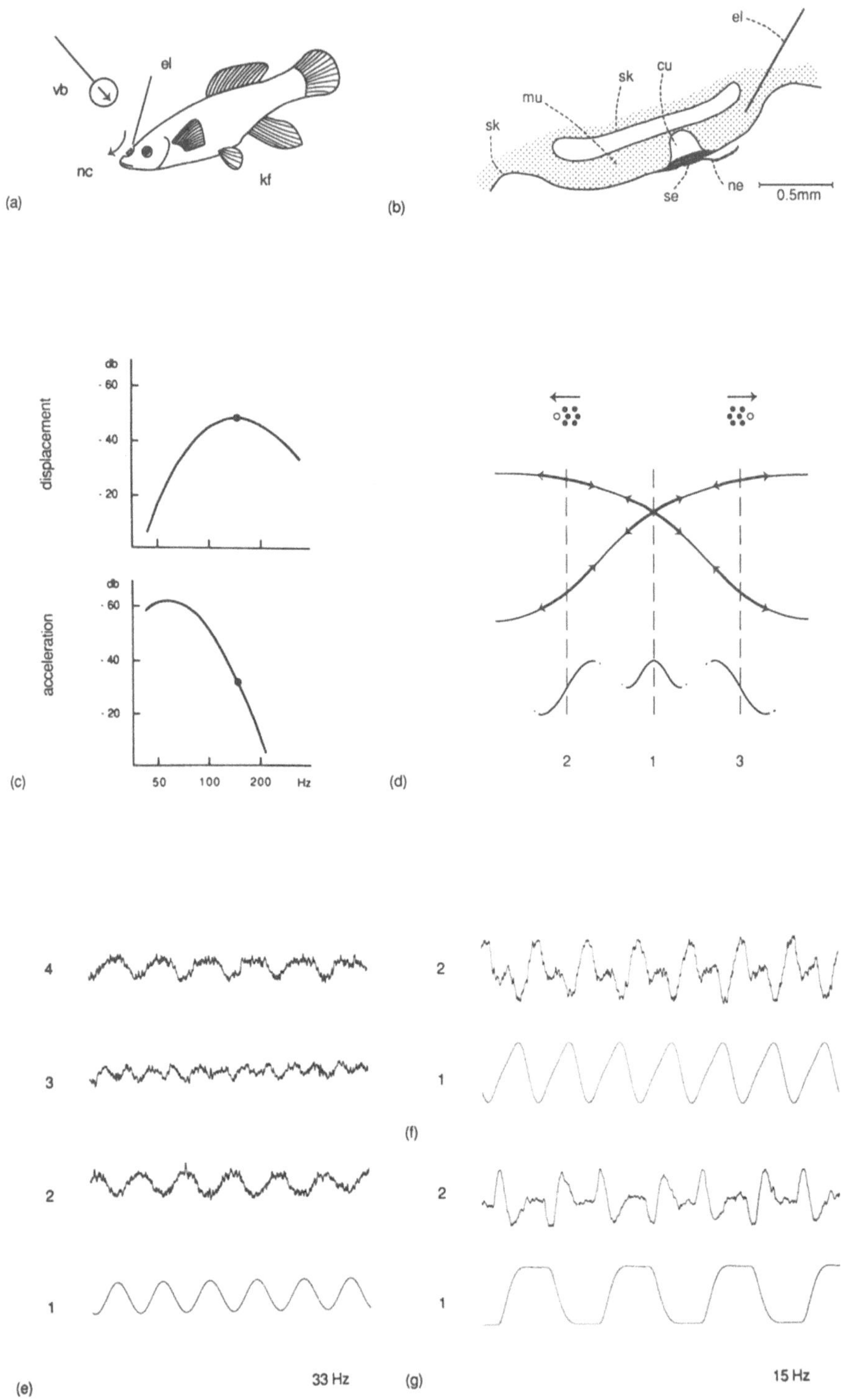
vb
el
nc
kf
(a)
el
sk
cu
mu
sk
se
ne
0.5mm
(b)
db
60
40
20
displacement
db
60
40
20
acceleration
50
100
200
Hz
(c)
2
1
3
(d)
4
3
2
1
(e)
33 Hz
2
1
(f)
2
1
(g)
15 Hz

fish with a nearby vibrating sphere while recording the extracellular microphonic potentials of single neuromasts. Plotted versus the displacement of the sphere, the results showed moderately sharp tuning curves with best frequencies between 140 and 180 Hz and phase angles ranging from about 30° to 210° over a frequency range of 33 to 333 Hz (Fig 9.3a–c).

Because the phase angles were reasonably close to zero at the lower frequencies, Harris and van Bergeijk concluded that the lateral line canal organs respond to the displacement, not the velocity or acceleration, of the source. In an attempt to explain a remaining 30° phase lag, they briefly considered the phase shifts expected from the viscous nature of the force driving the cupula and from the complex properties of the hydrodynamic boundary layer in the canals. However, disregarding the low Reynolds numbers, they opted to reject both ideas, as the resulting phase advances would increase, rather than diminish, the discrepancy.

Van Bergeijk and Alexander (1962) actually describe the canals as filled with a thick, viscous mucous and wider than the elongated cupulae. Yet, even if the cupulae effectively block the canals, the system does not respond to the displacement of the outside water at low frequencies (see previous section). Unfortunately, despite Jielof et al.'s (1952) compelling biophysical analysis and revealing microscopic observations, Harris and van Bergeijk appeared to be unaware that the amplitude and phase response curves in their 1962 paper were mutually inconsistent without invoking extraordinary mechanical or physiological phenomena.

The inconsistency is obvious. For the canal fluid to follow the inertia-controlled displacement of the outside water, and for the cupula to follow the displacement of the canal fluid, the motions of both the canal fluid and the cupula must be largely inertia controlled (which they are not). Any restraining viscous or stiffness forces would render the canal fluid sensitive to the velocity or acceleration of the outside water, or the cupula sensitive to the velocity or acceleration of the canal fluid. However, in the absence of restraining viscous or stiffness forces, Harris and van Bergeijk's tuning curves would be inexplicable.

A hint as to a possible solution of the issue is obtained by adding the phase shifts predicted from the viscous nature of the driving force, 90°, and from the complex properties of the boundary layer, from 45° to 90°, to the 30° that Harris and van Bergeijk reported. This gives a total of approximately 180° as expected from an acceleration detector. If the cupulae effectively block the canal, similar results are obtained. Might Harris and van Bergeijk have misjudged the phase of the response by 180°? They apparently did. Unfortunately, their mistake was not exposed until recently (Kalmijn 1988a).

In monitoring the phase of the vibrating sphere, Harris and van Bergeijk used a velocity transducer. Could they have added, rather than subtracted, the 90° phase advance of their device? Or did they lose track of the polarity of the recorded potentials? If undisclosed, either mistake would cause a phase error of 180°. Or might they have focused on the wrong peaks of the double-frequency microphonic potentials? Extracellularly, the summated receptor potentials of oppositely oriented hair cells produce double-frequency microphonics as a result of their nonlinear response characteristics (Fig. 9.3d).

To remove any lingering doubt in such a fundamental matter, the author and his son Jelger repeated Harris and van Bergeijk's experiments, using the same fish species and following essentially the same stimulation and recording techniques. The results were in close conformity with the original data, except that the microphonic potentials at the lower frequencies were nearly in phase with the acceleration, not the displacement, of the vibrating sphere (Fig. 9.3e). In response to other than purely sinusoidal stimuli, the microphonic potentials followed the acceleration of the sphere even more conspicuously (Fig. 9.3f,g).

3.2 Lateral Line Response Curves

Since the studies of Harris and van Bergeijk (1962), nearly all physiological data on the lateral line have been presented and interpreted in terms of displacement. Certainly, the displacement of a source is readily calibrated by optical means. Moreover, researchers often consider displacement conceptually easier to grasp than velocity or acceleration. Admittedly, the practice of plotting the data in terms of displacement is not necessarily wrong, but, historically, it has proved sufficiently misleading to cause serious misconceptions about the function and operation of the lateral line and inner ear (Kalmijn 1988a).

Plotted per unit of displacement, the physiological tuning curves usually show best frequencies well above 20 Hz and only weak responses at frequencies below 20 Hz (upper curves of Fig. 9.4a,b). Therefore, they do not seem to support the fishes' prevalent use of the lower frequencies shown behaviorally (Dijkgraaf 1963). Moveover, little evidence has been found for a *prevailing* significance of the higher frequencies to which the lateral line seems most sensitive. The problem consists in the use of the terms "most sensitive" and "best frequency" without proper justification of the stimulus parameter to which they refer.

A good practice would be to express the fields in nature and the physiological responses they elicit in terms of the stimulus parameter that the sensory system detects. Stating that the lateral line is most sensitive to, or has its best frequency at, 80 Hz is deceiving if the response is expressed in terms of displacement, whereas the system is designed to detect acceleration. The use of an arbitrary parameter can be justified only if the response of the system is directly referred to the natural stimulus (see Fig. 4.8 in Kalmijn 1988a). Even then it tends to obscure, rather than expose, the nature of the detection process.

Thus, when the data from the literature are replotted in terms of the proper stimulus parameter (per unit or acceleration of a combination of velocity and acceleration), the once familiar high-frequency tuning curves change into much broader curves exhibiting predominantly low-pass characteristics (lower curves of Fig. 9.4a,b). In other words, expressed in the stimulus parameter that the lateral line detects, the same experimental data reveal a much higher sensitivity to the lower frequencies and a much lower sensitivity to the higher frequencies than the original plots at first sight seem to indicate (Kalmijn 1988a).

Hence, the lateral line system does not appear particularly insensitive to the lower frequencies nor does it appear overly sensitive to the higher frequencies.

Instead, it shows a more nearly level response from the very low frequencies to about the former best frequencies, with or without a slight overshoot at the high-frequency end. The new curves fully support the behavioral responses to the common low-frequency flow fields of moving objects without excluding the detection of higher frequencies. These considerations do justice to both the behavioral and physiological data by removing their seeming incongruity.

Even though inherently a low-frequency system, the lateral line shows great diversity in the design of the peripheral sensory apparatus, suggesting important functional adaptations. For example, there are free neuromasts and canal organs; large and small, narrow longitudinal and broad transverse cupulae; watery and mucous canal cores; many and few external pores (see Coombs et al. 1988 for review). These structural features may be essential for an emphasis on either the lower or higher frequencies or to weigh the sensitivity of the sensory system against its susceptibility to damage by controlling the system's natural frequency, its damping, and the coupling to the outside world (see also Denton and Gray 1988, Chapter 11; Kalmijn 1988a; van Netten and Kroese, Chapter 12).

Montogomery and MacDonald (1987) recently reported that the lateral line of an antarctic fish shows best afferent responses to stimuli of about 40 Hz, which would be within the frequency range of the fields produced by the small crustaceans on which the fish feeds. However, when the response curves are replotted in terms of the proper stimulus parameter, the former best frequency degrades to a point far down the high-frequency slope. Furthermore, it is questionable whether a small crustacean tethered to a force transducer shows a similar frequency spectrum as the unrestrained animal (see also Montgomery, Chapter 28).

An evaluation of the frequency range ultimately depends on the fields that the animal detects in nature and on the physical constraints imposed on the system. Thus, the stimulus conditions are drastically different for fish that are in touch with the water surface, rest on the bottom, or move freely in the water column. Moreover, the fields produced by sources at the surface or on the bottom may greatly differ from the fields in midwater. Therefore, the frequency-dependent properties of the surface and bottom should be taken into account in presenting the data for fishes living near either of the two interfaces.

Note that the proper stimulus parameter depends further on whether the detection process is studied physiologically at the cell, the organ, or the system level,

---

►

FIGURE 9.4. Physiological frequency-response curves of lateral line and inner ear sensory systems. *Top row*: in terms of source displacement; *bottom row*: in terms of the proper stimulus parameter. The lower curves clearly show the low-pass nature of the two sensory systems. Note that by convention the lateral line curves give the amplitude of the response per unit of stimulus (in −dB), whereas the inner ear curves give the amplitude of the stimulus (in +dB) per unit of response. (a) Lateral line free neuromasts, *Xenopus laevis*, nerve recording (Kroese et al. 1978). (b) Lateral line canal organs, *Sarotherodon niloticus*, nerve recording (Muenz 1985). (c) Inner ear otolith organs, *Salmo salar*, cardiac response (Hawkins and Johnstone 1978). Dots in curves indicate best frequencies in top row and rolloff frequencies in bottom row. (After Kalmijn 1988a.)

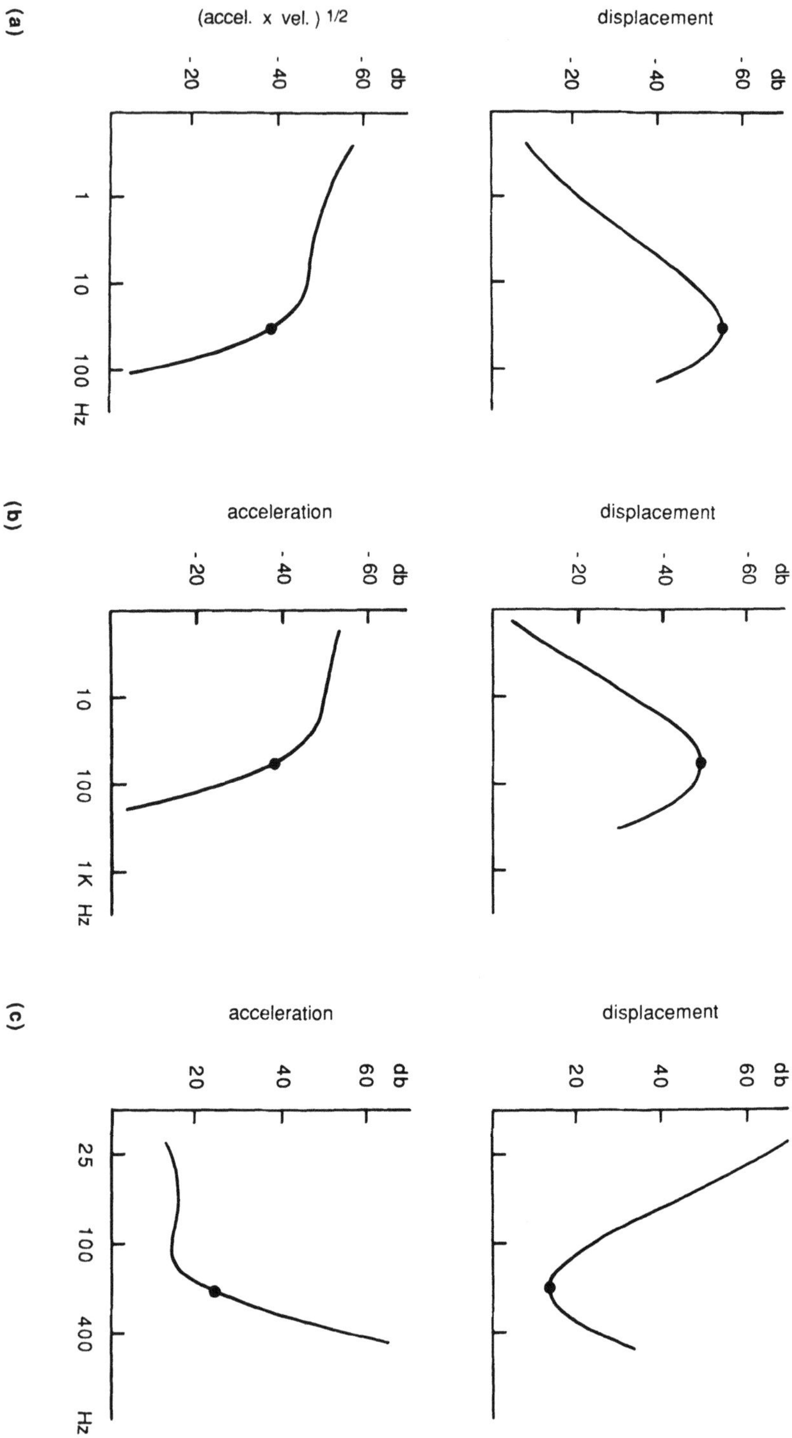
displacement
(accel. x vel.) 1/2
db
-20
-40
-60
1
10
100 Hz
(a)
acceleration
10
100
1K Hz
(b)
20
40
60
25
400
Hz
(c)

or behaviorally at the whole-animal level. Compromises or multiple plots may be necessary to discuss matters from different points of view or to compare the results of different authors (Kalmijn 1988a; see also Coombs and Janssen, Chapter 15; Denton and Gray, Chapter 11). The many discussions on the proper stimulus parameter in this volume bear witness to the importance of the issue.

3.3 Inner Ear Response Curves

In sensing the animal's whole-body motions, the otolith organs of the inner ear are generally regarded as acceleration detectors. Nevertheless, their frequency response curves have customarily been expressed in terms of displacement or, occasionally, in terms of velocity, rather than acceleration. Only Enger (1966) once noted that the accelerations required for a given level of response were remarkably independent of frequency. In terms of displacement per unit of response, however, the otolith organs yield sharp tuning curves with best frequencies ranging from 100 to 200 Hz (Fig. 9.4c, top).

If, to be consistent, the inner ear data from the literature are replotted in terms of acceleration per unit of response (Fig. 9.4c, bottom), the results once more reveal low-pass characteristics, rather than sharp, high-frequency tuning curves (Kalmijn 1988a). Whereas the arrays of free neuromasts feature roll-off frequencies at, say, 40 Hz and the system of canal organs at 80 Hz, the frequency range of the inner ear otolith organs extends to over 160 Hz. Even though this strongly suggests a progression towards an inclusion of higher frequencies, all three systems remain exquisitely sensitive to low frequencies as well.

Thus, in terms of acceleration, the inner ear, as well as the lateral line, appears sensitive to much lower frequencies than the original response curves seem to suggest. Because of the arbitrary method of plotting, the low-frequency sensitivity of the inner ear remained obscure for many years, even though the inertial nature of the sense organs was commonly appreciated. Not surprisingly, in hindsight, behavioral and cardiac-conditioning studies have in the meantime revealed a remarkable responsiveness to water motions of frequencies well below 20 Hz, both in sharks (Myrberg 1978) and in bony fishes (Olson 1976; Karlsen and Sand 1987).

In the detection of gravity, conspicuous responses are noted even at frequencies close to zero hertz. For the detection of moving objects, however, a sensitivity to such extremely low frequencies may not always be needed or desirable. Low-frequency filtering as an adaptation to the stimulus fields and noise levels in nature may take place in the peripheral sensory apparatus, in the process of sensory transduction, in the conversion of the graded receptor potentials to the sequences of discrete afferent nerve impulses, or in the central nervous system. Nevertheless, many vertebrates are sensitive to extremely low-frequency vibrations as well (Lewis 1984).

Even though the microphonic potentials may be attenuated at the higher frequencies as a result of capacitive shunting by the receptor epithelia, the receptor potentials and afferent nerve impulses do not suffer from such losses. Thus, the frequency range of the otolith organs may extend well beyond 160 Hz, certainly in fish that respond to acoustic pressure (see below and Platt et al., Chapter 32). In regard to prey too small to accelerate the whole recipient fish, the lateral line might conceivably also mediate meaningful responses at frequencies up to 80 Hz or perhaps even higher, although only from short distances.

4. Relevance of Acceleration Fields

The conclusions reached so far may be reviewed briefly. In fish, both the lateral line system and the inner ear otolith organs are essentially acceleration detectors, or nearly so. Some components respond to a lesser fractional derivative, as do the arrays of free neuromasts; others are perhaps more sensitive to a higher fractional derivative. In their capacity as acceleration detectors, the lateral line system and otolith organs may, at higher frequencies, operate in a velocity or displacement mode, or at any gradation in between.

Furthermore, the lateral line of unrestrained fish responds to the spatial differences in the flow fields about moving objects. Whereas uniform flows hardly stimulate the lateral line (since most fish are nearly neutrally buoyant), nonuniform flows are readily resolved in detail along the lengths of the sensory arrays (Sand 1981, 1984). In contrast, the otolith organs respond to the volume motions of the water and are unaffected by the nonuniformity of the field. The lateral line and inner ear are also stimulated by the motions of the fish themselves.

The above results pose several new, challenging questions. Which are the hydrodynamic fields that fish detect in the water column? What might be the advantages of detecting either the spatial derivative of the acceleration or the acceleration per se? What has driven the animals to attend to the higher source frequencies as well? How do we define hydrodynamic detection, and where do we begin to speak of sound and hearing? These issues will be addressed in the second part of this chapter, after a few introductory remarks.

The static acceleration of gravity and the dynamic accelerations initiated by the fish themselves play an important role in the control of equilibrium, posture, and locomotion (Platt 1988). Furthermore, the accelerations from remote sources, in particular the accelerations associated with surface waves or infrasound, offer plausible orientation cues (Karlsen and Sand 1987). The present chapter, though, focuses on the function of the lateral line and inner ear in the detection of moving objects in an endeavor to help clarify the origin of vertebrate hearing.

In the author's opinion, it seems perfectly reasonable to assume that the inner ear sense organs evolved from early forms of the lateral line neuromasts (cf. Jørgensen, Chapter 6). However, it would be a mistake to declare any one of the basic functions—the sense of equilibrium, the control of locomotion, or the detection of moving objects—as the oldest or the most important. On the contrary, it is more logical to suppose that the various functions evolved concurrently from an early stage on and, over time, gave rise to the present morphological diversity of the hair cell organs.

5. Fields of Moving Objects

The lateral line and inner ear sense organs evolved in response to the accelerations initiated by the fish themselves, as well as to accelerations originating in the natural environment. This section discusses the hydrodynamic and acoustic fields of moving objects, such as prey, because of their profound importance for the functional evolution of the two sensory systems.

5.1 Multipole Expansion of Local-Flow Field

To deal with the seemingly unforgiving complexity of the hydrodynamic flow fields about moving objects, one can expand their sources into infinite series of terms each consisting of a monopole, a dipole, a quadrupole, and octupole, and so forth. Since the contributions of successive terms fall off with increasing powers of the distance, sufficiently far from the source the field becomes rather simple, as the lowest nonvanishing term tends to dominate the remainder of the series (Harris 1964; Kalmijn 1988a). Near the source, however, additional, higher-order terms are needed for a detailed description of the field.

The monopole moment results from changes in the volume of the object, as in the instance of a pulsating sphere. The flow field about a pure monopole is radially symmetrical. The fluid accelerations fall off with the square of the distance and are proportional to the surface area of the source. However, the monopole moments of fish tend to be extremely weak, except where air sacs, if present, are driven into reverberation. Consequently, when viewed from a distance, the fields of quietly swimming fishes closely resemble those of dipole sources.

In front of a moving body the water is pushed ahead by the advancing surface; in the rear it fills the space vacated at the trailing edge, and to the sides it moves along wide arcs, giving rise to the familiar dipole configuration (Fig. 9.5a). The field of a pure dipole source, such as a rigid, vibrating sphere, is axially symmetrical. The fluid accelerations are not directed at the source, except along the dipole axis. They fall off with the cube of the distance and are proportional to the volume of the moving body.

This field description is correct only instantaneously. When the source moves through the water, the flow pattern moves with it, so that, as time passes, the same field lines are traced by different fluid particles. If the distance traveled by the source is small compared to the distance of observation, the displacement of the field-line pattern is usually negligible. However, the larger distances of travel are biologically highly relevant, as they enable stationary predators to detect not only accelerating but also steadily moving prey (see below).

In the vicinity of a dipole source, the flow field may, in a first approximation, be described as though the water were inviscid and incompressible, which

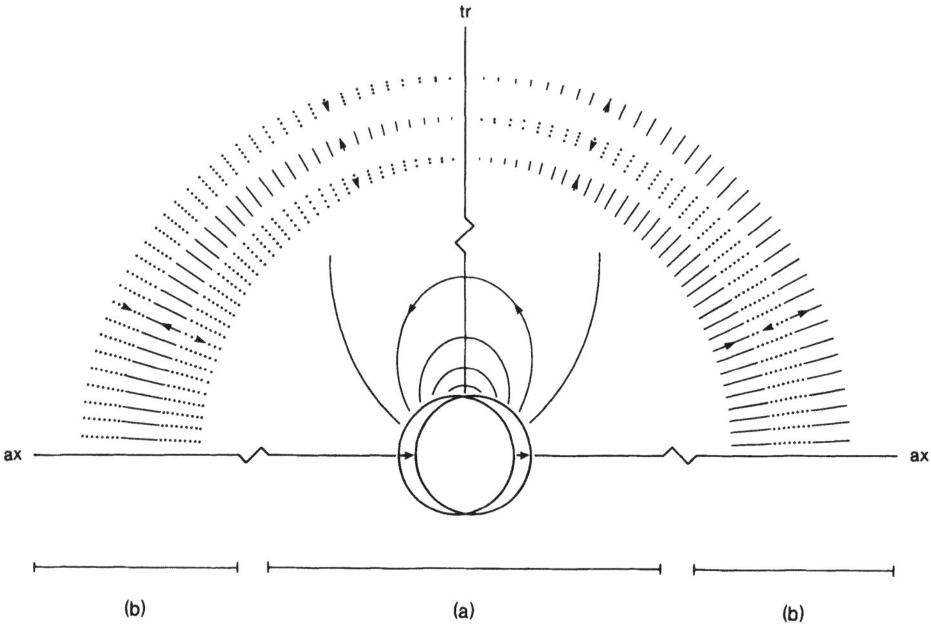

FIGURE 9.5. Dipole field of vibrating sphere. (a) Dipolar local flow. (b) Radial propagating sound wave. Arrowheads suggest particle velocities, heavy shading denotes compression, light shading denotes rarefaction. The dipolar and radial regions are separated by a wide, highly complex transition zone, not indicated here. ax) Axis of vibration; tr) transverse plane; not to scale. (After Kalmijn 1988a.)

defines the hydrodynamic *local flow.* The negative pressure gradients accelerating the fluid particles result directly from an elevated pressure at the advancing surface and from a reduced pressure at the trailing edge of the moving body. When referred to the acceleration of the source, the fluid accelerations, the pressures, and the pressure gradients of the local flow are independent of frequency.

To qualify as inviscid and incompressible, flow fields must be predominantly inertia-controlled. If the flow fields are also free of vorticity, as they are about the idealized sources discussed here, they can furthermore be expressed in terms of scalar potential fields. If not mentioned specifically, the viscous forces and their consequences are assumed to be negligible. Definitely, the wake of a fish is not of a potential nature. Thus, the next sections apply in first instance to the detection of moving objects in front or to the sides of the recipient fish, leaving vorticity and turbulence for future treatment.

In recent studies, Dr. Enger and the author recorded the hydrodynamic flow fields of quietly approaching and passing prey fish with a water-motion transducer (Kalmijn 1988a; Enger et al., Chapter 29). The triphasic wave forms obtained were consistent with the expected dipole configuration of the local-flow

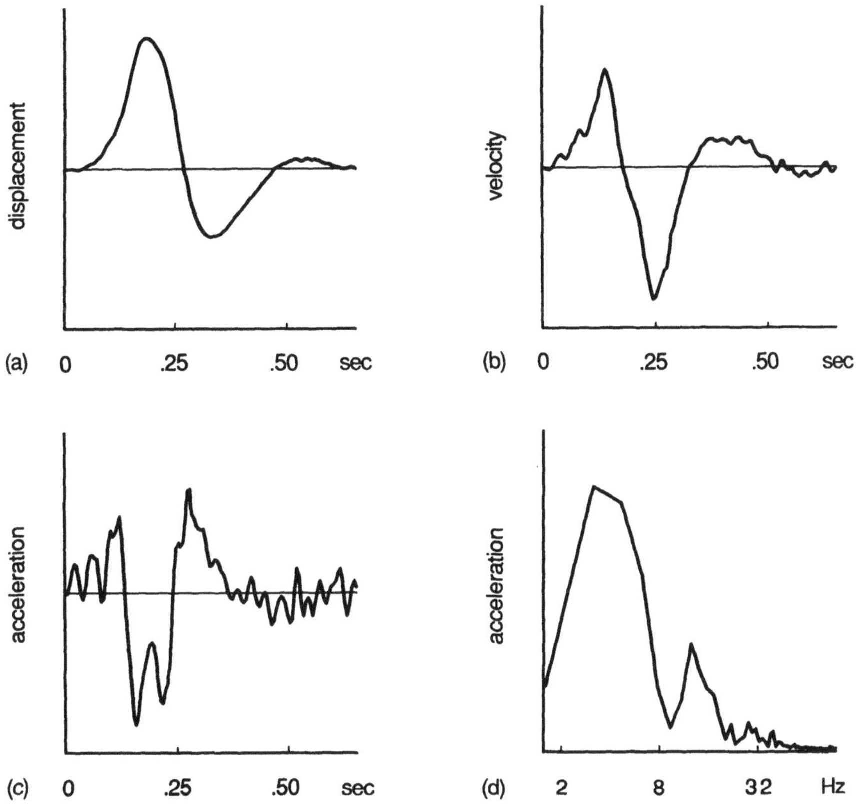

FIGURE 9.6. Flow field of a quietly approaching and passing goldfish, revealing dipole configuration and predominantly low-frequency spectral composition. (a) Displacement time series, (b) velocity time series, (c) acceleration time series, (d) acceleration amplitude spectrum, in relative units. The lower frequencies derive mainly from the steady (dc) motion of the source relative to the stationary receiver; the higher frequencies reflect the (ac) tail beat and the ensuing fluctuations in the animal's swimming velocity. (After Kalmijn 1988a; see also Enger et al., Chapter 29.)

field resulting from the fishes' whole-body motions (Fig. 9.6a–c; cf. Fig 6.3 in Kalmijn 1988b).

Moreover, the acceleration spectra of quietly moving fish revealed a striking predominance of frequencies below 20 Hz (Fig. 9.6d). Evidently, the lower frequencies derived mainly from the triphasic transient wave form caused by the animal's constant-velocity (dc) motion relative to the stationary receiver, whereas the highest (ac) frequencies were clearly due to the fish's tail beat and the ensuing fluctuations in its swimming speed. These observations form a cornerstone of the ideas presented in this chapter.

5.2 Propagating Sound Waves of Moving Objects

Besides giving rise to the pressure gradients that accelerate the fluid particles, the elevated pressure in front and the reduced pressure in the rear of a moving object inevitably lead to slight compressions and rarefactions of the water as well, even when the source moves at a constant velocity. Near the source, however, the fluid motions due to compression and rarefaction remain negligibly weak compared to the motions of the incompressible flow, since the water more readily escapes from the front to the back of the sphere than it is compressed.

When at larger distances from the source the local-flow field becomes rapidly weaker, and when at higher frequencies the pressure differences required to accelerate the water masses become significantly larger, the motions resulting from compression and rarefaction gain progressively in importance. The consequences are dramatic, for the compressed regions tend to relax by expanding on the neighboring parcels of water, which, after compression, in turn expand on the water neighboring those, and so on, thus giving rise to the familiar *propagating sound wave*, commonly identified with the sound we hear.

In homogeneous media, the propagating sound wave travels radially outward from the source (Fig. 9.5b). However, the field it produces need not be radially symmetrical, as it is about a monopole source. Indeed, about a dipole source, the sound wave is strongest along the dipole axis and vanishes in the zero plane perpendicular to it. The pressure fluctuations and longitudinal particle accelerations in the propagating sound waves of small objects fall off gently, with only the first power of the distance, independent of the source configuration.

Whereas the particle accelerations of the local-flow field are frequency-independent when referred to the source acceleration, in the propagating sound wave they gain in strength as frequency increases. About monopole sources, they are directly proportional to the frequency, whereas about dipole sources they are proportional to the frequency squared. Therefore, dipole sources are rather poor at producing low-frequency propagating sound waves compared to monopole sources such as drummed swim bladders and most man-made underwater sound projectors.

Monopoles must exhibit varying source velocities, whereas dipoles must exhibit varying source accelerations to produce propagating sound waves (see Morse and Ingard 1968). Thus, except perhaps for the most frantic jerking and struggling movements, the low-frequency whole-body motions of animals are not expected to produce appreciable propagating sound waves. Swim bladder sounds and the higher-frequency snapping, rubbing, and rushing noises made by fishes and various invertebrates are more likely to produce sound waves with particle motions detectable at some distance above ambient noise levels (Hawkins 1986).

The speed at which the sound wave travels is determined by the inertial properties and the compressibility of the medium. In natural waters, it is close to 1,500 m/s, or about $4^1/_2$ times the speed in air, depending on the temperature, salinity, and depth. Even though the local-flow field near the source does not propagate in a wavelike fashion, any hydrodynamic disturbance needs time to spread and does so with the speed of sound.

5.3 Dipolar and Radial Field Regions

One may be tempted to consider the hydrodynamic local flow, which is bound to the source, and the propagating sound wave, which radiates away from it, as two distinct physical phenomena. In appearance, they certainly are (Fig. 9.5). Nevertheless, it must be kept in mind that both the local flow and the propagating sound wave are governed by one and the same acoustic wave equation. The two are partial solutions, only the sum of which has real physical meaning, although either one may regionally dominate depending on the boundary conditions.

Because the particle accelerations in the propagating sound wave fall off more gently than in the local flow, at sufficient distances from the source the propagating sound wave prevails, which defines the acoustic *far field*. The region in which the propagating sound wave does prevail is the acoustic *near field*. These definitions are in keeping with engineering conventions. Note that the local flow prevails only in the inner part of the near field, whereas the remainder of the near field is ruled by both the local flow and the propagating sound wave.

For fish that rely on the configuration of the field to locate moving objects, it is more interesting to know where the fields are predominantly *dipolar* and where they are predominantly *radial*. Since the local flow is frequency-independent, but the propagating sound wave is not, the radial and dipolar regions are conveniently measured in units of the wavelength. Thus, the predominantly dipolar part of the field about a moving object extends to about $\frac{1}{2\pi}$ times the wavelength, whereas the predominantly radial part of the field does not fully develop until at least one wavelength from the source (Kalmijn 1988a).

Consequently, at source frequencies from zero to 20 Hz, the acceleration fields of fishes are predominantly dipolar up to distances of at least 12 m, whereas they do not become predominantly radial until distances of well over 75 m. Thus, the detection of quietly swimming prey is expected to take place mainly in the dipolar region of the local-flow field. The far-field particle motions produced by biological sound sources are probably significant only in special cases as, for example, in swim bladder drumming and similar vocalizations. The far-field acoustic pressure, however, offers more promise, as will be detailed below.

In a broad transition zone between the dipolar and radial regions, the field is highly complex. Here, the fluid particles trace ellipses of various orientations and aspect ratios, since the particle motions of the contributing local flow and propagating sound wave differ in both direction and phase. Furthermore, the dipolar region degenerates not only at larger distances, but also in the near vicinity of the source, where the descriptionof the local-flow field requires additional, higher-order multipole terms.

In summary, whereas in the immediate vicinity of moving objects the flow fields are rather complicated, farther away they are approximately dipolar. At large distances, beyond a broad, highly complex transition zone, the particle motions eventually become purely radial. Vorticity and turbulence are largely confined to the wake of the object. Monopole fields are radial at all distances from the source, but, with notable exceptions, they are biologically rare compared to the ubiquitous dipole flow fields of moving animals.

6. Lateral Line and Inner Ear Functions

The search for the proper lateral line and inner ear stimuli and for the biologically relevant hydrodynamic and acoustic fields in nature has resulted in compelling, circumstantial evidence as to the functions of the two sensory systems. Nevertheless, definite answers can be obtained only by presenting the animals with well-posed questions in thoughtfully designed behavioral experiments.

6.1 The Lateral Line

Since in the detection of moving objects by unrestrained fish the lateral line detects the spatial differences in the acceleration, rather than the acceleration per se, the distance of response is bound to be limited. Whereas the acceleration of the local-flow field about a dipole source falls off rather steeply, with the third power of the distance, the *differences* in acceleration along the lateral line arrays fall off even more steeply, with approximately the fourth power of the distance from the source.

Close to the source, the fluid accelerations vary greatly in strength and direction over only short distances, presenting the lateral line with the spatial nonuniformity it is most sensitive to. Farther from the source, the field is more nearly uniform, leaving little differential motion for the lateral line to detect. Thus, the passive, whole-body motions of the almost neutrally buoyant fish tend to deprive the lateral line of its physical stimulus (Fig. 9.7).

In the knowledge that the spatial differences in the fluid acceleration about a dipole source are proportional to the acceleration and the volume of the target but fall off with about the fourth power of the distance, and that the effective stimulus increases with the length of the sensory arrays, one may estimate that the lateral line is effectively stimulated by the fields of quietly moving prey only within distances of, at most, a few body lengths of the recipient fish (Denton and Gray, Chapter 11; Gray 1984; Kalmijn 1988a).

Aside from the limited detection range, the speed and precision that the lateral line offers in predation are impressive. Because the sense organs are arranged in long, linear arrays, both on the head and usually also along the sides of the trunk, the fish can simultaneously sample the distribution of the acceleration differences along almost the full length of the body. From the shape of the field, the animal may, by proper data processing, extrapolate the position of the target and strike at it in one fast, well-aimed move.

How the recipient fish derives the position of the source from the spatial configuration of the acceleration field detected by the lateral line is not precisely

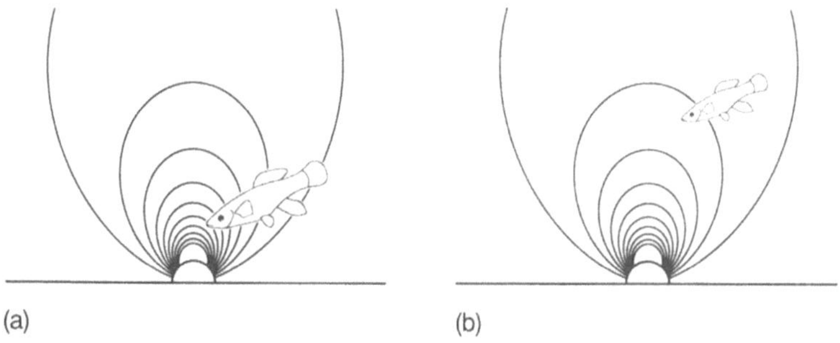

FIGURE 9.7. Detection of local-flow field produced by dipole source. (a) Near the source, the local-flow field is highly nonuniform; conspicuous spatial differences in the water motions over the surface of the fish present strong lateral stimuli. (b) Farther from the source, the field is more nearly uniform; motions of the whole fish largely deprive the lateral line of its physical stimulus, leaving the detection of the local-flow field mainly to the inner ear.

known. Although the fields of prey are mainly dipolar when observed from some distance, close to the source they feature appreciable higher-order moments and, often, vorticity as well. However, it is not unreasonable to assume that the mode and spread of the stimulus distribution along the lateral line arrays give the fish both the position and distance of a nearby target ("proximity detection"; see also Pumphrey 1950).

These considerations are consistent with the results of the recent behavioral studies conducted on the function of the lateral line in a midwater predatory fish, the bluegill *Lepomis macrochirus* (Enger et al., Chapter 29). When observed by use of a covert-infrared video system, the bluegills successfully attacked real and hydrodynamically simulated prey, evidently relying on the lateral line for the execution of their short-range strikes. Interestingly, only sources rich in frequency components below 10 Hz elicited credible attacks.

Hence, in prey detection, the lateral line enables the recipient fish to strike at its target with great speed and precision, although only from short distances. Near the source, both higher-order moments and vorticity render the flow field rather complicated, especially in the wake of the tail. The proximity method of detection evidently is not very demanding on the configuration of the flow field. At any rate, the lateral line functions as a hydrodynamic detector.

6.2 The Inner Ear

Since the imposed whole-body motions of the recipient fish tend to deprive the lateral line of its physical stimulus, one may expect that the recipient fish resorts to the detection of its imposed whole-body motions in order to detect its target. Because the imposed accelerations are proportional to the accelerations of the water properly averaged over the volume occupied by the fish, rather than to the spatial differences in the acceleration, the distance of response has the potential of being substantially larger than for the lateral line.

At some distance from a dipole source, but well within the local flow, the accelerations imposed on the recipient fish are proportional to the product of the accelerations and the volume of the moving object divided by the cube of the distance (Fig. 9.7b). Thus, the motions of the recipient fish reproduce the motions of the source, although on a much smaller scale and generally along differently oriented, slightly curvilinear axes. When the source and receiver are moving with respect to one another, these conclusions are valid only instantaneously.

To monitor the whole-body motions, fish have three pairs of otolith organs and three pairs of semicircular canals, all particularly well suited for a three-dimensional analysis of the common acceleration fields imposed on the animals. The otolith organs are highly sensitive to low-frequency accelerations, including those produced by quietly moving prey, as is the lateral line. Whether the inner ear is more effective in detecting the accelerations per se than the lateral line is in detecting their spatial differences is difficult to say.

Hence, by logical inference, it was recently hypothesized that the otolith organs of the inner ear are engaged in the detection of the local-flow fields about moving objects at distances beyond the reach of the lateral line (Kalmijn 1988a). In this capacity, the otolith organs would serve as hydrodynamic sensors, attending to different physical features of the same local-flow fields that the lateral line detects. Pursuing this line of reasoning, one may even conjecture that the detection of near-field local flows is one of the original, and still one of the main, functions of the fish's inner ear.

The inner ear has an advantage over the lateral line in that it detects the flow fields of moving objects at distances at which the higher-order moments are negligibly weak and the configuration of the field is nearly dipolar. However, since the otolith organs detect the whole-body accelerations averaged, in effect, with respect to only a single point, they do not mediate any direct orientational cues as to where the source might be. To derive the position of the source from the direction of the field, the fish has to know where in the field it is, for which it must know the position of the source.

This vicious circle may be broken by assuming a simple approach algorithm, based on the spatial configuration of the dipole field, guiding a predator to its prey, without requiring it to know the position of the source (Fig. 9.8). The algorithm is based on the observation that in the local-flow field any approach path that intersects the field lines at a constant angle leads to the source. Hence, the otolith organs would only have to detect the direction of the acceleration field with respect to the body axes. The approach paths must, however, lie in a plane through the dipole axis and the point from where the approach was initiated (for further details, see Kalmijn 1988a).

The approach algorithm was first conceived in an effort to explain the oriented responses in sharks to the bioelectric fields of prey. However, the infrared bluegill recordings also showed unmistakable evidence of deliberate approaches initiated at sizable distances from the moving targets, whether or not the lateral

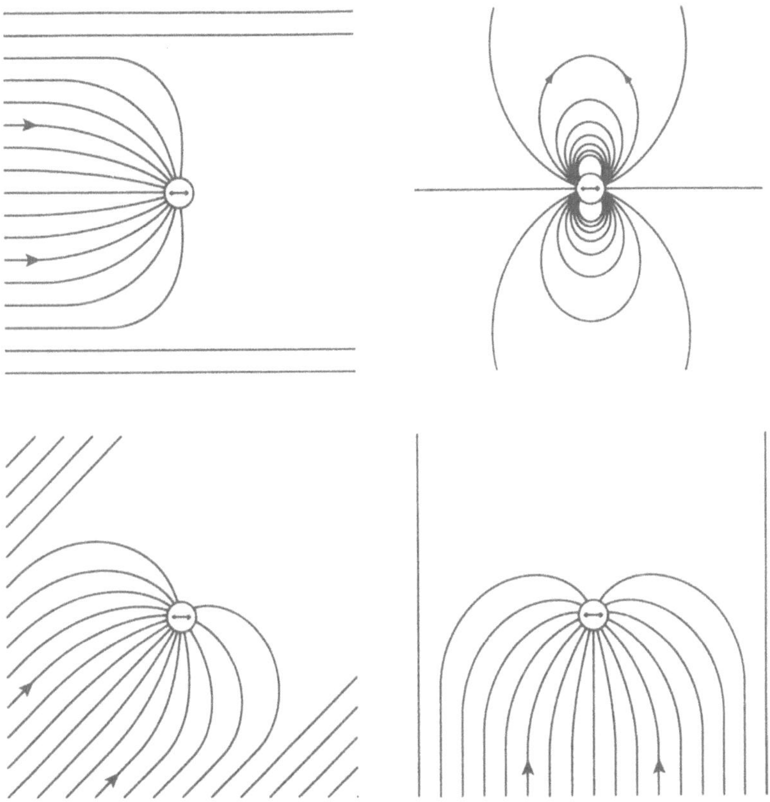

FIGURE 9.8. Approach paths predicted by proposed dipole detection algorithm. By maintaining a constant angle between the direction of the field and the body axes, the animal invariably arrives at the source of the field. *Top right*: dipole field. *Other*: approach paths for an animal entering field under angles of 0°, 45°, and 90° with respect to dipole axis; valid only for paths in planes through dipole axis. (From Kalmijn 1988b.)

line was blocked (Kalmijn, Enger, and Sand, unpublished data). The lateral line was needed only for the final attack. As the design of the experiments excluded the use of visual and chemical cues, the results were consistent with the hypothesis that the approaches were mediated by the inner ear.

This leads to the notion that a predator first detects the local-flow field of the prey at a distance by use of the inner ear, which then guides the animal into the vicinity of its target (by an approach algorithm). Subsequently, the predator detects the spatial differences in the local-flow field near the target by the lateral line, which then provides the information needed for the final strike (by a proximity method). Thus, the local flow field beyond the range of the lateral line must be regarded as the domain of the inner ear.

6.3 The Function of Hearing

The otolith organs of the inner ear typically respond to moving objects from larger distances, and are sensitive to higher frequencies, than the lateral line. A continuation of this trend may have led to detection of the far-field propagating sound wave. This idea is here referred to as the functional evolution of the lateral line and inner ear, even though, at this stage, it is merely a line of thinking to put our limited knowledge in order, instead of a well-documented theory concerning the historical course of events.

Since the far-field particle accelerations are radially directed, they offer convenient orientational cues. They also show good potential for long-distance responses by virtue of their gentle decline. However, the far-field particle accelerations of moving objects are extremely small, as the lower frequencies are highly ineffective in producing far-field sound, and the sources are usually very weak in the higher frequencies. On the other hand, the propagating sound wave dominates the local flow at higher frequencies closer to the source.

The usefulness of a particular frequency range depends furthermore on the ambient noise spectrum. After first falling off steeply at the lower frequencies, the noise levels generally increase to form a broad maximum at a few hundred hertz, before eventually falling off again (Wenz 1964). Nevertheless, the sounds of drummed swim bladders and the noises produced by the sudden motions of fishes when startled might contain sufficiently high frequencies and have large enough acceleration amplitudes to warrant detection of the particle motions in the far field (see Buwalda 1981; Hawkins 1986).

Thus, in addition to complementing the lateral line in the near field, the otolith organs of the inner ear may also have given the animal an early means of exploring the far field. In detecting the *motional* disturbances about moving objects, the lateral line and inner ear thus seemed to have reached their highest degree of perfection. In all fish, from the most primitive to the most advanced, we find the two sensory systems as well-balanced sets of particle-motion detectors, obviously playing and important role in the daily conduct of the animals.

So far, we have not attributed great significance to the *pressure* fluctuations in the fields about sound sources. Fish are not known to have suitable pressure detectors, except for the species equipped with swim bladders or similar air sacs. Whether by design or fortuitously, swim bladders tend to reverberate with the pressure fluctuation in both the near and the far field (von Frisch 1938; Dijkgraaf 1960). The ensuing tissue motions spread to the inner ear, with or without the aid of special ossicles, producing stimuli appreciably stronger than the particle accelerations of the inducing sound wave.

In bony fishes, the use of the swim bladder as a pressure-to-motion converter has given rise to excellent far-field hearing. It not only has increased the distance of response dramatically, but also has extended the frequency range from a few hundred to a few thousand hertz. Even though the acoustic pressure lacks direction, it has introduced the animals into a whole new world of sound, partially

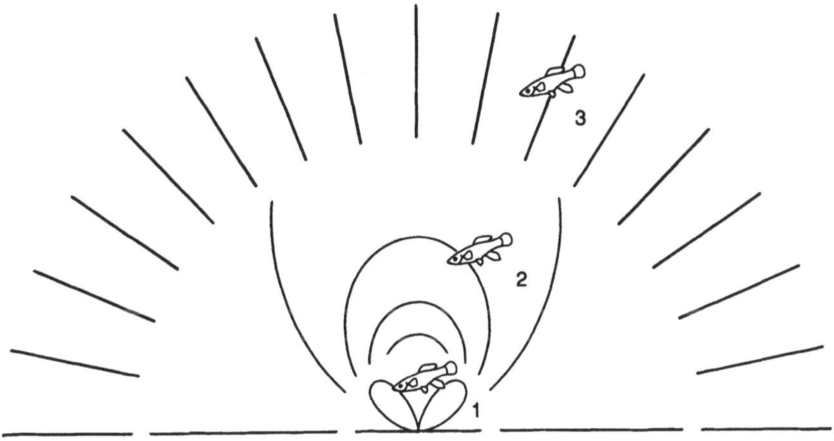

FIGURE 9.9. Functions of lateral line and inner ear in the detection of moving objects. 1. Near the source, the local-flow field is rather complicated; the lateral line detects the spatial differences in the particle acceleration field along the length of the body: hydrodynamic detection. 2. Farther from the source, the local-flow field is predominantly dipolar; the inner ear detects the particle acceleration field properly averaged over the volume of the body: near-field hearing. 3. At large distances from the source, the propagating sound wave dominates; the inner ear may, under favorable conditions, detect the radial particle acceleration field: far-field hearing. In the local flow, the accelerations detected by the lateral line and inner ear result from the acceleration field of the source and from the velocity field of the source as it moves relative to the receiver. With the swim bladder acting as a pressure-to-motion converter, where present, the acoustic pressure is detected by the inner ear both in the near and in the far fields. In higher animals, the detection of the far-field pressure has become the ruling principle of vertebrate hearing.

consisting of their own vocalizations. Ultimately, the sensitivity to acoustic pressure would become an essential feature of the advanced mechanisms of hearing in higher vertebrates (van Bergeijk 1967).

Undoubtedly, in detecting the acoustic pressure of the propagating sound wave, the fish's inner ear functions as an organ of hearing. Nevertheless, it has here been argued that most, if not all, fish primarily use the otolith organs in detecting the local-flow fields of moving objects, whether or not they employ the organs in the far field as well. Consequently, if the near-field local flows are not regarded as sound, we come to the disconcerting conclusion that in many fishes the inner ear often does not function as an organ of hearing, even though it mediates meaningful behavioral responses to the most common biological sources of sound.

This dilemma is readily resolved if the local flow is included in our definition of sound. After all, the acoustic wave equation not only governs the propagating sound wave but also demands the local-flow field for its solution. When the local flow, as an essential part of the acoustic field, is detected by the inner ear, it is certainly proper to speak of hearing. When the local flow, in its guise of an incompressible flow, is detected by the lateral line, it is equally proper to speak of hydrodynamic detection (Fig. 9.9).

7. Summary

After a review of the detection process, the neurophysiological data on the lateral line and inner ear are examined. It is shown that the two sensory systems respond in principle to acceleration or a combination of acceleration and velocity. The lateral line of free-moving fish detects the spatial differences in imposed flow fields, whereas the inner ear detects the flow fields per se. A reevaluation of the earlier data in terms of the proper stimulus parameter confirms the expected low-pass frequency characteristics of the lateral line and inner ear.

The dipole fields of moving objects are described; monopole sources are mentioned only briefly. The concepts of local flow, propagating sound wave, near field, and far field are clarified. The lateral line is considered a hydrodynamic detector, attending to the local-flow field in the vicinity of the source. The inner ear is thought to cover the remainder of the local-flow field and, in many fishes, to detect the propagating sound wave as well. It is inferred that the acoustic function of the inner ear originated in the near field.

Acknowledgments. This chapter is founded in the behavioral studies of Dijkgraaf (1963), the physics approach of Jielof, Spoor, and de Vries (1952), and the neurophysiological recordings of Harris and van Bergeijk (1962). The author has greatly enjoyed the collaboration of his colleagues Dr. Per S. Enger and Dr. Olav Sand, and of his wife Vera and his son Jelger. He is further particularly thankful for the inspiring discussions with Drs. Charles S. Cox, Sven Dijkgraaf, James T. Enright, Peter E.P. Görner, Bernd Jähne, J. Møup Jørgensen, Alfons B.A. Kroese, Sietse M. van Netten, and Gary J. Rose. Dr. Georgia H. Greaves kindly helped with the artwork. Dr. Angela K. Quinn and Ronald S. Kaufmann meticulously proofread the manuscript. Drs. Eric O. Hartwig, Christopher Platt, and Bernhard J. Zahuranec have regularly given the author the needed moral support. The work has been conducted under the auspices of the Office of Naval Research and the National Science Foundation.

References

Bergeijk WA van (1967) The evolution of vertebrate hearing. In: Neff WD (ed) Contributions to Sensory Physiology. New York: Academic Press, pp. 1–49.

Bergeijk WA van, Alexander S (1962) Lateral line canal organs on the head of *Fundulus heteroclitus* J. Morphol 110:333–346.

Blaxter JHS, Denton EJ, Gray JAB (1981) Acousticolateralis system in clupeid fishes. In: Tavolga WN, Popper AN, Fay RR (eds) Hearing and Sound Communication in Fishes. New York: Springer-Verlag, pp. 39–59.

Buwalda RJA (1981) Segregation of directional and nondirectional acoustic information in the cod. In: Tavolga WN, Popper AN, Fay RR (eds) Hearing and Sound Communication in Fishes. New York: Springer-Verlag, pp. 139–171.

Coombs S, Janssen J, Webb JF (1988) Diversity of lateral line systems: Evolutionary and functional considerations. In: Atema J, Fay RR, Popper AN, Tavolga WN (eds) Sensory Biology of Aquatic Animals. New York: Springer-Verlag, pp. 553–593.

Dale T (1980) Surface morphology of the acoustico-lateralis sensory organs in teleosts: Functional and evolutionary aspects. In: Ali MA (ed) Environmental Physiology of Fishes. New York: Plenum Press, pp. 387–401.

Denton EJ, Gray JAB (1983) Mechanical factors in the excitation of clupeid lateral lines. Proc R Soc Lond 218:1–26.

Denton EJ, Gray JAB (1988) Mechanical factors in the excitation of the lateral lines of fishes. In: Atema J, Fay RR, Popper AN, Tavolga WN (eds) Sensory Biology of Aquatic animals. New York: Springer-Verlag, pp. 595–617.

Dijkgraaf S (1960) Hearing in bony fishes. Proc R Soc Lond 152:51–54.

Dijkgraaf S (1963) The functioning and the significance of the lateral-line organs. Biol Rev 38:51–105.

Enger PS (1966) Acoustic threshold in goldfish and its relation to the sound source distance. Comp Biochem Physiol 18:859–868.

Flock A (1965) The ultrastructure of the lateral line canal organ. Acta Otolaryngol. (Suppl.) 199:7–90.

Flock A (1971) Sensory transduction in hair cells. In: Loewenstein WR (ed) Handbook of Sensory Physiology, Vol. I. New York: Springer-Verlag, pp. 396–441.

Flock A, Flock B, Murray E (1977) Studies on the sensory hairs of receptor cells in the inner ear. Acta Otolaryngol 83:85–91.

Frisch K von (1938) The sense of hearing in fish. Nature 141:8–11.

Gray JAB (1984) Interaction of sound pressure and particle acceleration in the excitation of the lateral-line neuromasts of sprats. Proc R Soc Lond 220:299–325.

Harris GG (1964) Considerations on the physics of sound production by fishes. In: Tavolga WN (ed) Marine Bio-Acoustics. Oxford; U.K.; Pergamon Press, pp. 233–247.

Harris GG, van Bergeijk WA (1962) Evidence that the lateral-line organ responds to near-field displacements of sound sources in water. J Acoust Soc Am 34:1831–1841.

Hawkins AD (1986) Underwater sound and fish behavior. In: Pitcher TJ (ed) The Behavior of Teleost Fishes. London: Croom Helm, pp. 114–151.

Hawkins AD, Johnstone ADF (1978) The hearing of the Atlantic salmon, *Salmo salar* J Fish Biol 13:655–673.

Hudspeth AJ (1982) Extracellular current flow and the site of transduction by vertebrate hair cells. J Neurosci 2:1–10.

Hudspeth AJ, Corey DP (1977) Sensitivity, polarity, and conductance change in the response of the vertebrate hair cells to controlled mechanical stimuli. Proc Natl Acad Sci USA 74:2407–2411.

Jielof R, Spoor A, de Vries H (1952) The microphonic activity of the lateral line. J Physiol 116:137–157.

Kalmijn AJ (1988a) Hydrodynamic and caustic field detection. In: Atema J, Fay RR, Popper AN, Tavolga WN (eds) Sensory Biology of Aquatic Animals. New York: Springer-Verlag, pp. 83–130.

Kalmijn AJ (1988b) Detection of weak electric fields. In: Atema J, Fay RR, Popper AN, Tavolga WN (eds) Sensory Biology of Aquatic Animals. New York: Springer-Verlag, pp. 151–186.

Karlsen HE, Sand O (1987) Selective and reversible blocking of the lateral line in freshwater fish. J. Exp Biol 133:249–262.
Kroese ABA, van der Zalm JM, van den Bercken J (1978) Frequency response of the lateral-line organ of *Xenopus laevis*. Pflügers Arch 375:167–175.
Kuiper JW (1967) Frequency characteristics and functional significance of the lateral line organ. In: Cahn PH (ed) Lateral Line Detectors. Bloomington: Indiana University Press, pp. 105–121.
Lewis ER (1984) Inertial motion sensors. In: Bolis L, Keynes RD, Maddrell SHP (eds) Comparative Physiology of Sensory Systems. Cambridge U.K.; Cambridge University Press, pp. 587–610.
Montgomery JC, MacDonald JA (1987) Sensory tuning of lateral line receptors in antarctic fish to movements of planktonic prey. Science 235:195–196.
Morse PM, Ingard KU (1968) Theoretical Acoustics. New York: McGraw-Hill.
Muenz H (1985) Single unit activity in the peripheral lateral line system of the cichlid fish *Sarotherodon niloticus* L.J Comp Physiol 157:555–568.
Myrberg AA (1978) Underwater sound – its effect on the behavior of sharks. In: Hodgson ES, Mathewson RF (eds) Sensory Biology of Sharks, Skates, and Rays. Washington: U.S. Government Printing Office, pp. 391–417.
Netten SM van, Kroese ABA (1987) Laser interferometric measurements on the dynamic behavior of the cupula in the fish lateral line. Hear Res 29:55–61.
Olson K (1976) Evidence for localization of sound by fish in schools. In: Schuijf A, Hawkins AD (eds) Sound Reception in Fish. Amsterdam: Elsevier, pp. 257–270.
Platt C (1988) Equilibrium in the vertebrates: Signals, senses, and steering underwater. In Atema J, Fay RR, Popper AN, Tavolga WN (eds) Sensory Biology of Aquatic Animals. New York: Springer-Verlag, pp. 783–809.
Pumphrey RJ (1950) Hearing. Symp Soc Exp Biol 4:3–18.
Sand O (1981) The lateral line and sound reception. In: Tavolga WN, Popper AN, Fay RR (eds) Hearing and Sound Communication in Fishes. New York: Springer-Verlag, pp. 459–480.
Sand O (1984) Lateral-line systems. In: Bolis L, Keynes RD, Maddrell SHP (eds) Comparative Physiology of Sensory Systems. Cambridge, U.K.; Cambridge University Press, pp. 3–32.
Wenz JM (1964) Curious noises and the sonic environment in the ocean. In: Tavolga WN (ed) Marine Bio-Acoustics. Oxford, U.K.; Pergamon Press, pp. 101–119.

10
Hydrodynamic Imaging of the Surroundings by the Lateral Line of the Blind Cave Fish *Anoptichthys jordani*

E.S. Hassan

1. Introduction

It is known that certain fish that have been blinded are still able to avoid collision with stationary objects such as the aquarium wall (Dijkgraaf 1962). Congenitally blind fish, like the blind cave fish *Anopthichthys jordani*, can swim around in their environment as surely as fish with vision.

Behavioral experiments on this fish (von Campenhausen et al. 1981; Weissert and von Campenhausen 1981) have shown that it is able to detect and recognize stationary objects, such as bars arranged in different patterns. It is proposed that this capability of the blind cave fish is achieved with the aid of the lateral line system. When the fish glides through the open water, a flow field is set up around its body due to the displacement of water at the head and suction in the tail region (Fig. 10.1). When the fish glides in the vicinity of an obstacle, the flow field will be altered in a characteristic way that depends on the geometry and dimensions of the obstacle. The alteration of the flow field will give rise to an alteration in the water displacement at the skin of the fish, and this can be perceived by the fish with the aid of its lateral line system.

The aim of this chapter is to discuss how these alterations in water displacement can create an image of an object—e.g., a cylindrical bar—on the skin of the moving fish and how this information could be spatially and temporally mapped by the lateral line system. These questions were investigated by determining mathematically the spatial and temporal course of water displacement on the skin of the fish when gliding in the vicinity of an obstacle. The mathematical treatment of this problem was done for the two-dimensional case. This case is applicable for a fish with a dorsoventrally flattened body shape and, in particular, for the midtrunk region along its side.

2. Mathematically Modeling Hydrodynamic Interaction Between Two Bodies

The flow field around a fish gliding in open water can be simulated by a constant flow with the same velocity as the fish but in the direction opposite to its

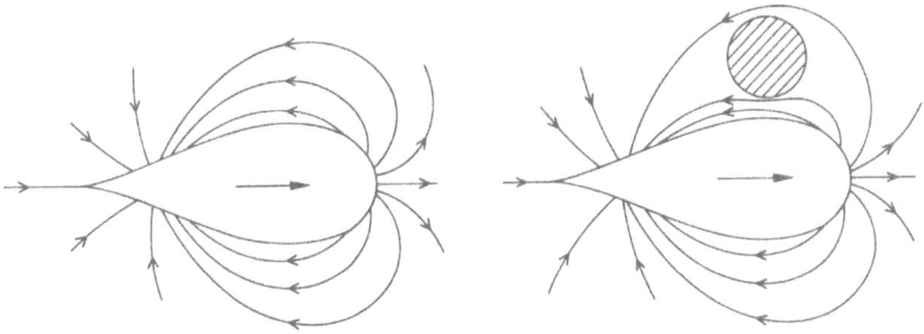

FIGURE 10.1. Schematic diagram of the flow field around a gliding fish. Left: in open water. Right: passing an obstacle. (From Hassan 1985.)

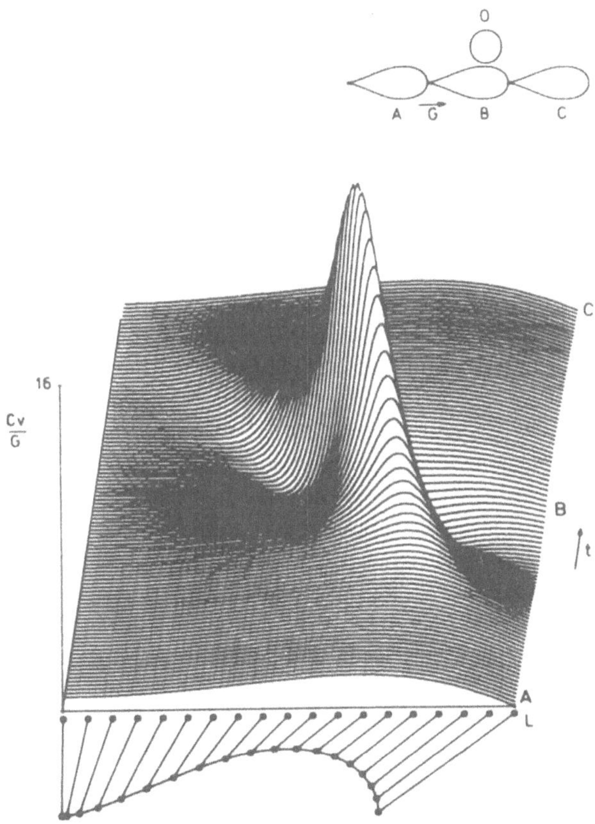

FIGURE 10.2. Flow velocity distribution on the surface of a fish gliding past a stationary obstacle upper inset. The vertical axis is the flow velocity (Cv) divided by the gliding velocity of the fish (G). Horizontal axis (L): the contour of the side of the fish facing the obstacle stretched to a straight line . Points on this axis connect to the corresponding points of the fish's side drawn below the axis. t: Time. (From Hassan 1985).

movement. The presence of the fish in this flow can be represented as the line distribution of singularities, such as sources and sinks, along the axis of the fish inside its body. In determining the line distribution of these singularities, the boundary condition of vanishing normal flow at the surface of the fish must be fulfilled. That is, the flow must be parallel to the surface of the fish. When the fish glides past an obstacle, the flow field must also fulfill the boundary condition at the surface of the obstacle. This can be achieved by introducing singularities to represent the obstacle. This then requires the introduction of additional singularities for the fish body. The result is a model of the hydrodynamic interaction between the two bodies–fish and obstacle. For the two-dimensional case, the fish can be modeled as an infinitely high cylinder with cross section the same as the horizontal cross section of a fish. The obstacle must then be modeled as an infinitely high cylinder with a circular cross section (see Hassan, 1985, for further details).

3. Velocity Profiles Across Time and Space

From this kind of mathematical model, one can determine the velocity profile across time and space that results when the fish glides past a cylinder. Figure 10.2 shows the results of such modeling. The vertical axis in this figure is the flow velocity on the skin of the fish normalized for the glide velocity. The horizontal axis represents the extended contour of the side of the fish facing the obstacle. Points on this axis are connected by lines to the corresponding points on the real contour of the fish's side. The third axis is time. Each curve in this figure plots the flow velocity distribution for a certain instant in time. The bottom curve (A) represents the flow velocity distribution when the fish is at position A (inset) relative to the obstacle and the top curve (C) for the position C (inset). These curves were calculated with arbitrary units of 100 for fish length and 40 for fish width. The obstacle has a diameter the same as the width of the fish. As can be seen, a wave forms in front of the fish as it approaches the obstacle. The wave reaches its maximum amplitude when the fish is opposite the obstacle, then dies away again at the tail end as the fish passes the obstacle.

Figure 10.3 shows the pressure distribution on the side of the fish. Figure 10.4 shows the pressure gradient distribution, which is presumed to be the stimulus for the canal system. Note here that the profile of pressure gradient is sharper than that of the flow velocity. This means that the frequency content of the stimulus to the canal system is higher than that to the free neuromasts. This can be correlated with the fact that the neuromasts of the canal organ respond to higher frequencies than do the free neuromasts, as been shown (Münz and Claas 1983; Münz, Chapter 14, Schellart and Kroese, Chapter 21).

Figure 10.5 shows the dependency of the flow velocity distribution on the diameter (given in arbitrary units) of the obstacle. Each curve here represents the distribution at the moment when the fish is opposite the obstacle, depending on the radius of the obstacle (bottom curve for obstacle radius of 0.1; top curve for

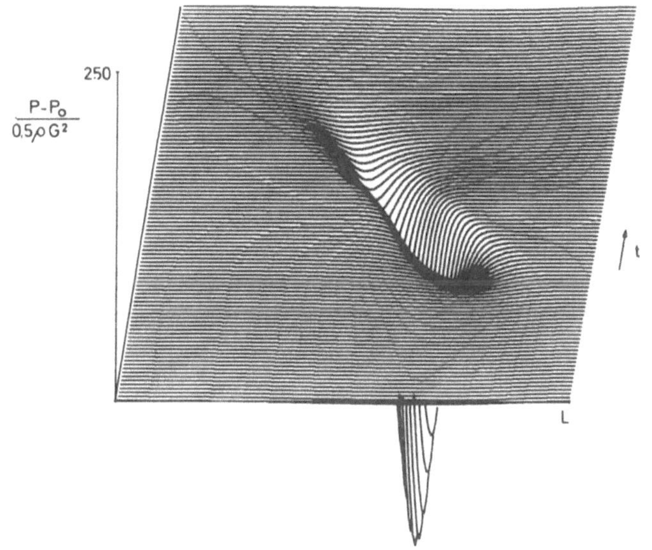

FIGURE 10.3. Pressure distribution on the surface of the gliding fish. (From Hassan 1985.)

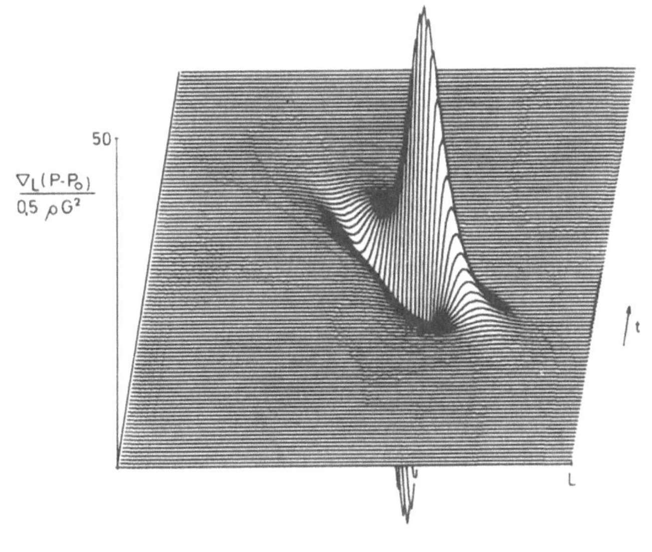

FIGURE 10.4. The pressure gradient on the surface of the gliding fish. (From Hassan 1985.)

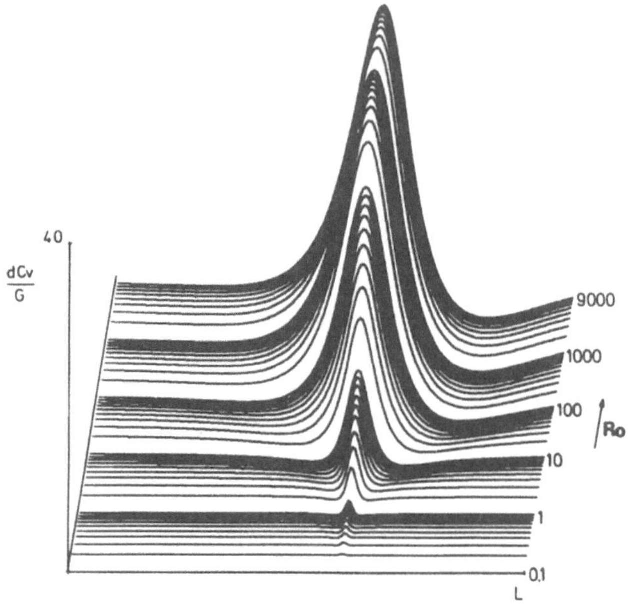

FIGURE 10.5. Plot of the flow velocity distribution as a function of the radius (Ro) of the stationary obstacle. (From Hassan 1985.)

obstacle radius of 9,000; i.e., for all practical purposes the obstacle is a wall). As can be seen, for small obstacles, the increase in amplitude of the flow velocity is small and confined to a small region. For larger obstacles, the amplitude and extent of the velocity distribution become larger. Figure 10.6 shows the dependency on the distance between the fish and the obstacle. Here, the closer the obstacle, the higher the peak amplitude, and the more sharply the spatial distribution of the flow velocity is tuned.

Now, for each obstacle at a certain distance, the distribution of the flow velocity can be characterized with two parameters—its peak amplitude (dCv/G), and its spatial extent (W), defined as the length of the contour range where dCv/G has positive values as a percentage of the total contour length. When these two parameters are made the axes of a coordinate system and the corresponding combinations of the radius (Ro) of the obstacle and its distance (D) are plotted as points in this coordinate system, with all points of the same distance value connected with a line, then one obtains a group of curves as shown in Figure 10.7. As can be seen in such a representation, different combinations of size and distance of the obstacle correspond to different points int the coordinate system. This means that there is an unequivocal combination of the size and the distance from the obstacle for every combination of dCv/G and W. For the fish, this means that information about both the size and the distance of objects can be acquired by analyzing the velocity distribution over its whole body surface. In

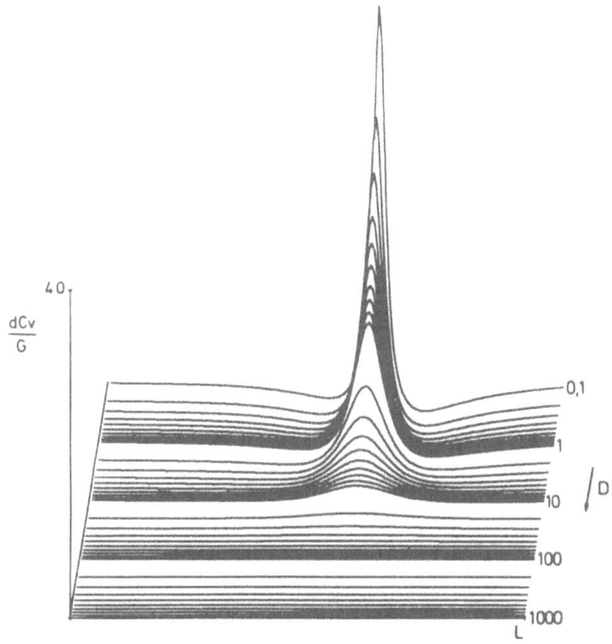

FIGURE 10.6. Plot of the flow velocity distribution as a function of the distance (D) of the stationary obstacle from the gliding fish. (From Hassan 1985.)

other words, a fish may not need a mechanism for size constancy. However, if it lacks such a mechanism, the fish must know its own velocity.

4. Spatial Discrimination by the Blind Cavefish

The results from the mathematical model show that certain obstacles produce well-defined spatial and temporal distributions of water flow at the skin of the fish. How far the blind cave fish can use these two aspects of the stimulus to the lateral line system was the basic question for the following behavioral experiments. In these experiments, the performance of the fish in discriminating spatial intervals was investigated. This was done in a training tank like that in Fig 10.8, in which the fish was trained to discriminate between two grids of bars with different spatial intervals between them as shown. The task of the fish was to associate each grid with the opening in its middle. When the fish swam through the one opening, it was rewarded with food. When the fish swam through the other opening, it was punished by a sudden jet of water at the opening. After 10 choices by the fish, the grid plate was rotated slowly through 180° to avoid place training.

In carrying out these experiments, the intervals between the bars of the learning grid (i.e., the one the fish has to learn to choose) were kept constant, and the

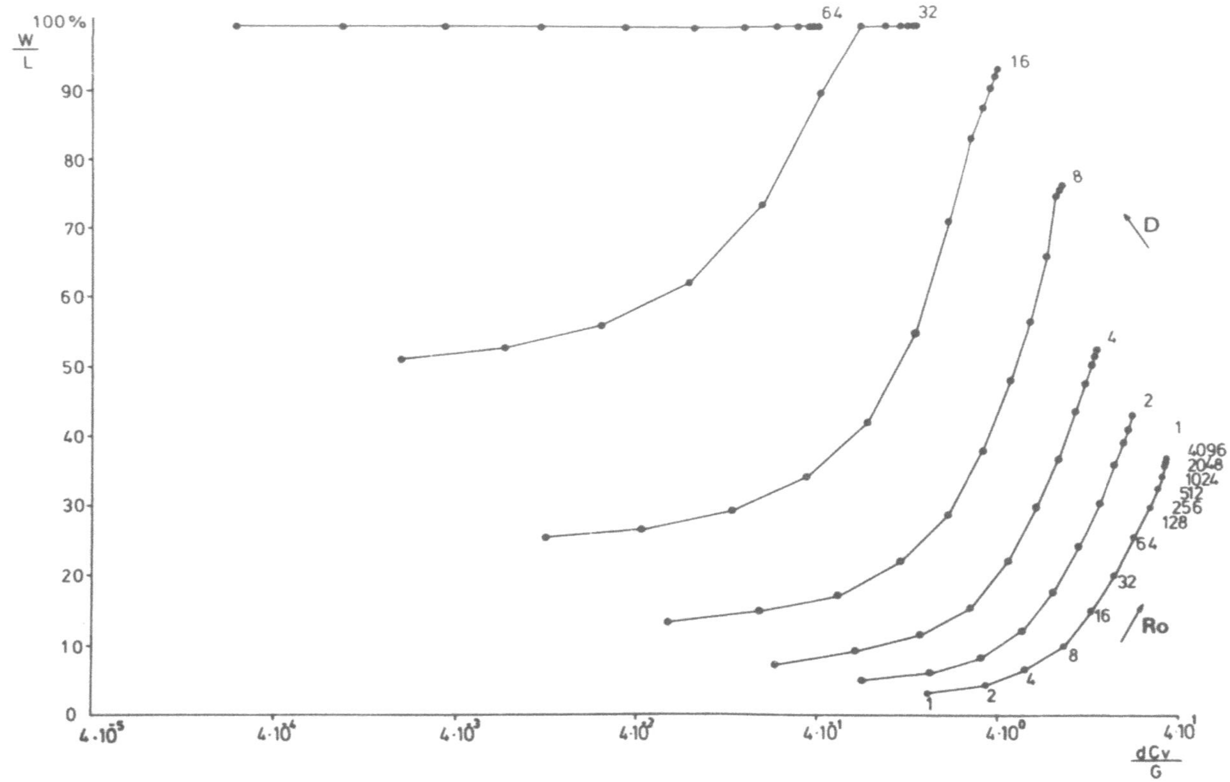

FIGURE 10.7. Plot of the parameters of the stationary obstacle, its radius (Ro), and its distance from the gliding fish (D) in a coordinate system established with the peak amplitude of the flow velocity increase (dCv) as the horizontal axis and the extent (W) of the area in which the increase occurs expressed as a percentage of the contour length (L) as the vertical axis. (From Hassan 1985.)

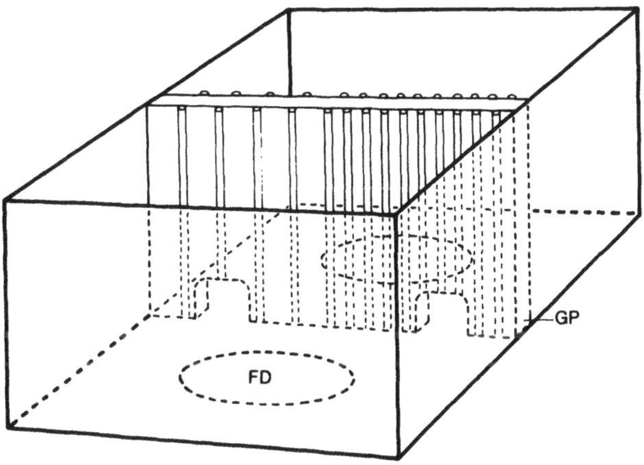

FIGURE 10.8. Schematic diagram of the experimental tank. PG Grid plate; FD; food depression. (From Hassan, 1986.)

intervals between the bars of the comparison grid (i.e., the one the fish was to avoid) were altered in steps so that they gradually became closer to those of the learning grid. The size of the intervals between the bars was limited both by the body length of the fish (5 to 7 cm) and by the diameter of the bars (1 mm). In this study, at any one time, the fish was alongside at least two bars. The lower limit was 5 mm to minimize the effect of unavoidable inaccuracies in the construction of the grids (see Hassan 1986 for further details).

Figure 10.9 shows the result of such experiments. In these particular experiments, the bars of the learning grid were 20 mm apart in (a) and 10 mm apart in (b). The horizontal axes of these graphs represent the interval between the bars of the comparison grid. The vertical axis represents the percentage of correct choices. The fish had to make at least 400 choices after each change to the interval between the bars of the comparison grid. As can be seen, the fish can detect a difference of 2.5 mm between the intervals of the learning and comparison grids with over 70% certainty. An average value of 1.5 mm was calculated as the smallest difference which the fish can detect with *just* 70% certainty. This value for the smallest detectable difference corresponds to a spatial frequency limen of 15% for the 10-mm interval and 7.5% for the 20-mm interval.

5. Summary and Conclusions

The results of the mathematical model show that the spatial and temporal pattern of the flow field surrounding a moving fish provides the fish's lateral line system with a stimulus for imaging its surroundings. The behavioral experiments on the blind cave fish showed that it can discriminate between spatial intervals with a difference limen of 15% at 10 mm and of 7.5% at 20 mm.

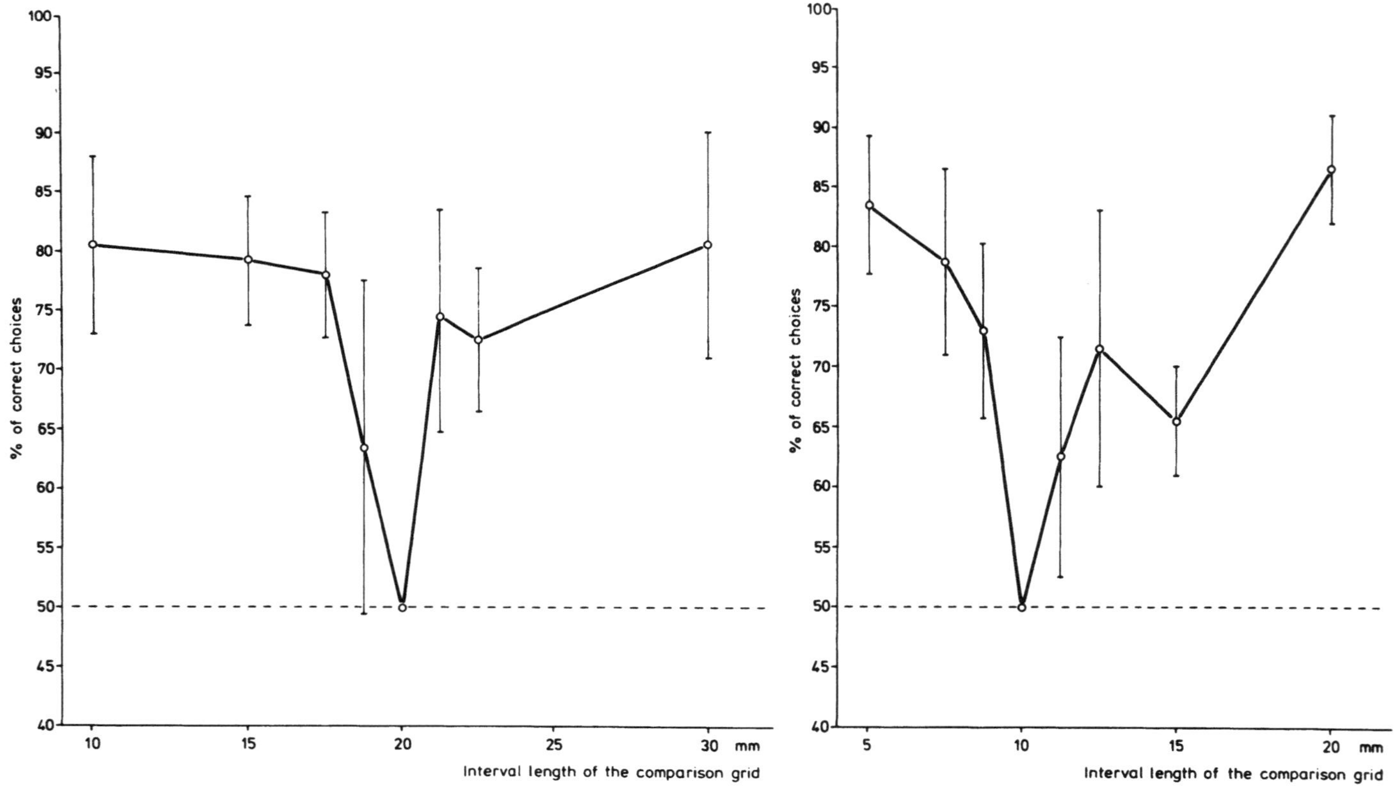

FIGURE 10.9. Discriminatory performance data of one fish; (a) for a learning grid with intervals of 20 mm between the bars; (b): for a learning grid with intervals of 10 mm. Each horizontal axis represents the interval between the bars of the comparison grid. The vertical axes represent the percentage of correct choices. No experiments were carried out with learning and comparison grids that had the same intervals as each other, so the point at 50% correct choices is actually hypothetical. (From Hassan 1986.)

These values of the difference limen for spatial intervals cannot be explained in terms of the temporal pattern of the stimulus alone. For a fish gliding past a grid, the temporal pattern will depend on the fish's velocity and on the interval between the bars of the grid. Measurement of the velocity of the blind cave fish gave a value of 16 to 25 cm/s (Weissert and von Campenhausen 1981). More recent measurements (Teyke 1985, 1987) revealed a value of 15 cm/s for a fish swimming in unfamiliar surroundings, which decreased to about 10 cm/s as the fish becomes familiar with its surroundings. Assuming a velocity of 20 cm/s, the stimulus frequency would be 20 Hz for grid intervals of 10 mm and 10 Hz for intervals of 20 mm. The frequency difference limens of the blind cave fish at these frequencies are 25% and 40%, respectively (Frühbeis 1984). These values were determined for vibrations produced by two beads oscillating simultaneously. When the beads oscillate alternately, the frequency difference limen become poorer (von Campenhausen, personal communication). Thus it can be concluded that improvement in the difference limen for spatial interval in the blind cave fish is the result of its ability to use the spatial distribution of the stimulus. However, for a moving fish, this spatial distribution will vary in time, so it can be argued that the moving fish must use the phase relationship of the stimulus along its body. Such an analysis of phase relationship requires a comparison between the temporal patterns of the stimulus at different points on the fish's surface. The precision of this analysis increases in proportion to the number of points compared—i.e., the number of neuromasts. This may explain why such a large number of neuromasts are found in the skin of the blind cave fish (Schemmel 1967).

Acknowledgments. I wish to thank Dr. Mark Tritsch for improving the English.

References

Campenhausen C von, Ries SI, Weissert R (1981) Detection of stationary objects by the blind cave fish *Anoptichthys jordani* (Characidae). J Comp Physiol 143:369–374.

Dijkgraaf S (1962) The functioning and significance of the lateral line organs. Biol Rev 38:51–105.

Fühbeis B (1984) Verhaltensphysiologische Untersuchungen zur Frequenzunterscheidung und Empfindlichkeit durch das Seitenlinienorgan des Blinden Höhlenfisches *Anoptichthys jordani* (Hubbs et Innes). Dissertation, Universität Mainz, Mainz, FRG.

Hassan ES (1985) Mathematical analysis of the stimulus of the lateral line organ. Biol Cybern 52:23–36.

Hassan ES (1986) On the discrimination of spatial intervals by the blind cave fish (*Anoptichthys jordani*). J Comp Physiol A 159:701–710.

Münz H, Claas B (1983) The functional organization of neuromasts in the lateral-line system of cichlid fish. Advances in Vertebrate Neuroethology (London, NATO Series A) 56:301–307.

Schemmel C (1967) Vergleichende Untersuchungen an den Hautsinnesorganen ober- und unterirdisch lebender *Astyanax* Formen. Z Morphol Tiere 61:255–316.

Teyke T (1985) Collision with and avoidance of obstacles by blind cave fish *Anoptichthys jordani* (Characidae). J Comp Physiol A 157:837–843.
Teyke T (1987) Zur Orientierung der Blinden Höhlenfische (*Anoptichthys jordani*)–Strömungsfeld und innere Landkarte. Dissertation Universität Mainz, Mainz, FRG.
Weissert R, von Campenhausen C (1981) Discrimination between stationary objects by the blind cave fish *Anopichthys jordani* (Characidae). J Comp Physiol 147:375–381.

11 Some Observations on the Forces Acting on Neuromasts in Fish Lateral Line Canals

E.J. DENTON and J.A.B. GRAY

1. Introduction

There are good reasons for measuring stimuli to lateral-line systems as pressure gradients (Denton and Gray 1988), the most likely useful natural stimuli to lateral line sense organs being local pressure gradients produced by a fish's own swimming movements and local pressure gradients produced by external sources. In what follows we shall assume that (1) the effective stimulus for a canal neuromast will be one that produces a movement of the cupula *relative* to the wall of the canal, (2) the force moving the cupula must be transmitted to the cupula through the liquid that the canal contains, and (3) displacements of such liquid and the cupula of the neuromast will arise from *net* pressure gradients along the short stretch of canal carrying the neuromast. This net gradient can be measured in terms of the local accelerations of the medium adjacent to the canal *relative* to the surface of the fish (Denton and Gray 1983). It follows that if the medium adjacent to the lateral line moves in exactly the same way as the surface of the fish that carries a particular "section" of lateral line, the neuromast in this section will not be stimulated. When considering an external source, the stimuli to lateral lines will often be proportional to the accelerations of the sonic surface (Gray 1984; Kalmijn 1988, Chapter 9).

It will be attempted here to relate the displacements of the cupulae of canal neuromasts to *net* pressure gradients, mainly considering alternating sinusoidal and transient stimuli. Measuring and calculating the forces acting on an obstacle, such as a cupula, when it is exposed to steady or alternating flows of liquid along a canal of uniform cross section, is difficult to do for a wide range of frequencies and amplitudes of particle displacement. Nonetheless this has been successfully done by van Netten and Kroese (1987, Chapter 12). It is even more difficult if the canal's cross section varies along its length and around the cupula as is very often the case (see, e.g., Jansson et al 1987; Denton and Gray 1988). The simplest kind of stimuli to consider are those of the flows inside lateral line canals produced by steady pressure gradients along the canals. Such pressure gradients might be found in head lateral lines when a fish is gliding forward and thus generating a field of pressures over the surfaces of its head (see Kuiper, 1967; DuBois et al.

1974; Hassan, Chapter 10). Such external pressures are proportional to the velocity of the fish squared, and they can be large. Thus, the maximum pressure on the snout of the sprat (*Sprattus sprattus*) gliding forward with a velocity of 0.5 ms^{-1} will be over 100 Pa, and high pressure gradients will exist over the fish's lateral line system. It is estimated, however, that for such a sprat the appropriate Reynolds numbers for flows inside the lateral line canals will be low. Thus a pressure gradient of 10 Pa/mm will, for a canal of diameter of 100 μm, give a flow of liquid along the canal corresponding to a Reynolds number of only about 0.2. (The Reynolds number is taken as $(\rho v l)/\eta$ where ρ is the density of sea water, v is the mean velocity of liquid inside the canal, l is the diameter of the canal, and η is the viscosity of seawater.) Under these conditions, it is expected that flow would be laminar and that the flow profiles would be parabolic across the canal, and viscous/shear forces would predominate over inertial forces.

With alternating pressure gradients, such as can be given either by a fish's own swimming movements or by external sources, the picture is more complicated, and inertial forces will become relatively more important as frequency rises. The relative importance of shear and inertial forces in the lateral line canal of a particular animal will depend on the structure and dimensions of the canal and on the range of frequencies that are useful to the animal. In some lateral line systems, low-frequency stimuli are thought to be important in nature (see Kalmijm, Chapter 9; Enger et al., Chapter 29). However Kroese and van Netten (Chapter 13) and van Netten and Kroese (Chapter 12) have shown by experiments on the ruffe and on theoretical grounds that inertial forces become significant, at least in some canals, at frequencies that are thought to be important to the fish. Our recordings from the neuromast nerves of sprat, ray (*Raja clavata*), and two sizes of ruffe (*Gymnocephalus cernua* = *Acerina cernua*) do show that velocity in the canal can be used as a reasonable indicator of stimulus strength in the frequency range likely to be of biological importance.

Although there are many types of lateral line canal organs (see Coombs et al. 1988; Denton and Gray 1988), this chapter focuses on two lateral line types shown diagrammatically in Fig. 11.1. One type, characteristic of clupeoid fishes, has cupulae that are very low or flattened and that fill only a small portion of the canal lumen. The flows in capillary tubes without obstacles and sprat lateral lines were shown by Denton and Gray (1983) to be very similar to each other in both amplitude and phase relations over a range of frequencies between about 5 and 200 Hz. The second type is one in which the cupulae completely, or almost completely, block the lumen of the canal. This appears to be the case for a number of fish e.g., *Lota vulgaris* (Flock 1967), *Coryphenoides rupestris* (Fänge et al. 1972), and *Trachurus trachurus* (our own observations). Cupulae almost block the structurally relatively simple head lateral line canals of the killifish *Fundulus heteroclitus* described and studied by van Bergeijk and Alexander (1962) and Harris and van Bergeijk (1962). If the cupula were to completely block a section of canal, we might regard the system as consisting of a spring, a mass, and a resistance in series, the spring being provided by the cupula, the mass being

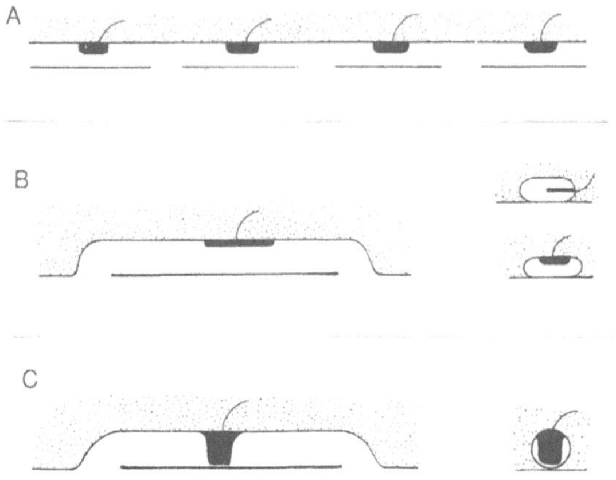

FIGURE 11.1. Diagrams of simple lateral lines, neuromasts including cupulae shown black. The tissues are shown as stippled areas, and the lines arising from neuromasts represent nerves. (A) Typical lateral line with neuromasts between successive openings to the external medium—a length of canal between openings is described as a section of lateral line. (B) Section of a typical clupeoid lateral line—e.g., that of an anchovy. The two small diagrams on the right show cross sections through canals at the positions of neuromasts. The upper diagram is for the sprat; its neuromasts are placed on the sides of the canal (see Fig. 1, Denton and Gray 1988). The lower diagram is for an anchovy. (C) A section of a type of lateral line for which the cupulae almost block the lumen of the canal. The small diagram shows a cross section through the canal at the position of a neuromast.

mainly that of the liquid in the canal, and the resistance the frictional forces acting against movements of liquid in the canal.

In this chapter, experiments on the sprat are described, and the results obtained on this animal are discussed. There is also a section in which results obtained by Harris and van Bergeijk on *Fundulus* are reanalyzed to show how the displacements of cupulae are related to pressure gradients along the canal. Surprisingly, over the working range of frequencies, the amplitude and phase relations of the two systems (sprat and *Fundulus*) are very like each other. In addition, some new experiments on the head lateral line of the ruffe are described. In this animal thin compliant membranes cover the lateral line canals, and these appreciably affect the responses of the system.

2. Measuring Displacements of Cupulae in Sprat

In earlier experiments, direct observations of the liquid flows within the lateral line canals of the sprat were made in relation to sinusoidal flows of liquid over the surface of the fish. The general approach was that of observing, through a dissecting microscope, the displacements of light scattering particles that had been introduced into the canal and which were also present in the medium. In the

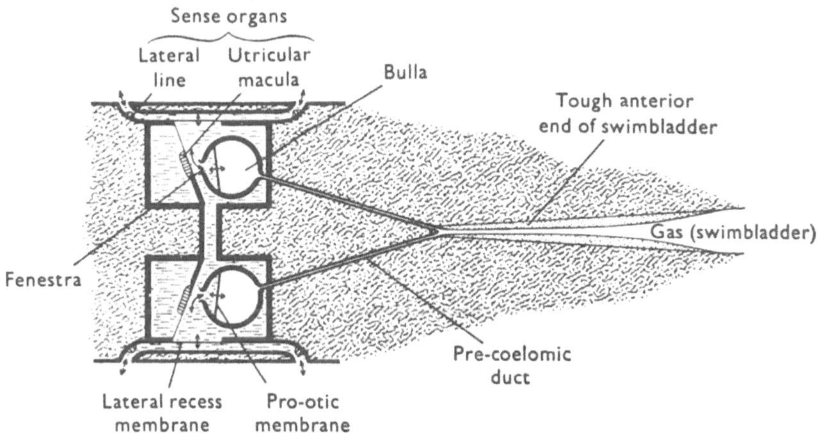

FIGURE 11.2. Diagram of the main mechanical components of the lateral line-bulla-swim bladder system of the sprat. A change in pressure causes a displacement of the pro-otic membrane and thus of the lateral recess membrane and movement of liquid along the lateral line canals and through the pores of the lateral line system. (After Denton and Blaxter 1976.)

sprat, the cupulae are thin plates oriented along the length of the canal (Fig. 11.1B), and it is easy to show that liquid can flow very readily on both sides of the cupulae.

In some experiments the cupulae were stained with Janus green B (this causes a small amount of shrinkage) and could then be seen sharply silhouetted against the highly silvered inner surface of the sprat's lateral line canal. At the magnifications used (×50) no movements of the cupulae were seen even when the liquid on either side of such cupulae was made to have large displacements. To measure the displacements of sprat neuromast cupulae, we took advantage of the fact that in the sprat, as in other clupeoids, the lateral line system is closely coupled to an auditory system that contains gas bubbles within two auditory bullae (Fig. 11.2). As a consequence of this special adaptation, changes of pressure applied to the whole fish can produce flows of liquid within the lateral line system and can produce such flows without causing significant displacements of the fish itself. The relations between alternating pressure applied to the fish and displacements of liquid in the lateral line system were found by Denton and Blaxter (1976), Gray and Denton (1979), and Denton et al. (1979). They showed that displacements of liquid are proportional to pressure change over a very wide range of pressures and that, for a given pressure change, such displacements vary relatively little with frequency up to 200 Hz.

The apparatus used to measure displacements of sprat cupulae is shown in Fig. 11.3. Fish were killed with anesthetic. A 1% solution of Janus green B in seawater was pipetted over the head of the fish for several minutes to stain the cupulae, and the excess stain was then washed out of the canals. The fish was placed

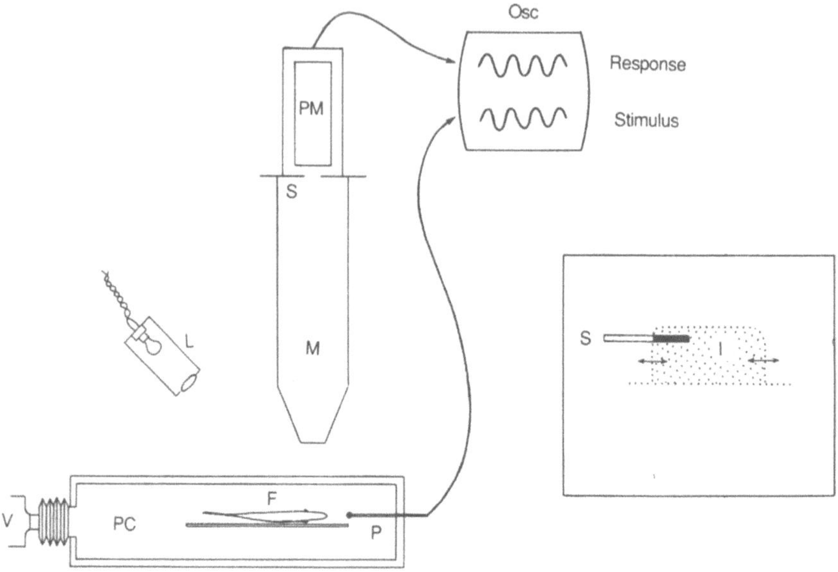

FIGURE 11.3. Diagram of apparatus used to measure displacements of cupulae in the lateral line of the sprat. V, vibrator; PC, pressure chamber; L, light source; F, fish; P, pressure transducer; M, microscope; S, slit; PM, photomultiplier; Osc, oscilloscope. Inset: Plan view of slit (S) showing the way in which the image (I) of the stained cupula was arranged so as to obscure part of the slit.

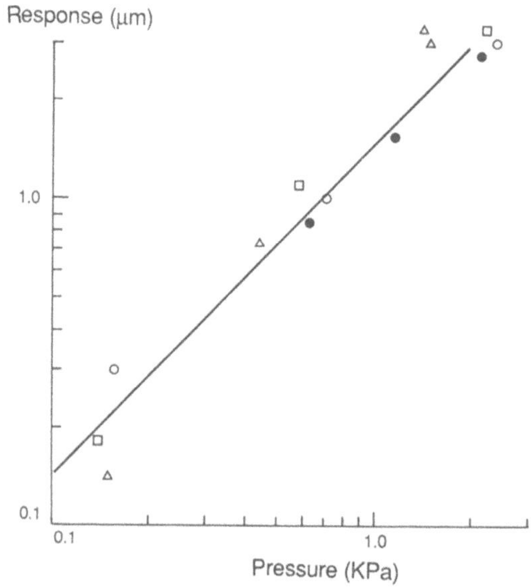

FIGURE 11.4. Sprat. Displacements of the cupula in response to pressure changes inside the apparatus shown in Fig. 11.3. These measurements were made at 160 Hz. Abscissa–peak-to-peak pressure in KPa. Ordinate–peak-to-peak displacement in microns. The different symbols relate to different specimens of sprat. The line indicates the relation expected if displacements are proportional to pressure.

TABLE 11.1. Sprat lateral line.

Frequency (Hz)	Displacement of cupula / Velocity of liquid in canal
40	0.07 nm/μm•s^{-1}
80	0.06 nm/μm•s^{-1}
160	0.12 nm/μm•s^{-1}
220	0.08 nm/μm• s^{-1}

in a closed chamber (PC) filled with seawater within which pressure changes of known amplitude and frequency could be produced. Using the microscope (M), an image of the edge of a stained cupula was focused on to a slit (s). Light passing through the slit fell on a photomultiplier (PM), the output of which was displayed on an oscilloscope (Osc) and recorded on tape. Sinusoidal changes in pressure applied to the fish produced sinusoidal changes of output from the photomultiplier. These were caused by displacements of the cupula with respect to the base, which did not move in response to pressure changes. It proved easy to calibrate the system so that absolute displacements of the cupula edge could be found.

Figure 11.4 shows that, at a given frequency, cupula displacements are linearly related to pressure change and, hence, linearly related to displacements of liquid within the lateral line canal. Maximum pressures of a few KPa were used–these are large but not excessive–about 80 to 100 dB above threshold for the lateral line. Table 11.1 shows that, over the range of frequencies studied, i.e., between 40 and 220 Hz, there was relatively little change in the ratio of cupular displacement to velocity of seawater in the canal. The cupula displacements for the parts of the cupula farthest from its base were around 0.08 nm for maximum velocities of liquid within the canal of 1 μm s^{-1}; this is probably two or three times threshold (Gray 1984). The displacements of the cupula became less as the base of the cupula was approached.

The cupula displacements given on Figure 11.4 are peak-to-peak values, and so the lower values correspond to displacements from a rest position of about 100 nm. The cupula displacements at its base will be appreciably less than this and so correspond to displacements of the hair bundles of the receptors within the range for which such displacements might be expected to be proportional to the applied force (Crawford and Fettiplace 1985).

3. The Forces on Cupulae

In an earlier publication (Denton and Gray 1983), it has been shown that in response to sinusoidally varying pressure gradients, the liquid flows along sprat lateral line canals are very similar to those found in straight capillary tubes of about the same cross-sectional areas. At low frequencies, the resistance of

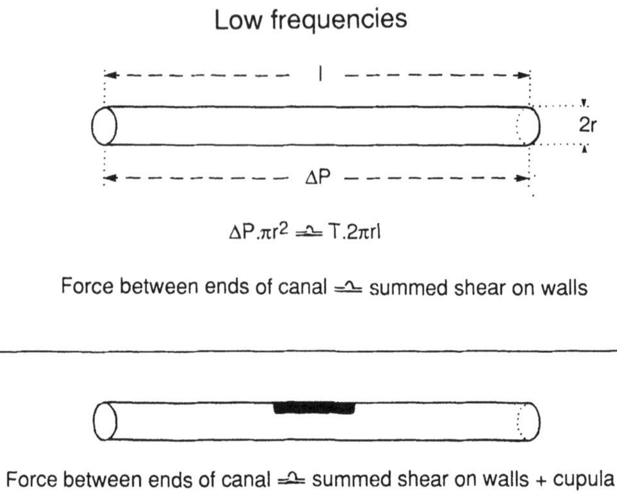

FIGURE 11.5. Top: Capillary tube. The force acting across the tube is $\Delta P \cdot \pi r^2$, where ΔP is the pressure between the ends of the tube and r is the radius of the tube. When Hagen-Poiseuille flows operate, this is equaled by the force $T \cdot 2\pi rl$, where T is the shear force on unit area of the internal surface of the capillary and l is the length of the capillary. Bottom: Diagram of a section of lateral line. Here, part of the force balancing the force $\Delta P \cdot \pi r^2$ is exerted by the shear forces on the walls of the cupula.

a canal can be calculated from the Hagen-Poiseuille equation for static flows. If the flow is little affected by compliances in the system, e.g., in the cupulae or in membranes covering the "openings" of the canals, this resistance will provide most of the impedance to alternating flows of liquid. For a capillary tube of constant internal diameter, the force applied between the ends of the capillary (which is the pressure difference between the ends of the capillary multiplied by the cross-sectional area of the tube) is equal to the summed shear forces over the internal area of the tube (Fig. 11.5). Shear forces against the wall of the tube are proportional to the velocity gradients against the wall. At low frequencies, when the transverse profile of velocities is parabolic, the shear force exerted on the wall can be estimated if the velocity of liquid flow through the tube is known. To estimate the forces acting on sprat cupulae, two assumptions were made: (1) Shear forces on the sides of a cupula can be calculated from the Hagen-Poiseuille relations, and (2) shear forces per unit area are twice those on the walls of a capillary tube of the same cross-sectional area as the lateral line canal. This factor is used because the cupula divides the lumen of the canal, and thus the transverse flow gradients (which will determine the magnitudes of the shear forces) will be approximately doubled. The value calculated in this way is likely to be a minimum value; this is because deviations from simple theory tend to increase the forces for a given flow (see Denton and Gray 1987).

Now the dimensions of the sprat lateral line canals and the cupulae studied are known (a typical cupula was about 250 μm long, 100 μm high, and 30 μm wide), and so the approximate forces on a cupula for given velocities of liquid flow through the canal can be estimated. Since the displacements of the cupulae for such velocities are also known (see Table 11.1), the stiffness (Sc) of a cupula—i.e., the force needed to give unit displacement—can be found. From the results of the experiments described above, the stiffness of a cupula can be calculated to be about 3.5×10^{-2} Nm^{-1} at 50 μm from the base of the cupula. The mass (Mc) of a cupula will be about 7.5×10^{-10} kg. This means that at lower frequencies the ration Sc:$Mc\omega^2$, where ω is 2π times frequency, is high (even at 100 Hz it will be about 100) and so, in the working range of the system, the movements of a cupula will be stiffness-controlled, and its displacements will be almost in phase and proportional to the shear forces exerted on its surface.

At low frequencies, the forces applied between the ends of a section of lateral line canal will be almost equaled by the shear forces on the walls of the canal and the cupula. As frequency rises, a greater and greater fraction of such forces will be used to accelerate and decelerate the liquid inside the canal. If it is assumed that at all frequencies the flow patterns across the canal are parabolic and everywhere in phase, it is possible to calculate the ratio:

$$\alpha = \frac{\text{shear force acting on the walls and cupula}}{\text{force acting between the ends of the section}}$$

for a given section of lateral line or capillary tube at any given frequency. For a canal of radius 100 μm, the ratio α remains substantially constant up to 80 Hz, and it then falls with increasing frequency. The ratio α will in practice fall less steeply at the higher frequencies than simple theory would predict. This is because (1) the actual velocities along the canal or tube are greater than predicted (see Fig. 11.6), and (2) transverse velocity gradients are steeper (see below). Such a mechanical system would be effective at allowing forces on the cupulae to follow transients, certainly for "rise" times slower than about 3 to 4 ms.

The assumption of Hagen-Poiseuille flow patterns is reasonable at low frequencies, but at higher frequencies the flow patterns are more complicated. In describing oscillatory flow patterns found in canals (pipes), Schlichting (1968) makes use of the dimensionless quality

$$k = \sqrt{\frac{\omega}{\nu}} \cdot r$$

where ω is 2π times frequency, ν is kinematic viscosity, and r is the radius of the canal. When $k \ll 5$, the flow patterns are similar to those found in static (Hagen-Poiseuille) conditions—i.e., everywhere in phase with a parabolic velocity distribution across the canal. Where $k \gg 5$, there is a marked boundary layer within which velocities increase with distance from the wall, while the main mass of liquid in the center of the canal will have, at any instant, a constant velocity independent of the distance from the wall. For the intermediate condition,

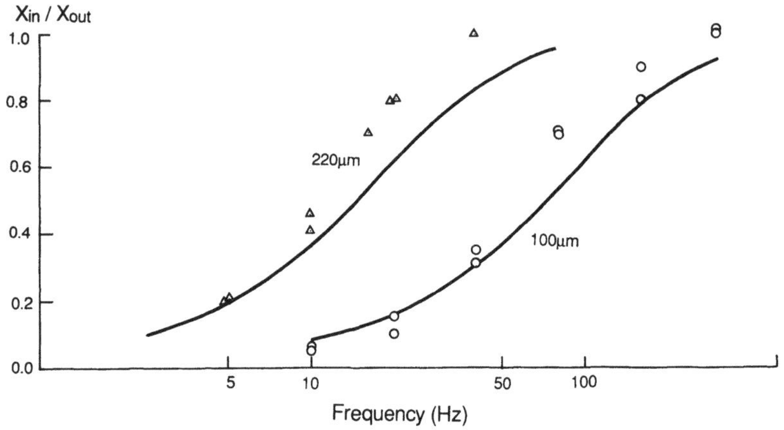

FIGURE 11.6. Amplitude responses in capillary tube. The ordinate shows amplitude of displacement inside a capillary (X_{in}) divided by that in the medium (X_{out}). The abscissa is frequency (Hz). The points are experimental results (two observers), △ for capillary of radius 220 μm, 0 for capillary of radius 100 μm. The smooth curves were calculated from the theoretical relation $(j\omega I)(j\omega I + R)$ (see text for explanation).

k around 5, the flow pattern across the tube is more complicated. There are phase differences in the flows across the whole width of the canal. There is the added complication that, at higher frequencies, maximum velocities of flow are not always in the center of the canal. All these conditions—i.e., $k \ll 5$, *k* around 5, and $k \gg 5$—exist in lateral line canals within the ranges of frequencies for which lateral lines are thought to operate. Thus, for a canal of radius 100 μm (many canals of approximately this radius exist—e.g., in sprat), $k = 5$ at 400 Hz, and 1 at 16 Hz. For canals of radius 1 mm (canals with this or greater radii are often found in deep-sea fish), $k = 5$ at 4 Hz, 1 at 0.16 Hz, and 50 at 100 Hz. For these larger canals, the transverse velocity gradients on which shear forces depend are, even for frequencies of only a few hertz, confined to the sides of the canal. This perhaps explains why very low cupulae are sometimes found in very large canals—e.g., in the deep-sea fish *Poromitra* sp. (Denton and Gray 1988).

It may be noted that for canals of different radii, corresponding frequencies (where the oscillatory flow patterns are the same) would be expected to be proportional to the reciprocals of the cross-sectional areas of the canals. This expectation is well borne out by earlier experiments (Denton and Gray 1983) in which the ratios of the displacements (X_{in}) inside sprat lateral lines and inside capillaries to those in the external media (X_{out}) were found over a range of frequencies. Results on two capillaries of radii, 100 μm and 220 μm, respectively, are shown on Fig. 11.6, where they may be compared with estimates based on the assumption that the ratios will be given by $(j\omega I)/(j\omega I + R)$, where I is the inertance (mass divided by the square of the cross-sectional area of the tube), R

is the viscous resistance for static (Hagen-Poiseuille) flow conditions, j is the complex operator, and ω is 2π times frequency. The lateral line of the sprat gives experimental results that agree well with those found for capillaries and show very similar deviations from simple theory. On the assumptions given above, the relative values for the quantity α would be given by dividing, for each frequency studied, the value of $(X_{in})/(X_{out})$ by the frequency.

One interesting conclusion that can be drawn from these results is that frequency-response curves for different sizes of canal (or capillary) are invariant if plotted with frequency times cross-sectional area of the canal as abscissa.

It is generally agreed that the responses of the receptors and their nerves are related to the displacement of the hair cells (see Flock 1971), though the relation is often nonlinear. It is, however, a reasonable, though not certain, assumption that a particular level of response is related to a particular displacement of the neuromast during a transient. If a constant level of a neural response, be it a measure of receptor activity or of nerve impulse discharges, is found at different frequencies (or the equivalent in transients), then it is likely that the force on the cupula is the same in each case, and this can be related to the total force across the section of canal. In the case of the sprat, as has just been explained, it was possible to measure the velocity in the canal. It was also possible to show that results of experiments with constant responses to transients using the neural discharge as indicators were consistent with measurements of velocity in the canal (data of Gray 1984, Fig. 8).

In subsequent sections, examples are considered in which there have been no direct measurements of velocity, and results from constant response experiments have been used to compare with theory.

4. *Fundulus heteroclitus*

Harris and van Bergeijk (1962) have, over a range of frequencies, measured the microphonic potentials produced by neuromasts in the head lateral line of the fish *Fundulus heteroclitus* in response to the stimuli produced by a sphere oscillating in the external medium. They plotted their results (their Fig. 12) as microphonic response against frequency for given amplitudes of mechanical stimulus measured as displacements of the sphere. These results are easily presented in terms of the acceleration of the source (Kalmijn 1988). In Fig. 11.7 the reciprocal of the pressure gradient required to give a constant microphonic response is plotted against frequency (see above). Harris and van Bergeijk noted that their results were consistent with a mechanical system that had a resonant frequency (f_n) of about 170 Hz and which was a little underdamped (the microphonic potential being taken as proportional to the displacement of the cupula). On Fig. 11.7, for comparison with their results (replotted to give the reciprocal of the pressure gradient for a constant microphonic response against frequency), there is a curve for critical damping for a simple resonant system with $f_n = 170$ Hz. These results show that although the coupling between movements of the cupula and the move-

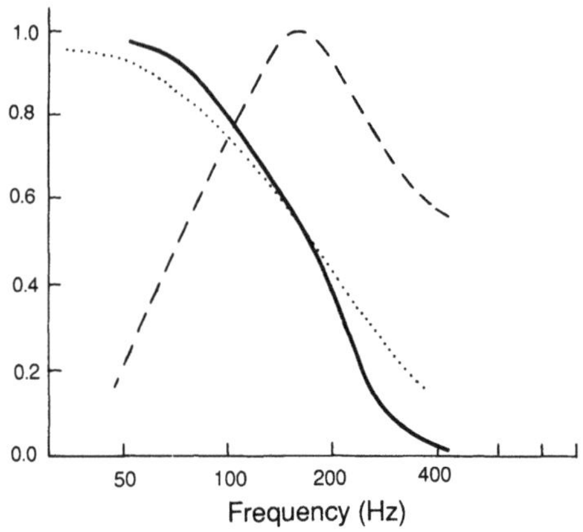

FIGURE 11.7. *Fundulus heteroclitus*. Dashed line: Curve given by Harris and van Bergeijk (1962), in which microphonic response of a lateral line neuromast, as ordinate, is plotted against frequency for a given *amplitude* of displacement of the source. Solid line: Data of Harris and van Bergeijk (1962) replotted so as to give as ordinate reciprocal of the *acceleration* of the source required to elicit constant microphonic response. Dotted line: Curve similar to that shown by solid line but for a simple mechanical system which has an undamped resonant frequency of 170 Hz and is critically damped. Maxima made equal to 1.

ments of the liquid within the canal is perhaps different from that found in the sprat, the constant response curves are similar (the similarity extends to phase relations; Kalmijn (1988, Chapter 9) has shown that Harris and van Bergeijk gave phase curves from *Fundulus* that were 180° in error). Both systems (sprat and *Fundulus*) might be expected to respond well mechanically to low frequencies and to follow transients well provided that the rise times involved were not faster than a few milliseconds.

5. *Gymnocephalus cernua* (the Ruffe)

As early as 1952, Jielof et al. made excellent measurements on the lateral line system of the ruffe. In particular they measured the stiffnesses of cupulae. They found the stiffness to increase for large displacements. For the smaller displacements of their experiments, they suggested values for stiffness of about 1.25×10^{-1} Nm^{-1} if the displacements are measured at the tips of the cupulae. The displacements close to the bases of the cupulae were about 30% of those at their tips. Van Netten and Kroese (Chapter 12) have shown, by very elegant techniques, that for small displacements, much smaller than those used by Jielof et al., the motion is the same from the tip of the cupula to very close to the base. They have measured the stiffness and obtained a value of 1.3×10^{-1} Nm^{-1}.

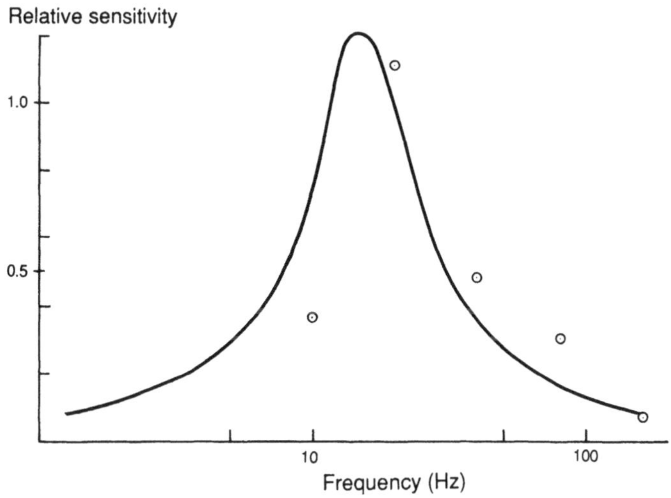

FIGURE 11.8. The ruffe. Curve: predicted reciprocal of acceleration of water outside the fish (relative to the fish) for a constant velocity around a neuromast in the middle of the infraorbital canal plotted against frequency (Hz). Circles: reciprocal of acceleration of water outside for a constant response recorded from a *nerve* from neuromasts in the same position.

Observations have been made on this fish whose canals consist of large chambers covered with compliant membranes, each with a small hole, separated by partitions with gaps in which the neuromasts lie. Measurements of these canals, using the simple theory previously described Denton and Gray (1988), have permitted predictions of velocities in different parts of the canal system in relation to net external pressure gradients. Using the constant response technique with nerve discharges as indicators (see above), estimates of equivalent forces on the cupula at different frequencies have been attempted. The results are consistent with the hypothesis that these forces are proportional to the velocity of the liquid around the neuromast (Fig. 11.8).

6. Displacements of Cupulae and of the Hair Bundles of Receptors

Displacements of liquid inside a section of a lateral line canal will be accompanied by forces acting over the whole surface area of the cupula of its neuromast. At low frequencies, it should be expected that all of these forces would be in phase with each other and almost in phase with the displacements of the cupula and the hair bundles of the receptors. If the system is to be efficient in stimulating the receptor cells, then these forces would have to be effectively transmitted to the

TABLE 11.2. Hair cell numbers.

Fish	Hair cell density (cells/μm²)	Approx. No. of receptors/neuromast (10-cm fish)
Sprat[a]	0.23	1000 (500 for 7 cm fish)
Anchovy	0.045	1000
Anoplogaster	0.034	800
Poromitra	0.046	10000
Ruffe	0.038	5400 (3400 for 8 cm fish)

[a] From Best and Gray (1982).

hair bundles, which would in turn provide the main restoring forces acting against displacements (see Denton and Gray 1988).

There is a good deal of evidence showing that, for a receptor cell, the bundle of stereocilia (hairs) behaves as a system of linked and inflexible rods pivoted about a point close to the cell surface (Flock et al. 1977; Crawford and Fettiplace, 1985). The pivotal stiffness of such bundles has been measured for several different preparations and found to be around 2×10^{-14} N m/rad (Ashmore 1984; Strelioff and Flock 1984; Crawford and Fettiplace 1985). This corresponds to a stiffness value of about 4×10^{-4} Nm^{-1} if we were to measure displacements close to the top of the bundle of stereocilia, typically about 6 to 7 μm from the surface of the cells.

We have found the numbers and densities (number per unit area) of hair cells for the neuromasts of a number of species of fish, and some of our results are given in Table 11.2. We have, in particular, made such measurements for sprat and for ruffe of the sizes used respectively in the experiments on sprat described above and those of Jielof et al. (1952) on ruffe. The stiffness for whole cupulae (Sc) can therefore be compared with those given for 7 μm above the surfaces of the hair cells by multiplying the stiffness of a single bundle of hairs (Sr) by the numbers (n) of receptors. These comparisons have been made. They show that for both the sprat and the ruffe, the value of $n \cdot$Sr, for the combined actions of the receptors, is in the range of four to 10 times that of Sc, for the cupula. It can be recalled, however, that the stiffness of a hair bundle and of the intact nueromasts, with their cupulae, of sprat and ruffe were estimated for different distances from the surfaces of the receptor cells (about which the hairs pivot). For a bundle of hairs alone, the stiffness is inversely proportional to the square of the height (above the cell surface) at which force is applied to the hair bundle (Crawford and Fettiplace 1985). If such an inverse square law were to apply to all parts of a cupula and the reactive force was exerted by the hair cells, the stiffness measured at the cupula would be expected to be much less than actually observed. This would imply that most of the reactive forces would not have been given by the hair cells. However, this is not thought to be the case, and there is evidence, at least for some cupulae, that the whole cupula slides over the hair cells surface

(Flock 1967; see also van Netten and Kroese, Chapter 12). The strong likelihood remains, therefore, that a large fraction of such forces can be effective as stimuli to receptor cells, and that a large fraction of the stiffnesses of cupulae can be accounted for by the numbers of hair cells, the stiffness of the hair bundles, and the nature of the immediate mechanical coupling between the hair bundles and the cupulae. In this brief discussion, the fact that the stiffness of a hair bundle is known to be under nervous control has not been taken into account.

7. Effects of Change of Scale

Pumphrey (1950) has discussed the relation between fish behavior and the properties of lateral lines on the assumption that the sensitivity of the lateral line remains at the same level regardless of the size of the fish. The arguments given above suggest that this assumption is unlikely to be correct. Some results, calculations, and ideas about the performance of lateral lines of very different sizes and designs have already been published (Denton and Gray 1988). Here a particular case, that of fish of identical (or similar) design but different size, is considered. Some results obtained suggest that, at least over some ranges of size, all dimensions of the lateral line system can increase approximately in proportion to the length of the fish. The mechanical sensitivity of a neuromast will be measured in terms of the summed forces exerted on its cupula for a given net pressure gradient along the section of lateral line canal in which the neuromast lies.

If two lateral lines are now taken, differing in linear scale by the factor L, then, for a given net pressure gradient, the force F_s acting along a section of lateral line is greater by the factor L^3 for the larger lateral line (the section is $L \times$ longer and the cross section is L^2 greater). At frequencies for which a similar fraction of F_s acts on the cupula, the conclusion is reached that the larger lateral line will be more "sensitive" by a factor L^3. When external sources of vibration are considered, this advantage will often not be the only one that the larger fish will have. Consider the simple case of a fish facing a vibrating source (Fig. 11.9). Since the whole fish behaves as a rigid body, all parts of the fish along its length will in a given situation vibrate in the same way (Denton and Gray 1983). A given section of the lateral line will only be stimulated when the vibration of the medium over this section of lateral line differs from that of the surface of the fish in which the lateral line lies. There will be a position along the fish's length, which might be called the neutral position, where there is no net stimulation because lateral line and medium move in the same way. Net stimulation will increase with distance from the neutral position, and corresponding distances are greater for larger fish. The magnitude of this important effect will change appreciably with the distance and the orientation of the fish with respect to the source, but it will often give an advantage in effective stimulus to the larger fish which is xL or greater.

When this is taken into account, it is concluded that the larger fish (with scaled up lateral lines) will, for a given-low frequency source, have a sensitivity advantage of roughly xL^4 on the basis of the mechanics. One might therefore detect a

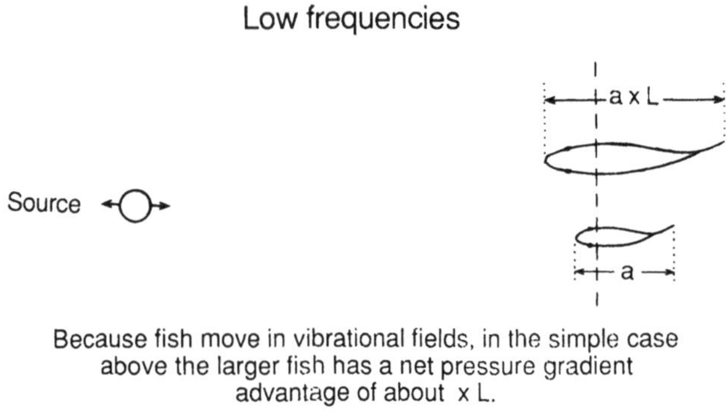

FIGURE 11.9. Two fish, one of which has all of its dimensions greater than the other by a factor L, face the same vibrating source. The fish will move in the vibrational field. The dashed vertical line indicates the positions where the fish and the medium will move in the same way, and so where the lateral line receptors will not be stimulated. Both ahead and behind this line, the corresponding points of the lateral line will be more strongly stimulated for the larger than for the smaller fish.In this simple case the larger fish has a net pressure gradient advantage of about $\times L$.

sphere vibrating at low frequencies at a distance that is about xL greater than would the smaller fish; the net stimulus to a fish facing a vibrating sphere falls off as L^4 (Denton and Gray 1983). If, however, the larger and smaller fish are thought of as predator and prey, respectively, then they must be considered as sources as well as detectors of vibration. If the larger and smaller fish were to make similar movements at a given frequency, the larger fish would be a source xL^4 greater (xL^3 because the stimulus given is proportional to the volume of the moving object and xL, because the amplitude of movement will for similar movements be xL greater).

From these simple arguments, the conclusion is reached that predator and prey will detect each other at the same distance. The true picture is evidently much more complicated. Thus, for example, (1) small fish often have relatively wider lateral lines so the advantage in sensitivity will not be so great as the above argument suggests; (2) in general, smaller fish make movements more quickly than larger fish, which will give an advantage to the larger fish; and (3) with faster transients, the sensitivity advantage to the larger fish is less great because the mechanical system of a larger lateral line will be less good at following rapid transients.

In the above analysis, we have not taken account of the fact that, in general, the number of hair cells increases with the size of the animal. If this number is proportional to L^2, then for a given force applied to the cupula, its displacement will be diminished by L^2. However, since the number of hairs displaced is increased by L^2, the larger animal could, depending on the ways in which the responses of the hair cells are combined, have, for a given displacement, advantages in sensitivity or discrimination.

8. Possible Behavioral Significance of the Head Lateral Line of the Ruffe

One reason for interest in the relation between water motion in the canal, the force on the cupula, and the signals sent to the brain is that if this is understood, the nerve activity can be predicted from a knowledge of the motion. A very simple method of obtaining a first approximation of the velocity of water in canals from a knowledge of the net acceleration of the outside water relative to the fish has already been proposed (Denton and Gray, 1988). This net acceleration can be calculated for positions around a known external source if it approximates a simple type. On this basis, calculations have been made to predict the excitation of all the neuromasts on a ruffe's head, with the head in varying positions in relation to a small source (a 1.5-mm diameter sphere with a peak acceleration of 8 $mm \cdot s^{-2}$; i.e., approx 1-mm displacement at 13 Hz). It is estimated that the fish could detect such a source at a distance of about 6 cm. In certain positions, the predicted change in the pattern of activity of the neuromasts on the head is very great for small changes of position relative to the source ($\sim$ 0.5 cm). More information is needed about sources, but from present knowledge it seems that the ruffe could detect and locate chironomid larvae, its common prey, in the top layers of the mud using its head canals. Kuiper (1967) showed that fish with some neuromasts denervated were unable to locate a rod held in the hand in the way they can with all the organs functional.

9. Summary and Conclusions

1. The displacements of neuromast cupulae in response to known alternating displacements of liquid inside the lateral line canal have been measured for the sprat (*Sprattus sprattus*). Over the likely working range of frequencies, the displacements of cupulae are approximately proportional to the velocities of liquid in the canal. The measurements allowed the stiffness of the cupula/hair cell complex to be estimated, and the conclusion was reached that, over this frequency range, the movements of the complex are stiffness-controlled. Recordings from neuromast nerves of sprat, ray (*Raja clavata*), and ruffe (*Gymnocephalus cernua*) show that velocity in the canal can be used as a reasonable indicator of stimulus strength in the relevant frequency range.
2. Some earlier results of Harris and van Bergeijk (1962) on the killifish, *Fundulus heteroclitus*, have been compared with those obtained on the sprat, expressing both sets of results in terms of the pressure gradients along a canal required to give constant response (see also Kalmijn 1988, Chapter 9). Although the coupling between movements of the cupula and movements of liquid within the canal seem to be very different for the two species, the constant response curves obtained are very similar. For lower frequencies (<100 Hz), the responses are approximately proportional to the pressure gradients – i.e., to accelerations of a source of vibration.

3. The numbers of hair cells/neuromast have been found for several species of fish. It is argued that the stiffness of the cupula-hair cell complex is largely determined by the hair cells.
4. The mechanical design of fish lateral lines is briefly discussed, some attention being given to responses to transient and oscillatory stimuli and to effects of change of scale.
5. For one fish, the ruffe, it has been shown that when a source of vibration is just ahead of the fish, the distribution of stimuli over the lateral line system changes greatly with small changes in the position of the source relative to the fish.

References

Ashmore JF (1984) The stiffness of the sensory hair bundles of frog saccular hair cells. J Physiol 350:20P.

Best ACG, Gray JAB (1982) Nerve fibre and receptor counts in the sprat utriculus and lateral line. J Mar Biol Assoc UK 62:200–213.

Bergeijk WA van, Alexander (1962) Lateral line canal organs on the head of *Fundulus heteroclitus* J Morphol 110:333-346.

Coombs S, Janssen J, Webb J (1988) Diversity of lateral line systems: Evolutionary and functional considerations. In: Atema J, Fay RR, Popper AN, Tavolga WN (eds) Sensory Biology of Aquatic Animals. New York: Springer-Verlag, pp. 553–593.

Crawford AC, Fettiplace R (1985) The mechanical properties of ciliary bundles of turtle cochlear hair cells. J Physiol 340:359–379.

Denton EJ, Blaxter JHS (1976) The mechanical relationships between the clupeid swimbladder, inner ear and lateral line. J Mar Biol Assoc UK 56:787–807.

Denton EJ, Gray JAB (1983) Mechanical factors in the excitation of clupeid lateral lines. Proc R Soc Lond B 218:1–26.

Denton EJ, Gray JAB (1988) Mechanical factors in the excitation of the lateral lines of fish. In: Atema J, Fay RR, Popper AN, Tavolga WN (eds) Sensory Biology of Aquatic Animals. New York: Springer-Verlag, pp 595–617.

Denton EJ, Gray JAB, Blaxter JHS (1979) The mechanics of the clupeid acousticolateralis system: Frequency responses. J Mar Biol Assoc UK 59:27–47.

DuBois AB, Cavagna GA, Fox RS (1974) Pressure distribution on the body surface of swimming fish. J Exp Biol 60:581–591.

Fänge R, Larsson Å, Lidman U (1972) Fluids and jellies of the acusticolateralis system in relation to body fluids in *Coryphaenoides rupestris* and other fishes. Mar Biol 17:180–185.

Flock Å (1967) Ultrastructure and function in the lateral line organs. In: Cahn PH (ed) Lateral-Line Detectors. Bloomington: Indiana University Press, pp. 163–197.

Flock Å (1971) Sensory transduction in hair cells. In: Loewenstein WR (ed) Handbook of Sensory Physiology, Vol. 1: Principles of Receptor Physiology. New York: Springer-Verlag, pp. 396–441.

Flock Å, Flock B, Murray E (1977) Studies on the sensory hair cells of receptor cells in the inner ear. Acta Otolaryngol 83:85–91.

Gray JAB (1984) Interaction of sound pressure and particle acceleration in the excitation of the lateral line neuromasts of sprat. Proc R Soc Lond B 220:299–325.

Gray JAB, Denton EJ (1979) The mechanics of the clupeid acoustico-lateralis system: Low frequency measurements. J Mar Biol Assoc UK 59:11–26.
Harris CG, van Bergeijk WA (1962)Evidence that the lateral line organ responds to near-field displacements of sound sources in water. J Acoust Soc Am 34:1831–1841.
Janssen J, Coombs S, Hoekstra D, Platt C (1987) Anatomy and differential growth of the lateral line system of the mottled sculpin, *Cottus bairdi* (Scorpaeniformes: Cottidae). Brain Behav Evol 30:210–229.
Jielof R, Spoor A, de Vries H (1952) The microphonic activity of the lateral line. J Physiol 116:137–157.
Kalmijn Ad J (1988) Hydrodynamic and acoustic field detection. In: Atema J, Fay RR, Popper AN, Tavolga WN (eds) Sensory Biology of Aquatic Animals. New York: Springer-Verlag, pp 83–130.
Kuiper JW (1967) Frequency characteristics and functional significance of the lateral line organ. In: Cahn PH (ed) Lateral Line Detectors. Bloomington: Indiana University Press, pp. 105–121.
Netten SM van, Kroese ABA (1987) Laser interferometric measurements on the dynamic behavior of the cupula of the fish lateral line. Hear Res 29:55–61.
Pumphrey RJ (1950) Hearing. Soc Exp Biol Sym 4:5–18.
Schlichting H (1968) Boundary Layer Theory. New York: McGraw-Hill.
Strelioff D, Flock Å (1984) Stiffness of sensory hair bundles in the isolated guinea pig cochlea. Hear Res 15:19–28.

12 Dynamic Behavior and Micromechanical Properties of the Cupula

SIETSE M. VAN NETTEN and ALFONS B.A. KROESE

12.1 Introduction

In the transmission and processing of mechanically coded information by organs of the acousticolateralis system of vertebrates, the transfer of vibrations via fluid to accessory structures of hair cells plays an important role (Pumphrey 1950). In the lateral line system, the hydrodynamic coupling between the cupula and the surrounding water together with the mechanical properties of the cupula filters the mechanical information that reaches the sensory hair cells.

The peripheral mechanical part of canal lateral line transduction is shown schematically in Fig. 12.1. The first step which is formed by the transmission of motion from external water to canal fluid, involves canal hydrodynamics and has recently been investigated experimentally by Denton and Gray (1983, Chapter 11), who showed that the canal acts as a high-pass filter with respect to fluid displacements of the surrounding water (see also Kroese and van Netten Chapter 13). In the present chapter recent experimental data (van Netten and Kroese 1987) on the subsequent two transmission steps, which concern the dynamic behavior of the cupula and the sensory hair cell bundles in response to canal fluid displacements, will be discussed. These steps involve the hydrodynamic and mechanical properties of the cupula in combination with the micromechanical properties of the sensory hair cells and their mechanical coupling to the cupula. The last step in transmission depicted in Fig. 12.1 comprises the mechanoelectrical transduction performed by the sensory hair cells and is dealt with by Kroese and van Netten in Chapter 13.

12.2 Fluid Forces Acting on the Cupula

The hydrodynamic excitation of lateral line cupulae has been discussed by many authors. Most conclusions about the fluid forces that drive the cupula are based on the analysis of electrical responses of the sensory hair cells rather than on direct mechanical measurements of the cupula itself. Based on these electrophysiological measurements and general considerations, the fluid forces acting

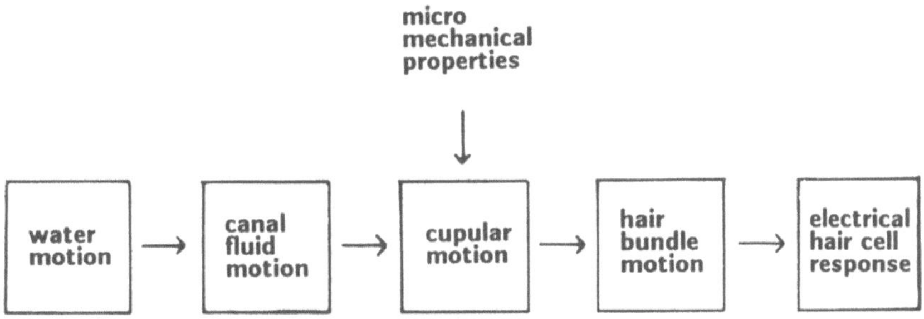

FIGURE 12.1. Schematic drawing of the peripheral mechanical part of lateral line transduction. In this chapter the step from canal fluid motion to cupular motion and the subsequent transmission to the sensory hair bundles are discussed.

on lateral line cupulae have generally been supposed to consist solely of viscous drag, implying that the excitatory fluid force is proportional to the relative velocity of the fluid flowing past the cupula (Görner 1963; Harris and Milne 1966; Flock 1971; Oman 1972; Schwartz 1974; Russell 1976; Strelioff and Honrubia 1978; Kroese et al. 1978, 1980; Denton and Gray 1983, Chapter 11). One of the main reasons for the lack of direct mechanical data on cupular motion is that under physiological conditions, displacement amplitudes are well below the resolving power of a light microscope (e.g., Hudspeth 1983). In a small number of studies lateral line cupular motion has been measured in the relatively large micrometer displacement range using light microscopic techniques in combination with stroboscopic illumination (Jielof et al. 1952; Kuiper 1956; Liff and Shamres 1972). From these mechanical measurements (Jielof et al. 1952; Oman 1972) and on the basis of theoretical studies of lateral line hydrodynamics (Kalmijn 1987, Chapter 9) it has been suggested that boundary layer effects, which are frequency-dependent, might play a role in the mechanical excitation of the cupula. This means that fluid forces other than viscous drag play a role in the mechanical excitation of the cupula. To what extent the dynamic behavior of the cupula is affected by boundary layer effects, however, has not been investigated previously. Obviously, the answer to this question is not only important in relation to the understanding of lateral line hydrodynamics, but it also enables to relate a well-defined stimulus of the sensory hair bundles to electrical responses of hair cells (see Fig. 12.1 and Kroese and van Netten Chapter 13).

With the use of specially developed techniques (van Netten 1988; Kroese and van Netten 1987), the dynamic behavior of the cupula of the lateral line in vivo in the submicrometer displacement range has been investigated (van Netten and Kroese 1987). Here, the measurements will be described and interpreted in terms of the physical and hydrodynamic properties of the lateral line system using a theoretical hydrodynamic description that includes a frequency-dependent boundary layer around the cupula (van Netten 1987). This description allows for variation of the relevant physical parameters—namely, the cupular radius, the

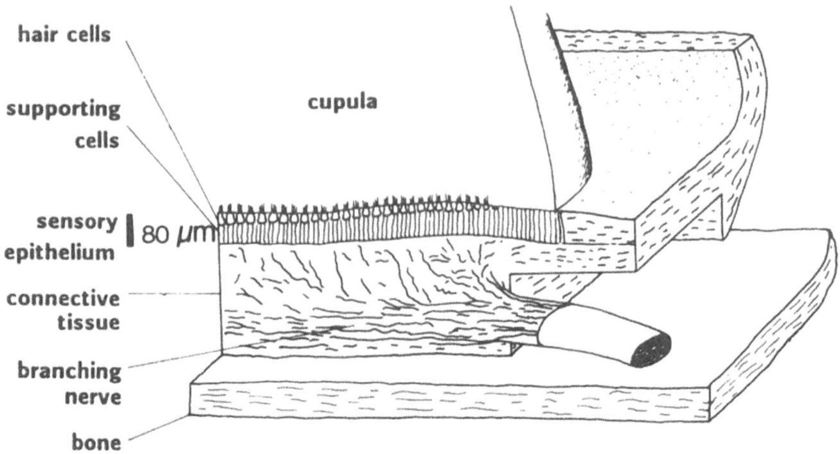

FIGURE 12.2. Schematic drawing of a neuromast in the lateral line canal of the ruff. After Jakubowski (1963).

amount of stiffness coupling between cupula and epithelium, and the viscosity of the fluid. In this way, the functional significance of these parameters for lateral line transduction is illustrated. It appears that only one dimensionless parameter, formed by a combination of the physical parameters of the system, determines whether the cupula detects fluid displacement, fluid velocity, or fluid acceleration in a certain frequency range.

12.3 Methods

12.3.1 Preparation

Experiments were performed on ruff (*Acerina cernua* L.) about 11 cm long. Fish were anesthetized with intraperitoneal injection (24 mg/kg) of Saffan (Glaxovet) (Oswald 1978) and held rigidly in place by head and body clamps, while artificial respiration was provided by a flow of tap water (0.25 l/min) with a temperature of about 15 °C. Canal organs in the supraorbital canal were revealed by carefully removing the overlying skin and bony bridge. A schematic longitudinal section of a neuromast is shown in Fig. 12.2, where the hair cells can be seen to be located in between the supporting cells, firmly sustained by connective tissue and the innervating nerve that rest on the bony bottom of the canal.

12.3.2 Mechanical Stimulation

Local sinusoidal water motions were produced by a glass sphere located in the canal about 4 mm from the cupula (see Fig. 12.3). The glass sphere had a diameter of 0.8 mm and was mounted on a piezoelectric driver. Over the range

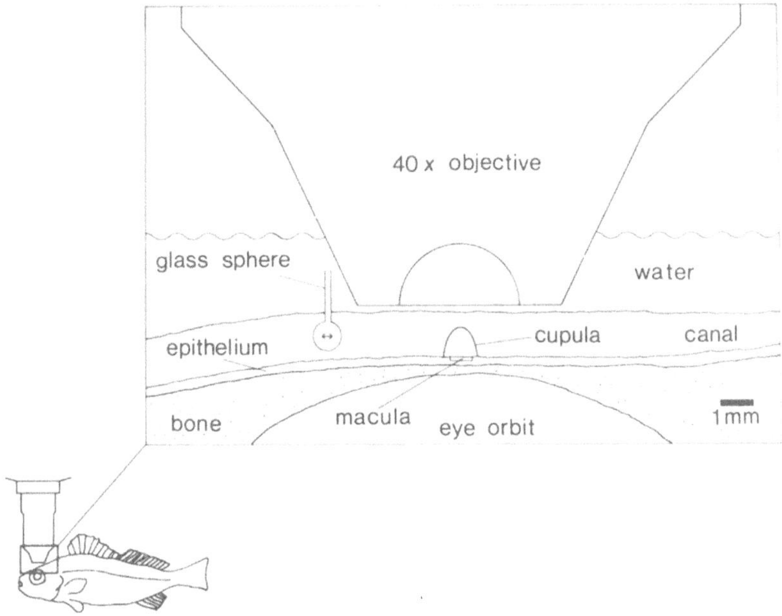

FIGURE 12.3. Schematic drawing of the lateral line preparation. The fish is placed below the water immersion objective of the microscope. Local sinusoidal water motions are produced by the glass sphere located in the canal. Optical access to the macula is possible because of the transparency of the cupula. From van Netten and Kroese (1987).

of frequencies used (15 to 500 Hz), the sphere had a constant displacement in response to a given voltage. The movement of the sphere was parallel to the longitudinal axis of the canal and in the direction in which the hair cells have their maximum sensitivity (Flock 1971, Kroese and van Netten 1987).

12.3.3 Visualization Technique (ILPM)

The organ in the canal was observed during an experiment with incident light polarization microscopy (ILPM), an imaging technique (magnification 1,000×) developed to visualize sensory hair cells and hair bundles under in vivo conditions (van Netten and Kroese 1986; Kroese and van Netten 1987). The method permits a clear view of the plane of the macula. Figure 12.4 shows a representative picture of the image obtained with the microscope. The apical surface of the hair cells (H), the hair bundles (B), and the supporting cells (S) can be easily identified. Furthermore, with the ILPM, clear images of the macula with a resolution of 0.5 μm can be obtained up to a depth of about 150 to 200 μm below the apical surface of the hair cells. This allows visualization of erythrocytes passing through the capillary vessels supplying the macula, thus providing a simple

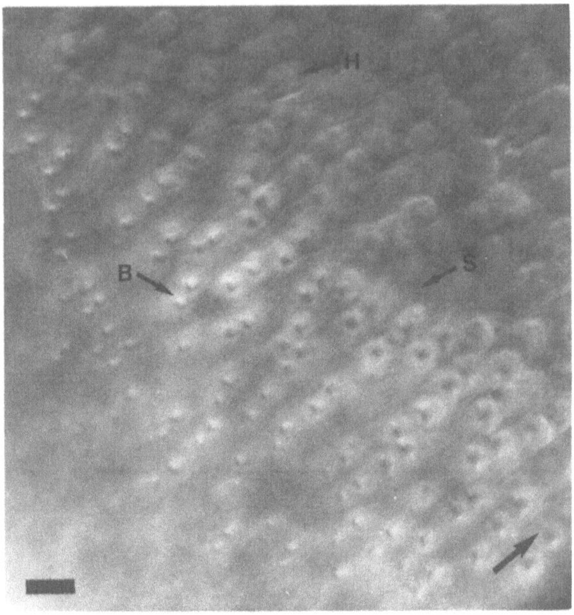

FIGURE 12.4. The sensory epithelium of a neuromast of the lateral line observed in vivo with an incident light-polarizing microscope (ILPM). Focal plane ranges from top of the hair bundles (bottom left) to just below the apical membrane of the hair cells (top right). H, hair cell; B, hair bundle; S, supporting cell. Arrow indicates the direction of the longitudinal axis of the canal. Bar: 10 μm. From Kroese and van Netten (1987).

means of monitoring the condition of the in vivo preparation (van Netten and Kroese 1986).

12.3.4 Submicromechanical Measuring Technique (LIM)

To be able to measure the submicroscopic displacements of the cupula, a specially developed laser interferometer microscope (LIM) was employed. The instrument will be described here only briefly. A detailed description can be found elsewhere (van Netten 1988). The LIM consists of a differential heterodyne laser interferometer that has been combined with the incident light-polarizing microscope (ILPM) described in the previous section. The instrument permits the measurement of nanometer vibrational displacements of a light-scattering object having microscopic dimensions. A schematic drawing of the optics of the LIM is shown in Fig. 12.5. Two laser beams, emerging from a beam splitter (HM1), are directed to the microscope and focused on the object (OB) under two different angles to produce an interference pattern at their cross section (ø 5 μm) at OB. The optical frequency of the two laser beams is shifted by

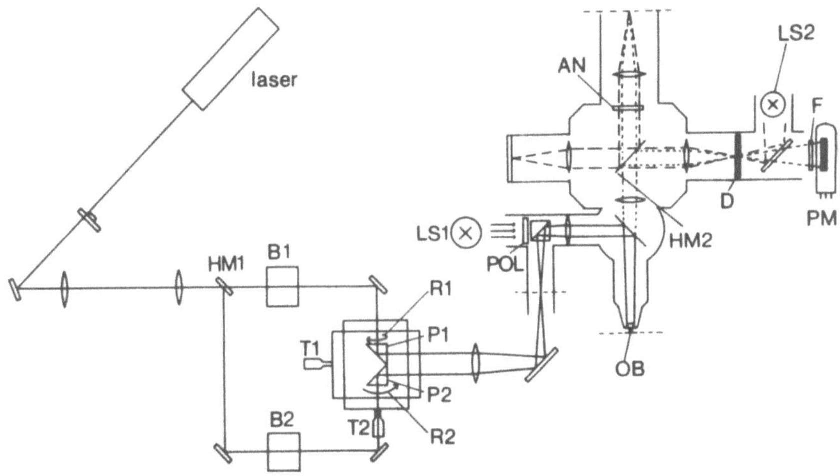

FIGURE 12.5. Optics (schematical) of the laser interferometer microscope (LIM). Light source LS1 together with the analyzer (AN) and polarizer (POL) are part of the incident light-polarizing microscope (ILPM). Two laser beams are directed via half-mirror HM1, prisms P1 and P2, and Bragg cells B1 and B2 to the microscopic object OB with precision rotators R1 and R2 and translators T1 and T2 under visual control of the microscope. The position of pinhole D can be adjusted while it is illuminated by LS2 and visualized through the microscope simultaneously with the image of the object. This ensures that only backscattered light from the object is detected by photomultiplier PM. F is an optical filter. For details on the laser interferometer microscope, see van Netten (1988).

Bragg cells (B1 and B2) by 80 and 79.6 MHz, respectively, resulting in an optical frequency difference f_{het} of 400 kHz between the two beams. Using this optical heterodyne technique causes the fringes of the interference pattern to move with a constant velocity across the cross section of the beams at OB. The light intensity scattered back by a moving microscopic object at OB is then given by Eq. 1 (van Netten 1988):

$$I(t) = A\{1 + \cos[2\pi(f_{het}t + 2n\sin\alpha\, x(t)/\lambda) + \phi_0]\} \quad (1)$$

where A is a constant, α is the angle of the laser beams with the optical axis, ϕ_0 is a phase-offset, n is the refractive index, λ is the wavelength of the laser light, and $x(t)$ is the displacement of the object OB as a function of time. The first term between square brackets results from the moving fringes, whereas the second term arises from the motion of the object OB. The intensity fluctuations of the backscattered light thus can be described by a carrier with frequency f_{het} that is phase-modulated in proportion to the displacement of the object. The backscattered light received by the microscope objective is directed to a photomultiplier (PM) with the use of beam splitter HM2. A pinhole (D) is used to ensure that only light scattered from the object is detected. For the cupular displacement measurements, the photomultiplier signal was demodulated with a frequency tracker that has an output proportional to the frequency of the input signal. For a slightly phase-modulated carrier (Eq. 1), this means that apart from a dc component arising from the heterodyne technique the tracker output is proportional to the velocity of the object. Calibration of the laser interferometer microscope is relatively simple, since it does not depend on the amplitude of the photomultiplier signal but involves only the determination of the geometrical factor $\sin\alpha$.

The advantage of combining an interferometer with a high numerical aperture microscope is that it permits focusing of the laser beams under visual control on an object with microscopic dimensions and, in addition, increases the amount of backscattered light that can be used for interferometry. The use of the optical heterodyne technique makes the instrument insensitive to slow background displacements that result from artificial respiration of the fish with tap water and from erythrocytes passing through the capillary vessels of the macula.

Since the cupula is transparent and has almost the same refractive index as water (Jielof et al. 1952), its reflectivity in aquatic media such as the canal fluid is very small. Therefore, cupular motion was measured via a small reflecting polystyrene bead (ø 1.15 μm) that was positioned under visual control of the incident light-polarizing microscope on a selected location in the cupula. Beads were delivered from a glass micropipette (ø 4 μm) by means of a voltage step. Micropipettes were bent to fit under the objective (Hudspeth and Corey 1978). For the cupular displacement measurements, the laser power incident on the preparation was about 100 μW, with an interference spot size of about 5 μm. For a polystyrene bead located in the interference pattern, this results in a backscattered light power of about 1 nW, which suffices to operate the laser interferometer microscope and the signal processing electronics.

The amplitude and phase of the tracker output were measured by means of two lock-in amplifiers referenced in quadrature to the stimulus signal. The accuracy of the amplitude measurements is within 1 dB, and the phase measurements have an accuracy within 10°. The experimentally determined lower limit for sinusoidal displacement measurements is inversely proportional to stimulus frequency and amounts to about 80 pm (rms) for 120 Hz, using a measuring time of 300 s.

The laser power that can be used for interferometry on a biological object is limited, because cells are susceptible to radiation damage. Irradiation of a lateral line hair cell and its hair bundle for 30 min with the laser interferometer microscope did not affect the resting membrane potential recorded intracellularly, nor did it have any visual effect on the hair bundles or the hair cells. Thus, the laser light used for the displacement measurements does not appear to cause immediate damage to the hair cells.

12.4 Submicromechanical Measurements of Cupular Motion

Figure 12.6 shows the amplitude of cupular displacement as measured via a bead located at the top of the cupula in response to a 110-Hz stimulation. The figure clearly demonstrates that the amplitude of cupular displacement is linearly

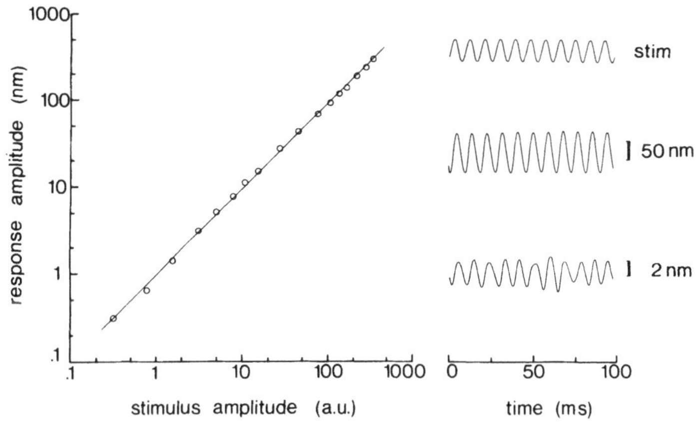

FIGURE 12.6. Amplitude (rms) of the displacement of the top of the cupula as a function of stimulus amplitude. One arbitrary unit (a.u.) corresponds to a glass sphere displacement of 20 nm. The right-hand side shows the sinusoidal stimulus of 110 Hz (upper trace) and the averages of 68 filtered traces of the responses with an amplitude (rms) of 42 nm (middle trace) and 1.4 nm (lower trace). From van Netten and Kroese (1987).

related to the stimulus amplitude from 0.3 nm up to 300 nm. The phase of the response showed little variation over this range, and there was no noticeable distortion of the wave form, implying that the system can be regarded as linear up to displacements of 300 nm. In Fig. 12.7, results are given from a typical experiment in which the amplitude and phase of cupular displacement were measured as a function of frequency of the sphere displacement. Cupular motion was measured at two different locations in the same cupula. One bead (solid circles) was placed 10 μm below the top of the cupula, which had a height of about 450 μm; the other bead (open circles) was located at the center of the base of the cupula, at only a few micrometers from the tip of a hair bundle. Since the amplitude and phase responses of the top and the base of the cupula are identical, it is concluded that the cupula is sliding over the sensory epithelium.

As shown in Fig. 12.7, the amplitude of cupular displacement increases with increasing frequency of stimulation until a maximum is reached at about 110 Hz (mean = 118 Hz, SD = 22 Hz, n=8). Q_{3dB} values related to the maximum were found in the range of 0.9 to 1.5. The slope of the amplitude curve increases gradually with frequency from 20 dB/dec at 20 Hz to about 40 dB/dec at about 80 Hz. At a frequency of about 110 Hz, the amplitude starts to decrease and approaches a constant value at higher frequencies. The phase of the response has a constant value of about 120° at low frequencies and drops to about 0° at higher frequencies. These gradual changes in the slope of the amplitude in combination with the phase drop can be accounted for by the frequency-dependent boundary layer around the cupula, as will be demonstrated in the next sections.

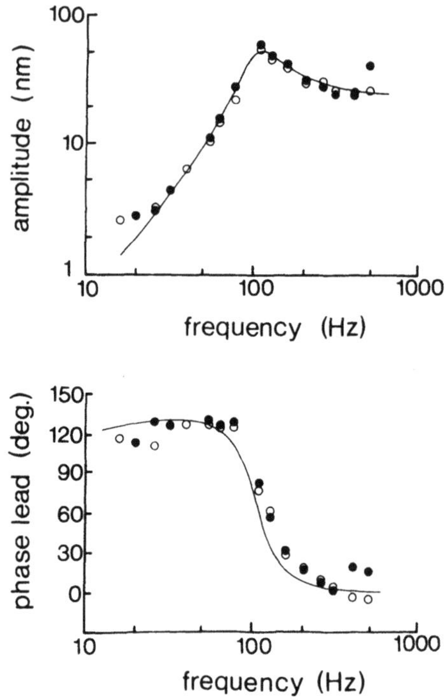

FIGURE 12.7. Amplitude (rms) and phase of cupular displacement as a function of stimulus frequency. Measurements were made at the cupular top (•) and at the cupular base (○). Phase is given with respect to glass sphere displacement. The solid curves show the model results (see text) for a sliding stiffness of 0.13 N/m and a viscosity of $5.1 \cdot 10^{-3}$ kg/(m·s). From van Netten and Kroese (1987).

12.5 Hydrodynamics of Cupular Motion

To quantify the effect of the boundary layer on the measured dynamic behavior of the cupula, a hydrodynamic description of the mechanical excitation of the cupula has been developed (van Netten 1987). In this description, the dome-shaped cupula is considered to be a rigid half-sphere which slides as a whole over a frictionless plate owing to the flow of a viscous Newtonian fluid past it. The flow is assumed to be oscillatory and to be spatially uniform, except for the region where the flow is disturbed by the presence of the cupula. The effect of the boundary layer of the canal wall can be neglected, since the canal diameter is relatively large (1 mm) (e.g., van Netten 1987). To account for the restoring forces, the cupular base is considered to be coupled elastically to the sensory epithelium. Note that in this description, the cupula is driven as well as damped by canal fluid forces.

For the lateral line, it has been shown (van Netten 1987) that under the experimental conditions used, the fluid flow is essentially linear and the canal fluid is incompressible. The incompressibility of the fluid implies that fluid motion around the cupula associated with sound can be neglected. Harris and van Bergeijk (1962) arrived at the same conclusion with respect to the lateral line by considering the response of neuromasts to a near field stimulus source (e.g., Sand 1981).

To calculate the response of the cupula to the oscillatory fluid displacements, all forces acting on the cupula are equated to cupular mass times acceleration:

$$M\frac{d^2X}{dt^2} = -SX + M^*\frac{d^2D}{dt^2} + F_{\text{fluid}} \tag{2}$$

Here X is the displacement of the cupula, t is time, M is the mass of the cupula, M^* is the mass equivalent to the mass of fluid with a volume of the cupula, S is the sliding stiffness, D is the fluid displacement, and F_{fluid} is the fluid force acting on the cupula. The first term on the right-hand side of Eq. 2 is the restoring force associated with the elastic coupling of the cupula to the epithelium. The second term accounts for the buoyancy force exerted on the cupula by the accelerating fluid. F_{fluid} includes all remaining fluid forces acting on the cupula and incorporates the forces of the frequency-dependent boundary layer. To calculate F_{fluid} an expression derived by Stokes has been used (Stokes 1851, Landau and Lifshitz 1959, Freeman and Weiss 1986). This expression is based on an exact solution of the Navier-Stokes equation for a (half-) sphere oscillating with a small amplitude in an incompressible fluid. The result is (using complex notation):

$$F_{\text{fluid}} = 6\,\pi a \mu V_{\text{rel}}\,[1 + (1+i)\,(\text{Re}/2)^{1/2} + i(\text{Re}/9)] \tag{3}$$

where a is the radius of the sphere, μ is the viscosity of the fluid, and V_{rel} is the velocity of the sphere relative to the fluid. Re is described by the following relation:

$$\text{Re} = a^2\omega\rho/\mu \tag{4}$$

where ρ is the density of the fluid and ω is the angular frequency. Re is the (ac linear) Reynolds number and gives a measure of the ratio of (linear) inertial to viscous fluid forces acting on an oscillating sphere (Batchelor 1967). Equation 3 shows clearly the transition of viscous (real) to inertial (imaginary) fluid impedance with increasing Reynolds number. By substituting Eq. 3 into Eq. 2 with $X = X_0\exp(i\omega t)$, $D = D_0\exp(i\omega t)$ and matching the density of the aqueous cupula to that of the fluid, X_0 can be solved in the frequency domain to obtain the frequency response of the cupula (van Netten 1987):

$$X_0 = \frac{i[f/f_t] + 1/2\,\sqrt{2}\,(i-1)[f/f_t]^{3/2} - 1/3\,[f/f_t]^2}{P_c + i[f/f_t] + 1/2\,\sqrt{2}\,(i-1)[f/f_t]^{3/2} - 1/3\,[f/f_t]^2}\,D_0 \tag{5a}$$

with:

$$f_t = \mu/(2\pi a^2\rho) \tag{5b}$$

$$f/f_t = f(2\pi a^2\rho)/\mu = \text{Re} \tag{5c}$$

$$P_c = Sa\rho/(6\pi\mu^2) \tag{5d}$$

It is evident from Eqs. 5 that the frequency response depends on only two independent parameters, f_t and P_c, which are determined by the four physical parameters of the model: S (sliding stiffness), ρ (fluid density), a (cupular radius), and μ (fluid viscosity). The parameter f_t appears in Eq. 5a exclusively in the ratio f/f_t. This dimensionless ratio is identical to the Reynolds number for periodic fluid flow (see Eqs. 4 and 5c). Thus, f_t indicates the transition frequency at which the inertial fluid force component starts to dominate the viscous component. This means that at frequencies below f_t the cupula detects the velocity, whereas at frequencies exceeding f_t it detects the acceleration of the canal fluid. With regard to the dynamic behavior of the cupula, f_t serves as a scale factor on the frequency axis.

P_c is the only parameter besides the scale factors f_t and D_0, implying that there exists only one family of frequency responses of which the forms are completely determined by the dimensionless number P_c. The numerical value of P_c determines whether a cupula detects displacement, velocity, or acceleration of the canal fluid in a certain frequency range. For $P_c \ll 1$, the response described by Eqs. 5 consists of a low-frequency part where cupular displacement is proportional to fluid velocity and a high-frequency part where the cupula detects the displacement of the canal fluid. No pronounced maximum is present in this case. For $P_c \gg 1$, Eq. 5 describes a response that has a velocity detection character at frequencies below f_t (Eq. 5b). For these values of P_c a pronounced maximum is present at a frequency f_m given by van Netten (1987):

$$f_m = f_t\,[3P_c]^{1/2}\ (P_c \gg 1) \tag{6}$$

At frequencies between f_t and f_m, the cupula detects the acceleration, whereas at frequencies exceeding f_m, it detects canal fluid displacement. The Q_{3dB} value of the maximum is also related to P_c (van Netten 1987), in agreement with its function as a form parameter:

$$Q_{3dB} = \tfrac{2}{3}[P_c/3]^{1/4}\ (P_c \gg 1) \tag{7}$$

12.6 Comparison of Theory and Experimental Data

To fit the calculated frequency response (Eq. 5) to the measured data, the stiffness and the viscosity were varied while the cupular radius was fixed to the average observed cupular base of 0.3 mm and the fluid density was set to that of water. The result for a sliding stiffness of 0.13 N/m and a viscosity of 0.0051 kg/(m·s) is added to Fig. 12.7 and fits the measured data very well. The scale-factor D_0, which is the amplitude of the canal fluid displacements, was 24 nm. These physical parameters result in $f_t = 9$ Hz and $P_c = 80$. The f_t of 9 Hz means that for the low-frequency part of the measured data, viscous as well as inertial fluid forces contribute to cupular excitation. At frequencies far below 9 Hz, the cupular displacement is completely determined by the viscous boundary layer around it and is therefore proportional to canal fluid velocity. At frequencies between f_t and f_m (Eq. 5b), the inertial component will dominate, meaning that at these frequencies the cupular displacement is proportional to canal fluid acceleration.

Equation 5 expresses the cupular response to water motion in the canal. It therefore does not include the first transmission step shown in Fig. 12.1. The effect of this step, however, can be neglected for the large diameter (1 mm) of the supraorbital canal of the ruff with respect to the frequency range used (van Netten 1987; see also Denton and Gray 1983). For neuromasts located in canals with smaller diameters and at lower frequencies, however, the first transmission step will contribute to the frequency response.

12.7 Dependence of Cupular Frequency Response on Physical Parameters

The calculated frequency response (Eq. 5) can also be used to examine the significance of the physical parameters in lateral line transduction. In Fig. 12.8, the dependence of the response on cupular radius is shown. Radii are expressed with respect to the radius (0.3 mm) of the supraorbital neuromasts of the ruff, which is fairly large in comparison with the size of neuromasts in other species. The shift of the maximum toward lower frequencies for increasing cupular radius is a consequence of the increasing vibrating mass. The phase responses at the low-frequency part ($f < f_m$), in combination with a slope inclining to 20 dB/dec for small cupular radii, indicate that a small cupula functions as a fluid velocity detector (slope of amplitude 20 dB/dec; phase lead 90°) rather than an acceleration detector (slope of amplitude 40 dB/dec; phase lead 180°) (e.g., Kroese et al. 1978, Kroese and Schellart 1987). It is also clear that the shape of the maximum, i.e. the Q_{3dB} value, changes with cupular radius, owing to the change in viscous drag (see also Eq. 7).

The dependence of cupular response on the sliding stiffness is shown in Fig. 12.9. The shift of the amplitude maximum to higher frequencies, together with a sharpening of the maximum with increasing stiffness, is evident.

The plateau reached at frequencies exceeding the resonance frequency results from the fact that at these frequencies, viscous as well as stiffness forces acting on the cupula are completely overruled by the inertial forces, with the result that the cupula is mechanically indistinguishable from the fluid and therefore follows exactly the fluid displacement.

12.8 Hair Bundle Stiffness and Coupling to the Cupula

The laser interferometric measurements of cupular motion as a function of frequency demonstrate clearly that the submicrometer motion of the top and the base of the cupula are identical. This means that there is no internal bending of the cupula itself during stimulation, but only a sliding motion over the sensory epithelium (e.g., Dijkgraaf 1963). Therefore, the stiffness forces involved in the dynamic behavior can originate only from regions near the base of the cupula. If the mechanical connection between the tips of the hair bundles and the cupular base is very tight, and if there is no other significant elastic coupling to the epithelium, the sliding stiffness of the cupula is completely determined by the pivoting stiffness of the individual hair bundles. Since the supraorbital macula of

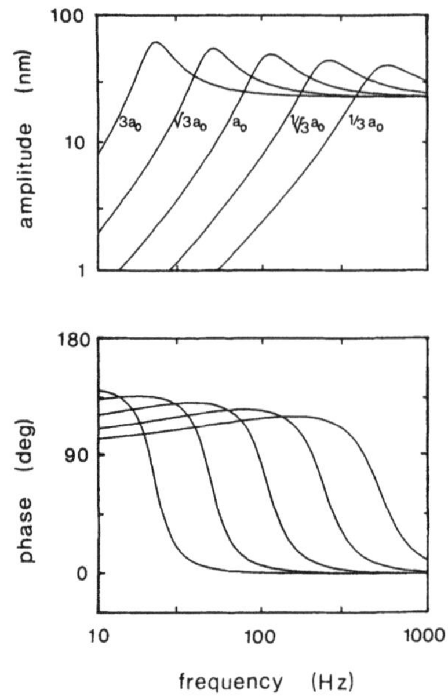

FIGURE 12.8. Dependence of the calculated frequency response upon cupular radius. Radius is varied around a_0 = 0.3 mm, the radius of the supraorbital canal neuromast of the ruff. From van Netten (1987).

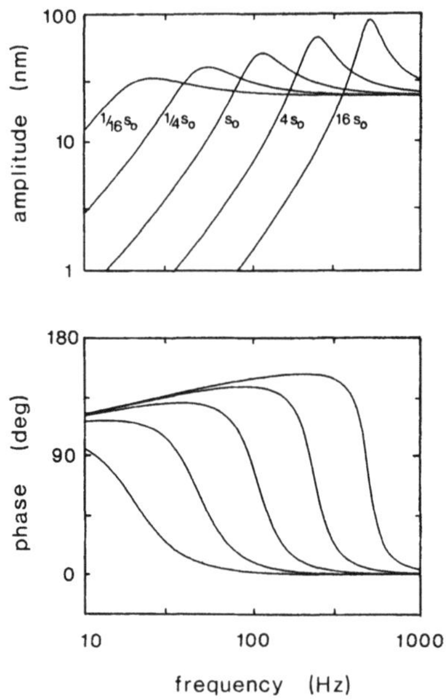

FIGURE 12.9. Dependence of the calculated frequency response upon cupular sliding stiffness. Stiffness is varied around s_0 = 0.13 N/m, the value determined for the supraorbital canal neuromast of the ruff. From van Netten (1987).

the ruff contains about 1,000 hair cells, each having a hair bundle with a length of approximately 15 μm (Kroese and van Netten 1987), the value of 0.13 N/m determined for the whole cupula would result in a pivotal stiffness of $2.9 \cdot 10^{-14}$ Nm/rad per lateral line hair bundle. This value is close to $2.2 \cdot 10^{-14}$ Nm/rad, the mean of values measured for cochlear hair cells in the turtle (Crawford and Fettiplace 1985), $2.2 \cdot 10^{-14}$ to $2.8 \cdot 10^{-14}$ Nm/rad, the range found for saccular hair cells in the frog (Howard and Ashmore 1986, Howard and Hudspeth 1987), and $1.5 \cdot 10^{-14}$ Nm/rad, the stiffness of hair bundles in the apical turn of cochlea in the guinea pig (Strelioff and Flock 1984). This comparison strongly suggests that the cupula is coupled mechanically to the sensory epithelium only by the hair bundles. There is evidence from electron microscopic studies for a network of microfibrils in the cupula of the lateral line concentrated around and in between the stereocilia, which could possibly serve for this coupling (Flock 1967, 1971). Another indication of a tight coupling of the base of the cupula to the tips of the hair bundles comes from experiments (van Netten and Kroese, unpublished) in which the cupula was given (large, 1 μm) dc displacements while observing the hair bundle tip displacements with the ILPM. Tip displacements (kinocilia) were

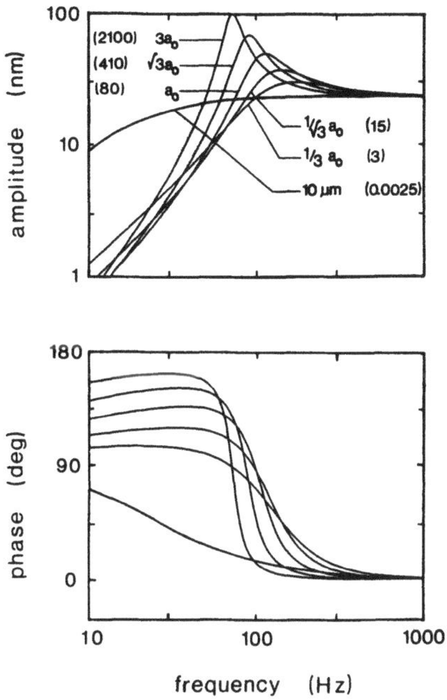

FIGURE 12.10. Calculated cupular frequency response as a function of cupular radius and sliding stiffness, which was taken proportional to the square of the radius. The hair cell density is therefore assumed to be the same for all curves. The radius is varied around a_0 = 0.3 mm, the radius of the supraorbital canal neuromast of the ruff. The curve marked with 10 μm is the limiting case of a one hair cell neuromast (see Relevance of Cupular Size). Figures indicated between brackets are the corresponding P_c numbers of the frequency responses. After van Netten (1987).

found to follow the motion of the cupular base regions at and above the level of the hair bundle tips.

12.9 Relevance of Cupular Size

The size and shape of neuromasts and canals vary considerably among bony fishes as summarized by Coombs et al. (1987; see also Coombs and Janssen Chapter 15). Also the number of hair cells in different neuromasts varies (e.g., Münz Chapter 14). If the sliding stiffness of the cupula can be attributed completely to the stiffness of the sensory hair bundles, then the sliding stiffness is proportional to the number of hair cells. As a first approximation it is reasonable to assume that the hair cell density (number of hair cells per surface area of the macula) is constant for different neuromasts. This would mean that the sliding stiffness is approximately proportional to the square of the radius of the organ. This assumption was made in Fig. 12.10, where the cupular frequency response is given for various radii. The main difference with Fig. 12.8, where the radius is also varied but stiffness is held constant, is that the maxima are confined to a limited region around 100 Hz. It is also obvious from the figure that the sensitivity increases for increasing cupular radius at these frequencies. The transition from velocity detection to acceleration detection for increasing cupular radius is even more pronounced than in Fig. 12.8. For comparison, also the limiting case of a single hair cell underlying a cupula with a diameter of 10 μm is shown in Fig.

12.10. The single hair cell cupula seems to have the advantage of a higher sensitivity at low frequencies. However, it can be calculated that the rms Brownian motion associated with such a cupula would be about 30 nm. A cupula like that of the ruff has a rms Brownian motion level on the order of 1 nm, and therefore the response to canal fluid motion has a much better signal-to-noise ratio.

12.10 Concluding Remarks

The hydrodynamic description of the excitation of the cupula shows that the combination of cupular sliding stiffness, cupular diameter, canal fluid viscosity and density, characterized mathematically by the P_c number, determines the resulting detection behavior. Small P_c numbers ($P_c \ll 1$) are associated with cupulae that detect fluid velocity at low frequencies and fluid displacement at high frequencies. For cupulae with $P_c > 1$, there is a frequency range below the transition frequency f_t, where the cupula is driven by viscous fluid forces (velocity detection), and a frequency range extending from f_t to f_m (Fig. 12.10), where the cupula detects the acceleration of the canal fluid. At higher frequencies ($f > f_m$), these cupulae also are displacement detectors. The investigated supraorbital neuromasts of the ruff have relatively large dimensions and large numbers of hair cells (i.e., large sliding stiffness) and therefore belong to the last ($P_c \gg 1$) group. It has therefore to be expected that lateral line neuromasts with smaller dimensions and smaller numbers of hair cells such as for instance superficial neuromasts will have smaller P_c numbers and a correspondingly stronger velocity detection behavior.

References

Batchelor GK (1967) An Introduction to Fluid Dynamics. Cambridge, U.K.: Cambridge University Press.

Coombs S, Janssen J, Webb JF (1988) Diversity of lateral line systems: Evolutionary and functional considerations. In: Atema J, Fay RR, Popper AN, Tavolga WN (eds) Sensory Biology of Aquatic Animals. New York: Springer-Verlag, pp. 553–593.

Crawford AC, Fettiplace R (1985) The mechanical properties of ciliary bundles of turtle cochlear hair cells. J Physiol 364:359–379.

Denton EJ, Gray JAB (1983) Mechanical factors in the excitation of clupeid lateral lines. Proc R Soc Lond B 218:1–26.

Dijkgraaf S (1963) The functioning and significance of the lateral-line organs. Biol Rev 38:51–105.

Flock Å (1967) Ultrastructure and function in the lateral line organs. In: Cahn PH (ed) Lateral Line Detectors. Bloomington: Indiana University Press, pp. 163–197.

Flock Å (1971) Sensory transduction in hair cells. In: Loewenstein WR (ed) Handbook of Sensory Physiology, Vol. I: Principles of Receptor Physiology. New York: Springer-Verlag, pp. 396–441.

Freeman DM, Weiss TF (1986) On the role of fluid inertia and viscosity in stereociliary tuft motion: Analysis of isolated bodies of regular geometry. In: Allen JB, Hall JL,

Hubbard A, Neely ST, Tubis A (eds) Peripheral Auditory Mechanisms. Berlin: Springer-Verlag, pp. 147–154.
Görner P (1963) Untersuchungen zur Morphologie und Electrophysiologie des Seitenlinienorgans vom Krallenfrosch (*Xenopus laevis* Daudin). Z Vergl Physiol 47:316–338.
Harris, GG, Bergeijk WA van (1962) Evidence that the lateral line organ responds to near field displacements of sound sources in water. J Acoust Soc Am 34:1831–1841.
Harris GG, Milne DC (1966) Input-output characteristics of the lateral-line sense organs of *Xenopus laevis*. J Acoust Soc Am 40:32–42.
Howard J, Ashmore JH (1986) Stiffness of sensory hair bundles in the sacculus of the frog. Hear Res 23:93–104.
Howard J, Hudspeth AJ (1987) Mechanical relaxation of the hair bundle mediates adaptation in mechanoelectrical transduction by the bullfrog's saccular hair cell. Proc Natl Acad Sci USA 84:3064–3068.
Hudspeth AJ (1983) Mechanoelectrical transduction by hair cells in the acousticolateralis sensory system. Annu Rev Neurosci 6:187–215.
Hudspeth AJ, Corey DP (1978) Controlled bending of high-resistance glass microelectrodes. Am J Physiol 234:C56–C57.
Jakubowski M (1963) Cutaneous sense organs of fishes. I. The lateral-line organs in the stone-perch (*Acerina cernua* L.). Acta Biol Cracoviensia Zool 6:59–82.
Jielof R, Spoor A, De Vries HL (1952) The microphonic activity of the lateral line. J Physiol 116:137–157.
Kalmijn AJ (1987) Hydrodynamic and acoustic field detection. In: Atema J, Fay RR, Popper AN, Tavolga WN (eds) Sensory Biology of Aquatic Animals. New York: Springer-Verlag, pp. 83–130.
Kroese ABA, Netten SM van (1987) The application of incident light polarization microscopy for the visualization of vertebrate sensory hair cells in vivo. J Microsc 145:309–317.
Kroese ABA, Schellart NAM (1987) Evidence for velocity and acceleration sensitive units in the trunk lateral line of the trout. J Physiol 394:13P.
Kroese ABA, Zalm JM van der, Bercken J van den (1978) Frequency response of the lateral line organ of *Xenopus laevis*. Pflügers Arch 375:167–175.
Kroese ABA, Zalm JM van der, Bercken J van den (1980) Extracellular receptor potentials from the lateral line of *Xenopus laevis*. J Exp Biol 86:63–77.
Kuiper JW (1956) The microphonic effect of the lateral line organ. Thesis, University of Groningen, The Netherlands. 159 pp.
Landau LD, Lifshitz EM (1959) Fluid Mechanics. London: Pergamon Press.
Liff HJ, Shamres S (1972) Structure and motion of cupulae of lateral line organs in *Necturus maculosus*. III. A technique for measuring the motion of free-standing lateral line cupulae. Q Progr Rep Res Lab Electronics MIT 104:332–336.
Netten SM van (1987) Laser interferometric study of the mechanosensitivity of the fish lateral line. Thesis, University of Groningen, The Netherlands.
Netten SM van (1988) Laser interferometer microscope for the measurement of nanometer vibrational displacements of a light-scattering microscopic object. J Acoust Soc Am 83:1667–1674.
Netten SM van, Kroese ABA (1986) Visualization of sensory hair cells in an in vivo preparation. In: Allen JB, Hall JL, Hubbard A, Neely ST, Tubis A (eds) Peripheral Auditory Mechanisms. Berlin: Springer-Verlag, pp. 377–384.
Netten SM van, Kroese ABA (1987) Laser interferometric measurements on the dynamic behaviour of the cupula in the fish lateral line. Hear Res 29:55–61.

Oman CM (1972) Structure and motion of cupulae of lateral line organs in *Necturus maculosus*. IV. Preliminary model for the dynamic response of the freestanding lateral-line cupula based on measurements of cupula stiffness. Q Progr Rep Res Lab Electronics MIT 104:336–342.

Oswald RL (1978) Injection anaesthesia for experimental studies in fish. Comp. Biochem Physiol 60C:19–26.

Pumphrey R (1950) Hearing. Symp Soc Exp Biol 4:3–18.

Russell IJ (1976) Amphibian lateral line receptors. In: Llinas R, Precht W (eds) Frog Neurobiology. Heidelberg: Springer-Verlag, pp. 513–550.

Sand O (1981) The lateral line and sound reception. In: Tavolga WN, Popper A, Fay RR (eds) Hearing and sound communication in fishes. New York: Springer-Verlag, pp. 459–480.

Schwartz E (1974) Lateral line mechano-receptors in fishes and amphibians. In: Fessard A (ed) Handbook of Sensory Physiology, Vol. III/3: Electroreceptors and Other Specialized Receptors in Lower Vertebrates. New York: Springer-Verlag, pp. 257–278.

Stokes GG (1851) On the effect of the internal friction of fluids on the motion of pendulums. Trans Camb Phil Soc 9:6–106.

Strelioff D, Flock Å (1984) Stiffness of sensory-cell hair bundles in the isolated guinea pig cochlea. Hear Res 15:19–28.

Strelioff D, Honrubia V (1978) Neural transduction in *Xenopus laevis* lateral line system. J Neurophys 41:432–444.

13 Sensory Transduction in Lateral Line Hair Cells

ALFONS B.A. KROESE and SIETSE M. VAN NETTEN

13.1 Introduction

Sensory hair cells are specialized epithelial cells, which function as mechano-electrical transducers in the acousticolateralis organs of vertebrates. Since early in the 1950s (Jielof et al. 1952), the lateral line has been extensively used as a model for studying the fundamental processes of hair cell physiology (for reviews see Dijkgraaf 1963; Flock 1971; Russell 1976; Sand 1984). In fact, most of the functional properties of hair cells, such as their directional sensitivity (Flock and Wersäll 1962; Harris et al. 1970) and their nonlinear input output function (Flock 1965), were discovered as a result of lateral line studies.

Within the past decade, however, on the subcellular level many investigations into the mechanism of mechanoelectrical transduction in hair cells have been performed on hair cells in other organs (see Hudspeth 1983, 1985; Roberts et al. 1988 for extensive reviews). Therefore, as an introduction, this chapter will first deal briefly with the transduction in sensory hair cells in general.

In the second part, current knowledge on the transformation in the lateral line organ from stimulus to electrical hair cell response will be summarized, with emphasis on recent data obtained by the authors on the mechanical and electrical properties of hair cells in the fish canal lateral line. Further, the contribution of the neuromast micromechanics and the hair cells to the frequency characteristic of the canal lateral line will be discussed.

The main conclusion of the present chapter is that the frequency-selective properties of the lateral line are predominantly determined by hydrodynamic cupular driving forces and mechanical properties of the neuromast, not by individual hair cell tuning.

13.2 Sensory Transduction in Hair Cells

13.2.1. Hair Bundle Motion

The most striking morphological specialization of the hair cell is the hair bundle, which consists of a single kinocilium and a number of stereocilia arranged in an hexagonal array on the apical surface of the cell. Hair cells are excited by displacements of their stereocilia (Hudspeth and Jacobs 1979). The stereociliar core contains a large number of actin filaments (Flock and Cheung 1977), which make a stereocilium behave like a rigid rod (de Rosier et al. 1980). Direct microscopic observation (Flock and Orman 1983) and measurements of the stiffness of hair bundles by use of the force-fiber technique (Strelioff and Flock 1984; Crawford and Fettiplace 1985; Howard and Ashmore 1986; Howard and Hudspeth 1987; Ohmori 1987) have shown that a displacement of the tip of the hair bundle causes a pivoting motion of the stereocilia.

13.2.2 Mechanoelectrical Transduction Channels

There is strong evidence now that mechanically sensitive ionic channels in the hair cell membrane are involved in the process of mechanoelectrical transduction (for reviews see Hudspeth 1985; Roberts et al. 1988). Opening of the transduction channels appears to be induced by movement of the hair bundle into the direction of the kinocilium and causes an increase in membrane conductance (c.f. Davis 1965); for movement in the opposite direction, the reverse holds (Hudspeth and Corey 1977).

The linearity of the current-voltage relation of the transduction channels (Corey and Hudspeth 1979; Ohmori 1985, 1987; Crawford and Fettiplace 1981) demonstrates that the average opening and closing of the transduction channels are controlled by the hair bundle position and are not dependent on the membrane potential. From measurement of the transduction current in voltage-clamped hair cells (Corey and Hudspeth 1979) and membrane-noise analysis (Holton and Hudspeth 1986), it has been estimated that there are about 50 to 300 transduction channels per hair bundle.

The type of ion carrying the transduction current is unknown, although the reversal potentials measured point to K^+ ions as the principal charge carrier (Corey and Hudspeth 1979; Crawford and Fettiplace 1981). The finding that the transduction channels are not very selective for any particular monovalent cation species (Corey and Hudspeth 1979; Ohmori 1985) also indicates that K^+ is the most likely ion to carry the transduction current, because of the (in general) high K^+ concentration at the apical surface of hair cells. On the other hand, it has been demonstrated that Ca^{2+} ions are necessary at the apical surface of the cell for the transduction mechanism to function (Sand 1975; Hudspeth and Corey 1977; Jørgensen 1984; Ohmori 1985) and that the calcium concentration affects properties of the transduction process (e.g., Eatock et al. 1987). Further, the results of experiments on the lateral line (Jørgensen 1984) and on vestibular hair cells (Ohmori 1985) indicate that at least some Ca^{2+} might enter the cell through the transduction channels.

13.2.3 Mechanoelectrical Transduction

Whether the coupling between stereociliar motion and the transduction channels is accomplished through structural connections or directly via a molecular force cannot be decided (Hudspeth 1985; Holton and Hudspeth 1986; Howard and Hudspeth 1987). The only available information on the location of the transduction channels in the apical part of the cell, obtained by use of an extracellular current probe, suggests that the transduction channels may be on the tips of the stereocilia (Hudspeth 1982). The ototoxic aminoglycosides selectively block the transduction channels in sensory hair cells (Hudspeth and Kroese 1983; Ohmori 1985, Kroese et al. 1989), but their affinity to the channels is too low to enable localization experiments.

An interesting notion, based on indirect evidence, is that the small fibrous filaments between the stereocilia (Pickles et al. 1984; Neugebauer and Thurm 1987) transfer shear between adjacent stereocilia directly to the transduction channels (Holton and Hudspeth 1986; Howard and Hudspeth 1987). Involvement of these filaments in transduction could explain the hair cell's highly directional response properties (Flock 1971; Shotwell et al. 1981).

13.2.4 Receptor Potentials

Accompanying the opening of the transduction channels during hair bundle motion in the direction of the kinocilium are an increase in hair cell membrane conductance and an inward current flow. The small potential change induced by this flow causes the activation of several (at least three) types of voltage- and ion-dependent ionic channels in the basal hair cell membrane (Lewis and Hudspeth 1983; Art and Fettiplace 1987; Hudspeth and Lewis 1988). The electrical transduction signal is amplified by the activation of these channels and is then called the receptor potential. Such receptor potentials have been recorded in hair cells of many different organs, including the lateral line.

13.2.5 Afferent and Efferent Innervation

At the basal side of the hair cell, the receptor potential causes a modulation of the synaptic activity by changing the transmitter release and thereby the postsynaptic electrical activity of the afferent nerve endings (Furukawa 1986; Flock et al. 1973b; Sand et al. 1975). The nature of the afferent transmitter is not known for sure, but there is some evidence in support of L-glutamate or a related compound to be the afferent transmitter substance in the lateral line organs (e.g., Bobbin et al. 1985).

Hair cells in most organs are also innervated by efferent nerve fibers, activation of which causes a hyperpolarization of the hair cell membrane and a subsequent reduction in the sensitivity of the cell for mechanical stimulation (Flock and Russell 1976; Art et al. 1985). In the cochlea of the turtle *Pseudomys scripta elegans*, the efferents acting on the electrically tuned hair cells are able to modulate sensitivity as well as frequency selectivity of the cells (Art et al. 1985). In the lateral line, the efferent synapse has properties characteristic for cholinergic synapses (Russell 1971), and the efferent fibers are active just before and during active movement of the animal (see Roberts and Meredith Chapter 22).

13.2.6 Frequency Selectivity of Hair Cells

In all organs of the acousticolateralis system, some frequency filtering specific for the organ takes place in accessory structures that drive the hair bundle (Pumphrey 1950). It has recently become clear that the hair cells themselves may also possess an intrinsic tuning mechanism.

One such mechanism of hair cell frequency selectivity is that found in those auditory organs of lower vertebrates in which the hair bundles are freestanding and directly driven by fluid motion. In the auditory papilla of the alligator lizard *Gerrhonotus multicarinatus*, for instance, tuning is thought to be associated with the mechanics of the coupling between the overlying fluid and the stereocilia (Holton and Weiss 1983). In these organs the hair bundle length varies systematically from cell to cell, and ciliary bundles of different length are tuned to different characteristic frequencies (Weiss and Leong 1985). The activity of these cells is thus dominated by the physical properties of the hair bundles.

Another tuning mechanism found in lower vertebrate hair cells is electrical tuning—that is, filtering in the hair cell that takes place subsequent to the transducer current as a consequence of the electrical properties of the cell membrane (for review see Fettiplace 1987). The hair cell membrane is tuned by an electrical resonance (Crawford and Fettiplace 1981) which arises from an interaction between a voltage dependent Ca^{2+} conductance and a Ca^{2+}-activated K^+ conductance (Lewis and Hudspeth 1983; Art and Fettiplace 1987; Hudspeth and Lewis 1988). In the turtle cochlea, the kinetics of the membrane conductance of the K^+ channel varies with the location of the hair cell on the papilla to produce a distribution of resonant frequencies covering the auditory range (Art and Fettiplace 1987). Tuning in some organs of lower vertebrates is thus at least partly determined by the electrical properties of the hair cell membrane. Interestingly, a similar mechanism has been proposed to explain the tuning of electroreceptors in weakly electric fish (Hopkins 1976), organs that are closely related to the lateral line organ.

13.3 Sensory Transduction in Lateral Line Hair Cells

13.3.1 Pathway from Stimulus to Lateral Line Nerve

The lateral line neuromasts transform information on water motion into afferent nerve activity (for reviews see Flock 1971, Russell 1976). The physical properties of the relevant water motion around the animal have recently been experimentally and theoretically studied by Kalmijn (1987, Chapter 9) and Hassan (Chapter 10). For the canal lateral line, the next transmission step is the filtering caused by the canal fluid mechanics and the geometry of the canal (see Denton and Gray 1983, Chapter 11).

Cupular motion, which for the canal organ has been quantified by van Netten and Kroese (Chapter 12), results in a motion of the hair bundles of the sensory cells. The receptor potential modulates the activity of the afferent nerve fibers via the chemical afferent synapse (Flock et al. 1973b; Sand et al. 1975; Russell 1976). Further information on synaptic transmission and impulse generation in the lateral line afferents can be found elsewhere (Russell 1976; Pabst 1977). Characteristics of afferent nerve activity have been studied in the lateral line of *Xenopus* (see e.g., Russell 1976; Kroese et al. 1978; Elepfandt and Wiedemer 1987) as well as of fish (e.g., Münz 1985, Chapter 14, Kroese and Schellart 1987; Coombs and Janssen, Chapter 15; Montgomery, Chapter 28).

In the present chapter, the transmission steps that occur in the canal lateral line between motion of the cupula and generation of the hair cell receptor potential are considered.

13.3.2 Lateral Line Hair Cells

In the lateral line, the sensory epithelium consists of a single class of hair cells from whose apical ends a kinocilium and a bundle of about 50 stereocilia protrude into the overlaying cupula. The dimensions of the lateral line hair cells and hair bundles vary between organs in different species (Flock 1971; Russell 1976). In the canal organ of the ruff *Acerina cernua* L., the hair cells are pear shaped and about 15 μm high. The diameter of the cells is at the basal side about 10 μm and at the apical side only about 7 μm. The height of the hair bundle is about 15 μm (Kroese and van Netten 1987). The hair cells are separated and surrounded by a large number of supporting cells. Each neuromast contains two populations of hair cells with opposite orientation, both with their most sensitive direction parallel to the length axis of the canal (Flock and Wersäll 1962) and a separate afferent innervation. In canal organs the hair cells of these two groups are intermingled, apparently in a random fashion (Flock and Wersäll 1962; Russell 1976; Kroese and van Netten 1987).

The first intracellular recordings of hair cell receptor potentials in response to mechanical stimulation were made from hair cells in the lateral line of *Necturus* (Harris et al. 1970), and similar recordings have been obtained in the lateral line by others (Flock et al. 1973b; Sand et al. 1975; Flock and Russell 1976). The amplitude of the receptor potentials measured in response to stimuli of intensities approaching the upper physiological range suggested a quite low mechanical sensitivity for the lateral line hair cells in comparison to that of hair cells in other organs (Crawford and Fettiplace 1985; Russell et al. 1986; Ohmori 1987; Hudspeth 1983).

13.3.3 Sensitivity of Lateral Line Hair Cells

We have recently made routine intracellular recordings from hair cells in the supraorbital lateral line canal (neuromast No. 4; Jakubowski 1963) of the anesthetized ruff. The setup and the procedures used have been described in detail in the previous chapter and elsewhere (Kroese and van Netten 1987; van Netten and Kroese 1987). The recordings were made with high impedance (250 to 300 MΩ) glass microelectrodes filled with 3 M potassium acetate, which were

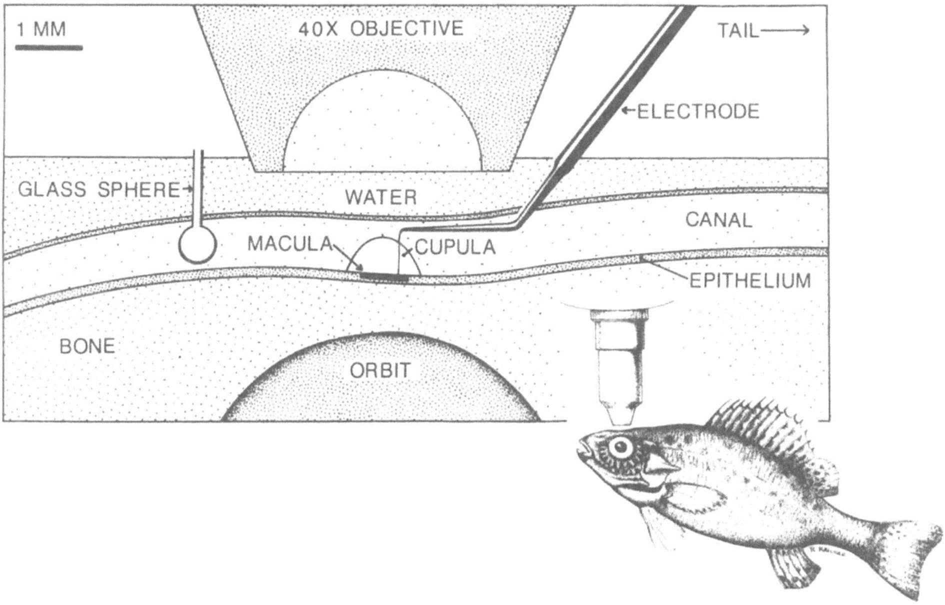

FIGURE 13.1. Schematic drawing of the lateral line preparation. The fish is placed below the water immersion objective of the microscope. Underneath the water immersion objective and just below the skin of the head runs the lateral line canal. In the canal the neuromast and the position of the glass sphere and of the electrode for intracellular recording are shown. Optical access to the macula is possible because of the transparency of the cupula.

bent twice to fit below the objective (see Fig. 13.1). Visual control by incident light polarization microscopy (ILPM), a new neuronal imaging technique (Kroese and van Netten 1987), permitted electrode penetration of the hair cells at the edge of the apical surface opposite to the kinocilium location. In this way any damage to the cuticular plate of the hair cell or to the epithelial integrity by the recording electrode could be avoided.

In these experiments, the potential recorded during the visually controlled placement of the electrode into the cupula and during the descent of the electrode toward the epithelium did not deviate from zero until the electrode touched the epithelium (see Fig. 13.2). Thus, there was no evidence for a positive potential in the cupula adjacent to the apical surface of the cells, such as the endocupular potential that has been reported in the *Xenopus* lateral line (Russell and Sellick 1976; McGlone et al. 1979).

An example of a stable, intracellularly recorded resting membrane potential of a hair cell with a duration of more than 30 min is shown in Fig. 13.2. Experiments on 28 fish yielded a resting membrane potential of -59 ± 6 mV (mean $\pm$ SD; $n = 105$). The membrane impedance of the hair cells was 110 ± 40 MΩ (mean $\pm$ SD; $n = 52$) (cf. Flock et al. 1973a), and the time constant of the membrane

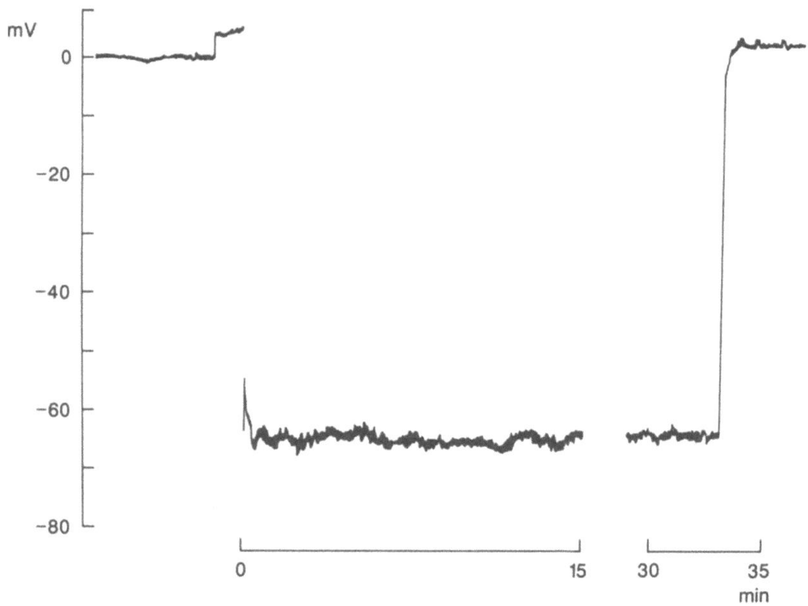

FIGURE 13.2. Recording of the potential during penetration of a lateral line hair cell. From left to right: The electrode was under visual control slowly advanced through the cupula (potential 0 mV) until it touched the apical surface of the hair cell (small positive potential). Penetration of the cell was accomplished by applying an oscillating voltage to the electrode tip by overcompensating the capacitance neutralization and signaled by a sudden change in dc potential. A stable resting membrane potential of the hair cell (−64 mV) was recorded for about 33 min and disappeared when the electrode was moved 1 μm upward.

appeared to be smaller than that of the recording system (0.7 ms), as concluded from current injections.

Intracellular recordings of receptor potentials were made from hair cells in neuromasts that were mechanically stimulated with sinusoidal water motion evoked by a small glass sphere in the canal, located about 4 mm from the neuromast (see Fig. 13.1 and van Netten and Kroese, Chapter 12). An example of receptor potentials recorded from two different hair cells is shown in Fig. 13.3. Receptor potentials were only recorded when the electrode was in an intracellular hair cell position, under visual control. Stimulation with water motions that induced cupular displacements larger than 10 to 20 nm usually resulted in abrupt termination of the intracellular recording. Presumably such terminations were caused by mechanical damage to the hair cell or by the loss of sealing of the electrode. The susceptibility of the hair cell preparation to mechanical damage under these conditions is not surprising. During mechanical stimulation the cupula is sliding over the sensory epithelium (Dijkgraaf 1963; van Netten and Kroese 1987) in a direction perpendicular to the electrode, which penetrates over a distance of about 400 to 600 μm through the cupula. Therefore, intracellular

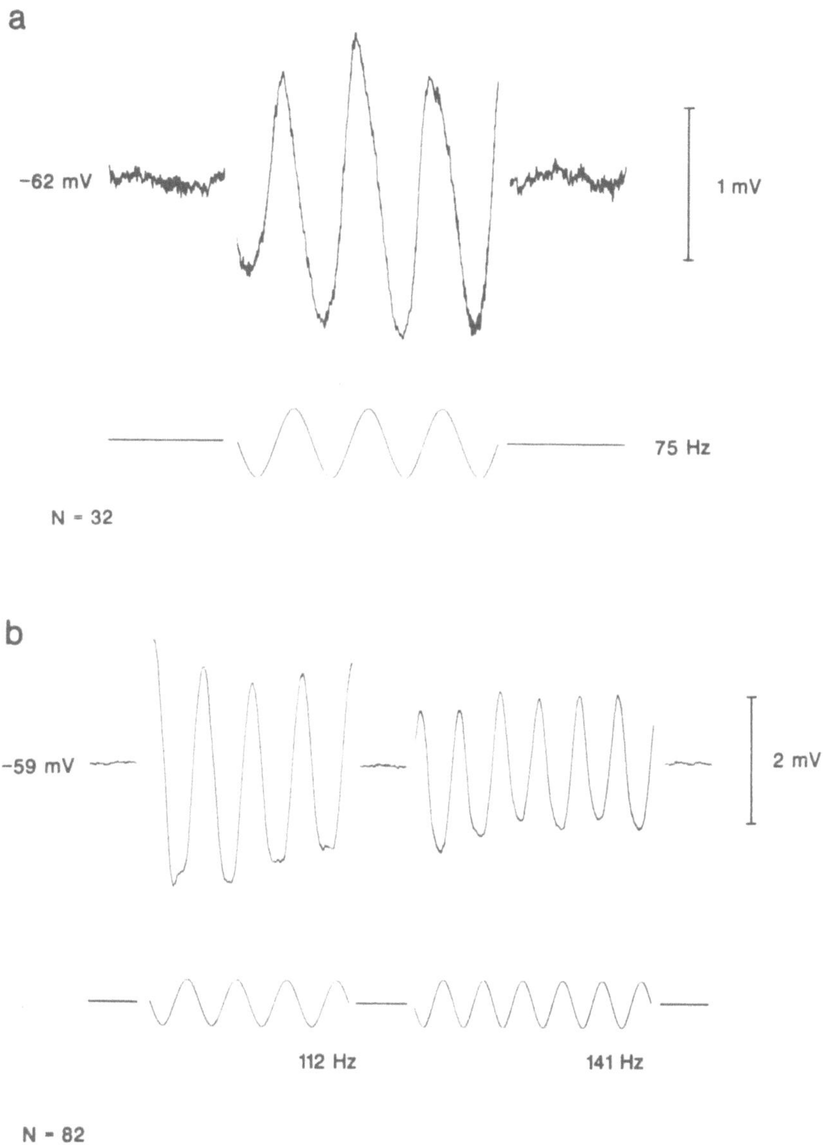

FIGURE 13.3. Average receptor potentials for small displacements of the cupula of two hair cells (a and b) in different neuromasts. Lower traces: for each example the tone burst stimulus (frequency indicated below) was delivered by the glass sphere in the canal. Upper traces: steady-state part of the response to the tone burst. N = number of averaged responses. Estimated cupular displacement was 4 nm (pp) for cell a and 10 nm (pp) for cell b (both frequencies). Resting membrane potential of the cells is indicated.

recordings of receptor potentials were only made during stimulation with small amplitude stimuli.

Since the relation between the movement of the glass sphere and the cupular motion has been quantified before by measurement of cupular motion (van Netten and Kroese, Chapter 12), the absolute sensitivity of the hair cells to cupular motion could be estimated. The amplitude of the receptor potentials of five hair cells in different neuromasts was on average 0.4 mV (± 0.2 mV) per nanometer cupular motion.

An important conclusion of these results is that the sensitivity of the lateral line hair cells is of the same order of magnitude as that reported for cochlear hair cells, measured by direct stimulation of the hair bundles (Crawford and Fettiplace 1985; Russell et al. 1986). That the intracellular recordings were made under visual control may explain the high resting membrane potentials and the high mechanical sensitivity of the lateral line hair cells compared to earlier reports (see Section 13.3.2).

The measured hair cell sensitivity is in fair agreement with the sensitivity of afferent nerve fibers and behavioral lateral line responses, although these cannot be compared directly. The most sensitive canal lateral line afferents respond to water motion in the nm range (see e.g., Münz 1985; Coombs and Janssen, Chapter 15), in which range also the thresholds for behavioral responses of fish are found (e.g., Kuiper 1956; Bleckmann 1980; Bleckmann and Bullock, Chapter 19; Coombs and Janssen, Chapter 15).

13.3.4 Input-Output Relation of Hair Cells

Because of the limitations imposed on the intracellular recording of receptor potentials described above, in most experiments the electrical hair cell response was measured by recording the receptor potentials extracellularly. Extracellular receptor potentials had been recorded before in canal lateral line organs of fish (Jielof et al. 1952; see also Sand 1984) and in the superficial lateral line of *Xenopus* (Kroese et al. 1980), but up to now they have been related only to water motion induced by a glass sphere, and not directly to cupular motion.

In the ruff, extracellular receptor potentials were recorded by means of a silver wire electrode (diameter 100 μm) located in the canal close to and at the caudal side of the neuromast. Hydrodynamical distortion of the fluid flow due to the presence of the electrode was checked by varying the position of the electrode and found to be negligible. Because of the directional sensitivity of the hair cells, their alternating opposite orientation in the macula, and their nonlinear input-output relation, the extracellular receptor potentials show a frequency doubling with respect to the stimulus (Flock 1965; Corey and Hudspeth 1983). Amplitude and phase of the extracellular receptor potentials were measured with two lock-in amplifiers referenced to the stimulus signal. It is important to note that the phase depicted in the figures is the phase of the receptor potentials of one of the two populations of hair cells in the neuromast. Simultaneously with the extracellular receptor potentials, the nanometer displacement of the cupula was measured

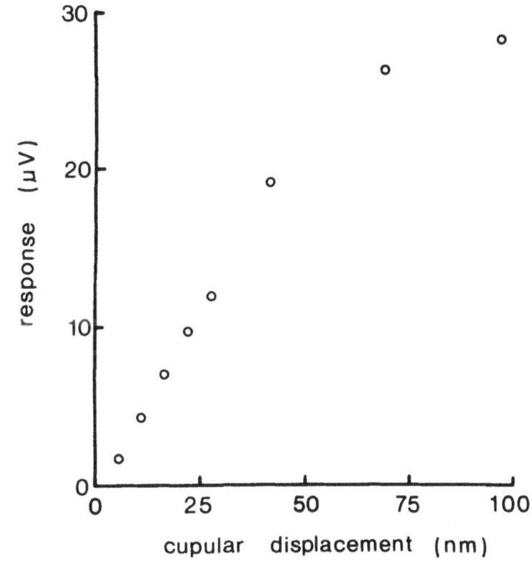

FIGURE 13.4. Relationship between cupular displacement and electrical response of a typical neuromast stimulated with sinusoidal fluid motions with a frequency of 110 Hz. Abscissa gives the displacement amplitude (rms) of the cupula. Ordinate is the amplitude (rms) of the extracellular receptor potential measured simultaneously with the cupular displacements. The phase of the response remained constant (± 4°) over the range of stimulus levels used.

with a heterodyne laser interferometer coupled to the microscope (see van Netten and Kroese Chapter 12; van Netten 1987, 1988).

Figure 13.4 shows that the amplitude of the extracellular receptor potentials increases linearly with cupular displacement up to about 50 nm and, above that value, slowly saturates. The cupular displacement of about 100 nm at which the response saturates compares well with the hair bundle displacements at which the depolarizing component of intracellularly recorded potentials in cochlear hair cells saturates (Crawford and Fettiplace 1985; Russell et al. 1986). The displacement level of saturation of the extracellular receptor potential is also in accordance with the intracellularly measured sensitivity of the lateral line hair cells (see Section 13.4.1) of 0.4 mV/nm cupular motion. For the smallest stimulus amplitudes (up to about 6 nm), the receptor potential amplitude increased with the square of cupular motion, as should be expected because of the origin of these potentials (Boston 1980).

13.3.5 No Intrinsic Frequency Selectivity in Single Lateral Line Hair Cells

In general (see e.g., Münz 1985), frequency response characteristics of lateral line organs measured on the afferent nerve fibers always show a maximum in sensitivity at frequencies between 20 Hz (superficial organs) and 200 Hz (canal organs). In the supraorbital canal neuromast of the ruff, the afferent sensitivity to water displacement shows a maximum at about 80 to 100 Hz (Wubbels 1988).

It should be pointed out here that the occurrence of a maximum in the sensitivity of a neuromast results from the way in which the frequency responses are plotted, namely with amplitude and phase with respect to water displacement (Kalmijn 1987). Experimental (Denton and Gray 1983; Münz 1985; Kroese and Schellart 1987) and theoretical (Kalmijn 1987) studies have shown that for canal organs the adequate stimulus is the acceleration and not the displacement of the water around the fish. From plotting the frequency response with respect to water acceleration it appears that the canal lateral line functions as a low-pass filter for water acceleration (Kalmijn 1987, Chapter 9), which rightly stresses the high relevance of low-frequency water motion for the lateral line organ. Measurement of cupular motion and hydrodynamic calculations (van Netten and Kroese Chapter 12, van Netten 1987) have shown, however, that the canal lateral line does not function exactly and exclusively as an acceleration detector over the whole relevant frequency range.

Measurement of cupular motion in the physiological range has demonstrated that during mechanical stimulation the cupula as a whole is sliding over the macula in the preferred direction (van Netten and Kroese Chapter 12). For this reason it should be expected that during cupular motion all hair bundles in a neuromast will receive an identical mechanical stimulus. It was also found from observation with ILPM that the hair bundles of the majority of hair cells in the macula have about the same length and the same number of stereocilia (Kroese and van Netten 1987). An exception is formed by the hair cells at the border of the macula where, probably owing to postembryonic production of hair cells (Corwin 1985), maturizing hair cells with divergent hair bundles and sizes can be distinguished (Kroese and van Netten, unpublished observation). Further, in the lateral line the hair bundles are embedded in the cupula by a matrix of fibrous material that intertwines among them (Flock 1971, Jørgensen and Flock 1973), so it seems unlikely that individual hair cells in a lateral line neuromast are tuned to distinct frequencies by differences in mechanical properties of their hair bundle or in coupling to the cupula.

To investigate the possible presence of an electrical tuning mechanism in hair cells of the lateral line, a spectral analysis of stable recordings of the resting membrane potential of 54 hair cells (mean rmp −59 mV), each with a duration of more than 10 min, was made. In the averaged spectra no evidence of any spontaneous oscillations in the resting membrane potential was found. Such oscillations are characteristic for electrically tuned hair cells (for review see Fettiplace 1987). Furthermore, current injections into the hair cells, causing depolarizing changes in resting membrane potential of up to 30 mV, failed to induce any oscillations in resting membrane potential (see Fig. 13.5). Such current injections are known to cause damped oscillations in resting membrane potential in the electrically tuned turtle cochlear (Crawford and Fettiplace 1981) and frog saccular (Ashmore 1983; Lewis and Hudspeth 1983) hair cells. Thus, these experiments

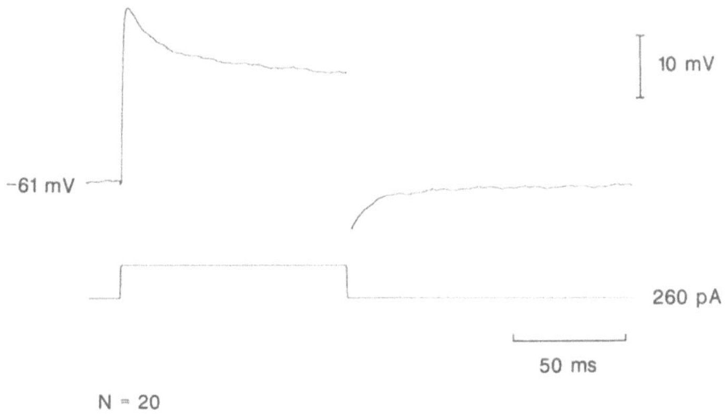

FIGURE 13.5. Change in membrane potential of a hair cell evoked by a current step (260 pA; lower trace) injected through the recording electrode. The upper trace shows the membrane potential response, after balancing out of the voltage drop across the electrode resistance with a bridge circuit. Note that there are no oscillations in membrane potentials after onset and offset of current injection.

do not provide any evidence for an electrical tuning mechanism in hair cells of the lateral line.

There is thus no evidence for tuning in individual lateral line hair cells by hair bundle properties or by electrical properties of the hair cell membrane. This conclusion is in accordance with the finding in a limited number of studies that the afferent frequency response characteristics of different fibers from a single lateral line neuromast (or functional unit) appear to be quite similar (Kroese et al. 1978; Kroese and Schellart 1987). Differences between the hair cells in different neuromasts, however, in length, diameter, and stiffness of hair bundles may, as will be shown below, have a strong influence on the frequency response of the neuromasts.

13.3.6 Frequency Characteristics of Cupular Motion and Hair Cell Response

To determine the contribution of the several transmission steps to the frequency response properties of the neuromast, cupular motion, the direct input for the hair cells, and the electrical response of the cells as a function of stimulus frequency have been measured. The extracellular receptor potentials in these experiments were evoked by cupular displacements with amplitudes between 6 and 50 nm, which is in the linear range (see Section 13.3.4). As has been argued above, it is assumed that in the lateral line individual hair cells have identical transfer properties, and extracellularly recorded receptor potentials thus depict single hair cell properties.

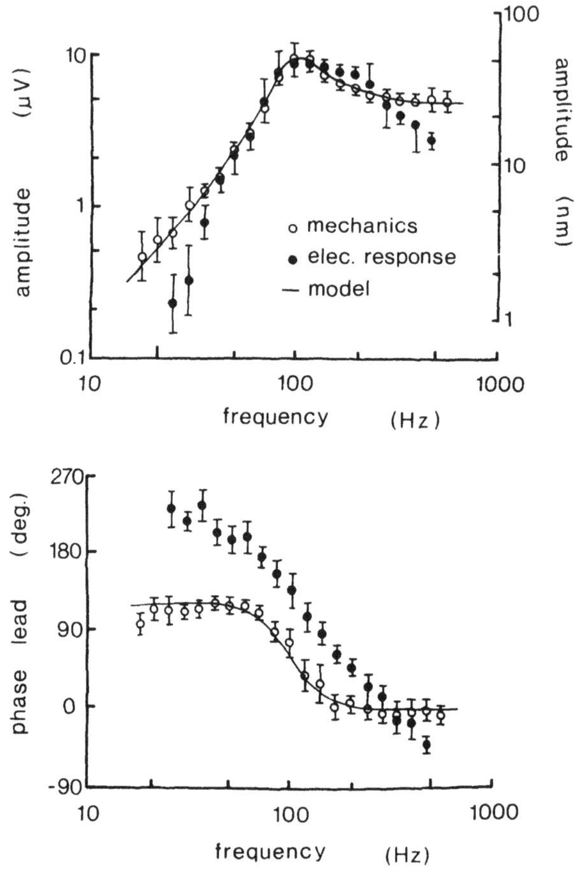

FIGURE 13.6. Amplitude (rms ±SD) and phase (± SD) of the extracellular receptor potentials (solid symbols) and cupular displacements (open symbols) as a function of the frequency of the displacement of the glass sphere. The phase of the electrical response represents the phase of receptor potentials in one of the two oppositely oriented hair cell populations in the neuromast. Results were obtained from measurements on five (receptor potential) and eight (cupular displacement) neuromasts. For each group, the amplitude curves, which had the same shape per group, were shifted along the frequency axis as to have their maxima located at the mean maximum, which was 114 ± 14 Hz for the receptor potentials and 118 ± 22 Hz for the cupular displacements. Subsequently the amplitude curves were shifted along the vertical axis to be superimposed, from which the SD indicated in the figure was calculated. Phase curves were shifted accordingly along the frequency axis. The solid line shows the result of a hydrodynamic model describing cupular displacement in terms of the relevant physical parameters (see van Netten and Kroese Chapter 12).

The frequency responses of cupular motion and of the receptor potentials presented in Fig. 13.6 have similar band-pass characteristics with a maximum amplitude at about 115 Hz. The solid line in the figure is a fit to the measured cupular frequency response predicted by hydrodynamic calculations of the cupular driving forces (van Netten and Kroese Chapter 12). From the conformity between both characteristics it can be concluded that the frequency selectivity of the lateral line with respect to water displacements resides predominantly in the mechanical tuning properties of the cupula, and not primarily in the hair cells.

In canal lateral line organs, the mechanical tuning properties of the cupula are controlled by the viscosity of the canal fluid, the cupular and canal geometry and dimensions, and the stiffness coupling of the cupula to the canal epithelium (van Netten and Kroese Chapter 12). The geometry and dimensions critical to cupular movement show large variations between lateral line organs of different species and between organs of individual fish (for review see Coombs et al. 1988). From the hydrodynamic calculations it appears that the location of the maximum and the shape of the frequency response of a neuromast are strongly influenced by cupular sliding stiffness. Since the stiffness coupling between the cupula and the canal is most probably brought about by the hair bundles (van Netten and Kroese 1987; van Netten and Kroese 1989), the present results lead to the conclusion that the mechanical properties and also the length and number of hair bundles will strongly influence the frequency selectivity of the organ. Thus, notwithstanding the lack of individual hair cell tuning reported above, the hair bundles contribute definitely to frequency selectivity of the neuromast.

13.3.7 Transmission Step from External Water to Canal Fluid

Denton and Gray (1983) have demonstrated that in the lateral line canal of the sprat (diameter 140 μm) *Sprattus sprattus* L., the transmission step from external water motion to water motion in the canal shows a high pass characteristic up to 80 Hz. However, because of the relatively large diameter of the supraorbital canal in the ruff (1 mm), the transmission in this canal is independent of frequency over the whole range of frequencies where measurements were made. From the results of measurements on capillaries by Denton and Gray (1983), it has been estimated that in the ruff this canal effect plays a role only below frequencies of about 15 Hz. In control experiments, the skin overlaying the canal was left intact, and the neuromast's frequency response was measured by recording extracellular receptor potentials in response to a stimulus in the water exterior to the fish. The frequency responses measured in this way did not differ essentially from those measured under open canal conditions with local stimulation. Thus, in the present experiments on the ruff prepared with a partly exposed canal, all measurements can be considered as having been performed under conditions where the stimulus is applied in the water surrounding the fish.

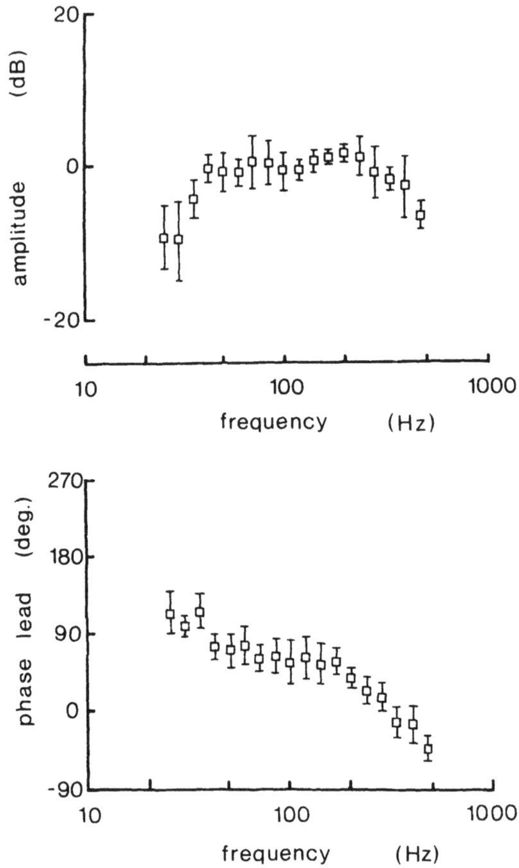

FIGURE 13.7. Transfer function (± SD) of the sensory hair cells, obtained by dividing the receptor potentials by the cupular displacements displayed in Fig. 13.6. Amplitude is given with respect to an arbitrary reference level.

13.3.8 Transfer Function of the Lateral Line Hair Cell

The contribution made by intrinsic hair cell properties to the frequency selectivity of the lateral line organ was further examined by quantification of the transfer function of the lateral line hair cells. The transfer function of the hair cells depicted in Fig. 13.7 was obtained by dividing the electrical response by the cupular displacement. The plateau in the amplitude characteristic shows that the sensitivity of the hair cells to cupular motion is constant over a broad frequency range (see Hudspeth and Corey 1977) of 30 to 300 Hz. This finding is in agreement with the conclusion made above that the hair cells are not tuned individually to specific frequencies. However, especially the phase response shows that the hair cells are not purely detectors of cupular displacement. At low frequencies the phase lead of the hair cell response of about 90° and the slope in the amplitude characteristic of about 20 dB/dec point to a differentiating filter action by the hair cells. The hair cells thus function as a high-pass filter of cupular motion. The attenuation of the amplitude of the hair cell response at frequencies above 300 Hz is attributed to the time constant of the hair cell membrane.

The dependence of the hair cell responses at low stimulus frequencies on cupular velocity rather than on cupular displacement (Fig. 13.7) is not likely the result of a viscous coupling between the hair bundles and the cupula, since the cupula is tightly coupled to the sensory hair bundles by filaments (see van Netten and Kroese Chapter 12).

A plausible explanation of the high-pass filter characteristic of the lateral line hair cell might be an adaptation of the cell comparable to that reported in hair cells of the bullfrog sacculus (Corey and Hudspeth 1983; Eatock et al. 1987). The adaptation in bullfrog hair cells involves a relatively slow shift in the displacement-response curve along the displacement axis during a static displacement of the hair bundle. The time course of this adaptation indicates that the response of the hair cells is high-pass filtered, with a corner frequency of 5 to 20 Hz (Eatock et al. 1987), which frequency is very well comparable to that found here (Fig. 13.7, about 30 Hz) for the lateral line hair cells. It has been suggested recently that this high-pass filtering action of the saccular hair cells is the result of position-dependent changes in the mechanical properties of the hair bundles (Howard and Hudspeth 1987, 1988).

The adaptation tends to hold the hair cell in the sensitive range of its displacement-response function (Hudspeth 1983) and thereby increases the operational range of the organ considerably. The dynamic range of the superficial lateral line organ of *Xenopus*, for instance, has been shown to be at least 80 dB (Elepfandt and Wiedemer 1987). The undesirable reduction of the lateral line neuromast's sensitivity to low frequencies as a result of the high-pass filtering by the hair cell is in the ruff canal neuromast compensated by the different (see Fig. 13.6) mechanical behavior of the cupula at low frequencies.

13.4 Summary and Conclusions

The results of the present experiments demonstrate that frequency selectivity in the lateral line is mainly controlled by hydrodynamics and cupular mechanics, and not by specific tuning mechanisms in the hair cells. Yet the hair cells are in another way strongly involved in frequency selectivity of the organ, since cupular mechanics depends on the hair bundle stiffness coupling of the cupula to the sensory epithelium. In this respect, the lateral line shows a remarkable parallel with the mammalian cochlea, in which the hair cells probably make an important contribution to the mechanics of the cochlear partition and thereby to frequency selectivity (e.g., Flock and Strelioff 1984). Further, the mechanical sensitivity of lateral line hair cells is comparable to that of cochlear hair cells.

On the basis of these results it could be expected that for lateral line neuromasts with differences in size (i.e., with differences in cupular dimensions and/or the number of hair cells per neuromast), in the length of the hair bundles, or in the stiffness of the bundles, the frequency responses will be different. There is, however, too striking a lack of firm data on frequency response characteristics of lateral line neuromasts with quantitatively described hair cell morphology to confirm this prediction.

Most likely, hydrodynamic conditions and cupular mechanics play a predominant role in bringing about the response of superficial neuromasts to water velocity and of the canal neuromasts to water acceleration. The degree to which the remarkable difference in the number of hair cells between superficial and canal neuromasts (e.g., Münz and Claas 1983; Münz Chapter 14; Coombs and Janssen Chapter 15) contributes to the different response characteristics of both types of neuromasts remains to be elucidated.

Acknowledgments. We are grateful to Dr. Nico A.M. Schellart for critically reading the manuscript. This work was supported by the Netherlands Organization for Scientific Research (NWO).

References

Art JJ, Fettiplace R (1987) Variation of membrane properties in hair cells isolated from the turtle cochlea. J Physiol 385:207–242.

Art JJ, Crawford AC, Fettiplace R, Fuchs PA (1985) Efferent modulation of hair cell tuning in the cochlea of the turtle. J Physiol 360:397–421.

Ashmore JF (1983) Frequency tuning in a frog vestibular organ. Nature 304:536–538.

Bleckmann H (1980) Reaction time and stimulus frequency in prey localization in the surface feeding fish *Aplocheilus lineatus*. J Comp Physiol A 140:163–172.

Bobbin RP, Bledsoe SC Jr, Winbery SL, Jenison GL (1985) Actions of putative neurotransmitters and other relevant compounds on *Xenopus laevis* lateral line. In: Drescher DG (ed) Auditory Biochemistry. Springfield, IL: Thomas, pp. 102–122.

Boston JR (1980) A model of lateral line microphonic response to high-level stimuli. J Acoust Soc Am 67:875–881.

Coombs S, Janssen J, Webb JF (1988) Diversity of lateral line systems: Evolutionary and functional considerations. In: Atema J, Fay RR, Popper AN, Tavolga WN (eds) Sensory Biology of Aquatic Animals. New York: Springer-Verlag, pp. 553–595.

Corey DP, Hudspeth AJ (1983) Analysis of the microphonic potential of the bullfrog's sacculus. J Neurosci 3:942–961.

Corey DP, Hudspeth AJ (1979) Ionic basis of the receptor potential in a vertebrate hair cell. Nature 281:675–677.

Corwin JT (1985) Perpetual production of hair cells and maturational changes in hair cell utlrastructure accompany postembryonic growth in an amphibian ear. Proc Natl Acad Sci USA 82:3911–3915.

Crawford AC, Fettiplace R (1981) An electrical tuning mechanism in turtle cochlear hair cells. J Physiol 306:79–125.

Crawford AC, Fettiplace R (1985) The mechanical properties of ciliary bundles of turtle cochlear hair cells. J Physiol 364:359–379.

Davis H (1965) A model for transducer action in the cochlea. Cold Spring Harbor Symp Quant Biol 30:181–190.
Denton EJ, Gray JAB (1983) Mechanical factors in the excitation of clupeid lateral lines. Proc R Soc Lond B 218:1–26.
Dijkgraaf S (1963) The functioning and significance of the lateral-line organs. Biol Rev 38:51–105.
Eatock RA, Corey DP, Hudspeth AJ (1987) Adaptation of mechanoelectrical transduction in hair cells of the bullfrog's sacculus. J Neurosci 7:2821–2836.
Elepfandt A, Wiedemer L (1987) Lateral-line responses to water surface waves in the clawed frog, *Xenopus laevis*. J Comp Physiol A 160:667–682.
Fettiplace R (1987) Electrical tuning of hair cells in the inner ear. Trends Neurosci 10:421–425.
Flock Å (1965) Electron microscopic and electrophysiological studies on the lateral line canal organ. Acta Otolaryngol Suppl 199:1–90.
Flock Å (1971) Sensory transduction in hair cells. In: Loewenstein WR (ed) Handbook of Sensory Physiology. New York: Springer-Verlag, pp. 396–441.
Flock Å, Cheung HC (1977) Actin filaments in sensory hairs of inner ear receptor cells. J Cell Biol 75:339–343.
Flock Å, Orman S (1983) Micromechanical properties of sensory hairs on receptor cells of the inner ear. Hear Res 11:249–260.
Flock Å, Russell IJ (1976) Inhibition by efferent nerve fibres: Action on hair cells and afferent synaptic transmission in the lateral line canal organ of the burbot *Lota lota*. J Physiol 257:45–62.
Flock Å, Strelioff D (1984) Graded and nonlinear mechanical properties of sensory hairs in the mammalian hearing organ. Nature 310:597–599.
Flock Å, Wersäll J (1962) A study of the orientation of the sensory hairs of the receptor cells in the lateral line organs of fish, with special reference to the function of the receptors. J Cell Biol 15:19–27.
Flock Å, Jørgensen JM, Russell IJ (1973a) Passive electrical properties of hair cells and supporting cells in the lateral line canal organ. Acta Otolaryngol 76:190–198.
Flock Å, Jørgensen M, Russell IJ (1973b) The physiology of individual hair cells and their synapses. In: Moller AR (ed) Basic Mechanisms of Hearing. New York: Academic Press, pp. 273–306.
Furukawa T (1986) Sound reception and synaptic transmission in goldfish hair cells. Jpn J Physiol 36:1059–1077.
Harris GG, Frishkopf LS, Flock Å (1970) Receptor potentials from hair cells of the lateral line. Science 167:76–79.
Holton T, Hudspeth AJ (1986) The transduction channel of hair cells from the bull-frog characterized by noise analysis. J Physiol 375:195–227.
Holton T, Weiss TF (1983) Frequency selectivity of hair cells and nerve fibres in the alligator lizard cochlea. J Physiol 345:241–260.
Hopkins CD (1976) Stimulus filtering and electroreception: Tuberous electroreceptors in three species of gymnotoid fish. J Comp Physiol 111:171–207.
Howard J, Ashmore JF (1986) Stiffness of sensory hair bundles in the sacculus of the frog. Hear Res 23:93–104.
Howard J, Hudspeth AJ (1987) Mechanical relaxation of the hair bundle mediates adaptation in mechanoelectrical transduction by the bullfrog's saccular hair cell. Proc Natl Acad Sci USA 84:3064–3068.
Howard J, Hudspeth AJ (1988) Compliance of the hair bundle associated with gating of

mechanoelectrical transduction channels in the bullfrog's saccular hair cell. Neuron 1:189–199.
Hudspeth AJ (1982) Extracellular current flow and the site of transduction by vertebrate hair cells. J Neurosci 2:1–10.
Hudspeth AJ (1983) Mechanoelectrical transduction by hair cells in the acousticolateralis sensory system. Ann Rev Neurosci 6:187–215.
Hudspeth AJ (1985) The cellular basis of hearing: The biophysics of hair cells. Science 230:745–752.
Hudspeth AJ, Corey DP (1977) Sensitivity, polarity, and conductance change in the response of vertebrate hair cells to controlled mechanical stimuli. Proc Natl Acad Sci USA 74:2407–2411.
Hudspeth AJ, Jacobs R (1979) Stereocilia mediate transduction in vertebrate hair cells. Proc Natl Acad Sci USA 76:1506–1509.
Hudspeth AJ, Kroese ABA (1983) Voltage-dependent interaction of dihydrostreptomycin with the transduction channels in bullfrog sensory hair cells. J Physiol 345:66P.
Hudspeth AJ, Lewis RS (1988) Kinetic analysis of voltage- and ion-dependent conductances in saccular hair cells of the bull-frog, *Rana catesbeiana*. J Physiol 400:237–274.
Jakubowski M (1963) Cutaneous sense organs of fishes. I. The lateral-line organs in the stone-perch (*Acerina cernua* L.). Acta Biol Cracoviensia Zool 6:59–82.
Jielof R, Spoor A, De Vries Hl (1952) The microphonic activity of the lateral line. J Physiol 116:137–157.
Jørgensen F (1984) Influence of Ca^{2+} on the voltage dependent mechanosensitivity of the hair cells in the lateral line organs of *Xenopus laevis*. Acta Physiol Scand 120:481–488.
Jørgensen JM, Flock A (1973) The ultrastructure of lateral line sense organs in the adult salamander *Ambystoma mexicanum*. J Neurocytol 2:133–142.
Kalmijn AJ (1988) Hydrodynamic and acoustic field detection. In: Atema J, Fay RR, Popper AN, Tavolga WN (eds) Sensory Biology of Aquatic Animals. New York: Springer-Verlag, pp. 83–130.
Kroese ABA, van Netten SM (1987) The application of incident light polarization microscopy for the visualization of vertebrate sensory hair cells *in vivo*. J Microsc 145:309–317.
Kroese ABA, Schellart NAM (1987) Evidence for velocity- and acceleration-sensitive units in the trunk lateral line of the trout. J Physiol 394:13P.
Kroese ABA, van der Zalm JM, van den Bercken J (1978) Frequency response of the lateral-line organ of *Xenopus laevis*. Pflugers Arch 375:167–175.
Kroese ABA, van der Zalm JM, van den Bercken J (1980) Extracellular receptor potentials from the lateral line organ of *Xenopus laevis*. J Exp Biol 86:63–77.
Kroese ABA, Das A, Hudspeth AJ (1989) Blockage of the transduction channels of hair cells in the bullfrog's sacculus by aminoglycoside antibiotics. Hear Res (in press).
Kuiper JW (1956) The microphonic effect of the lateral line organ. Thesis, University of Groningen, The Netherlands.
Lewis RS, Hudspeth AJ (1983) Voltage- and ion-dependent conductances in solitary vertebrate hair cells. Nature 304:538–541.
McGlone FP, Russell IJ, Sand O (1979) Measurement of calcium ion concentrations in the lateral line cupulae of *Xenopus laevis*. J Exp Biol 83:123–130.
Münz H (1985) Single unit activity in the peripheral lateral line system of the cichlid fish *Sarotherodon niloticus* L. J Comp Physiol A 157:555–568.
Münz H, Claas B (1983) The functional organization of neuromasts in the lateral-line system of a cichlid fish. In: Ewert JP, Capranica RR, Ingle DJ (eds) Advances in Vertebrate Neuroethology. New York: Plenum, pp. 301–307.

van Netten SM (1987) Laser interferometric study of the mechanosensitivity of the fish lateral line. Thesis, University of Groningen, The Netherlands.

van Netten SM (1988) Laser interferometer microscope for the measurement of nanometer vibrational displacements of a light-scattering microscopic object. J Acoust Soc Am 83:1667–1674.

van Netten SM, Kroese ABA (1987) Laser interferometric measurements on the dynamic behaviour of the cupula in the fish lateral line. Hear Res 29:55–61.

van Netten SM, Kroese ABA (1989) Hair cell mechanics controls the dynamic behaviour of the lateral line cupula. In: Wilson JP, Kemp DT (eds) Cochlear Mechanisms: Structure, Function and Models. New York: Plenum Press, pp. 47–55.

Neugebauer D-Ch, Thurm U (1987) Surface charges of the membrane and cell adhesion substances determine the structural integrity of hair bundles from the inner ear of fish. Cell Tissue Res 249:199–207.

Ohmori H (1985) Mechano-electric transduction currents in isolated vestibular hair cells of the chick. J Physiol 359:189–217.

Ohmori H (1987) Gating properties of the mechano-electrical transducer channel in the dissociated vestibular hair cell of the chick. J Physiol 387:589–609.

Pabst A (1977) Number and location of the sites of impulse generation in the lateral-line afferents of *Xenopus laevis*. J Comp Physiol A 114:51–67.

Pickles JO, Comis SD, Osborne MP (1984) Cross-links between stereocilia in the guinea pig organ of Corti, and their possible relation to sensory transduction. Hear Res 15:103–112.

Pumphrey RJ (1950) Hearing. Symp Soc Exp Biol 4:3–18.

de Rosier DJ, Tilney LG, Egelman E (1980) Actin in the inner ear: The remarkable structure of the stereocilium. Nature 287:291–296.

Roberts WM, Howard J, Hudspeth AJ (1988) Hair cells: Transduction, tuning and transmission in the inner ear. Ann Rev Cell Biol 4:63–92.

Russell IJ (1971) The pharmacology of efferent synapses in the lateral line system of *Xenopus laevis*. J Exp Biol 54:643–658.

Russell IJ (1976) Amphibian lateral line receptors. In: Llinas R, Precht W (eds) Frog Neurobiology. Heidelberg: Springer-Verlag, pp. 513–550.

Russell IJ, Sellick POOM (1976) Measurement of potassium and chloride ion concentrations in the cupulae of the lateral lines of *Xenopus laevis*. J Physiol 257:245–255.

Russell IJ, Richardson GP, Cody AR (1986) Mechanosensitivity of mammalian auditory hair cells in vitro. Nature 321:517–519.

Sand O (1975) Effects of different ionic environments on the mechanosensitivity of lateral line organs in the mudpuppy. J Comp Physiol A 102:27–42.

Sand O (1984) Lateral-line systems. In: Bolis L, Keynes RD, Maddrell SHP (eds) Comparative Physiology of Sensory Systems. London: Cambridge University Press, pp. 3–32.

Sand O, Ozawa S, Hagiwara S (1975) Electrical and mechanical stimulation of hair cells in the mudpuppy. J Comp Physiol A 102:13–26.

Shotwell SL, Jacobs R, Hudspeth AJ (1981) Directional sensitivity of individual vertebrate hair cells to controlled deflection of their hair bundles. Ann NY Acad Sci 374:1–10.

Strelioff D, Flock Å (1984) Stiffness of sensory-cell hair bundles in the isolated guinea pig cochlea. Hear Res 15:19–28.

Weiss TF, Leong R (1985) A model for signal transmission in an ear having hair cells with free-standing stereocilia (III micromechanical stage). Hear Res 20:157–174.

Wubbels RJ (1988) Frequency response of single unit afferents innervating the lateral line system of *Acerina cernua*. In: Duifhuis H, Horst JW, Wit HP (eds) Basic Issues in Hearing. London: Academic Press pp. 164–168.

14 Functional Organization of the Lateral Line Periphery

HEINRICH MÜNZ

14.1 Introduction

In the lateral line, as in any other sensory system, the function of the system depends on the peripheral receptive structures. These receptive structures, which are the neuromasts in the mechanosensory lateral line system, determine the character of the information available to higher brain centers. Parameters related to the functional organization of the lateral line periphery include the spatial arrangement and the innervation of neuromasts and the presence of different morphological types of neuromasts which may have different physiological properties. Additionally, the centrifugal control of the periphery mediated by the efferent innervation has to be considered.

This chapter will focus on the differences between the two types of sense organs, superficial and canal neuromasts, and the implications of these differences for the function of the system. The distribution and spatial arrangement of neuromasts have been described for many teleostean (Devillers 1958; Disler 1971; Coombs et al. 1988; Northcutt Chapter 3; Webb Chapter 4) and amphibian species (Escher 1925; Lannoo 1987a,b). However, information on both the morphological and functional differences of neuromasts and on the innervation of single neuromasts or groups of neuromasts (functional units) is less detailed. The classification of a neuromast as either a superficial or canal neuromast is usually based on its position. Superficial neuromasts are situated in the epidermis. Their apical structures protrude from the body surface into the surrounding water. Canal neuromasts are completely recessed in the skin or in the bony tissue of the skull or scales. Their apical structures extend into a tubular, fluid-filled chamber which in the majority of species communicates with the surrounding water via pores of various shape. This classification does not consider the many morphological variations within each type of neuromast or in the structure of canals (reviewed by Coombs et al. 1988). Up to now it is not clear whether all neuromasts in each category are truly homologous. Canal neuromasts are formed only in cartilaginous and bony fish. Amphibians do not have canallike structures in their lateral line. Most of the data presented here are obtained from teleosts and especially from our own studies on cichlid fish of the genus *Sarotherodon* (syn. *Tilapia* or *Oreochromis*) (Münz 1979, 1985, 1986).

14.2 Differences in the Ontogenetic Development of Superficial and Canal Neuromasts

During ontogeny, all neuromasts first appear in a superficial position. Thus they can be regarded, at least for a certain time, as superficial neuromasts. Behavioral studies with fish indicate that the neuromasts function efficiently already in early embryonic stages of the individual (Blaxter and Fuiman Chapter 24, Disler 1971). When the animal grows, some of the superficially situated neuromasts recede into canal tubes and become canal neuromasts (Dijkgraaf 1952). For cichlid fish, the ontogenetic development of individual neuromasts has been followed from their first appearance until canals are formed (Münz 1986). In these fish—and probably in many other species as well—the fate of a neuromast as either a superficial or canal neuromast in the adult seems to be determined by the time it appears. Usually the first neuromasts to develop in the fish embryo will later become canal neuromasts and are therefore called presumptive canal neuromasts (Peters 1973, Münz 1979, 1986). The number of these presumptive canal neuromasts is almost identical with the number of scales finally bearing canal structures. A similar invariance has also been described for other teleost fish (Disler 1971) and has been often used as a taxonomic character. In several teleostean species, the increase in the number of scales and the development of canal structures end at a certain stage in the fish's development (Disler 1971, Münz 1986). The limitation of the number of canal structures in these species may be coupled to the termination of new scale formation.

In contrast, at least in cichlids, the number of superficial neuromasts appears to increase throughout the whole life of the animal independent of scale development (Peters 1973, Münz 1986). Whereas changes in the number of neuromasts have been described for larval stages of several species (for review, see Disler 1971), postlarval changes have been quantitatively documented in only a few species (Peters 1973, Münz 1986, Janssen et al. 1987). Nevertheless, postlarval changes should be considered as more widely distributed among teleosts.

In the cichlid fish *Sarotherodon*, both the start of development of the canal structures and the number of superficial neuromasts correlate with the body length of the fish and not with the age of the animal (Münz 1986). In cyprinid fish, the numbers of superficial neuromasts on the trunk scales that later become canal neuromasts differ between species and among individuals of the same species (Dijkgraaf 1952, 1963). Presumptive canal neuromasts of one species remain in a superficial position in others. To explain differences among individuals of one species, Dijkgraaf (1952) proposed that the development of the lateral line may be directly influenced by environmental hydrodynamic factors. This hypothesis was tested by raising young *Sarotherodon* in either fast-moving or still water. The results showed that the different environmental conditions did not change the number of canal neuromasts (Münz 1986), although this may be different in cyprinid species.

The increase in size of both types of neuromasts during postlarval growth has been directly compared for *Sarotherodon* (Fig. 14.1) and the mottled sculpin

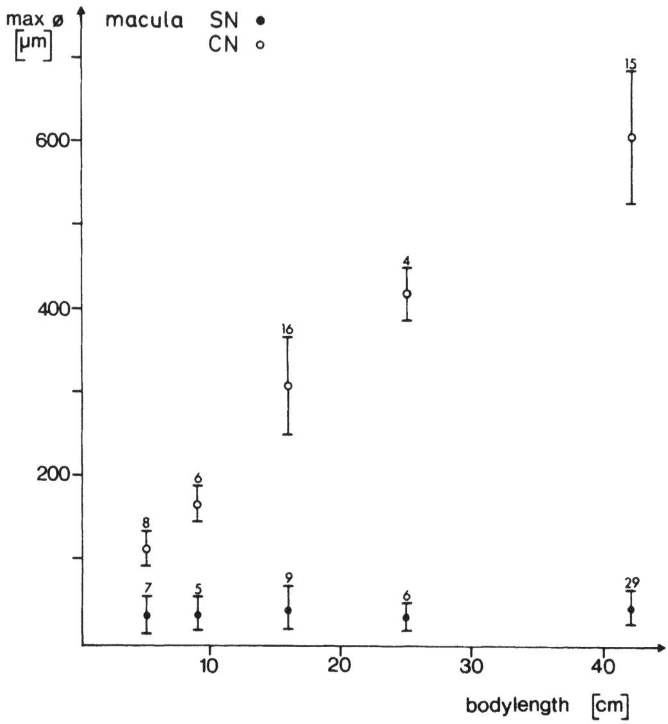

FIGURE 14.1. Maximum diameters (max ∅) of the maculae of superficial (SN) and canal neuromasts (CN) in correlation with the body length of *Sarotherodon niloticus*. Neuromasts on several scales of the upper trunk lateral line were counted. Each measurement represents the mean value of neuromasts of two animals. Bars, standard deviation; small numbers, number of neuromasts.

(Janssen et al. 1987). The macula diameter (maximum diameter of the area covered by hair cells) of a canal neuromast increases with the body size, whereas superficial neuromasts seem to stop growing once they reach a diameter of about 100 μm. The ranges of maximum macula diameters for canal and superficial neuromasts obtained from several species are depicted in Fig. 14.2. Data are taken from unpublished studies of this author and from the literature.

In summary, these data indicate that in many teleostean fish, the size of canal neuromasts exceeds the size of superficial neuromasts. Counts of the hair cells in amphibian neuromasts (Lannoo 1987b) and in the maculae of the fish ear (for review see Corwin 1981, Popper and Hoxter 1984) led to the hypothesis that the size of a macula correlates with the number of associated hair cells. For the lateral line, this relationship would imply that canal neuromasts have far more hair cells than superficial neuromasts, as has been demonstrated for the mottled sculpin (Janssen et al. 1987).

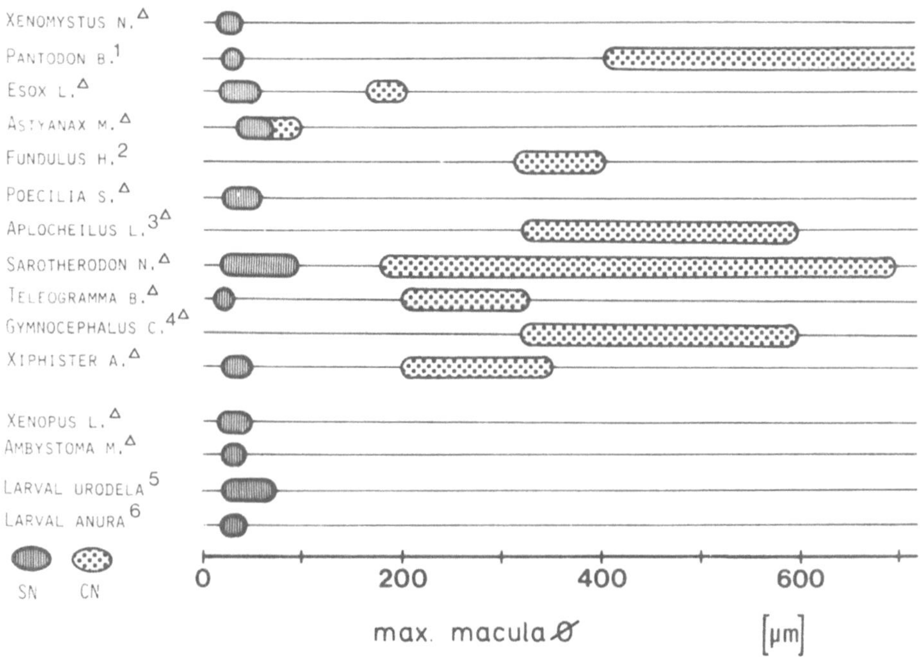

FIGURE 14.2. Range of maximum diameters (max macula Ø) of the maculae of superficial (SN, hatched columns) and canal neuromasts (CN, stippled columns) of different species. Data are taken from the author's unpublished studies (△) and from the literature: [1]Müller and Tittel 1984; [2]van Bergeijk and Alexander 1962; [3]Schwartz 1970; [4]Kuipers 1956, [5]Lannoo 1987b; [6]Lannoo 1987a.

Although developmental differences between superficial and canal neuromasts may be tied to morphological constraints, the functional consequences of these differences are still unclear. The postlarval increase in the number of superficial neuromasts may be an adaptation for maintaining a constant density of sense organs as the animal grows and the body surface expands. Another possibility is an enhancement of sensitivity by increasing the total number of hair cells on distinct points of the body surface. Enhancement of sensitivity may also explain the postlarval increase in the size of canal neuromasts. Although either mechanism – increasing the size of the macula or adding new neuromasts – may make the system more sensitive, the two have different functional consequences. An increase in the number of hair cells, which presumably underlies an increase in macula size, is accompanied by changes in the shape of the cupula. This in turn leads to changes in the dynamic properties of stimulus transduction (see Denton and Gray Chapter 11; van Netten and Kroese Chapter 12). In summary, the results from ontogenetic studies of canal and superficial neuromasts demonstrate that the peripheral organization of the lateral line continues to change in the postlarval life of the fish.

14.3 Morphological Differences Between Canal and Superficial Neuromasts

All neuromasts are composed of three cell types: the sensory cells (hair cells), the supporting cells between the sensory cells, and the mantle cells, which separate the neuromast from the surrounding epidermal tissue. The absolute number and the ratio of cell types in a neuromast may differ both within an individual and between species. Although all superficial neuromasts are similar with respect to their superficial location in the epidermis, differences in associated structures such as dermal hillocks, pits, and dermal ridges are very common (for review, see Coombs et al. 1988). Canal structures also vary considerably between and within species.

These differences in the accessory structures certainly modify the transduction properties of the neuromasts and must be taken into account in functional descriptions of typical superficial or canal neuromasts. The transformation of a superficial neuromast into a canal neuromast is often accompanied by a considerable change in the form of the cupula. Whereas the cupulae of superficial neuromasts are long and slender and may be either oval or round in cross section, the cupulae of canal neuromasts commonly have an oval cross section and are shorter and more massive, reflecting the larger size of the macula and the limitations by the tubular cavity (for differences in canal neuromast cupulae, see Denton and Gray, Chapter 11). As mentioned above, the different shapes of the cupulae, together with other accessory structures, modify the stimulus transduction and surely change the functional properties of the neuromasts (for details, see Denton and Gray, Chapter 11; van Netten and Kroese Chapter 12).

In the preceding section, the size differences between superficial and canal neuromasts were mentioned. In addition, canal and superficial neuromasts differ with respect to their innervation. In *Sarotherodon*, for example, canal neuromasts are innervated by more fibers than are superficial neuromasts (Münz 1979). The ratio varied between 2:1 and 5:1. Fiber counts in adult fish from a number of taxonomic groups indicate that canal neuromasts are often innervated by far more fibers than superficial neuromasts (Fig. 14.3). Unfortunately, most of the studies do not distinguish between afferent and efferent fibers. An efferent innervation has been demonstrated for all canal neuromasts studied so far (see Roberts and Meredith Chapter 22). Whether efferent terminals are present in each superficial neuromast of all species and on each hair cell is still an open question.

Another morphological difference is found in the type of innervation. In many teleosts, the canal neuromasts are innervated by fibers that retain their myelination within the sensory epithelium up to the synaptic area (Flock 1965; Münz 1979), whereas in superficial neuromasts of teleosts and amphibians, the myelination of all fibers terminates at the basal membrane (Flock 1967; Münz 1979). Whether this morphological difference corresponds to a functional difference between afferent lateral line fibers is still uncertain.

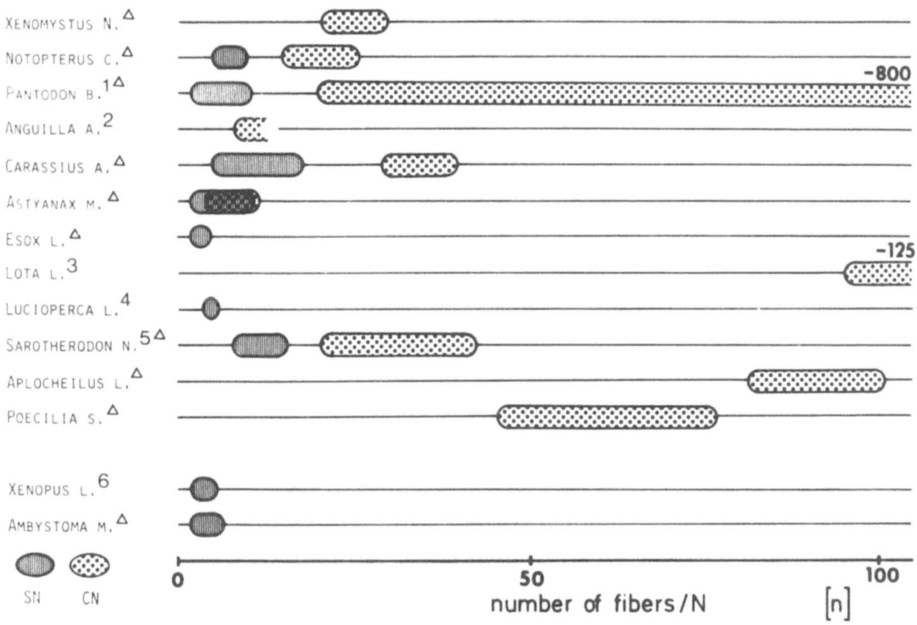

FIGURE 14.3. Numbers of fibers below superficial (SN, hatched areas) and canal neuromasts (CN, stippled areas) of different species. Data are taken from the author's unpublished studies (△) or from the literature: [1]Müller and Tittel 1984; [2]Alnaes 1973; [3]Flock 1965; [4]Jakubowski 1966; [5]Münz 1965; [6]Görner 1963, Flock 1967.

One important characteristic of the peripheral organization is the innervation pattern of neuromasts. The spatial distribution and the orientation of the neuromasts innervated by the same afferent fiber determine the spatial resolution and the directional sensitivity of that fiber. In the trunk lateral line of the African clawed toad, *Xenopus*, all superficial neuromasts in a given row (stitch) are innervated by two afferent fibers (Görner 1963; Flock 1967). With the exceptions of developmental stages of neuromasts or in some superficial neuromasts of the blind cave fish *Astyanax mexicanus* (unpublished results of the author), such a simple innervation pattern is not known for superficial neuromasts in the lateral line of teleost fish (Münz 1985). In cichlid fish, superficial neuromasts of one row are innervated by six to 10 afferent fibers, and these show a complicated branching pattern. The number of neuromasts innervated by the same afferent fiber varies within one row and from row to row (Münz 1985). In cichlid fish, canal and superficial neuromasts are innervated by completely separate afferent fiber populations. Hence, the information from canal and superficial neuromasts is carried separately to the brain. Whether this is a general feature in teleosts has yet to be proved. Single fiber recordings demonstrate that each fiber innervates only hair cells with the same directional orientation, even when the fiber innervates more than one neuromast (Münz 1985). Whether the input from

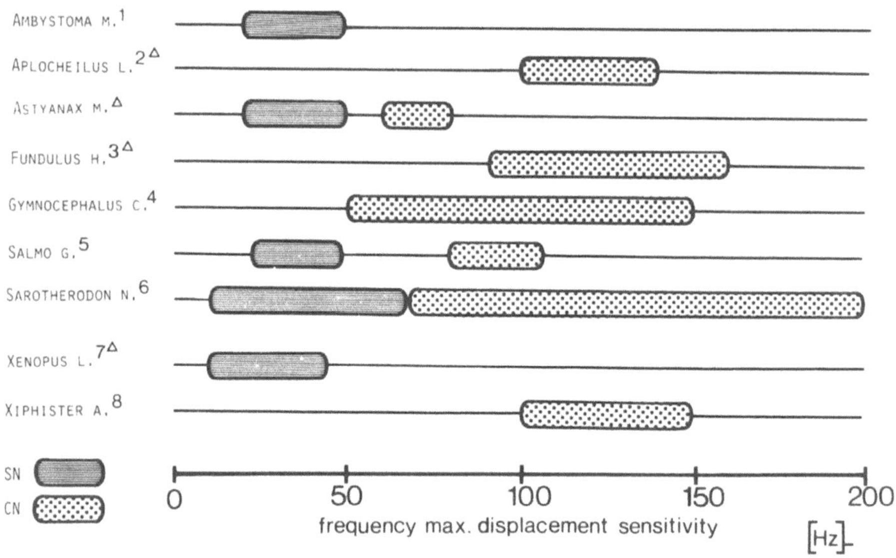

FIGURE 14.4. Range of the frequencies of the maximal displacement sensitivity of superficial (SN, hatched areas) and canal neuromasts (CN, stippled areas) of different species. Data are taken from the author's unpublished studies (△) and from the literature: [1]Münz et al. 1984; [2]Bleckmann and Topp 1981; [3]Harris and Van Bergeijk 1962; [4]Kuiper 1965, 1967; [5]Kroese and Schellart 1987; [6]Münz 1985; [7]Kroese et al. 1978, Strelioff and Honrubia 1978; [8]Bleckmann and Münz 1988.

canal and superficial neuromasts is separately processed in the CNS is still an open question.

In summary, the superficial and canal neuromasts have an identical cellular composition but differ considerably in accessory structures and innervation. In fish studied so far, information from canal and superficial neuromasts is carried by separate afferent fibers. In many teleosts, a canal neuromast may have a stronger and different influence on the CNS, since it is innervated by more afferent fibers than a superficial neuromast.

14.4 Physiological Differences Between the Two Types of Neuromasts

The transduction properties of superficial and canal neuromasts have been directly compared for only two teleost species (Fig. 14.4) (Münz 1985; Kroese and Schellart 1987; Schellart and Kroese Chapter 21). In cichlids and in the salmonid trout (*Salmo gairdneri*), the two types of neuromasts respond to water displacement of different frequencies, but the sensitivity of canal neuromasts extends to higher frequencies than that of superficial neuromasts. Whereas in cichlid fish both types of neuromasts respond to water acceleration, only the canal neuromasts do so in the salmonid fish. Superficial neuromasts respond to water velocity.

The few data available from the literature (Fig. 14.4) support the assumption that an extension of the response to higher frequencies may be a feature of canal neuromasts. One of the reasons for these differences in frequency response may be the different sizes of the sensory maculae and number of hair cells (Denton and Gray Chapter 11; van Netten and Kroese Chapter 12). To date, it is not clear whether the differences in the frequency range are of adaptive significance in the sense that all animals use the higher frequencies for behavior. From theoretical considerations (Kalmijn Chapter 9) and studies of prey-catching behavior of the sunfish (Enger et al. Chapter 29), it can be assumed that frequencies below 10 Hz are important for the lateral line. But lateral line mediated, unconditioned responses in the mottled sculpin to stimulus frequencies from 10 to over 100 Hz indicate that higher frequencies may also be of functional significance (see Coombs and Janssen Chapter 15).

In surface-feeding fish, detection of the higher-frequency components of water waves is biologically important (Bleckmann and Schwartz 1982; Hoin-Radkovsky et al. 1984). In these specialized animals, the frequency components of a surface wave are determined by canal neuromasts or modified canal neuromasts on the head. The behavioral experiments have shown that the animals use the higher frequencies for localizing the prey at a short distance. Not much is known about the frequency content of water movements in the natural environment of fishes that do not hunt at the water surface. In the future, data about ambient water levels and the flow dynamics around individuals in a fish school or during sudden changes of direction may lead to new interpretations of the broader frequency response of canal neuromasts.

Interspecies comparison reveal a correlation between the relative abundance of canal versus superficial neuromasts and the animals' life-style or the hydrodynamic conditions of their surroundings (for review, see Dijkgraaf 1963). Fishes that either live in flowing water or are active swimmers possess a well-developed canal system, whereas closely related species living in calm water have few or no canal neuromasts and many superficial neuromasts. The functional interpretation was that the canal structure develops to "protect" the neuromast from constant stimulation caused by environmental factors or by the animal's own movement (Dijkgraaf 1963). Some counterexamples were also mentioned by Dijkgraaf (1963), so it can be concluded that other factors besides "protection" may also contribute to the expression of canal neuromasts. One of these factors may be the broader frequency range of canal neuromasts and its ability to extract signals from high noise levels. To what extent both types of neuromasts are stimulated by the "noise" in the natural surroundings and how the boundary layer modifies the excitation of the neuromasts, as well as the influence of the efferent activity, remain to be studied. In summary, the physiological differences raise the distinct possibility that the canal and the superficial neuromasts serve different purposes.

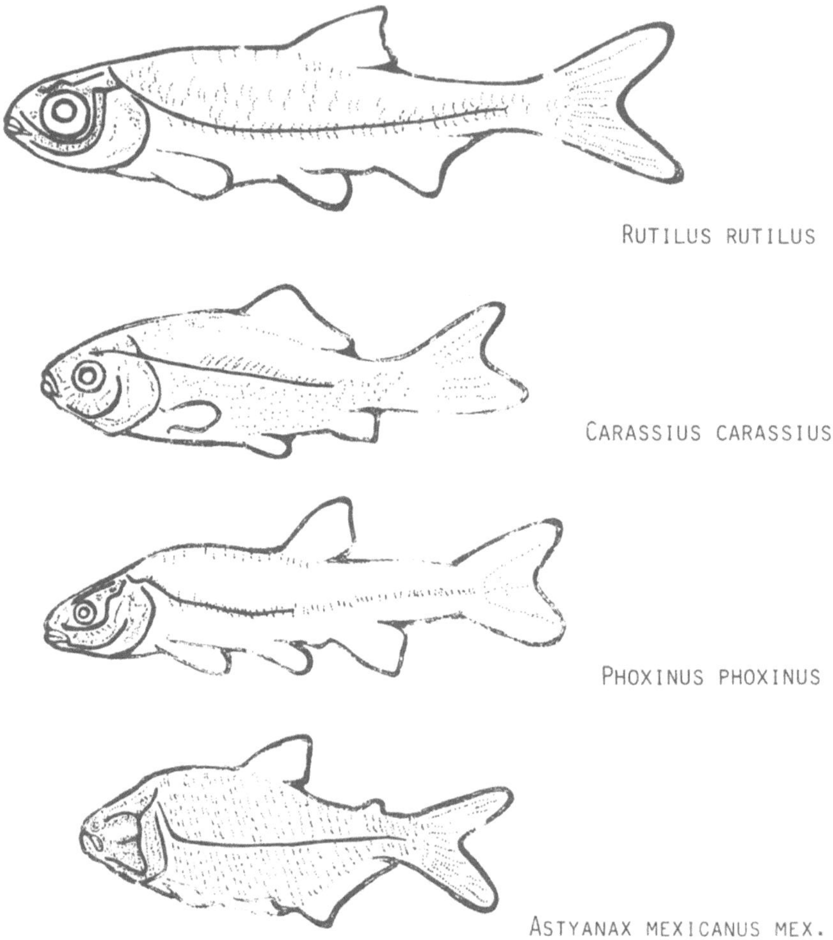

FIGURE 14.5. Arrangement of superficial neuromasts (black dots) and canal lines (black lines) in four cypriniform species. Data obtained from Disler 1971, Dijkgraaf 1934, Schemmel 1967.

14.5 Spatial Arrangement of Neuromasts

The spatial arrangement of the lateral line structures has been described for many fishes and amphibians (reviewed by Devillers 1958; Dijkgraaf 1934, 1963; Disler 1971; Coombs et al. 1988; Northcutt Chapter 3; Webb Chapter 4). In summary, these studies demonstrate that the lateral line system does not simply consist of a bilateral trunk canal: it can have a rather complex, species-specific organization (Coombs et al. 1988). The spatial arrangement and the number of neuromasts vary considerably even between species of the same family (Fig. 14.5). Additionally, the arrangement can vary within certain limits among individuals of the same species.

The reason for these heterogeneous and complex patterns of neuromast arrangement is still an open question: do they represent functional adaptations, or are they determined by developmental and other morphological constraints? (See Northcutt Chapter 3 for further discussion.) In general, the distribution pattern for canal neuromasts is far more conservative across extant fish groups than that for superficial neuromasts. A basic "bauplan" for neuromast distribution upon which variation may be imposed has been proposed for actinopterygian fish (Coombs et al. 1988).

To evaluate the published observations on the spatial arrangement of neuromasts for functional or evolutionary considerations is difficult. Many publications describe the distribution of only one type of lateral line sense organ, the canal neuromast. Because of difficulties in observing superficial neuromasts in vital or unstained material, many studies omit a description of their distribution. In contrast, the pores or elevated structures associated with the canal neuromasts can often be detected quite easily and are often the only macroscopically visible parts of the lateral line. Nevertheless, in many species, the superficial neuromasts represent the majority of lateral line receptor organs. Furthermore, a description of lateral line distribution is complicated by the fact that not all tubular structures in the fish integument belong to the lateral line. In several extinct forms (Denison 1966, 1979), a "pore canal" system can be found and in some living species (Walters 1963), a subdermal canal system has been described. These systems can be easily distinguished from true lateral line canals on the basis of its morphology and the lack of neuromasts.

The pore canal system may be phylogenetically related to the lateral line (Denison 1966), but its function is still speculative. Recent studies indicate that even in some teleost fishes, canal structures exist that resemble typical lateral line canals but do not enclose any neuronal structures such as neuromasts or lateral line nerves (Münz and Bleckmann 1988). Whether these structures belong to the lateral line system or are related to the pore canal system or to other tubular structures in the fish integument is still uncertain. A third problem in generalizing from published data on the arrangement of neuromasts results from changes in the distribution of both types of neuromasts during larval and postlarval development, which was mentioned above. In nearly all of the published distributions of neuromasts, these factors are insufficiently considered and must be clarified for functional interpretations. (However, see Northcutt Chapter 3 and Webb Chapter 4.)

Nevertheless there exist examples in which the lateral line arrangement apparently reflects functional adaptation (Dijkgraaf 1963, Chapter 2). In fishes in which the pectoral fins insert quite dorsally on the side of the body, the trunk canal neuromasts are displaced from their "normal" midbody position. This displacement protects the neuromasts from disturbances caused by the beating of the pectoral fins. The causal relationship between the course of the lateral line and the fin position has been demonstrated in a flatfish species with unilateral pectoral fins. On the side with a pectoral fin, the trunk canal line is displaced, whereas on the side that lacks this fin, the course of the canal line is in the midbody position (Dijkgraaf 1963, Chapter 2). Unpublished observations on several perciform fish also indicate that superficial neuromasts are absent in the area of the pectoral fins.

In several species of blind cave fish, where the lateral line is the only sensory system available for detecting obstacles at a distance, the number of superficial neuromasts is increased (Schemmel 1967; Poulsen and White 1969), whereas canal neuromasts are absent or partly reduced (Dijkgraaf 1963; Wilkens 1977). The lateral line detects obstacles by measuring the modifications of the water flow along the body as the fish swims (Campenhausen et al. 1981; Hassan Chapter 10). The temporal course of the stimulus at different points on the surface of the fish is measured and compared. The precision of this analysis depends on the number of sampled points. The superficial neuromasts seem to be the important neuromast type for this task, because their number can be increased extensively to cover the whole body surface.

To summarize, the two types of neuromasts differ in their spatial arrangement. Canal neuromasts seem to be restricted to a characteristic set of lines on the head and trunk of the animals probably based on a "Bauplan" (Coombs et al. 1988; Northcutt Chapter 3). Although the arrangement of superficial neuromasts may follow a basic Bauplan, it is much more variable. They can be located exclusively close to the canal lines or can cover the whole body surface and the tail fin, as shown in some characid (Schemmel 1967) and cyprinid species (Disler 1961; Pudrowski unpublished) (Fig. 14.5).

14.6 Conclusion

The peripheral organization provides the lateral line centers in the brain with information on the hydrodynamic conditions by measuring amplitude, time, and direction of water movements at numerous points on the body surface. Variations in the morphological, physiological, and spatial features are presumably adaptations to accomplish this task.

References

Alnaes EP (1973) Two types of lateral line afferents in the eel (*Anguilla anguilla*). Acta Physiol Scand 87:535–548.

Bergeijk WA van, Alexander S (1962) Lateral line canal organs on the head of *Fundulus heteroclitus*. J Morphol 110:333-346.

Bleckmann H, Schwartz E (1982) The functional significance of frequency modulation within a wave train for prey localization in the surface-feeding fish *Aplocheilus lineatus* (Cyprinodontidae). J Comp Physiol 145:331-339.

Bleckmann H, Topp G (1981) Surface wave sensitivity of the lateral line organs of the topminnow. *Aplocheilus lineatus*, Naturwissenschaften 68:624–625.

Bleckmann H, Münz H (1988) The physiology of lateral line mechanoreceptors in *Xiphister atropurpureus*, a teleost with multiple lateral lines. Brain Behav Evol (submitted).

Campenhausen C, Riess I von, Weissert R (1981) Detection of stationary objects by the blind cave fish *Anoptichthys jordani* (Characidae), J Comp Physiol 143:369–374.
Coombs S, Janssen J, Webb J (1988) Diversity of lateral line system: Evolutionary and functional considerations. In: Atema J, Fay RR, Popper AN, Tavolga WN (eds) Sensory Biology of Aquatic Animals. Heidelberg: Springer-Verlag, pp. 553–593.
Corwin JT (1981) Postembryonic production and aging of inner ear hair cells in sharks. J Comp Neurol 201:541–553.
Denison AB (1966) The origin of the lateral-line sensory system. Am Zool 6:369–370.
Denison RH (1979) Acanthodii. In: Schultze HP, Kuhn O (eds) Handbook of Paleoichthyology, Vol. 2. Stuttgart: G. Fischer Verlag.
Devillers C (1958) Le système latéral. In: Grasse PP (ed) Traite de Zoologie, Vol. 13. Paris; Masson, pp. 940–1032.
Dijkgraaf S (1934) Untersuchungen über die Funktion der Seitenorgane an Fischen. Z Vergl Physiol 20:162–214.
Dijkgraaf S (1952) Bau and Funktionen der Seitenorgane und des Ohrlabyrinths bei Fischen. Experientia 8:205–216.
Dijkgraaf S (1963) The functioning and significance of the lateral-line organs. Biol Rev 38:51–105.
Disler N (1971) Lateral line sense organs and their importance in fish behavior. Isr Prog Sci Transl III:-328.
Escher K (1925) Das Verhalten der Seitenlinienorgane der Wirbeltiere und ihrer Nerven beim Übergang zum Landleben. Acta Zool 6:307–414.
Flock Å (1965) Electron microscopic and electrophysiological studies on the lateral line canal system. Acta Otolaryngol (Suppl) 199:1–90.
Flock Å (1967) Ultrastructure and function in the lateral line organs. In: PH Cahn (ed) Lateral Line Detectors. Bloomington: Indiana University Press, pp. 163–197.
Görner P (1963) Untersuchungen zur Morphologie und Elektrophysiologie des Seitenlinienorgans vom Krallenfrosch (*Xenopus laevis* Daudin). Z Vergl Physiol 47:316–338.
Harris G, van Bergeijk WA (1962) Evidence that the lateral line organ responds to near-field displacements of sound sources in the water. J Acoust Soc Am 34:1831–1841.
Hoin-Radkovsky I, Bleckmann H, Schwartz E (1984) Determination of source distance in the surface-feeding fish *Panthodon buchholzi*. Anim Behav 32:840–851.
Jakubowski M (1966) Cutaneous sense organs of fishes. IV. The lateral line organs in the perch-pike (*Lucioperca lucioperca* L.) and (*Perca fluviatilis* L.), their topography, innervation, vascularization, and structure. Acta Biol Cracoviensia Zool 9:138–149.
Janssen J, Coombs S, Hoekstra D, Platt C (1987) Anatomy and differential growth of the lateral line system of the mottled sculpin, *Cottus bairdi* (Scorpaeniformes: Cottidae). Brain Behav Evol 30:210–229.
Kroese ABA, van der Zalm JM, van den Bercken J (1978) Frequency response of the lateral line organ of *Xenopus laevis*. Pflugers Arch 375:167–175.
Kroese ABA, Schellart NAM (1987) Evidence for velocity and acceleration sensitive units in the trunk lateral line of the trout. J Physiol 393:28P.
Kuiper JW (1956) The Microphonic Effect of the Lateral Line Organ. Groningen: Publication of the Biophysical Group of the "natuurkundig Laboratorium", Groningen, Netherlands.
Kuiper JW (1967) Frequency characteristics and functional significance of the lateral line organ. In: PH Cahn (ed) Lateral Line Detectors. Bloomington: Indiana University Press, pp. 105–122.
Lannoo MJ (1987a) Neuromast topography in anuran amphibians. J Morphol 191:115–129.

Lannoo MJ (1987b) Neuromast topography in urodele amphibians. J Morphol 191:247-263.
Müller U, Tittel G (1984) Vergleichende Anatomie der Seitenlinien von *Aplocheilus lineatus* (Cyrpinodotidae) und *Panthodon buchholzi* (Panthodontidae) und ihr Bezug zu physiologischen Unterschieden. Verh Dtsch Zool Ges 77:311.
Münz H (1979) Morphology and innervation of the lateral line system in *Sarotherodon niloticus*. Zoomorphology 93:73-86.
Münz H (1985) Single unit activity in the peripheral lateral line system of the cichlid fish *Sarotherodon niloticus* L., J Comp Physiol A 157:555-568.
Münz H (1986) What influences the development of canal and superficial neuromasts? Ann Kon Mus Mid Afr Zool Wetensch 251:85-89.
Münz H, Bleckmann H (1988) Non-sensory lateral lines in a teleost fish? Naturwissenschaften (submitted).
Münz H, Claas B, Fritzsch B (1984) Electroreceptive and mechanoreceptive units in the lateral line of the axolotl *Ambystoma mexicanum*. J Comp Physiol A 154:33-44.
Peters HM (1973) Anatomie und Entwicklungsgeschichte des Lateralissystems von Tilapia (Pisces, Cichlidae). Z Morphol Tiere 74:89-161.
Popper AN, Hoxter B (1984) Growth of a fish ear. 1. Quantitative analysis of hair and ganglion cell proliferation. Hear Res 15:133-142.
Poulsen TL, White WB (1969) The cave environment. Science 165:971-981.
Schemmel C (1967) Vergleichende Untersuchungen an den Hautsinnesorganen ober- und unterirdisch lebender *Astyanax*-Formen. Z Morphol Tiere 61:255-316.
Schwartz E (1970) Ferntastsinnesorgane von Oberflächenfischen. Z Morphol Tiere 67:40-57.
Strelioff D, Honrubia V (1978) Neural transduction in *Xenopus laevis* lateral line system. J Neurophysiol 41:432-444.
Walters V (1963) The trachipterid integument and an hypothesis on its hydrodynamic function. Copeia 1963:260-270.
Wilkens H (1977) Die Rudimentation des Rumpfkanals bei kavernikolen Populationen des Astyanax (Characidae, Pisces). Exerientia 33:604.

15
Peripheral Processing by the Lateral Line System of the Mottled Sculpin (*Cottus bairdi*)

SHERYL COOMBS and JOHN JANSSEN

15.1 Introduction

Although there is considerable information on the peripheral anatomy of lateral line systems (see Dijkgraaf 1962; Russell 1976; Coombs et al. 1988 for reviews), relatively little is known about how sensory information is encoded by the nervous system, either at the periphery or more centrally. To answer questions of this nature, it is necessary to have information not only on the peripheral anatomy of the system and its connections to and within the central nervous system but also on the response properties of neurons at various levels of the nervous system. This kind of information must then be evaluated in terms of the response of the entire system–i.e., the behavioral output or sensory capabilities of the animal.

Although there is a growing body of information on peripheral and central processing in the mechanosensory lateral line system of the amphibian, *Xenopus laevis* (see chapters by Görner and Mohr, Elepfandt, Class et al., and Will), comparable data on teleost fishes are fewer. Both spontaneous and evoked activity in peripheral lateral line nerves has been measured for a number of different teleost fishes (Hoagland 1932, 1933a,b, 1934; von Schriever 1935; Katsuki et al. 1950, 1951; Suckling and Suckling 1950, 1964; Suga 1967; Suckling 1967; Alnaes 1973; Spath and Schweickert 1977; Hensel et al. 1975). But only more recent studies have systematically examined the way in which signals are encoded by peripheral nerve fibers, particularly in terms of their frequency response (Sand 1981; Bleckmann and Topp 1981; Topp 1983; Münz 1985; Montgomery and Macdonald 1987; Kroese and Schellart 1987; Schellart and Kroese Chapter 21; Wubbels 1988). The relationship between nerve fiber response and the sensory capabilities of the animal, however, has only been touched upon for the topminnow, *Aplocheilus lineatus* (Bleckmann and Topp 1981; Topp 1983) and the antarctic fish, *Pagothenia borchgrevinki* (Montgomery and Macdonald 1987).

The Lake Michigan mottled sculpin, *Cottus bairdi*, has several advantages for studying sensory processing by the mechanosensory lateral line system. One is that stimulation of the lateral line system in this fish results in a naturally occurring, stereotyped feeding behavior (Hoekstra and Janssen 1985, 1986) that can be used in the lab as a behavioral indicator of sensory capabilities. A second is that the sculpin freezes its body movements, including respiratory movements of the gills, while "listening" with its lateral line. Thus, the natural state of the animal when using its lateral line for feeding approximates the unnatural state of the artificially immobilized animal for which physiological measures of neural responses can be made. A third advantage is that the morphology of the system, although relatively simple (e.g., canal and superficial neuromasts show none of the elaborate specializations found in many bony fishes; see Coombs et al. 1988 for review), changes as a function of position on the body (Janssen et al. 1987). This makes it possible to probe different portions of the system in order to study the relationship between structure and function. The major goals of this chapter will be (1) to present preliminary behavioral and physiological results on stimulus detection and encoding by the peripheral lateral line system of the mottled sculpin, and (2) to place these results in the larger context of sensory processing by vertebrate hair cell systems.

15.2 Gross Anatomy of the Mottled Sculpin Lateral Line System

The lateral line system of the mottled sculpin, like that of most bony fishes, has both canal and superficial neuromasts that are distributed on the head and trunk of the fish (Fig. 15.1). Canal and superficial neuromasts differ in a number of features, but the most obvious—the overall size and shape of neuromasts—are a consequence of differential growth (Janssen et al. 1987). When measured in terms of number of hair cells and total area of the sensory and neuromast surface, neuromasts from the preopercular, infraorbital, and supraorbital canals are relatively homogeneous in their size and growth, but as a group they grow significantly faster than trunk canal neuromasts and slower than mandibular canal neuromasts, the largest neuromasts in the system (Fig. 15.2). In addition, the trunk canal differs from head canals by being enclosed by nonossified tissue.

In contrast, superficial neuromasts, regardless of their position on the fish, are significantly smaller than all canal neuromasts. Moreover, their surface area growth, though negligible relative to canal neuromasts, is symmetrical as opposed to the asymmetrical elongation of canal neuromasts (Janssen et al. 1987). This difference in symmetry is also accompanied by a concomitant difference in the size and shape of the sensory surface and its overlying cupula and most likely has a significant bearing on the different response properties of superficial and canal neuromasts (Janssen et al. 1987; Kalmijn Chapter 9; Münz 1985, Chapter 14; Kroese and Schellart 1987; Kroese and van Netten Chapter 13).

As summarized in Fig. 15.1, the mottled sculpin has a very convenient spatial separation of neuromast types, whereby the largest canal neuromasts in the

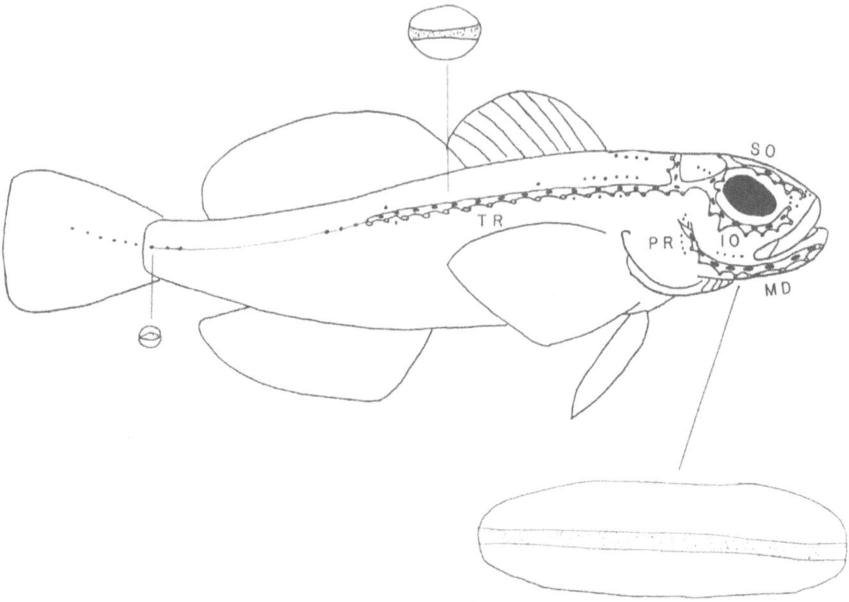

FIGURE 15.1. Schematic showing the distribution of canal and superficial neuromasts on the body of the mottled sculpin. Enlarged drawings outline the dorsal surface of canal neuromasts from the mandible and trunk and of superficial neuromasts on the tail to indicate the extent to which neuromasts vary in size and shape. Stippled areas represent the hair cell sensory surface. MD, mandibular canal; SO, supraorbital canal; IO, infraorbital canal; PR, preopercular canal; TR, trunk canal.

system are located on the head of the fish in bony canals and the smallest canal neuromasts are located on the trunk in nonossified canals. Finally, the very different superficial neuromasts, though found in conjunction with both head and trunk canals, are the only end organs found in the tail region of the mottled sculpin. The experiments described below were designed to stimulate these three different regions—head, trunk, and tail—to determine if the sensitivity of the animal to live prey or the frequency response of the system to artificial stimuli changed as a function of stimulus position. In addition, behaviorally measured thresholds from the trunk of the sculpin were compared to physiologically measured thresholds from afferent fibers innervating trunk lateral line organs.

15.3 Methods

15.3.1 Stimulus Generation and Measurement

The stimulus for both behavioral and physiological experiments was generated with the same dipolar source. A plastic sphere, 6 mm in diameter, was attached

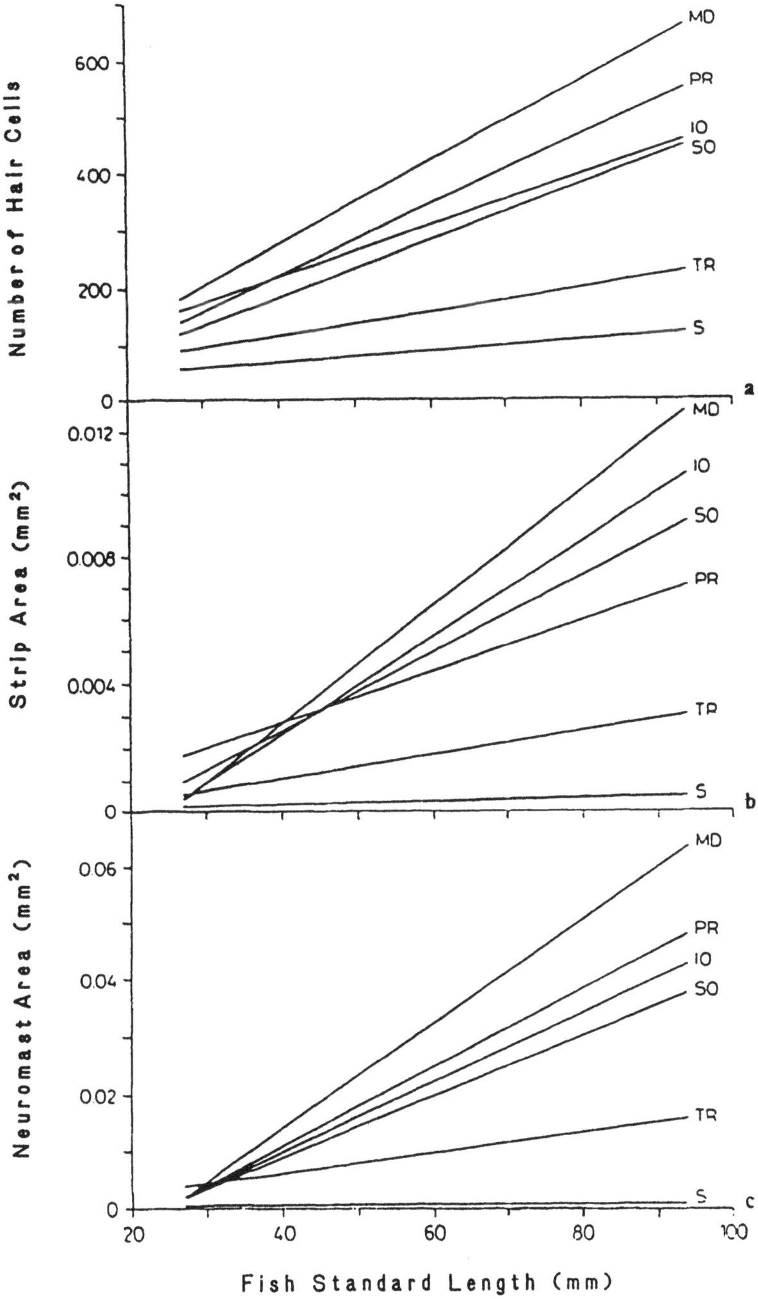

FIGURE 15.2. Growth functions based on linear regressions of hair cell number (a), sensory strip area (b), and neuromast area (c) for various canal neuromasts (see Fig. 15.1 for canal abbreviations) and for superficial (S) neuromasts. (From Janssen et al. 1987. Used with permission from S. Karger AG.)

to the tip of a 16-gauge, 12-cm-long syringe needle which was vibrated by a B&K minishaker. Electronic signals to the minishaker were digitally synthesized and played out by a function generator of a modular hardware system interfaced with an IBM XT. Signals were low-pass filtered (400 Hz cutoff frequency), manually and programmably attenuated, and amplified before being sent to the minishaker. Water velocity levels created by the vibrating sphere were measured with a hot-film anemometry system (TSI, Inc.) (Coombs and Janssen, submitted). The sensing element of the anemometer probe was oriented to respond maximally to flow along the axis of sphere vibration and was placed at the same radial distance from sphere center as fish positioned in behavioral and physiological experiments. Thus, signal levels reported here were measured in the absence of the fish and do not take into account possible influences of the fish's body on the signal at the fish surface.

15.3.2 Behavioral Methods

Behavioral experiments were of two types. In one set of experiments, freely moving sculpin were videotaped while feeding on live *Daphnia*, and the threshold distance at which they detected *Daphnia* was measured as a function of prey position. In a second set of experiments, sculpin were trained to keep fixed positions, and the threshold level at which they detected a vibrating sphere was measured as a function of sphere position and vibration frequency. For both experiments, 5- to 10-cm (standard length; SL) mottled sculpin (*Cottus bairdi*) were collected from Lake Michigan and blinded by enucleation (Hoekstra and Janssen 1986) several days before experiments began.

For the first set of experiments, methods nearly identical to those used by Hoekstra and Janssen (1986) were used. Freely moving, blinded sculpin were videotaped while feeding on live *Daphnia* (10 to 25 in an area 25 cm square and 20 cm deep), and the feeding response was used as an indicator of detection. The prey-to-fish distance and the relative position of *Daphnia* (1.5 to 2 mm in length) were determined with video tracings. Threshold distances were determined by assigning a value of +1 for those distances at which prey were detected and −1 for those not detected. Distances were ranked in order of increasing value, and cumulative scores of assigned values were plotted as a function of distance rank (Barlow et al. 1972). The distance at which the cumulative score was highest was used as an estimate of the distance at which the probability of detection was 0.5.

For the second set of experiments, sculpin were trained to rest on a glass plate which could be moved to a number of different positions around the bottom of a tank measuring 80 cm long, 47 cm wide, and 36 cm high. Water level was maintained at 7 cm above the glass plate, and the center of the vibrating sphere was 12 mm above the glass plate with its driving shaft (syringe needle) perpendicular to it. Two concentric circles visible through the glass plate marked fixed radial distances of 1.5 and 3 cm from sphere center. The glass plate was moved to position the fish with either the base of its tail (caudal peduncle), the middle of its trunk canal (at the junction between first and second dorsal fins), or the middle of its head (at the eye socket) 1.5 cm away from sphere center.

Once the fish was positioned, the experimenter initiated a computer-controlled sequence which determined a random intertrial interval length (from 5 to 15 s) followed by a 5-s trial. Thirty percent of all trials resulted in no vibration of the sphere (blank trials), whereas 70% resulted in sinusoidal vibration of the sphere in the vertical axis. The sculpin's response to the trial was scored by the experimental observer as either a detection response (orientation toward or actual strike at the sphere) or a nondetection response (no response or movement away). If the fish moved before the trial began, he was repositioned, and a new trial sequence was begun.

An adaptive tracking technique was used in which the fish's response to a trial determined the amplitude of the subsequent trial, such that a detection response resulted in decreasing the amplitude of the next trial by 5 dB, and a nondetection response resulted in increasing the amplitude of the next trial by the same amount. The signal level midway between that for each consecutive yes and no response was defined as a transition threshold. Approximately three to six transition thresholds were obtained daily, and if the rate of response to blank trials was greater than 20%, transition thresholds for that day were thrown out. Preliminary data reported here are based on 20 transition thresholds from each of two fish at each vibration frequency and stimulus position.

15.3.3 Physiological Methods

Methods for recording evoked activity in single cells from the posterior lateral line nerve were very similar to those used in recording from saccular fibers of the goldfish auditory system (Fay 1978). Fish of the same size range (7 to 10 cm SL) were anesthetized with MS-222 (200 mg/L), immobilized with intramuscular injections of Flaxedil (2 to 4 μg/g body weight), and clamped onto a respirator tube which provided a continuous flow of aerated water through the mouth and over the gills.

The fish and respirator assemblage were then lowered into a vibration-isolated, water-filled test tank so that only the dorsal surface of the head remained above the water surface. The trunk canal and most of the end organs supplied by the posterior lateral line nerve were submerged. A hole was made in the dorsal cranium to expose the brain, and overlying fluids were aspirated away to reveal where the posterior lateral line nerve entered the medulla. Micropipettes filled with 3 M KCl (tip impedance ranging from 10 to 50 Ω) were positioned with a micromanipulator and advanced through the nerve with a motorized microdrive. Pulsed, low-pass noise signals were used to search for single fibers and spike activity, and stimuli were acoustically and visually monitored. The output of the electrode was amplified within a 300- to 3,000-Hz band width, and spikes from single fibers were discriminated and converted to TTL pulses by a voltage-level discriminator. The voltage-level discriminator was interfaced with an IBM XT through a modular hardware system (Modular Instruments, Inc.) to accumulate on-line peristimulus time and period histograms. Spike activity and stimulus waveforms were also stored on hard disk for subsequent data analysis. The

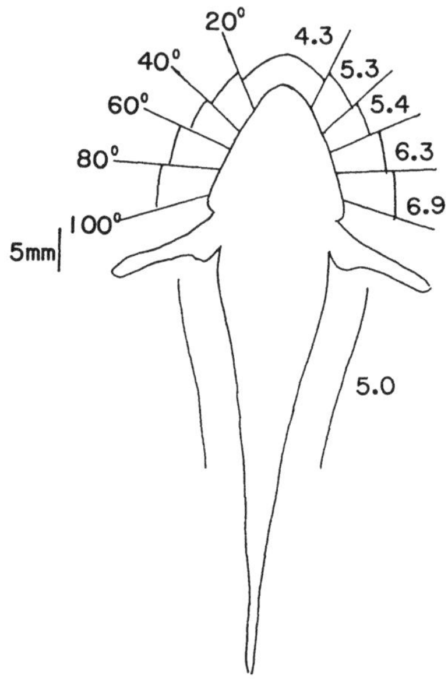

FIGURE 15.3. Threshold distances at which sculpin responded to live *Daphnia* are delineated for angular zones around the head and along the trunk. A two-factor ANOVA (independent variable = threshold distance, dependent variables = angular zone and fish) showed a significant angle effect ($F = 26.5$; $p < .01$; 5, 21 dF) and no fish effect ($F = 2.4$; $p > .03$; 3, 21 dF). A subsequent Tukey's multiple-comparisons test showed that threshold distances were significantly less for the snout and trunk than for the side of the head ($p < .01$).

vibrating sphere was positioned at the same angle and distance (1.5 cm) from the fish as was done for the trunk position in behavioral experiments.

The spike activity evoked by 10 repetitions of a 10-cycle, sinusoidal burst was analyzed in the form of peristimulus time and period histograms for a wide range of stimulus intensities and frequencies. Both average rate and phase-locking (synchronization coefficient) (Anderson 1973) measures were plotted as a function of intensity. The signal intensities at which spike rate began to rise above spontaneous levels and at which the synchronization coefficient was 0.4 were defined as thresholds. Independent estimates of threshold sensitivity using both criteria were possible under most conditions, and these rarely differed by more than 5 dB. In general, thresholds based on rate criteria were used most frequently in constructing tuning curves. However, synchronization criteria were occasionally used at low frequencies in spontaneously active units when it was difficult to drive the unit above spontaneous levels of activity.

15.4 Behavioral Results and Discussion

Results from both sets of behavioral experiments show that the mottled sculpin is not uniformly sensitive along its body, despite the fact that lateral line organs extend nearly the entire length of the fish (Fig. 15.1). The threshold distance for detection of *Daphnia* positioned directly in front of the snout (20° to either side of the midline) was significantly less than for prey detected elsewhere along the head (Fig. 15.3). Likewise, the threshold distance for detection along the trunk was significantly less than the adjacent head regions (60° to 100° off the midline) and essentially unmeasurable at the tail. In general, threshold distances of *Daphnia* detection are very small (within 7 mm) and indicate how close the operating range of this system is for small invertebrate prey. Enger et al. (Chapter 29) report a somewhat longer range (within 2 to 3 cm) for the detection of vertebrate prey by the lateral line system of the bluegill (*Lepomis macrochirus*).

The mottled sculpin results reported here are quite similar to those reported by Hoekstra and Janssen (1986) for which average reactive distances to *Daphnia*, rather than threshold distances, were measured. They are also consistent with results from the second set of behavioral experiments for which sensitivity to a vibrating sphere was measured as a function of stimulus frequency and position (Fig. 15.4). Threshold sensitivity for frequencies below 150 Hz was best for stimuli delivered near the head, intermediate for those near the trunk, and worst for tail stimulation. Best sensitivity for the head was in the nanometer range (−186 dB re: 1 M) at 100 Hz and tail sensitivity was 15 to 20 dB less than that. Although a sample size of two animals is not adequate for a full statistical analysis of these preliminary results, a 3-factor ANOVA shows that these position effects are significant ($p < 0.01$).

Although the physiological and anatomical bases for these sensitivity differences have not been experimentally demonstrated, sensitivity seems to be related to the distribution of different-size canal neuromasts on the mottled sculpin. The tail-to-headward increase in sensitivity is correlated with an increase in canal neuromast size and distribution density. Similarly, reduced sensitivity at the tip of the snout (Fig. 15.3) (see Hoekstra and Janssen 1986 for further details) and the base of the tail is correlated with the absence of canal neuromasts. The correlation between increased sensitivity and canal neuromast size is in accordance with the results of van Netten and Kroese (Chapter 12) on measured and calculated response properties of different-size cupulae.

In general, the overall shapes of the behavioral functions change very little as a function of stimulus position (Fig. 15.4), although there may be some frequency-specific effects occurring at the tail. What is striking about the basic shape of these functions is that when plotted in terms of acceleration, they show a relatively flat frequency response out to around 100 Hz. Although there appears to be a moderate increase (around 10 to 15 dB) in sensitivity from 10 to 40 Hz, this may be a consequence of masked thresholds at lower frequencies. Ambient levels of water motion measured in the experimental tank decline from around −58 dB re: 1 M/s^2 at 10 Hz to levels that were immeasurable at 50 Hz.

►

FIGURE 15.4. Mean behavioral thresholds plotted in terms of displacement (top), velocity (middle), and acceleration (bottom) for the mottled sculpin. Behavioral thresholds from two other species are plotted on the bottom panel for the topminnow (*Aplocheilus lineatus*) in response to surface waves (data from Bleckmann 1980) and for the percid fish *Acerina cernua* in response to electromagnetic vibration of the skin overlying lateral line canals on the head (data from Kuiper 1967).

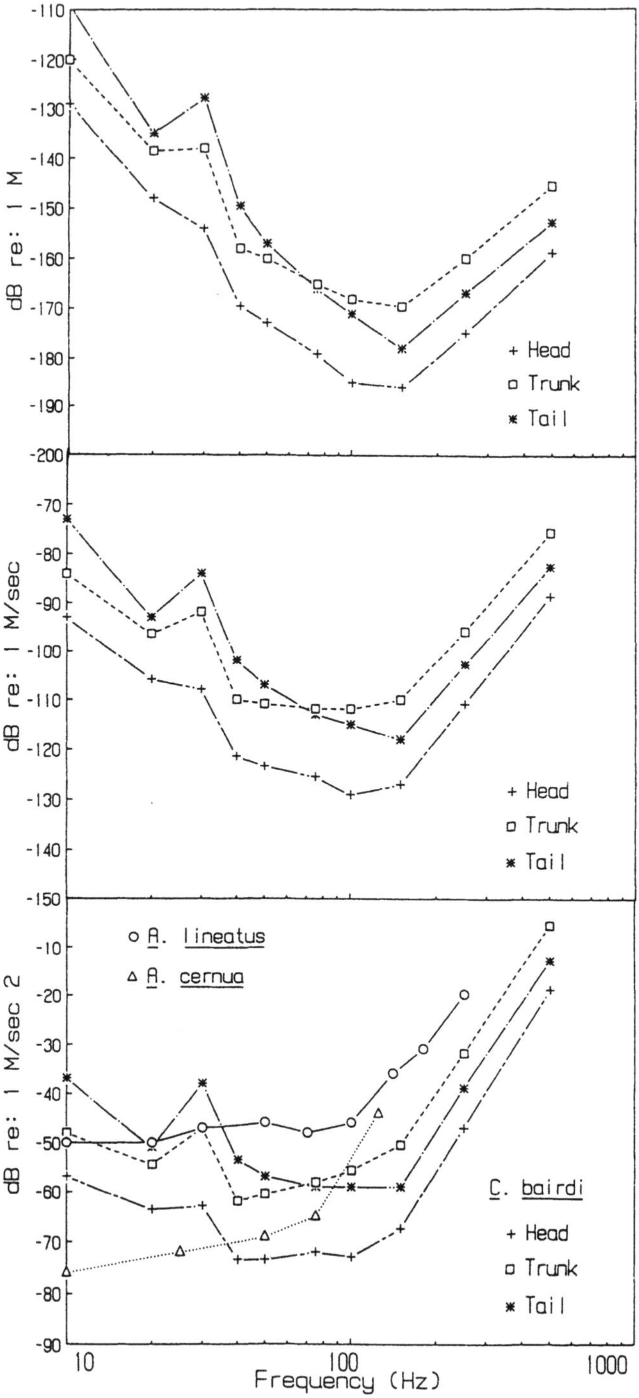
dB re: 1 M
+ Head
□ Trunk
✱ Tail
dB re: 1 M/sec
+ Head
□ Trunk
✱ Tail
dB re: 1 M/sec 2
○ A. lineatus
△ A. cernua
C. bairdi
+ Head
□ Trunk
✱ Tail
Frequency (Hz)

Behavioral thresholds from two other fish species, the topminnow (*Aplocheilus lineatus*) and the ruff (*Acerina cernua*) (Fig. 15.4), are also nearly independent of frequency when plotted as acceleration thresholds. It appears as if the behavioral responses of these fish and the mottled sculpin are governed predominantly by acceleration sensitivity. These results are consistent with both theoretical (Kalmijn Chapter 9) and measured (Gray 1984; Münz 1985; Kroese and Schellart 1987; Denton and Gray Chapter 11; van Netten and Kroese Chapter 12) properties of lateral line canal organs.

15.5 Physiological Results and Discussion

Tuning curves were collected from 40 single cells in 13 fish from the posterior lateral line nerve. Spontaneous activity from these fibers was of three types: (1) silent or < 10 spikes per second, (2) bursting (bimodal interspike interval histogram), and (3) irregular (Poisson-like distribution of intervals). Average spontaneous rates ranged from one to 120 spikes per second. Bursting fibers accounted for over 68% of all units encountered. Spontaneous activity types were similar to those reported for lateral line fibers in other teleost fishes (Münz 1985) and for eighth nerve (saccular) fibers in fish (Enger 1963; Fay 1981; Moeng and Popper 1984) with the exception of types showing regular activity (unimodal interspike interval histogram) which have not yet been encountered in sculpin posterior lateral line fibers. To what extent spontaneous activity may have been evoked by ambient noise levels is unclear, but cessation of respiratory water flow (one potential source of ambient noise) caused no immediate reductions in spontaneous activity. Reductions in spontaneous activity could be observed several minutes after respiratory flow had been turned off, but these were presumably due to anoxia.

In terms of tuning curves, units fell into two basic classes. Forty-three percent of tuning curves were relatively flat when plotted in terms of acceleration, most of them beginning to lose sensitivity by about 75 to 100 Hz (Fig. 15.5). Another 42% of fibers appeared to be more or less tuned in this frequency region, although tuning was very broad ($Q_{10} < 1$) (Fig. 15.6). Taken together, these fibers appear to form more of a gradual continuum than two distinct fiber types, with the high-frequency cutoff of flat units corresponding to best frequency in band-pass units (Fig. 15.7). Average spontaneous rates for these fibers ranged from 0 to 100 spikes per second.

In contrast, a distinctly different class of fibers (15% of total population) showed low-pass characteristics when plotted in acceleration coordinates, with best sensitivity around 30 Hz (Fig. 15.8). As with the other class of fibers, some of these fibers were more highly tuned than others. Mean spontaneous activity for this class was always below 40 spikes per second.

Lateral line fibers in other fish species show similar properties (Fig. 15.9). In general, fibers seem to fall into two categories based on whether responses are nearly independent of frequency when plotted in terms of acceleration (e.g., *A.*

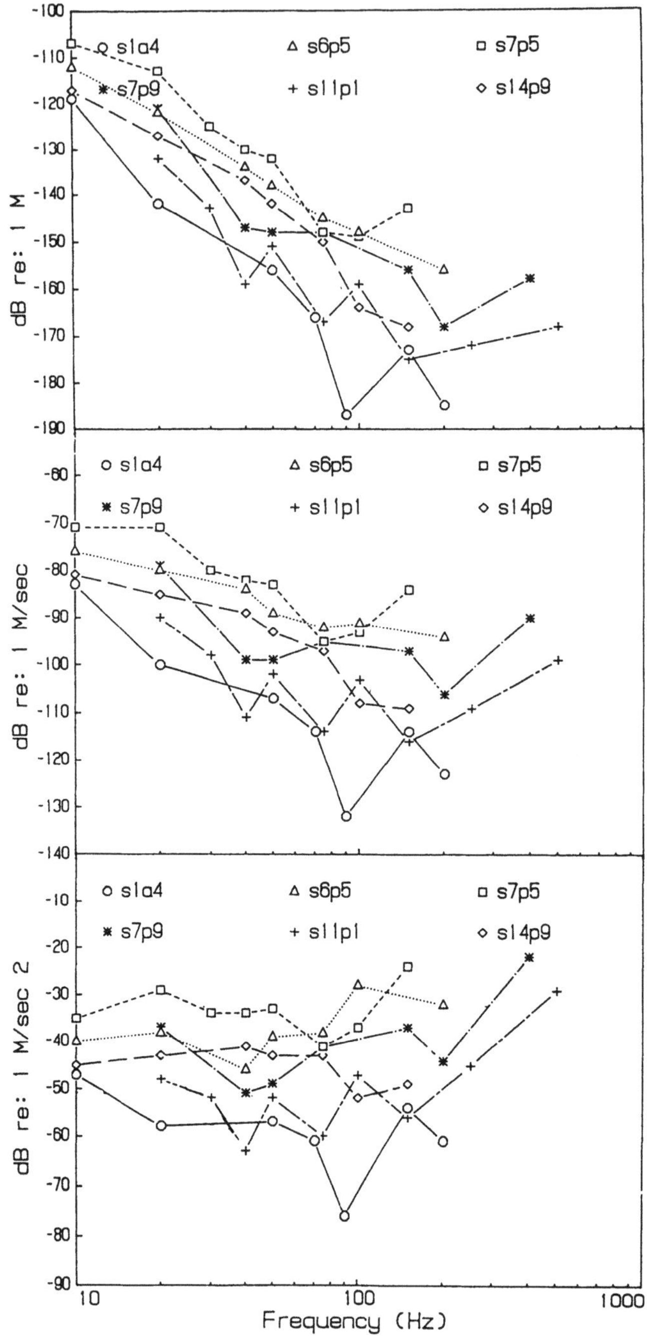

FIGURE 15.5. Tuning curves from posterior lateral line fibers plotted in terms of displacement (top), velocity (middle), and acceleration (bottom). Tuning curves in this figure represent fibers that are relatively untuned when expressed in terms of acceleration.

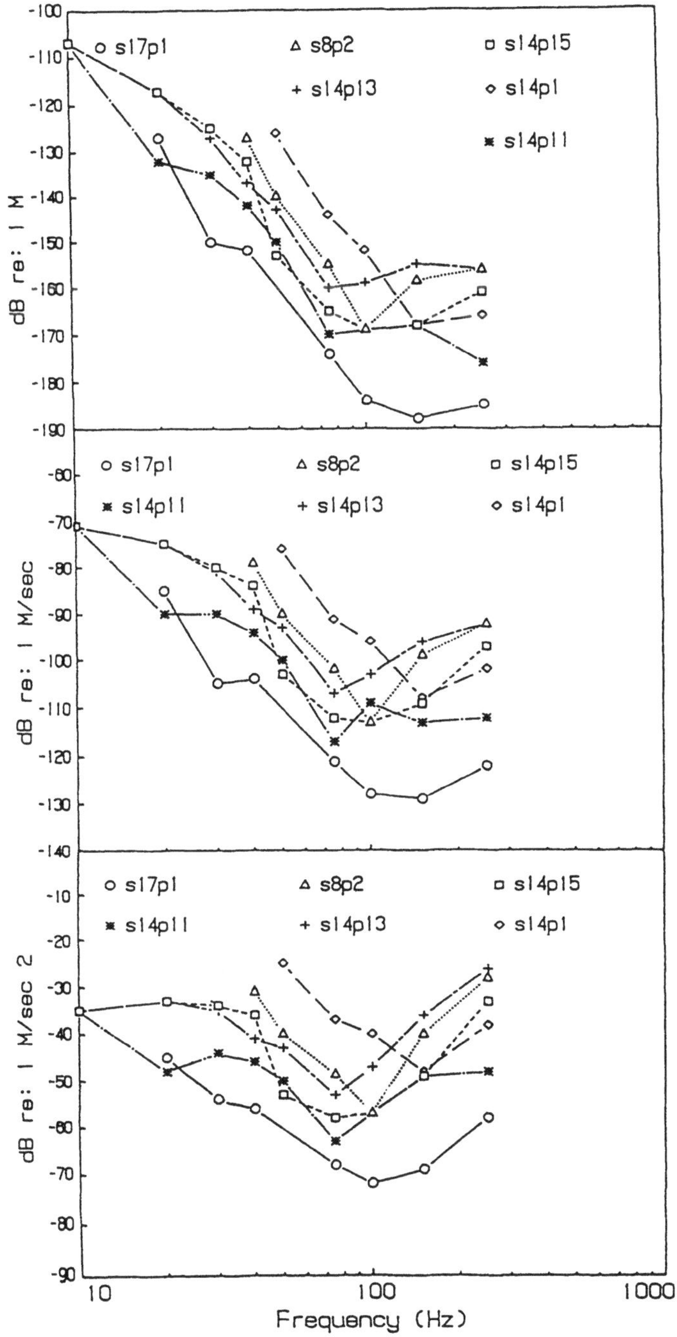

FIGURE 15.6. Representative tuning curves having band pass characteristics when plotted in terms of acceleration.

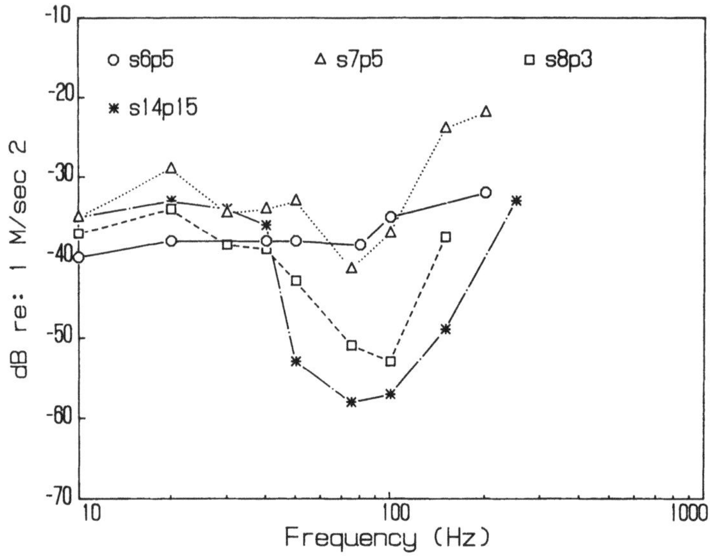

FIGURE 15.7. Tuning curves from both flat (Fig. 15.5) and band-pass (Fig. 15.6) fibers plotted together to show how these two types form a continuum with the high-frequency cutoff of flat fibers corresponding to best frequency in band-pass fibers.

lineatus and sculpin fiber s7p5) or in terms of velocity (e.g, *R. rutilus* and fiber s15p16 from the sculpin). Fibers innervating supraorbital organs of *A. lineatus* have subsequently been shown to respond to the acceleration component of surface wave stimulation (Topp 1983), and evidence for a velocity-sensitive population of fibers innervating superficial neuromasts and an acceleration-sensitive population innervating canal neuromasts has recently been provided for the trout (*Salmo gairdneri*) by Kroese and Schellart (1987). These results are also somewhat consistent with results of comparative studies by Münz (1985, Chapter 14) in which afferent fibers from superficial neuromasts can be distinguished from those of canal neuromasts on the basis of their frequency response. Although the current experimental approach makes it impossible to determine the origin of differently tuned fibers in the mottled sculpin, it is quite reasonable to suggest that low-pass fibers (Fig. 15.8) innervate superficial neuromasts and that flat and band-pass fibers (Figs. 15.5 to 15.7) innervate canal neuromasts.

Because the anesthetic (MS-222) used in this study is known to have pronounced effects on the activity of lateral line efferent and afferent fibers (Spath and Schweickert 1977; Hensel et al. 1975), the question of whether this drug had any effect on the reported results should be raised. The possibility that MS-222 had any significant effects on the results reported here, however, is unlikely, because (1) the flow of the respiratory water containing the anesthetic was not directly over the trunk lateral line system, (2) the experimental tank was continuously flushed with fresh water, (3) spontaneous and evoked activity was robust, (4) the sensitivity of afferent fibers was as good as that measured

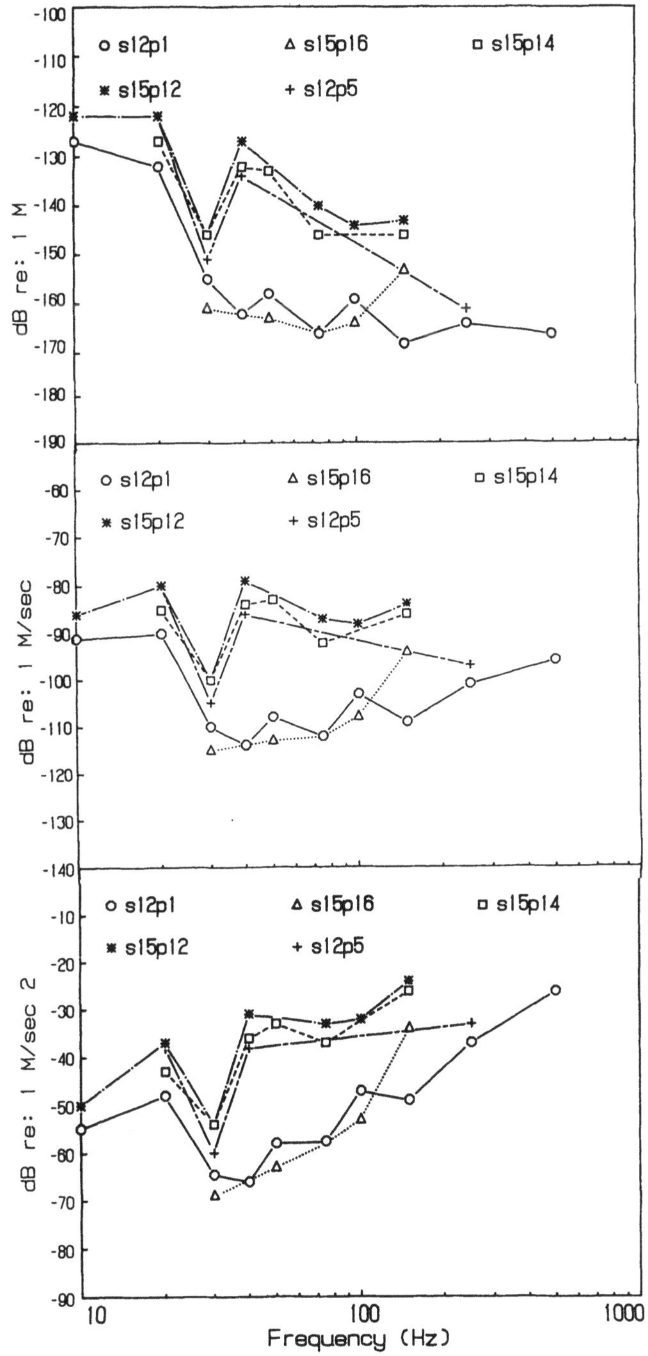

FIGURE 15.8. Representative tuning curves from fibers having low-pass characteristics when plotted in terms of acceleration.

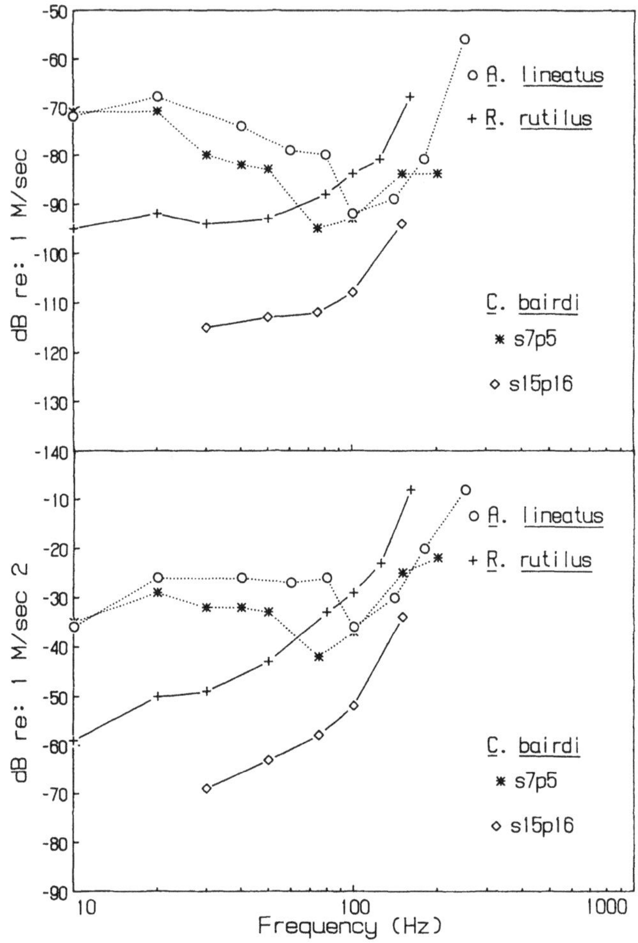

FIGURE 15.9. Tuning curves from lateral line fibers in three fish species plotted in terms of velocity (top) and acceleration (bottom). Data from the topminnow, *A. lineatus*, are from Bleckmann and Topp (1981); those for the roach, *R. Rutilus*, are from Sand (1981).

behaviorally in unanesthetized animals, and (5) MS-222 has much smaller effects on afferent fibers than on efferent fibers (Spath and Schweickert 1977).

15.6 General Discussion

In the mottled sculpin lateral line system, there are clearly single afferent fibers that are as sensitive and as broadly tuned as behavioral functions obtained under the same stimulus conditions (Fig. 15.10). A similar close correspondence between behavioral and physiological functions has been reported for the

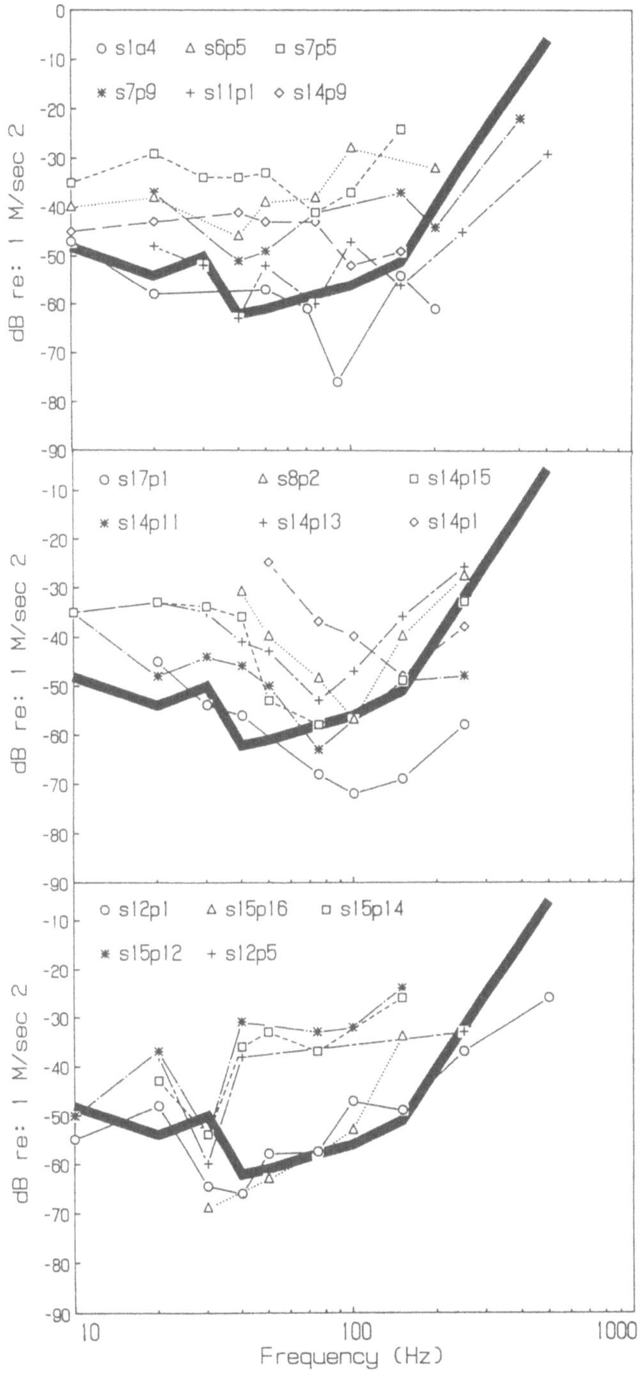

FIGURE 15.10. Behavioral thresholds for the trunk (heavy solid line, from Fig. 15.4) plotted with representative tuning curves from three classes of posterior lateral line fibers (from Figs. 15.5, 15.6 and 15.8). All data are plotted in terms of acceleration.

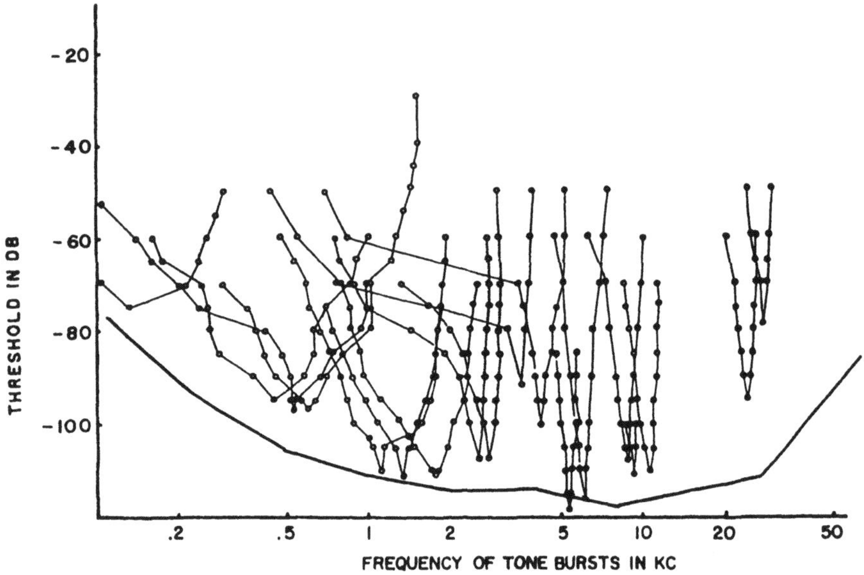

FIGURE 15.11. Collective tuning curves from single eighth nerve fibers in the cat (from Kiang 1965; used with permission from MIT Press) plotted with behavioral thresholds from the same animal (from Heffner and Heffner 1985). Neural tuning curves are plotted in dB re: 4 V input to the earphone and behavioral thresholds have been normalized to the tip of the most sensitive tuning curve (−118 dB).

topminnow (Bleckmann and Topp 1981). This is in marked contrast to the relationship between neural and behavioral measures of sensitivity in most vertebrate auditory systems. In most terrestrial vertebrates that have been studied, there is no one-to-one correspondence between a tuning curve from a single auditory fiber and the behavioral function. Rather, tuning curves from eighth nerve fibers have a continuum of best frequencies that span the frequency range of the behavioral function. Thus, the behavioral audiogram is best described by an array of overlapping tuning curves, each more narrowly tuned than the behavioral function, and each with a different best frequency (Fig. 15.11). For animals such as fish, however, which are restricted to low-frequency hearing (see Fig. 15.12), there may be single fibers that are as broadly tuned as the audiogram (Fay and Ream 1986). These comparisons indicate that integration across differently tuned fibers is not necessary to account for the behavioral sensitivity of the lateral line system and perhaps for other low-frequency systems as well.

Although the lateral line is clearly at the lowest end of the frequency continuum of vertebrate hearing (Fig. 15.12; see also Fig. 32.2 in Chapter 32 by Platt et al.), it rivals most other hair cell systems in terms of best sensitivity. Furthermore, when both behavioral and physiological measures of absolute sensitivity and frequency range of response for the lateral line are compared with those of the fish auditory system, the degree of similarity can be (depending on fish species) quite

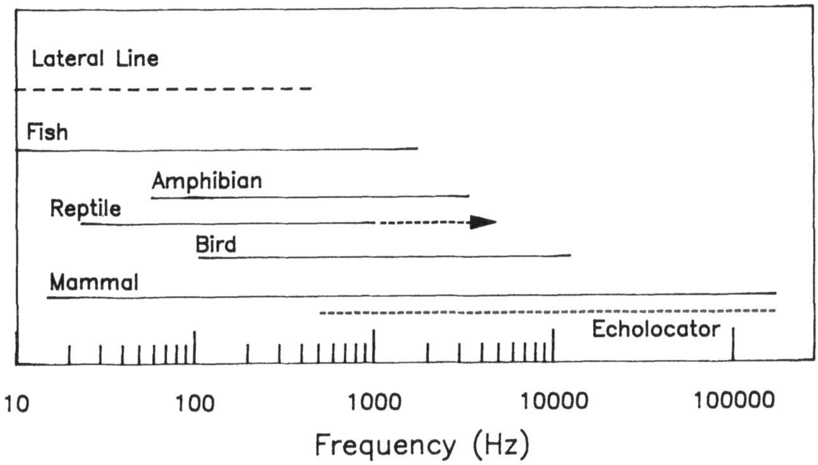

FIGURE 15.12. Frequency range of detection for various vertebrate auditory systems (solid lines) and fish lateral line systems (dashed line). Ranges were determined at 60 dB above best sensitivity. Data from Fay (1988).

striking (Platt et al. Chapter 32). Although there may be some overlap between the two systems in terms of sensitivity and frequency response, it should be emphasized that the two systems are probably quite different in the way in which they are stimulated and with respect to the biologically relevant stimuli to which they respond (Dijkgraaf Chapter 2; Kalmijn Chapter 9; Enger et al. Chapter 29; Platt et al. Chapter 32).

15.7 Summary and Conclusions

The peripheral lateral line system of the mottled sculpin, like that of most species for which it has been studied, it adapted for the detection of low-frequency (below 200 and 300 Hz) water movements. Although the sensitivity of the system to sinusoidal water movements varies as a function of stimulus position, the overall band width of response is similar. Best sensitivity (−185 dB re: 1 M, −73 dB re: 1 M/s^2) near the head is correlated with the large canal neuromasts found in head canals, especially the mandibular canal, intermediate sensitivity with small canal neuromasts on the trunk, and worst sensitivity with the absence of canal neuromasts on the tail. Since behavioral measures of sensitivity are largely independent of frequency when plotted in terms of acceleration, the feeding response of the mottled sculpin may be driven primarily by acceleration sensitivity.

The sensitivity and frequency range of one class of single afferent fibers innervating trunk neuromasts are sufficient to account for behavioral sensitivity measured at the trunk. A second class of afferent fibers from the trunk has a frequency response profile that is distinctly different from that of other fibers and less similar to behavioral measures of sensitivity. From their frequency response it is suggested that these fibers respond to the velocity component of water motion, that they innervate superficial neuromasts, and that their contribution to the particular behavioral response measured here is minimal.

Behavioral sensitivity of the lateral line system of the mottled sculpin and other fish species is nearly as good as that measured for other vertebrate auditory systems. Unlike auditory systems in terrestrial vertebrates, however, integration across differently tuned afferent fibers does not appear to be necessary to account for behavioral sensitivity. In terms of overall sensitivity and band width of response, the lateral line system is most similar to other low-frequency hair cell systems that are acceleration-sensitive, like the otolithic end organs of the fish ear.

Acknowledgments. Support for this research came from the National Institutes of Health (Program Project Grant P01NS2191) and from the Max Goldenberg Foundation. Many thanks to Shirley Pride for enduring "fishpan" hands during the tedious hours of collecting behavioral data and to A. Kroese for many insightful and helpful comments on an earlier draft of this chapter.

References

Alnaes E (1973) Two types of lateral line afferents in the eel. Acta Physiol Scand 87:535–548.

Anderson D (1973) Quantitative model for the effects of stimulus frequency upon synchronization of auditory nerve discharges. J Acoust Soc Am 54:361–364.

Barlow RE, Bartholomew DJ, Bremmer JM, Brunk HD (1972) Methods of Statistical Inference under Order Restrictions. New York: Wiley.

Bleckmann H (1980) Reaction time and stimulus frequency in prey localization in surface-feeding fish *Aplocheilus lineatus*. J Comp Physiol 140:163–172.

Bleckmann H, Topp G (1981) Surface wave sensitivity of the lateral line organs of the top-minnow *Aplocheilus lineatus*. Naturwissenshaften 68:624–625.

Coombs S, Janssen J, Webb JF (1988) Diversity of lateral line systems: Evolutionary and functional considerations. In: Atema Y, Fay RR, Popper AN, Tavolga WN (eds) Sensory Biology of Aquatic Animals. New York: Springer-Verlag, pp. 553–595.

Coombs S, Janssen J (submitted) Hot film anemometry for measuring lateral line stimuli. J Acous Soc Am.

Denton EJ, Gray JAB (1982) The rigidity of fish and patterns of lateral line stimulation. Nature 297:679–681.

Dijkgraaf S (1962) The functioning and significance of the lateral line organs. Biol Rev 38:51–105.

Enger PS (1963) Single unit activity in the peripheral auditory system of a teleost fish. Acta Physiol Scand (Suppl 3) 59:9–48.

Fay RR (1978) Coding of information in single auditory nerve fibers of the goldfish. J Acoust Soc Am 63:136–146.

Fay RR (1981) Coding of acoustic information in the eighth nerve. In: Tavolga WN,

Popper AN, Fay RR (eds) Hearing and Sound Communication in Fishes. New York: Springer-Verlag, pp. 189–222.

Fay RR (1988) Hearing in Vertebrates: A Psychophysics Databook. Winnetka, IL: Hill-Fay Associates, 628 pp.

Fay RR, Ream TJ (1986) Acoustic response and tuning in saccular nerve fibers of the goldfish (*Carassius auratus*). J Acoust Soc Am 79:1883–1895.

Gray JAB (1984) Interaction of sound pressure and particle acceleration in the excitation of the lateral-line neuromasts of sprats. Proc R Soc Lond B 220:299–325.

Heffner RS, Heffner HE (1985) Hearing range of the domestic cat. Hear Res 19:85–88.

Hensel H, Bromm B, Nier K (1975) Effect of ethyl m-aminobenzoate (MS 222) on ampullae of Lorenzini and lateral line organs. Experentia 31:958–961.

Hoagland H (1932) Impulses from sensory nerves of catfish. Proc Natl Acad Sci USA 18:701–705.

Hoagland H (1933a) Electrical responses from the lateral line nerve of catfish. I. J Gen Physiol 16:695–714.

Hoagland H (1933b) Quantitative analysis of responses from lateral line nerves of fishes. II. J Gen Physiol 16:715–732.

Hoagland H (1934) Electrical responses from the lateral line nerves of catfish. III. J Gen Physiol 17:77–82.

Hoekstra D, Janssen J (1985) Non-visual feeding behavior of the mottled sculpin, *Cottus bairdi*, in Lake Michigan. Environ Biol Fish 12:111–117.

Hoekstra D, Janssen J (1986) Lateral line receptivity in the mottled sculpin (*Cottus bairdi*). Copeia 1986:91–96.

Janssen J, Coombs S, Hoekstra D, Platt C (1987) Anatomy and differential growth of the lateral line system of the mottled sculpin, *Cottus bairdi* (Scorpaeniformes: Cottidae). Brain Behav Evol 30:210–229.

Katsuki Y, Yoshino S, Chen J (1950) Action currents of the single lateral line nerve of fish. I. On the spontaneous discharge. Jpn J Physiol 1:87–99.

Katsuki Y, Yoshino S, Chen J (1951) Action current of the single lateral line nerve fiber of fish. II. On the discharge due to stimulation. Jpn J Physiol 1:179–194.

Kiang N (1965) Discharge Patterns of Single Fibers in the Cat's Auditory Nerve. Cambridge, MA: MIT Press, 154 pp.

Kroese ABA, Schellart NAM (1987) Evidence for velocity-and acceleration-sensitive units in the trunk lateral line of the trout. J Physiol 394:13P.

Kuiper JW (1967) Frequency characteristics and functional significance of the lateral line organ. In: Cahn P (ed) Lateral Line Detectors. Bloomington: Indiana University Press, pp. 105–122.

Moeng RS, Popper AN (1984) Auditory response of saccular neurons of the catfish, *Ictalurus nebulosus*. J Comp Physiol A 155:615–624.

Montgomery JC, Macdonald JA (1987) Sensory tuning of lateral line receptors in antarctic fish to the movements of planktonic prey. Science 235:195–196.

Münz H (1985) Single unit activity in the peripheral lateral line system of the cichlid fish *Sarotherodon niloticus* L. J Comp Physiol A 157:555–568.

Russell I (1976) Amphibian lateral line receptors. In: Llinas R, Precht W (eds) Frog Neurobiology. New York: Springer-Verlag, pp. 513–550.

Sand O (1981) The lateral line and sound reception. In: Tavolga WN, Popper AN, Fay RR (eds) Hearing and Sound Communication in Fishes. New York: Springer-Verlag, pp. 459–480.

von Schriever H (1935) Aktionspotentiale des N. Lateralis bei Reizung der Seitenorgane von Fischen. Pflugers Arch 235:771–784.

Spath M, Schweickert W (1977) The effect of metacaine (MS222) on the activity of the efferent and afferent nerves in the teleost lateral line system. Pharmacology 297:9–16.
Suckling EE, Suckling JA (1950) The electrical response of the lateral line system of fish to tone and other stimuli. J Gen Physiol 34:1–8.
Suckling EE, Suckling JA (1964) Lateral line as a vibration receptor. J Acoust Soc Am 36:2214–2216.
Suckling JA (1967) Trunk lateral line nerves: Some anatomical aspects. In: Cahn PH (ed) Lateral Line Detectors. Bloomington: Indiana University Press, pp. 45–52.
Suga N (1967) Electrosensitivity of canal and free neuromast organs in gymnotid electric fish. J Comp Neurol 131:453–458.
Topp G (1983) Primary lateral line response to water surface waves in the topminnow *Aplocheilus lineatus* (Pisces, Cyprinodontidae). Pflugers Arch 397:62–67.
Wubbels RJ (1988) Frequency response of single unit afferents innervating the lateral line system of *Acerina cernua*. Eighth International Symposium on Hearing, Rijksuniversiteit Groningen, Groningen, The Netherlands.

Part IV
Central Processing: Neuroanatomy and Physiology

16
Central Mechanosensory Lateral Line Centers and Pathways Among the Elasmobranchs

ROBERT L. BOORD and JOHN C. MONTGOMERY

The central neuronal pathways and connections required to process mechanosensory lateral line information probably arose, concomitantly with the peripheral receptors, in the first vertebrates. The cartilaginous fishes (Chondrichthyes) share the mechanosensory component of the lateral line system with all other anamniotes, although there are vast differences in the central nervous systems of those vertebrates that possess lateral line sensory capabilities. Even among the elasmobranchs, there are different levels of brain organization and profound variations within each level (Northcutt 1978). It is important to identify mechanosensory lateral line centers and their fiber relations with other octavolateralis centers among those elasmobranchs that possess different patterns of brain organization, in order to establish homologies and to better understand the evolution of mechanoreception in chondrichthyans. However, our knowledge of the central mechanosensory lateral line pathways is limited to only a few elasmobranchs, e.g., the batoids (thornback guitarfish *Platyrhinoidis triseriata*, clearnose skate *Raja eglanteria*, and the little skate *Raja erinacea*), the squalomorph sharks (spiny dogfish *Squalus acanthias*), and the galeomorph sharks (carpet shark *Cephaloscyllium isabella*). Central lateral line pathways have been studied mostly in skates that are considered advanced elasmobranchs in terms of brain organization (Northcutt 1978), but some information is available in the spiny dogfish and carpet shark which, in contrast to skates, are more primitive in terms of brain organization. The purpose of this chapter is to delineate some of the central mechanosensory lateral line centers and their connections at each level of the neuraxis and to establish homologies in these few species of cartilaginous fishes.

16.1 Lateral Line Nerves

McCready and Boord (1976) showed, by gross dissection of the superficial roots of the fifth, seventh, eighth cranial nerve complex in the smooth dogfish *Mustelus*, that the lateral line nerves enter the medulla separately. Although they did not trace nerve fibers from the peripheral receptors, they speculated that the ventral root of the anterior lateral line nerve and the posterior lateral line nerve

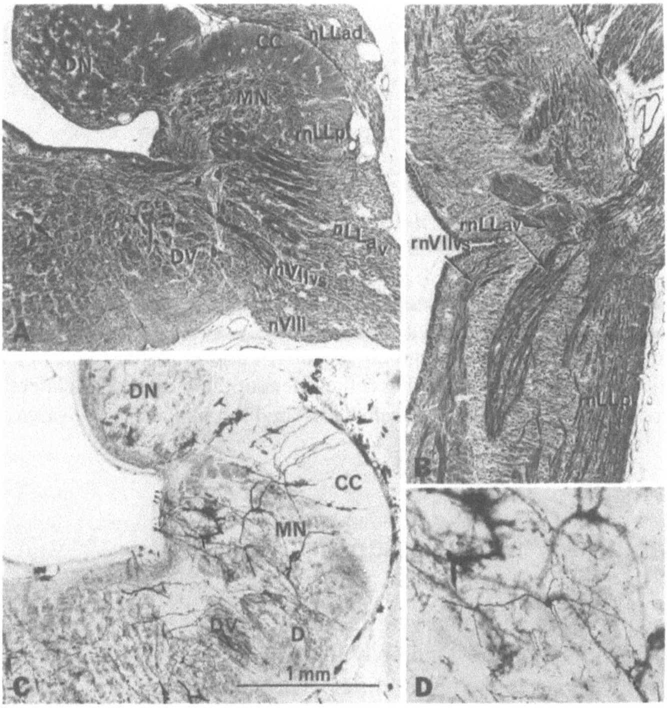

FIGURE 16.1. (A) Cross section through the medulla of the carpet shark *Cephaloscyllium isabella* at a level of the ventral root of the anterior lateral line nerve showing the topographic relations of the medial octavolateralis nucleus. Silver stain. Bar scale shown in C = 1.0 mm. (B) Horizontal section through the medulla of the carpet shark showing the root fibers of the mechanosensory lateral line nerves within nucleus medialis. Silver stain. Bar scale shown in C = 1.0 mm. (C) A Golgi-stained section showing large multipolar neurons and a small multipolar neuron within the medial nucleus of the carpet shark. Bar scale = 1.0 mm. (D) Enlargement of the small multipolar neuron of Fig. 16.1C.

were pure special somatic sensory nerves that innervated mechanoreceptors of the head and the remainder of the body, respectively. Even lateral line fibers that innervate those sense organs of that segment of the body supplied by the glossopharyngeal nerve enter the medulla with the posterior lateral line nerve. This has subsequently been verified by electrophysiological experiment in the thornback guitarfish; e.g., the ventral root of the anterior lateral line nerve consists of fibers that innervate head mechanosensory neuromasts (Bodznick and Northcutt 1980). There is no evidence that neurons innervating free neuromasts enter the medulla as a separate root, but since free neuromasts occur predominantly in the skin of the body and tail, the majority are carried by the posterior lateral line nerve. There are relatively few free neuromasts on the head of elasmobranchs, and these are innervated by fibers of the external mandibular branch of the anterior lateral line nerve (Tester and Kendall 1967).

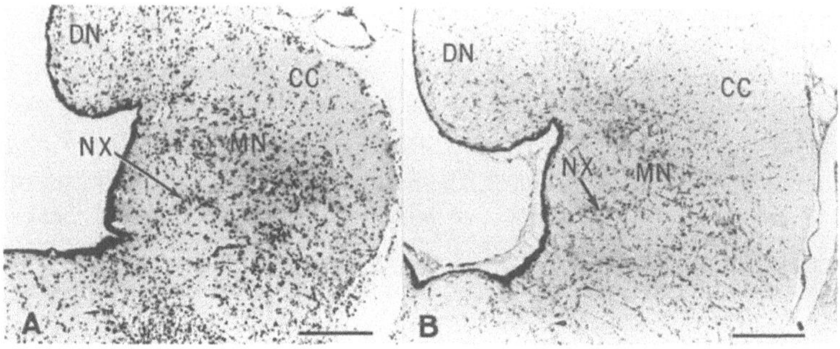

FIGURE 16.2. Cross sections through the medulla at a level rostral to the ventral root of the anterior lateral line nerve showing cell clusters of the medial octavolateralis nucleus of the carpet shark *Cephaloscyllium isabella* (A) and the spiny dogfish *Squalus acanthias* (B). Cell stain. Bar scale = 0.5 mm.

16.2 Medullary Mechanosensory Centers

The lateral line fibers that innervate mechanoreceptors in all cartilaginous fishes, as well as all other nonbony and bony fishes, terminate within a column of cells derived from the alar plate of the rhombencephalon, termed the medial octavolateralis nucleus. The medial octavolateralis nucleus occupies a similar topographic position in all anamniotes and is homologous among all anamniotic vertebrate radiations (Northcutt 1984).

The medial octavolateralis nucleus and its overlying cerebellar crest are termed the posterior lateral line lobe (Fig. 16.1A). The most prominent cells of the medial nucleus are large multipolar neurons whose ventral dendrites arborize extensively among the incoming fibers and whose dorsal dendrites extend into the cerebellar crest and ramify among the parallel fibers (Paul and Roberts 1977). In addition to the large multipolar neurons of diverse sizes and shapes, small interneurons occur among the large multipolar cells (Fig. 16.1C,D). Although the large neurons occur throughout the nucleus, they are most dense immediately beneath the molecular layer of the cerebellar crest, where they occur in clusters (Fig. 16.2). There is no evidence that these clusters of neurons are functionally related to particular classes of peripheral receptors—i.e., neuromasts versus pit organs—but there is a somatotopic arrangement of first-order mechanosensory afferents within the medial nucleus (Koester 1983, Bodznick and Schmidt 1984). The ascending and descending roots of the anterior and posterior lateral line nerves and their terminal fields are confined to the medial two-thirds and lateral one-third of the medial nucleus, respectively (Fig. 16.1A,B). The central terminal field of the anterior lateral line nerve is divided into dorsal and ventral portions by a plate of triangular and ovoid neurons, situated within the ventromedial part of the medial nucleus and termed nucleus X (Northcutt 1978). Mechanosensory afferents of each ramus of the anterior lateral line nerve terminate, with considerable overlap, on either side of nucleus X (Bodznick and Schmidt 1984).

On the basis of [^{14}C]2-deoxyglucose autoradiography and evoked potential recordings following acoustic stimulation, Corwin and Northcutt (1982) believe that nucleus X is a component of an auditory lemniscal pathway in the thornback guitarfish *Platyrhinoidis triseriata*. Since no increase in metabolic activity or responses to acoustic stimuli are detectable in known projection sites of lateral line nerves within the medial nucleus, they suggest that the lateral line may not be involved in the processing of auditory information. However, Bodznick and Schmidt (1984) offer evidence that the cells of nucleus X receive a mechanosensory lateral line input and suggest that auditory and lateral line inputs may converge at this level.

The medial octavolateralis nucleus also receives primary octaval afferents from each inner-ear end organ, particularly from the macula of the saccule (Barry 1987). Octaval root fibers invade all parts of the medial nucleus and therefore necessarily overlap mechanosensory lateral line terminal fields, but whether they terminate on secondary neurons of the medial nucleus has not been anatomically demonstrated. There is anatomical evidence that the dorsal dendrites of the large neurons of the medial nucleus receive lateral line as well as other somatosensory input from the vestibulolateral lobe of the cerebellum via the granule cell-parallel fiber route, the details of which have been reported by Schmidt and Bodznick (1987) in the little skate. The vestibulolateral lobe of the cerebellum is divided into a pars medialis and a pars lateralis. Pars medialis, which is sometimes called the lower lip, bridges the roof of the fourth ventricle and consists of a typical cerebellar organization—i.e., Purkinje cell, granule cell, and molecular layers. Pars lateralis (the auricle of the cerebellum) is composed predominantly of populations of granule cells which are continuous with the granule cell layer of pars medialis.

The granule cells of the vestibulolateral lobe of the cerebellum are the source of parallel fibers to the molecular layer (cerebellar crest) that overlies nucleus medialis (as well as the electrosensory dorsal nucleus). In skates, pars lateralis consists of a dorsal granular ridge and a lateral granular layer that are the source of parallel fibers to the molecular layers of the electrosensory dorsal nucleus and the mechanosensory medial nucleus, respectively. The granule cells of pars medialis also supply parallel fibers to the molecular layer of the medial nucleus but not to the dorsal nucleus. The lateral granule cell layer receives an input from the octaval nerve, mechanosensory lateral line nerves, lateral funicular nucleus, anterior octaval nucleus, contralateral lateral granular layer, and the reticular formation. The medial division of the vestibulolateral lobe receives an input from the octaval and mechanosensory lateral line nerves, descending octaval nuclei, lateral funicular nuclei, and the nucleus of the medial longitudinal fasciculus (Barry 1987; Schmidt and Bodznick 1987; Koester 1983).

Similar inputs to the medial nucleus from the vestibulolateral lobe of the cerebellum of sharks can be expected but have not been proved. Montgomery (1981) reports that in the carpet shark (*Cephaloscyllium isabella*), vestibular information is supplied to the medial nucleus via parallel fibers from the granule cells of both the medial (upper leaf of the auricle) and lateral (lower leaf of the auricle) parts of the vestibulolateral lobe of the cerebellum. The variety of large multipolar cells, perhaps arranged in functional clusters, interconnecting interneurons, plethora of synaptic endings (Paul et al. 1977), and proprioceptive, octaval, and mechanosensory lateral line inputs strongly suggest that there is an integration of information even at the level of the primary mechanosensory medial nucleus. The integration of mechanosensory lateralis with other somatosensory modalities (octaval, proprioceptive, tactile) may provide the animal with information on body and fin position (Barry 1987; Schmidt and Bodznick 1987). However, multimodal inputs to anatomically specified large multipolar neurons of the medial nucleus have not been directly demonstrated (Bleckmann and Bullock, Chapter 19).

16.3 Mesencephalic Mechanosensory Centers

Efferents of the medial nucleus are the axons of the large multipolar neurons, which relay information to toral levels of the midbrain and give rise to extensive commissural connections. In *Raja*, where the wall of the mesencephalon is greatly expanded and the ventricle much reduced relative to that of *Squalus* and *Cephaloscyllium*, a large nuclear complex, termed the lateral mesencephalic nucleus, consists of lateral, mediodorsal, medioventral, (Fig. 16.3A), and anterior subdivisions (Boord and Northcutt 1982). Of these, the mediodorsal nucleus is the main recipient of ascending mechanosensory lateral line fibers, as shown in *Raja* by anatomical methods (Boord and Northcutt 1982) and in *Platyrhinoidis* by electrophysiological methods (Bleckmann et al. 1987). The latter authors show a rostrocaudal point-to-point somatotopy within the mediodorsal nucleus, with the head represented rostrally.

In contrast to batoids (Fig. 16.3A), the lateral mesencephalic nucleus of *Cephaloscylium* (Fig. 16.3B) and *Squalus* consists of populations of cells that are not readily separated into individual nuclei. Though at any particular level of the lateral mesencephalic nucleus aggregations of cells may be recognized, these cell aggregations are not cytologically homogeneous and cannot be identified as being the recipient of ascending mechanosensory lemniscal fibers from any particular part of the medial nucleus. Field potentials recorded from the contralateral lateral mesencephalic nucleus of *Cephaloscyllium* in response to either electric fields or to shock to the posterior lateral line nerve reveal that the peaks of mechanosensory field potentials tend to be less deep than those of the electrosensory fields (Fig. 16.4). Whereas there is a small separation between electrosensory and mechanosensory field potentials, there is much overlap between these two lateral line modalities.

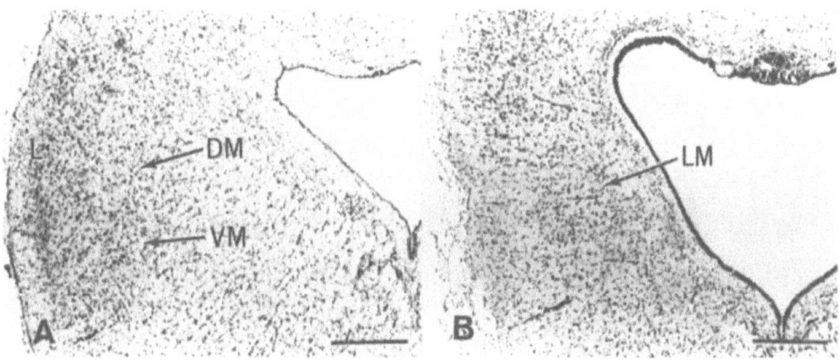

FIGURE 16.3. Cross section through the mesencephalon, at a level of the sixth nerve roots, of the clearnose skate *Raja eglanteria* (A) and the carpet shark *Cephaloscyllium isabella* (B) showing the lateral mesencephalic nuclear complex. Bar scale = 0.5 mm.

The ascending pathways of the medial octavolateralis nucleus of *Cephaloscyllium*, as revealed by the transport of horseradish peroxidase, are shown in Fig. 16.5. HRP injections were restricted to the caudal pole of the medial nucleus in order to minimize the uptake of HRP by nonmechanosensory lateral line axons (Boord and Northcutt 1988). Successful uptake of the enzyme by the large projectional neurons of nucleus medialis is indicated by extensive labeling of their dorsal dendrites that radiate within the cerebellar crest. Efferents from MON include extensive commissural connections (Figs. 16.5C–E, 16.6B) and ipsilateral and contralateral lemnisci (Fig. 16.5).

Efferents from the medial nucleus join the arcuate system at medullary levels and course to the medial nucleus of the opposite side (Figs. 16.5E, 16.6B) where they terminate. Numerous labeled cells in the contralateral medial nucleus indicate that commissural connections are extensive. Labeled cells also occur within the reticular formation, within the contralateral anterior and descending octaval nuclei, and along the ascending lemniscal pathways. No labeled neurons occur within a column of small to medium-size cells of various shapes that is termed nucleus B (Smeets and Nieuwenhuys 1976; Smeets et al. 1983). Nucleus B appears to receive an ipsilateral projection from the medial nucleus in *Cephaloscyllium* (Fig. 16.5D). A second-order mechanosensory input to nucleus B is present in *Raja* (Bodznick unpublished) and probably in all elasmobranchs, although Boord and Northcutt (1988) could not unequivocally demonstrate this projection in *Squalus*. Nucleus B, at caudal levels, is situated immediately adjacent to the descending trigeminal tract and extends rostrad to the level of the cerebellar peduncle. It receives substantial afferent connections from the electrosensory dorsal octavolateralis nucleus and occupies a feedback position in the descending electrosensory pathways by projecting back to the dorsal nucleus (Bodznick and Boord 1986). In this respect, nucleus B is comparable to the teleostean nucleus praeeminentialis that consists of mechanosensory and electrosensory subdivisions (Sas and Maler 1983, Finger and Tong 1984),

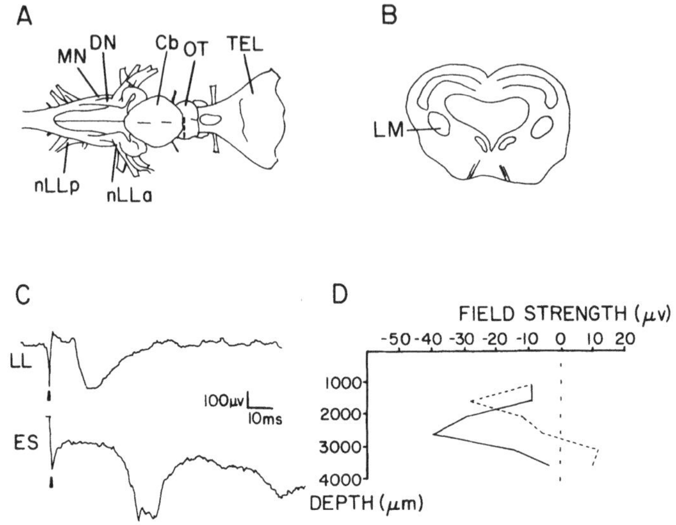

FIGURE 16.4. (A) Diagram of a dorsal view of the brain of the carpet shark *Cephaloscyllium isabella*. The transverse dotted line on the surface of the midbrain (OT) shows the position of the electrode penetrations. (B) Transverse section at the level indicated by the dotted line in A showing the position of the lateral mesencephalic nucleus (LM). (C) An example of lateral line mechanosensory (LL) and electrosensory (ES) field potentials recorded at one electrode location. The stimulus occurred at the time shown by the arrow. (D) Vertical distribution of field potentials in one electrode penetration through the central part of the lateral mesencephalic nucleus showing the peak of the electrosensory field at a greater depth than that of the mechanosensory lateral line field.

but functional subdivisions within nucleus B of elasmobranchs have not been demonstrated.

Only a small proportion of the fibers of the lemnisci reach midbrain levels to terminate within the lateral mesencephalic nucleus (Fig. 16.5A), which indicates that a large proportion terminate at lower levels of the brain stem. Those labeled fibers that enter the caudoventral pole of the lateral mesencephalic nucleus orient parallel with respect to the long axes of the fusiform, triangular, and polygonal cells and form a distinct terminal field (Fig. 16.6D). The placement of the injection precludes mapping the rostrocaudal distribution of the mechanosensory terminal field within the lateral mesencephalic nucleus of *Cephaloscyllium*. In *Squalus*, the mechanosensory lemniscus terminates within the ventromedial part (Boord and Northcutt 1988). Clearly, the lateral mesencephalic nucleus of *Squalus* and *Cephaloscyllium* is a homologue of the lateral mesencephalic complex of *Raja*; however, the mediodorsal nucleus of *Raja* cannot be homologyzed with a particular subdivision of the lateral mesencephalic nucleus of sharks. Horseradish peroxidase injections into the wall of

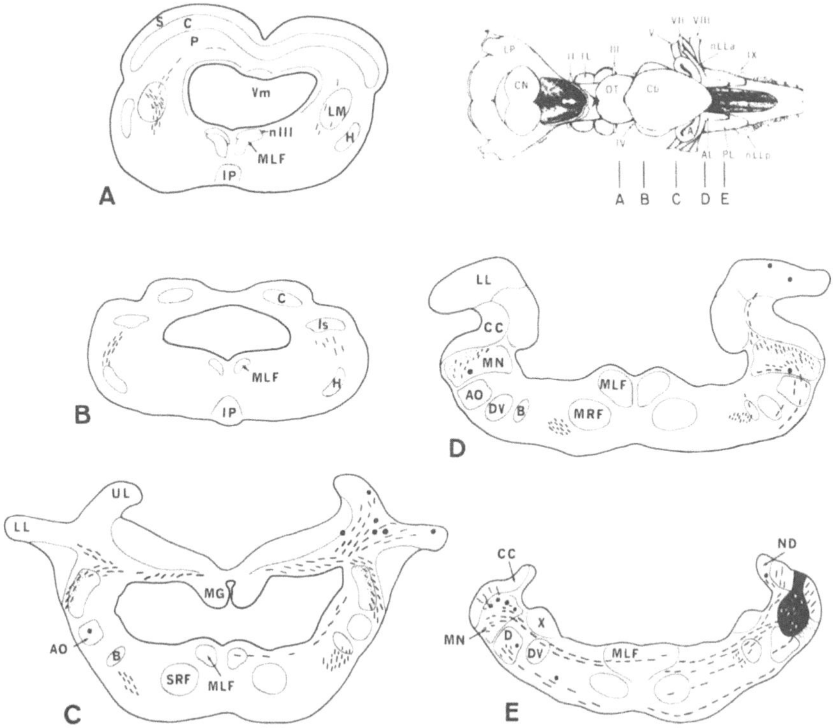

FIGURE 16.5. Diagrams of transverse sections through the brainstem of the carpet shark, at the levels indicated in the diagram of the dorsal view of the brain, showing the course of labeled axons (dashes) and terminals (dots) following an HRP injection in the medial octavolateralis nucleus. Large, solid circles represent retrogradely labeled cells.

the midbrain including the lateral mesencephalic nucleus result in the retrograde labeling of large numbers of the large multipolar cells of the contralateral medial nucleus (Fig. 16.7A,B). They also include labeled cells of the dorsal nucleus, nucleus B, nucleus X, anterior octaval nucleus, and an occasional cell of the medullary reticular formation. This precludes speculation as to what parts of the lateral mesencephalic nucleus in *Cephaloscyllium* receive connections from different medullary octavolateralis centers, as is the case in *Raja*.

The anterior nucleus of the lateral mesencephalic complex of *Raja* does not appear to receive a direct projection from the medullary medial octavolateralis nucleus (Boord and Northcutt 1982) but does receive a minor input from the dorsomedial nucleus. Neither are mechanosensory lateral line fibers of the second order traceable rostral to the lateral mesencephalic nucleus in *Squalus* or *Cephaloscyllium*. HRP injections into the anterior nucleus of *Raja* reveals no labeled cells in the medullary medial nucleus. A homologue of the anterior nucleus of skates has not been identified in sharks.

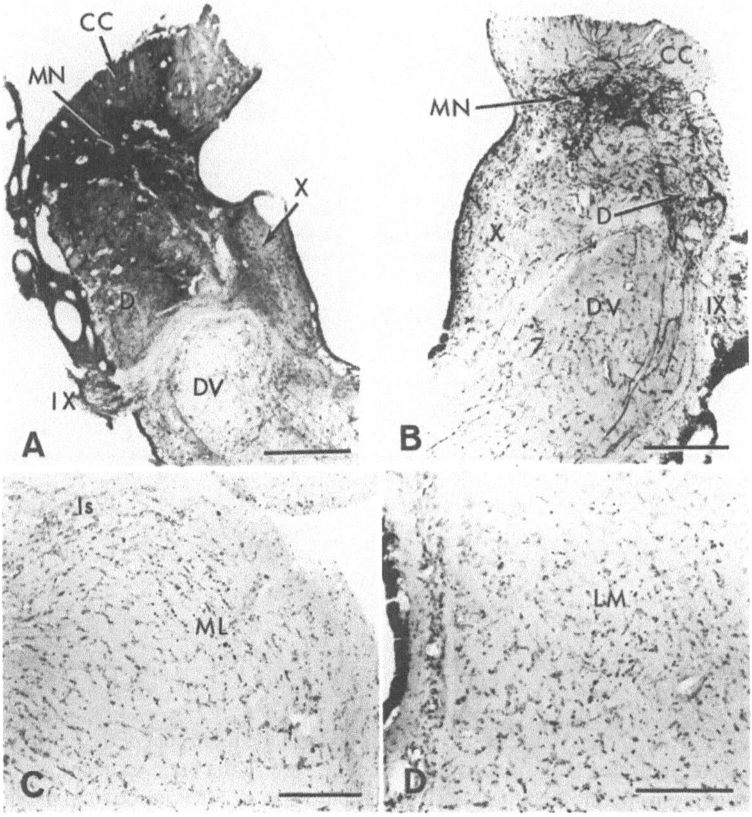

FIGURE 16.6. (A) Transverse section through the caudal pole of the medial mechanosensory octavolateralis nucleus of the carpet shark showing the HRP injection site (cf. Fig. 16.5E). DAB stain. Bar scale = 0.5 mm. (B) Transverse section through the contralateral medial octavolateralis nucleus showing TMB-labeled neurons and fibers (cf. Fig. 16.5E). Bar scale = 0.5 mm. (C) Transverse section at isthmic levels showing labeled contralateral ascending lemniscal fibers (cf. Fig. 16.5B). Bar scale = 0.25 mm. (D) Transverse section at a midbrain level showing labeled fibers within a restricted part of the lateral mesencephalic nucleus (cf. Fig. 16.5A). TMB stain. Bar scale = 0.25 mm.

A certain proportion of ascending mechanosensory lemniscal fibers course through the lateral mesencephalic nucleus to terminate in the central tectal zone and probably about the cells of the peripheral tectal zone. This has been shown in all elasmobranchs thus far studied. Moreover, studies of tectal efferents indicate that the lateral mesencephalic nucleus also receives an input from the tectum in *Raja* (Northcutt and Boord 1981) and in *Raja* and *Scyliorhinus* (Smeets 1981), but this is uncertain because of the possibility that tectal lesions may have interrupted commissural fibers between the lateral mesencephalic nuclei of either side.

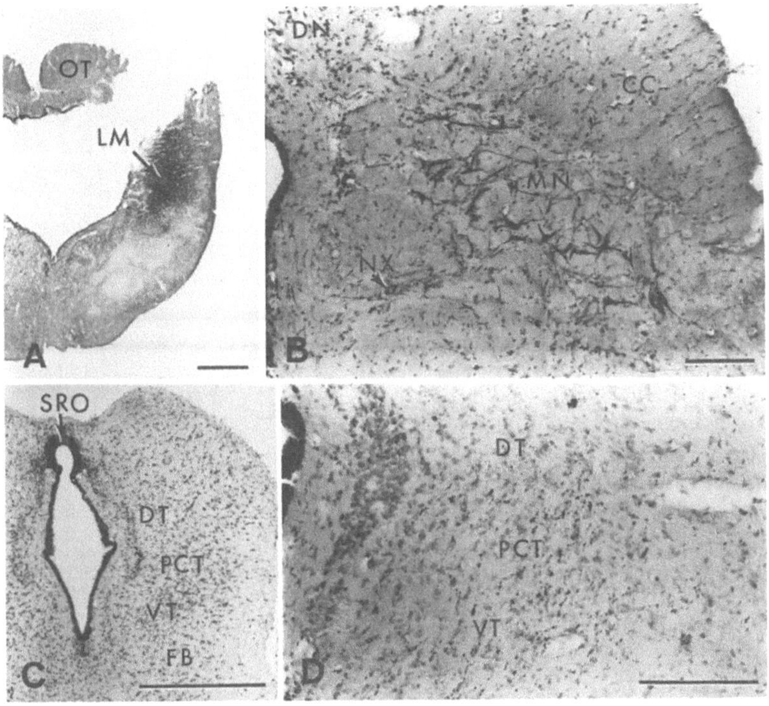

FIGURE 16.7. (A) Transverse section through the mesencephalon of the carpet shark showing an HRP injection confined to the lateral mesencephalic nucleus. A portion of the tectum was removed, and the HRP was directly pin-injected into the lateral mesencephalic nucleus. DAB stain. Bar scale = 1.0 mm. (B) Transverse section through the medial octavolateralis nucleus of the carpet shark showing labeled large multicellular neurons and labeled cells of nucleus X resulting from retrograde transport of HRP from the contralateral injection site of A. TMB stain. Bar scale = 0.5 mm. (C) Transverse section through diencephalon of the carpet shark showing posterior thalamic nuclei. Cresyl violet stain. Bar scale = 1.0 mm. (D) Labeled fibers within the posterior central thalamic zone resulting from anterograde transport from the HRP injection site of A. TMB stain. Bar scale = 0.25 mm.

16.4 Forebrain Centers

Little is known of the thalamic and telencephalic mechanosensory pathways and connections among the elasmobranchs. Bleckmann et al. (1987) recorded mechanosensory responses from the lateral tuberal nucleus of the hypothalamus and the posterior central thalamic nucleus of the thornback guitarfish *Platyrhinoidis*. There is support for this finding in the clearnose skate, where small electrolytic lesions confined to the dorsomedial mesencephalic nucleus reveal ventral projections to the posterior central thalamic nucleus (mostly ipsilateral) and the lateral posterior thalamic nucleus (mostly contralateral) (Boord and

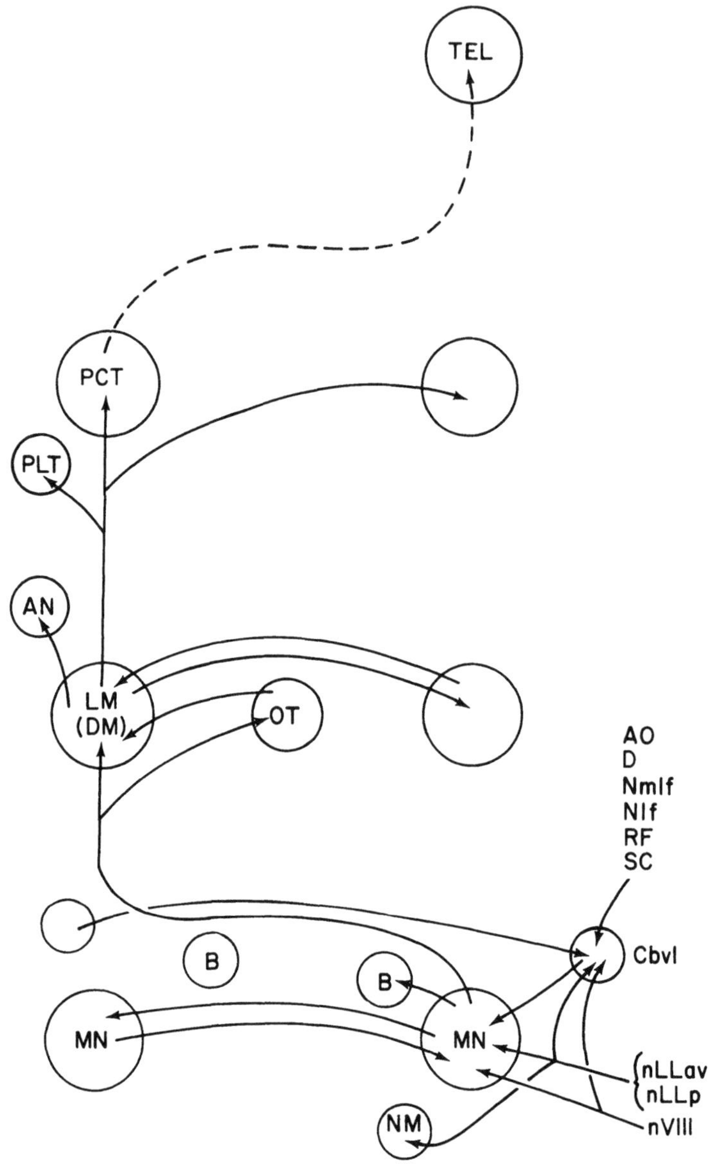

FIGURE 16.8. Generalized diagram showing the ascending circuitry of the elasmobranch mechanosensory lateral line system. It is based on anatomical and physiological evidence of various authors cited in the text. The dashed line represents a pathway that has not been substantiated by experimental anatomical evidence. The ipsilateral pathways are sparse and are not shown in the diagram.

Northcutt, unpublished observations). There also appears to be a minor projection to the contralateral anterior nucleus and commissural connections between the dorsomedial mechanosensory nuclei. Large HRP projections into the lateral mesencephalic nucleus of the carpet shark *Cephaloscyllium* also reveal a ventral projection to the posterior central zone of the thalamus (Fig. 16.7C,D), which corresponds to the position of the posterior central thalamic nucleus of skates—i.e., slightly ventral to the level of the medial thalamic nucleus and sulcus. The cells of this zone are smaller than those of the dorsal and ventral thalamic nuclei that border it dorsally and ventrally, respectively. This projection can reasonably be interpreted as lateral line, although electrosensory and mechanosensory components cannot be distinguished from each other. A few tectal fibers may course through the posterior central nucleus en route to the dorsal thalamic and pretectal nuclei, but Smeets (1981) and Northcutt and Boord (1981) do not report tectal projections to the posterior central thalamic nucleus in either sharks or skates.

Mechanosensory lateral line responses have been recorded in the telencephalon, as has activity from all octavolateralis modalities (Bullock and Corwin 1979; Bodznick and Northcutt 1984). Bleckmann et al. (1987) were the first to record telencephalic-evoked potentials and unit responses following direct stimulation of the posterior lateral line nerve of the thornback skate. They confined mechanosensory (and electrosensory) responses to the deep portion of the medial pallium. The diencephalic nuclei that receive a mechanosensory input are positioned to relay information to the telencephalon, but their efferent projections have not been elucidated by axonal tracing methods.

The pathways and connections of the ascending lateral line mechanosensory system, from medullary through telencephalic levels, are summarized in Fig. 16.8. Some of the pathways shown are equivocal, and certainly there are other centers and tracts yet to be revealed.

16.5 Concluding Remarks

The octavolateralis system of elasmobranchs consists of octaval (vestibular and acoustic), electrosensory lateral line, and mechanosensory lateral line components. There is convincing evidence from evoked potential and multiple and/or single-unit studies in a number of cartilaginous fishes that auditory and lateral line components are represented at all levels of the neuraxis, from medulla to telencephalon, including the cerebellum (Platt et al. 1974; Bullock and Corwin 1979; Bullock 1982; Corwin and Northcutt 1982; Bodznick and Boord 1986; Schweitzer 1986; Bleckmann et al. 1987). Correlative anatomical studies, required in order to determine the afferent and efferent connections of octavolateralis centers, are essential to compare and establish homologies among elasmobranchs and between elasmobranchs and other vertebrates. With regard to the mechanosensory component, some of the neural circuitry is known at lower levels of the brainstem, but forebrain centers and connections are highly speculative. The chief problem is anatomically distinguishing the mechanosensory lateral line pathways from other octavolateralis modalities.

Input to the primary medullary electrosensory dorsal nucleus and mechanosensory medial nucleus differs (Schmidt and Bodznick 1987) and is relayed to the midbrain via separate lemnisci. Among batoids, separate midbrain mechanosensory and electrosensory nuclei are present and are undoubtedly derived from parts of the lateral mesencephalic nucleus of more primitive elasmobranchs. The latter nucleus has not been subdivided into cytologically distinct functional areas in sharks, but subdivisions probably exist. The mechanosensory lateral line lemnisci among teleosts and other bony fishes terminate in separate zones of the torus semicircularis (Knudsen 1977; Bell et al. 1981; McCormick Chapter 17; Echteler 1984; Finger and Tong 1984), which are homologous to the mesencephalic mechanosensory fields of elasmobranchs.

The thalamus of elasmobranchs is known to receive input from a number of sensory sources, and thalamotelencephalic projections are known (Ebbesson and Schroeder 1971; Schroeder and Ebbesson 1974; Luiten 1981). Anatomical and physiological evidence suggests that lateral line information is relayed to thalamic centers (Schweitzer 1983; Boord and Northcutt 1983; Bleckmann et al. 1987), and it is tempting to predict that the posterior central thalamic nucleus and the posterior lateral thalamic nucleus are the main mechanosensory and electrosensory relay nuclei, respectively. Bodznick and Northcutt (1984) and Smeets and Northcutt (1987) show by the retrograde transport of horseradish peroxidase that the posterior lateral thalamic nucleus projects to the medial pallium in *Raja* and *Squalus*. The pallial area from which Bleckmann et al. (1987) recorded mechanosensory (and electrosensory) responses in *Platyrhinoidis* and which they interpret as deep medial pallium does not appear to coincide with the electrosensory area of *Raja*. This suggests that two pallial areas receive lateral line information, but each may serve as a multimodal integration center. Thus far, however, no pallial area has been cytoarchitectonically delineated as being electrosensory or mechanosensory. Moreover, horseradish peroxidase results have not been tested with regard to thalamotelencephalic lateral line projections; i.e., presumed lateral line pathways have not been traced in both directions between thalamus and telencephalon.

Elasmobranchs are believed to be a monophyletic group in which the lateral line system is well established, and the central mechanosensory and electrosensory pathways and centers can therefore be expected to be similar. Anatomical studies are required to identify homologous mechanosensory and electrosensory nuclei among representative of different elasmobranch radiations in order to better understand the evolution and neural basis of mechanoreception and electroreception.

Acknowledgments. Experiments on the carpet sharks (*Cephaloscyllium isabella*) were conducted at the Leigh Marine Laboratory of the University of Auckland,

New Zealand. We are grateful to Dr. Bob Creese, Director, for providing the required facilities, and to Mr. John Lowe, a commercial fisherman, who has a lively interest in sharks and those who study them, for supplying the specimens. This work was supported by USPHS grant NS11272.

Abbreviations in Figures

A	auricle of the cerebellum
AL	anterior lateral line lobe
AN	anterior nucleus of lateral mesencephalic nuclear complex
AO	anterior octaval nucleus
B	nucleus B (Smeets et al. 1983)
C	central tectal zone
Cb	cerebellum
Cbvl	vestibulolateral lobe of the cerebellum
CC	crista cerebellaris
CN	central nucleus of the telencephalon
D	descending octaval nucleus
DM	dorsomedial nucleus of the lateral mesencephalic nuclear complex
DN,ND	dorsal octavolateralis nucleus
DT	dorsal thalamus
DV	descending root of the trigeminal nerve
ES	electrosensory field potential
FB	forebrain bundle
H	nucleus H (Smeets et al. 1983)
OT	optic tectum
II	optic nerve
III	oculomotor nerve
IL	inferior lobe of the hypothalamus
IP	interpeduncular nucleus
Is	isthmic nucleus
IV	trochlear nerve
IX	glossopharyngeal nerve or root
L	lateral nucleus of the lateral mesencephalic nuclear complex
LL	lower leaf of vestibulolateral lobe of the cerebellum
LM	lateral mesencephalic nucleus
LP	lateral pallium of the telencephalon
MG	medial granular layer of the vestibulolateral lobe of the cerebellum
ML	mechanosensory lateral line lemniscus
MLF	medial longitudinal fasciculus
MN	medial octavolateralis nucleus
MRF	medial reticular formation
nIII	oculomotor nucleus
Nlf	lateral funicular nucleus

Nmlf	nucleus of the medial longitudinal fasciculus
nLLa	anterior lateral line nerve
nLLad	dorsal root of the anterior lateral line nerve
nLLav	ventral root of the anterior lateral line nerve
nLLp	posterior lateral line nerve
nVIII	octaval nerve
NM	nucleus magnocellularis
NX	nucleus X
OT	optic tectum
P	periventricular tectal zone
PCT	posterior central thalamus
PL	posterior lateral line lobe
PLT	posterior lateral thalamic nucleus
RF	reticular formation
rnLLav	ventral root of the anterior lateral line nerve
rnLLp	root of posterior lateral line nerve
rnVIIvs	visceral sensory root of the facial nerve
S	superficial tectal zone
SC	spinal cord
SRF	superior reticular formation
SRO	subcommissural organ
TEL	telencephalon
UL	upper leaf of vestibulolateral lobe of the cerebellum
V	trigeminal nerve
VII	facial nerve
VIII	octaval nerve
VM	ventromedial nucleus of the mesencephalic nuclear complex
VT	ventral thalamus
Vm	mesencephalic nucleus of the trigeminal nerve
X	vagal lobe (visceral sensory column)

References

Barry MA (1987) Afferent and efferent connections of the primary octaval nuclei in the clearnose skate, *Raja eglanteria*. J Comp Neurol 266:457–477.

Bell CC, Finger TE, Russell C (1981) Central connections of the posterior lateral line lobe in mormyrid fish. Exp Brain Res 42:9–22.

Bleckmann H, Bullock TH, Jørgensen JM (1987) The lateral line mechanoeceptive mesencephalic, diencephalic, and telencephalic regions in the thornback ray, *Platyrhinoidis triseriata* (Elasmobranchii). J Comp Physiol A 161:67–84.

Bodznick D, Boord RL (1986) Electroreception in Chondrichthyes: Central anatomy and physiology. In: Bullock TH, Heiligenberg W (eds) Electroreception. New York: Wiley, pp. 225–256.

Bodznick D, Northcutt RG (1980) Segregation of electro-and mechanoreceptive inputs to the elasmobranch medulla. Brain Res 195:313–321.

Bodznick D, Northcutt RG (1984) An electrosensory area in the telencephalon of the little skate *Raja erinacea*. Brain Res 298:117–124.
Bodznick D, Schmidt AW (1984) Somatotopy within the medullary electrosensory nucleus of the little skate, *Raja erinacea*. J Comp Neurol 225:581–590.
Boord RL, Northcutt RG (1982) Ascending lateral line pathways to the midbrain of the clearnose skate, *Raja eglanteria*. J Comp Neurol 207:274–282.
Boord RL, Northcutt RG (1983) Diencephalic and mesencephalic electrosensory centers of the clearnose skate. Am Zool 23:927.
Boord RL, Northcutt RG (1988) Medullary and mesencephalic pathways and connections of lateral line neurons of the spiny dogfish *Squalus acanthias*. Brain Behav Evol 32:76–88.
Bullock TH (1982) Electroreception. Annu Rev Neurosci 5:121–170.
Bullock TH, Corwin JT (1979) Acoustic evoked activity in the brain of sharks. J Comp Physiol 129:223–234.
Corwin JT, Northcutt RG (1982) Auditory centers in the elasmobranch brainstem: Deoxyglucose autoradiography and evoked potential recording. Brain Res 236:261–273.
Ebbesson SOE, Schroeder D (1971) Connections of the nurse shark's telencephalon. Science 173:254–256.
Echteler SM (1984) Connections of the auditory midbrain in a teleost fish, *Cyprinus carpio*. J Comp Neurol 230:536–551.
Finger TE, Tong S-L (1984) Central organization of eighth nerve and mechanosensory lateral line systems in the brain stem of ictalurid catfish. J Comp Neurol 229:129–151.
Knudsen EI (1977) Distinct auditory and lateral line nuclei in the midbrain of catfishes. J Comp Neurol 173:417–431.
Koester DM (1983) Central projections of the octavolateralis nerve of the clearnose skate, *Raja eglanteria*. J Comp Neurol 221:199–215.
Luiten PGM (1981) Two visual pathways to the telencephalon in the nurse shark (*Ginglymostoma cirratum*). II. Ascending thalamo-telencephalic connections. J Comp Neurol 196:539–548.
McCready PJ, Boord RL (1976) The topography of the superficial roots and ganglia of the anterior lateral line nerve of the smooth dogfish, *Mustelus canis*. J Morphol 150:527–538.
Montgomery JC (1981) Origin of the parallel fibers in the cerebellar crest overlying the intermediate nucleus of the elasmobranch hindbrain. J Comp Neurol 202:185–191.
Northcutt RG (1978) Brain organization in the cartilaginous fishes. In: Hodgson ES, Mathewson RF (eds) Sensory Biology of Sharks, Skates, and Rays. Washington: U.S. Government Printing Office, pp. 117–193.
Northcutt RG (1984) Evolution of the vertebrate central nervous system: Patterns and processes. Am Zool 24:701–716.
Northcutt RG, Boord RL (1981) Efferent projections of the optic tectum in the clearnose skate *Raja eglanteria*. Anat Rec 199:185A.
Paul DH, Roberts BL (1977) Studies on a primitive cerebellar cortex. I. The anatomy of the lateral-line lobes of the dogfish, *Scyliorhinus canicula*. Proc R Soc Lond B 195: 453–466.
Paul DH, Roberts BL, Ryan KP (1977) Comparisons between the lateral-line lobes of the dogfish and the cerebellum: An ultrastructural study. J Hirnforsch 18:335–343.
Platt CJ, Bullock TH, Czéh G, Kovacevic N, Konjevic D, Gojkovic MK (1974) Comparison of electroreceptor, mechanoreceptor and optic evoked potentials in the brain of some rays and sharks. J Comp Physiol 95:323–355.

Sas E, Maler L (1983) The nucleus praeeminentialis: A Golgi study of a feedback center in the electrosensory system of gymnotid fish. J Comp Neurol 221:127–144.
Schmidt AW, Bodznick D (1987) Afferent and efferent connections of the vestibulolateral cerebellum of the little skate, *Raja erinacea*. Brain Behav Evol 30:282–302.
Schroeder D, Ebbesson SOE (1974) Nonolfactory telencephalic afferents in the nurse shark (*Ginglymostoma cirratum*). Brain Behav Evol 9:121–155.
Schweitzer J (1983) The physiological and anatomical localization of two electroreceptive diencephalic nuclei in the thornback ray, *Platyrhinoidis triseriata*. J Comp Physiol A 153:331–341.
Schweitzer J (1986) The neural basis for electroreception in elasmobranchs. In: Uyeno T, Arai R, Taniuchi T, Matsuura K (eds) Indo-Pacific Fish Biology: Proceedings of the Second International Conference. Tokyo: Ichthyological Society of Japan, pp. 392–407.
Smeets WJAJ (1981) Efferent tectal pathways in two chondrichthyans, the shark *Scyliorhinus canicula* and the ray *Raja clavata*. J Comp Neurol 195:13–23.
Smeets WJAJ, Nieuwenhuys R (1976) Topological analysis of the brain stem of the sharks *Squalus acanthias* and *Scyliorhinus canicula*. J Comp Neurol 165:333–368.
Smeets WJAJ, Northcutt RJ (1987) At least one thalamotelencephalic pathway in cartilaginous fishes projects to the medial pallium. Neurosci Lett 78:277–282.
Smeets WJAJ, Nieuwenhuys R, Roberts BL (1983) The Central Nervous System of Cartilaginous Fishes. Berlin: Springer-Verlag.
Tester AL, Kendall JI (1967) Innervation of free and canal neuromasts in the sharks *Carcharhinus menisorrah* and *Sphyrna lewini*. In: Cahn PH (ed) Lateral Line Detectors. Bloomington: Indiana University Press, pp. 53–69.

17
Central Lateral Line Mechanosensory Pathways in Bony Fish

CATHERINE A. MCCORMICK

17.1 Introduction

The lateral line mechanosensory system originated early in vertebrate history. It was apparently present in at least some extinct agnathous forms and inherited from these vertebrates by the early gnathostome fishes. So far as is known, the system is present in petromyzontid agnathans, in all extant cartilaginous and bony fishes, and in most larval and many postmetamorphic amphibians (Boord and Montgomery Chapter 16; Fritzsch Chapter 5; Northcutt Chapter 3). By all indications, the lateral line mechanosensory system evolved in early agnathans and was retained in subsequent radiations; it is thus by definition homologous among those vertebrates that possess it. This homology is reflected both in the anatomical similarity of the neuromasts across groups and in the similarity in the overall pattern of central connections of the system, at least among gnathostomes. This chapter will summarize our current understanding of these central anatomical pathways in bony fish, pointing out variations within this group and similarities to cartilaginous fishes and amphibians.

17.2 Organization of First-Order Fibers and Nuclei

17.2.1 Lateral Line Nerves

In bony fishes, as in other species, information from neuromasts on the head is carried into the central nervous system by the anterior lateral line nerve (ALLN), whereas information from more caudal regions is relayed via the posterior lateral line nerve (PLLN). Each of these nerves is composed of bipolar neurons with cell bodies forming a peripheral ganglion closely apposed to but distinct from either the trigeminal-facial complex or the vagal complex (Fig. 17.1a,b). So far as is known, the lateral line nerves carry information only from lateral line organs—neuromasts, electroreceptors, or the spiracular organ (Barry and Bennett Chapter 30). Neuromast information is always carried by both nerves (when both are present), whereas electroreceptor information may be confined to portions of the anterior lateral line nerves (e.g., amphibians, nonteleost electroreceptive fishes, and some gymnotids) or may be carried by both lateral line nerves (most electroreceptive teleosts). When electroreceptors are present, lateral line nerve fibers innervating them terminate in nuclei different from those receiving mechanosensory lateral line information (Fig. 17.1a; for reviews, see McCormick 1982; Bullock et al. 1983). Mechanoreceptive lateral line fibers terminate in both the medulla and the cerebellum (Fig. 17.1c).

17.2.2 Organization of Medullary Lateral Line Areas

The dorsal portion of the medulla in bony fishes is devoted to processing first-order information mainly from the lateral line system and the inner ear. Figure 17.1a,b illustrates the arrangement of these sensory regions in two nonteleost species: the sturgeon *Scaphirhynchus* (Chondrostei; New and Northcutt 1984)

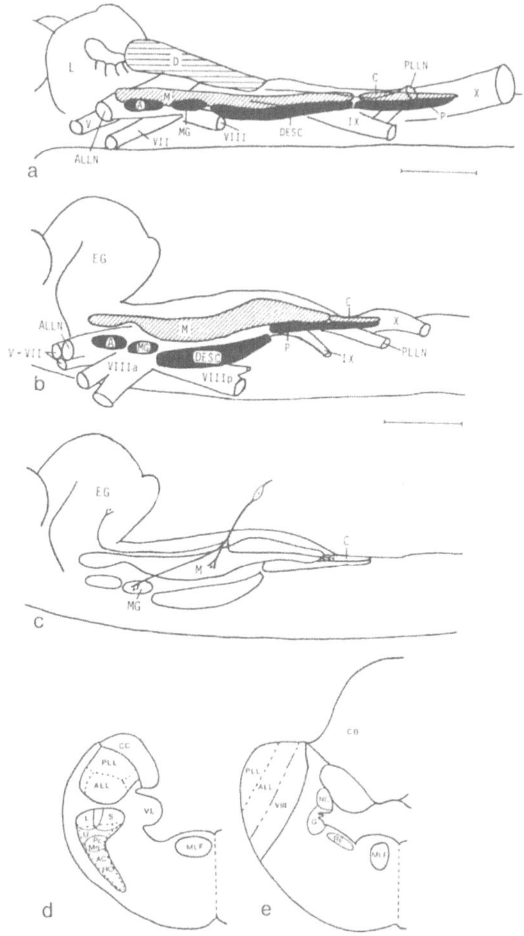

and the bowfin *Amia* (Halecomorphi; McCormick 1981). The dorsalmost area of the medulla in fishes contains the first-order nuclei of the lateral line mechanosensory system and, when present, the electrosensory system (Fig. 17.1a-d). The mechanosensory nuclei form a column which lies lateral or ventrolateral to the first-order electrosensory nucleus (nucleus dorsalis) in nonteleost electrosensory species (Fig. 17.1a) but medial or ventromedial to the electrosensory area (electrosensory lateral line lobe) of electroreceptive teleosts (Fig. 17.2). The mechanosensory column receives its major first-order input from lateral line mechanosensory fibers, although other first-order inputs are also present (see below). The principal component of this column is nucleus medialis or intermedius, which occupies at least the rostral two-thirds of the column. A much smaller nucleus caudalis lies caudally in the column (Fig. 17.1a–c).

Ventral, or largely ventral, to the lateral line columns are the nuclei which receive input from the inner ear via the eighth nerve. These nuclei comprise the octavus column (Fig. 17.1a–d).

◀

FIGURE 17.1. Basic organization of lateral line areas of the medulla and cerebellum. (a) Lateral view of the caudal portion of the brain of the sturgeon *Scaphirhynchus*. Nuclei receiving mainly electrosensory (horizontal lines), mechanosensory (diagonal lines), or inner-ear (black) inputs have been projected onto the lateral aspect of the brain. Bar scale = 1 mm. After New and Northcutt (1984). (b) Lateral view of the caudal portion of the brain of the bowfin *Amia*. Nuclei receiving mainly mechanosensory (diagonal lines) or inner ear (black) inputs have been projected onto the lateral aspect of the brain. Bar scale = 1 mm. After McCormick (1981). (c) Lateral view of the brain of *Amia* schematically showing the regions that receive first-order lateral line mechanosensory input. (d) Transverse hemisection through the medulla of *Amia*. Shown schematically are the projection areas of the posterior (PLL) and anterior (ALL) lateral line nerve within nucleus medialis, as well as the main projection areas of the inner ear end organs within the descending octaval nucleus. (e) Transverse hemisection through the eminentia granularis of the cerebellum in *Amia* schematically showing the projection areas of the anterior (ALL) and posterior (PLL) lateral line nerves and the eighth nerve (VIII). Abbreviations: A, anterior octaval nucleus; Ac, projection field of the anterior semicircular canal; ALL, projection field of the anterior lateral line nerve; ALLN, anterior lateral line nerve; C, nucleus caudalis; CB, cerebellum; D, nucleus dorsalis; DESC, descending octaval nucleus; EG, eminentia granularis; G, secondary gustatory nucleus; Hc, projection field of the horizontal semicircular canal; IX, glossopharyngeal nerve; L, projection field of the lagena; LL, lateral lobule of the cerebellum; M, nucleus medialis; MG, nucleus magnocellularis; MLF, medial longitudinal fasciculus; Mn, projection field of the macula neglecta; NC, nucleus cerebelli; P, posterior octaval nucleus; Pc, projection field of the posterior semicircular canal; PLL, projection field of the posterior lateral line nerve; PLLN, posterior lateral line nerve; RF, reticular formation; S, projection field of the saccule; U, projection field of the utriculus; V, trigeminal nerve; VII, facial nerve; V-VII, trigeminal facial complex; VIII, eighth nerve; VIIIa, anterior ramus of the eighth nerve; VIIIp, posterior ramus of the eighth nerve; VL, vagal lobe; X, vagus nerve.

17.2.3 Primary Afferent Input to the Mechanosensory Column

As lateral line fibers enter the medulla, they form ascending and descending bundles. Both the anterior and posterior lateral line nerves supply nucleus medialis and nucleus caudalis throughout their entire rostrocaudal extent (Fig. 1c). In all bony fishes, fibers distribute such that head neuromasts are represented medially or ventromedially, and trunk neuromasts laterally or dorsolaterally (Fig. 17.1d; Claas and Münz 1981; McCormick 1981, 1983a; New and Northcutt 1984; Finger and Tong 1984; Meredith 1984; Meredith et al. 1987; Carr and Matsubara 1981; Braford 1986; Puzdrowski 1987, personal communication; Song 1987, personal communication). However, a precise body map of neuromast position has not been demonstrable. A similar situation is present in amphibians (Altman and Dawes 1983; Lowe and Russell 1983; Will et al. 1985). This is in contrast to the very orderly body maps of the electroreceptors present in both teleosts and chondrichthyans (Bell and Russell 1978; Carr et al. 1982; Heiligenberg and Dye 1982; Bodznick and Schmidt 1984) and is curious in light of the proposed spatial mechanism for processing mechanosensory input (Hassan Chapter 10). However, it has been recently shown in *Xenopus* that the optic tectum is the first central site of directional localization in the lateral line system (Claas et al. Chapter 20; Zittlau et al. 1986); perhaps a similar situation exists in fishes.

The lateral line mechanosensory system in many bony fishes contains both free neuromasts and canal neuromasts (Münz Chapter 14; Northcutt Chapter 3). These two receptor subclasses conceivably provide the organism with different types of information (Münz 1985, Chapter 14; Schellart and Kroese Chapter 21). There are as yet no studies that specifically address the possibility of any differential mapping of these two types of lateral line mechanoreceptors. Perhaps such studies would reveal that there is a more orderly organization of the periphery in nucleus medialis and nucleus caudalis than is currently believed.

Nucleus medialis and nucleus caudalis differ from one another not only in size but also in the types of cells present, relationship with the cerebellum, and second-order connections. So far as is known, nucleus caudalis, which contains only small granule cells, does not contribute fibers to the lateral line lemniscus which ascends to isthmal and midbrain levels (Bell 1981a). In *Amia*, only projections to the contralateral nucleus caudalis are present. It is also not in contact with descending cerebellar fibers coursing in the cerebellar crest. The cerebellar crest is primarily a layer of axons whose cell bodies lie in the eminentia granularis of the cerebellum. Nucleus medialis is overlain by these cerebellar crest fibers, and the dorsalmost cells of this nucleus—the "crest," or Purkinje-like cells—receive cerebellar input onto the dorsal dendrite that they extend into this fiber layer. The ventral dendrite of the crest cells receives first-order lateral line input, as do the polygonal and granule cells in more ventral regions of the nucleus (McCormick 1981). Since first-order mechanosensory input also reaches the eminentia granularis, crest cells in nucleus medialis are influenced by both primary and processed mechanosensory input (also see Caird 1978).

Nucleus medialis is also the only primary mechanosensory nucleus that contributes axons to the ascending lateral line lemniscus. The axons of crest cells form this tract in all bony fishes (Bell 1981a; Braford 1982; Finger 1982; Finger and Tong 1984; McCormick personal observations); in *Amia* polygonal cells of nucleus medialis also contribute to this pathway. Reciprocal connections between the left and right medial nuclei are mediated primarily by the crest cells.

17.2.4 Primary Afferent Input to the Cerebellum

Some fibers in the ascending mechanosensory bundles course into the cerebellum and synapse in various regions (Fig. 17.1c,d). In all bony fishes, the eminentia granularis—a subdivision of the vestibulolateral lobe (Bass 1982)—receives the majority of such input. In many species, mechanosensory lateral line fibers are seen only ipsilaterally in the eminentia granularis (Luiten 1975; McCormick 1981, 1983a; Zottoli and Van Horne 1983; Finger and Tong 1984; Meredith 1984; Meredith et al. 1987). Bilateral projections to the eminentia granularis are present in the sturgeon *Scaphirhynchus* (New and Northcutt 1984) and mormyrids (Bell and Russell 1978) and are probably independent specializations (New and Northcutt 1984). A separate region of the eminentia receives first-order electrosensory input in species that have such connections.

In most species that have been studied, the anterior and posterior lateral line nerves terminate in largely separate regions within the eminentia granularis (Fig. 17.1e). The exception occurs in mormyrids, where the two lateral line nerves extensively overlap in their terminal fields (Bell and Russell 1978; Bell 1981b).

Yet another variation in first-order mechanoreceptor input to the cerebellum concerns the termination of afferents in areas outside the vestibulolateral lobe. In some fish, these fibers enter the cerebellar corpus ipsilaterally (Bell 1981b; Zottoli and Van Horne 1983; Meredith 1984; Blubaum-Gronau and Schwartz 1987, personal communication) or bilaterally (Claas and Münz 1981). Projections to the cerebellar valvula have also been reported (Claas and Münz 1981).

17.2.5 Relationship Between Primary Mechanosensory and Inner Ear Inputs

Based on the studies generated in the first half of this century, the lateral line and auditory inner-ear systems have been traditionally described as sharing common central processing areas. For example, nucleus medialis was thought to be the major first-order processing center for both of these systems (Larsell, 1967). However, recent anatomical and physiological studies generally show that although there may be some central regions in which the two systems converge, the majority of the information is processed in regions specific to each system, at least below the level of the diencephalon. This finding probably reflects the fact that the sources of stimuli for lateral line and inner-ear sensory systems differ and that the behavioral functions served by the two systems are also likely different (see Dijkgraaf Chapter 2; Enger et al. Chapter 29; Kalmijn Chapter 9; Platt et al. Chapter 32) By and large, then, first-order lateral line fibers synapse in nuclei within the dorsal mechanosensory column which relay the information to mechanosensory regions of the torus semicircularis, whereas first-order inner-ear fibers synapse in the nuclei of the octaval column which relay the information either to auditory regions of the torus semicircularis or to structures involved in vestibular processing. Likewise, fibers from the lateral line and inner-ear systems terminate in many species in separate, though adjoining, areas of the vestibulolateral lobe. In mormyrid fish, however, no such segregation in the cerebellum is present (Bell 1981b).

Nevertheless, there are first-order structures in which segregation of these inputs is not maintained. One such site of convergence is the Mauthner cell, onto which synapses from the eighth and lateral line nerves have been clearly established (Zottoli and Danielson Chapter 23). Another consistent finding has been the termination of some fibers of both of the lateral line nerves in an octaval nucleus—nucleus magnocellularis (Bell and Russell 1978; Carr and Matsubara 1981; McCormick 1981, 1983a; Zottoli and Van Horne 1983; New and Northcutt 1984; Meredith 1984; Meredith et al. 1987; Braford 1986; Kitch et al. 1987; Puzdrowski 1987, personal communication; Song 1987, personal communication). Nucleus magnocellularis does not appear to contribute fibers to the ascending lemniscal systems associated with either lateral line mechanoreceptors or the inner ear (Bell 1981a; Echteler 1984; Finger and Tong 1984; McCormick personal observations), and the significance of the overlap is unknown. It has also been suggested by some that some of the cells at the dorsal boundary of the descending octaval nucleus may be a point of convergence (Carr and Matsubara 1981; Finger and Tong 1984); this suggestion may stem partly from the absence of a clear boundary between the descending and overlying medial nuclei in some species.

Eighth-nerve fibers have been reported in or near the dorsolateral line column in a number of bony fishes (Bell 1981b; McCormick 1981, 1983b; Meredith and Butler 1983). Some of these fibers do appear to terminate in nucleus medialis, but their numbers are usually small relative to those synapsing in the octaval column. However, Meredith et al. (1987) describe in *Anguilla* a light octaval input to all but the most rostral portion of nucleus medialis, which receives a much heavier projection. Likewise, Bell (1981b) also noted a significant octaval projection to nucleus medialis in mormyrids. The significance of these connections is unknown, but it is noteworthy that in both bony and cartilaginous fishes (Barry 1987) they originate from all inner-ear end organs (with the possible exception of the macula neglecta). Thus, these eighth-nerve terminations in nucleus medialis do not support the traditional idea that nucleus medialis is the primitive first-order auditory nucleus. Along these same lines, it is known that the dendritic fields of cells in both the octaval and the lateral line columns can be extremely large (Larsell 1967; Meredith 1984); the distal dendrites of one column might thus extend down into another and receive inputs from that column's primary source. Inputs on these distal dendrites would presumably not influence the cell

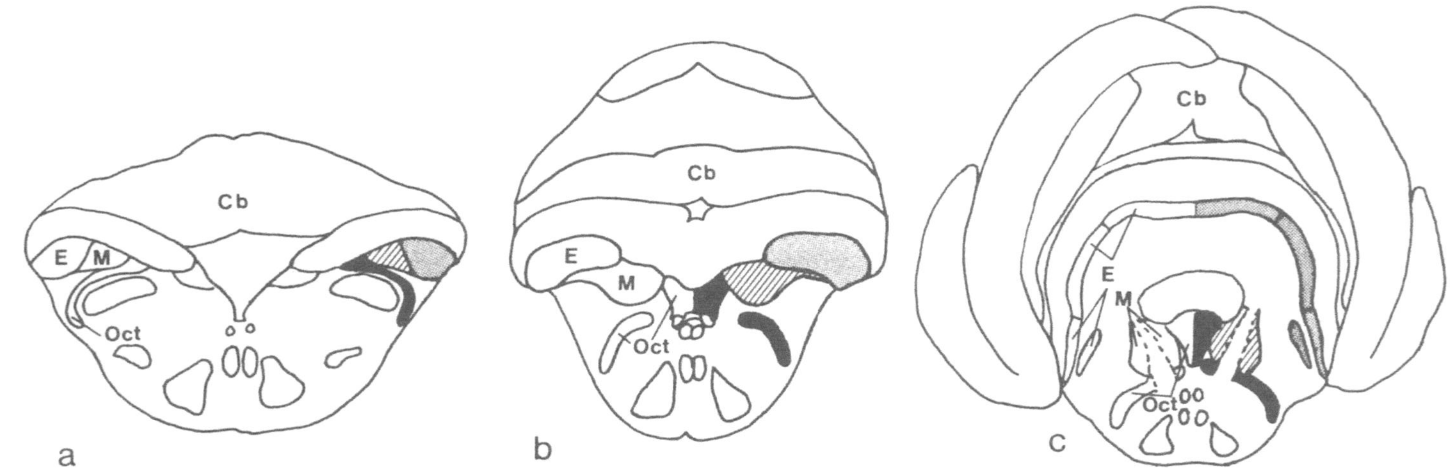

FIGURE 17.2. Transverse sections through the brains of three teleost species showing the positions of electrosensory (E, stippled), mechanosensory (M, diagonal lines) and octaval (Oct, black) nuclei. After McCormick and Braford (1988). (a) The ostariophysan *Ictalurus*; (b) the osteoglossomorph *Xenomystus*; (c) the osteoglossomorph *Gnathonemus*.

as strongly or regularly as more proximal inputs; it is therefore probably misleading to interpret this type of convergence as indicating the absence of separate lateral line and inner-ear central pathways.

Other eighth-nerve fibers that course dorsally actually terminate on cells that appear to be dorsomedially directed extensions of the underlying octaval cell column. These dorsomedial octaval cells, which are located medial to nucleus medialis, have been reported in some closely related osteoglossomorphs (Fig. 17.2; Bell 1981a; Braford 1982; McCormick and Braford 1988), although in the paper by Bell these cells were included within the boundaries of the lateral line mechanosensory zone. Finger and Tong (1984) reported a dorsomedial octavus zone in a euteleost, the ostariophysan *Ictalurus*. Based on normal analysis, a similar dorsomedial area is present in two other ostariophysans – *Carassius* and *Eigenmannia* – although eighth-nerve input to these cells in *Eigenmannia* has not been established (McCormick and Braford 1988). These osteoglossomorph and ostariophysan species have in common the fact that they possess accessory peripheral auditory structures specialized for sound pressure sensitivity, although these structures evolved independently in the two groups (for a review, see Fay and Popper 1980). Perhaps the dorsomedial extensions of the octavus column are related to these peripheral specializations and their demand for specialized processing capabilities. In any case, the possible existence of octaval structures in dorsal areas of the medulla should be considered before claims of first-order central overlap between the lateral line and inner ear sensory systems are made.

17.3 The Ascending Pathway Between Medulla and Midbrain

As previously noted, ascending mechanosensory connections appear to originate only from the crest and polygonal neurons of nucleus medialis; nucleus caudalis appears to have only commissural connections. Crest cells in addition project to the contralateral nucleus medialis via a tract that courses through the medulla in a more dorsal location than does the lateral line lemniscus (Fig. 17.3a).

Apart from such commissural connections, the axons of nucleus medialis terminate in medullary, isthmal, and midbrain structures. These ascending axons collectively form the lateral line lemniscus which courses bilaterally, with contralateral predominance. The largest portion of this pathway is that which synapses in the midbrain torus semicircularis and optic tectum; those portions that supply lower levels of the neuraxis are not well understood in most species.

▶

FIGURE 17.3. Transverse sections showing the connections of nucleus medialis in *Amia* as determined using HRP. In the illustrated case, HRP was deposited in the right (labeled M) nucleus medialis. Triangles = HRP-labeled cell bodies; dashed lines = HRP-labeled fibers; dots = terminal fields.

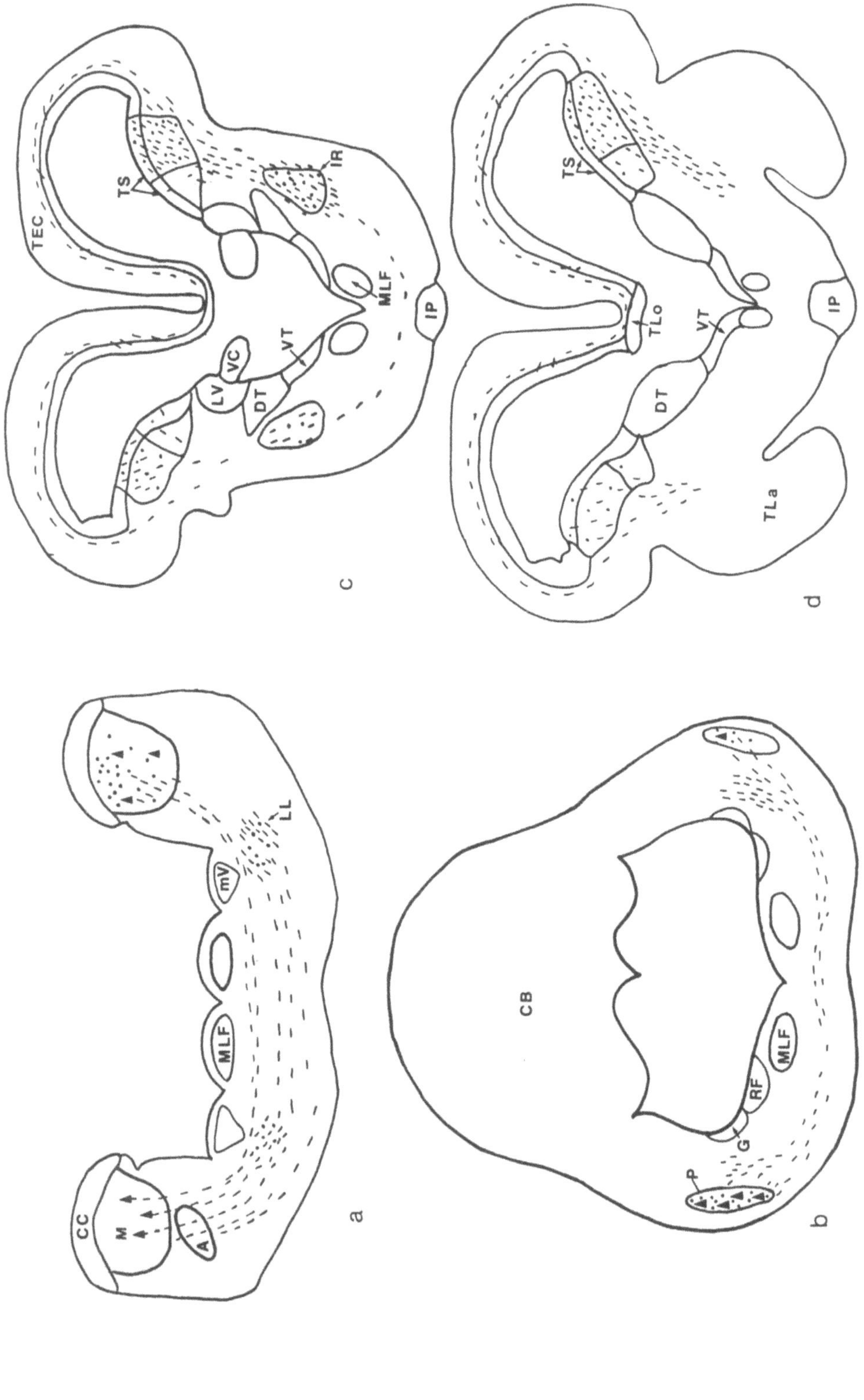
CC
M
A
MLF
mV
LL
a
CB
P
G
RF
MLF
b
TEC
TS
IR
LV
VC
DT
VT
MLF
IP
c
TS
TLo
VT
DT
IP
TLa
d

17.3.1 Projections of Nucleus Medialis to Other Medullary Structures

In some bony fishes, one or more cell groups known to project to the torus semicircularis are located within or in the vicinity of the medullary reticular formation near the level of entry of the anterior lateral line nerve (Braford and McCormick 1979; Bell 1981b; Finger and Tong 1984; Echteler 1984). These cells have been labeled the superior olive based on their location and their efferent projection to the torus semicircularis, but its other connections are incompletely known, and it has not been rigorously characterized in terms of the sensory modalities it processes. In *Amia* and *Ictalurus*, lateral line lemniscus fibers synapse in this group bilaterally (Fig. 17.3a), and in *Ictalurus* axons originating from it course to mechanosensory but not auditory portions of the torus (Braford and McCormick 1979, Finger and Tong 1984). However, in *Cyprinus*, a species closely related to *Ictalurus*, similar cells have been classified physiologically as auditory in nature (Fig. 17.6a; Echteler 1985). It therefore remains to be determined whether there are auditory and lateral line subgroups within the superior olive or whether some or all of these cells are bimodal for the two classes of sensory input.

17.3.2 Projections to the Isthmus

It is firmly established in all electroreceptive teleosts that an isthmal structure, nucleus praeeminentialis, is involved in processing electrosensory and mechanosensory lateral line information (for reviews see Finger 1986; Carr and Maler 1986; Braford 1986; Bell and Szabo 1986). Braford (1986) pointed out the close relationship between this nucleus and the eminentia granularis, with which it is in cellular continuity in *Xenomystus* and at least some other species. Although electroreception is believed to have been reinvented at least twice among teleosts (Bullock 1974), the nucleus praeeminentialis of all of these species is similar in that there is a dorsal subdivision interconnected with electrosensory structures, and a ventral subdivision interconnected with mechanosensory structures. Each subdivision receives descending input from the functionally corresponding ipsilateral region of the torus semicircularis and is bilaterally reciprocally interconnected in most species with the functionally corresponding primary medullary nucleus. In the case of the mechanosensory system, these bilateral reciprocal interconnections are seen in all electroreceptive teleosts. Interconnections between the electrosensory portion of nucleus praeeminentialis and the vestibulolateral lobe are also known in *Ictalurus* and *Eigenmannia* (Finger 1986; Carr and Maler 1986; Bell and Szabo 1986).

Since nucleus praeeminentialis does have a mechanosensory portion, it would be surprising if this portion of the nucleus did not exist in other bony fishes, though such a scenario is not impossible. Finger (1982, 1986) has reported the existence of a nucleus praeeminentialis in the nonelectrosensory species *Carassius auratus*. It is reciprocally connected to nucleus medialis and receives descending input from the ipsilateral torus semicircularis. In *Amia*, a group of cells that is probably nucleus praeeminentialis also is present (Fig. 17.3b). It begins caudally as a small group of granule cells at the base of, and in cellular continuity with, the eminentia granularis and expands rostrally to occupy the lateral edge of the brain in its isthmal region. Its reciprocal connections with nucleus medialis show ipsilateral predominance. This is also the case in the electroreceptive catfish *Ictalurus* but not in *Xenomystus*, which exhibits contralateral predominance. In *Amia*, most lateral line lemniscus fibers that course ipsilaterally terminate in the presumed nucleus praeeminentialis, such that the continuation of this ipsilateral tract to the midbrain is quite small. Descending fibers from the ipsilateral torus semicircularis in *Amia* course just lateral to nucleus praeeminentialis, but there is no clear evidence of terminations. In the absence of further anatomical and comparative data, it should only be tentatively suggested that the nucleus praeeminentialis of *Amia* and *Carassius* is homologous to the ventral division of nucleus praeeminentialis of electroreceptive teleosts. The specific functions of nucleus praeeminentialis for either mechano- or electroreceptivity are unknown in any fish.

17.3.3 Projections to the Midbrain: Isthmoreticular Nucleus and Perilemniscal Nuclei

The most caudal midbrain structure that may receive mechanosensory fibers is the isthmoreticular, or perilemniscal, nucleus. In *Amia*, the lateral line lemniscus passes directly over, and most likely terminates in, this nucleus (Fig. 17.3c). The isthmoreticular nucleus was first described in *Amia* (Braford and McCormick 1979) and was named after a similar nucleus in anurans which is now considered to be like the nucleus of the lateral lemniscus of amniotes. In *Amia*, horseradish peroxidase injected into the torus semicircularis will retrogradely label these cells bilaterally with contralateral predominance, a pattern of connections similar to that of the anuran and mammalian nuclei of the lateral lemniscus. In *Amia*, the isthmoreticular nucleus is also retrogradely labeled after HRP injections in the tectum. The isthmoreticular nucleus is similar in general location to the more rostral perilemniscal cells which are afferent to the torus in teleosts (Echteler 1984, Finger and Tong 1984). In teleosts, there is evidence that there may be two functional populations of perilemniscal cells. Echteler (1984) provided evidence that at least some of these cells in *Cyprinus* belong to the ascending auditory system. In *Ictalurus*, Finger and Tong (1984) reported large perilemniscal cells morphologically like the isthmoreticular cells in *Amia* that were retrogradely labeled after HRP injections in the mechanosensory portion of the torus. They appear to distinguish these cells from other perilemniscal cells associated with the auditory system. It remains to be determined in *Amia* whether the isthmoreticular nucleus is really the same cell group as the perilemniscal nucleus of teleosts and whether or not it can be subdivided into separate auditory and lateral line portions.

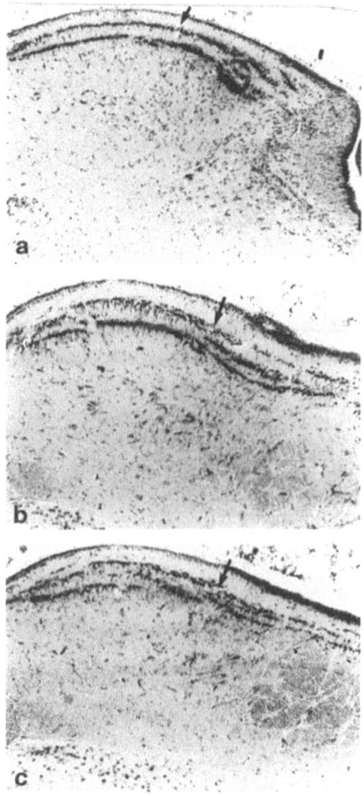

FIGURE 17.4. Photomicrographs of Nissl-stained transverse sections through the torus semicircularis of *Amia*; Fig. 17.4a is the most caudal section. Arrows indicate the boundaries between the lateral mechanosensory and medial octaval regions.

17.3.4 Projections to the Midbrain: Torus Semicircularis

The torus semicircularis is the principal midbrain termination site of mechanosensory lateral line afferents. The subdivisions of the torus were first described by Knudsen (1977) in *Ictalurus*, where, in a lateral to medial direction, there are largely separate electrosensory (nucleus lateralis pars lateralis), mechanosensory (nucleus lateralis pars medialis), and auditory (nucleus centralis) regions (Fig. 17.5b). These subdivisions in *Ictalurus* were further described, along with additional ones, by Finger and Tong (1984). Separate toral sensory areas are also present in gymnotids, although in these electroreceptive fishes the electrosensory region forms the large, laminated dorsal torus which overlies the lateral mechanosensory and medial auditory regions of the ventral torus (Fig. 5c; for a review see Carr and Maler 1986). The related nonelectroreceptive ostariophysan *Cyprinus* also exhibits a laterally positioned mechanosensory region (nucleus lateralis) and a medially positioned auditory region (nucleus centralis) (Figs. 17.5a, 17.6b; Echteler 1984, 1985). It is likely that all nonelectroreceptive ostariophysan fishes have a lateral to medial organization of the torus semicircularis.

The electroreceptive osteoglossomorphs also exhibit largely segregated electrosensory, mechanosensory, and auditory toral areas. The arrangement of the three toral areas differs in several respects, however. In mormyrids, the elec-

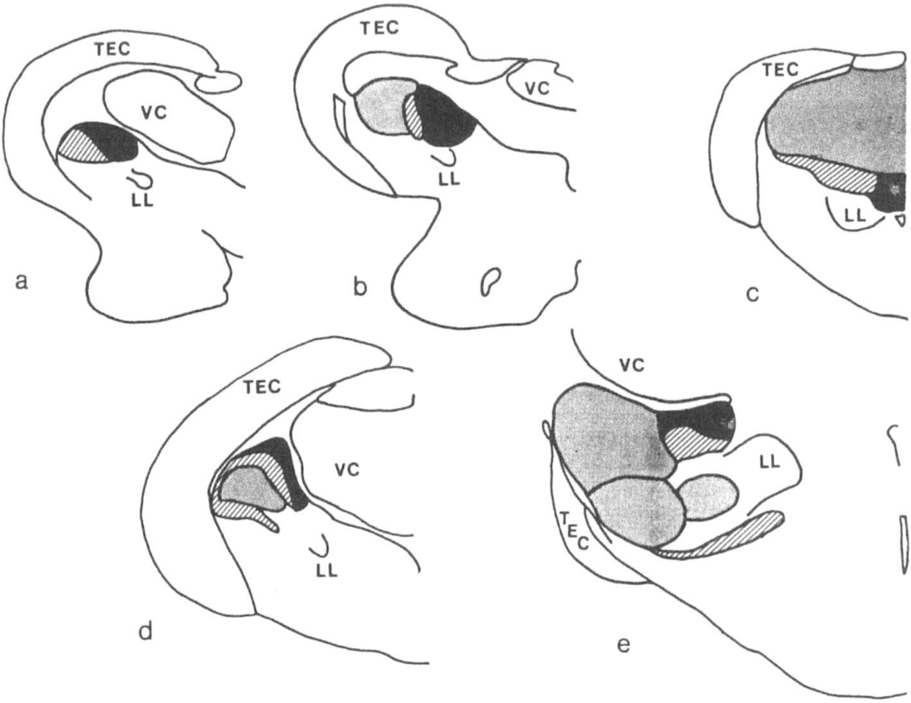

FIGURE 17.5. Transverse hemisections through the midbrain of various teleosts showing the positions of electrosensory (stippled), mechanosensory (diagonal lines), and octaval (black) areas. After McCormick and Braford (1988). (a) The ostariophysan *Cyprinus*; (b) the ostariophysan *Ictalurus*; (c) the ostariophysan *Eigenmannia*; (d) the osteoglossomorph *Xenomystus*; (e) the osteoglossomorph *Gnathonemus*.

trosensory structures, which have a nuclear rather than a laminated organization, lie lateral to the mechanosensory zone at some levels, but at others, one of these nuclei subdivides the mechanosensory zone (Fig. 17.5e; Bell 1981b; for a review, see Bell and Szabo 1986). The entire electrosensory nucleus in the related species *Xenomystus* also interrupts the mechanosensory zone, resulting in one portion lying dorsal and one ventral to the electrosensory nucleus (Fig. 17.5d; Braford 1986). In mormyrids and *Xenomystus*, horseradish peroxidase injections in the dorsal medulla which include first-order saccular neurons suggest that the auditory toral area is in close proximity to the mechanosensory area, but the exact position has not been determined (Bell 1981b, Braford 1982).

In *Amia*, the torus semicircularis exhibits medial and lateral subdivisions similar to those in the nonelectroreceptive ostariophysan *Cyprinus*, based on afferent fiber connections and subtle cytoarchitectonic differences (Figs. 17.3d, 17.4). Both of these subdivisions have a superficial laminated portion and a deeper, nonlaminated portion. The medial subdivision receives fibers from the dorsal half of the descending octaval nucleus and may thus be auditory in nature. The

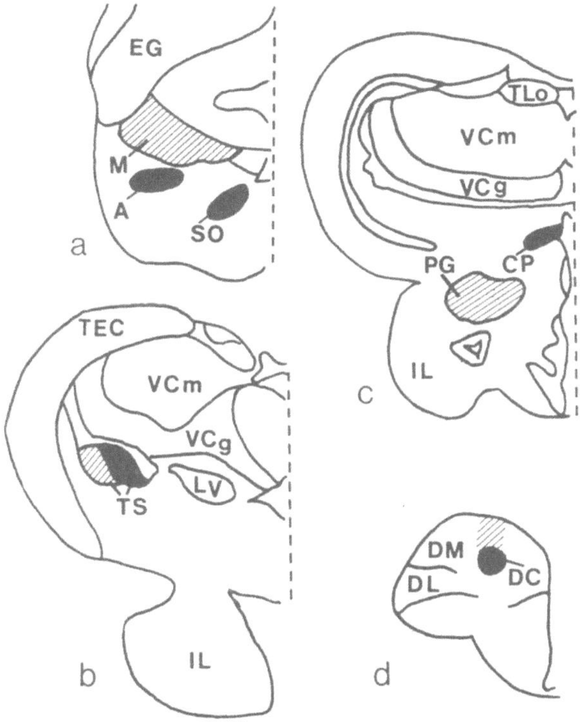

FIGURE 17.6. Transverse hemisections through the brain of *Cyprinus carpio* showing the position of mechanosensory (diagonal lines) and acoustic (black) centers identified using anatomical and/or electrophysiological methods. After Echteler (1984, 1985).

medial subdivision reaches its largest size in the caudal half of the torus semicircularis and is absent in the rostral one-third of the torus. The lateral division of the torus receives axons from nucleus medialis and extends almost the entire length of the torus.

The torus semicircularis is considered to be part of the ascending lemniscal pathway for mechanoreception and other octavolateralis modalities. However, it also contains multimodal units that integrate these modalities, as described by Schellart and Kroese (Chapter 21). As with the cases on convergence in the medulla, the existence of these multimodal units in the torus does not necessitate abandoning the notion that a large proportion of the information from lateral line and inner-ear receptors is processed by nuclei which form lemniscal pathways specific to each system.

17.3.5 Projections to the Midbrain: Optic Tectum

The deep layers of the optic tectum are a site of integration of various sensory modalities. Such integration is well known in the deep layers of the superior colliculus of mammals (Brodal 1981). Some axons from nucleus medialis course directly to it in cartilaginous and bony fishes (Fig. 17.3c,d; Boord and Northcutt 1982, 1988; Finger 1982; Finger and Tong 1984; McCormick personal observations; Braford personal communication). In electroreceptive species, electrosensory information reaches it via the electrosensory regions of the torus semicircularis (bony fishes) or directly from nucleus dorsalis (cartilaginous fishes) (for reviews see Bodznick and Boord 1986, Carr and Maler 1986, Bell and Szabo 1986). It is known that in at least some electroreceptive teleosts, electrosensory input maps topographically in the optic tectum such that it is in spatial register with the fish's visual map and that there are also many cells that are multimodal for vision and the electric sense (Bastian 1982). Similarly detailed studies on mechanosensory input to the torus are not available, and it is therefore not known how this information is used by the optic tectum in any of the fishes. However, it was recently shown that in *Xenopus* the optic tectum is the probable first site of directional coding in the mechanosensory system (Zittlau et al. 1986; Claas et al. Chapter 20).

17.4 Representation of Mechanoreception in the Diencephalon and Telencephalon

The final portion of the basic ascending pathway for mechanoreception involves the relay of information from midbrain to thalamus, and then from thalamus to telencephalon. Although this part of the circuitry is not well understood, a number of forebrain structures are known to be involved with processing octavolateralis information. The nonelectrosensory portions of the torus semicircularis project as a whole to at least two thalamic structures: the central posterior nucleus and the preglomerular complex (Fig. 17.7a,b; Braford and McCormick 1979; Echteler 1984; Murakami et al. 1986). In teleosts, the thalamic ventromedial nucleus and the anterior tuberal nucleus of the hypothalamus also receive toral connections (Echteler 1984). In electroreceptive teleosts, electroreceptive toral efferents supply the preglomerular complex and at least in some cases other thalamic structures; these data are discussed elsewhere (Finger 1986; Carr and Maler 1986; Wulliman and Northcutt 1987). The nonelectrosensory projections of the torus to the above-mentioned thalamic groups have been studied and interpreted in different ways by different investigators, and there is disagreement as to the extent to which separation exists between thalamic mechanosensory and auditory inputs.

Echteler's studies (1984, 1985) in the nonelectroreceptive ostariophysine *Cyprinus carpio* focused primarily on the auditory pathways but also contributed information about mechanosensory lateral line projections. He concluded that the central posterior nucleus, which is in the dorsal thalamus, processes auditory information (Fig. 17.6c). In this species, it receives projections from the medial (auditory) portion of the torus semicircularis and responds to auditory but not

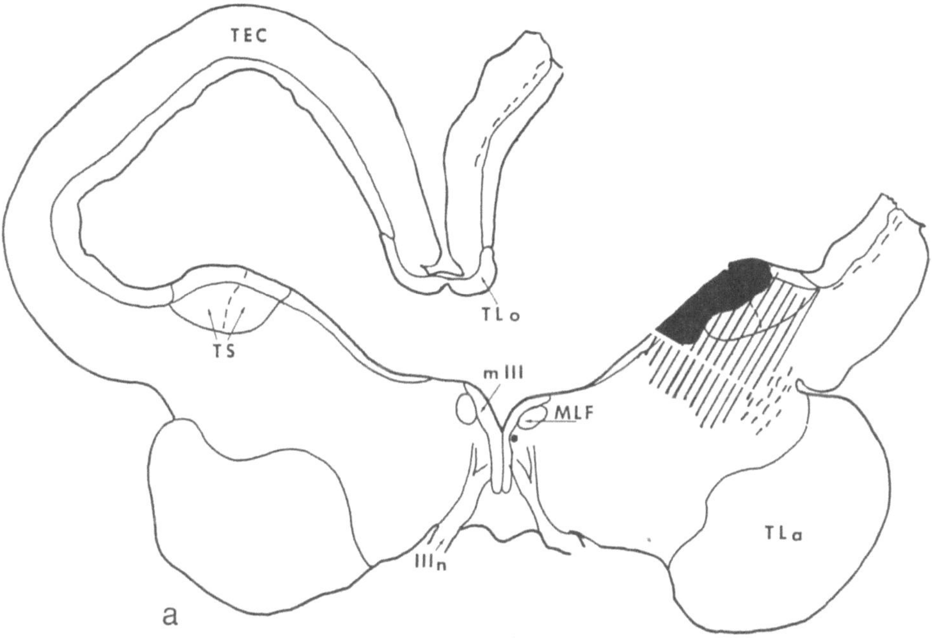

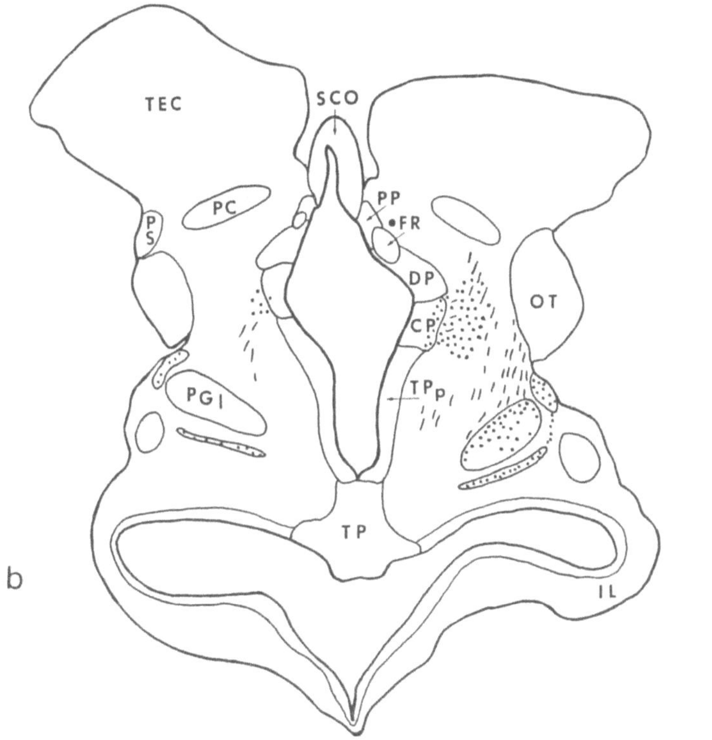

FIGURE 17.7. Projections of the torus semicircularis to the thalamus in *Amia* determined using HRP. The black area is the focus of the HRP injection site; the diagonal lines are the extent of spread of the injection. Dashes = HRP-labeled fibers; dots = terminal fields. From Braford and McCormick (1979).

lateral line stimuli. His anatomical study also revealed that the central posterior nucleus projects back upon the medial torus, that the right and left medial tori are reciprocally connected, and that the medial torus also projects to the anterior tuberal nucleus of the hypothalamus and sparsely to a major cerebellar thalamic target, the ventromedial nucleus. He concluded that the central posterior nucleus is an important component of the auditory lemniscal pathway. This is an attractive conclusion from a comparative standpoint, since the central posterior nucleus of fish is in a location similar to a major acoustic thalamic nucleus in anurans, the central nucleus (Neary 1974, 1988; Wilczynski 1978, 1981). (Another similarity with anurans is that midbrain auditory input is relayed to the hypothalamus (Echteler 1984; Neary 1988).) The efferent connections of the central posterior nucleus of *Cyprinus* were not determined anatomically. However, electrophysiological recording indicated the area dorsalis pars centralis as the first telencephalic acoustic processing zone. A second auditory zone is located in the caudal area dorsalis pars medialis near the zone sensitive to lateral line stimuli (Fig. 17.6d).

Echteler's studies imply that mechanosensory input is processed in portions of the preglomerular complex and demonstrate electrophysiologically that mechanosensory information is relayed in some way to the caudal portion of the telencephalic area dorsalis pars medialis (Fig. 17.6c,d). The preglomerular complex is a migrated portion of the posterior tuberculum and is made up of nuclear subpopulations which vary in number in different species. In *Amia*, for example, there are four major groups (McCormick personal observations), whereas in *Xenomystus* there are six (Braford personal communication). These subpopulations have not been studied comparatively, and their connections are largely unknown. Large toral injections of horseradish peroxidase result in efferent projections to only some portions of the ipsilateral preglomerular complex in at least some species. In *Cyprinus*, such connections were seen only when these injections included the lateral, mechanosensory portion of the torus. Additionally, no auditory responses were reported in this complex in this species (Echteler 1985). In *Cyprinus*, toral efferents supply only the lateral portion of the preglomerular complex. In *Amia* such fibers supply only two of the caudal subpopulations, the lateral and the central (Fig. 17.7).

Electrophysiological recordings in the telencephalon of *Cyprinus* revealed a mechanosensory processing area in the caudal area dorsalis pars medialis (Echteler 1985). Although this area was in close proximity to an auditory zone, best responses to auditory and lateral line stimuli were recorded at different depths. This strongly suggests that a subpopulation within the area dorsalis pars medialis represents the end point of the lemniscal pathway for mechanoreception. Braford has shown in *Xenomystus* that injections of horseradish peroxidase in the caudal area dorsalis pars medialis retrogradely label cells in those portions of the preglomerular complex that receive toral efferents (Braford 1982, personal communication). However, no electrophysiological studies have been done in *Xenomystus* that would identify with certainty these thalamic and telencephalic areas as mechanosensory.

In summary, the studies cited above conclude or suggest strongly that the mechanosensory lemniscal pathway involves portions of the thalamic preglomerular complex and caudal area dorsalis pars medialis of the telencephalon, whereas the lemniscal auditory pathway involves the thalamic central posterior nucleus, portions of the area dorsalis pars centralis, and perhaps portions of the area dorsalis pars medialis different from the lateral line zones. Two other sets of studies provide conflicting evidence or interpretations of the above scenario.

An electrophysiological study by Finger and Bullock (1982) in a species closely related to *Cyprinus*, the catfish *Ictalurus nebulosus*, revealed both auditory and mechanosensory units in a diencephalic area similar in position to the preglomerular complex of other fishes. They referred to this area as the mechanoreceptive thalamic zone (MTZ) and demonstrated some units that were unimodal for a given octavolateralis modality and others that were bimodal. Finger (1980) showed that MTZ cells project to restricted portions of the caudal area dorsalis pars medialis in *Ictalurus* but did not comment on whether these thalamic cells were confined to a given subpopulation within the MTZ. Neither of these two studies mentioned the central posterior nucleus. Finger and Bullock (1982) concluded that some segregation between lateral line and acoustic modalities exists in the MTZ.

Murakami et al. (1986) concluded in an anatomical study that the preglomerular nucleus of *Sebasticus marmoratus*, a nonelectroreceptive scorpenid teleost, is the major thalamic relay for both lateral line and acoustic modalities. The basis of their conclusion relies on their assertion that the thalamic central posterior nucleus is not a distinct entity but is instead part of a multimodal thalamic complex, the nucleus ventromedialis thalami, that receives input from a variety of sources. They therefore consider the preglomerular nucleus to be the only likely candidate for the thalamic portion of both the acoustic and lateral line lemniscal systems. However, they also conclude on the basis of connectional differences that the preglomerular nucleus of *Sebasticus* is not homologous to the preglomerular complex of other species. They traced efferent fibers from the preglomerular nucleus to both area dorsalis pars medialis and area dorsalis pars dorsalis.

Clearly, the complexity of forebrain structures, different interpretations of them, and possibly species differences have resulted in a confusing picture of the rostral portion of the ascending lateral line and acoustic pathways. Future anatomical and physiological studies must in particular make an effort to better characterize those forebrain structures that are composed of a number of subpopulations with potentially different connections and functions.

17.5 Summary and Conclusions

There is clearly a great deal that remains elusive concerning the representation of neuromast information in the brains of bony fishes. It is reasonably clear that bony fish have separate lemniscal pathways for electroreception and mechanoreception that extend at least to the level of the diencephalon and probably to the telencephalon. It is also reasonably clear that separate lemniscal pathways for hearing and mechanoreception extend to the level of the torus semicircularis, although there are known (e.g., magnocellular octaval nucleus; Mauthner cell) or possible (e.g., superior olive; perilemniscal cells) sites in which information from these two modalities converges. The identity of the acoustic and neuromast lemniscal pathways beyond the level of the midbrain is not yet agreed upon. Several important questions need to be addressed. First, if neuromast and acoustic information is processed in structures currently recognized under a single name (i.e., the preglomerular complex), to what extent are separate subpopulations processing the two modalities versus cells that are influenced by both? Additionally, whatever the extent of segregation between these two modalities, what functions are subserved by populations in which extensive convergence occurs versus populations specific for a given modality?

Finally, an important issue that should be considered is the degree to which we should expect the lateral line mechanosensory pathways to be similar among verbebrate classes, and even within a given class. Various lines of evidence point to the origin of the mechanosensory system in the Agnatha. It is reasonable to assume that this system arose only once, and is thereby (1) homologous and (2) anatomically similar peripherally and centrally among anamniotes that possess it. Lateral line mechanosensory information is carried along an ascending lemniscal system which is basically similar among anamniotes and is also similar to the lemniscal pathways for the electrosensory and auditory systems. However, it is possible that each vertebrate class may exhibit a range of modifications of specific features of the mechanosensory pathway, including (1) the degree to which mechanosensory, electrosensory, and inner-ear information converge upon a given nucleus or cell; (2) the precise number of nuclei in the mechanosensory pathway; (3) the nature of the descending systems; and (4) the degree to which the nuclei of the octavolateralis system as a whole are differentiated from each other at each level of the neuraxis.

For example, Boord and Montgomery (Chapter 16) discuss the substantial differences in central lateral line organization in primitive versus advanced elasmobranchs and point out the need to study more representatives of the various radiations in order to generate a more accurate picture of how central pathways have evolved within this single vertebrate class. The organization of the lateral mesencephalic nucleus (the probable homologue of the torus semicircularis) is a case in point. In the advanced elasmobranch *Raja*, it is a well-differentiated complex with distinct termination zones for three of the octavolateralis modalities (electrosensory, mechanosensory, and presumed auditory), whereas in two squalomorphs (*Squalus* and *Cephaloscyllium*), it is a poorly differentiated zone in which extensive overlap between the three modalities occurs. Interestingly, the torus semicircularis of osteichthyans is also variable in its degree of differentiation. Whereas the torus semicircularis of electroreceptive teleosts is well defined cytoarchitonically, its subdivisions in *Amia* and various nonelectroreceptive teleosts are much less distinct. Unlike elasmobranchs, however, all bony fish studied thus far appear to have largely separate toral areas for the octavolateralis modalities.

There may likewise be differences in the degree to which inner-ear fibers project to nucleus medialis, but these differences cannot be interpreted until we have a clearer idea as to the role various otic end organs play in hearing versus the vestibular sense; the latter also likely differs among various species.

Without belaboring discussion of the four potential sources of pathway variability listed above, it is clear that more studies of the mechanosensory and other octavolateralis pathways need to be undertaken in order to (1) establish the primitive pattern of organization of these pathways at least for each of the three gnathostome anamniote classes, (2) determine how these primitive patterns of organization reflect the functional properties of the components of the octavolateralis system (i.e., are these primitive functional properties similar among the classes?), and (3) get some idea of the range of variation in central organization within each class. This variation can then be looked at from a functional standpoint (i.e., what new capabilities, if any, do they reflect?) as well as from an evolutionary standpoint (have similar solutions been arrived at independently among different classes?).

Acknowledgments. I thank Mark R. Braford Jr., Robert L. Boord, and Sheryl Coombs for their critical readings of this manuscript. The technical assistance of Susan Sutterfield and Virginia Kunch is also gratefully acknowledged. This work was supported by NSF grant BNS 19378.

References

Altman JS, Dawes EA (1983) A cobalt study of medullary sensory projections from lateral line nerves, associated cutaneous nerves, and the VIIIth nerve in adult *Xenopus*. J Comp Neurol 213:310–326.

Barry MA (1987) Afferent and efferent connections of the primary octaval nuclei in the clearnose skate, *Raja eglanteria*. J Comp Neurol 266:457–477.

Bass AH (1982) Evolution of the vestibulolateral lobe of the cerebellum in electroreceptive and nonelectroreceptive teleosts. J Morphol 174:335–348.

Bastian J (1982) Vision and electroreception: Integration of sensory information in the optic tectum of the weakly electric fish *Apteronotus albifrons*. J Comp Physiol 147: 287–297.

Bell CC (1981a) Some central connections of medullary octavolateral centers in a mormyrid fish. In: Tavolga WN, Popper AN, Fay RR (eds) Hearing and Sound Communication in Fishes. New York: Springer-Verlag, pp. 383–392.

Bell CC (1981b) Central distribution of octavolateral afferents and efferents in a teleost (Mormyridae). J Comp Neurol 195:391–414.

Bell CC, Russell CJ (1978) Termination of electroreceptor and mechanical lateral line afferents in the mormyrid acousticolateral area. J Comp Neurol 182:367–382.

Bell CC, Szabo T (1986) Electroreception in mormyrid fish: Central anatomy. In: Bullock TH, Heiligenberg W (eds) Electroreception. New York: Wiley, pp. 375–421.

Blubaum-Gronau E, Schwartz E (1987) Topological representation of primary afferents from different parts of the lateral line system in the butterflyfish *Pantodon buchholzi*. Poster presented at Neurobiology and Evolution of the Lateral Line System, Bielefeld, FRG, Aug. 31–Sept. 4, 1987.

Bodznick D, Boord RL (1986) Electroreception in Chondrichthyes: Central anatomy and physiology. In: Bullock TH, Heiligenberg W (eds) Electroreception. New York: Wiley, pp. 225–256.

Bodznick D, Schmidt AW (1984) Somatotopy within the medullary electrosensory nucleus of the little skate, *Raja erinacea*. J Comp Neurol 225:581–590.

Boord RL, Northcutt RG (1982) Ascending lateral line pathways to the midbrain of the clearnose skate, *Raja eglanteria*. J Comp Neurol 207:274–282.

Boord RL, Northcutt RG (1988) Medullary and mesencephalic pathways and connections of lateral line neurons of the spiny dogfish *Squalus acanthias*. Brain Behav Evol 32:76–88.

Braford MR Jr (1982) Electroreceptive and mechanoreceptive afferents of the torus semicircularis in the notopterid fish, *Xenomystus nigri*. Soc Neurosci Abstr 8:764.

Braford MR Jr (1986) African knifefishes: The Xenomystines. In: Bullock TH, Heiligenberg W (eds) Electroreception. New York: Wiley, pp. 453–464.

Braford MR Jr, McCormick CA (1979) Some connections of the torus semicircularis in the bowfin, *Amia calva*: A horseradish peroxidase study. Soc Neurosci Abstr 5:139.

Brodal A (1981) Neurological Anatomy. New York: Oxford University Press.

Bullock TH (1974) An essay on the discovery of sensory receptors and the assignment of their functions together with an introduction to electroreceptors. In: Fessard A (ed) Handbook of Sensory Physiology, Vol. III/3. New York: Springer-Verlag, pp. 1–12.

Bullock TH, Bodznick D, Northcutt RG (1983) The phylogenetic distribution of electroreception: Evidence for convergent evolution of a primitive vertebrate sense modality. Brain Res Rev 6:25–46.

Caird DM (1978) A simple cerebellar system: The lateral line lobe of the goldfish. J Comp Physiol 127:61–74.

Carr CE, Maler L (1986) Electroreception in gymnotiform fish: Central anatomy and physiology. In: Bullock TH, Heiligenberg W (eds) Electroreception. New York: Wiley, pp. 319–373.

Carr CE, Matsubara J (1981) Central projections of the octavolateralis nerves in gymnotiform fish. Soc Neurosci Abstr 7:84.

Carr CE, Maler L, Sas E (1982) Peripheral organization and central projections of the electrosensory nerves in gymnotiform fish. J Comp Neurol 211:139–153.

Claas B, Münz H (1981) Projection of lateral line afferents in a teleost's brain. Neurosci Lett 23:287–290.

Echteler S (1984) Connections of the auditory midbrain in a teleost fish, *Cyprinus carpio*. J Comp Neurol 230:536–551.

Echteler S (1985) Organization of central auditory pathways in a teleost fish, *Cyprinus carpio*. J Comp Physiol A 156:267–280.

Fay RR, Popper AN (1980) Structure and function in teleost auditory systems. In: Popper AN, Fay RR (eds) Comparative Studies of Hearing in Vertebrates. New York: Springer-Verlag, pp. 1–42.

Finger TE (1980) Non-olfactory sensory pathway to the telencephalon in a teleost. Science 210:671–673.

Finger TE (1982) Connections of the lateral line lobe (n. medialis) in goldfish, *Carassius auratus*. Soc Neurosci Abstr 8:763.

Finger TE (1986) Electroreception in catfish: Behavior, anatomy, and electrophysiology. In: Bullock TH, Heiligenberg W (eds) Electroreception. New York: Wiley, pp. 287–317.
Finger TE, Bullock TH (1982) Thalamic center for the lateral line system in the catfish, *Ictalurus nebulosus*: Evoked potential evidence. J Neurobiol 13:39–48.
Finger TE, Tong SL (1984) Central organization of eighth nerve and mechanosensory lateral line systems in the brainstem of ictalurid catfish. J Comp Neurol 229:129–151.
Heiligenberg W, Dye J (1982) Labelling of electroreceptive afferents in a gymnotoid fish by intracellular injection of HRP: The mystery of multiple maps. J Comp Physiol 148:287–296.
Kitch RD, Highstein SM, Baker R (1987) Central projections of the primary octavolateralis afferents in the toadfish. Soc Neurosci Abstr 13:133.
Knudsen EI (1977) Distinct auditory and lateral line nuclei in the midbrain of catfishes. J Comp Neurol 173:417–431.
Larsell O (1967) The Comparative Anatomy and Histology of the Cerebellum from Myxinoids Through Birds. Jansen J (ed) Minneapolis: University of Minnesota Press.
Lowe DA, Russell IR (1982) The central projections of lateral line and cutaneous sensory fibers (XII and X) in *Xenopus laevis*. Proc R Soc Lond B 216:279–297.
Luiten PGM (1975) The central projections of the trigeminal, facial, and anterior lateral line nerves in the carp (*Cyprinus carpio* L.). J Comp Neurol 160:399–418.
McCormick CA (1981) Central projections of the lateral line and eighth nerves in the bowfin, *Amia calva*. J Comp Neurol 197:1–15.
McCormick CA (1982) The organization of the octavolateralis area in actinopterygian fishes: A new interpretation. J Morphol 171:159–181.
McCormick CA (1983a) Central projections of the lateral line and eighth nerves in the pike cichlid, *Crenicichla lepidota*. Brain Res 265:177–185.
McCormick CA (1983b) Central projections of inner ear endorgans in the bowfin, *Amia calva*. Am Zool 23:895.
McCormick CA, Braford MR Jr (1988) Central connections of the octavolateralis system: Evolutionary considerations. In: Atema J, Fay RR, Popper AN, Tavolga WN (eds) Sensory Biology of Aquatic Animals. New York: Springer-Verlag, pp. 733–756.
Meredith GE (1984) Peripheral configuration and central projections of the lateral line system in *Astronotus ocellatus* (Cichlidae): A nonelectroreceptive teleost. J Comp Neurol 288:342–358.
Meredith GE, Butler AB (1983) Organization of eighth nerve afferent projections from individual endorgans of the inner ear in the teleost *Astronotus ocellatus*. J Comp Neurol 220:44–62.
Meredith GE, Roberts BL, Maslam S (1987) Distribution of afferent fibers in the brainstem from end organs in the ear and lateral line in the European eel. J Comp Neurol 265:507–520.
Münz H (1985) Single unit activity in the peripheral lateral line system of the cichlid fish *Sarotherodon niloticus* L. J Comp Physiol A 157:555–568.
Murakami T, Fukuoka T, Ito H (1986) Telencephalic ascending acousticolateral system in a teleost (*Sebasticus marmoratus*), with special reference to the fiber connections of the nucleus preglomerulosus. J Comp Neurol 247:383–397.
Neary T (1974) Diencephalic efferents of the torus semicircularis in the bullfrog, *Rana catesbeiana*. Anat Rec 178:425.
Neary T (1988) Forebrain auditory pathways in ranid frogs. In: Fritzsch B, Ryan MJ, Wilczynski W, Hetherington TE, Walkowiak W (eds) The Evolution of the Amphibian Auditory System. New York: Wiley, pp. 233–252.

New JG, Northcutt RG (1984) Central projections of the lateral line nerves in the shovel-nose sturgeon. J Comp Neurol 225:129–140.
Puzdrowski RL (1987) The distribution, innervation and central projections of canal and superficial neuromasts in *Carassius auratus*. Poster presented at Neurobiology and Evolution of the Lateral Line System, Bielefeld, FRG, Aug. 31–Sept. 4, 1987.
Song J (1987) Peripheral distribution and central projection of the lateral line nerves in the garfish (*Lepisosteus platyrhincus*). Poster presented at Neurobiology and Evolution of the Lateral Line System, Bielefeld, FRG, Aug, 31–Sept. 4, 1987.
Wilczynski W (1978) Connections of the midbrain auditory center in the bullfrog, *Rana catesbeiana*. Doctoral Thesis, University of Michigan, Ann Arbor.
Wilczynski W (1981) Afferents to the midbrain auditory center in the bullfrog, *Rana catesbeiana*. J Comp Neurol 198:421–433.
Will U, Luhede G, Gorner P (1985) The area octavo-lateralis in *Xenopus laevis*. I. The primary afferent projections. Cell Tissue Res 239:147–161.
Wulliman MF, Northcutt RG (1987) Telencephalic and retinal projections allow reinterpretation of the diencephalon in mormyrids. Soc Neurosci Abstr 13:130.
Zittlau KE, Claas B, Münz H (1986) Directional sensitivity of lateral line units in the clawed toad *Xenopus laevis* Daudin. J Comp Physiol A 158:469–477.
Zottoli SJ, Van Horne C (1983) Posterior lateral line afferent and efferent pathways within the central nervous system of the goldfish with special reference to the Mauthner cell. J Comp Neurol 219:100–111.

Abbreviations for Figures 17.2 to 17.7

A	anterior octaval nucleus
Cb	cerebellum
CC	cerebellar crest
CP	central posterior nucleus
DC	area dorsalis pars centralis of the telencephalon
DL	area dorsalis pars lateralis of the telencephalon
DM	area dorsalis pars medialis of the telencephalon
DT	dorsal tegmental nucleus
E	electrosensory nucleus (nuclei) of teleosts
EG	eminentia granularis of the cerebellum
FR	fasciculus retroflexus
G	secondary gustatory nucleus
IIIn	oculomotor nucleus
IL	inferior lobe
IP	nucleus interpeduncularis
IR	isthmoreticular nucleus
LL	lateral line lemniscus
LV	nucleus lateralis valvulae
M	nucleus medialis
mIII	oculomotor nucleus
MLF	medial longitudinal fasciculus
mV	trigeminal motor nucleus
Oct	octaval nucleus (nuclei)
OT	optic tract
P	nucleus praeeminentialis

PC	central pretectal nucleus
PG	nucleus preglomerulosus
PGl	nucleus preglomerulosus pars lateralis
PP	paraventricular pretectal nucleus
PS	superficial pretectal nucleus
SCO	subcommissural organ
SO	superior olive
TEC	optic tectum
TLa	torus lateralis
TLo	torus longitudinalis
TS	torus semicircularis
TP	posterior tubercle
TPp	paraventricular nucleus of the posterior tubercle
VC	valvula cerbelli
VCg	granular layer of the valvula cerebelli
VCm	molecular layer of the vavul cerebelli
VT	ventral tegmental nucleus

18 Central Mechanosensory Lateral Line System in Amphibians

UDO WILL

18.1 Introduction

In recent years neurobiological research on the lateral line system has produced an impressive body of knowledge not the least of which is evidenced by the present volume. In this chapter, present data on the central organization of the amphibian lateral line system will be reviewed. Commencing with the central termination of lateral line afferents, it will successively consider higher nuclear levels, taxonomic differences being discussed at each level. Yet, already at the medullary level, comparative neuroanatomical material exists mainly for only urodeles and anurans. As for the third amphibian order, gymnophionans, several studies have dealt with the peripheral lateral line system (Taylor 1970; Hethrington and Wake 1979; Fritzsch et al. 1985; Wahnschaffe et al. 1985). Knowledge about central connections, however, is still limited to a very few data. Furthermore, there are hardly any data on the physiology of the lateral line at the medullary level. Neuroanatomical data on higher projections, except for the projection to the mesencephalon in urodeles and gymnophionans, are available only for anurans. However, far from being complete, they are limited to the mesencephalon and its connections.

Despite all these gaps in our knowledge of the amphibian lateral line system, the data presently available allow for a preliminary assessment of some basic features of the system and their functional implications. Focusing on "lower" levels, these will be discussed in a separate section. Lower levels of sensory systems are often thought of as being more or less simple, serially connected relay stations within an open-loop neuronal network. From this review, however, a picture as complex as that of the auditory system (Will 1988) will arise for the amphibian lateral line system, with divergent connections and recursive organization being salient features. Besides functional implications, the organization of the mechanosensitive lateral line system and its relation to the other parts of the octavolateralis systems also have implications for a discussion of the evolution of the whole system. The identification of a nucleus saccularis and the analysis of overlap and convergence patterns may yield some new aspects for the octavolateralis discussion. These problems will be addressed in the last section.

18.2 Central Organization

As in other anamniotes, afferent fibers from the peripheral lateral line organs in amphibians enter the alar plate (i.e., the sensory zone of the dorsal rhombencephalon) via two separate nerves dorsorostral and dorsocaudal to the eighth nerve. The posterior lateral line nerve, supplying trunk organs, is closely associated with the vagus nerve, whereas the anterior lateral line nerve, supplying head organs, consists of two distal components which are associated with the trigeminal and the facial nerve, respectively (Escher 1925). If electroreceptive lateral line organs are present, fibers from these organs enter the brain via the dorsal part or a separate dorsal branch of the anterior lateral line nerve (Fritzsch 1981; Fritzsch et al. 1985). Before describing the afferent terminal pattern, a short characterization of the alar plate nuclei will be given.

18.2.1 Nuclei of the Alar Plate

Out-group comparison with other anamniotes suggests that the pattern of nuclear arrangement of the alar plate in urodeles is the simplest and least derived among amphibians. Within the periventricular gray of the dorsal rhombencephalon in urodeles, three periventricular cell masses can be delineated on the basis of topographical relations to sulci and afferent fiber courses in addition to criteria like cell size and density of cell populations (Fig. 18.1a).

The dorsal nucleus (DN; nucleus dorsalis areae octavolateralis of Opdam and Nieuwenhuys 1976) is a group of small-size cells adjacent to the taenia of the choroid plexus. Rostrocaudally, it extends from the fifth to the tenth nerve root. The neuropil lateral to this nucleus, the "dorsal island of Kingsbury" (Herrick 1948), receives primary afferents from the electroreceptive ampullary organs (Fritzsch 1981; Münz et al. 1984).

The intermediate nucleus (IN; nucleus intermedius areae octavolateralis of Opdam and Nieuwenhuys 1976) is a group of small- to medium-size cells ventral to the dorsal nucleus. It stretches from the tip of the lateral recess down to the caudal root of the tenth nerve. The ventral border of this nucleus is at the level of the sulcus intermediodorsalis. The neuropil lateral to this nucleus receives mechanoreceptive lateral line afferents (Fritzsch 1981; Münz et al. 1984) and, ventrally, octavus afferents (Will and Fritzsch 1988).

The first two nuclei are not present in the family Plethodonthidae, which completely lack a lateral line system (Fritzsch et al. 1987).

The ventral zone (VZ; nucleus ventralis areae octavolateralis of Opdam and Nieuwenhuys 1976) is a group of medium- to large-size neurons, its dorsal and ventral borders being marked by the sulcus intermediodorsalis and the sulcus limitans, respectively. It extends from the lateral recess to the obex region. This zone and the laterally adjacent neuropil receive primary afferents from the inner ear (Will and Fritzsch 1988).

Gymnophionans and anurans show specific variations of this basic pattern. In gymnophionans with a free larval stage, a DN, an IN, and a VZ can be delineated

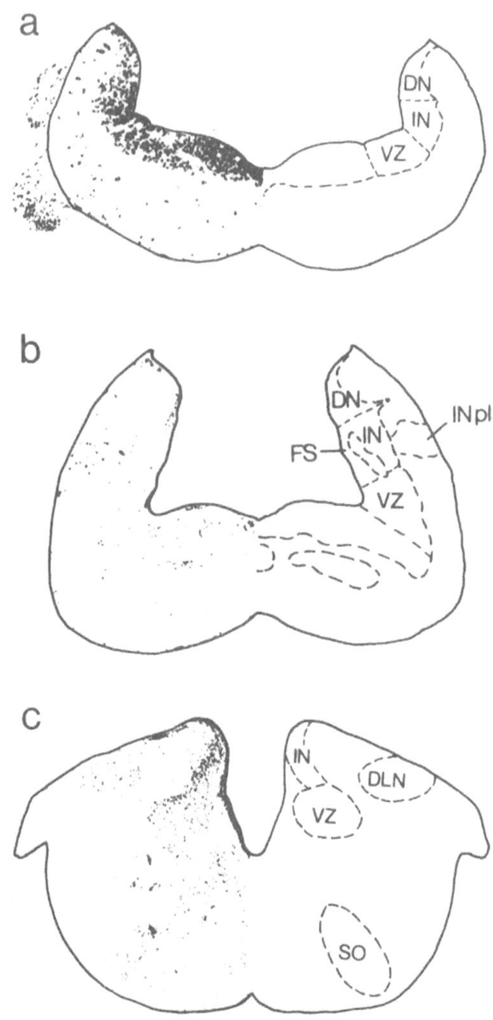

FIGURE 18.1. Cell masses (left) and nuclei of the alar plate in a cross section of the rhombencephalon at the level of the eighth nerve in *Ambystoma mexicanum* (a), larval *Ichthyophis kohtaoensis* (b), and adult *Xenopus laevis* (c). DN, dorsal nucleus; IN, intermediate nucleus; IN pl, intermediate nucleus pars lateralis; DLN, dorsolateral nucleus; VZ, ventral zone; FS, solitary fascicle; SO, superior olive.

as in urodeles (Fig. 18.1b). Gymnophionans without a free larval stage either show only a continuous periventricular gray or possess a dorsal nucleus and a ventral zone (*Thyphlonectes*; McCormick and Braford 1987). Anurans lack an electroreceptive sensory system (Fritzsch et al. 1984) and thus do not possess a DN. If the lateral line system is present (as in larval anurans and in adult members of Discoglossidae, Pipidae, Leptodactylidae, and Ranidae; Fritzsch et al. 1987), the dorsomedial periventricular cell mass (Fig. 18.1c) and its adjacent neuropil receive mechanoreceptive lateral line afferents and thus correspond to the IN and the lateral line neuropil of urodeles and gymnophionans.

Four nuclei have been described in the anuran VZ (Will and Fritzsch 1988). Two of them, the anterior nucleus (AN) and the caudal nucleus (CN), also receive lateral line afferents and therefore belong to the lateral line zone as well.

In addition to these variations, both anurans and gymnophionans exhibit laterally positioned cell groups which, however, are not homologous. In anurans (Fig. 18.1c), these cells form the dorsolateral nucleus (DLN) which receives afferents from the basilar papilla, the amphibian papilla, and the lagena of the inner ear. In gymnophionans, the lateral cell group is somewhat confluent with the ventral IN and the dorsal VZ (Fig. 18.1b) and receives lateral line afferents dorsally and saccular afferents ventrally. These cells have been described as the dorsal octavus nucleus of gymnophionans in homology to the anuran dorsolateral octavus nucleus (Kuhlenbeck 1967–1978) or as the IN proper because of their lateral line afferents (Fritzsch et al. 1985). However, they resemble more the ventrocaudal IN of urodeles and anurans, because they receive both lateral line and saccular afferents, and some of these neurons project to the spinal cord. Therefore, these cells have been named intermediate nucleus pars lateralis (IN pl), without implying that they are purely lateral line neurons (Will and Fritzsch 1988).

18.2.2 Lateral Line Afferents

Upon entering the alar plate, lateral line afferents bifurcate into ascending and descending branches, form longitudinal fascicles lateral to the nuclei, and send off numerous beaded collaterals (Altmann and Dawes 1983; Fritzsch 1981; Fritzsch et al. 1984, 1985; Herrick 1948; Lowe and Russell 1982; Will et al. 1985a; Will 1986). The longitudinal fascicles are confined to the ipsilateral medulla and do not enter the spinal cord or the corpus cerebelli. The terminal pattern implies that activity in any afferent fiber reaches all the various parts of the medullary lateral line zone. Physiologically, this has been indicated by Plassmann (1980) in an analysis of pre- and postsynaptic medullary potentials following electrical stimulation of the posterior lateral line nerve in *Xenopus*.

No simple topographical relations between peripheral organs and their central projections have been identified anatomically. All afferents seem to terminate in a similar pattern. However, in *Xenopus* there are slight differences for the anterior and posterior nerve fibers in the center of the terminal field, with the collaterals and boutons of the anterior lateral line being closer to the somata of IN neurons than those of the posterior lateral line (Will et al. 1985).

With respect to fascicular organization and collateral branching, a urodelian/gymnophionan and an anuran pattern of central afferent organization can be distinguished. The urodelian/gymnophionan pattern can be characterized as follows. Electroreceptive afferents, entering via the dorsal root of the anterior lateral line nerve, terminate with numerous collaterals which branch all over the dorsal island (Fig. 18.2a,b). The fascicle they form is the shortest of all lateral

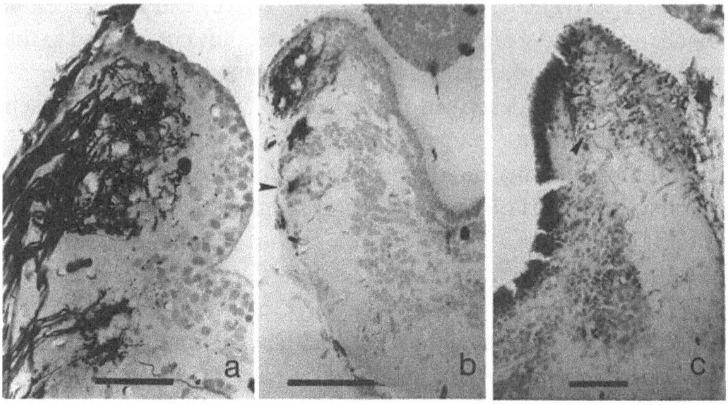

FIGURE 18.2. (a) Microphotograph of a cross section through the alar plate in *Ambystoma tigrinum* with HRP-labeled fibers of the anterior lateral line nerve. Electroreceptive fibers terminate dorsal to the two fascicles (arrowheads) of mechanoreceptive fibers from the trigeminal part of the anterior LL nerve. Courtesy of B. Fritzsch. (b) Cross section through the left alar plate in larval *Ichthyophis kohtaoensis* in which afferents of the anterior lateral line nerve were labeled by HRP. Electroreceptive fibers terminate in the dorsal island. One of the two fascicles of mechanoreceptive afferents (arrowhead) course in between the two parts of the IN pl. (c) Cross section through the right alar plate in an adult *Ichthyophis kohtaoensis* specimen following HRP application to the saccular branch of the peripheral eighth nerve. Arrowheads point to afferent fibers terminating in the medial IN. Bars are 100 μm.

line fascicles. Mechanoreceptive afferents terminate lateral to the IN. Beaded collaterals of these fibers are most prominent at the level of entrance of the lateral line nerves. However, compared with the collaterals of the electroreceptive afferents, they are very short and mainly confined to the immediate vicinity of the fascicles (Fig. 18.2a,b).

The longitudinal fascicles of the mechanoreceptive fibers are arranged in a stacked manner. Fibers from the posterior lateral line form the outer two fascicles, and fibers from the trigeminal part of the anterior lateral line form the innermost two fascicles (Fig. 18.2a). Fibers from the facialis part of the anterior lateral line form two fascicles in between the inner and outer fascicles. In gymnophionans (larval *Ichthyophis*; Firtzsch et al. 1985), one of the two fascicles from each part of the lateral line afferents courses close to the dorsal cell group of the IN pl, while the other runs in between the two IN pl cell groups (Fig. 18.2b). The fibers give rise to short, beaded collaterals, and some of the boutons appear to establish contacts with cells of the IN pl. It is not known whether collaterals contact cells of the medial, periventricular part of the IN.

In the anuran pattern there are no electroreceptive fibers, and the ascending and descending branches of the mechanoreceptive afferents give rise to a dense network of richly beaded collaterals within the neuropil as well as within the IN (Fig. 18.3a). (In adult *Bombina* (Fig. 18.3b), however, the collateral branching

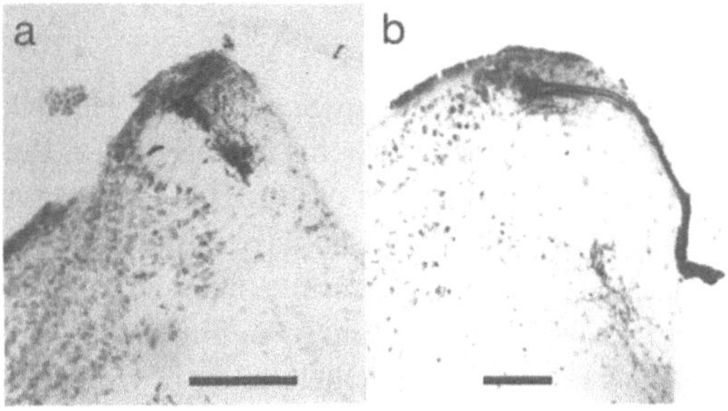

FIGURE 18.3. Terminal patterns of LL afferents in anurans. In adult specimen of both *Hymenochirus curtipes* (a) and *Bombina orientalis* (b), the posterior LL afferents are labeled by HRP application to the peripheral LL nerves. In *Hymenochirus* the ventromedial, nonlabeled part of the neuropil is occupied by fibers from the anterior lateral line nerve, whereas adult *Bombina* has lost this nerve completely. Note that the labeled fibers in *Hymenochirus* form two subfascicles. Bars are 100 μm.

pattern resembles that of urodeles much more than that of pipids.) The afferent fibers form two adjacent fascicles, a dorsal and a ventral one, which are supplied by fibers from the posterior and the anterior lateral line, respectively. The fibers of each fascicle are topographically organized according to the main line of the peripheral lateral line nerves. Will et al. (1985) have suggested that the fascicular organization is not determined functionally, because it is not replicated at the level of the terminating collaterals. Instead, it might be determined by ontogenetic processes—i.e., the way in which the afferents invade the alar plate.

18.2.3 Non-Lateral-Line Afferents

In all amphibians, some afferent fibers from the inner ear project to the ventral lateral line zone. Notably, this zone is also the major protection field of saccular afferents. They terminate in the lateral line neuropil, ventral (urodeles) or lateral (anurans) to the lateral line fascicles (Will and Fritzsch 1988). In species that lose their lateral line system during metamorphosis, secondary saccular neurons (see below) appear to constitute a major part of the remnant of the larval IN. In *Ichthyophis*, for example, which loses the lateral line system during metamorphosis, the postmetamorphic alar plate appears to be reduced as compared to larval forms. As in some urodeles, reduction seems to affect mainly the dorsal island and the DN but hardly the IN, which is the main recipient of saccular afferents (Fig. 18.2c). They terminate predominantly in the ventral IN pl, and several fibers even reach medial IN cells (arrowheads in Fig. 18.2c). In anurans that lose the lateral line system, the "nucleus saccularis" (i.e, the remnant of the larval IN) is part of what has been described as the nucleus caudalis (Opdam et al. 1976).

18.2.4 Medullary Organization

In his monograph "The Brain of the Tiger Salamander," Herrick (1948) holds the view that "at the first (central) synapse this specificity (of the peripheral sense organs) is no longer preserved in terms of localized gray centers or pathways of conduction." His interpretation was essentially based on the identification of a type of alar plate neurons that seem to receive input simultaneously from lateral line afferents, from octavus afferents, and from fibers of the descending trigeminal tract. In contrast to Herrick's view, the studies of Gomez Segade (1980) and Gomez Segade and Carrato Ibanez (1981) on *Salamandra* and *Chioglossa* indicate a variety of specific and nonspecific inputs onto secondary alar plate neurons. Applying classical neuroanatomical methods, they describe at least four different types of neurons for the lateral line zone of the alar plate: (1) DN neurons with dendrites arborizing within the dorsal island neuropil and tract "a"; (2) dorsal IN neurons, which send dendrites to all fascicles of lateral line afferents except for the dorsal island and tract "a"; (3) mediolaterally oriented, ventral IN neurons contacting mechanoreceptive lateral line afferents; and (4) dorsal VZ neurons that receive lateral line input via dorsal dendrites and input from trigeminal afferents and the reticular formation via ventral dendrites. Four different types of neurons are also described for the VZ (Gomez Segade 1980). Dorsal VZ neurons contact octavus afferents or octavus and lateral line afferents. Some small cells in the caudal VZ are reported to contact afferent fibers other than octavus afferents. The other two types of neurons receive input either selectively from octavus afferents or from octavus and trigeminal afferents and from the reticular formation.

The results of Gomez Segade and Carrato Ibanez are corroborated by HRP studies on *Ambystoma* and *Cynops* of this author. HRP injections into the dorsal mesencephalic tegmentum labeled the following types of lateral line neurons within the alar plate (Fig. 18.4). The first type consisted of elongated, dorsoventrally oriented neurons in the lateral aspects of the DN or the IN with numerous lateral dendritic processes (Fig. 18.4a,c). DN neurons contact various electroreceptive as well as mechanoreceptive lateral line afferents, whereas IN neurons contact various mechanoreceptive afferents only. These neurons are predominantly contralateral to the injection site and probably correspond to the type of dorsal alar plate neuron demonstrated by Herrick (1948), although dendrites extending to the descending trigeminal tract were not found in the present study. Some of the ventralmost dendrites might, however, establish contacts with octavus afferents. A second type of DN and IN neurons with restricted dendritic fields seems to get selective input from only a limited number of either electro- or mechanoreceptive lateral line afferents (Fig. 18.4b,d). A third type of neuron in the ventral IN extends its dendrites selectively into the dorsal terminal field of octavus afferents (mainly saccular fibers) and probably receives only octavus input (arrowhead in Fig. 18.4b). A fourth type of neuron is located in the dorsal VZ and exhibits two main dendritic branches, one extending to the ventral,

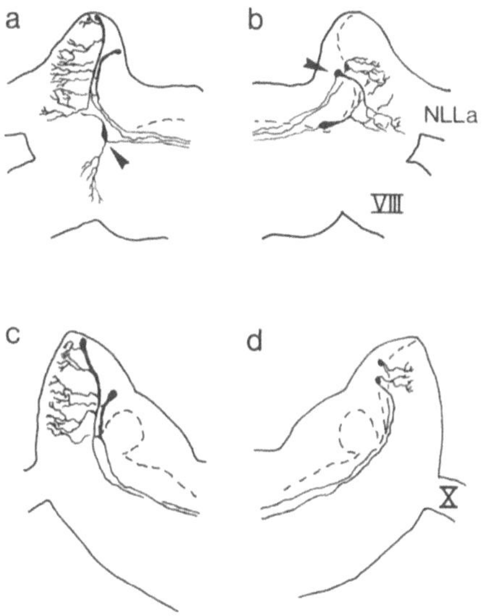

FIGURE 18.4. Reconstruction of alar plate neurons projecting to the mesencephalon in *Ambystoma mexicanum* at the level of the eighth (a,b) and the tenth nerve (c,d). Cells were retrogradely labeled following HRP micropipette application to the dorsal tegmentum (for arrowheads see text). NLLa: anterior lateral line nerve.

mechanoreceptive part of the lateral line neuropil, the other branching within octavus afferents (arrowhead in Fig. 18.4a).

A similar spectrum of cell morphology has been described for anurans (Will et al. 1985b). Hence, in amphibians, there is a variety of input relations for alar plate neurons, ranging from specific input of a subgroup of fibers of one modality to "unspecific" input from at least three modalities via lateral line, octavus, and trigeminal afferents. These different degrees of convergence contribute to the complexity of processing at the medullary level, as do the connections within the alar plate. As stated above, responses of lateral line afferents will affect the different parts of the medullary lateral line zone in parallel. In urodeles, further connections within the alar plate are probably formed by the correlation tracts "a" and "b" (Kingsbury 1895; Herrick 1948). Tract "a" is thought to be related to the lateral line system, because in adult urodeles in which the lateral line system is atrophied tract "a" also disappears (Kreht 1930; Gomez Segade 1980). Tract "b" seems to be related primarily to the octavus system. There are no physiological data on the correlation tracts. Herrick (1948) suggests that they are involved in some kind of specific filtering from the merged sense qualities of the alar plate. Some fibers of both tracts have been described as being commissural, connecting the respective areas on both sides (Herrick 1948).

In anurans, the dorsal AN gives rise to an ipsilateral fiber bundle running caudalward within the lateral line neuropil and a bilateral longitudinal fiber tract, part of which decussates within the cerebellum and runs within the contralateral lateral line neuropil (Will et al. 1985b). All these fibers may establish contacts with lateral line afferents or with dendrites of IN and CN neurons or with both. They resemble tract "a" fibers in urodeles. Other connections of AN neurons have not yet been identified, and the relation between the AN and the cerebellum remains unclear. The CN exhibits reciprocal bilateral connections with the AN (dominantly ipsilateral) and dominantly ipsilateral projections to the cerebellar auricle (Will 1988).

Extensive commissural connections exist between the lateral line zones on both sides of the medulla (Herrick 1948; Gomez Segade 1980; Will 1988). In anurans, commissural fibers of the IN and the CN arise as internal arcuate fibers and either terminate contralaterally as internal arcuate fibers or, predominantly, turn to the external arcuate fiber tract after traversing the medullary midline and terminate as external fibers. Most of the AN commissural fibers, however, do not change their position in the arcuate fiber system. IN commissural fibers terminate mainly within the LL neuropil and thus may contact afferents as well as dendrites of secondary neurons. These commissural fibers are probably responsible for a suppressive action on contralateral IN neurons which have been demonstrated by Plassmann (1980).

18.2.5 Projections to the Spinal Cord

In all amphibians the alar plate projection to the spinal cord is bilateral with ipsilateral dominance and originates mainly in the VZ. In urodeles two groups of alar plate neurons project directly to the spinal cord (Will 1988). Several cells of the rostral group, which are located in the VZ between the rostral pole of the eighth nerve and the first vagus root (Fig. 18.5), extend their dendrites into the areas of the octavus and the lateral line afferents. Some dorsal neurons, however, have dendrites exclusively in the area of the lateral line afferents (arrowhead in Fig. 18.5b). It should be noted that these dendrites are restricted to the terminal field of the mechanoreceptive fibers: there seems to be no direct DN projection to the spinal cord. The other group is located caudal to the vagus nerve, dorsomedial to the solitary fascicle. Their dendrites are confined to the caudal IN, which, however, receives lateral line as well as octavus (primarily saccular) afferents (Will 1988). Besides these direct spinal cord projections, tracts "a" and "b," as well as alar plate neurons, have been reported to establish connections with descending neurons of the medullary basal plate (Herrick 1948, Gomez Segade 1980).

In adult *Ichthyophis*, as in other amphibians, most alar plate neurons projecting to the spinal cord are located in the ventral zone. Some, however, are located in the ventral IN with dendrites ramifying within the nucleus and the adjacent neuropil (Fig. 18.6). In anurans the CN and the IN (Ten Donkelaar et al. 1981; Will et al. 1985b; Will 1988) project to the spinal cord. Neurons receiving lateral line input are located in the ventral IN and in the adjacent dorsal VZ. Several cells have dendrites contacting both octavus and lateral line afferents. Some of these neurons in *Bombina* are shown in Fig. 18.7a. Another pathway to spinal cord motor centers is formed by the Mauthner cell, which receives lateral line input via the dorsolateral dendrite (Will 1986b). This input may be only modulatory, however, as it has not been demonstrated that input from lateral line afferents alone is sufficient to fire the Mauthner cell (see also Zottoli and Danielson Chapter 23).

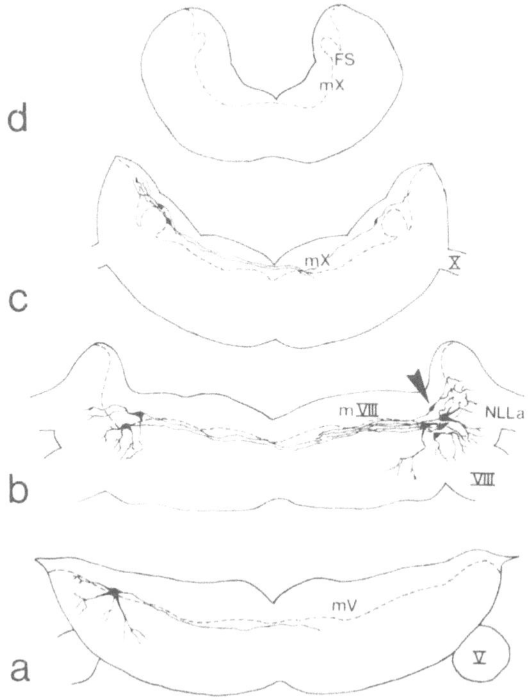

FIGURE 18.5. Reconstruction of alar plate neurons projecting to the spinal cord in *Ambystoma mexicanum* following HRP injection into the right spinal cord at the level of the second spinal segment. Arrowhead in (b) indicates a neuron of the ventral zone extending its dendrites exclusively into the intermediate nucleus and the adjacent neuropil. Roman numbers indicate cranial nerves. NLLa, anterior lateral line nerve; FS, solitary fascicle; mX, mVII, mV, motor nuclei X, VII, and V. (From Will 1988.)

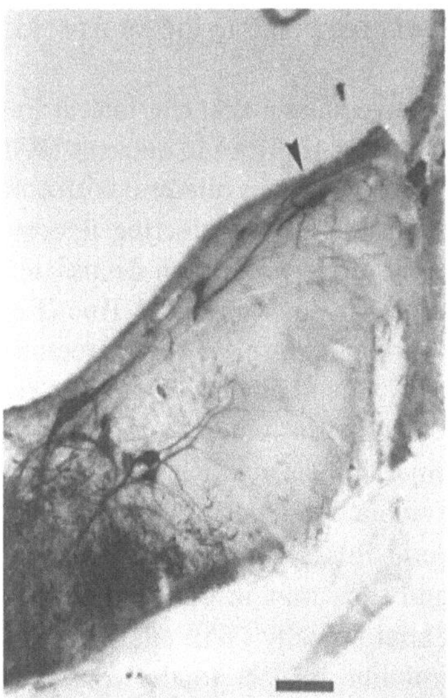

FIGURE 18.6. Microphotograph of a cross section through the right medulla in an adult *Ichthyophis kohtaoensis*. Needle application of HRP was performed to the rostral spinal cord ipsilaterally. Arrowhead points to labeled neuron in the IN. Bar is 100 μm.

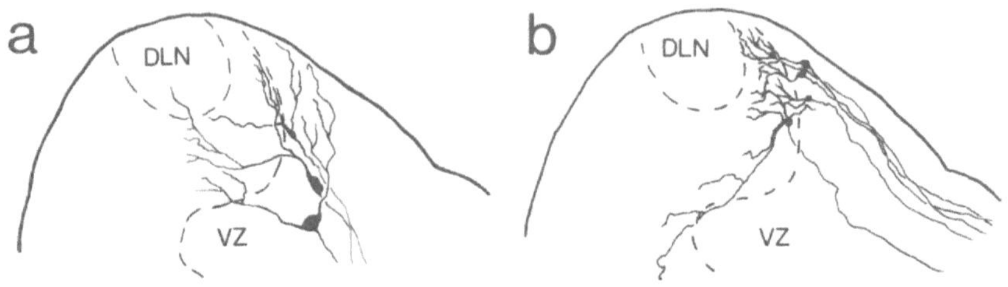

FIGURE 18.7. (a) Reconstruction of alar plate neurons projecting to the spinal cord in *Bombina orientalis*. Neurons were labeled by HRP injection into the ipsilateral rostral spinal cord. (b) Reconstruction of labeled alar plate neurons following HRP injection into the caudomedial torus of *Bombina orientalis*. The ventrolateral dendrite of the ventral neuron extends into the spinotrigeminal tract.

18.2.6 Projections to the Mesencephalon

Fibers from the IN ascend, via arcuate fibers, with the lemniscus bulbaris to the mesencephalon (Kreht 1930; Herrick 1948). In urodeles they have been reported to turn dorsally at the isthmus and to ramify within the optic tectum (Herrick 1948; Gomez Segade 1980). However, HRP preparations by this author indicate that IN neurons project bilaterally to the dorsal mesencephalic tegmentum with contralateral dominance (Will 1988). In adult *Ichthyophis*, too, HRP injections into the dorsal tegmentum result in labeling of IN neurons, and injections into the IN label terminal structures mainly in the dorsal tegmentum. Therefore, the dorsal tegmentum seems to be the main target of the ascending IN projection in urodeles and gymnophionans, and this projection corresponds to the toral projection of the IN in anurans.

Anatomical and electrophysiological studies have shown that the lateral line system projects to the magnocellular nucleus of the torus (Tmc) in anurans (Will et al. 1985b; Lowe 1986; Zittlau et al. 1985). The projection is bilateral with contralateral dominance, the ratio of ipsi- and contralaterally projecting neurons being 1:4 in *Xenopus*. Several fibers of the contralateral projection decussate a second time, within the torus and terminate in the ipsilateral medial Tmc (Fig. 18.8). Several IN neurons that project to the torus also project to the tectum. Axon collaterals branch off ventral to the Tmc, ascend further rostrally, and terminate in layers 2 to 7 of the rostrolateral tectum. In *Xenopus*, the ratio of labeled IN neurons following tectal and toral injections is about 1:20.

Lowe (1986) provides electrophysiological evidence for spatially distinct representations of the anterior and posterior lateral line system in the Tmc, the former dominating in the rostrolateral Tmc and the latter in the mediocaudal Tmc. Such a representation was not identified anatomically (Will et al. 1985b): fibers of the anterior and the posterior lateral line project to the rostral as well as the caudal IN. The rostal IN projects to the lateral Tmc, and the caudal IN projects to the lateral Tmc as well as to the medial Tmc. Hence, the toral

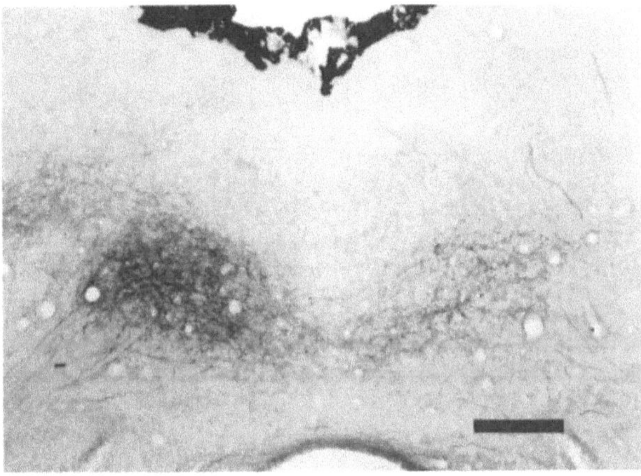

FIGURE 18.8. Labeled terminal field in the medial Tmc of *Xenopus laevis* following HRP injection into the right IN. Note that several fibers decussate a second time within the torus and terminate in the ipsilateral Tmc. Bar is 200 μm.

projections from both parts of the IN mediate responses from the anterior as well as from the posterior lateral line nerve, and the separate toral representation of the anterior and posterior lateral line system may, at least in part, be due to toral processing.

Besides pure lateral line neurons (see Fig. 18.9b), there are other types of neurons projecting to the mesencephalon with similar trajectories (Fig. 18.4 for urodeles and Fig. 18.7b for anurans. These include (1) neurons of the ventral IN and the dorsal VZ with dendrites in the terminal fields of lateral line and octavus afferents, (2) neurons similar to those of the first group but with additional extended lateral dendrites reaching trigeminal afferents, and (3) secondary saccular neurons of the ventrocaudal IN which project to the mediocaudal Tmc (Will 1988). In *Xenopus*, the dendrites of these neurons course around the lateral line fascicles and branch extensively in dorsal saccular terminal field (Fig. 18.9a).

In urodeles, some of the ascending lemniscal fibers have been reported to branch, one collateral continuing further into the dorsal thalamus (Herrick 1948). Collateral branching of alar plate neurons has also been found at medullary levels (Herrick 1948). In anurans, several ascending axons of IN neurons branch in the vicinity of the seventh motor nucleus. They give rise to collaterals that terminate in the reticular formation, in the vicinity of octavolateralis efferents and the ventromedial dendrite of the Mauthner cell (Fig. 18.9b).

18.2.7 Mesencephalic Connections

Except for the lateral line projection to the dorsal mesencephalic tegmentum, nothing is known about ascending projections in urodeles and gymnophionans. In anurans, too, diencephalic and telencephalic pathways are yet to be revealed.

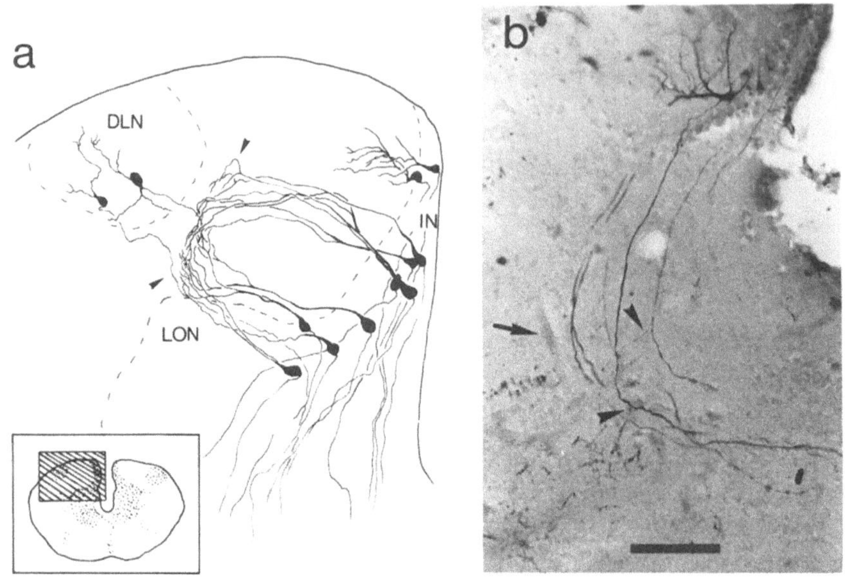

FIGURE 18.9. (a) Reconstructed saccular neurons in the IN of *Xenopus laevis* labeled by HRP injection into the medial Tmc. The dendrites of these neurons course around the descending limb of LL afferents and branch in the terminal field of saccular afferents (arrowheads). The inset indicates the location of the drawing. LON, lateral octavus nucleus; DLN, dorsolateral nucleus. (b) Microphotograph of a medullary cross section in *Xenopus laevis*. Ascending axons of IN neurons give rise to collaterals (arrowheads) which branch in the reticular formation. Axons were retrogradely labeled by toral HRP injection. Arrow indicates position of the Mauthner cell. Bar is 100 μm.

However, there is some, albeit incomplete, information about the anuran mesencephalic connections. The Tmc and the Tp are the main sensory input stations: the IN (secondary lateral line and saccular neurons) projects to the Tmc, the Tp receives an octavus projection (from the dorsolateral nucleus and the superior olive) (Wilczynski 1981; Pettigrew 1981; Will et al. 1985b; Feng 1986b), and the Tmc receives a somatosensory projection bordered by the lateral line projection laterally and the acoustic octavus projection medially (Will and Luhede 1984). This topography is corroborated by the results of an evoked potential study by Lowe (1986) in which lateral line evoked activity was mapped lateral and caudolateral to acoustically evoked activity. In frogs without a lateral line system, the somatosensory projection to the torus is extended laterally and caudally and appears to cover the whole area of the Tmc (Comer and Grobstein 1981).

Within the torus, three regions appear to contact all sensory "input stations." Neurons of the laminar nucleus (Tl) extend dendrites to various parts of the Tmc and the Tp. Lateral line activity that has been recorded from the Tl (Lowe 1986) might reach this nucleus via the extended dendrites of its cells. A similar pathway has been suggested for the auditory input to the Tl (Northcutt 1980). However,

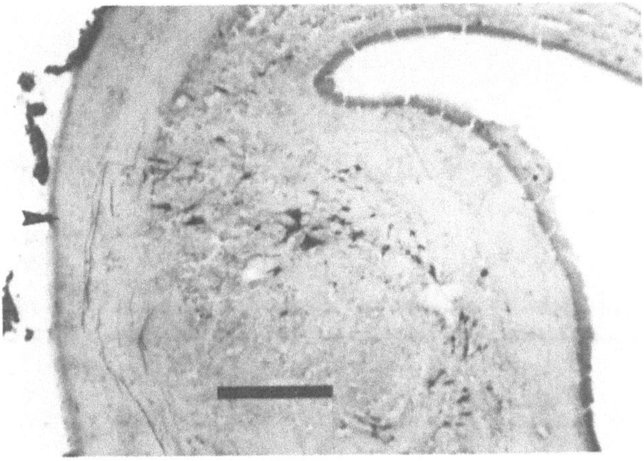

FIGURE 18.10. Microphotograph of mesencephalic coronal section from *Xenopus laevis* following HRP injection into the left ventral funiculus at the obex region. Arrowhead indicates the ventral border of the outer tectal layers. Bar is 200 μm.

some neurons of the Tmc and the Tp seem to project to the Tl. The dorsolateral torus, a subtectal zone where the main toral nuclei merge, exhibits reciprocal connections with the Tl and receives projections from various parts of the Tmc and the Tp. Similarly, the ventrolateral Tmc receives projections from various parts of the Tmc, the Tp, and the dorsolateral torus (Will and Luhede 1984).

The torus shows ascending projections to the thalamus and to the tectum opticum. On the basis of latency measurements, Lowe (1986) has already suggested the existence of a torotectal lateral line projection besides the direct tectal projection from the IN. Zittlau et al. (1986) revealed a functional significance of this torotectal projection: in response to lateral line stimulation with surface waves, the directional specificity was considerably greater in tectal than in toral units (for lateral line processing in the tectum, see Claas et al., Chapter 20).

A descending projection from the dorsal mesencephalon originates in the laminar nucleus (Tl) of the torus and the ventral tectum (Ten Donkellaar et al. 1981; Will and Luhede 1984). Following HRP injections into the inferior reticular formation or into the ventral funiculus of the rostral spinal cord, the labeled efferent cells form a continuous band (Fig. 18.10), some single cells being scattered throughout the tectum. The projection is bilateral with contralateral dominance, and the ratio of ipsi- and contralaterally labeled neurons is 1.8:1 in *Xenopus*. Despite their continuity, the descending efferents of the torus and the ventrolateral tectum obviously perform different functions, as can be concluded from their different input relations, from their different ipsi-contralateral distribution, and from lesion experiments (Comer and Grobstein 1981a,c).

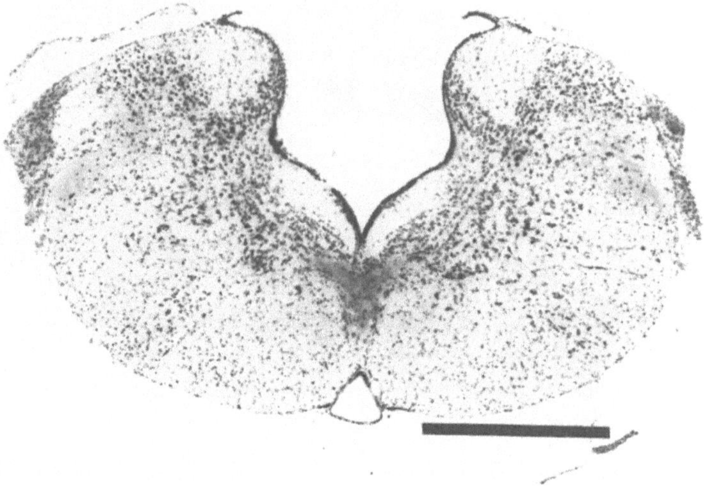

FIGURE 18.11. Microphotograph of a Nissl-stained coronal section through the medulla in *Xenopus laevis*. The preparation was made 1 year after destruction of all peripheral LL organs by cauterization. No LL nerve was identified in the preparation. The IN seems hardly to be affected, whereas the area of the lateral line neuropil is reduced owing to the degeneration of the primary afferents (compare with Fig. 18.1c). Bar is 1 mm.

18.3 Some Basic Principles of Organization

From the data presented above, several basic principles of organization can be deduced. The first concerns the IN and its role within the whole of the lateral line system. If IN cells were only serially connected relay stations, we would expect them to be seriously affected if disconnected from primary afferents. However, this is not the case. Figure 18.11 shows a cross section through the medulla in an *X. laevis* specimen 1 year after the destruction of all peripheral lateral line organs. The primary afferents have disappeared, but the IN hardly seems to be affected. The survival of the medullary nucleus following the destruction of peripheral organs was also reported for the VZ in *Rana* (Will et al. 1988). These results strongly suggests that, at least for the octavolateralis system, the medullary input stations are incorporated in circuitries that are more or less independent of the connections with primary afferents.

For the anuran lateral line zone, this circuitry consists of four parts—the AN, the rostral IN, the caudal IN, and the CN. AN and CN are connected reciprocally, and both have connections with the other parts of the lateral line zone via parallel fibers originating in the AN. The alar plate nuclei are connected reciprocally with the reticular formation. Finally, there is an elaborated commissural fibers system that mutually connects the circuits on both sides of the medulla.

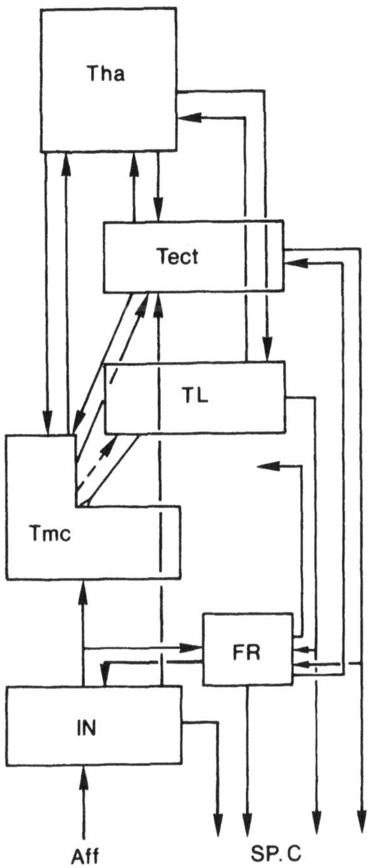

FIGURE 18.12. Schematic diagram of the connections known within the anuran LL system. (AN, CN, and contralateral connections are not considered.)

The maintenance of these circuits in deafferented animals indicates that they still perform vital functions. Maybe their "resting" activity provides essential reference points, as does the VZ resting activity for the vestibular system (Flohr et al. 1981).

Another important feature is that most of the nuclei show reciprocal connections with their nearest neighbors. The ascending IN projections to the torus and the tectum, however, are not reciprocal. The existence of reciprocal connections at different levels results in a complex hierarchical arrangement of feedback loops of different order. Feedback loops and recursive organization have recently been proposed as essential elements for basic behavior control systems (Powers 1973). They seem to be, indeed, important constituents even on the sensory side of such systems.

It should be emphasized that the collateral branching of IN axons implies that IN neurons project divergently to more than one brain area. Collateral branchings seem to be more widespread than previously thought and constitute an essential element in the wiring of neuronal networks. In octavus and lateral line systems, collateral branching, has been reported for alar plate neurons in urodeles (Herrick 1948), second-order vestibular neurons in the cat (McCrea et al. 1980), and secondary acoustic neurons in anurans (Will et al. 1985b; Feng 1986; Will 1988). The importance of this is that output at one level may act as input at several other levels of the network, resulting in complex feedback loops. For example, reticular formation response to input from the IN is under the influence of input from mesencephalon which itself receives input from the IN. In urodeles, even a direct IN projection to the thalamus has been reported (Herrick 1948), which might constitute a third-order feedback loop via the thalamotoral connection.

The main connections within the lateral line system are summarized in Fig. 18.12. The picture that emerges is that of a complex hierarchical network in which feedback, diverging connections, and parallel processing play important roles. In this network the hierarchy is not constituted by a simple-to-complex-to-hypercomplex, pyramidlike type of convergent processing. Instead, it emerges from the different levels of feedback loops involved. Such networks have recently been proposed as models for the organization of sensory systems in analyses of the visual system (Stone et al. 1979) and the somatosensory system (Dykes 1983) and on theoretical grounds (Powers 1973). A more detailed modeling of the lateral line system, however, seems premature, because essential physiological data are still lacking. A further discussion of the lateral line network wirings on anatomical grounds will nevertheless be a useful tool to guide further electrophysiological studies.

18.4 Relationship Between the Octavolateralis Subsystems

The structural relations between the octavolateralis subsystems as described in this report have some importance for an understanding of the function as well as the evolution of the whole system. From the turn of the century until the late 1960s, it was generally believed that all octavolateralis afferents terminate in the same medullary nucleus (e.g. Larsell 1967). Experimental studies during the 1970s advanced evidence for distinct projections of afferents of different modalities (for review see Northcutt 1981). Studies in the 1980s, however, stressed that although the main terminal fields are distinct for the three types of octavolateralis afferents, there is indeed some overlap of afferents. The extent of overlap, however, varies considerably among species (Barry 1987; Bell 1981; Carr and Matsubara 1981; McCormick 1981; Meredith and Butler 1983; Meredith 1984, 1985; Northcutt 1978a). In amphibians, overlap occurs in the center of the afferent terminal zone—i.e., among ventral lateral line, dorsal vestibular, and acoustic afferents. The dorsal (electro- and dorsal mechanoreceptive) and ventral (vestibular) terminal zones appear to be monomodal.

Afferent overlap per se does not necessarily imply that there is convergence of modalities at the level of second-order neurons. Excitation from "misled" afferents might reach the "proper" nucleus via extended dendrites from this nucleus. The dendritic structures of secondary neurons in amphibians and, in the case of urodeles, the limited extension of afferent collaterals, however, strongly support the idea that there is considerable convergence and overlap in the following cases: (1) DN neurons with dendrites in the electroreceptive and in the mechanoreceptive lateral line afferent terminal field, a few of which might receive additional octavus input via their ventralmost dendrites; (2) IN neurons with dendrites in the mechanoreceptive lateral line and the octavus afferent terminal field; (3) dorsal VZ neurons with dendrites in the octavus and the mechanoreceptive afferent terminal field; (4) dorsal VZ neurons with dendrites in the zone of mechanoreceptive lateral line, octavus, and trigemus afferents; and (5) VZ neurons with dendrites in the terminal field of octavus afferents and the descending trigeminal tract. All five types project to the mesencephalon, and the main types of multimodal convergence have been identified electrophysiologically in the torus of pipids (Zittlau et al. 1985). Neurons of types 3–5 were also found to project to the spinal cord. Interestingly, spinal cord projecting neurons that receive mechanoreceptive lateral line input are dominantly located in the dorsal VZ.

Hence, at the medullary level there is considerable anatomical evidence for various kinds of convergence of different degrees which subsequently are relayed to the spinal cord, the mesencephalon, and the reticular formation. For the latter, Herrick (1948) has already suggested that this convergence is involved in an unspecific activation of the motor zone, resulting in mass movements. On the other hand, input comparison from the lateral line and the octavus has also been suggested for more specific reactions in clupeids (Denton et al. 1979) and pipids (Görner 1973, 1976), the latter being able to orient toward surface waves even after destruction of all lateral line receptors.

The close relation between the mechanoreceptive lateral line and the octavus system is also stressed by the fact that the amphibian IN is not a pure lateral line nucleus (Will 1988): it contains secondary saccular neurons that project to the torus with trajectories similar to those of secondary lateral line neurons. There are striking similarities between this "nucleus saccularis" and the auditory nucleus within the mechanosensitive medial nucleus in elasmobranchs (nucleus X of Northcutt 1978; nuclei C1 and C2 of Smeets et al. 1983). Although its input has not yet been revealed, it has been described to project to the torus (Barry 1987; see also Boord and Montgomery, Chapter 16). Probably the "nucleus saccularis" represents one of the "old" acoustic pathways in vertebrates.

The data available do not give any substantial hints as to which of the octavolateralis subsystems preceded the others (for a discussion of the octavolateralis hypothesis, see Northcutt 1980). Therefore, a far more appropriate question to ask is, "What did the old system look like?" One might argue that overlap as described here for the amphibians is unique and reflects an adaptation to a particular life-style. However, at least overlap between the mechanoreceptive lateral line and the octavus system, with features similar to those in amphibians, is also found in other anamniotes. Furthermore, if it can be demonstrated that elasmobranchs possess a nucleus homologue to the amphibian nucleus saccularis, then both overlap and nucleus saccularis, with its direct mesencephalic projection, can be considered plesiomorphic characters of the octavolateralis system.

This author is therefore tempted to suggest that the octavolateralis system probably arose as a nonspecific hair cell system, a kind of alert or arousal system that responded to characteristics of both water and sound waves (see also Kalmijn, Chapter 9). Subsequent processes might then have led to a differentiation into separate subsystems. Of these, the pure vestibular and the electroreceptive subsystems, after their early emergence, have remained remarkably constant in their basic features during phylogeny, whereas the "intermediate" subsystems of the overlap zone show enormous variability and considerable evolutionary potential. Hence the separate subsystems are to be considered as derived characters of an ancestral octavolateralis system, of which afferent overlap zone, convergence of modalities, and structural similarities of the hair cells may be the closest, though faint reminiscences.

References

Altmann JS, Dawes EA (1983) A cobalt study of medullary sensory projections from the lateral line nerves, associated cutaneous nerves and the VIIIth nerve in adult *Xenopus*. J Comp Neurol 213:310–326.

Barry MA (1987) Afferent and efferent connections of the primary octaval nuclei in the clearnose skate, *Raja eglanteria*. J Comp Neurol 266:457–477.

Bell CC (1981) Central distribution of octavolateral afferents and efferents in a teleost (Mormyridae). J Comp Neurol 195:391–414.

Carr C, Matsubara J (1981) Central projections of the octavolateralis nerve in a gymnotiform fish. Neurosci Abstr 7:84.

Comer C, Grobstein P (1981a) Tactually elicited prey acquisition behavior in the frog, *Rana pipiens*, and a comparison with visually elicited behavior. J Comp Physiol 142:141–150.

Comer C, Grobstein P (1981b) Involvement of midbrain structures in tactually and visually elicited prey acquisition behavior in the frog, *Rana pipiens*. J Comp Physiol 142:151–160.

Comer C, Grobstein P (1981c) Organization of sensory inputs to the midbrain of the frog, *Rana pipiens*. J Comp Physiol 142:161–168.

Denton EJ, Gray JAB, Blaxter JHS (1979) The mechanics of the clupeid acoustico lateralis system: Frequency responses. J Mar Biol Assoc UK 59:27–47.

Dykes RW (1983) Parallel processing of somatosensory information: A theory. Brain Res Rev 6:47–115.

Escher K (1925) Das Verhalten der Seitenlinienorgane der Wirbeltiere und ihrer Nerven beim Uebergang zum Landleben. Acta Zool Bd 6:307:414.

Feng AS (1986) Afferent and efferent innervation patterns of the cochlear nucleus (dorsal medullary nucleus) of the leopard frog. Brain Res 367:183–191.

Flohr H, Bienhold H, Abeln W, Masckovics I (1981) Concepts of vestibular compensation. In: Flohr H, Precht W (eds) Lesion Induced Neuronal Plasticity in Sensorimotor Systems. Berlin: Springer, pp. 153–172.

Fritzsch B (1981) The pattern of lateral line afferents in urodeles. Cell Tiss Res 218:581–594.

Fritzsch B, Nikundiwe AM, Will U (1984) Projection patterns of lateral line afferents in anurans: A comparative HRP study. J Comp Neurol 229:451–469.
Fritzsch B, Wahnschaffe U, Crapon de Caprona M-D, Himstedt W (1985) Anatomical evidence for electroreception in larval *Ichthyophis kohtaoensis*. Naturwissenschaften 72:102–104.
Fritzsch B, Drewes RC, Ruibal R (1987) The retention of the lateral line nucleus in adult anurans. Copeia 1987 1:127–135.
Gomez Segade LA (1980) Morphology and evolution of the acoustic-lateral area in the rhombencephalon of Salamandridae. I. *Salamandra salamandra*. Trab Inst Cajal Invest Biol Madrid 71:37–55.
Gomez Segade LA, Carrato Ibanez A (1981) Morphology and evolution of the acoustic-lateral area in the rhombencephalon of Salamandridae. II. *Chioglossa lusitanica*. Trab Inst Cajal Invest Biol Madrid 72:111–119.
Görner P (1973) The importance of the lateral line system for the perception of surface waves in the claw toad, *Xenopus laevis* D. Experientia 29:295–296.
Görner P (1976) Source localization with the labyrinth and the lateral line in the clawed toad, *Xenopus laevis*. In: Schuif A, Hawkins AD (eds) Sound Reception in Fish. Amsterdam: Elsevier, pp. 171–183.
Herrick CJ (1948) The Brain of the Tiger Salamander. Chicago: University of Chicago Press. 409 pp.
Hetherington TE, Wake MH (1979) The lateral line system in larval *Ichthyophis* (Amphibia: Gymnophiona) Zoomorphologie 93:209–225.
Kingsbury BF (1895) On the brain of *Necturus maculatus*. J Comp Neurol 5:139–205.
Kreht H (1930) Ueber die Faserzuege im Zentralnervensystem von *Salamandra maculosa* L. Z Mikrosk Anat Forsch 23:239–320.
Kuhlenbeck H (1967–1978) The Central Nervous System of Vertebrates. Basel: Karger.
Larsell O (1967) The Comparative Anatomy and Histology of the Cerebellum from Myxinoids Through Birds. Minneapolis: University of Minnesota Press.
Lowe DA (1986) Organisation of lateral line and auditory areas in the midbrain of *Xenopus laevis*. J Comp Neurol 245:498–513.
Lowe DA, Russell IR (1982) The central projections of lateral line and cutaneous sensory fibers (XII and X) in *Xenopus laevis*. Proc R Soc Lond B 216:279–297.
McCormick CA (1981) Central projections of the lateralis and eighth nerves in the bowfin, *Amia calva*. J Comp Neurol 197:1–16.
McCormick CA, Braford MR (1987) Central connections of the octavolateralis system: Evolutionary Considerations In: Atema J, Fay RR, Popper AN, Tavolga WN (eds) Sensory Biology of Aquatic Animals. New York: Springer.
McCrea RA, Yoshida K, Berthoz A, Baker R (1980) Eye movement related activity and morphology of second order vestibular neurons terminating in the cat abducens nucleus. Exp Brain Res 40:468–473.
Meredith GE (1984) Peripheral configuration and central projections of the lateral line system in *Astronotus ocellatus* (Cichlidae): A nonelectroreceptive teleost. J Comp Neurol 228:342–358.
Meredith GE (1985) The distinctive central utricular projections in the herring. Neurosci Lett 55:191–196.
Meredith GE, Butler AB (1983) Organization of the eighth nerve afferent projections from individual endorgans of the inner ear in the teleost, *Astronotus ocellatus*. J Comp Neurol 220:44–62.

Münz H, Claas B, Fritzsch B (1984) Electroreceptive and mechanoreceptive units in the lateral line system of the axolotl *Ambystoma mexicanum*. J Comp Physiol A 154: 33–44.

Northcutt RG (1978a) Primary projections of VIII nerve afferents in a teleost, *Gillichthys mirabilis*. Anat Rec 193:638.

Northcutt RG (1978b) Brain organization in cartilaginous fishes. In: Hodgson ES, Mathewson RF (eds) Sensory Biology of Sharks, Skates, and Rays. Arlington, VA: Office of Naval Research, pp. 117–193.

Northcutt RG (1980) Central auditory pathways in anamniotic vertebrates. In: Popper AN, Fay RR (eds) Comparative Studies of Hearing in Vertebrates. New York: Springer-Verlag, pp. 79–118.

Northcutt RG (1981) Audition and the central nervous system of fishes. In: Tavolga WN, Popper AN, Fay RR (eds) Hearing and Sound Communication in Fishes. New York: Springer-Verlag, pp. 331–355.

Opdam P, Nieuwenhuys R (1976) Topological analysis of the brainstem of the axolotl *Ambystoma mexicanum*. J Comp Neurol 165:285–306.

Opdam P, Kemali M, Nieuwenhuys R (1976) Topological analysis of the brainstem of the frogs *Rana esculenta* and *Rana catesbeiana*. J Comp Neurol 165:307–332.

Pettigrew AG (1981) Brainstem afferents to the torus semicircularis of the queensland cane toad (*Bufo marinus*). J Comp Neurol 202:59–68.

Plassmann W (1980) Central neuronal pathways in the lateral line system of *Xenopus laevis*. J Comp Physiol 136:203–213.

Powers WT (1973) Feedback: Beyond behaviorism. Science 179:351–356.

Smeets WJAJ, Nieuwenhuys R, Roberts BL (1983) The Central Nervous System of Cartilaginous Fishes. Berlin: Springer.

Stone J, Dreher B, Leventhal A (1979) Hierarchical and parallel mechanisms in the organization of visual cortex. Brain Res Rev 1:345–394.

Taylor EH (1970) The lateral line sensory system in the caecilian family Ichthyophiidae (Amphibia Gymnophiona). Univ Kansas Sci Bull 48:861–868.

Ten Donkelaar HJ, de Boer van Huizen R, Schouten FTM, Eggen SJH (1981) Cells of origin of pathways descending to the spinal cord in the clawed toad. Neuroscience 6(11):2297–2312.

Wahnschaffe U, Fritzsch B, Himstedt W (1985) The fine structure of the lateral line organs of larval *Ichthyophis* (Amphibia, Gymnophiona). J Morphol 186:369–377.

Wilczynski W (1981) Afferents to the midbrain auditory center in the bullfrog, *Rana catesbeiana*. J Comp Neurol 198:421–433.

Will U (1986a) Das Seitenliniensystem bei *Bombina* (Discoglossidae, Anura). Verh Dtsch Zool Ges 79:302.

Will U (1986b) Mauthner neurons survive metamorphosis in anurans. J Comp Neurol 244:111–120.

Will U (1988) Organization and projections of the area octavolateralis in amphibians. In: Fritzsch B et al. (eds) The Evolution of the Amphibian Auditory System. New York: Wiley, pp. 185–208.

Will U, Fritzsch B (1988) The octavus nerve of amphibians. Peripheral and central distribution. In: Fritzsch B et al. (eds) The Evolution of the Amphibian Auditory System. New York: Wiley, pp. 159–183.

Will U, Luhede G (1984) Organization of sensory input to the torus semicircularis of *Xenopus laevis*. Verh Dtsch Zool Ges 77:345.

Will U, Luhede G, Görner P (1985a) The octavo-lateralis area in *Xenopus laevis*. I. The primary afferent projections. Cell Tiss Res 239:147–161.

Will U, Luhede G, Görner P (1985b) The octavo-lateralis area in *Xenopus laevis*. II. Second order projections and cytoarchitecture. Cell Tiss Res 239:163–175.

Will U, Kortmann H, Flohr H (1988) HRP study on structural changes in the commissural fiber system of *Rana temporaria* following labyrinthectomy. In: Flohr H (ed) Lesion Induced Neuronal Plasticity. New York: Springer-Verlag, pp. 345–355.

Zittlau KE, Claas B, Münz H, Görner P (1985) Multisensory interaction in the torus semicircularis of the clawed toad *Xenopus laevis*. Neurosci Lett 60:77–81.

Zittlau KE, Claas B, Münz B (1986) Directional sensitivity of lateral line units in the clawed toad, *Xenopus laevis* Daudin. J Comp Physiol 158:469–477.

19 Central Nervous Physiology of the Lateral Line, with Special Reference to Cartilaginous Fishes

HORST BLECKMANN and THEODORE H. BULLOCK

19.1 Introduction

Fishes and aquatic amphibians use the mechanoreceptive lateral line to detect weak water currents (Dijkgraaf 1963; Bleckmann 1986; Kalmijn 1988) and water surface waves (Schwartz 1971; Bleckmann 1988). Electrophysiological studies have shown that the pattern of impulses carried by primary lateral line afferents encodes information about the nature of the peripheral stimulus with respect to duration, amplitude, frequency, and phase (see Münz Chapter 14). If the activity of several neuromasts, which may differ with respect to the alignment of their most sensitive axis, is integrated over time and space, the additional information of stimulus direction and, perhaps, stimulus distance may be obtained. Thus the peripheral lateral line provides the brain with all cues necessary to evaluate a complex wave stimulus with respect to stimulus origin, stimulus duration, and stimulus type.

In contrast to the considerable attention directed to the peripheral physiology and anatomy of the lateral line, relatively little has been done with the central pathway of the lateral line, especially with respect to higher brain centers. The present chapter reviews the central physiology of the mechanoreceptive lateral line, particularly in cartilaginous fishes (for teleosts see also Schellart and Kroese, Chapter 21). We will describe the central physiology of the lateral line pathway from medulla to telencephalon. Besides more general comments we will discuss the following subheadings under each level of the neuraxis: threshold and dynamic range, receptive field (RF) organization, and multimodality.

The topics of efferent control of lateral line input, of Mauthner cell function with respect to lateral line physiology, and of the central physiology of the amphibian lateral line system will not be discussed here. For these aspects the reader is referred to Roberts and Meredith (Chapter 22), Zottoli and Danielson (Chapter 23), and Claas et al. (Chapter 20).

19.2 Central Organization

Earlier studies claimed extensive overlap between central terminations of the primary afferents of the lateral line and the eighth nerves (e.g., Mayser 1882; Herrick 1897). More recent studies, however, report little overlap in the primary projections of these nerves, except for the magnocellular octaval nucleus (e.g., Bodznick and Northcutt 1980; Northcutt 1981; McCormick 1981; Finger and Tong 1984; New and Bodznick 1985; McCormick and Braford 1988). Some exceptions do exist, however (cf. Bell 1981). In lampreys, cartilaginous fishes, and primitive bony fishes, the primary target of the mechanoreceptive ventral root of the anterior lateral line nerve and the exclusively mechanoreceptive posterior lateral line nerve is the medial or intermediate octavolateralis nucleus. The anterior and posterior lateral line lobes of electroreceptive teleosts are thought to be homologous to parts of the medial nucleus in the foregoing groups (McCormick 1981). Recent anatomical and physiological studies indicate that the mechanoreceptive lateral line of teleost fishes and aquatic amphibians is represented at all levels of the neuraxis–i.e., from medulla to telencephalon (Echteler 1985; Northcutt and Plassmann personal communication). The same holds true for cartilaginous fishes (Bleckmann et al. 1987, 1989) (Fig. 19.1).

19.3 Central Physiology

19.3.1 Medulla

Electrical potentials from the brain of goldfish (*Carassius auratus*) and perch (*Perca fluviatilis*) in response to stimuli known or believed to be mediated by the lateral line were first seen by Hoagland (1935). Alnaes (1973a,b), Russell (1974, 1976), Paul and Roberts (1977a,b), and Caird (1978) pioneered the study of second-order lateral line neurons. In the common eel, *Anguilla anguilla* (Alnaes 1973a,b), and the goldfish, *Carassius auratus* (Caird 1978), a direct electric shock to the posterior lateral line nerve (pLLN) leads to an evoked potential (EP) in both the ipsilateral and, to a lesser degree, the contralateral lateral line lobe. Paul and Roberts (1977a,b) failed to find such a contralateral response to electrical stimulation of the lateral line nerves in the dogfish *Scyliorhinus canicula*, but their recordings were made only from the surface of the hindbrain. Peak latencies of ipsilateral evoked potentials vary between 5.7 ms (goldfish) and 9.1 ms (eel). Excitatory activity from the contralateral lateral line nerve impinges with an added latency of 0.8 ms on the same neurons, and bilateral summation occurs when the ipsilateral input leads the contralateral by less than 10 ms (Alnaes 1973b). In both the goldfish and the dogfish, EP decrease in amplitude if two successive nerve shocks are separated by less than 100 to 150 ms (Caird 1978;Paul and Roberts 1977a). In the goldfish at an interpulse interval of 10 to 30 ms, a weak summation can be observed (Caird 1978).

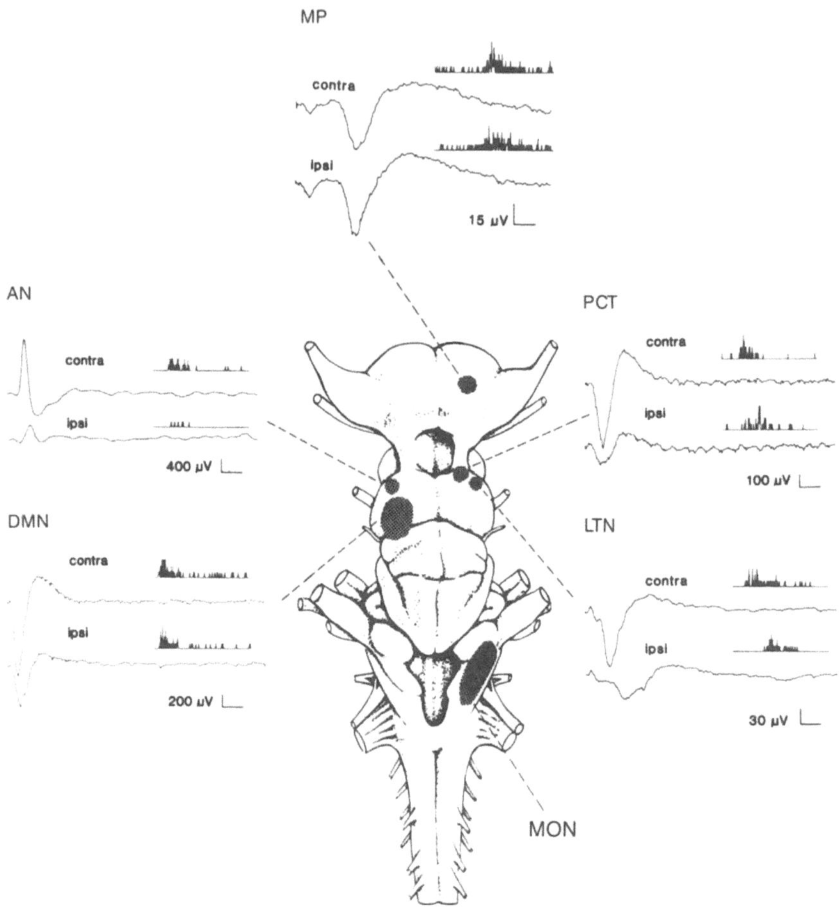

FIGURE 19.1. Dorsal view of the brain of *Platyrhinoidis triseriata*. The approximate extension of the different areas that respond to a pLLN-shock or a quasinatural stimulus are indicated by stipples. **AN**, anterior nucleus of the mesencephalic nuclear complex; **DMN**, dorsal nucleus of the mesencephalic nuclear complex; **LTN**, lateral tuberal nucleus of the diencephalon; **PCT**, posterior central thalamic nucleus of the diencephalon; **MON**, medial octavolateralis nucleus of the medulla; **MP**, medial pallium of the telencephalon. The responses recorded in the **MON** are not shown. The original drawing of the brain was provided by R. G. Northcutt.

In cartilaginous and teleost fishes, at least three cell types can be distinguished in the mechanosensitive lateral line lobe (Alnaes 1973b; Paul and Roberts 1977a; Caird 1978):

1. Spontaneously active, constant-latency units which follow a pLLN shock up to at least 100 Hz. These short-latency units are assumed to represent primary afferents.

2. Units whose spontaneous discharge pattern does not change after peripheral nerve stimulation.
3. Long-latency units that do not follow a stimulus repetition rate of 100 Hz. The latter units are assumed to represent secondary cells. At least two types can be distinguished: (a) subcristal units that respond to a nerve shock with a single spike. These units have only been reported for the goldfish and are called gray cells by Caird (1978). (b) Subcristal units that are often spontaneously active and which follow a pLLN shock with a single spike or a burst of spikes. The latency of these cells varies extensively, but latencies shorter than 1.2 ms were never observed. Cells of type 3b, misleadingly called primitive Purkinje cells by Caird (1978), integrate a large number of disparate afferent input channels (Alneas 1973b; Caird 1978).

In the eel *A. anguilla*, the spontaneous activity of secondary subcristal units, unlike that of primary afferents, is not coupled to gill movements. The observed suppression of the spontaneous discharge or of responses evoked by a contralateral lateral line nerve shock (Alnaes 1973b) during one phase of opercular movement may be due to common-mode rejection, as has been reported for the electroreceptive lateral line (Montgomery 1984). The mechanism of common-mode rejection implies primary projections to the contralateral medial nucleus, as has been recognized only in petromyzontids (Ronan and Northcutt 1987). Inhibitory mechanisms can also be observed in the goldfish medulla. Spontaneous and excited Purkinje-like cells show a strong depression of activity for up to 250 ms after caudal lobe stimulation (Caird 1978). A weak degree of stellate cell-type inhibition may be mediated by cristal interneurons onto the Purkinje-like cells. A second inhibitory mechanism is present in the lateral line lobe. Direct electrical stimulation of the pLLN results not only in a single spike but also in a suppression of spontaneous activity in both anterior and posterior lateral line secondary cells (Fig. 19.2A–C). This suppression follows the initial response for up to 50 ms. Inhibitory postsynaptic potentials following pLLN stimulation show this to be postsynaptic inhibition (Fig. 19.2E,F). The central sensory inhibition mechanisms in the lateral line lobe could increase, for instance, directional information by enhancing time of arrival or intensity differences between different organs (Caird 1978).

Central aspects of the mechanoreceptive lateral line of the thornback guitarfish, *Platyrhinoidis triseriata*, have been studied (Bleckmann et al. 1987, 1989). Both EP and unit responses recorded in the medullary medial octavolateralis nucleus (MON) are tonic or phasic-tonic. As quasinatural stimulus, a vibrating sphere (diameter 17.6 mm), placed 4 to 5 mm above the infraorbital lateral line canal, was used. Like acoustic "frequency following responses" recorded in the hindbrain of different vertebrate taxa (Bullock 1986), MON-EPs often show a delayed, tonic wave form that maintains the fundamental of the stimulus frequency and may add the second harmonic (Fig. 19.3). MON units are phase-coupled to the stimulus, firing up to one spike per stimulus cycle (Fig. 19.4). MON responses recorded in the hindbrain of the thornback did not require a low

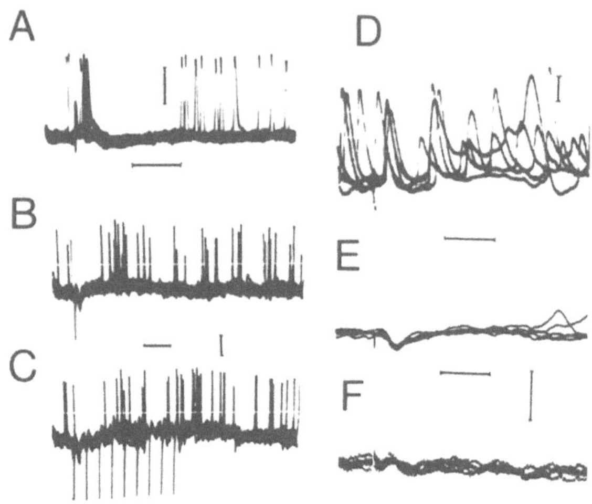

FIGURE 19.2. Inhibition in secondary cells. (**A**) Nuclear cells recorded from the lateral line lobe of the goldfish fire a single spike on stimulating the pLLN, and this response is followed by a 10- to 60-ms inhibition of spontaneous activity. (**B**) This inhibitory effect is widespread, and spontaneously active nuclear cells are also inhibited by pLLN stimulation. (**C**) This inhibition is not increased when multiple pLLN stimuli are given. The anterior lateral line unit shown in (**B**) and (**C**) was penetrated, and (**D**) shows the injury discharge produced in this unit. Approximately 20 nA of hyperpolarizing current produced the blocked inhibitory postsynaptic potential (ipsp) shown in (**F**). The current was then switched off, and the cell stayed silent, showing a large ipsp (**E**). Calibration bars, 20 ms, 10 mV (after Caird 1978).

stimulus repetition rate. Even at one stimulus per second (stimulus duration 500 ms), only a weak response decrement could be observed (Bleckmann et al. 1989).

19.3.1.1 Threshold and Dynamic Range

If driven with a vibrating sphere, primary and secondary fibers of the goldfish's lateral line lobe show a tonic response that is phase-coupled to the stimulus. At 30 Hz, the stimulus threshold is 2 to 8 μm p-p water displacement (Caird 1978). If the stimulus amplitude is increased, primary lateral line afferents and secondary cells may show either no saturation (up to 42 dB over threshold), saturation, or even inhibition (Caird 1978). Some cells display nonmonotonic curves; that is, after a period of saturation or even inhibition, the response increases again if the stimulus amplitude is further increased. Primary fibers give up to 10 spikes per cycle, whereas secondary cells rarely exceed three spikes under the same stimulus conditions. Secondary cells thus show some compression compared to the afferent fibers, but because of the small sample size and the amplitudes tested, it is not clear whether this leads to a reduced dynamic range of secondary cells (Caird 1978). If the stimulus probe is moved away from the

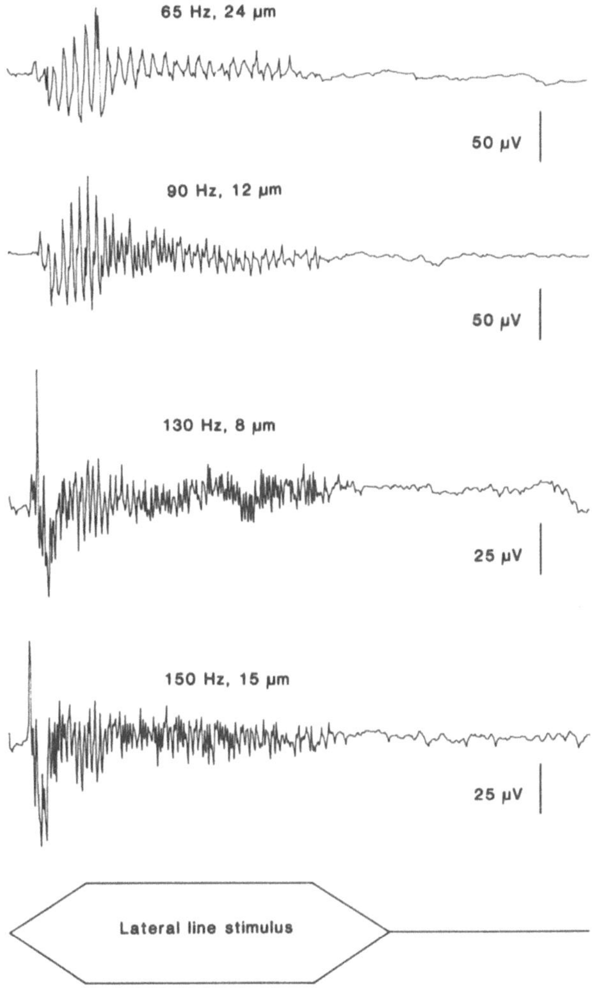

FIGURE 19.3. EPs (n = 16) recorded in the ipsilateral MON. Note that the tonic response component follows and may finally double the stimulus frequency. Stimulus duration was 500 ms, with rise and fall times of 100 ms, respectively. Stimulus amplitudes are given in the figure as p-p displacement.

fish, the response of primary afferents decreases much more sharply than the response of secondary cells. The secondary cells must therefore have, in addition to their punctate excitatory fields, an inhibitory projection from adjacent lateral line organs (Caird 1978).

Without exception, units recorded in the MON of *Platyrhinoidis* have minimal displacement thresholds in the range of 75 to 150 Hz. Within this range, a p-p displacement of 0.02 μm is sufficient to generate a neural response in the most

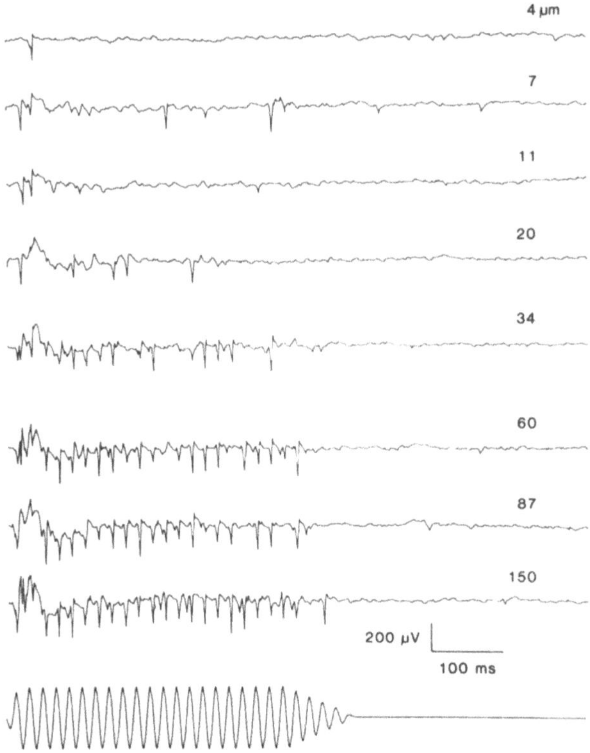

FIGURE 19.4. Example of a single unit recorded in the ipsilateral MON in response to a 50-Hz vibrating sphere stimulus. The stimulus amplitudes are given to the right of each recording as p-p displacement calculated at the skin. Note that the unit is phase-coupled to the stimulus and shows saturation if the stimulus displacement exceeds 87 µm.

sensitive preparations and recording sites (Bleckmann et al. 1989). In the thornback, the dynamic range of MON responses, determined for the frequencies 25, 50, and 100 Hz, varies at least between 30 and 75 dB. For instance, at a stimulus frequency of 100 Hz, a unit may show saturation if the stimulus amplitude exceeds 6 µm p-p displacement, or it may encode the stimulus amplitude up to at least 150 µm, the highest amplitude tested. Whether units with a small dynamic range represent primary afferents and those with a large dynamic range represent secondary cells is not known. If a given response to a 100-Hz stimulus is taken as a reference, 50-Hz and 25-Hz responses are usually weaker for the same stimulus amplitude. In general, the increase in neural response per increase in stimulus displacement decreased with stimulus frequency. It cannot be ruled out that some of the variation in sensitivity among units or preparations may be due to the general condition of the experimental animals (Fields and Ellisman 1985).

19.3.1.2 RF Organization

If the trunk lateral line of the goldfish is stimulated with a vibrating probe, primary lateral line afferents and some secondary cells show a very sensitive punctate RF, which usually corresponds to a single canal organ located ipsilateral to the recording site. A second type of subcristal units is driven by a large number of primary afferents from parts of the trunk lateral line separated by up to 20 cm (Alnaes 1973b; Caird 1978). Thus, no simple somatotopic organization is observed within the lateral line lobe. Given the small RF of primary afferents, the high convergence of lateral line input on secondary cells is surprising. However, as in other modalities, it is very likely that subsystems exist in parallel that preserve high spatial resolution along the neuraxis or sacrifice it for other functions. In *P. triseriata*, for instance, midbrain lateral line units still can have a small, well-defined RF (Bleckmann et al. 1987). In *A. anguilla*, approximately 33% of the secondary neurons investigated are influenced by contralateral lateral line stimuli. In most cases contralateral stimulation induces smaller responses than ipsilateral stimulation. No attempt was made to assess the extension of contralateral peripheral RFs (Alneas 1973b).

The nucleus praeeminentialis is a group of cells located deep in the lateral part of the brainstem. This nucleus is treated by different authors as mesencephalic or metencephalic. The nucleus praeeminentialis receives in its small ventral portion input from the mechanosensory lateral line lobe and responds to water disturbances with high sensitivity (Tong 1982). By using a small brush to touch the skin, Tong (1982) found that each of nine tested units had RFs contralaterally: six on the head, three on the trunk. In bimodal neurons, the mechanosensitive RF is congruent with the electrosensitive RF.

19.3.1.3 Multimodality

Units recorded in the area acousticolateralis of the catfish *Ictalurus nebulosus* are reported to be sensitive to both electric field and weak lateral line stimulation (Andrianov and Ilyinski 1973). Although these authors conclude that electro- and mechanoreceptors converge, all present anatomical data indicate that these two modalities terminate in different hindbrain areas, and other workers have not seen bimodal units in the medulla. This claim should be further investigated. There are some indications that Purkinje-like cells in the lateral line lobe of the goldfish integrate lateral line information with other sorts of information. However, although auditory and tactile Purkinje-like cells could be identified, no multimodal sensory Purkinje-like cells were found (Caird 1978).

In *Ictalurus*, about 10% of the units recorded in the nucleus praeeminentialis are bimodal, responding to both weak electric and mechanical stimuli, such as that caused by a water drop falling from a height of 10 cm into the bath 15 cm from the fish (Tong 1982). The restricted RFs of these units indicate that they were driven by lateral line but not by eighth nerve input.

19.3.2 Cerebellum

In gnathostomes, a part of the primary lateralis projection to the medial nucleus commonly continues into the eminentia granularis of the vestibulolateral cerebellum (Bell and Russell 1978; Claas and Münz 1981; New and Northcutt 1984). Physiological studies extend and localize the response area. Lee and Bullock (1984) applied electrical stimulation either to the medial octavolateralis nucleus or to the pLLN in *Ictalurus*. Direct electric shocks to the pLLN elicit triphasic spikelike activity with a peak latency of about 4 to 6 ms in the molecular layer overlying the eminentia granularis. The lateral eminentia granularis is dominated by a large negative wave with an onset latency of 16 ms and a peak latency of 25 ms followed by a positive wave peaking at about 75 ms. Direct electric shocks applied to the medial nucleus give rise of field potential responses in the ipsilateral and contralateral eminentia granularis (lateral part) and, in addition, in the valvula. The field potential response in each active site consists mainly of a fast positive-negative spikelike wave followed by a larger negative wave with maximal amplitude in the granular layer. These authors found a total of 52 units, 26 in the lateral eminentia granularis and 26 in the valvula, 44 of which were Purkinje cells with simple spikes.

In the catfish, the response of cerebellar lateral line units can be divided into two categories—those with higher rates of spontaneous discharge, which are depressed even at low repetition rate, and those of lower spontaneous discharge, which are excited. In the latter, a single stimulus may elicit one or several bursts of activity, depending on the intensity, and repetition as high as 8/s elicits response following. Besides the Purkinje cells, eight assumed eurydendroid cells were found in the lateral eminentia granularis. These respond either to natural stimuli (a water jet applied to the skin) or to direct brain stimulation, with minimum latencies slightly longer than the onset of the inhibitory phase of the Purkinje cells that respond to the same stimulus (Lee and Bullock 1984).

Callens et al. (1967) are the first to report cerebellar (probably Purkinje cell) responses to natural lateral line stimuli. Although they do not define the stimuli they use, they state that in the goldfish cerebellar neurons show outburst reactions at the beginning or the end of the presumed lateral line stimulus. In the lateral part of the corpus cerebelli of the goldfish, a large number of Purkinje cells respond to water currents directed to different parts of the body. By cutting the lateral line nerves, it can be shown that these responses are mediated by the lateral line system (Kotchabhakdi 1976). In intact fish a water current directed toward different parts of the body produces an increase in the frequency of ongoing activity. The initial excitatory response may be followed by excitatory-inhibitory sequences much like those observed in response to stimulation of other sensory modalities that are mediated by the mossy fiber system (Kotchabhakdi 1976).

In the thornback *Platyrhinoidis*, direct electrical shock of the pLLN results in spike responses in the auricle bilaterally (Tong and Bullock 1982). The onset latency of the response recorded in the ipsilateral auricle is 8 to 9 ms; the onset latency of the response recorded in the contralateral auricle 10 to 11 ms. EPs recorded from the surface of the ipsilateral auricle show a negative wave that peaks at 25 ms. This wave reaches a maximum at a depth of about 750 μm and reverses polarity at 1,750 μm. These authors did not succeed in getting cerebellar responses to a falling water drop or to gentle agitation of the water by a thin rod. However, Fiebig (1988), who studied the connections and some physiological aspects of the corpus cerebelli in the same species, reports that responses to a falling water drop can be recorded from the ipsi- and contralateral granule cell layer of the caudal half of the anterior lobe of the cerebellar corpus and throughout the posterior lobe of the cerebellar corpus. Optimal responses were recorded about 1,100 μm deep from all recording sites. Most of the recording sites that responded to water drops falling on the water surface also responded to touch with multiunit activity (MUA). Whether responses to water drops were mediated by touch receptors or lateral line organs were activated by the tactile stimulus is not yet clear. Bleckmann (unpublished observation) found Purkinje cells in the anterior lobe of the hornshark, *Heterodontus francisci*, that responded during a vibrating ball stimulus (100 Hz, 130 μm p-p displacement) with an inhibition of the spontaneous activity, followed by a weak off-response at stimulus end.

19.3.2.1 RF Organization

In the corpus and valvula cerebelli of the goldfish, units that respond to stimulation of the lateral line organs in the tail or body region are found in a more posterior part than those that respond to stimulation of the head region. Most responses are recorded on the ipsilateral side, although a few responding cells are scattered on the contralateral side. RFs are usually large, diffuse, and overlapped with one another, and no pattern of specific topographic projection can be observed (Kotchabhakdi 1976).

The lateral line units recorded in the eminentia granularis of *Ictalurus* primarily respond to ipsilateral stimuli (Lee and Bullock 1984). Again, the RFs of the mechanoreceptive units are large and difficult to delimit. They overlap with one another and do not appear to fall into any recognizable pattern. However, units in the valvula, which may or may not be multimodal (see below), have their RFs confined to the head region (Lee and Bullock 1984).

Mapping of tactile and/or lateral line cerebellar RFs in *Platyrhinoidis* again reveals a complex somatotopy. The RFs are large and include ipsi- and contralateral body areas with considerable overlap among adjacent recording sites. However, caudal body areas, such as the tail and anal fins, are mostly represented in a relatively small region in the caudal tip of the posterior lobe of the cerebellar corpus; more rostral body areas (pectoral fins, trunk, and head area) are represented in the rest of the posterior lobe of the cerebellar corpus and the caudal part of the posterior lobe of the cerebellar corpus (Fiebig 1988).

19.3.2.2 Multimodality

In the valvula of the catfish cerebellum, Lee and Bullock (1984) found that about 20% of the large units were multimodal. Four of these units responded to visual and lateral line mechanosensory inputs (or equivalent electrical stimulation of the pLLN and the MON, respectively), and three units responded to tactile or proprioceptive and lateral line mechanoreceptive input (or equivalent electrical stimulation). The response of these units, as in the unimodal units, consists of inhibition which may or may not be preceded by excitation. Units in the corpus were all unimodal. Many Purkinje cells recorded from the goldfish cerebellum are multimodal. However, multimodal Purkinje cells that respond to lateral line stimuli as well are not reported (Kotchabhakdi 1976).

19.3.3 Midbrain

Anatomical studies and electrophysiological experiments have shown that in teleosts (Schellart et al. 1984) and cartilaginous fishes (Boord and Northcutt 1982), ascending MON efferents mainly terminate in the contralateral torus semicircularis and tectum. In the electroreceptive teleost *Ictalurus*, the mechanoreceptive lateral line region is intercalated between the medial auditory and the lateral electroreceptive region of the torus (Knudsen 1977). The supposed lateral line torus equivalent in cartilaginous fishes is the dorsal nucleus of the mesencephalic nuclear complex (DMN; Boord and Northcutt 1982). In *Platyrhinoidis*, the anterior nucleus of the mesencephalic nuclear complex (AN), which borders the DMN rostrolaterally, also receives input from the mechanoreceptive lateral line (Schweitzer 1986; Bleckmann et al. 1987). EP and unit responses recorded in the DMN and AN are phasic or phasic tonic, respectively, often depending on stimulus intensity. A phase coupling to the stimulus is usually not observed but may occur at high stimulus intensities. In general, response latencies were significantly longer than those observed in the MON. The longest response latencies of 24 to 80 ms, depending on stimulus frequency and stimulus amplitude, were recorded in the midbrain AN.

Although very few studies exist, the tectum of fishes is probably involved in the processing of lateral line information (Callens et al. 1967). In the thornback *P. triseriata*, EPs and multiunit activity can be recorded after a pLLN shock in both the contra- and ipsilateral central tectal zone (Bleckmann et al. 1987). Ipsilateral responses are usually weaker than contralateral responses.

19.3.3.1 Threshold and Dynamic Range

Units in the mechanoreceptive region of the catfish torus often are highly sensitive to water displacements, generated by an oscillator-driven solenoid, which moves a 0.5 cm^2 plastic plate through the water (Knudsen 1977). A step displacement of 25 μm is sufficient to evoke a neural response, even if the distance between the probe and the fish is 3 cm (Knudsen 1977). Recently, Bleckmann and

Münz (1989) studied the lateral line system of *Xiphister atropurpureus* (Stichaeidae, Perciformes), a teleost with multiple trunk lateral line canals. The few midbrain recordings done in this animal show a maximal displacement sensitivity in the frequency range 75 to 150 Hz. Within this range, a p-p displacement of 0.5 μm is sufficient to generate a neural response. Lateral line responses recorded in the DMN and AN of *Platyrhinoidis* are highly sensitive, with a minimal displacement threshold of 0.005 μm (DMN) and 0.08 μm (AN) in the frequency range of 75 to 150 Hz (Bleckmann et al. 1989). Units that differed from each other remarkably in terms of the shape of the frequency response curve were not found. At a frequency of 100 Hz, AN responses often show saturation if the stimulus amplitude exceeds 10 to 75 μm p-p displacement.

The observed saturation is often followed by a high-intensity decrease in response. For a given stimulus intensity, the response magnitude may or may not decrease with decreasing stimulus frequency. Because the determination of the dynamic range of neural responses requires the application of high stimulus intensities, control experiments were done with animals whose eighth nerve was cut bilaterally. These animals gave responses similar in sensitivity to those of intact preparations, thus making it unlikely that the responses recorded in intact fish were affected by eighth nerve input.

19.3.3.2 RF Organization

Lateral line units in the torus of the catfish receive input from a restricted portion of the contralateral body surface. The rostrocaudal position of the recording site corresponds to the relative rostrocaudal position of the RF (Knudsen 1977). The topographic representation of body surface in the mechanoreceptive region of the torus is in close register with the topographic projection of RFs to the electroreceptive region; that is, mechanoreceptive and electroreceptive units at the same rostrocaudal level in the torus receive input from coincident body areas (Knudsen 1976, 1977). A similar somatotopic organization is found in the DMN of *Platyrhinoidis*. Again, more rostral RFs map on more rostral midbrain areas (Bleckmann et al. 1987).

19.3.3.3 Multimodality

In *Ictalurus*, the activity of many units throughout the torus can be influenced to a limited extent by electric field, lateral line, acoustic, and visual stimuli. However, one of the modalities clearly dominates at any given recording site (Knudsen 1977); Nederstigt and Schellart (1986) recorded single units in the trout torus semicircularis. They found units that were (1) exclusively sensitive to frequencies higher than 125 Hz (33%, $n = 67$), (2) exclusively sensitive to frequencies lower than 125 Hz (12%), and (3) broad-band units that responded to both low and high frequencies (55%). The low-frequency units are believed to get input from the lateral line. Nederstigt and Schellart (1986) speculate that the different behavior of some broad-band units may reflect convergence from acousticolateral subsystems with distinct properties. It may be added that 22%

and 35%, respectively, of the low-frequency and broad-band units show visual sensitivity too (for further information, see Schellart and Kroese, Chapter 21).

In the tectum of the goldfish, some visually activated cells show a change in their response pattern if the period of illumination or obscurity is accompanied by lateral line stimulation (Callens et al. 1967). The influence of the lateral line can be facilitatory, suppressive, or even selectively facilitatory and suppressive on the off-burst or sustained activity. Control experiments were not done, however, and it is uncertain whether the responses were partly or exclusively mediated by eighth nerve input.

19.3.4 Diencephalon

Until recently, many investigators implied that the optic tectum is the major cephalic target for lateral line information and receives its input directly from the torus semicircularis (Kappers et al. 1936; Meek 1983). Consequently, there are only a few papers addressing the question of whether lateral line information reaches and is processed in diencephalic or telencephalic brain areas. Finger and Bullock (1982) were the first to demonstrate that an electrical stimulus applied to the lateral line nerve leads to EPs and multiunit hash in a mechanoreceptive thalamic zone, which had already been described anatomically (Finger 1980). The peak latency of the main thalamic response (EP) is 30 ms. If two nerve shocks are given in quick succession, i.e., in a shock-pair paradigm, the thalamic response shows only minor depressive or facilitatory effects within the 70-ms time frame tested. However, if several nerve shocks are given in quick succession, i.e., in a shock-train paradigm, the thalamic response decays to 50% of the single shock level at a repetition rate of 0.75/s (Finger and Bullock 1982).

In the diencephalon of *Platyrhinoidis*, two distinct areas responding to either ipsi- or contralateral pLLN shock have been found. EP and unit responses that can be recorded in the caudal area of the diencephalon are restricted to the posterior central thalamic nucleus (PCT) (Bleckmann et al. 1987). The latency of the initial negative peak is 22 ms; spikes can sometimes be evoked that have a latency of 17 ms. A second diencephalic area involved in the processing of lateral line information is slightly ventral to the posterior lateral thalamic nucleus, in an area that corresponds to the hypothalamic lateral tuberal nucleus (LTN). In this area, neural responses following a contralateral pLLN shock occur with a latency of 19 ms for MUA and 28 ms for EPs (Bleckmann et al. 1987). The responses recorded in the LTN and PCT can be clearly separated by their dynamic characteristics. In the LTN a response can hardly be evoked if the iteration rate of the stimulus exceeds 2 Hz. A similar degree of response decrement is reached in the PCT at about 6 Hz (Bleckmann et al. 1987).

19.3.4.1 Threshold and Dynamic Range

The only fish in which diencephalic responses to a quasinatural lateral line stimulus (sphere of 17.6 mm diameter vibrating with a frequency of 6.5 to 200 Hz) have been recorded is the guitarfish, *Platyrhinoidis* (Bleckmann et al. 1989). Here, potentials recorded in the diencephalic LTN are often purely phasic, especially at low stimulus intensities. Minimal displacement thresholds are again in the frequency range of 75 to 150 Hz; in some preparations, a p-p displacement of 0.2 μm is sufficient to generate a phasic neural response. At higher intensities, an additional tonic and/or late response component may appear. The nature of the response depends on the stimulus repetition rate. When the stimulus interval was 55 s, both components appeared as soon as the p-p displacement surpassed 25 μm (25 Hz), 12 μm (50 Hz), and 10 μm (100 Hz). At an interval of 7 s, the tonic component appeared at 88 μm (50 Hz) and 50 μm (100 Hz), whereas the 25-Hz stimulus caused no response, even at 150 μm (Bleckmann et al. 1989).

19.3.4.2 RF Organization

Almost no information is available on the diencephalic RFs. Bleckmann et al. (1989) recorded few unit responses in the LTN of the thornback. In one case, the RF was mapped; it was restricted to the anterior half of the body, was large and complex, and seemed to follow the course of the infraorbital and body lateral line canals (Fig. 19.5).

19.3.4.3 Multimodality

In some penetrations, the mechanoreceptive thalamic center of the catfish gave multiple unit local responses to both a lateral line nerve shock and an acoustic stimulus (Finger and Bullock 1982). Two observations argue, however, against a common interaction between the thalamic acoustic and lateral line pathway in the catfish. First, in many thalamic loci, responses can be evoked with an acoustic stimulus but not with a lateral line nerve shock, and vice versa. Second, in a stimulus pair test, the acoustic response revealed long-lasting depression unlike that of the lateral line response, and the physiology of the acoustic pathway was clearly distinct from that of the lateral line pathway. No responses to electric field and light stimuli could be recorded in this center (Finger and Bullock 1982). As in the catfish, many hot-spot recording sites in the diencephalon of the thornback did not show any obvious interactions between visual (light flash), weak acoustic, electroreceptive, and mechanoreceptive lateral line responses. One recording site, histologically confirmed to be in the PCT, gave clear multiunit responses to waves caused by an approaching plastic strip only if the eye facing the strip was uncovered. Controls showed that responses were not caused by visual stimuli alone (Bleckmann et al. 1987).

In *Platyrhinoidis*, some recording sites in the LTN responded to both a vibrating ball and an electric field stimulus (Bleckmann et al. 1987). However, in most cases an interaction in terms of facilitation or inhibition could not be observed. In a few preparations and recording sites, an electric field stimulus clearly inhibited the response to a lateral line stimulus, and vice versa (Bleckmann et al. 1987).

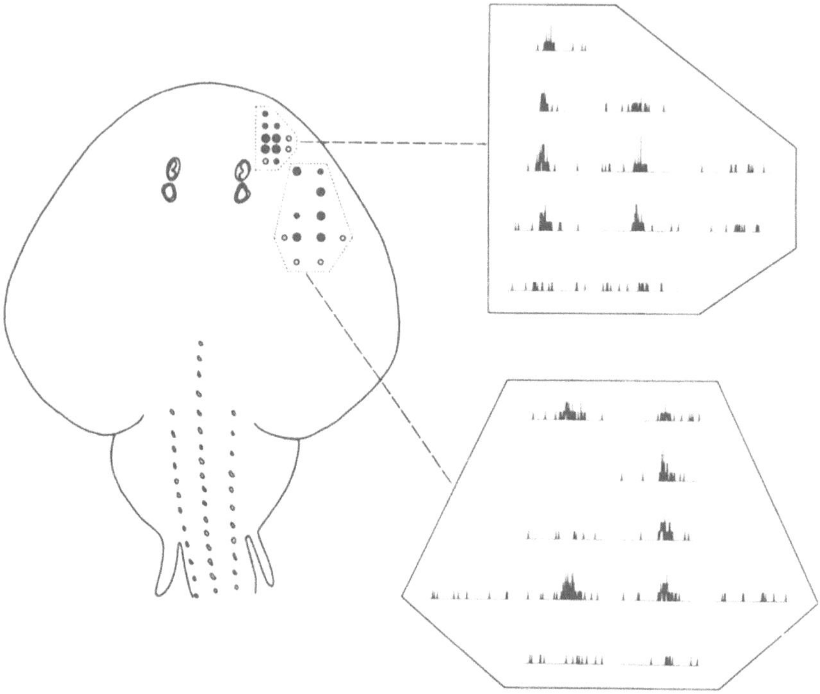

FIGURE 19.5. Top view of *P. triseriata* showing the positions of the vibrating sphere (dots). Depending on the site of stimulation, recordings, made in succession from the same recording site within the contralateral LTN, differ with respect to the number of spikes evoked (expressed as peristimulus time histogram shown to the right). For the rostral stimulation site (borders of stimulation sites are drawn by dotted lines), RFs were mapped using a stimulus frequency of 100 Hz and a stimulus of 2 μm p-p displacement. For the more caudal stimulation site, RFs were mapped again using a stimulus frequency of 100 Hz but a stimulus displacement of 7.5 μm p-p. Weaker stimuli failed to give any responses at this body area. Dot size represents 0 to 15 spikes (○), 16 to 30 spikes (•), and 30 to 65 spikes (●) within the first 150 ms of stimulation. Responses $<$ 15 spikes are believed to be not distinguishable from spontaneous activity. Good stimulation sites follow approximately the route of the lateral line canal system (not shown in the figure). All peristimulus time histograms are based on the average of eight responses.

19.3.5 Telencephalon

There are some indications that the mechanoreceptive lateral line pathway of teleosts reaches the telencephalon (Echteler 1985). From recent anatomical studies we know that the majority of the output from the torus semicircularis does not project to the tectum but instead ends on a specific diencephalic nucleus (Braford and McCormick 1979; Finger 1980), described in the catfish as a mechanoreceptive thalamic zone (Finger and Bullock 1982). In *Ictalurus*, this zone, which receives acoustic and lateral line information, is known to project to the telencephalon (Finger 1980).

Echteler (1985) studied the organization of the central auditory pathways in the carp, *Cyprinus carpio*, and found units responding to lateral line stimuli (a water jet applied from a glass pipette against lateral line neuromasts) within the posterior dorsomedial telencephalon, an area that responds to acoustic stimuli as well. Based on multiunit recordings, it remains unclear whether the sensitivity to both stimuli results from bimodal neurons. The best responses to lateral line stimuli occur, however, within 200 μm of the surface of the telencephalon, whereas maximum multiunit responses to acoustic stimulation are found at depths between 100 and 400 μm (Echteler 1985).

Until recently, the elasmobranch telencephalon was believed to receive only primary and secondary olfactory fibers (Backstrom 1924; Kappers et al. 1936; Aronson 1963). New experimental studies, however, have shown that it receives visual (Duff and Ebbesson 1973; Cohen et al. 1973; Veselkin and Kovačević 1973; Luiten 1981a,b), auditory (Bullock and Corwin 1979; Corwin and Northcutt 1982), and electroreceptive (Platt et al. 1974; Bullock 1979; Bodznick and Northcutt 1984) information. Thus, it is not surprising to find that the mechanosensory lateral line of selachians also maintains a distinct lemniscal pathway. In *Platyrhinoidis*, the telencephalic area that responds to an ipsi- or contralateral pLLN shock is slightly dorsal to the pallio/subpallial boundary and may correspond to the medial pallium (Bleckmann et al. 1987). Pallial lateral line responses, evoked by an ipsilateral pLLN shock, have a peak latency of 87 ms (EPs) and 49 to 61 ms (MUA).

In this fish, telencephalic field potentials can also be evoked with a vibrating sphere placed 5 mm above the infraorbital lateral line canal. The horizontal extension of the pallial area responsive to a vibrating sphere (100 Hz, 30 μm p-p displacement) is about 0.8 mm in the mediolateral and 0.8 to 1 mm in the anterioposterior direction. Good responses are found at a depth of about 2.8 to 3.6 mm. Despite intensive efforts, single unit or MUA to a vibrating sphere has not yet been obtained. Telencephalic lateral line responses can also be recorded in the hornshark, *Heterodontus* (Bleckmann et al. 1989).

The latency of telencephalic responses in the thornback increases with stimulus rise time but hardly changes with stimulus duration. Frequency-modulated wave stimuli (65 to 135 Hz within 500 ms) do not lead to better responses. Besides the long latencies, the most striking feature of telencephalic responses is a remarkable response decrement evident at an iteration rate of 2/min (Bleckmann et al. 1987, 1989).

19.3.5.1 Threshold and Dynamic Range

In the thornback, telencephalic responses are not as sensitive or as sharply "tuned" to a vibrating sphere stimulus as medullary or midbrain responses. At 100 Hz, EPs can be generated with a p-p water displacement of 1.9 μm. At a frequency of 12.5 Hz, p-p threshold displacement rises to 26 μm in the most

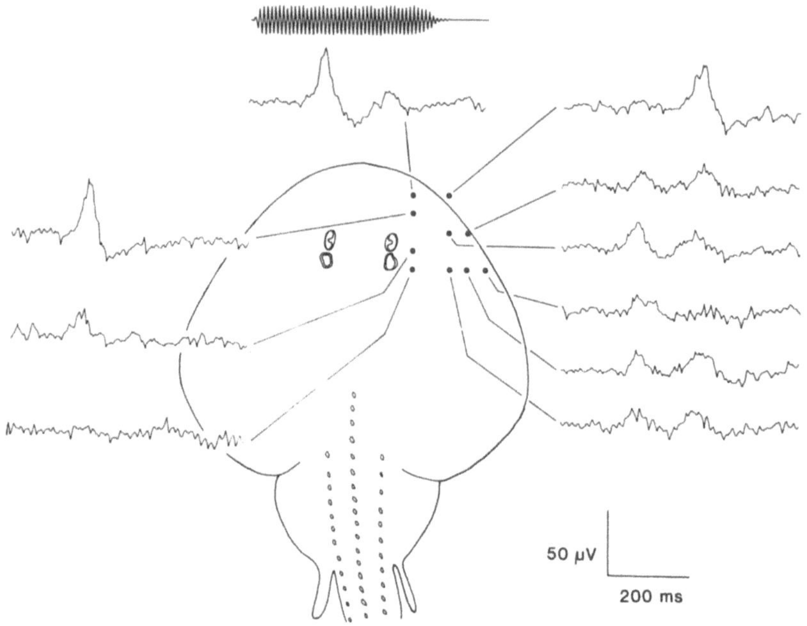

FIGURE 19.6. EPs recorded in the contralateral telencephalon in response to a vibrating ball stimulus (100 Hz, 30 μm p-p displacement). Recording depth in all cases was 4.6 mm. Note that the EPs consist of two long-latency, low-frequency waves whose size depends on the ball location (black dots on fish surface). In each case four responses were averaged.

sensitive preparations (Bleckmann et al. 1989). In contrast to midbrain and hindbrain lateral line areas, EPs may saturate shortly above threshold; dynamic range may be as small as 31.4 dB (50-Hz stimulus) and 11.8 dB (100-Hz stimulus). First and second waves of the response can differ with respect to dynamic range and/or threshold (Bleckmann et al. 1989).

19.3.5.2 RF ORGANIZATION

Telencephalic EPs following a vibrating ball stimulus are large and may differ with respect to the polarity of the response and the number and latency of the peaks, different preparations and recording sites, different repetition rates, and different positions of the vibrating ball (Bleckmann et al. 1989). One example is shown in Fig. 19.6.

19.4 Conclusion

As has been shown, the mechanoreceptive lateral line of fishes is present at all levels of the neuraxis, from medulla to telencephalon (Echteler 1985; Bleckmann et al. 1987; Northcutt and Plassmann unpublished). In addition, elec-

physiological results indicate that there are subsystems that may ascend in parallel (Bleckmann et al. 1987). On of the next important steps in lateral line research will be to switch from pLLN shocks to more natural stimuli in order to uncover which information transmitted to the brain by primary afferents is of special interest to the medulla, the cerebellum, the midbrain, and the forebrain.

The only fish whose lateral line centers from medulla to telencephalon have been investigated is the thornback guitarfish, *P. triseriata*. Surprisingly, in this animal we find little difference along the neuraxis with respect to absolute sensitivity, frequency and intensity response curves, and RF organization (Bleckmann et al. 1989). In part this may be due to the small sample size, but more likely the vibrating sphere, used in this study as a stimulus source, does not produce water motions well suited to drive specialized higher-order lateral line neurons. For instance, forebrain units that respond selectively to certain water movements were not found. Telencephalic units that can be driven selectively by natural stimuli are well known, for example, from the acoustic system of higher vertebrates (Bonke et al. 1979). For further electrophysiological investigations of lateral line centers, it may be wise to change to more complex, natural stimuli or at least to stimuli that are believed to be more natural. That means, one of the challenges to our ingenuity is the design of stimuli that are interesting to higher brain centers, including movement and cross-modal stimuli. One expectation—drawn from our knowledge in other sensory systems—is that the properties and preferences of responses in successive lateral line centers might define submodalities and parameter discrimination for which the system is specialized and hence suggest the ethologically significant stimuli for which the lateral line is important.

The use of complex stimuli leads to another problem that faces the experimenter who wants to study the sensory neurophysiology of the lateral line: the possibility that the neurons recorded from are multimodal or, even worse, are not driven by lateral line input at all. For instance, the inner ear of many fishes is a very sensitive sound detector which partly operates in the same frequency range as the lateral line (e.g. Hawkins 1981). Hence a stimulus believed to be a "lateral line stimulus"—like a falling water drop or a vibrating sphere—is often well suited to stimulate both lateral line and inner-ear receptors. In addition, spinal and trigeminal mechanoreceptors in the skin need to be assessed for their relative contribution (e.g. Nier 1976), perhaps with stronger stimuli. Ways to distinguish between lateral line and inner-ear responses include the following: to determine the RFs of units encountered, to use stimuli whose strength and/or frequency content is not suited to stimulate more than one sensory system, to cut the eighth nerve bilaterally to make sure that neural input cannot be contaminated by the inner ear, or to block lateral line function by use of $CoCl_2$ (see Enger et al. chapter 29). In many of the studies reported in this chapter, these precautions were not taken. Hence, some of the responses claimed in the literature as lateral line responses may have been caused by other sensory systems. Another precaution should be mentioned: in many electrophysiological experiments, Flaxedil is used to immobilize the animal. Although intramuscular injections of Flaxedil in concentrations just sufficient to cause immobilization also block efferent inhibition of lateral line activity (Russell and Roberts 1972), we should keep in mind that our experimental work is usually done without recognition of the presence of an efferent system and the possible complications efferent action might impose on our results.

With the use of quasinatural stimulus combinations and the precautions described above, we should be able to collect valuable information regarding the interaction between different end organs of the octavolateralis and visual system. Our present knowledge probably provides a good starting point for an "in-depth" neurophysiological exploration with dye-filled microelectrodes so that the electrical properties may be precisely correlated with morphologically specified neurons. This step should lead to a better understanding of the neuronal circuits that bridge between lateral line input and observed motor activity.

Besides asking more and more refined questions about the function of the lateral line of a "standard" bottom or midwater teleost, we should also investigate taxa highly specialized with respect to this sensory modality—e.g., fish with multiple lateral lines like *Xiphister atropurpureus*; surface-feeding fish like *Aplocheilus lineatus*; and fish that have a close peripheral contact between the lateralis and otic system, like some clupeids (see Denton and Gray Chapter 11) and mormyrids etc. We should look not only for the function of the lateral line system but also for major evolutionary changes, for example, by comparing one or more chondrichthian taxa. This means that instead of confining our attention to a few common species, we should also look at others to try to uncover (1) the ones with extremely specialized brain centers, comparable to species with the best vision or electroreception, and (2) commonalities or differences between more recent and more ancient taxa.

Acknowledgments. We thank Drs. Sheryl Coombs, Bernd Fritzsch, and especially Niko Schellart, for their thorough and helpful comments on an earlier draft of this chapter. The original research was supported by grants to H.B. from DFG (B1 242) and to T.H.B. from NSF and NIH.

References

Alnaes E (1973a) Lateral line input to the crista cerebellaris in the eel. Acta Physiol Scand 88:49–61.

Alnaes E (1973b) Unit activity of ganglionic and medullary second order neurones in the eel lateral line system. Acta Physiol Scand 88:160–174.

Andrianov GN, Ilyinsky OB (1973) Some functional properties of central neurons connected with the lateral-line organs of the catfish (*Ictalurus nebulosus*. J Comp Physiol A 86:365–376.

Aronson LR (1963) The central nervous system of sharks and bony fishes with special reference to sensory and integrative mechanisms. In: Gilbert PW (ed) Sharks and Survival. Boston: Heath, pp. 165–241.

Backstrom K (1924) Contribution to the forebrain morphology in selachians. Acta Zool 5:123–240.

Bell CC (1981) Central distribution of octavolateral afferents and efferents in a teleost (Mormyridae). J Comp Neurol 195:391–414.

Bell C, Russell CJ (1978) Termination of electroreceptor and mechanical lateral line afferents in the mormyrid acousticolateralis area. J Comp Neurol 182:367–382.

Bleckmann H (1986) Role of the lateral line in fish behaviour. In: Pitcher TJ (ed) The Behaviour of Teleost Fishes. London: Croom Helm, pp. 177–202.

Bleckmann H (1988) Prey identification and prey localization in surface-feeding fish and fishing spiders. In: Atema J, Fay RR, Popper AN, Tavolga W (eds) Sensory Biology of Aquatic Animals. New York: Springer, pp. 619–641.

Bleckmann H, Münz H (1988) The physiology of lateral line mechanoreceptors in *Xiphister atropurpureus*, a teleost with multiple lateral lines (submitted).

Bleckmann H, Bullock TH, Jørgensen JM (1987) The lateral line mechanoreceptive mesencephalic, diencephalic, and telencephalic regions in the thornback ray, *Platyrhinoidis triseriata* (Elasmobranchii). J Comp Physiol A 161:67–84.

Bleckmann H, Weiss O, Bullock TH (1989) The physiology of medullary, mesencephalic, diencephalic, and telencephalic lateral line areas in *Platyrhinoides triseriata*. J Comp Physiol A (in press).

Bodznick DA, Northcutt RG (1980) Segregation of electro-and mechanoreceptive inputs to the elasmobranch medulla. Brain Res 195:313–322.

Bodznick DA, Northcutt RG (1984) An electrosensory area in the telencephalon of the little skate, *Raja erinacea*. Brain Res 298:117–124.

Bonke D, Scheich H, Langner G (1979) Responsiveness of units in the auditory neostriatum of the guinea fowl (*Numida meleagris*) to species-specific calls and synthetic stimuli. I. Tonotopy and functional zones of field L. J Comp Physiol A 132:243–255.

Boord RL, Northcutt RG (1982) Ascending lateral line pathways to the midbrain of the clearnose skate *Raja eglanteria*. J Comp Neurol 207:274–282.

Braford M, McCormick CA (1979) Some connections of the torus semicircularis in the bowfin, *Amia calva*: A horseradish peroxidase study. Soc Neurosci Abstr 5:139.

Bullock TH (1979) Processing of ampullary input in the brain: Comparison of sensitivity and evoked responses among elasmobranchs and siluriform fishes. J Physiol (Paris) 75:297–407.

Bullock TH (1986) Interspecific comparison of brainstem auditory evoked potentials and frequency following responses among vertebrate classes. In: Cracco RQ, Bodis-Wollner I (eds) Evoked Potentials. New York: Alan R. Liss, pp. 155–164.

Bullock TH, Corwin JT (1979) Acoustic evoked activity in the brain of sharks. J Comp Physiol A 129:223–234.

Caird DM (1978) A simple cerebellar system: The lateral line lobe of the goldfish. J Comp Physiol A 127:61–74.

Callens M, Vandenbusche E, Greenway PH (1967) Convergence of retinal and lateral line stimulation on tectum opticum and cerebellar neurones. Arch Int Physiol Biochem 75:148–150.

Claas B, Münz H (1981) Projection of lateral line afferents in a teleost's brain. Neurosci Lett 23:287–290.

Cohen DH, Duff TA, Ebbesson SOE (1973) Electrophysiological identification of a visual area in shark telencephalon. Science 182:492–494.

Corwin JT, Northcutt RG (1982) Auditory centers in the elasmobranch brain: Deoxyglucose localization and evoked potential recording. Brain Res 236:261–273.

Dijkgraaf S (1963) The functioning and significance of the lateral line organs. Biol Rev 38:51–105.

Duff TA, Ebbesson SOE (1973) Electrophysiological identification of a visual area in shark telencephalon. Science 182:492–494.
Echteler SM (1985) Organization of central auditory pathways in a teleost fish, *Cyprinus carpio*. J Comp Physiol A 156:267–280.
Fiebig E (1988) Connections of the corpus cerebelli in the thornback guitarfish, *Platyrhinoidis triseriata* (Elasmobranchii). A study with WGA-HRP and extracellular granule cell recording. J Comp Neurol 268:567–583.
Fields RD, Ellisman MH (1985) Synaptic morphology and differences in sensitivity. Science 228:197–199.
Finger TE (1980) Nonolfactory sensory pathway to the telencephalon in a teleost fish. Science 210:671–673.
Finger TE, Bullock TH (1982) Thalamic center for the lateral line system in the catfish *Ictalurus nebulosus*: Evoked potential evidence. J Neurobiol 13:39–47.
Finger TE, Tong SL (1984) Central organization of eighth nerve and mechanosensory lateral line system in the brainstem of ictalurid catfish. J Comp Neurol 229: 129–151.
Hawkins AD (1981) The hearing abilities of fish. In: Tavolga WN, Popper AN, Fay RR (eds) Hearing and Sound Communication in Fish. New York: Springer, pp. 109–133.
Herrick CJ (1897) The cranial nerve components of teleosts. Anat Anz 13:425–431.
Hoagland H (1935) Electrical responses from the lateral line nerves of fishes. V. Responses in the central nervous system. J Gen Physiol 18:89–91.
Kalmijn Ad. J (1988) Hydrodynamic and acoustic field detection. In: Atema J, Fay RR, Popper AN, Tavolga W (eds) Sensory Biology of Aquatic Animals. New York: Springer, pp. 83–130.
Kappers CU Ariens, Huber G, Crosby EC (1936) The Comparative Anatomy of the Nervous System of Vertebrates, Including Man. Reprinted 1960. New York: Hafner, pp. 18–45.
Knudsen EI (1976) Midbrain responses to electroreceptive input in catfish: Evidence of orientation preferences and somatotopic organization. J Comp Physiol A 106:51–67.
Knudsen EI (1977) Distinct auditory and lateral line nuclei in the midbrain of catfishes. J Comp Neurol 173:417–432.
Kotchabhakdi N (1976) Functional organization of the goldfish cerebellum. Information processing of input from peripheral sense organs. J Comp Physiol A 112:75–93.
Lee LT, Bullock TH (1984) Sensory representation in the cerebellum of the catfish. Neuroscience 13:157–169.
Luiten PGM (1981a) Two visual pathways in the telencephalon of the nurse shark (*Ginglymostoma cirratum*). I. Retinal projections. J Comp Neurol 196:531–538.
Luiten PGM (1981b) Two visual pathways in the telencephalon of the nurse shark (*Ginglymostoma cirratum*). II. Ascending thalamotelencephalic connections. J Comp Neurol 196:539–548.
Mayser P (1882) Vergleichend anatomische Studien über das Gehirn der Knochenfische mit besonderer Berücksichtigung der Cyprinoiden. Z Wiss Zool 36:259–364.
McCormick CA (1981) Comparative neuroanatomy of the octavolateralis area of fishes. In: Tavolga WN, Popper AN, Fay RR (eds) Hearing and Sound Communication in Fishes. New York: Springer, pp. 375–381.
McCormick CA, Braford MR (1988) Central connections of the octavolateralis system: Evolutionary considerations. In: Atema J, Fay RR, Popper AN, Tavolga W (eds) Sensory Biology of Aquatic Animals. New York: Springer, pp. 733–756.
Meek J (1983) Functional anatomy of the tectum mesencephali of the goldfish. An

explorative analysis of the functional implications of the laminar structural organization of the tectum. Brain Res Rev 6:247–297.
Montgomery (1984) Noise cancellation in the electrosensory system of the thornback ray: Common mode rejection of input produced by the animals own ventilatory movement. J Comp Physiol A 155:103–111.
Nederstigt LJA, Schellart NAM (1986) Acousticolateral processing in the torus semicircularis of the trout *Salmo gairdneri*. Pflügers Arch 406:151–157.
New JG, Northcutt RG (1984) Central projections of the lateral line nerves in the shovelnose sturgeon. J Comp Neurol 225:129–140.
New JG, Bodznick D (1985) Segregation of electroreceptive and mechanoreceptive lateral line afferents in the hindbrain of chondrostean fishes. Brain Res 336:89–98.
Nier K (1976) Cutaneous sensitivity to touch and low-frequency vibration in selachians. J Comp Physiol A 109:345–355.
Northcutt RG (1981) Audition and the central nervous system of fishes. In: Fay RR, Popper A, Tavolga W (eds) Hearing and Sound Communications in Fishes. New York: Springer, pp. 331–355.
Paul DH, Roberts BL (1977a) Studies on a primitive cerebellar cortex. II. The projection of the posterior lateral line nerve to the lateral line lobes of the dogfish brain. Proc R Soc Lond B 195:467–478.
Paul DH, Roberts BL (1977b) Studies on a primitive cerebellar cortex. III. The projection of the anterior lateral line nerve to the lateral line lobes of the dogfish brain. Proc R Soc Lond B 195:479–496.
Platt CJ, Bullock TH, Czéh G, Kovačević N, Konjević Dj, Gojković M (1974) Comparison of electroreceptor, mechanoreceptor, and optic evoked potentials in the brain of some rays and sharks. J Comp Physiol A 95:323–355.
Ronan M, Northcutt RG (1987) Primary projections of the lateral line nerves in adult lampreys. Brain Behav Evol 30:62–81.
Russell IJ (1974) Central and peripheral inhibition of lateral line input during the startle response in goldfish. Brain Res 80:517–522.
Russell IJ (1976) Central inhibition of lateral line input in the medulla of the goldfish by neurones which control active body movements. J Comp Physiol A 111:335–358.
Russell IJ, Roberts BL (1972) Inhibition of spontaneous lateral line activity by efferent nerve stimulation. J Exp Biol 57:77–82.
Schellart NAM, Zweijpfenning RCVJ, Nederstigt LJA (1984) Convergence of auditory and lateral line processing in the torus semicircularis of the trout, *Salmo gairdneri*. Neurosci Lett 42:39–44.
Schwartz E (1971) Die Ortung von Wasserwellen durch Oberflächenfische. Z Vergl Physiol A 74:64–80.
Schweitzer J (1986) Functional organization of the electroreceptive midbrain in an elasmobranch (*Platyrhinoidis triseriata*): A single unit study. J Comp Physiol A 158:43–58.
Tong SL (1982) The nucleus praeeminentialis: An electro-and mechanoreceptive center in the brainstem of the catfish. J Comp Physiol A 145:299–303.
Tong SL, Bullock TH (1982) The sensory functions of the cerebellum of the thornback ray, *Platyrhinoidis triseriata*. J Comp Physiol A 148:399–410.
Veselkin VP, Kovačević N (1973) Non-olfactory telencephalic afferent projections in elasmobranch fishes. Zh Evol Biokhim Fiziol 9:585–592 (in Russian).

20
Direction Coding in Central Parts of the Lateral Line System

BARBARA CLAAS, HEINRICH MÜNZ and KLAUS E. ZITTLAU

20.1 Introduction

Some predatory fishes and aquatic amphibians are able to use surface waves to find prey, such as struggling insects trapped on the water surface (for review, see Bleckmann 1986, 1987; Bleckmann et al. Chapter 25). The sensory system mainly responsible for the detection of surface waves is the mechanosensory lateral line system. The sensory organs of the lateral line system, the neuromasts, are stimulated by the water motion induced by the propagating surface waves. Sensory processing in the lateral line system provides the animals with information about the direction and, at least in some surface feeding fish, the distance to the source of a surface wave (Bleckmann and Schwartz 1982).

In some species, quite accurate and reproducible orienting reactions toward the source of a surface wave can be elicited even under laboratory conditions (Schwartz 1965; Görner 1976, Elepfandt 1982; Görner et al. 1984). This behavioral response to lateral line stimuli – in particular the ability to detect the direction of a wave source – has been extensively studied in the African clawed toad *Xenopus laevis* (Görner 1976; Elepfandt 1982; Görner et al. 1984). Therefore, *Xenopus* offers a suitable model for combined behavioral and physiological approaches to the study of neuronal mechanisms underlying behavior guided by the lateral line.

In *Xenopus*, the coding of stimulus directions has been studied with electrophysiological methods at different levels of the lateral line system (Zittlau et al. 1985, 1986). Directional tuning has been determined for primary fibers, ascending medullary units, and toral units as well as for units in the tectum opticum. Only in the tectum opticum were units driven by the lateral line sharply tuned to characteristic directions on the water surface (Zittlau et al. 1985, 1986).

The present study represents a first attempt to analyze how the directional information is computed from the stimulus parameters encoded in the periphery. For this purpose, the response of lateral line units in the tectum to different surface wave stimuli is described. Second, the question of whether the tectum opticum belongs to the sensorimotor interface for behavior guided by the lateral line is addressed. For this purpose, the orienting reaction of *Xenopus* was quantitatively analyzed before and after lesions of various sizes were made in the tectum.

20.2 Material and Methods

The electrophysiological recordings were carried out on 35 African clawed toads, *Xenopus laevis* (Daudin), from the Bielefeld laboratory colony. Animals of both sexes with body lengths of 3.5 to 4.0 cm were used. The animals were first anesthetized by immersion in a 0.1% aqueous solution of tricaine methane sulfonate (Sigma) and immobilized by an intralymphatic injection of Alloferin (Roche) (2 μg/g bodyweight). Next, the skin and bone covering the midbrain and then the dura mater were removed. At the recording site, a small cut was made in the pia mater. To exclude visual input, 14 animals were blinded by cutting the optic nerves prior to the recordings.

20.2.1 Electrophysiological Recordings

Recordings were made with glass micropipettes filled either with a 2.5 M NaCl solution or with metal (Lettvin et al. 1959). Electrode resistances were 6 to 12 megohm for electrodes filled with electrolyte and 0.4 megohm for metal-filled electrodes. The neuronal activity was amplified, monitored, and stored for final off-line processing using conventional equipment. The activity recorded from single or multiple units was passed through a window discriminator to obtain standard pulses (0.05 ms, 5 V) for further processing by computer (Data General DG20). The responses of single or multiple units were quantified by counting the number of spikes per stimulus.

For stimulation with surface waves, the animal was fixed on a thin ridge of clay and placed in the center of a round perspex tank (diameter 34 cm) filled with tap water at ~ 20°C. Surface waves were produced by blowing a short air pulse (100 ms) onto the water surface. The air pulse was produced by a modified loudspeaker connected to a glass tube with a tip diameter of 1 mm. The loudspeaker was driven by a function generator. The glass tube, ending 1 cm above the water surface, was moved around the animal in circles of different diameters in steps of 30°. The central point of these circles was between the forelimbs of the animal at the midline of the trunk. Surface waves induced by clicks contain several frequencies. Because of physical properties of water surface, surface waves are strongly modified as they propagate; the wave amplitude decreases while the duration of the wave increases (Fig. 20.1). Since the attenuation of the amplitude increases with the frequency of the surface wave, the spectral composition of the wave also changes with the distance traveled.

For a second series of experiments, sinusoidal surface waves were produced by a small sphere (8 mm) vibrating on the water surface. The sphere was driven by a vibrator at frequencies from 5 to 100 Hz. The sphere was moved around the animal in circles that had the same central point as described above. The surface waves were simultaneously recorded near the head using the method originally described by Rudolph (1967), in which the resistance between a small silver wire immersed in the water (receiver electrode) and a ground electrode is measured.

Any change of the water level produces a change in resistance which is proportional to the amplitude of the water wave.

20.2.2 Behavioral Experiments

For 24 *Xenopus*, the reactions to surface wave stimuli were determined before and after a small area in the tectum opticum was electrolytically lesioned. Prior to the operation, the animals were anesthetized as described above. The skull was opened and a current (~ 5 μA/1–5 s) was passed through a tungsten steel electrode into the tissue. The animals' reactions to surface wave stimuli were retested 4 to 6 h after recovery from anesthesia. All animals had been blinded early in life and were kept in tanks with a water depth of 10 cm. Since the animals were swimming around in the experimental tank, it proved to be more convenient to produce the surface waves by dipping a glass rod (2 mm in diameter) into the water by hand. The waves did not differ from those obtained by blowing air on the water surface. The reactions were recorded on videotape. Stimulus and reaction angles were determined using a digitizer (Apple IIe) for single-frame analysis.

After the behavioral experiments were completed, the animals were killed with an overdose of tricaine methane sulfonate. The brains were dissected, and the size and the location of the lesion were determined on the basis of 40-μm frozen sections counterstained with cresyl violet. The lesions were nearly spherical with diameters between 100 and 400 μm. They reached from the superficial layers to the periventricular gray.

20.3 Results

In the intact animal, most units in the superficial layers of the tectum opticum reacted to photic stimuli, such as illumination of the whole eye. When a blinking LED was moved around the animal by hand, most of the units had distinct receptive fields. Scattered among the visual units in more ventral parts of the tectum were other units, which reacted exclusively to water surface waves. These units were also found in animals in which the optic nerves were cut prior to the recordings. Thus, they were considered to be driven by lateral line input, and a total of 186 of these were recorded. In the central and rostral parts of the tectum opticum, they were found exclusively at depths of 190 to 210 μm below the tectal surface. In more lateral areas this band lay between 300 and 700 μm below the surface. The recording sites correspond to lamina 6 (Levine 1980).

Eighty percent of units encountered had no resting activity; the rest had mean spike rates of less than 0.2 imp/s. Surface waves induced by air pulses were used as search stimuli to aid in recognizing lateral line units. Recordings from individual units lasted up to 2.5 h. During this time the responses to stimuli sometimes varied considerably. For single units, the number of spikes ranged

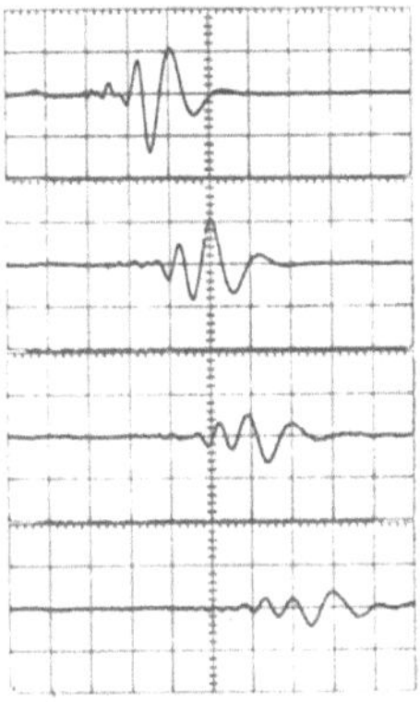

FIGURE 20.1. Changes in the pattern of a surface wave induced by an air pulse as a function of the distance traveled. The distances to the wave centers are given at the right margin of the figure. Horizontal divisions (time): 100 μs; vertical divisions (wave amplitude): 60 μm.

from one to more than 10 spikes per stimulus produced by an air pulse at a distance of 8 cm (stimulus duration 450 μs; see Fig. 20.1). In multiple-unit recordings, up to 86 spikes per stimulus could be elicited, depending on the number of units present in the recording. Spikes were elicited from only a few stimulus positions; units were sharply tuned to the most effective stimulus direction (Fig. 20.2c). Latencies varied between 80 and 500 ms, depending on the stimulus distance and the receptive field of the unit. For frontal or lateral receptive fields, the shortest latencies (80 ms for 6 cm distance) were obtained for the most effective stimulus direction. For units with more caudal receptive fields, the most effective stimulus direction did not coincide with the stimulus direction producing the shortest latencies. The latencies were shortest for stimulus directions between 180° and 135°.

As the recording site was varied from caudomedial to rostrolateral positions in the tectum, the receptive fields shifted from caudal to rostral on the contralateral water surface. Ipsilateral receptive fields were found to be represented in the same order in frontal parts of the tectum opticum. In the central tectum no lateral line units were found. Figure 20.2 schematically depicts the position of lateral line units in the tectum and their most sensitive directions. When optic and lateral line units were recorded in the same electrode track, the receptive field of the optic unit was pointing in the same direction as the receptive field of the lateral line unit. Thus, it appears that the visual map and the lateral line map of directions on the water surface are in register.

Receptive fields for surface waves produced by air pulses were compared in three units to fields induced by sinusoidal stimuli of 20 Hz (amplitude 40 μm; 200 ms duration). Although the number of spikes per stimulus was greater for sinusoidal stimuli, the form and the direction of the receptive fields were not significantly different. When different frequencies (5 Hz; 10 to 50 Hz in steps of 10 Hz) of equal amplitude and duration were presented, the response of the unit was the same.

Sensitivity to different stimulus amplitudes was also tested in three units by reducing the amplitude of 40 Hz sinusoidal waves. The stimulus was positioned

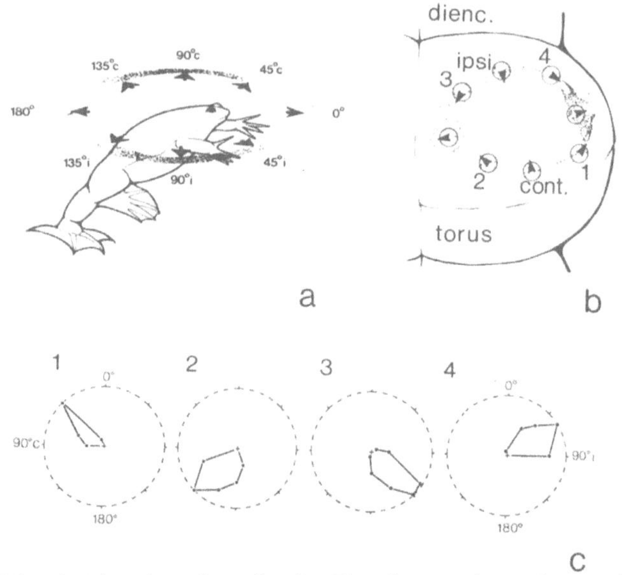

FIGURE 20.2. Directional tuning of tectal units. The diagram shows the way in which directions on the water surface are represented in the tectum opticum. The stippled arrows indicate how the 360° around the frog (a) are mapped onto a circle in the tectum (b). The surface of the tectum opticum and the torus semicircularis (torus) is drawn as seen from above. For several specific stimulus directions (solid arrowheads in (a), the corresponding arrows in (b) indicate the locations of units tuned to these directions. In (c) the receptive fields of four units recorded in two animals are shown. Each point is the mean of the responses to 10 consecutive stimuli. Responses are given in relative units. The response in the most sensitive direction corresponds to 100%.

in the most effective direction of the unit. The units were relatively insensitive to differences in the amplitude; reducing the amplitude of the surface wave from 80 μm to 4 μm reduced the response by about 25%. In a second series of experiments ($n = 12$ units), the distance to the center of a pulse-induced surface wave was increased from 6 to 16 cm in steps of 2 cm. This decreases the amplitude but increases the duration of the stimulus owing to the dispersion of the surface waves. The reactions of lateral line units were heterogeneous. One unit responded to stimuli only at a distance of 6 cm, and three units responded best for intermediate distances (8 and 12 cm). For most units, the number of impulses per stimulus increased with increasing distance.

In *Xenopus* the probability of a behavioral response to surface waves is not always constant. Therefore, it was necessary to test the proclivity of each animal to respond prior to each behavioral experiment. This was done by gently touching the animal with a small piece of meat held with tweezers. When the animal is motivated (i.e., likely to respond), it immediately turns to the stimulated side and catches the meat. Following a positive response to this test, the reactions to surface wave stimuli were recorded.

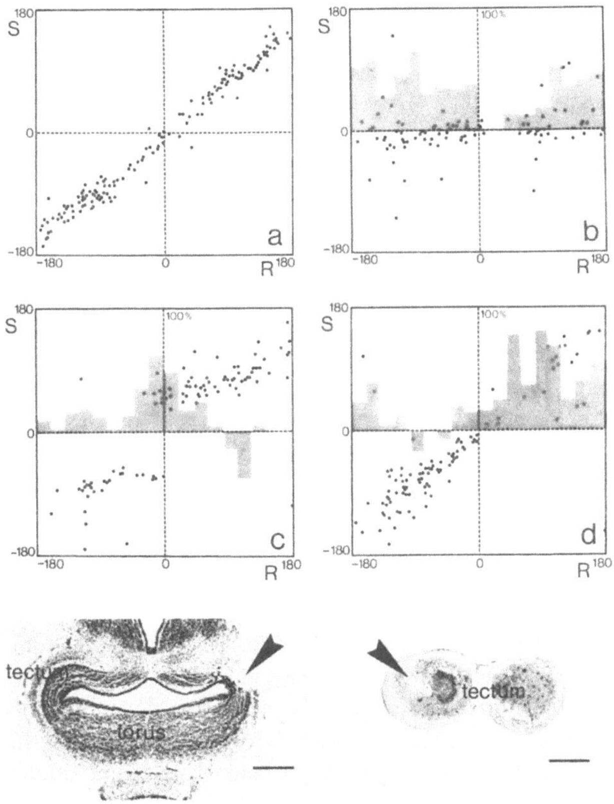

FIGURE 20.3. Reactions of three *Xenopus* to surface wave stimuli before and after lesions in the tectum opticum. (a) An example of the reaction by an intact animal. For 157 reactions the turning angles (R) are plotted against the stimulus angles (S). After a lesion of the entire tectum, the animal reacted as shown in (b). The two other animals had lesions of the rostrolateral part of the right tectum (c) and of the central area of the left tectum (d). The corresponding lesions, indicated by the arrowheads, are shown in Nissl-stained horizontal sections through the tectum. The shaded histograms show the percentage of reduction in response frequency after lesioning. Bin widths of histograms: 20°; bar: 400 μm.

The turning behavior of *Xenopus* is described in terms of stimulation and reaction angles as defined by Görner (1984). Stimulation angles were measured when the water surface was touched by the glass rod, and reaction angles were determined after the turning, when the toad remained motionless in the new position for several seconds. The stimulus distances varied between 5 and 10 cm.

For all intact animals ($n = 24$), the relation between the stimulus angle and the reaction angle was linear (Fig. 20.3a). The regression line had a slope of ~ 45°, and the scatter of the measured values was small. The stimulus distances had no effect on the turning angles. The frequencies with which the animals responded to stimuli varied from 70% to 100%. The reactions of animals with small lesions in the tectum differed from those of intact animals in two respects. For stimulus directions corresponding to the receptive fields of the destroyed neuronal population, the response frequency of the animal was reduced (see shaded histograms in Fig. 20.3b–d). The response angles to stimulus directions corresponding to the area lesioned differed from the stimulation angle, and sometimes the animal even turned away from the stimulus. When the whole tectum was destroyed bilaterally, only one of three animals responded to surface waves. The response frequency decreased to 36% of that from the intact animal. All stimulus directions were equally affected. Responses that occurred, however (a stereotyped behavior consisting of a forward movement with a few hasty swimming strokes), clearly indicated that the animal detected the presence of the surface wave but could not localize the source. In contrast, lesions rostral to the tectum (thalamus) had no effect on the ability to determine the direction of a wave source, but they reduced the frequency with which animals responded to both lateral line and somatosensory stimuli.

20.4 Discussion

The lateral line units, which were recorded in layer 6 of the tectum, probably represent third-order neurons, which get their input from the torus semicircularis (Zittlau et al. 1988). These units differed from those found at other levels of the system (Zittlau et al. 1985, 1986) in showing a sharp tuning for stimulus direction. The receptive fields changed gradually with the position of the unit in the tectum; that is, lateral line units appear to be arranged topographically according to their receptive fields. They form a map of directions on the water surface around the animal (Zittlau et al. 1986). Multiunit recordings did not differ from single-unit recordings, since neighboring units have similar directional preferences.

In the visual system, maps are based on a topological projection from retinal ganglion cells to the tectum (Gaze et al. 1974; Toth et al. 1980), but a comparable topological projection does not exist in the lateral line system. Each surface wave stimulus excites many receptors. Therefore, information from many receptors must be integrated to compute the stimulus direction based on primary lateral line information. Information about the stimulus direction can be encoded in the afferent activity in several ways. First, the animal can make use of the directional sensitivity of the neuromasts. Theoretically, each direction on the water surface can be unequivocally coded by a single pair of neuromasts with different hair cell orientations. This problem is discussed in detail by Görner and Mohr (Chapter 27).

Second, the surface wave produces a complex pattern of activity, referring to the response rates across many afferent lateral line fibers. This pattern contains information about the direction of wave propagation. Because of the relatively slow propagation velocity of the surface wave, some parts of the body are stimulated before others. As the wave travels over the body, its amplitude is attenuated owing to the physical properties of surface waves. Furthermore, the amplitude can be increased or decreased locally, owing to interference of the waves following different paths around the body.

This time and intensity pattern can be used in two ways. On the one hand, the activity of stitches or groups of stitches in different lines or on different parts of the body can be measured simultaneously and compared. On the other hand, the temporal changes in the activity of groups of stitches or lines can be measured. Neuroanatomical (Will et al. 1985) and electrophysiological studies (Plassmann 1980) have shown that interactions between the inputs from both sides of the body exist already at the first relay station of the lateral line system in the medulla, the nucleus lineae lateralis. It has been hypothesized that this interaction contributes to the decision to turn left or right (Plassmann 1980).

To date it is not clear how many inputs from the lateral line periphery are needed to determine the stimulus direction. Behavioral experiments have shown that *Xenopus* is able to orient to surface waves with unchanged accuracy when all but four stitches of the occipital line have been destroyed (Elepfandt 1982; Görner et al. 1984). The latency measurements here can be interpreted in the same way. Under the present experimental conditions, the shortest latencies occurred for the caudal stimulus position (180°), because the distance between the pipette and the first stitches reached by the surface wave (in this case those on the caudal part of the trunk) was shorter than for all other stimulus angles. This is caused by the fact that *Xenopus* has an elongated body and the stimulation device was centered on the midline of the body between the forelegs. Thus, the response of the tectal neurons to caudal stimulus positions in fact precedes the activation of stitches on the head which in turn means that the latter are not involved in the computation of the receptive fields. This corresponds to results of Lowe (1986), who found that separate electrical stimulation of individual branches of either trunk or head lateral line nerve led to a predominance of inputs from the head in the lateral tectum, whereas those from the trunk were larger in caudal areas of the tectum. It would clearly be useful to study the receptive fields of tectal neurons in animals with defined lesions in the lateral line periphery to answer this question.

The results of the single-unit recordings show that lateral line units in the tectum are unresponsive to changes in the amplitude of the stimulus, which in the present experiments ranged from 4 to 80 μm. These amplitudes are within the physiological range of the lateral line periphery, and it is assumed that the tectal cells are not saturated with the smaller amplitudes.

When the stimulus amplitude was decreased by increasing the stimulus distance, eight units increased their response. This response characteristic is probably due to the increase in the duration of the stimulus with increasing distance (see Fig. 20.1). Only three units had a response optimum at an intermediate distance. Thus, it seems that most tectal units are not tuned for distinct stimulus distances but rather respond to all stimuli within their receptive fields which are above threshold. However, for surface wave stimuli, information about the distance is present in the duration of the response, but it seems that the animal makes no use of this information. Behavioral experiments have shown that *Xenopus* is not able to localize the center of the wave source exactly, which requires determining both the direction and the distance. At best, *Xenopus* can probably distinguish stimuli that are very close (2.5 cm) from those far away (7.5 cm) (Görner and Mohr Chapter 27).

As revealed by the recordings reported here and by comparison with studies on the representation of the visual world in *Xenopus* (Udin and Keating 1981), the lateral line map is in register with the map of the visual surroundings. In those tectal areas where no lateral line activity was found, the receptive fields of visual units were above the water surface.

It seems to be a common feature that sensory systems like the visual and the lateral line system in *Xenopus* map the animals' surroundings on the superficial layers of the tectum opticum (or colliculus superior). In each species studied so far (including *Xenopus*; Gaze et al. 1974), the most dorsal layers of the tectum, which receive input from retinal ganglion cells, are retinotopically organized. Analogous to the lateral line cells directly beneath the visual layers in *Xenopus*, cells responding to nonvisual stimuli such as somatosensory or acoustic stimuli are present beneath the optic layer in a variety of animals, including mammals. These cells are arranged in maps; their receptive fields are in register with the receptive fields of the visual system (for review, see Sparks 1988).

To date, the biological significance of this alignment is not fully understood. In contrast to the visual or somatosensory systems in which the topological projection might arise as a consequence of the developmental growth, the lateral line map in *Xenopus* is a computational map. Thus alignment probably offers advantages for sensory processing. One can hypothesize that an alignment of independent sensory maps may help to facilitate the perception of multimodal signals all arising from the same source. A second reason for aligning sensory maps might be to permit cues from all sensory modalities to access shared maps of motor space.

Small unilateral lesions in those tectal areas where the contralateral surroundings are represented led to characteristic deficits in the orienting behavior of *Xenopus*; the reaction frequencies for the corresponding stimulus directions were reduced, and the response directions differed from the stimulus directions. Thus, the lesion experiments show that in *Xenopus* the tectum is necessary for generating the orienting behavior. When the whole tectum was destroyed bilaterally, the orienting reaction no longer occurred (two animals did not react to surface wave stimuli at all, and one animal responded by moving forward without turning). In contrast to the results here, Elepfandt (Chapter 26) found no impairment of the orienting reaction after ablation of the tectum. Up to now, this difference cannot be explained, but it might be that the lesions in the two experiments were not identical, since it is difficult to destroy the tectum entirely without damaging the torus semicircularis.

Because of the directional tuning of the tectal cells and the finding that the tectum is necessary for the generation of the orienting response to surface waves, it can be assumed that the tectum belongs to the sensorimotor interface. In a variety of animals, the tectum opticum (colliculus superior) seems to be an important part of the neuronal substrate responsible for the generation of orienting responses, such as moving the head or the external ears.

In the deep layers of the monkey colliculus superior, neurons have been recorded that are active before the onset of saccadic eye movements and which induce eye movements when electrically stimulated (Sparks 1978). This activity is almost perfectly correlated with the occurrence of a saccade. These experiments suggest that in the monkey the entire sensorimotor interface is located inside the colliculus, which contains both the sensory map in the superficial layers and the motor map in the deeper layers.

In *Xenopus*, electrical stimulation of the tectal lateral line units has not yet been performed. Such experiments could show whether or not these units are related to movement. However, the frequency with which tectal neurons respond to surface stimuli was greater than the response frequency shown by *Xenopus* in behavioral experiments. Time periods in which neurons were unresponsive to stimuli, a situation that could be compared to a behaviorally "unmotivated" *Xenopus*, did not occur. Thus, it is assumed that the lateral line map probably represents the sensory side of the sensorimotor interface but is not part of the motor output side.

20.5 Summary and Conclusions

In the tectum of *Xenopus laevis*, units were recorded that responded to stimulation of the lateral line with surface waves. These units were tuned to characteristic stimulus directions. In the tectum, the lateral line units are arranged topographically according to their receptive fields; thus, they form a map of directions on the water surface. This map is in register with the map of the visual surroundings.

After lesions in the tectum, both the accuracy and the frequency of the orienting reactions decreased for stimulus directions corresponding to the receptive fields of the units in the lesioned area. When the tectum was destroyed entirely, the orienting reaction no longer occurred. It is concluded that the lateral line map in the tectum represents the sensory side of the sensorimotor interface for the orienting reaction toward the source of surface waves.

References

Bleckmann H (1986) Role of the lateral line in fish behaviour. In: Pitcher TJ (ed) The Behaviour of Teleost Fishes. London: Croom Helm, pp. 177–202.

Bleckmann H (1988) Prey identification and prey localization in surface-feeding fish and fishing spiders. In: Atema J, Fay RR, Popper AN, Tavolga WN (eds) Sensory Biology of Aquatic Animals. New York: Springer-Verlag, pp. 619–641.

Bleckmann H, Schwartz E (1982) The functional significance of frequency modulation within a wave train for prey localisation in the surface feeding fish *Aplocheilus lineatus*. J Comp Physiol 145:331–339.

Elepfandt A (1982) Accuracy of taxis response to water waves in the clawed toad (*Xenopus laevis* D.) with intact or with lesioned lateral line system. J Comp Physiol 148:535–545.

Gaze RM, Keating MJ, Chung SH (1974) The evolution of the retinotectal map during development in *Xenopus*. Proc R Soc Lond B 185:301–330.

Görner P (1976) Source localization with labyrinth and lateral line in the clawed toad (*Xenopus laevis*). In: Schuijf A, Hawkins AD (eds) Sound Reception in Fish. Amsterdam: Elsevier, pp. 171–184.

Görner P, Moller P, Weber W (1984) Lateral-line input and stimulus localization in the African clawed toad *Xenopus* sp. J Exp Biol 108:315–328.

Lettvin JY, Maturana HR, McChulloch WS, Pitts WH (1959) What the frog's eye tells the frog's brain. Proc Inst Radio Eng NY 47:1940–1951.

Levine RL (1980) An autoradiographic study of the retinal projections in *Xenopus laevis* with comparison to *Rana*. J Comp Neurol 189:1–29.

Lowe DA (1986) Organization of lateral line and auditory areas in the midbrain of *Xenopus laevis*. J Comp Neurol 245:498–513.

Plassmann W (1980) Central neuronal pathways in the lateral line system of *Xenopus laevis*. J Comp Physiol 136:203–213.

Rudolph P (1967) Zur Ortungsverhalten von *Gyrinus substraticus* Steph. (Taumelkäfer). Z Vergl Physiol 50:341–361.

Schwartz E (1965) Bau und Funktion der Seitenlinie des Streifenhechtlings (*Aplocheilus lineatus* Cuv. u. Val.). Z Vergl Physiol 50:55–87.

Sparks DL (1978) Functional properties of neurons in the monkey superior colliculus: Coupling of neuronal activity and saccade onset. Brain Res 156:1–16.

Sparks DL (1988) Neural cartography: Sensory and motor maps in the superior colliculus. Brain Behav Evol 31:49–56.

Toth P, Lazar GY, Görcs T (1980) Retinal projections in the adult *Xenopus laevis*: A study with cobalt filling technique. Acta Morphol Acad Sci Hung 28:365–374.

Udin SB, Keating MJ (1981) Plasticity in the central nervous pathway in *Xenopus*: Anatomical changes in the isthmo-tectal projection after larval eye rotation. J Comp Neurol 203:575–594.

Will U, Luhede G, Görner P (1985) The area octavo-lateralis in *Xenopus laevis*. II. Second order projections and cytoarchitecture. CTR 239:163–175.

Zittlau KE, Claas B, Münz H (1985) Directional sensitivity of lateral line neurons of the clawed toad *Xenopus laevis* D. Neurosci Lett (Suppl) 22:435.

Zittlau KE, Claas B, Münz H (1986) Directional sensitivity of lateral line units in the clawed toad *Xenopus laevis* Daudin. J Comp Physiol 158:469–477.

Zittlau KE, Claas B, Münz H (1988) An HRP study of tectal afferents in *Xenopus laevis* with special emphasis on their relationship to the lateral line system. Brain Behav Evol 32:208–219.

21
Interrelationship of Acousticolateral and Visual Systems in the Teleost Midbrain

NICO A.M. SCHELLART and ALFONS B.A. KROESE

21.1 Introduction

It is well established that the mesencephalic torus semicircularis of fish obtains input from the octavus, lateral line, and visual systems. In electroreceptive teleosts and elasmobranchs, electroreceptive information reaches the torus as well. The modalities mentioned are processed separately not only in unimodally sensitive units but also in multimodal neurons.

After a brief introduction about the neuroanatomy of the torus and its connections with other brain structures, the characteristics of acousticolateral responses in nonelectroreceptive teleosts will be discussed. Attention will be paid to excitatory and inhibitory processes, acousticolateral frequency coding, temporal features of the responses, and the extent to which the acoustic and lateral line systems are intermingled in the torus. Acousticolateral processing in the torus area has to be compared with that in more peripheral parts of the acousticolateral systems to understand more completely the information processing along the acousticolateral pathway.

Current knowledge of central acousticolateral processing in teleosts will be reviewed, with the emphasis on the topographical organization of the modalities and on their response features. The characteristics of visually elicited unit responses in the tegmentum will be dealt with in order to evaluate the topographic intermingling of the acousticolateral and visual processes and the various kinds of interactions between the modalities at the level of single cells.

21.2 Anatomy and Connections of the Torus Semicircularis

The torus semicircularis is a longitudinal ridge forming the floor of the mesencephalic ventricle. In electroreceptive teleosts, the torus is well developed and occupies nearly the whole ventricular space, with the tectum much thinner than seen in other fish. The large number of well-stratified granular and plexiform layers of these tori is most remarkable (e.g., Carr and Maler 1986). The cytoarchitecture of these tori strongly deviates from those of nonelectroreceptive fish. The tori of only a few species of the latter group of teleosts have been examined neuroanatomically (Ito 1974). They occupy only the caudal region of the ventricular floor, spreading laterally to the lateral and intercalated part. The superficial layer of these tori is subependymal, containing many thin and hardly any thick fibers. Underneath this layer are generally two (sometimes one) patchy, irregular soma layers with increasing distance to the ventricle in the medial direction. The ventral part of the torus consists of a single layer, which is as thick as all other layers together. This lower layer contains scattered medium-size and large neurons as well as the radiation of the fasciculus longitudinalis lateralis (fll). In the trout, *Salmo*, this layer is split in a dorsal and ventral part by the tractus isthmotectalis. The ventrolateral part of the torus is bound by the tractus tectospinalis and the nucleus dorsolateralis tegmenti (DLT), a secondary visual structure, obtaining efferents from the tectum (Grover and Sharma 1979; de Wolf et al. 1983). Medial to the torus, the nucleus lateralis valvulae (NLV) is found. The torus, probably only the pars ventralis, is ipsilaterally innervated by type XII neurons of Meek and Schellart (1978) in the stratum griseum centrale of the tectum (Grover and Sharma 1981; Echteler 1984; Murakami et al. 1986) and type XIII neurons in the stratum album centrale (Grover and Sharma 1981; Echteler 1984). Visual information from the tectum probably reaches the torus also via the DLT (de Wolf et al. 1983). In the trout, the tectal projection mainly reaches the ventral and dorsorostral torus.

The torus has a large number of afferent and efferent connections (Fig. 21.1). Ascending afferent pathways come mainly from the octavolateral area in the medulla. The principal ascending pathway is via the fll. However, a minority of the ascending auditory fibers innervate the torus not directly, but bilaterally via the nucleus oliva superior (NOS) and the nucleus lemniscus lateralis (e.g., de Wolf et al. 1983; Meredith and Roberts 1986). In nonelectroreceptive teleosts, the NOS is generally poorly developed. However, in the electroreceptive catfish *Ictalurus*, the main acoustic pathway is via the NOS (Finger and Tong 1984). The medullar ascending projections have been reviewed by McCormick (1983). The connections between the distinct octavolateral nuclei in the hindbrain and the various parts of the torus have been described in detail for *Ictalurus* (Finger and Tong 1984). In the salmonid, *Oncorhynchus*, an ascending spinal projection has

►

FIGURE 21.1. Afferent and efferent connections of the teleost torus semicircularis. Heavy arrows indicate the direct visual and acousticolateral pathways. The efferent structures are indicated by dashed boxes. Abbreviations: DLT, nucleus dorsolateralis tegmenti; LHRH, tegmental nucleus with neurons containing luteinizing hormone-releasing hormone; NCH, nucleus cerebellaris hypothalamus; NLL, nucleus lemniscus lateralis; NOI, nucleus oliva inferior; NOS, nucleus oliva superior; NP, nucleus praeeminentialis; NPG, nucleus preglomerulosus; NT; nucleus tuberalis; NVMT, nucleus ventromedialis thalami. The connections are based on the papers of Echteler (1984), Finger and Tong (1984), Grover and Sharma (1979, 1981), Ito et al. (1986), Meredith and Roberts (1986), Münz et al. (1981), Murakami et al. (1983, 1986), Oka et al. (1986), and de Wolf et al. (1983).

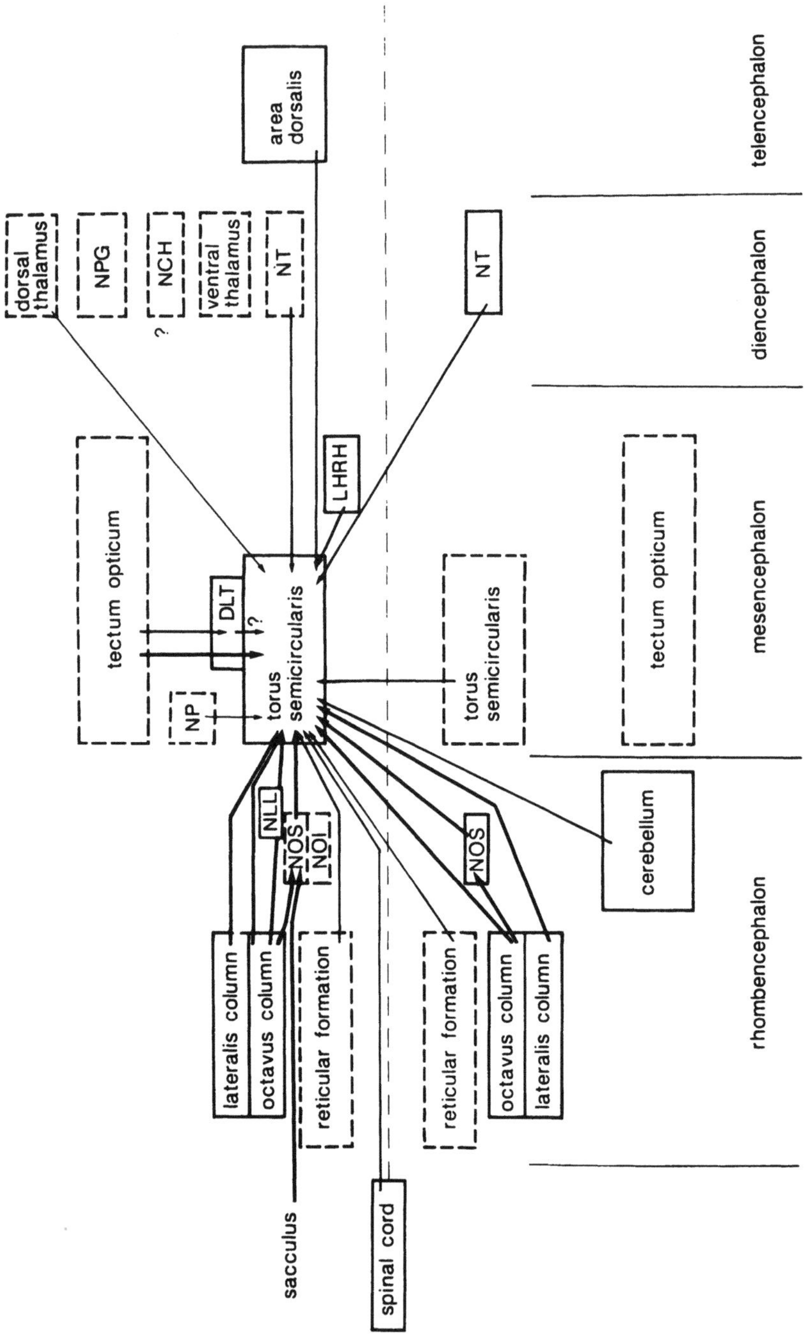
telencephalon
diencephalon
mesencephalon
rhombencephalon
area dorsalis
dorsal thalamus
NPG
NCH
?
ventral thalamus
NT
NT
LHRH
tectum opticum
DLT
?
torus semicircularis
NP
torus semicircularis
tectum opticum
NLL
NOS
NOL
NOS
cerebellum
lateralis column
octavus column
reticular formation
reticular formation
octavus column
lateralis column
sacculus
spinal cord

recently been described (Oka et al. 1986). In the trout torus there are connections with thick spinal fibers, the afferent or efferent character of which is unknown (de Wolf et al. 1983). Other afferent, mostly small projections to the torus come from the ipsi- and contralateral nucleus ventromedialis thalami (Ito et al. 1986), the nucleus tuberalis (Echteler 1984), the ipsi- and contralateral NOS as far as present (de Wolf et al. 1983; Echteler 1984), the cerebellum, and the telencephalic area dorsalis (Finger 1980; Murakami et al. 1983). Finally, there is a heavy projection from the contralateral torus via the commissura transversa (e.g., de Wolf et al. 1983; Echteler 1984).

The torus projects to both tectal halves, the nucleus cerebellaris hypothalami (Wolf et al. 1983), the nucleus preglomerulosus (Murakami et al. 1986), the central part of the dorsal thalamus (Echteler 1984), the reticular formation (de Wolf et al. 1983; Echteler 1984), the ipsilateral NOS (Echteler 1984), and, in *Ictalurus*, to the nucleus praeeminentalis in the hindbrain (Finger and Tong 1984).

21.3 Characteristics of Acousticolateral Torus Responses

21.3.1 Multimodality in the Mesencephalon

In the lower midbrain, Acousticolateral units, Visual units, and Bimodal units (i.e., units sensitive to both of these modalities) can easily be found (e.g., Page and Sutterlin 1970; Schellart 1983). In electroreceptive teleosts, electroreceptive torus units are also found (e.g., Knudsen 1977). However, in the trout tectum opticum, acousticolateral and bimodal unit activity is difficult to find (Schellart et al. 1987). In the tectum of other nonelectroreceptive teleosts, the occasional occurrence of auditory (Niida 1973), visuolateral line (Galand and Liege 1975), and visuotactile units (Galand and Liege 1975; O'Benar 1976) has been reported. In other nuclei projecting onto the torus, bimodal activity is not known to occur. Therefore, the bimodal units are most probably intrinsic torus neurons. On the basis of spike wave form and response properties as described in the various publications (see Table 21.1), the torus Acousticolateral and Visual units recorded from are also thought to be intrinsic neurons. However, recordings from afferents cannot be excluded, such as those from incoming fll fibers and fll fibers just outside the torus. In some studies a minority of the recordings originates from the DLT too.

21.3.2 Frequency Range of Acousticolateral Units

In the goldfish, *Carassius* (Page 1970), and the trout, *Salmo gairdneri* (Nederstigt and Schellart 1986), the frequency ranges over which torus units showed a response to a tone burst of a moderate strength have been determined. In both the goldfish and trout experiments, the stimulus was generated by an oscillating membrane mounted in one of the walls of the experimental water-filled tank. A statistically detectable change (with respect to the spontaneous activity) of the

TABLE 21.1. Acousticolateral response features of single- and multiunit recordings in the teleost torus.

	Page 1970, Page and Sutterlin 1970	Echteler 1985	Grözinger 1967	Horner et al. 1980	Nederstigt and Schellart 1986, Schellart 1983, Schellart et al. 1987	Knudsen 1977	Plassmann 1985
Animal	Goldfish	Carp	Tench	Cod	Trout	Catfish	Catfish
Anesthetic	Tricaine		G 29505	None	None	Tricaine	Tricaine
Paralyzer	Tubocur.	Flaxedil	Flaxedil	Tubocur.	Pancuronium	Tubocur.	Tubocur.
Stimulus generated in	Water[a]	Air[b]	Water[a]	Air[b]	Water[a,c] Air[b]	Water[c] Air[b]	Water[b]
Feature							
Single/multi	s[d]	m	m, s	s[d]	s[d]	s	s
Freq. range	90[e]–2,000		200[f]–3,500		40[f]–700		100–1,800
Best freq.	140–700		450–800		150–350		100–1,800
No. best freq.	Rarely > 1				Often > 1		
Phase locking	None		Often	Weak	Often		
Tonic/phasic	Both				Both		
Habituation	Often				Often		
Latency (ms)	4[g]	7[h]	2.7		12[g]		7[h]
I behavior	Perhaps		Some	Some	Often		
E⁺ behavior	Some				Some		
Threshold (Pa)	0.5–2[i]	⩽0.2	0.003–0.04	⩽0.1	0.005–0.2		0.001–0.03
SpA (spikes/s)	0–1		Low	<20	0–50		
Modality	Ac, V, AcV	Ac	Ac	Ac	L[j], Ac, AcL[j], V, Bi	L, AcL?, Ac	Ac

Abbreviations: Ac, acoustic; AcL, acousticolateral; AcV, acousticovisual; Bi, bimodal; E⁺, see Fig. 21.2; I, inhibitory; L, lateral line; SpA, spontaneous activity; V, visual.
[a] Oscillating membrane.
[b] Loudspeaker.
[c] Vibrating submerged plate.
[d] Including recordings from the vicinity of the torus (e.g., NDT and fll).
[e] Artificial limit.
[f] Some units may be sensitive to lower frequencies.
[g] Estimated by applying tone bursts.
[h] Estimated by applying clicks.
[i] Estimated with a high criterion.
[j] Only indirect evidence.

number of spikes elicited by the tone burst was considered a response. These responses were used to determine the frequency range of responsiveness to a stimulus of fixed intensity.

In the torus of the trout (Schellart 1983; Nederstigt and Schellart 1986), the frequency response range was determined with a stimulus of 2.5 Pa. The acoustic displacement waves accompanying this stimulus had amplitudes of several microns in the 40- to 100-Hz range. Since the submerged fish was loosely restrained, the fish body made oscillations of 0.2 μm at most (Schellart, unpublished acousticomechanical measurements), enough to stimulate the otolith organs directly. In the nonostariophysine trout, most units show responsiveness between 150 and 300 Hz. Only 11% of the units show responsiveness above 500 Hz (Nederstigt and Schellart 1986). These findings are in agreement with the frequency range of the behaviorally determined sound pressure audiograms of salmonids (e.g., Hawkins and Johnstone 1978). The torus units of the ostariophysine goldfish were found to respond to a 7-Pa stimulus from at least 140 Hz to at most 2,000 Hz. In contrast to the trout, 50% of the units are responsive above 500 Hz (Page 1970).

The low-frequency nature of salmonid swimbladder-mediated hearing as compared to that of the goldfish is thought by us to be related to the exceptionally large size of the swimbladder, which shows a low frequency tuning (Schellart and de Munck 1987), and to the lack of specialized ossicular structures transmitting swimbladder pulsations to the ears.

21.3.3 Types of Acousticolateral Units

On the basis of their frequency ranges in the trout, three types of acousticolateral sensitive units (Fig. 21.2) can be distinguished: H (high-frequency), L (low-frequency), and B (broad-band) units (Nederstigt and Schellart 1986; Schellart et al. 1987). As a criterion, 125 Hz is used to distinguish L and H units. The best frequency of the posterior lateral line organs of the trout seldom exceeds a frequency of 110 Hz (Kroese and Schellart 1987), and the same holds for other fish and amphibians (see Münz 1985 for references). Therefore, the H units are thought to receive input exclusively from the otolith systems. On the other hand, teleost auditory nerve fibers rarely show best frequencies below 125 Hz and always respond up to at least 175 Hz (e.g., Coombs and Fay 1987; Fay 1981; Horner et al. 1981). This seems to exclude dominant auditory input to L units. These are apparently driven mainly by the lateral line system. For frequencies below 125 Hz, some B units may have obtained input from the lateral line system too, since in the experimental tank the displacement movements of the fish body were smaller than those of the surrounding water. It is assumed that vestibular input does not occur above 40 Hz. The classification of B, H, and L units is generally independent of stimulus strength between 0.25 and 25 Pa.

The frequency ranges of the torus units in the goldfish do not cluster into various types (Page's Fig. 12, 1970). Most units have a broad-band character, and none of them has a frequency response range smaller than half an octave. The

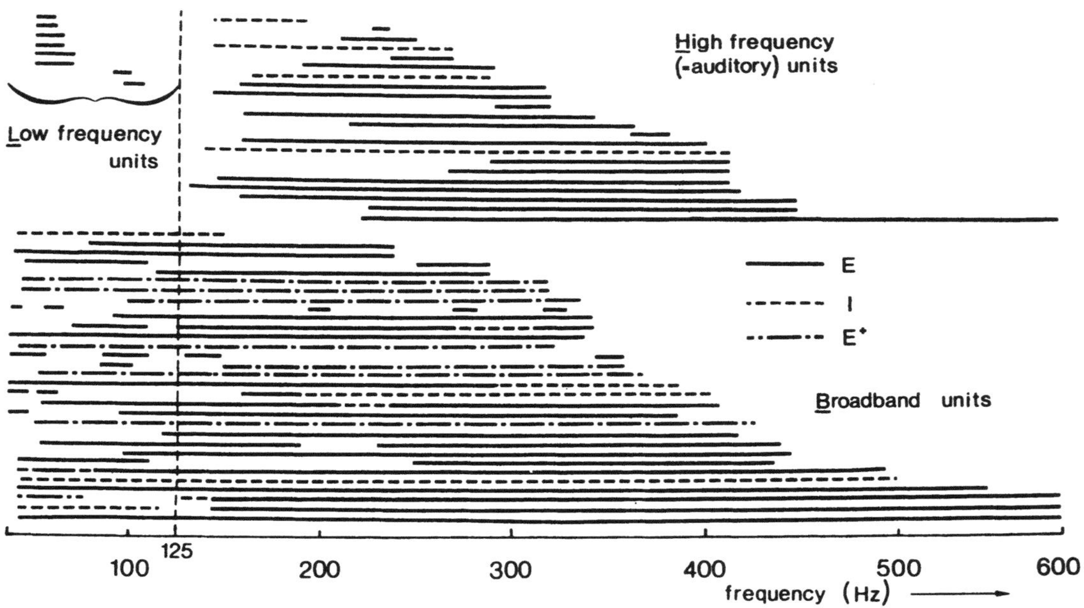

FIGURE 21.2. "Band widths" of L, H, and B units of the trout torus semicircularis. Each horizontal bar, whether or not interrupted, presents for one particular unit the range of frequencies that gave a just perceptible response to a 2.5-Pa tone burst of 720-ms duration, irrespective whether the response was excitatory (E), inhibitory (I), or excitatory with extension after cessation of the tone burst (E^+). The lowest frequencies applied were 40, 45, or 50 Hz (from Nederstigt and Schellart 1986).

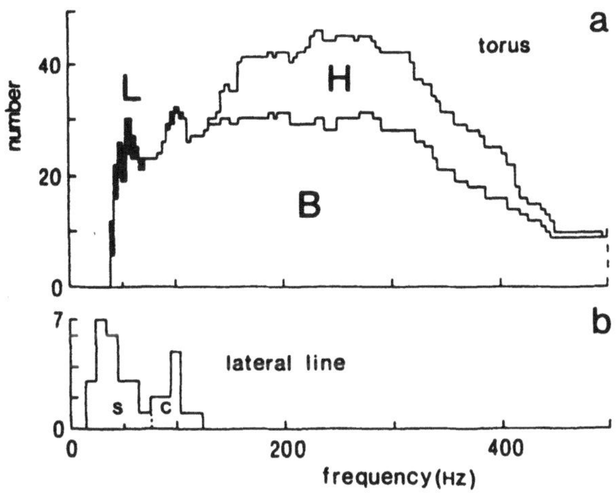

FIGURE 21.3. The distribution of "band widths" of the acousticolateral units of the torus (a), related to the best frequency of trunk lateral line units (b), both recorded in the trout. The upper histogram gives the distribution of the frequency ranges of the L, H, and B units (from Nederstigt and Schellart 1986). The lower histogram gives the best frequency of 23 S and 11 C trunk lateral line units measured during local stimulation of the trunk lateral line.

responses of the goldfish units are thought to be elicited exclusively by input from the otolith systems.

Recently, the authors determined the linear frequency characteristics of afferent lateral line nerve fibers during local stimulation of the trunk lateral line of the trout (Kroese and Schellart 1987). Individual lateral line units were stimulated with small-amplitude, sinusoidal, local water displacements (4 to 400 Hz), produced by a 3-mm sphere at a distance of 1 to 4 mm from the skin. Linear frequency response analysis revealed that two types of fibers can be distinguished: one with a best frequency around 40 Hz (S units), the other of about 90 Hz (C units). The S type is thought to innervate superficial neuromasts, and the C type the canal neuromasts. In the cichlid *Sarotherodon*, a similar distinction has been found: the best frequency of superficial neuromasts ranges from 10 to 70 Hz, and that of canal neuromasts from 30 to 200 Hz with a mean of approximately 120 Hz (Münz 1985). On the basis of the amplitude and phase characteristics of both types in the trout, it is concluded that S units respond to water velocity and C units to water acceleration (Kroese and Schellart 1987). The occurrence of the best frequency, plotted as a function of frequency, yielded two distributions coinciding with the two small, low-frequency peaks in the distribution of the "band widths" of the trout Acousticolateral-sensitive torus units (Fig. 21.3). This strengthens the suggestion that the L units and part of the B units are fed by the lateral line system.

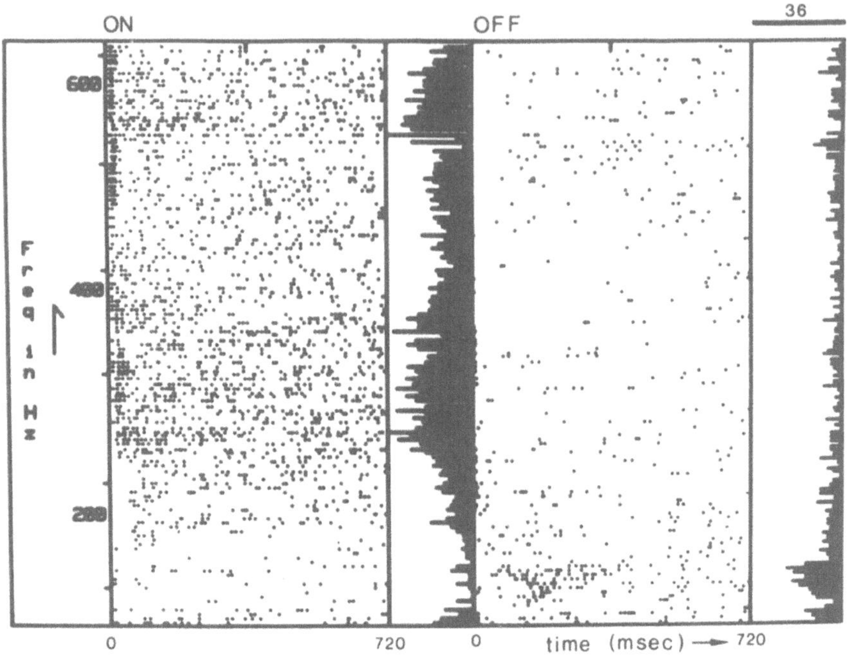

FIGURE 21.4. Frequency-time dot display of a spontaneously active B unit without phase locking. The dot display contains the responses of the unit to about 220 tone bursts (left panel, ON) of 720 ms, each followed by a silent interval (right panel, OFF). Each dot represents a spike. The frequency of the tones in the bursts increases from 60 Hz (bottom) to 610 Hz (top). In the two histograms the number of spikes elicited in each 720-ms period is given. The horizontal bar in the right top corner gives the scaling of the histograms. In the lateral line frequency range, the unit exhibits inhibitory behavior.

21.3.4 Frequency Response Curves

For units lacking phase locking (i.e., synchronization of the spikes to the individual stimulus cycles), only spike rate is available for determining the responsiveness of the unit. For these units, the mean spike rate at a given intensity, or the intensity necessary to evoke a criterion rate response (tuning curve), can be determined to obtain information about frequency behavior. Frequency characteristics based on the spike rate of single units in the lower midbrain have mainly been studied in the goldfish (Page 1970) and the trout (Nederstigt and Schellart 1986). All kinds of characteristics have been encountered, but this has not led to a classification more precise than the L, H, and B designation.

Figure 21.4 illustrates a frequency-time dot display of a trout unit, obtained with tone bursts between 60 and 620 Hz at a constant intensity. The middle vertical histogram gives the number of spikes elicited during the tone bursts of 720-ms duration, and the right histogram gives the number of spikes fired during the intermittent silent intervals. The highest spike rates are found between 230 and 350 Hz and between 530 and 580 Hz. The frequency response curves, measured in terms of either the spike rate or the sensitivity, often show a complex pattern of various peaks and troughs, as can be seen in the example of Fig. 21.5a. In this figure, both histograms of the dot display are replotted as spike density (vertical) versus the logarithm of the frequency (horizontal). B units especially often appeared to have more than one maximum, a phenomenon occasionally also observed in the goldfish (Page 1970). Since constant pressure as a function of frequency does not guarantee constant kinetics of the sound stimulus, only gross features of the shapes of the frequency response curves could be evaluated. To rule out the possibility that the peaks at several frequencies are due to insufficient control of the complex tank acoustics, the curves were also determined with a stimulus of constant accleration for some units. The acceleration stimulus yielded similar peaks in the frequency response rate curves as obtained with constant pressure. In other experiments featuring airborne sound pressure generated by a loudspeaker, whether the fish was submerged or not, the same results have been obtained. Therefore, multiple peaks in the frequency response curves are a physiological phenomenon.

21.3.5 Phase Locking

In addition to frequency ranges and frequency response curves determined with rate information, the torus units of the goldfish and trout were also inspected for the occurrence of phase locking. In the torus of the trout, phase locking has been observed for only 9% of the units. This is in sharp contrast to the behavior of primary afferent fibers of lateral line and auditory nerves, which all show phase locking to the stimulus. The trout B units with phase locking were all light-insensitive, spontaneously active, and of the B type (Schellart et al. 1987). Figure 21.5b gives the index of synchronization as a function of frequency together with the phase characteristic (Fig. 21.5c) of a well-responding B unit. For various frequencies, this unit gives nearly perfect phase locking. The few L units recorded from lack phase locking and bimodality. A majority (72%) of the B units fail to phase-lock. A part of them, nevertheless, respond to a tone

►

FIGURE 21.5. Spike density (a) phase locking (b,c) of a trout unimodal, spontaneously active B unit. Panel (a) shows the spike density during the tone burst and during the intermittent silent intervals as a function of the frequency of the tone burst. The spike density gives the mean spike density of the spikes elicited during each tone burst and the subsequent silent interval. The spike densities of the burst (upper part) are depicted upward, and those of the silent intervals downward (lower part). Since this unit showed inhibitory behavior for intermediate frequencies, the frequency response curve is comprised of a low- and a high-frequency part. The tuning curve—i.e., the strength of the phase locking expressed as the index of synchronization—is depicted in (b). The accompanying phase characteristic is presented in (c).

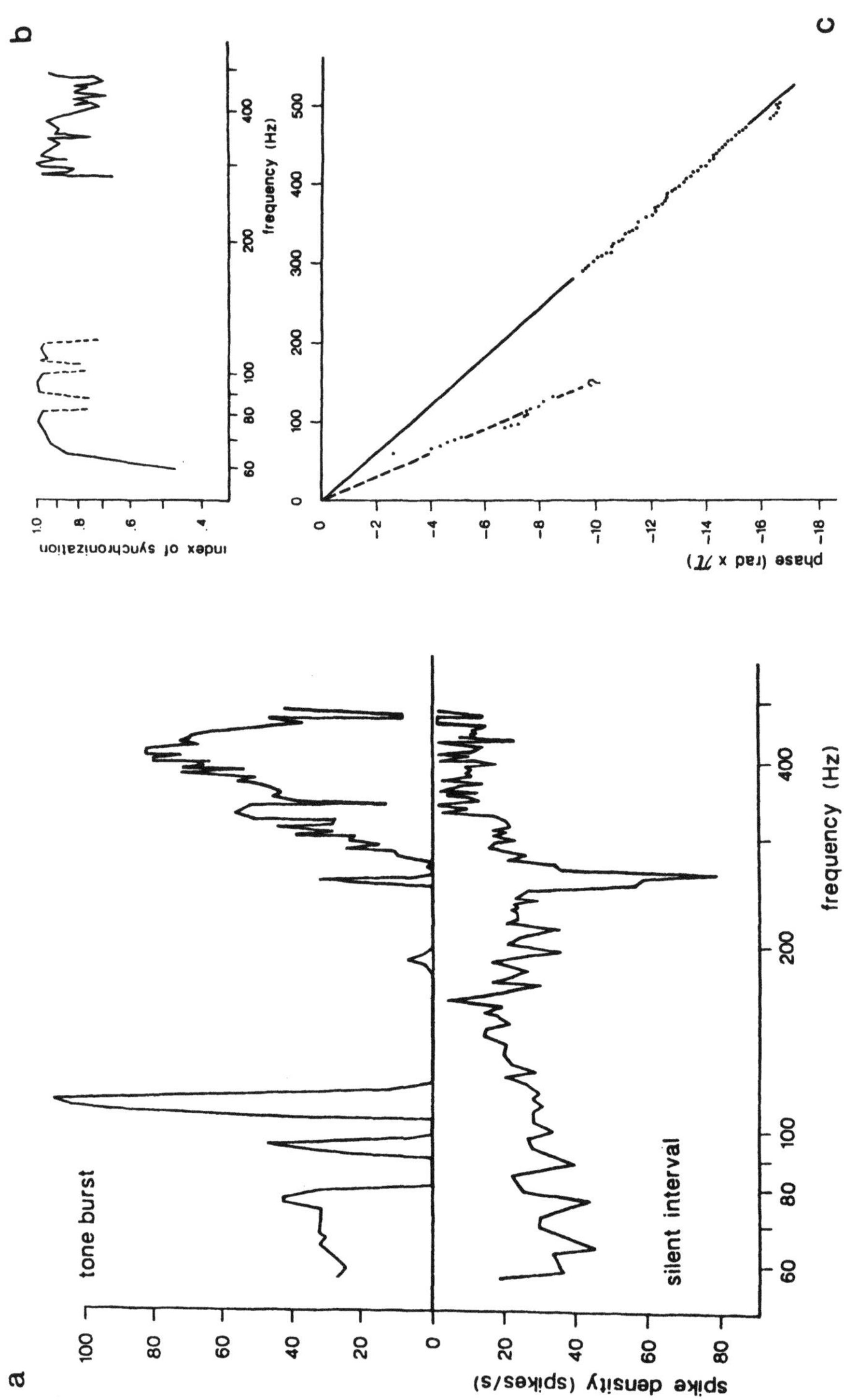
a
tone burst
silent interval
spike density (spikes/s)
frequency (Hz)
b
index of synchronization
frequency (Hz)
c
phase (rad x π)

burst with high spike firing levels, like the unit illustrated in Fig. 21.4. In the tench, *Tinca*, phase locking, studied by recording multiunit activity and slow potentials, has been shown to decrease in strength from the eighth nerve to the medulla and finally to the torus (Grözinger 1967). In the cod, *Gadus*, weakly developed phase locking is present in the torus (Horner et al. 1980), but in the tench's relative, the goldfish, phase locking has been found only in the medulla, not in the torus (Page 1970). Thus in general, phase locking appears to diminish along the acousticolateral pathway.

21.3.6 Excitatory and Inhibitory Responses

In the torus, in contrast to primary fibers, excitation (higher mean spike rate than spontaneous rate) *and* inhibition (lower rate) occur. Therefore, this duality of the response is an additional criterion for classification.

Excitatory units, which show an increased rate to all frequencies eliciting a response, are most common in trout (Fig. 21.2) and goldfish (Page 1970). In the trout, excitatory behavior is generally accompanied by a suppression of the spontaneous rate during the first 0.5 to 1 s after cessation of the tone burst. Two types of inhibitory behavior are distinguished: one shows suppression of firing during the tone burst with normal spontaneous rates during interburst intervals; the other also shows suppression during the bursts, but with a spike rate higher than spontaneous after cessation of the burst. In addition to excitatory and inhibitory units, there are also units showing frequency-dependent excitatory and inhibitory behavior (excitatory/inhibitory units). The unit of Figs. 21.4 and 21.5a shows the various behaviors. Finally, some units show an increase of spike density during both the bursts and a period of ~ 1 s after cessation. They respond with a very tonic discharge. In Fig. 21.2 they are denoted by E^+. A more detailed description of these types has been given elsewhere (Nederstigt and Schellart 1986).

Nearly all H units and the L units have an excitatory behavior (Fig. 21.2). The E^+ behavior occurs most frequently in Bimodal units. Generally, the accompanying visual response also has a tonic character. The more complex excitatory and inhibitory behavior of the B units does not imply bimodality; B units are more often unimodal than H units. B units with phase locking are never E^+ units or, of course, inhibitory units. In the goldfish, inhibitory, excitatory/inhibitory, and E^+ units appear to occur too, but on the basis of Page's data this distinction is less easy to make. Evidence for inhibitory behavior has been reported for the torus of the cod (Horner et al. 1980). Since E^+ behavior has not been described for recordings made from the medulla, it seems to be typical for the torus. This does not hold for inhibitory behavior, since in the eel, *Anguilla*, one-third of the medullar second-order neurons are either excited or inhibited by the contralateral lateral line nerve. However, the ipsilateral lateral line nerve gives excitation only at this level (Alnaes 1973).

21.3.7 Adaptation

It is generally accepted that auditory nerve fibers can follow sinusoidal amplitude modulation (SAM) of an acoustic carrier frequency. In the catfish torus, two types of units process this information. One type, the tonic one, follows the SAM tone bursts rather well, but the other only responds to the onset of a SAM input signal (Plassmann 1985). This latter type is more typical for central processing. In the torus, units that adapt rapidly to a tone burst (phasic behavior) as well as units that hardly adapt (tonic behavior) can be found. Mixed behavior also occurs. Tonic and phasic behavior can be frequency-dependent (Page 1970; Nederstigt and Schellart 1986).

21.3.8 Latency

The shortest latencies of the first spike of teleost torus responses to acousticolateral stimulation (clicks or tone bursts) range from 2.7 to 12 ms (Table 21.1). However, latency may amount to hundreds of milliseconds (Page 1970; Nederstigt and Schellart 1986). The lower limit is some milliseconds higher than that of unit responses in the medulla (Grözinger 1967; Page 1970; Echteler 1985). The large range observed in the torus is typical for acousticolateral responses in this midbrain structure. In the trout, the latency to low frequencies is often longer than to high frequencies, even if the response strength to the higher frequencies is smaller. The increase of the latency with decreasing frequency is much more than the increase of the stimulus period length. Medullar lateral line responses in fish (Alnaes 1973) and lateral line responses of the torus of *Xenopus* (Zittlau et al. 1985) have longer latencies than acoustic responses. Therefore, the longer latencies to lower frequencies found in the trout may be caused by lateral line input.

The latency can also be estimated from the phase characteristics of the linear frequency behavior of units with phase locking. Since the phase shift caused by the latency strongly surpasses the shift due to the dynamics, the latency can be measured directly from the phase plot. In Fig. 21.5c the slope of this plot yields for frequencies between 280 and 500 Hz a latency of 15 m (solid line). For the lower frequencies (60 to 140 Hz), the slope is probably steeper (dashed line), resulting in a longer latency. This agrees with the longer latency for the lower frequencies as observed in the frequency-time dot display (not shown).

21.3.9 Threshold Sensitivity and Responsiveness

The acousticolateral responsiveness of torus units varies considerably among individuals and species but also within a single animal. For stimulation strengths on the order of 1 to 25 Pa, the responses to acoustical stimuli can amount to several tens of spikes per second (Nederstigt and Schellart 1986), but Page and Sutterlin (1970) only rarely observed more than two impulses per second in the ventral torus. Among the various studies, the lowest pressure threshold ranges from 1 to 5 mPa, except for the study of the goldfish (Table 21.1). These values correspond with displacement thresholds in the nanometer range for frequencies around 100 Hz. The thresholds found are comparable with behavioral thresholds (e.g., Hawkins and Johnstone 1978). In the trout, B units were generally most and L units least responsive. The dynamic range is generally limited to 30 dB. In trout and goldfish, a minority of the units shows a clear decrease of response strength for the highest intensities. This nonmonotonic behavior is more seldom observed in earlier stages along the acousticolateral pathway (e.g., Page 1970).

21.3.10 Evaluation of Acousticolateral Response Features

Various temporal and dynamic response features observed in the torus appear, as noted before, to be typical for this structure. For the H units, these are the often occurring sensitivity to a visual stimulus and the narrow range of frequencies (the "band widths" of Fig. 21.2) to which they respond. For the B units, these features include the frequency-dependent excitatory/inhibitory behavior, the E^+ behavior, and frequency response curves (and tuning curves) with multiple best frequencies. The features of the B type indicate that the responses of B units are produced by interaction of inputs from various acousticolateral subsystems with different frequency ranges. Further, B units are less often sensitive to a visual stimulus than H units and more frequently show high responsiveness to acousticolateral stimulation. Together with the phase locking, these properties make the B units more similar to medullar units and, in some respects, to primary fibers than H units. It is therefore speculated that the B units may drive the H units.

21.4 Characteristics of Visual Torus Responses

Visual processing in the torus has mainly been studied in goldfish (Page and Sutterlin 1970) and trout (Schellart et al. 1987) by single-unit recordings. Like the goldfish, the trout has a well-developed visual system, as is obvious from its trichromatic retina, the size and precise stratification of its tectum, and its good visual acuity.

Visually evoked single-unit activity in the torus distinguishes itself in many ways from that observed in the superficial layers of the tectum. The torus units of the goldfish proved to be very phasic. Moreover, they often show rapid habituation to repeated stimulation. In the trout, the habituation of the response to an on-off flashing, immobile spot is the most striking response feature. Many units reproducibly habituate with time constants of approximately 1 to tens of seconds for a square-wave-modulated light stimulus of 0.35 to 2.8 Hz. Habituation is least pronounced when an on-off flashing spot was moved randomly. In the goldfish, visually habituating units have been found in the ventral torus and the DLT (Page and Sutterlin 1970). In this fish, as well as in the trout, many visually habituating torus units did not show habituation to auditory stimuli. This different behavior to both modalities suggests that visual habituation arises in the pathway between the superficial tectal layers and the Bimodal units in the torus. The visual responses to square-wave-modulated light are either phasic, tonic, or mixed phasic-tonic. In this respect they are similar to those of the tectum.

One of the most typical tectal response features is directional selectivity (see Schellart 1988), but in the torus this feature is rare. Visual responses of the torus are, unlike those of the superficial tectum, often erratic. As in the tectum, the majority of the visually sensitive units shown *on* behavior, whereas *off* and *on/off* behavior occur less frequently.

The receptive fields are generally larger than 80° and never smaller than 25° and lack well-defined borders. In the goldfish torus, the receptive fields are very large too. However, when a light or dark spot of some degrees in diameter is moved over a distance of only 2°, a response will be obtained. Repeated movement, like all kinds of repetitions, often results in habituation. Obviously, the very large size of the receptive fields hampers the existence of a well-developed retinotopy. The receptive fields always lack a center-surround antagonism, even when an adapting central spot is used. Wavelength coding has been found only once. The above data indicate that processing of wavelength, wave shape and contrast, precise position, and direction of movement is not a task of the torus. The generally occurring habituation to visual stimuli suggests that the torus provides the animal with the ability to observe sudden—also small—changes in the visual scenery.

21.5 Spontaneous Activity of Torus Units

In the trout, the majority of units are spontaneously active: most units show a resting activity of more than one impulse per second. However, this rate is the upper limit for the spontaneous activity in the anesthetized (MS 222) goldfish (Page 1970). The spontaneous discharge varies among the trout units from the less often observed regular firing to the more frequently found highly irregular and burstlike behavior. The spontaneous units are 1.4 ($p < .05$, $n = 95$) times more numerous in the rostral than the caudal part of the trout torus.

The fraction of units demonstrating a spontaneous discharge was larger for Bimodal units (76%) than for Acousticolateral and Visual units. This may reflect the multiinput nature of Bimodal units, the spontaneous activity being the result of neuronal "noise" from the sensory afferents. Mean spontaneous rates for Visual, Acousticolateral, and Bimodal units are 4.7, 8.8, and 12.3 spikes per second, respectively. These values suggest that the higher spontaneous activity of the Bimodal units may be the summation of the spontaneous activity of both unimodal types. Alnaes (1973) studied the spontaneous discharge of second-order lateral line neurons in the medulla of the eel. Firing rates up to 50 spikes per second have been observed with a population mean of 12.5. In the trout trunk lateral line, the population mean is 27.4 spikes per second. Spontaneous activity thus appears to diminish along the lateral line pathway, and the same seems to hold for the acoustic and visual pathways. A high acousticolateral sensitivity in the torus generally accompanies a high spontaneous activity. This implies that the highest spontaneous rates are found among B units and the lowest among L units. Among the Acousticolateral and Visual units, the excitatory and the *on* types are less often spontaneous than those with other response properties.

21.6 Topographic Organization of the Torus

21.6.1 Topography of Modalities

In goldfish, recordings have been made from Bimodal units in an area of the tegmentum adjacent to the fll (Page and Sutterlin 1970) and corresponding to the ventral torus and the DLT. In the trout, the Bimodal units, the least numerous type, occur in the deep as well as superficial part of the torus, with the highest occurrence in the rostral part (Schellart et al. 1987). In the goldfish, 80% of the units located in the ventral torus and DLT are purely visual, whereas only a few unimodal acoustic units can be found. In the trout, Visual units, the most common type, do not show a topographic preference, but the Acousticolateral units show preference for the middle and caudal parts. The dorsal torus of the goldfish appears to be reserved for acoustic neurons (Page 1970).

Besides the rough rostrocaudal segregation in the trout, a fine structure with respect to modality has also been found (Schellart et al. 1987). Neighboring units frequently have the same modality, resulting in the occurrence of separate clusters of Visual, Acousticolateral, and, less frequently, Bimodal units. The clusters are oblong and are oriented perpendicular to the torus surface. The vertical lengths of the clusters, possibly columns or slabs, are about 0.2 and 0.4 mm for Acousticolateral and Visual units, respectively.

21.6.2 Acoustic and Lateral Line Processing in the Torus: Intermingled or Not?

In the carp, *Cyprinus*, it has been found that the medial part of the torus processes acoustical but not lateral line information (Echteler 1984, 1985). The acoustic responses of the carp torus have probably not been mixed up with lateralis responses, since the applied airborne sound was too weak (< 1 Pa) to stimulate directly, or indirectly via the swimbladder (see Sand 1984), the lateral line organs. In contrast to the carp, goldfish acoustic responses have been recorded from the medial as well as lateral torus (Page's Fig. 2, 1970). Page's acoustic responses have not been contaminated by lateral line responses either. This is concluded from experiments in which evoked potentials to sound stimuli have been recorded from animals with a destroyed peripheral lateral line system and from animals with a swimbladder collapse. The data of Echteler (1985) and Page (1970) cannot be compared well, since the applied stimuli are insufficiently defined. Generally speaking, the other studies of acousticolateral central processing in teleosts suffer from the same drawback. From the trout data, it is concluded that acoustical information is processed in an area that covers at least 70% of the width of the torus, comprising the medial and intercalated part (Schellart et al. 1987). In these experiments the water displacements of the stimulus were up to 2 μm at 100 Hz in the vicinity of the lateral line, measured along the length axis of the fish. The oscillations of the fish body were about 10 times smaller. Recordings of the trunk lateral line units of the trout yielded thresholds of 50 to 500 nm at 100 Hz (Kroese and Schellart 1987). So, there is indirect evidence that the trout B units may receive input from the lateral line system, and the same holds for the L units. The correspondence of the best frequencies of lateral line S and C units and the peaks in the distribution of band widths of torus units (Fig. 21.3) strengthen this evidence.

These data are in line with the results of a reexamination of HRP track-tracing material of the trout torus (de Wolf et al. 1983). This material shows that the superficial (700-μm), medial, and intercalated parts receive projections from the medullar nucleus octavus magnocellularis, the nucleus octavus descendens, *and* the medullar lateral line nucleus—the nucleus intermedius octavolateralis. Recently, it has been found by HRP track-tracing in *Sebastiscus* that the lateral line nuclei mainly project to the lateral part of the torus and that the medial torus receives mainly projections from the octavus nuclei (Murakami et al. 1986). These findings are in agreement with those from the carp. On the other hand, the partial overlap of both systems, which is strongest in the intercalated zone of the torus, is in line with the observation that already in the medulla of other teleosts there is more (*Anguilla*: Meredith et al. 1987) or less (*Astronotus*: Meredith 1984) overlap of the primary projections of the otolith systems and the lateral line system. The HRP study of the trout also shows that the most rostral part of the torus does not receive medullary input (Schellart unpublished data). However, this part as well as the ventral torus does receive tectal projections. The remaining dorsal part lacks direct tectal input. Moreover, the ventral torus has afferent as well as efferent connections with the dorsal torus. In catfish, electric, lateral line, and acoustic stimuli are thought to be processed in the lateral, intercalated, and medial torus, respectively (Knudsen 1977). Although the lateral line and acoustic stimuli applied in this study were physically poorly defined, there is strong neuroanatomical evidence that acoustic and lateral line information is processed in more or less distinct areas of the catfish torus (Finger 1980; Finger and Tong 1984). For the remaining studies of Table 21.1, the method of stimulation and the intensities and frequencies applied probably rule out the possibility that the acoustic responses have been mixed up with lateral line responses.

In conclusion, unquestionable lateralis responses have been recorded in the medulla. In the catfish torus, it is quite probable that pure lateral line responses have been recorded, but for the trout there exists only indirect evidence. Mainly on the basis of neuroanatomical data in *Ictalurus*, there appears to be a segregated processing of acoustical and lateral line information, but in nonelectroreceptive teleosts, there is some (cyprinids) or considerable (trout) overlap.

21.6.3 Frequency Selectivity and Topography of Acousticolateral Units

A well-developed tonotopic organization, so firmly established for the colliculus inferior of higher vertebrates, has not been demonstrated for the fish torus. The small overall range of frequency sensitivity, hardly four octaves, of the trout torus and the small fraction of well-tuned units makes it difficult to reveal tonotopy in this teleost fish. However, segregation of B and H units has been demonstrated (Schellart et al. 1987). The B units occur preferentially in the middle and caudal parts of the torus, whereas in the rostral torus, H units are encountered nearly exclusively. The presence of units with a narrow band width is, by definition, a condition for the existence of genuine tonotopy. The B units do not have the appropriate characteristics to be such elements. In addition to their broad-band nature, they generally have two or three best frequencies and often show a mixed excitatory and inhibitory behavior. The B units may, however, still form part of the frequency-detecting mechanism. Such a hypothetical mechanism, for example, is composed of converging B units with different frequency properties. In this way the narrow frequency range of the H units – about half of them have a "band width" less than half an octave (Fig. 21.2) – can be realized. An alternative possibility is selective filtering of the output of a phase-locked B unit. Whether the H units are tonotopically arranged remains to be elucidated.

The functional significance of the correlated occurrence of visual sensitivity and of narrow-band H units is not clear. In the goldfish, visuoauditory units showed good frequency selectivity with, as in the trout, best frequencies between 150 and 350 Hz (Page and Sutterlin 1970). The given hypothesis (broad-band units providing the input of narrow-band units without phase locking) to realize tonotopy deviates strongly from what is found for most quadrupeds. In these animals the labyrinth itself has a tonotopic organization, and this tonotopy is preserved in central auditory structures. However, in fish, well-developed frequency selectivity is rare in more peripheral stages of the auditory pathway (e.g., Coombs and Fay 1987; Horner et al. 1981).

21.7 Visuoacousticolateral Interaction in the Torus

The significance of multisensory integration is improvement of detection, localization (orientation), and recognition. Improvement of detectability, leading to easier perception of an object, may be of importance under near-threshold conditions. This can be reached by two different mechanisms: algebraic interaction (e.g., dendritic summation) or multiplicative interaction (e.g., presynaptic facilitation). Improvement of localization (orientation) is reached by sensory spatiotopic mapping in register. Visuoacoustic spatiotopy has been established for various higher vertebrates, and a visuolateralis spatiotopy has recently been described for the tectum of *Xenopus* (Chapter 20, Claas et al.). Such maps have not been found in the torus of fish and amphibians. In ostariophysines, however, they cannot be excluded completely, since there are indications that features of the direction of a sound source are processed in the torus (Fay et al. 1982).

Interaction between the visual and acousticolateral systems has first been reported for the goldfish (Page and Sutterlin 1970) and more recently for the trout (Schellart 1983). Because of the often rapid habituation to visual stimulation, the frequently erratic nature of the visual responses, saturation, and instationarity of response strength, interaction experiments are difficult to perform. Therefore, the underlying mechanism of the interaction, be it basically algebraic or multiplicative, is difficult to unravel by manipulating the response strength with a sufficient number of stimulus strengths of both modalities. To overcome the habituation, on-off flashing spots were applied, which moved (pseudorandomly) in the visual field of the fish eye.

In the trout, interaction has been examined for Bimodal units as well as for unimodal units, since the ineffective modality may still influence the response to the other modality. When an on-off flashing, moving spot and a tone burst are given simultaneously and in phase, the strength of the response of most Bimodal units (14 of 19 units) is approximately the sum of the two unimodally elicited responses. This algebraic mode of interaction is also found for units encountered in the deep torus in the vicinity of the fll of the goldfish (Page and Sutterlin 1970). The majority of the unimodal Visual and Acousticolateral units do not change their response by simultaneous stimulation with an acoustic (Visual units) or on-off flashing light (Acousticolateral units). However, two of the 15 Acousticolateral units showed strong enhancement of the response with an additional visual stimulus, whereas the unit was not sensitive to any visual stimulation whatsoever. Similar behavior was found for two of 31 visual units.

Bimodal units can also show a strong deviation from the algebraic mode. Four units exhibited strong enhancement, and one showed a strong attenuation. Similar results were found in the goldfish torus (Page and Sutterlin 1970) and also between the acoustic and lateral line systems in the torus (Zittlau et al. 1985) and tectum of *Xenopus* (Lowe 1987). The responses of one of these units to uni-and bimodal stimulation are presented in Fig. 21.6. The three panels show in dot display fashion the response to a stationary, on-off flashing stimulus, which evoked a weak, already habituated response (left); the response to a tone burst, which was increased in frequency each time it was presented (middle); and the response to both stimuli presented simultaneously with the on and off sets synchronous (right). The response to light shows the common large latency (here ~ 220 ms), whereas the response to sound stimulation starts within 24 ms. The right panel shows a response approximately two times stronger than the sum of both individual responses. The increase in spike density is found from the instant at which the light response is expected. The right vertical histogram lacks a frequency-independent "background" of spike firing. Moreover, the shapes of the vertical middle and right histograms are alike for frequencies above 190 Hz. In the right panel below this frequency, some new peaks have arisen. The increase of spike firing for the whole "band width" and the appearance of the new peaks can most

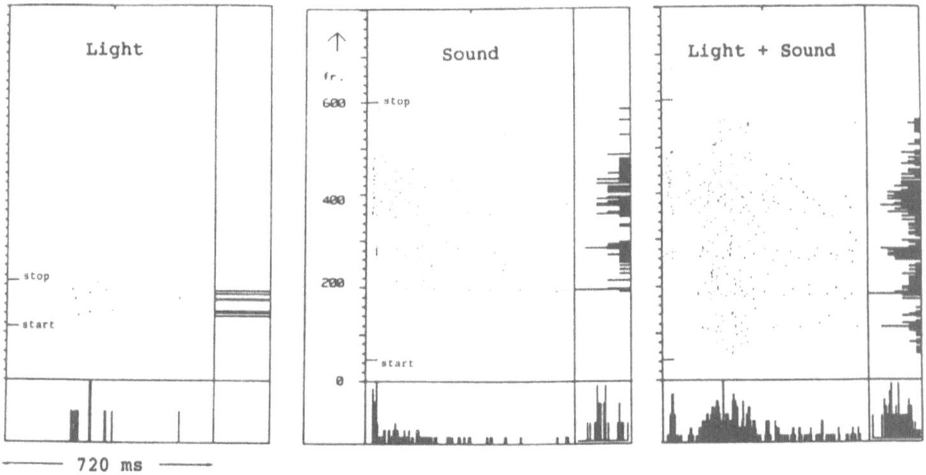

FIGURE 21.6. Visuoacousticolateral nonlinear interaction in a bimodal, nonspontaneous, non-phase-locked torus unit of the trout. The spike responses to tone bursts and a flickering spot of light are depicted in dot display fashion. The parts of the dot displays belonging to the silent and dark intervals are not shown. The subsequent stimulus periods are depicted along the ordinate. The lower small horizontal line (start) at the ordinate indicates the first stimulus period (a flickering spot of light or a tone burst alternated with silent periods). The upper small line (stop) indicates the last period presented. In the left panel only 40 periods of flicker were presented. The vertical histograms denote the number of spikes elicited during a single light-on or tone burst-on period. The horizontal histograms denote the number of spikes elicited in the subsequent time bins along the abscissa, added for all periods together. The histograms in the squares are the interspike interval histograms.

easily be understood as an enhancement of the acousticolateral sensitivity by the light stimulus – e.g., by presynaptic facilitation of this bimodal torus neuron. The right lower histogram indicates that this enhancement, which can be explained by a multiplicative type of interaction, has a very instantaneous nature. From the functional point of view this active visuoacousticolateral interaction can be interpreted as visually induced acousticolateral arousal with switchover times of the order of 0.1 s. Since the torus projects to the reticular formation, this fast arousal behavior may possibly play a role in sensorimotor control.

21.8 Concluding Remarks

As to complex processing of visual information, visually sensitive units in the torus of teleosts perform quite poorly in comparison to tectal units. This makes it improbable that the torus plays an important role in color and contrast vision and in the detection of the direction of movement. The generally occurring habituation to visual stimuli suggests that the torus is more involved in the detection of—even small—sudden changes in the visual scenery, such as a moving object.

The acousticolateral responsiveness is much more pronounced than responsiveness to visual stimuli. Acousticolateral units can be distinguished with respect to their frequency range, occurrence of phase locking, nature of the input (acoustic and/or lateral line), excitatory/inhibitory behavior of the response, sensitivity to visual stimulation, adaptation and habituation to tone bursts, and topographic preferences. The occurrence of some of these features is correlated.

Units that process both visual and acousticolateral input generally seem to act by summing algebraically the responses to individual stimulus modalities. A minority of the units show a more complex type of interaction which may be based on presynaptic facilitation.

In summary, the torus seems to be a most important center for higher-order acousticolateral information processing with an instantaneous visual alerting mechanism, which lowers the threshold for acousticolateral stimulation.

Acknowledgment. We thank Dr. R.J.A. Buwalda for comments on the manuscript. This work has been supported by the Netherlands Organization for the Advancement of Pure Research (ZWO-BION).

References

Alneas E (1973) Unit activity of ganglionic and medullary second order neurons in the eel lateral line system. Acta Physiol Scand 88:160–174.

Carr CE, Maler L (1986) Electroreception in gymnotiform fish. Central anatomy and physiology. In: Bullock TH, Heiligenberg W (eds) Electroreception. New York: Wiley, pp. 319–373.

Coombs S, Fay RR (1987) Response dynamics of saccular fibers: Effects of stimulus frequency and intensity on fibers with tuning, sensitivity and spontaneous activity. J Acoust Soc Am 81:1025–1035.

Echteler SM (1984) Connections of the auditory midbrain in a teleost fish, *Cyprinus carpio.* J Comp Neurol 230:536–551.

Echteler SM (1985) Organization of central auditory pathways in a teleost fish, *Cyprinus carpio.* J Comp Physiol 156:267–280.

Fay RR (1981) Coding of acoustic information in the eighth nerve. In: Tavolga WN, Popper AN, Fay RR (eds) Hearing and Sound Communication in Fishes. Berlin: Springer-Verlag, pp. 189–219.

Fay RR, Hillery CM, Bolan K (1982) Presentation of sound pressure and particle motion in the midbrain of goldfish. Comp Biochem Physiol 71:181–191.

Finger TE (1980) Non-olfactory sensory pathway to the telencephalon in a teleost fish. Science 210:671–673.

Finger TE, Tong S-L (1984) Central organization of eighth nerve and mechanosensory lateral line systems in the brainstem of Ictalurid catfish. J Comp Neurol 229:129–151.

Galand G, Leige B (1975) Responses visuelles unitaires chez la truite. In: Ali AM (ed) Vision of Fishes, New Approaches in Research. New York: Plenum, pp. 127–135.

Grover BG, Sharma SC (1979) Tectal projections in the goldfish (*Carassius auratus*). A degeneration study. J Comp Neurol 184:435–454.

Grover BG, Sharma SC (1981) Organization of extrinsic tectal connections in goldfish (*Carassius auratus*). J Comp Neurol 196:471–488.
Grözinger B (1967) Elektro-physiologische Untersuchungen an der Hörbahn der Schleie (*Tinca tinca* [L.]). Z Vergl Physiol 57:44–76.
Hawkins AD, Johnstone ADF (1978) The hearing of the atlantic salmon, *Salmo salar*. J Fish Biol 13:655–673.
Horner K, Sand O, Enger PS (1980) Binaural interaction in the cod. J Exp Biol 85:323–331.
Horner K, Hawkins AD, Fraser PJ (1981) Frequency characteristics of primary auditory neurons from the ear of the cod, *Gadus morhua* L. In: Tavolga WN, Popper AN, Fay RR (eds) Hearing and Sound Communication in Fishes. Berlin: Springer-Verlag, pp. 223–241.
Ito H (1974) Fine structure of the torus semicircularis of some teleosts. J Morphol 142:137–152.
Ito H, Murakami T, Fukuoka TK, Kishida R (1986) Thalamic fiber connections in a teleost (*Sebastiscus marmoratus*): Visual, somatosensory, octavolateral, and cerebellar relay region to the telencephalon. J Comp Neurol 250:215–227.
Knudsen EI (1977) Distinct auditory and lateral line nuclei in the midbrain of catfishes. J Comp Neurol 173:417–432.
Kroese ABA, Schellart NAM (1987) Evidence for velocity- and acceleration-sensitive units in the trunk lateral line of the trout. J Physiol 394:13P.
Lowe DA (1987) Single-unit study of lateral line cells in the optic tectum of *Xenopus laevis*: Evidence for bimodal lateral line/optic units. J Comp Neurol 257:396–404.
McCormick CA (1983) Organization and evolution of the octavolateral area of fishes. In: Northcutt RG, Davis RE (eds) Fish Neurobiology, Vol 1. Ann Arbor: University of Michigan Press, pp. 179–213.
Meek J, Schellart NAM (1978) A Golgi study of the goldfish optic tectum. J Comp Neurol 182:89–121.
Meredith GE (1984) Peripheral configuration and central projections of the lateral line system in *Astronotus ocellatus*(Cichlidae), a nonelectroreceptive teleost. J Comp Neurol 228:342–358.
Meredith GE, Roberts BL (1986) The relationship of saccular efferent neurons to the superior olive in the eel, *Anguilla anguilla*. Neurosci Lett 68:69–72.
Meredith GE, Roberts BL, Maslam S (1987) The distribution of afferent fibers in the brain stem from end organs in the ear and lateral line in the european eel. J Comp Neurol 265:507–520.
Münz H (1985) Single unit activity in the peripheral lateral line system of the cichlid fish *Sarotherodon niloticus* L. J Comp Physiol A 157:555–568.
Münz H, Stumpf WE, Jennes L (1981) LHRH systems in the brain of platyfish. Brain Res 221:1–13.
Murakami T, Morita J, Ito H (1983) Extrinsic and intrinsic fiber connections of the telencephalon in a teleost, *Sebastiscus marmoratus*. J Comp Neurol 216:115–131.
Murakami T, Fukuoka T, Ito H (1986) Telencephalic ascending acousticolateral system in a teleost (*Sebastiscus marmoratus*), with special reference to the fiber connections of the nucleus preglomerulosus. J Comp Neurol 247:383–397.
Nederstigt LJA, Schellart NAM (1986) Acousticolateral processing in the torus semicircularis of the trout *Salmo gairdneri*. Pflugers Arch 406:151–157.
Niida A (1973) Visual responses from ipsilateral optic tectum of crucian carp. J Fac Sci Hokkaido Univ Ser VI Zool 19:50–57.

O'Benar (1976) Electrophysiology of neural units in goldfish optic tectum. Brain Res Bull 1:529–541.

Oka Y, Satou M, Ueda K (1986) Ascending pathways from the spinal cord in the Hime salmon (landlocked red salmon, *Oncorhynchus nerka*). J Comp Neurol 245:104–112.

Page CH (1970) Electrophysiological study of auditory responses in the goldfish brain. J Neurophysiol 33:116–128.

Page CH, Sutterlin AM (1970) Visual-auditory responses in the goldfish tegmentum. J Neurophysiol 33:129–136.

Plassmann W (1985) Coding of amplitude-modulated tones in the central auditory system of catfish. Hear Res 17:209–217.

Sand O (1984) Lateral-line systems. In: Bolis L, Keynes RD, Maddrell SHP (eds) Comparative Physiology of Sensory Systems. Cambridge, U.K.: Cambridge University Press, pp. 3–32.

Schellart NAM (1983) Acoustical and visual processing and their interaction in the torus semicircularis of the trout, *Salmo gairdneri*. Neurosci Lett 42:39–44.

Schellart NAM (1989) The visual pathways and central non-tectal processing. In: Douglas RH, Djamgoz MBA (eds) The Visual System of Fish. London: Croom Helm (in press).

Schellart NAM, Munck HC de (1987) A model for directional and distance hearing in swimbladder-bearing fish based on the displacement orbits of hair cells. J Acoust Soc Am 82:822–829.

Schellart NAM, Kamermans M, Nederstigt LJA (1987) An electrophysiological study of the topographical organization of the multisensory torus semicircularis of the rainbow trout. Comp Biochem Physiol 88A:461–469.

Wolf FA de, Schellart NAM, Hoogland PV (1983) Octavolateral projections to the torus semicircularis in the trout, *Salmo gairdneri*. Neurosci Lett 38:209–213.

Zittlau KE, Claas B, Münz H, Görner P (1985) Multisensory interaction in the torus semicircularis of the clawed toad *Xenopus laevis*. Neurosci Lett 60:77–81.

22 The Efferent System

Barry L. Roberts and Gloria E. Meredith

22.1 Introduction

Our understanding of lateral line operation cannot be complete until the significance of the efferent innervation is understood. This innervation is a fundamental component of hair cell function: it is found not only in canal and superficial neuromasts of the lateral line but also in the inner ear. This efferent system, in common with the efferent innervation to the retina and muscle spindle, provides the opportunity for a dynamic control of sense organ properties, but how this is used and organized, in the case of octavolateralis efferent neurons, is unknown.

A common interpretation of efferent function is that it provides a feedback regulation of sensory activity. An alternative view is that these neurons belong to a feedforward system. This means that the activity of efferent neurons is determined not by sensory stimulation but by circuits within the central nervous system. In particular, efferent operation seems to be directly related to the motor state of the animal. We shall consider here the evidence from anatomical and physiological studies that supports this idea. Although the focus is on the lateral line, we shall, of necessity, refer to the more extensive literature that is available for the efferent supply to the ear.

22.2 Peripheral Organization of the Efferent System

An efferent innervation of octavolateralis sense organs, although widespread, is not universally present, nor is it always developed to the same degree. An efferent innervation is completely absent from the inner ear and lateral line of the hagfish *Eptatretus burgeri* (Amemiya et al. 1985; Kishida et al. 1987), the lateral line of lampreys (Yamada 1973; Fritzsch et al. 1984), and the basilar papilla of the bullfrog (Flock and Lam 1974). Efferent endings occur in restricted portions of the lagenar neuroepithelium of the perciform fish *Colisa* (Wegner 1982) and are absent from freestanding regions of the basilar papilla of several lizards (see Mulroy and Oblak 1985). There is a rich efferent supply to the head canals of the eel, *Anguilla*, whereas trunk canal organs are sparsely innervated (Yamada and

Hama 1972; Meredith and Roberts 1987). Such differences in the density of the efferent innervation are becoming increasingly evident as more and more species are studied in detail, but the significance of this variability is unclear.

Efferent endings have now been observed in electron microscopic studies of the lateral line organs of many species (see, e.g., Hama 1965, 1978; Yamada and Hama 1972; Yamada 1973). As in other portions of the octavolateralis system, in lateral line neuromasts, these endings typically form boutons, 2 to 4 μm in diameter, on the bases of hair cells and contain translucent and, sometimes, dense-core vesicles (Yamada and Hama 1972; Nakajima and Wang 1974; Sans and Highstein 1984). Efferent synapses may also occur en passant, for efferent fibers may innervate both canal organs and superficial neuromasts (Münz 1985). Efferent fibers also make contact with afferent fibers. Such efferent-afferent synapses have been seen in the inner ears of the goldfish, *Carassius auratus* (Hama 1969), and the toadfish, *Opsanus tau* (Sans and Highstein 1984), and in lateral line superficial neuromasts (Hama 1978).

The efferent endings are thought to produce acetylcholine as the neurotransmitter. Acetylcholine and its synthesizing enzyme, choline acetyltransferase, have been demonstrated biochemically in the lateral line organ (Flock and Lam 1974), and the effects of efferent nerve stimulation on hair cell activity can be blocked by acetylcholine antagonists (Russell 1971b; Flock and Russell 1973). However, acetylcholine may not be the only neurochemical involved, for a number of other neuroactive substances such as amino acids, various peptides, and γ-aminobutyric acid (GABA) have been reported to be present in efferent terminals in the mammalian cochlea (see Table 2 in Aschoff and Ostwald 1987).

The first ideas about the function of the efferent innervation came from studies on the cochlea (Galambos 1956) and lateral line (Russell 1971a), which showed spontaneous and evoked afferent activity to be reduced by efferent nerve stimulation. However, since 1980 this simple view of an inhibitory efferent system has been complicated by the finding that afferent activity of the vestibular system is sometimes unaffected, incompletely inhibited (Dieringer et al. 1977; Hartmann and Klinke 1980; Rossi et al. 1980), or even excited (Goldberg and Fernandez 1980; Rossi et al. 1980; Hartmann and Klinke 1980; Highstein and Baker 1985) by efferent nerve stimulation. In the lateral line, also, there is some evidence for an excitatory efferent action. In their study on the lateral line of the burbot, *Lota*, Flock and Russell (1973) reported, in addition to efferent-induced inhibition, an unexpected increase in afferent nerve activity when the lateral line nerve was stimulated in preparations treated with gallamine. Increases in afferent nerve activity have been observed in the lateral line of *Xenopus* to follow the application of calcitonin gene-related peptide. As this neuroactive substance is localized to small axons within neuromasts, this excitatory effect may involve the efferent system (Adams et al. 1987). Thus, it seems that efferent nerve stimulation can result either in inhibition or excitation of sensory activity (Fig. 22.1).

Electrophysiological studies on neuromasts have given further information about the peripheral action of the efferent system. The microphonic and summating potentials recorded extracellularly show changes in amplitude, shape, and

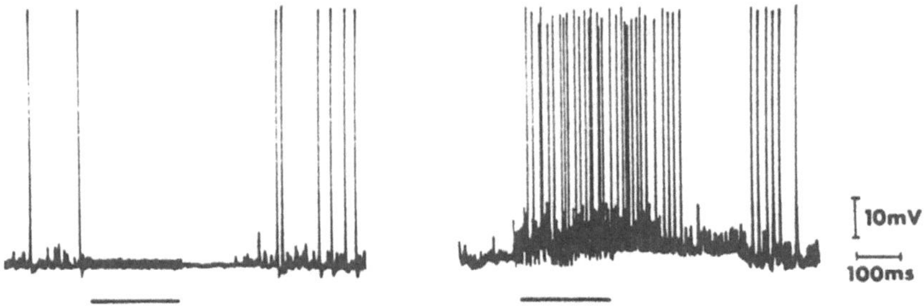

FIGURE 22.1. Efferent nerve action. The excitatory and inhibitory impact of efferent nerve stimulation seen in records from vestibular nerve fibers in the frog (from Rossi et al. 1980). A short stimulus train (200 ms long, 200 imp/sec) was applied to the vestibular nerve to activate the efferent axons. In the record of the afferent fiber on the left, the impact was inhibitory; for the fiber on the right, the activity was enhanced.

phase (Fig. 22.2B) during efferent nerve stimulation (Russell and Lowe 1983). Intracellular recordings from hair cells of the lateral line (Fig. 22.2C) (Flock and Russell 1976), sacculus (Ashmore and Russell 1982), and cochlea (Art et al. 1982) show that inhibitory postsynaptic potentials (IPSPs) are generated in hair cells, following efferent nerve stimulation. Such stimulation decreases the amplitude of excitatory postsynaptic potentials (EPSPs; Fig. 22.2A) and suppresses spontaneous miniature EPSPs (Flock and Russell 1976; Furukawa 1981) recorded from afferent fibers. These events, which are all abolished by the cholinergic blocker gallamine (Flock and Russell 1976), are the basis for the efferent inhibitory action on afferent nerve activity.

The excitatory action of efferent nerve stimulation reported for the vestibular system, however, does not seem to depend on membrane changes of hair cells. Highstein and Baker (1985) showed in the toadfish that efferent nerve stimulation produces EPSPs in vestibular fiber terminals which, if sufficiently large, evoke action potentials, thus enhancing the afferent discharge. The excitation is absent after the efferent nerves have been severed (Highstein and Baker 1985) and is unaffected by gallamine (Hartmann and Klinke 1980). An electron microscopic study (Sans and Highstein 1984) has shown that efferent-afferent fiber connections are present in the toadfish, and these may be the sites of efferent excitation (Highstein and Baker 1985).

Recent studies have shown that the mechanical properties of hair cells may also be influenced by the efferent system. Thus the stiffness of the stereocilia changes after a short tetanic shock is given to the frog saccular nerve (Ashmore 1984), and cochlear hair cells maintained in vitro expand and contract in response to current injection or iontophoretic application of acetylcholine (Brownell et al. 1985). The mechanism for the movement is unknown, although the actin and myosin present in hair cells are probably not responsible (Ashmore 1987).

As research on the efferent system at the periphery progresses, it is becoming clear that there is greater variability in synaptic location, content, and impact of

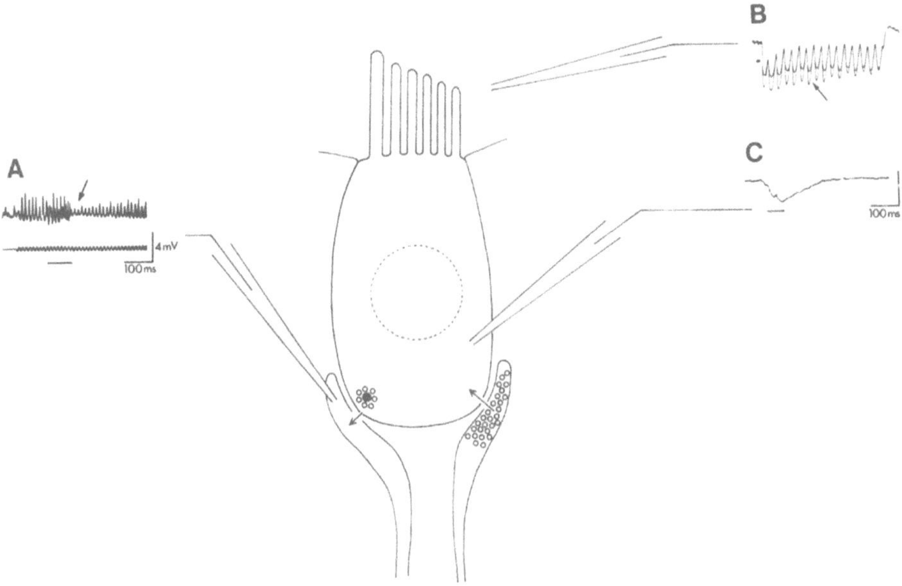

FIGURE 22.2. Efferent nerve inhibition. The diagram shows sample recordings from an afferent fiber (A); from the dorsal surface of a neuromast (B); and from within a hair cell (C) during efferent nerve stimulation. In (A), the decline in EPSP amplitude (arrow) is seen to follow efferent nerve stimulation (horizontal line). In (B), extracellular receptor potentials are shown in response to 25-Hz stimulation of the sense organ, recorded with (lower trace, arrow) and without efferent nerve stimulation. The trace in (C) shows the IPSP generated by efferent nerve stimulation (horizontal line). (A) and (C) from Flock and Russell (1976); (B) from Russell and Lowe (1983).

the efferent system than was previously thought. To give this structural and functional diversity some biological significance, an understanding is needed of how the efferent circuitry is organized and activated.

22.3 Central Organization of the Efferent System

The introduction of neuroanatomical tracers has enabled investigators during the last decade to locate the sources of the octavolateralis efferent innervation in a range of sense organs and species. In all vertebrate groups, the efferent supply originates from neurons that lie in the rhombencephalic tegmentum. In mammals, the neurons supplying the ear are separated into specific nuclei that are associated with cochlear or vestibular end organs (Warr 1975). Such distinctions have not been recognized in fishes, because a sharp separation of roles cannot be made for the octavolateralis end organs except for the semicircular canals, which have a vestibular function. The neurons that innervate the ear and lateral line of fishes and amphibians are located in a single rhombencephalic nucleus, the octavolateralis efferent nucleus (OEN; Meredith and Roberts 1986), which may, in some species, contain anatomical subdivisions (Zottoli and Van Horne 1983; Metcalfe et al. 1985; Highstein and Baker 1986). In addition, there is, in certain euteleosts, a minor lateral line supply which originates in the diencephalon (Zottoli and Van Horne 1983; Metcalfe et al. 1985). The efferent nucleus has a close association with the branchiomotor column and sometimes overlaps with the facial motor nucleus (see Meredith 1989). For this reason, we consider the efferent neurons to be a subset of the branchiomotor column that innervates neuroepithelium rather than muscle. This interpretation is reinforced by the fact that the presumed neurotransmitter, both for motoneurons and for efferent neurons, is acetylcholine and by the observation that the efferent axons follow a pathway, along with facial motoneuron axons, that is separate from octavolateral afferent fibers (Bell 1981; Pelligrini et al. 1985; Meredith and Roberts 1986, 1987; Zottoli and Danielson Chapter 23). Unlike branchiomotor neurons, however, efferent neurons in most species provide a bilateral innervation of the target organs, although this is ipsilateral in anuran and gymnophionian amphibians (Will 1982; Fritszch and Crapon de Caprona 1984).

Figure 22.3 illustrates the location of OEN in three vertebrate groups. This nucleus spans the midline in bony fishes (Claas and Münz 1980) but is situated more laterally in cartilaginous fishes (Koester 1983; Meredith and Roberts 1986) and aquatic amphibians (Will 1982; Fritszch and Crapon de Caprona 1984). In these vertebrate groups OEN cannot be readily subdivided functionally into auditory and vestibular groupings, and neurons supplying the ear do not lie apart from those innervating the lateral line. Neurons of the branchiomotor nuclei are arranged in a linear sequence that reflects the configuration of their targets (Withington-Wray et al. 1986). Such a sharp topography is not evident in the efferent nucleus, although a tendency is seen in some species for neurons that supply the head lateral line to concentrate in the rostral portion of the nucleus and for those that supply the body lateral line to lie more caudally (Highstein and Baker 1986; Meredith and Roberts 1987). This head-rostral and body-caudal organization is seen particularly clearly in anuran amphibians (Will 1982). Perhaps the absence of a sharp topography within OEN reflects the fact that some neurons innervate more than one target (Claas et al. 1981; Fritzsch and Wahnschaffe 1987). This multiple innervation has the significant functional implication that end organs of different location and modality are affected by a common efferent input.

There are few estimates of the number of efferent neurons in relation to target hair cells, but it is evident that the ratio of efferent neurons to hair cells is low in amphibians and fishes (Münz 1985; Meredith and Roberts 1986, 1987; Meredith 1989). For example, each efferent neuron supplies approximately 400 hair cells in the body lateral line of the dogfish, *Scyliorhinus*, if it is assumed that the efferent fibers supply all hair cells (Meredith and Roberts 1986). In the utriculus of the eel, *Anguilla*, one efferent neuron innervates 900 hair cells (see Meredith 1989). Such large "motor units," of course, imply a relatively crude level of control.

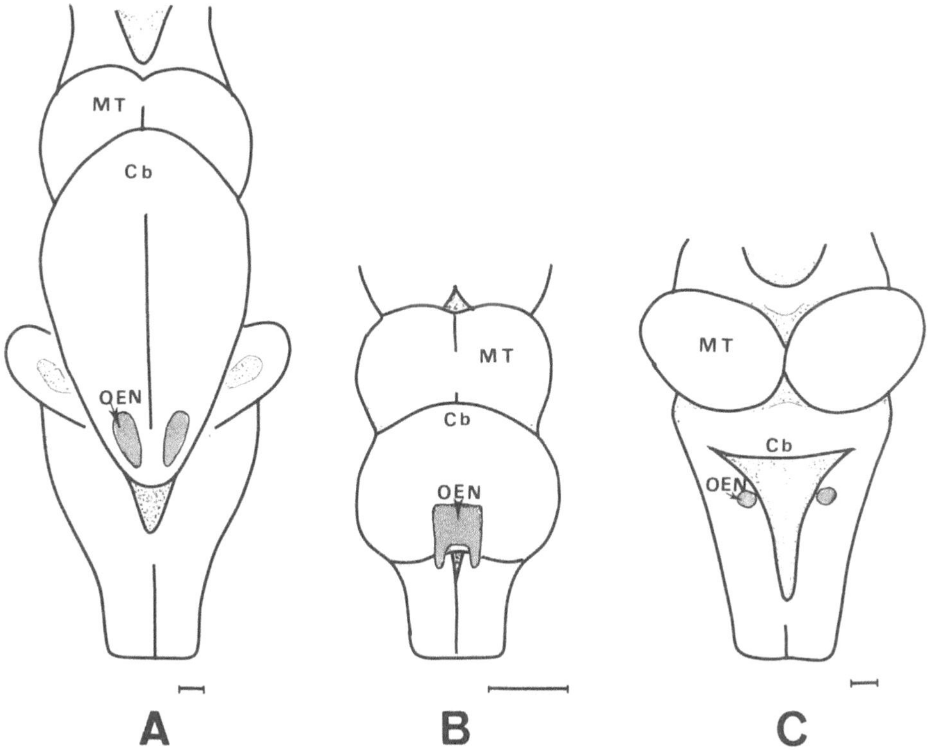

FIGURE 22.3. Efferent neuron location. The position of the octavolateralis efferent nucleus in the medulla as seen in dorsal view of the elasmobranch (A), teleost (B), and anuran (C) hindbrain. Abbreviations: Cb, cerebellum; MT, mesencephalic tectum; OEN, octavolateralis efferent nucleus. Scale bars equal 1 mm each. (A) is based on Meredith and Roberts (1986), (B) on Meredith and Roberts (1987), and (C) on Pelligrini et al (1985).

The efferent neurons of different species vary in size and form. They are small in mammals, e.g., diameter reaching 25 μm for auditory efferent neurons in the rat (Aschoff and Ostwald 1987), but are relatively large in fishes and amphibians, e.g., diameter reaching 88 μm for lateral line efferent neurons in *Scyliorhinus* (Meredith and Roberts 1986). They have extensive dendrites (Fig. 22.4) (Highstein and Baker 1986) which branch often and course through the inferior subdivision of the medial reticular formation. The dendrites in *Opsanus* extend even into the octavolateralis sensory nuclei (Highstein and Baker 1986). In cartilaginous fishes, the dendritic spread is confined to the same side of the tegmentum as the parent cell somata (Meredith and Roberts 1986), but in teleost fishes, cell bodies that lie close to the midline usually have bilateral dendritic trees (Fig. 22.4), whereas the more laterally placed somata have unilateral configurations. Indeed, in *Opsanus*, these differences in dendritic spread may form the basis for functional subdivisions within OEN (Highstein and Baker 1986). Certainly, the

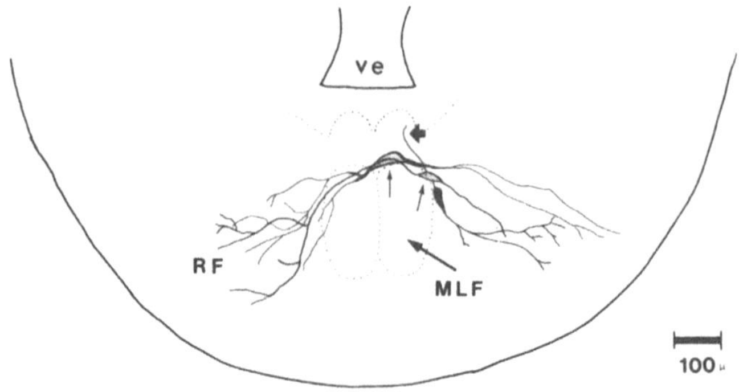

FIGURE 22.4. Efferent neuron morphology. This drawing illustrates, in a transverse section of the rhombencephalon of *Anguilla*, three efferent neurons from the octavolateralis efferent nucleus that have been labeled by applying HRP to their axons outside the brain. Note the bilateral spread of the dendrites of two neurons (small arrows) and the unilateral dendritic arrangement of a more laterally situated neuron (blackened soma). The thick arrow points to the axon of the latter neuron as it enters the axon bundle. MLF, medial longitudinal fasciculus; RF, reticular formation; ve, fourth ventricle.

subset of efferent neurons whose dendrites extend contralaterally will be influenced by bilateral sources, and this could have important implications for activation of the efferent system.

Most neurons in OEN stain for acetylcholinesterase (AChE; Meredith and Roberts 1987) and are immunoreactive for choline acetyltransferase (Zottoli and Danielson Chapter 23), thus indicating that they are cholinergic. There are neurons in OEN of the eel, however, that stain weakly or not at all for AChE. It may be that these neurons utilize another transmitter, because in mammals, where many efferent neurons are also thought to be cholinergic, others have been shown to react immunocytochemically for GABA and various peptides (see Table 2 of Aschoff and Ostwald 1987).

The morphological studies give the impression that OEN is homogeneous, at least in terms of neuronal morphology, topography, and transmitters. However, this does not mean that this nucleus functions uniformly, because different portions could receive inputs from different sources. Whether they do so, however, is unknown, because at present, there is very little information about the activation of the efferent system.

22.4 Activation of the Efferent System

The inhibitory and excitatory impact of the efferent system on afferent activity has been revealed in experiments in which efferent fibers were electrically stimulated. Efferent action in the living animal, however, may be more subtle, because the natural firing patterns of efferent neurons are probably complex.

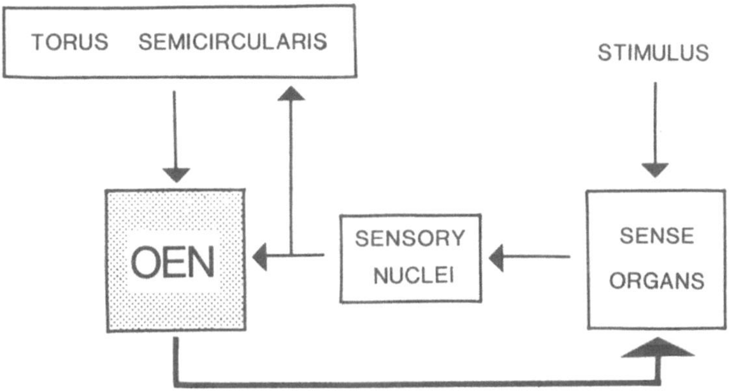

FIGURE 22.5. Efferent nerve function. A schematic diagram showing the octavolateralis efferent nucleus (OEN) as a component of a feedback circuit controlling sense organ properties. Input from sense organs, via sensory nuclei (nucleus intermedius for lateral line, the octavus nuclei for ear), may activate OEN which feeds back to the sense organs. A feedback pathway may also be provided via the torus semicircularis.

An obvious view of efferent function is that it provides a feedback inhibitory regulation of hair cell sensitivity; in this circuit (Fig. 22.5), strong stimulation of the hair cells will provoke efferent activity and so reduce hair cell output. There have been several reports that are consistent with this hypothesis. Thus, efferent neurons supplying the trunk lateral line of the eel become active when the lateral line is stimulated (Hashimoto et al. 1970), and vestibular efferent neurons are driven by vestibular stimulation (Schmidt 1963). The problem with these observations is that the effective stimulus to the efferent system is not sharply defined. The stimuli used to activate the lateral line (direct touch or water jets) may also stimulate other sense organs, and vestibular stimulation will, of course, evoke compensatory body movements and activate motor systems.

In a direct attempt to test the feedback hypothesis, Roberts and Russell (1972) measured efferent activity in response to various stimuli. They found that natural stimulation of the lateral line of the dogfish, *Scyliorhinus*, did not activate the efferent neurons, whereas other stimuli, such as touch, were effective. In contrast to this consistent finding with natural stimulation, synchronous electrical stimulation of the lateral line afferent fibers has been shown to evoke a "reflex" discharge from lateral line efferent neurons (Roberts and Russell 1972; Paul and Roberts 1977), a result that suggests that there is coupling between the afferent and efferent systems. This coupling is unlikely to be monosynaptic, not only because of the long latency of the responses but also because anatomical tracing studies indicate that primary lateral line fibers project well away from the efferent nucleus (Meredith and Roberts 1986; Meredith et al. 1987). However, a bisynaptic pathway may be present, for the axons of output neurons in the primary lateralis region (i.e., nucleus intermedius) course through the OEN region as they project to the contralateral lateralis zone or ascend to the torus semicircularis (unpublished observations). Ultrastructural studies are required to establish whether, indeed, they contact the efferent neurons.

Another source of sensory driving for OEN could be the torus semicircularis, an important sensory integrating center in the midbrain, which receives lateral line, auditory, and visual information (Schellart and Kroese Chapter 21). Little is known of its outputs, but it has been reported as projecting to the region of OEN in amphibians (Jacoby and Rubinson 1984; Will Chapter 18).

It is clear from several studies that stimulation of sensory receptors other than those involved in the ear and lateral line can activate the efferent system. Tactile stimulation of the head and body, for example, was found to be particularly effective in *Scyliorhinus* (Roberts and Russell 1972), and in some species visual stimuli evoke efferent discharges (Späth and Schweickert 1975; Hartmann and Klinke 1980; Highstein and Baker 1985; Münz and Claas 1986). Although these sensory modalities may directly drive OEN neurons, it is more likely that central circuits, integrating a variety of sensory modalities, are involved, because effective stimuli are known to be "arousing" and to provoke movement (Roberts and Russell 1972).

A correlation between efferent activity and the motor state of an animal was first reported for the vestibular system (Schmidt 1963; Görner 1967; Klinke and Schmidt 1970). Then Russell (1971a) showed, by recording simultaneously from efferent fibers and motor nerves in *Xenopus*, that the stimuli that drive the lateral line efferent system also cause body movement. This correlation was examined further by Roberts and Russell (1972), who recorded strong activity from efferent axons in the dogfish *Scyliorhinus* when the white musculature used for rapid locomotion was active. They also showed that there is an accompanying weak efferent output when the dogfish swims steadily using the tonic red muscle system. As an efferent output was not seen when the body was bent passively in a manner that would drive body proprioceptors, and so spinal cord circuits (Roberts and Williamson, 1983), it must arise from within brainstem motor circuits. Moreover, the correlation with motor systems extends to nonlocomotory movements, for efferent discharges have been seen in association with fin movements in *Scyliorhinus* (Roberts and Russell 1972), eye movements in *Opsanus* (Highstein and Baker 1985), and respiratory movements in *Scyliorhinus* (Paul and Roberts 1977) and *Xenopus* (Art and Kroese 1982). It seems that any movement of the animal involves a simultaneous activation of the octavolateralis efferent system. The efferent neurons can be thought of, therefore, as a component within some general activating circuit (Fig. 22.6), with the function of providing feed-forward control of sense organ sensitivity.

To fill in the anatomical features of the diagram in Fig. 22.6 requires a far greater understanding of OEN connections than we now have as well as knowledge of how locomotion is initiated, executed, and controlled. In the case of *Scyliorhinus*, we do know that the basic features of the locomotory program are produced within the spinal cord under the control of systems descending from the brainstem (Roberts and Williamson 1983); these systems (e.g., reticulospinal and

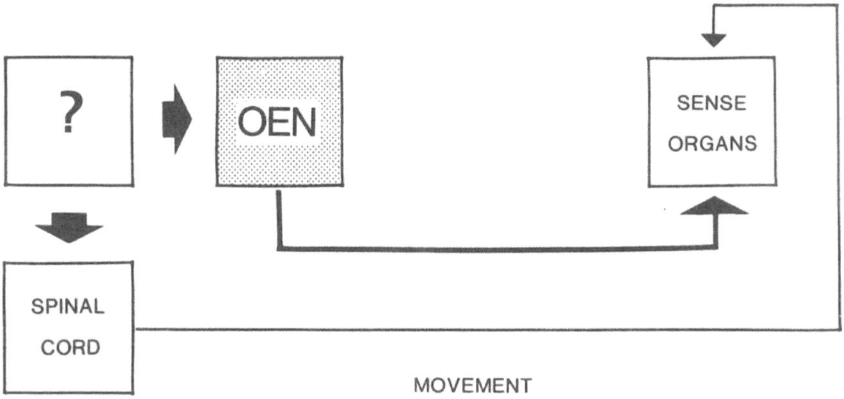

FIGURE 22.6. Efferent nerve function. A schematic diagram showing the octavolateralis efferent nucleus (OEN) as a component of a feed-forward circuit that controls sense organ properties. Postulated, and as yet unknown, centers (?) in the brain, involved in arousal and movement control, activate OEN and locomotor centers. The resulting movement will stimulate the sense organs.

vestibulospinal neurons) may also drive efferent neurons. Electrical stimulation of the caudal spinal cord in *Xenopus* (Russell 1971a) drives the efferent neurons synchronously, which suggests that neurons in OEN connect with either ascending spinal pathways or collaterals of descending systems. The latter interpretation was favored by Russell (1974, 1976), because he found, in the goldfish, a good correlation between efferent activity and the startle response. He showed that current injected intracellularly into the Mauthner cell (and also into reticulospinal and vestibulospinal neurons) would drive the efferent system. Moreover, a close anatomical relationship between efferent neurons and the Mauthner cell has been reported (Zottoli and Van Horne 1983; Metcalfe et al. 1985; Zottoli and Danielson Chapter 23). If, as we suggest, OEN is a subset of the lateral motor column of the brainstem, then it is possible that the relay neurons, identified by Hackett and Faber (1983) as being interposed between the Mauthner cell and cranial motoneurons, could also form a connection between brainstem–spinal cord pathways and the efferent system (see Zottoli and Danielson Chapter 23).

22.5 Functional Significance of the Efferent System

The association of the efferent neurons with brainstem motor systems and the coupling of efferent activity to locomotion raise questions about the significance of this relationship. During locomotion, lateral line sense organs are stimulated by body movements (Roberts 1972), yet the information that could be derived from this is unnecessary, as it is available from other sources. The strong stimulation that would result from fast, large body movements would be expected to deplete the excitatory neurotransmitter produced by the hair cell and so desensitize the afferent synapse (Roberts 1978). One role, therefore, for the efferent output would be to reduce hair cell sensitivity and to protect hair cells from such overstimulation. This idea satisfactorily explains the extensive spread of the innervation, the peripheral location of the efferent synapse, its inhibitory action, and the coupling with locomotion, and it is supported by the observation that lateral line sensitivity is relatively lower in moving than in stationary dogfish (Russell and Roberts 1974). However, this protective efferent task may not be relevant for the labyrinth, and even for the lateral line there are some features that do not fit this simple theory. Thus, some hair cells lack an efferent supply; the inhibitory impact is not maintained with successive stimulation; and the low discharge rate recorded from efferent fibers during steady locomotion would not exert a strong inhibition (see Russell and Roberts 1972). The excitatory effect of efferent nerve stimulation is also at variance with the idea that the efferent action is protective. Finally, the efferent fibers that supply the cochlea are sharply tuned (Robertson and Gummer 1985) and, when stimulated, change the tuning of cochlear hair cells (Art and Fettiplace 1984), which implies that the efferent system may have an active role in auditory processing. So although the efferent innervation may indeed serve to protect hair cells from overstimulation, it evidently has a wider role.

22.6 Summary and Conclusions

This survey has shown the efferent innervation of the lateral line to have important features in common with that supplying the sense organs of the ear. At the sense organ, efferent nerve fibers make synaptic connection with hair cells and afferent fibers. When activated, the efferent system inhibits and/or excites sensory nerve activity. The efferent fibers arise from neurons that lie in a single nucleus in the rhombencephalic tegmentum. The homogeneous composition of this nucleus, the small number of neurons it contains compared to the number of target hair cells, the bilateral innervation of sense organs that may be of more than one modality, are universal features of the central organization of the efferent system.

The efferent system is tied closely to brainstem mechanisms that are involved in motor control. Although the efferent system may be a component of a feedback loop, driven by sensory stimulation, it also belongs to a feed-forward circuit that controls sense organ sensitivity. The general significance of such a circuit may be to update continually sense organ performance in anticipation of changes in internal responsiveness or in external stimulus conditions that arise from an animal's actions. If this is so, the efferent system would be expected to be driven by those brain centers that determine "arousal." Such a conclusion has important implications for the way we think about the initiation and control of behavior. For although there is no doubt that behavior is influenced by sensory stimulation, we now have to recognize that sensory responsiveness will, in turn, be determined by the behavioral state, via the action of the efferent system.

References

Adams JC, Mroz EA, Sewell WF (1987) A possible neurotransmitter role for CGRP in a hair-cell sensory organ. Brain Res 419:347–351.

Amemiya F, Kishida R, Goris RC, Onishi H, Kusunoki T (1985) Primary vestibular projections in the hagfish, *Eptatretus burgeri*. Brain Res 337:73–78.

Art JJ, Fettiplace R (1984) Efferent desensitization of auditory nerve fiber responses in the cochlea of the turtle, *Pseudemys scripta elegans*. J Physiol 356:507–523.

Art JJ, Kroese ABA (1982) Effects of efferent activity during respiration on *Xenopus laevis* lateral-line afferent responses. J Physiol 332:21–22.

Art JJ, Crawford AC, Fettiplace R, Fuchs PA (1982) Efferent regulation of hair cells in the turtle cochlea. Proc R Soc Lond B 216:377–384.

Aschoff A, Ostwald J (1987) Different origins of cochlear efferents in some bat species, rats, and guinea pigs. J Comp Neurol 264:56–72.

Ashmore JF (1984) The stiffness of the sensory hair bundle of frog saccular hair cells. J Physiol 350:20P.

Ashmore JF (1987) A fast motile response in guinea-pig outer hair cells: The cellular basis of the cochlear amplifier. J Physiol 388:323–349.

Ashmore JF, Russell IJ (1982) Effect of efferent nerve stimulation on hair cells of the frog sacculus. J Physiol (Lond) 329:25–26P.

Bell CC (1981) Central distribution of octavolateral afferents and efferents in a teleost (Mormyridae). J Comp Neurol 195:391–414.

Brownell WE, Bader CR, Bertrand D, De Ribaupierre Y (1985) Evoked mechanical responses of isolated cochlear outer hair cells. Science 227:194–196.

Claas B, Münz H (1980) Bony fish lateral line éfferent neurons identified by retrograde axonal transport of horseradish peroxidase (HRP). Brain Res 193:249–253.

Claas B, Fritzsch B, Münz H (1981) Common efferents to lateral line and labyrinthine systems in aquatic vertebrates. Neurosci Lett 27:231–235.

Dieringer N, Blanks RHI, Precht W (1977) Cat efferent vestibular system: Weak suppression of primary afferent activity. Neurosci Lett 5:285–290.

Flock Å, Lam DMK (1974) Neurotransmitter synthesis in inner ear and lateral line sense organs. Nature 249:142–144.

Flock Å, Russell IJ (1973) The postsynaptic action of efferent fibres in the lateral line organ of the burbot *Lota lota*. J Physiol 235:591–605.

Flock Å, Russell IJ (1976) Inhibition by efferent nerve fibres: Action on hair cells and afferent synaptic transmission in the lateral line canal organ of the burbot, *Lota lota*. J Physiol 257:45–62.

Fritzsch B, Crapon de Caprona D (1984) The origin of centrifugal inner ear fibers of gymnophions (Amphibia): A horseradish peroxidase study. Neurosci Lett 46:131–136.

Fritzsch B, Wahnschaffe V (1987) Electron microscopical evidence for common inner ear and lateral line efferents in urodeles. Neurosci Lett 81:48–52.

Fritzsch B, Crapon de Caprona M-D, Wächtler K, Körtje K-H (1984) Neuroanatomical evidence for electroreception in lampreys. Z Naturforsch 39C:856–858.

Furukawa T (1981) Effects of efferent stimulation on the saccule of goldfish. J Physiol 315:203–215.

Galambos R (1956) Suppression of auditory nerve activity by stimulation of efferent fibres to cochlea. J Neurophysiol 19:424–437.

Goldberg JM, Fernández C (1980) Efferent vestibular system in the squirrel monkey: Anatomical location and influence on afferent activity. J Neurophysiol 43:986–1025.

Görner P (1967) Independence of afferent activity from efferent activity in the lateral line organ of *Xenopus laevis* (Daudin). In: Cohen PH (ed) Lateral Line Detectors. Bloomington: Indiana University Press, pp. 199–214.
Hackett JT, Faber DS (1983) Mauthner axon networks mediating supraspinal components of the startle response in the goldfish. Neuroscience 8:317–331.
Hama K (1965) Some observations on the fine structure of the lateral line organ of the Japanese sea eel *Rhynozymba nystromi*. J Cell Biol 24:193–210.
Hama K (1969) A study of the fine structure of the saccular macula of the goldfish. Z Zellforsch 94:155–171.
Hama K (1978) A study of the fine structure of the pit organ of the common Japanese sea eel *Conger myriaster*. Cell Tissue Res 189:375–388.
Hartmann R, Klinke R (1980) Efferent activity in the goldfish vestibular nerve and its influence on afferent activity. Pflügers Arch 388:123–128.
Hashimoto T, Katsuki Y, Yanagisawa K (1970) Efferent system of lateral-line organ of fish. Comp Biochem Physiol 33:405–421.
Highstein SM, Baker R (1985) Action of the efferent vestibular system on primary afferents in the toadfish, *Opsanus tau*. J Neurophysiol 54:370–384.
Highstein SM, Baker R (1986) Organization of the efferent vestibular nuclei and nerves of the toadfish, *Opsanus tau*. J Comp Neurol 243:309–325.
Jacoby J, Rubinson K (1984) Efferent projections of the torus semicircularis to the medulla of the tadpole, *Rana catesbeiana*. Brain Res 292:378–381.
Kimmel CB, Metcalfe WK, Schabtach ET (1986) Reticular interneurons: A class of serially repeating cells in the zebrafish hindbrain. J Comp Neurol 233:365–376.
Kishida R, Goris RC, Nishizawa H, Koyama H, Kadota T, Amemiya F (1987) Primary neurons of the lateral line nerves and their central projections in hagfishes. J Comp Neurol 264:303–310.
Klinke R, Schmidt CL (1970) Efferent influence on the vestibular organ during active movements of the body. Pflügers Arch 318:325–332.
Koester DM (1983) Central projections of the octavolateralis nerves of the clearnose skate *Raja eglanteria*. J Comp Neurol 221:199–215.
Meredith GE (1989) Comparative view of the central organization of afferent and efferent circuitry for the inner ear. Acta Biologica Hung (in press).
Meredith GE, Roberts BL (1986) Central organization of the efferent supply to the labyrinthine and lateral line receptors of the dogfish. Neuroscience 17:225–233.
Meredith GE, Roberts BL (1987) Distribution and morphological characteristics of efferent neurons innervating end organs in the ear and lateral line of the European eel. J Comp Neurol 265:494–506.
Meredith GE, Roberts BL, Suharti Maslam (1987) Distribution of afferent fibers in the brainstem from end organs in the ear and lateral line in the European eel. J Comp Neurol 265:507–520.
Metcalfe WK, Kimmel CB, Schabtach E (1985) Anatomy of the posterior lateral line system in young larvae of the zebrafish. J Comp Neurol 233:377–389.
Mulroy MJ, Oblak TG (1985) Cochlear nerve of the alligator lizard. J Comp Neurol 233:463–472.
Münz H (1985) Single unit activity in the peripheral lateral line system of the cichlid fish *Sarotherodon niloticus* L. J Comp Physiol A 157:555–568.
Münz H, Claas B (1986) Activity of efferent neurons in the lateral line system. Neurosci Lett (Suppl) 26:S375.
Nakajima Y, Wang DW (1974) Morphology of afferent and efferent synapses in hearing organ of the goldfish. J Comp Neurol 156:403–416.

Paul D, Roberts BL (1977) The location and properties of the efferent neurons of the head lateral-line organs of dogfish. J Comp Physiol 116:177–127.
Pelligrini M, Ceccotti F, Magherini P (1985) The efferent vestibular neurons in the toad (*Bufo bufo* L.): Their location and morphology. A horseradish peroxidase study. Brain Res 344:1–8.
Roberts BL (1972) Activity of lateral-line organs in swimming dogfish. J Exp Biol 56:105–118.
Roberts BL (1978) Mechanoreceptors and the behaviour of elasmobranch fishes with special reference to the acoustico-lateralis system. In: Hodgson ES, Matthewson RF (eds) Sensory Biology of Sharks, Skates and Rays. Washington: U.S. Government Printing Office, pp. 331–390.
Roberts BL, Russell IJ (1972) The activity of lateral-line efferent neurones in stationary and swimming dogfish. J Exp Biol 57:435–448.
Roberts BL, Williamson RM (1983) Motor pattern formation in the dogfish spinal cord. In: Roberts A, Roberts B (eds) Neural Origin of Rhythmic Movements. London: Cambridge University Press, pp. 331–350.
Robertson D, Gummer M (1985) Physiological and morphological characterization of efferent neurones in the guinea pig cochlea. Hear Res 20:63–77.
Rossi ML, Prigioni I, Valli P, Casella C (1980) Activation of the efferent system in the isolated frog labyrinth: Effects on the afferent epsps and spike discharge recorded from single fibres of the posterior nerve. Brain Res 185:125–137.
Russell IJ (1971a) The role of the lateral-line efferent system in *Xenopus laevis*. J Exp Biol 54:621–641.
Russell IJ (1971b) The pharmacology of efferent synapses in the lateral line system of *Xenopus laevis*. J Exp Biol 54:643–658.
Russell IJ (1974) Central and peripheral inhibition of lateral line input during the startle response in goldfish. Brain Res 80:517–522.
Russell IJ (1976) Central inhibition of lateral line input in the medulla of the goldfish by neurones which control active body movements. J Comp Physiol (A) 111:335–358.
Russell IJ, Lowe DA (1983) The effect of efferent stimulation on the phase and amplitude of extracellular receptor potentials in the lateral line system of the perch (*Perca fluviatilis*). J Exp Biol 102:223–238.
Russell IJ, Roberts BL (1972) Inhibition of spontaneous lateral-line activity by efferent nerve stimulation. J Exp Biol 57:77–82.
Russell IJ, Roberts BL (1974) Active reduction of lateral-line sensitivity in swimming dogfish. J Comp Physiol 94:7–15.
Sans A, Highstein SM (1984) New ultrastructural features in the vestibular labyrinth of the toadfish, *Opsanus tau*. Brain Res 308:191–195.
Schmidt RS (1963) Frog labyrinthine efferent impulses. Acta Otolaryngol 56:51–64.
Späth M, Schweickert W (1975) Lateral-line efferents to mechanical and visual stimuli. Naturwissenschaften 62:S579–S580.
Warr BW (1975) Olivocochlear and vestibular efferent neurons of the feline brainstem: Their location, morphology and number determined by retrograde axonal transport and acetylcholinesterase histochemistry. J Comp Neurol 161:159–182.
Wegner N (1982) A qualitative and quantitative study of a sensory epithelium in the inner ear of a fish (*Colisa labiosa*; Anabantidae), Act Zool (Stockh) 63:133–146.
Will U (1982) Efferent neurons of the lateral-line system and the VIII cranial nerve in the brainstem of anurans. Cell Tissue Res 225:673–685.

Withington-Wray DJ, Roberts BL, Taylor EW (1986) The topographical organization of the vagal motor column in the elasmobranch fish, *Scyliorhinus canicula* L. J Comp Neurol 248:95–104.
Yamada Y (1973) Fine structure of the ordinary lateral line organ. I. The neuromast of lamprey, *Entosphenus japonicus*. J Ultrastruct Res 43:1–17.
Yamada Y, Hama K (1972) Fine structure of the lateral-line organ of the common eel, *Anguilla japonica*. Z Zellforsch 124:454–464.
Zottoli S, Van Horne C (1983) Posterior lateral line afferent and efferent pathways within the central nervous system of the goldfish with special reference to the Mauthner cell. J Comp Neurol 219:100–111.

23
Lateral Line Afferent and Efferent Systems of the Goldfish with Special Reference to the Mauthner Cell

STEVEN J. ZOTTOLI and PAUL D. DANIELSON

23.1 Introduction

The Mauthner cells (M cells) are a pair of neurons found in the hindbrain of most fishes and amphibians (Stefanelli 1951; Zottoli 1978; Will 1986). These exceptionally large cells (up to 100 μm in cross-sectional diameter) are identifiable by their distinctive morphology, their position within the medulla oblongata, and their electrophysiological response characteristics. All of these features combine to make M cells an easily accessible "model system" for studying a number of synaptic phenomena, including electrotonic transmission (Furshpan 1964; Faber et al. 1980), synaptic integration (Bartelmez 1915; Bodian 1937; Chang et al. 1987; Faber and Korn 1978; Korn and Faber 1975a; Lin et al. 1983; Zottoli and Faber 1979; Zottoli et al. 1987a), and transmitter release mechanisms (Faber and Korn 1982; Korn et al. 1982, 1984, 1986; Korn and Faber 1987).

The M cell receives input from a variety of sensory sources such as the auditory, vestibular, visual, and lateral line systems. Axons from some of these systems are known to synapse onto restricted portions of the soma and major dendrites. For example, large club endings from saccular afferents synapse on the distal portion of the lateral dendrite (Bartelmez 1915; Bodian 1937; Lin et al. 1983). This monosynaptic sensory pathway from the sacculus to the M cell is thought to subserve the auditory-evoked startle response (Furshpan 1964; Furukawa 1966; Zottoli 1977). In general, while some input systems are clearly capable of exciting the M cell beyond threshold (e.g., auditory and visual), others (e.g., posterior lateral line) may rather contribute to the general excitatory state of that cell.

Electrical stimulation of the posterior lateral line nerve (NPLL) afferents results in a mixture of excitatory and inhibitory PSPs recorded in the M cell (Korn and Faber 1975a). The inhibitory components appear to be mediated in part by an identifiable subclass of interneurons (PHP neurons; Korn and Faber 1975b) which synapse on the M cell axon hillock and adjacent areas of the soma as unmyelinated club endings and small vesicle boutons (Nakajima 1974; Nakajima and Kohno 1978; Triller and Korn 1982; Tuttle et al. 1987). The PHP neurons mediate a two-component (electrical and chemical) inhibition (Korn and Faber 1976).

EPSPs evoked by electrical stimulation of NPLL are subthreshold unless the inhibitory components are blocked (i.e., the application of strychnine; Korn and Faber 1975a). However, there is no direct evidence that M cell action potentials can be elicited by "natural" lateral line stimuli, such as local water currents (however, see Blaxter and Fuiman Chapter 24). For example, a jet of water applied to the body PLL organs and free neuromasts of restrained goldfish (*Carassius auratus*) results in subthreshold depolarization of the M cell (Korn et al. 1974).

Excitation of the M cell above threshold activates an inhibitory network which feeds back onto both the M cell and auditory (Lin and Faber 1988) and lateral line afferents, the latter through lateral line efferent neurons (Russell 1974). Thus both afferent and efferent components of the posterior lateralis system (only one study has been conducted on the relationship of the anterior lateralis system to the M cell – Russell 1974) may modulate the level of M cell excitation (i.e., control the membrane potential relative to threshold).

The M cell initiates the first stage of a rapid startle response which is thought to be involved in escape from predators (Eaton and Hackett 1984; Eaton et al. 1988; DiDomenico et al. 1988, Nissanov and Eaton 1988). The relative importance of any one sensory input in activating the M cell during the attack of a predator probably depends on a number of factors, including (1) the major sensory systems used by the animal in detecting predators and the patterns of their excitatory and inhibitory connections with the M cell; (2) whether the predator is approaching from the air or the water; (3) the position of the prey relative to the predator in its environment (e.g., depth in the water, stationary vs. moving, and proximity to potential obstacles); and (4) the "excitatory state" of the M cell.

The M cell is one of the few vertebrate neurons that are morphologically and physiologically identifiable. This identifiability has allowed studies on the relationship of both synaptic inputs and outputs to a specific behavior, the startle response. This review focuses on the lateral line system, whose role in regulating M cell excitability is not yet fully understood. It is our hope to stimulate further experiments which will help clarify the role of the lateralis system in the startle response of fish.

23.2 Afferent Input from the Posterior Lateral Line

Information about the localization of NPLL input onto the M cell can be obtained by electrically stimulating the nerve and measuring the postsynaptic potentials that are generated along the cell's dendrites. M cell dendrites do not normally support action potentials (Furshpan and Furukawa 1962; Zottoli et al. 1987a; Chang et al. 1987), and maximal voltages of PSPs should be recorded closest to the source of current (i.e., the synaptic location of afferent fibers; Chang et al. 1987; Furshpan 1964; Korn and Faber 1975a; Zottoli and Faber 1979; Zottoli et al. 1987a). EPSPs (unmasked by the application of strychnine; Fig. 23.1A) evoked by stimulation of either NPLL were maximal in the soma and proximal dendritic portions of the M cell (Fig. 23.1B), and subthreshold EPSPs added to other excitatory input (Fig. 23.1C,D).

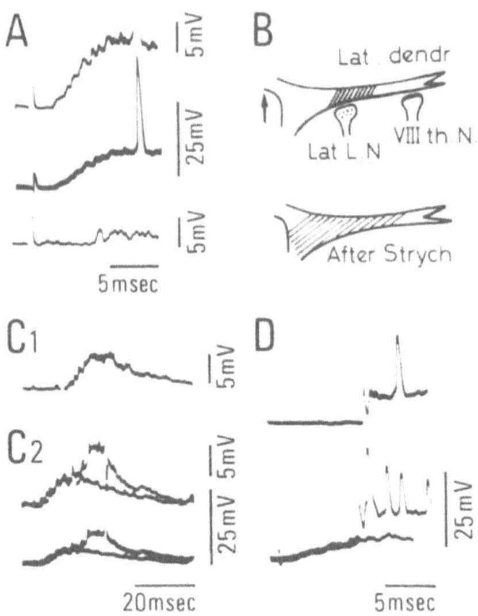

FIGURE 23.1. Excitatory NPLL inputs to the goldfish M-cell. All recordings were made 60 min after the injection of strychnine. (A) An EPSP evoked by contralateral NPLL electrical stimulation and recorded in the M cell lateral dendrite 100 μm away from the soma. Note the late firing of the M cell. High-gain ac (upper trace), low-gain dc (middle trace), and the corresponding extracellular field (lower trace). (B) Distribution of NPLL input determined electrophysiologically before (upper drawing) and after the application of strychnine. The largest EPSPs were restricted to the proximal portion of the lateral dendrite (hatched areas), just medial to terminal region of club endings before strychnine was applied. After strychnine application, the area over which maximal EPSPs could be recorded broadened to include the soma. (C, D) Facilitatory properties of PLNN input. (C_1) Subthreshold EPSP evoked by electrical stimulation of the ipsilateral NPLL. (C_2) When the stimulus in C_1 was paired with a conditioning stimulus of the contralateral NPLL, which also generated a subthreshold EPSP, summated resulted in the generation of a spike. Traces were superimposed with and without stimulation of the ipsilateral NPLL. Upper and lower traces, high-gain ac, and low-gain dc, respectively. (D) Stimulation of the eighth nerve root results in an EPSP that brings the M cell to threshold (upper trace). Pairing this stimulation with a conditioning stimulus of the contralateral NPLL results in a shortening of the latency of the first spike and repetitive discharge of the M cell. Superimposed recordings with and without eighth nerve stimulation. A, C_1, C_2, and D are from different experiments. (From Faber and Korn 1978. Used with permission from Raven Press, New York.)

The initial phase of some EPSPs had latencies consistent with monosynaptic transmission (range 0.8 to 3.1 ms for ipsi- and contralateral NPLL stimulation—Korn et al. 1974; Korn and Faber 1975a; Faber and Korn 1975). Results from these types of studies have suggested that some excitatory NPLL inputs synapse directly on the soma and proximal lateral dendrite of the goldfish M cell. This hypothesis was tested by applying horseradish peroxidase (HRP) onto the NPLL of goldfish and analyzing the projection patterns of lateral line afferents to the

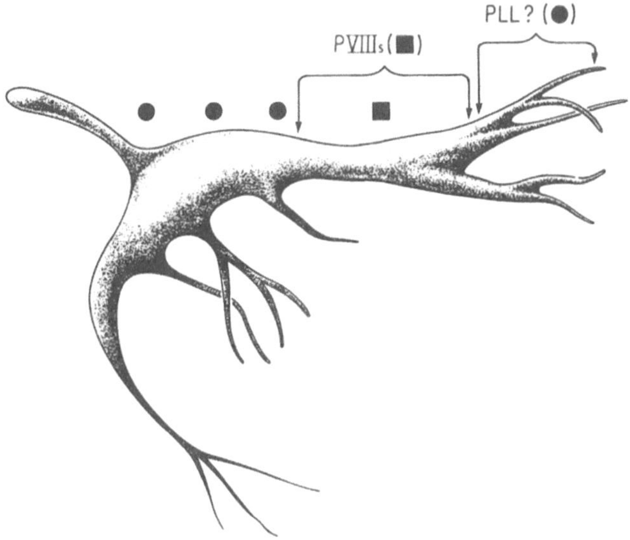

FIGURE 23.2. Termination zones of NPLL and saccular eighth (PVIIIs) nerve afferents onto a right goldfish M cell, as determined by physiological and histological techniques. The zone of morphological termination is delineated by brackets, and localization of the largest EPSP elicited by stimulation of the sensory nerve is designated by symbols. HRP application to the cut end of the saccular branch of the eighth nerve results in a distinct staining of the large club endings extending from 225 to 475 μm out on the lateral dendrite (the distance is taken relative to the axon hillock (Lin et al. 1983). The highest concentration of these terminals is 350 μm lateral, a distance at which the maximal EPSP has been recorded (Furshpan 1964). This correspondence between morphological and physiological results does not hold for NPLL. Specifically, the largest EPSPs are distributed over the soma and proximal lateral dendrite (Korn and Faber 1975a), and HRP application to the cut NPLL afferent fibers can be followed close to the distal portion of the M cell lateral dendrite. It is not known whether these fibers terminate monosynaptically on the dendrite (?).

M cell (Zottoli and Van Horne 1983). Surprisingly, no afferent projections were localized to the soma or the proximal lateral dendrite of the M cell. The discrepancy between morphological and physiological results is presented schematically in Fig. 23.2.

The excitatory PLL input to the goldfish M cell soma and proximal lateral dendrite is probably di- or polysynaptic in contradiction to the electrophysiological evidence of an initial monosynaptic component. If monosynaptic input exists, it is confined to the dorsal portion of the distal lateral dendrite, as has been hypothesized to be the case in the zebrafish (*Brachydanio rerio*; Metcalfe et al. 1985) and axolotl (*Ambystoma maculatum*; Kimmel et al. 1977). Perhaps the multineuronal pathway responsible for the initial portion of the NPLL-evoked EPSPs contains an electronic junction that would be difficult to discern from latency measurements (see Zottoli and Faber 1980 for another example of the problems associated with predicting morphological pathways from physiological latency measurements).

23.3 PLL Efferent Somata and Their Axonal Projections

23.3.1 PLL Efferent Somata Location and Morphology

PLL efferent somata have been found in the medulla oblongata of all teleost fishes studied (see Roberts and Meredith Chapter 22). Although there are species-specific differences in their rostrocaudal position and the arrangement of the cells within the medulla, the somata generally lie near the fasciculus longitudinalis medialis (flm) and are bordered caudally by the ninth cranial motor nucleus and rostrally by the entry of the eighth nerve (cichlid, *Sarotherodon niloticus*–Claas and Münz 1980; goldfish–Zottoli and Van Horne 1983; toadfish, *Opsanus tau*–Highstein and Baker 1986; however, see Bell 1981 for differences in efferent somata position in mormyrids relative to these teleosts). In the goldfish and larval zebrafish, the medullary efferent somata are divided into rostral (ren, goldfish; ROLE, zebrafish) and more caudal (cen, goldfish; RELL, zebrafish) nuclei. In addition to the hindbrain efferent somata, diencephalic efferent somata have been described in larval zebrafish (Metcalfe and Kimmel 1982; Metcalfe et al. 1985) and adult goldfish (Zottoli and Van Horne 1983). The arrangement of these three nuclei in the larval zebrafish is presented in Fig. 23.3.

Goldfish PLL efferent somata are fusiform or rounded and are either bi- or multipolar. Few of the cells in teleosts lie contralateral to the NPLL they supply (Strutz et al. 1980; Bell 1981; Zottoli and Van Horne 1983; Metcalfe et al. 1985; Meredith and Roberts 1987). Their dendritic morphology is difficult to discern on the basis of retrograde transport of HRP from the NPLL, since many neurons are usually filled in close proximity to one another (i.e., the dendrites of one cell are hard to distinguish from those of another), and all dendritic ramifications are not typically filled with HRP.

However, some efferent somata have dendrites that extend bilaterally; one dendrite extends ipsilaterally in a ventrolateral direction, and another travels across the midline and to the contralateral side of the brain. Although the initial direction of these dendrites has been characterized, the extent and end point of the dendritic ramifications are not known in the goldfish. PLL efferent neurons with dendrites that extend bilaterally are not unique to the goldfish and have been described in zebrafish (RELL cells; Metcalfe et al. 1985), *Sarotherodon niloticus* (Claas and Münz 1980), and eel (*Anguilla anguilla*; Meredith and Roberts 1987).

Efferent vestibular neurons in the toadfish have been intracellularly filled with HRP to determine the full extent of their dendritic ramifications (Highstein and Baker 1986). Two types were described on the basis of their size, position, and dendritic projections. Neurons in a lateral group were pyramidal in shape and had dendrites that extended only in a unilateral direction (Fig. 23.4A), and

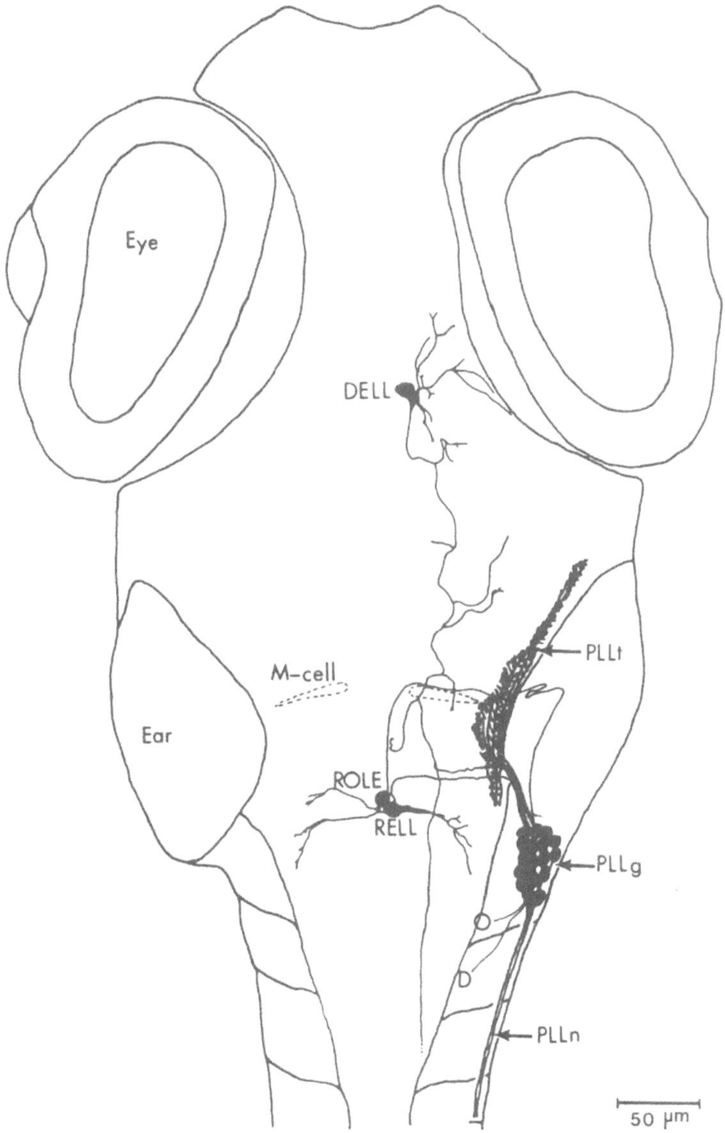

FIGURE 23.3. Schematic diagram of afferent and efferent components of the zebrafish NPLL. HRP was applied to the cut end of the right midbody branch of NPLL. The NPLL ganglion (PLLg) and afferent terminal fields (PLLt) can be seen in the hindbrain on the right edge of the drawing. NPLL efferent somata are situated in three nuclei (only one neuron in each is drawn for clarity of axonal and dendritic projections). The most rostral nucleus is located in the diencephalon (DELL), and the other two nuclei are found in the medulla (ROLE and RELL) caudal to the two M cells (dashed lines). This drawing is a composite of two preparations sectioned in the horizontal plane. (From Metcalfe et al. 1985.)

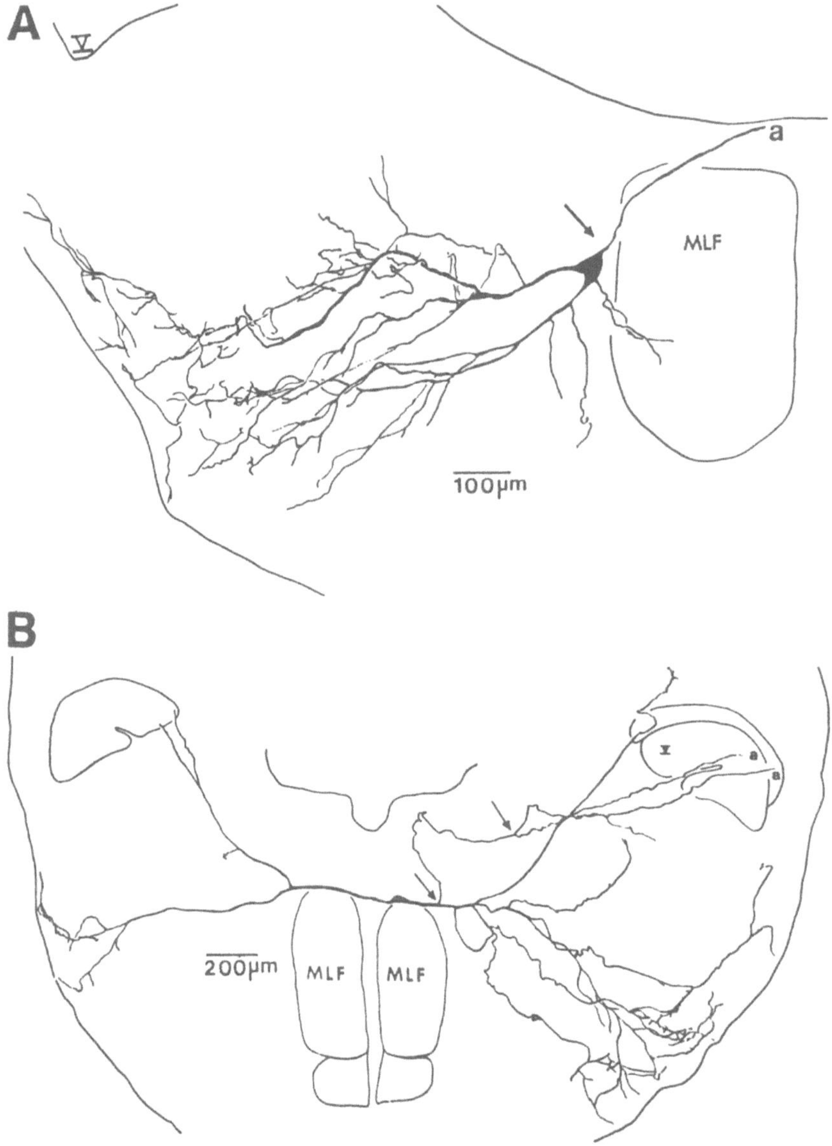

FIGURE 23.4. Reconstructions of vestibular efferent neurons of the toadfish intracellularly injected with HRP. (A) A neuron found in the lateral subgroup of vestibular efferent cells. The soma is pyramidal in shape, and its dendrites are exclusively unilateral. The arrow indicates the origin of the axon (the axon is also designated with an "a"). V, fifth nerve complex; FLM, fasiculus longitudinalis medialis. (B) A neuron found in the dorsal subgroup with a fusiform-shaped soma and bilaterally projecting dendrites. The arrows denote the dendritic origin of the axon and its collateralization (more lateral arrow). The axon branches were traced to the edge of the brain (a). (Modified from Highstein and Baker 1986.)

neurons in a dorsal group had fusiform somata with extensive bilateral dendritic trees (Fig. 23.4B; Highstein and Baker 1986). Whether a similar distinction exists for PLL efferent neurons in other species awaits experimentation with intracellular techniques.

23.3.2 Localization of Choline Acetyltransferase to PLL Efferent Somata

Efferent cells of the lateralis system are thought to be cholinergic, since their inhibitory effect on lateral line afferents is blocked by cholinergic antagonists. Stimulation of PLL efferent axons in the eel (*Astroconger myriaster*; Hashiomoto et al. 1970) or burbot (*Lota lota*; Flock and Russell 1973a) resulted in inhibition of spontaneous and evoked afferent activity recorded extracellulary in the NPLL. In addition, stimulation of lateral line efferent axons of the burbot resulted in a reduction in the amplitude of afferent EPSPs elicited by mechanical displacement of the cupula (Flock and Russell 1973a,b, 1976). This latter inhibitory effect was blocked by d-tubocurarine or gallamine triethiodide, suggesting that the efferent neuronal transmitter might be acetylcholine. Recently, choline acetyltransferase immunoreactivity (ChAT-IR) has been localized with immunohistochemical techniques to somata of NPLL efferent neurons that contain HRP retrogradely transported from NPLL (Danielson et al. 1988; see Rhodes et al. 1986; Zottoli et al. 1987b for techniques used in staining goldfish brains with AB8 antibody to ChAT). The presence of ChAT within some PLL efferent somata supports the view that these neurons are cholinergic. However, all diencephalic PLL efferent somata and some medullary PLLN efferent somata do not appear to contain ChAT, which may indicate that these neurons do not use acetylcholine as their transmitter.

23.3.3 Efferent Axonal Pathways Within the CNS

The axons of the PLL efferents exit the brain in two pathways (goldfish – Zottoli and Van Horne 1983; larval zebrafish – Metcalfe et al. 1985; see Fig. 23.3). One group of efferent axons extends laterally at the level of the rostral efferent nucleus (ren; ROLE; zebrafish) to the edge of the brain and combines with the NPLL root. The second group of axons continues rostrally. This bundle turns laterally at the level of the M cell and travels to the edge of the brain near the eighth nerve root (Zottoli and Van Horne 1983).

ChAT-IR staining of the goldfish brain has shown that the octavolateralis efferent axons extend within the central nervous system as a fascicle, along with axons originating from neurons within the ninth and seventh cranial motor nuclei. The tract lies dorsolateral to the flm and on the ventrolateral edge of the fourth ventricle. Specifically, the tract originates with axons of the ninth cranial motor nucleus. As these axons extend rostrally, they are joined by axons of the cen, ren, some unidentified neurons on the reticular formation, and the caudal and rostral portions of the seventh motor nucleus. Whether axons of a particular origin are

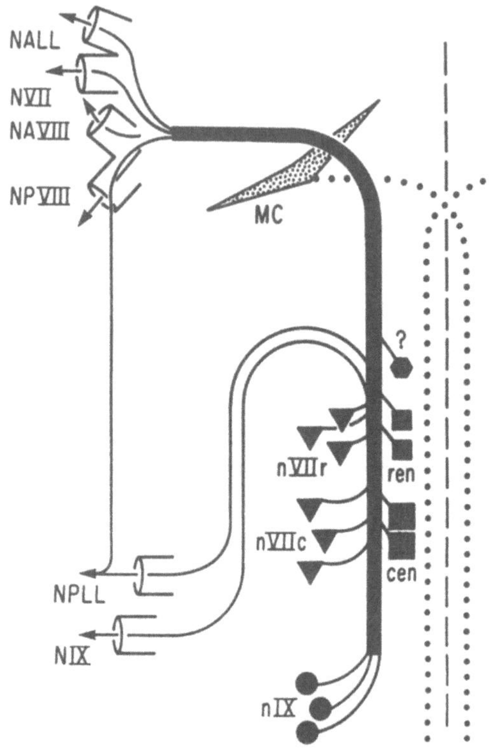

FIGURE 23.5. A ChAT-IR pathway in the goldfish central nervous system which includes octavolateralis efferent axons. A horizontal view of the left side of the hind brain. The midline is represented by the dashed line, the left M cell by stippling, and both M axons by dotted lines (note how the M axons decussate). The ninth (NIX), posterior lateral line (NPLL), anterior (NAVIII) and posterior (NPVIII) branches of the eighth nerve, and facial (NVII) and anterior lateral line nerve (NALL) roots are on the left side of this schematic diagram. The ChAT-IR pathway in transverse section is located dorsolateral to the flm just beneath the fourth ventricle (see Fig. 23.6). The ChAT-IR pathway (shown here as a thick, black line) originates caudally (bottom of drawing) with axons of the glossopharngeal motor neurons. More rostrally are axons of facial motoneurons (both the caudal and rostral divisions), axons of as yet unidentified neurons (?), and axons of octavolateralis efferent neurons join. The ninth motor axons exit near the rostral division of the facial nucleus, as do some of the lateralis efferent axons. The rest of the ChAT-IR axons extend rostrally to the level of the M axon, where they travel laterally toward the eighth cranial nerve root. At this point, the axons diverge and extend to the appropriate cranial nerves.

segregated within this tract is not known. ChAT-IR axons exit the fascicle as shown in Fig. 23.5. Photomicrographs of the ChAT-IR tract and some of the neurons that contribute to it are presented in Fig. 23.6. A similar pathway containing at least axons of the seventh cranial motor nucleus and axons of octavolateralis efferent neurons has been described in many other fish (Berkelbach van

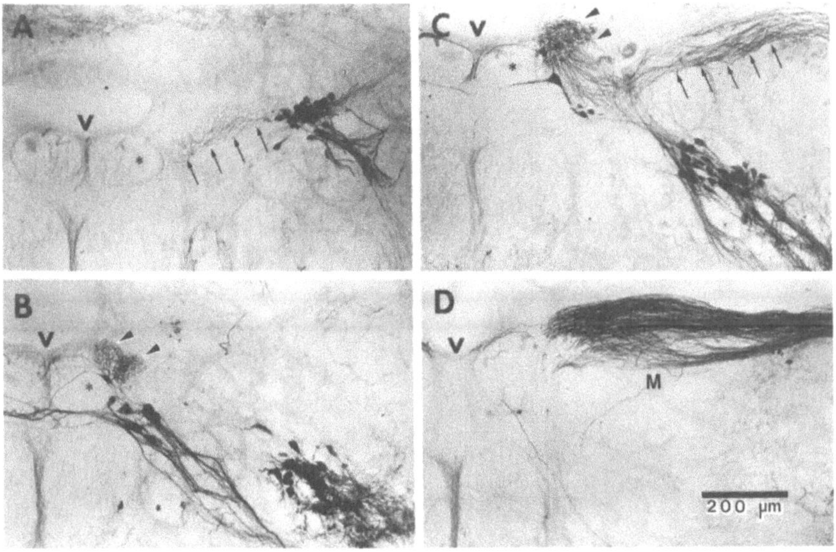

FIGURE 23.6. ChAT-IR somata that contribute to efferent pathways described in Fig. 23.5. (A–D) Transverse setions (80 µm) of the hindbrain from caudal to rostral. V, ventricle; the base of the V points ventrally to the midline. Dorsal is up, and lateral to the right. (A) ChAT-IR glossopharyngeal motoneurons and their axons (arrows), which extend toward the midline and form a tract (designated by arrowheads in B and C). (B) ChAT-IR facial (caudal division) motoneurons laterally and octavolateralis efferent neurons of the caudal efferent nucleus more medially. These cells contribute axons to the tract (arrowheads). (C) ChAT-IR facial neurons (lower right) contribute axons to the tract (arrowheads); the glossopharyngeal motor axons extend laterally toward the ninth motor root (arrows). (D) The ChAT-IR tract extends laterally over the M cell (M), which is not visible, since it does not stain with this antibody (Rhodes et al. 1986).

der Sprenkel 1915; Van der Horst 1918 in Ariëns Kappers et al. 1936; Bell 1981; Barry 1985; Meredith and Roberts 1986, 1987).

23.3.4 PLL Efferent Neurons and the M cell

The somata of many PLL efferent neurons lie in close proximity to the Mauthner axon, and it has been hypothesized that the M axon may form synaptic contacts with the efferent somata at this level (Zottoli and Van Horne 1983). However, electron microscopy has revealed no synapses between the M axon and efferent somata in the larval zebrafish (Metcalfe et al. 1985). Furthermore, the latency between M cell firing and the efferent inhibition of auditory afferent fibers in the goldfish is too long for a monosynaptic connection (Lin and Faber 1988). Therefore, the identity and number of neurons that are interposed between the M cells and efferent neurons remains unknown (dashed line and ? of Fig. 23.7).

A group of identifiable interneurons, cranial relay neurons (CRN), form a part of the M cell feedback inhibitory network (Fig. 23.7; M cell–CRN–INT B–

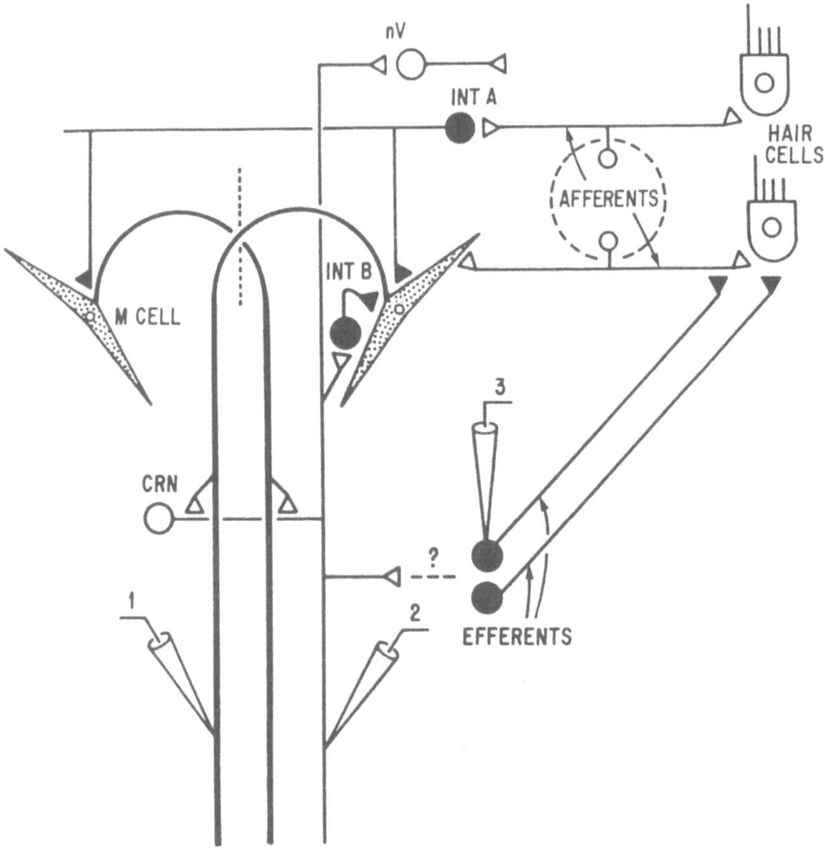

FIGURE 23.7. Schematic diagram of part of the M cell's feedback inhibitory network and its relation to excitatory input from hair cells of the octavolateralis system. Darkened cell bodies and terminals indicate inhibitory action. Excitation of either M cell (stippled) results in activation of a cranial relay neuron, which excites a number of cells including, at least, (1) cranial motoneurons (the fifth motor nucleus, nV); (2) interneurons inhibitory to the M cell (INT B, PHP neurons, which are responsible for the collateral IPSP); and (3) efferent neurons. The excitation of efferent neurons presumably contains a di- or polysynaptic synaptic pathway, the elements of which are not known (dashed line with question mark). Efferent neurons are known to inhibit both hair cells and afferent fibers in the goldfish inner ear. Since afferent fibers are known to excite another class of neurons inhibitory to the M cell (INT A, another class of PHP neurons; Zottoli et al. 1980), it is entirely possible that efferent neurons may synapse on and influence the excitability of hair cells and/or afferents projecting to these interneurons (connections not shown). Electrodes 1, 2, and 3 indicate a recording paradigm that might help elucidate the central connectivity of efferent neurons in relation to the M cell. All the neurons in this diagram except the efferents have been identified and recorded from in the goldfish. (Modified from Lin and Faber 1988; © 1988, Pergamon Journals Ltd.)

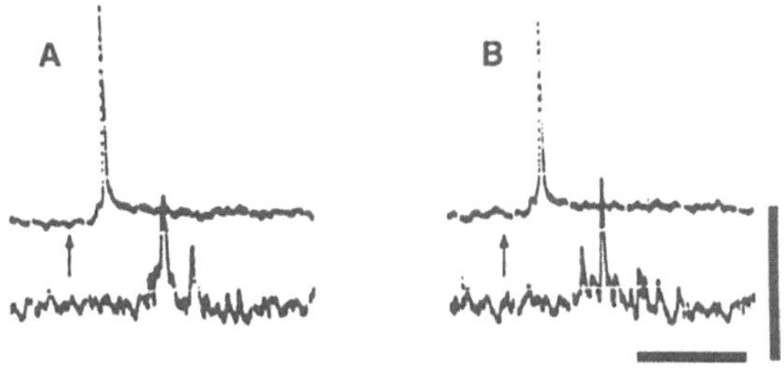

FIGURE 23.8. M cell activation of NPLL efferents. The M cell responds to a bench tap (arrows) in two separate trials (A and B, upper trace). The M cell impulse is followed in time by an efferent volley recorded in the proximal stump of a right NPLL (lower trace). Horizontal calibration, 20 ms. Vertical calibration, 50 mV and 0.5 mV for the upper and lower traces, respectively. (From Russell 1976.)

M cell) and a part of the excitatory pathway to cranial motoneurons responsible for the head level aspects of the startle response (Hackett and Faber 1983a,b). Lin and Faber (1988) have suggested that CRNs are interposed between the M cells and efferent neurons (Fig. 23.7; M cells–CRN–?–efferents) based on the following two results. (1) An action potential in either M cell was capable of eliciting the efferent inhibition (recorded as a hyperpolarization) of saccular afferents on one side. The CRNs are excited by both M cells and could therefore be the first element in the pathway between the M cells and efferent neurons (Hackett and Faber 1983a,b). (2) Fluctuations of efferent IPSPs recorded in saccular afferents paralleled fluctuations of collateral IPSPs recorded simultaneously from the M cell. This suggests that the collateral inhibitory and efferent inhibitory pathways share a common neuron such as the CRN (as shown in Fig. 23.7). Neurons homologous to the CRNs have been described in the hatchetfish (*Gasteropelecus sternicla*, giant interneurons–Auerbach and Bennett 1969) and the larval zebra fish (T cells–Kimmel et al. 1985).

Efferent neurons activated by M cells may exert their inhibitory effects at the level of the hair cells or on the afferent fibers innervating them. Activation of the goldfish M cell is followed 12 to 14 ms later by an efferent axon volley recorded in the PLLN (Fig. 23.8). The resultant inhibition of afferent activity can last for 50 ms (Russell 1974). This inhibition is most likely mediated by direct inhibitory action of efferent neurons on hair cells (Flock and Russell 1976), where they are known to terminate (Hama 1965). However, direct connections of efferent axons onto afferents (goldfish–Nakajima and Wang 1974; toadfish–Sans and Highstein 1984) have been shown to result in modulation of afferent activity in the auditory (Lin and Faber 1988) and vestibular (Highstein and Baker 1985) organs of fishes. For example, action potentials evoked by current injection in goldfish saccular afferent fibers were inhibited (Fig. 23.9, A_1) or delayed (A_2) by the activation of the M cell (Lin and Faber 1988). Furthermore, stimulation of efferent neurons by activation of the M cell resulted in an IPSP recorded in saccular affer-

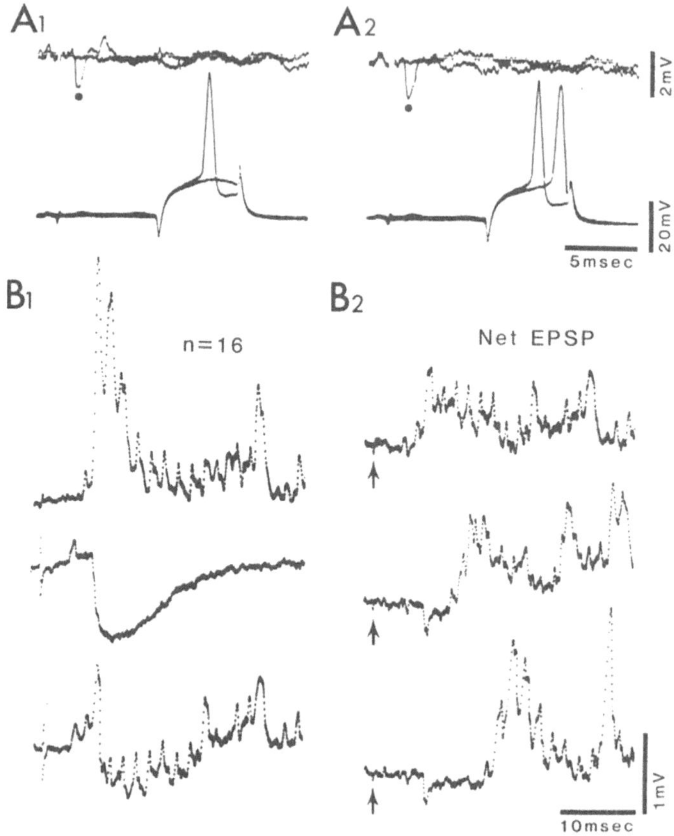

FIGURE 23.9. Inhibitory effect of an efferent PSP as recorded in an auditory afferent fiber. (A_1, A_2) Direct activation of a saccular afferent fiber (lower traces of A_1 and A_2) is inhibited (A_1) or delayed (A_2) when preceded by antidromic activation of the M cell. The M cell was activated (dot in upper traces) in one of the two traces as shown in the extracellular recordings from the M cell axon cap. (B_1, B_2) Interactions between an efferent IPSP and auditory-evoked EPSPs as recorded in an afferent fiber. All traces are an average of 16 sweeps. (B_1) Upper trace: Sound-evoked EPSP (400-Hz sound pulse with a duration of 25 ms). Middle trace: IPSP evoked by activation of the M cells. Lower trace: when the M cell stimulation was paired with the sound stimulus so that the initial phases of the EPSP and IPSP coincided, the excitation was attenuated. (B_2) The net EPSP (i.e., the difference between the paired response and the IPSP alone) was determined with the timing paradigm in B_1 (upper trace) and when the sound pulses were delayed by 5 ms (middle trace) and 10 ms (lower trace). Arrows designate when the spinal cord stimulation was applied. The net EPSP is less than control (B_1, upper trace) (i.e., the inhibitory effect is more than that predicted by linear summation of the two potentials). Note that the greatest reduction of the EPSP occurs when it is paired during the falling phase of the IPSP. (From Lin and Faber 1988; © 1988, Pergamon Journals Ltd.)

ent fibers, and this IPSP was capable of decreasing sound-evoked EPSPs (Fig. 23.9B; however see Highstein and Baker 1985 for excitatory effects of efferent activity on afferents).

23.4 Summary and Conclusions

Since the discovery of efferent innervation of the vestibular system in the cat (Rasmussen and Gacek 1958), there has been an active interest in elucidating the action of efferent neurons on their peripheral targets and the behavioral significance of efferent activation. The octavolateralis efferent system of fishes has many features that make it particularly attractive to study in this regard, including (1) the ability to intracellularly record from primary afferent fibers peripherally and efferent somata within the central nervous system, (2) the ability to stimulate efferent neurons independently of afferents, and (3) the ability to relate efferent activity to particular behaviors (e.g., the startle response).

The ability to conduct physiological studies on neuronal elements, including the M cell and efferent neurons, involved in the startle response provides the opportunity to elucidate the behavioral significance of efferent neurons. The duration of the efferent inhibitory influence (50 ms) on afferent activity following an M cell impulse would allow a reduction in sensitivity of PLL hair cells during the M cell-initiated stage of the startle response (i.e., efferents would block the "noise" produced by the movement of fish; Russell 1974; see Roberts and Meredith Chapter 22). This "noise reduction" might protect the hair cell–afferent synapse and allow it to be fully responsive after the massive movement during the initial phases of the startle response. However, the function(s) of efferent neurons are certainly more complex, as is indicated by the recent work of Lin and Faber (1988). Specifically, efferent inhibitory action on auditory afferents precedes a M cell-initiated C bend, indicating that the main function of efferent neurons may be the termination of auditory-evoked inputs before movement. Perhaps silencing of the afferents before the first stage of the startle response (presumably M cell-initiated) is important for parallel or subsequent components of this complex behavior (Eaton et al. 1988; DiDomenico et al. 1988; Lin and Faber 1988; Nissanov and Eaton 1988).

The ability to record intracellularly from the M cell, CRNs (Hackett and Faber 1983a,b), octavolateralis afferents (Highstein and Baker 1985; Lin and Faber 1988), and efferent somata (Highstein and Baker 1985) provides a unique opportunity to determine the circuitry (see the recording paradigm presented in Fig. 23.7) involved in the central activation of efferents and ultimately the relationship of that circuitry to the M cell-initiated startle response.

Acknowledgments. The authors are most grateful to Dr. Bruce Wainer for supplying the monoclonal antibody used in many of the experiments described in this paper. In addition, we thank Don Faber for critically reading an earlier version of this manuscript. Partial support for this research was supported by NSF grant BNS 86-06607 (S.J.Z.), NSF's College Science Instrumentation Program grant CSI-8650488, and NINCDS grant NS26032.

References

Ariëns Kappers CU, Huber GC, Crosby EC (1936) The Comparative Anatomy of the Nervous System of Vertebrates Including Man. New York: Hafner.

Auerbach AA, Bennett MVL (1969) Chemically mediated transmission at a giant fiber synapse in the central nervous system of a vertebrate. J Gen Physiol 53:183–210.

Barry MA (1985) Afferent and efferent connections of the primary octaval nuclei in the clearnose skate, *Raja eglanteria*. PhD Dissertation, University of Delaware.

Bartelmez GW (1915) Mauthner's cell and the nucleus motorius tegementi. J Comp Neurol 25:87–128.

Bell CC (1981) Central distribution of octavolateral afferents and efferents in a teleost (Mormyridae). J Comp Neurol 195:391–414.

Berkelbach van Der Sprenkel H (1915) The central relations of the cranial nerves in *Silurus glanis* and *Mormyrus caschive*. J Comp Neurol 25:5–65.

Bodian D (1937) The structure of the vertebrate synapse. A study of the axon endings on Mauthner's cell and neighboring centers in the goldfish. J Comp Neurol 68:117–159.

Chang YT, Lin JW, Faber DS (1987) Spinal inputs to the ventral dendrite of the teleost Mauthner cell. Brain Res 417:205–213.

Claas B, Münz H (1980) Bony fish lateral line efferent neurons identified by retrograde axonal transport of horseradish peroxidase (HRP). Brain Res 193:249–253.

Danielson PD, Zottoli SJ, Corrodi JG, Rhodes KJ, Mufson EJ (1988) Localization of choline acetyltransferase to posterior lateral line efferent somata of the goldfish. Brain Res 448:158–161.

DiDomenico R, Nissanov J, Eaton RC (1988) Lateralization and adaptation of a continuously variable central program following lesions of a reticulospinal command neuron. Brain Res (in press).

Eaton RC, Hackett JT (1984) The role of the Mauthner cell in fast-starts involving escape in teleost fishes. In Eaton RC (ed) *Neural Mechanisms of Startle Behavior.* New York: Plenum, pp. 213–266.

Eaton RC, DiDomenico R, Nissanov J (1988) Flexible body dynamics of the goldfish C-start: Implications for reticulospinal command mechanisms. J Neurosci 8:2758–2768.

Faber DS, Korn H (1975) Inputs from the posterior lateral line nerves upon the goldfish Mauthner cell. II. Evidence that the inhibitory components are mediated by interneurons of the recurrent collateral network. Brain Res 96:349–356.

Faber DS, Korn H (1978) Electrophysiology of the Mauthner cell: Basic properties, synaptic mechanism, and associated networks. In Faber DS, and Korn H (eds) Neurobiology of the Mauthner Cell. New York: Raven Press, pp. 47–131.

Faber DS, Korn H (1982) Transmission at a central inhibitory synapse. I. Magnitude of the unitary postsynaptic conductance change and kinetics of channel activation. J Neurophysiol 48:654–678.

Faber DS, Kaars C, Zottoli SJ (1980) Dual transmission at morphologically mixed synapses: Evidence from postsynaptic cobalt injections. Neuroscience 5:433–440.

Flock Å, Russell IJ (1973a) Efferent nerve fibers: Postsynaptic action on hair cells. Nature 243:89–91.

Flock Å, Russell IJ (1973b) The postsynaptic action of efferent fibers in the lateral line organ of the burbot *Lota lota*. J Physiol 235:591–605.
Flock Å, Russell I (1976) Inhibition by efferent nerve fibers: Action on hair cells and afferent synaptic transmission in the lateral line canal organ of the burbot *Lota lota*. J Physiol 257:45–62.
Furshpan EJ (1964) Electrical transmission at an excitatory synapse in a vertebrate brain. Science 144:878–800.
Furshpan EJ, Furukawa T (1962) Intracellular and extracellular responses of the several regions of the Mauthner cell of the goldfish. J Neurophysiol 25:732–771.
Furukawa T (1966) Synaptic interaction at the Mauthner cell of goldfish. Prog Brain Res 21A:46–70.
Hackett JT, Faber DS (1983a) Mauthner axon networks mediating supraspinal components of the startle response in the goldfish. Neuroscience 8:317–331.
Hackett JT, Faber DS (1983b) Relay neurons mediate collateral inhibition of the goldfish Mauthner cell. Brain Res 264:302–306.
Hama K (1965) Some observations on the fine structure of the lateral line organ of the Japanese sea eel, *Lyncozymba nystromi*. J Cell Biol 24:193–210.
Hashimoto T, Katsuki Y, Yanagisawa K (1970) Efferent system of lateral-line organ of fish. Comp Biochem Physiol 33:405–422.
Highstein S, Baker R (1985) Action of the efferent vestibular system on primary afferents in the toadfish, *Opsanus tau*. J Neurophysiol 54:370–384.
Highstein SM, Baker R (1986) Organization of the efferent vestibular nuclei and nerves of the toadfish, *Opsanus tau*. J Comp Neurol 243:309–325.
Kimmel CB, Metcalfe WK, Schabtach E (1985) T-reticular interneurones: A class of serially repeating cells in the zebrafish hindbrain. J Comp Neurol 233:365–376.
Kimmel CB, Schabtach E, Kimmel RJ (1977) Developmental interactions in the growth and branching of the lateral dendrite of Mauthner's cell (*Ambystoma mexicanus*). Dev Biol 55:244–259.
Korn H, Faber DS, Mariani J (1974) Existence de projections des nerfs postérieurs de la ligne latérale sui la cellule de Mauthner; leur effet antagoniste sur l'activation de ce neurone par les afférences vestibulaires. C R Acad Sci Paris 279:413–416.
Korn H, Faber DS (1975a) Inputs from the posterior lateral line nerves upon the goldfish Mauthner cell. I. Properties and synaptic localization of the excitatory component. Brain Res 96:342–348.
Korn H, Faber DS (1975b) An electrically mediated inhibition in goldfish medulla. J Neurophysiol 38:452–471.
Korn H, Faber DS (1976) Vertebrate central nervous system: Same neurons mediate both electrical and chemical inhibitions. Science 194:1166–1169.
Korn H, Faber DS (1987) Regulation and significance of probabilistic release mechanisms at central synapses. In: Edelman GM, Gall WE, Cowan WM (eds) Synaptic Function. New York: John Wiley, pp. 57–108.
Korn H, Mallet A, Triller A, Faber DS (1982) Transmission at a central synapse. II. Quantal description of release with a physical correlate for binomial n. J Neurophysiol 48:679–707.
Korn H, Faber DS, Burnod Y, Triller A (1984) Regulation of efficacy at central synapses. J Neurosci 4:125–130.
Korn H, Faber DS, Triller A (1986) Probabilistic determination of synaptic strength. J Neurophysiol 55:402–421.
Lin J-W, Faber DS (1988) An efferent inhibition of auditory afferents mediated by the goldfish Mauthner cell. Neuroscience 24:829–836.

Lin J-W, Faber DS, Wood MR (1983) Organized projection of the goldfish saccular nerve onto the Mauthner cell lateral dendrite. Brain Res 274:319–324.

Meredith GE, Roberts BL (1986) Central organization of the efferent supply to the labyrinthine and lateral line receptors of the dogfish. Neuroscience 17:225–223.

Meredith GE, Roberts BL (1987) Distribution and morphological characteristics of efferent neurons innervating end organs in the ear and lateral line of the European eel. J Comp Neurol 265:494–506.

Metcalfe WK, Kimmel CB (1982) Three types of posterior lateral line efferent neurons in larval zebrafish. Proc Soc Neurosci 8:763 (abstract).

Metcalfe WK, Kimmel CB, Schabtach E (1985) Anatomy of the posterior lateral line system in young larvae of the zebrafish. J Comp Neurol 233:377–389.

Nakajima Y (1974) Fine structure of synaptic endings on the Mauthner cell of the goldfish. J Comp Neurol 156:375–402.

Nakajima Y, Kohno K (1978) Fine structure of Mauthner cell: Synaptic topography and comparative study. In Faber DS, and Korn H (eds) Neurobiology of the Mauthner cell. New York: Raven Press, pp. 133–166.

Nakajima Y, Wang DW (1974) Morphology of afferent and efferent synapses in hearing organ of fish. J Comp Neurol 156:403–416.

Nissanov J, Eaton RC (1989) Reticulospinal control of rapid escape turning maneuvers in fishes. Am Zool 29: (in press).

Rasmussan GL, Gacek RR (1958) Concerning the question of an efferent fiber component of the vestibular nerve of the cat. Anat Rec 130:361–362.

Rhodes KJ, Zottoli SJ, Mufson EJ (1986) Choline acetyltransferase immunohistochemical staining in the goldfish (*Carassius auratus*) brain: Evidence that the Mauthner cell does not contain choline acetyltransferase. Brain Res 381:215–224.

Russell I (1976) Central inhibition of lateral line input in the medulla of the goldfish by neurons which control active body movements. J Comp Physiol 111:335–358.

Russell IJ (1974) Central and peripheral inhibition of lateral line input during the startle response in goldfish. Brain Res 80:517–522.

Sans A, Highstein SM (1984) New ultrastructure features in the vestibular labyrinth of the toadfish, *Opsanus tau*. Brain Res 308:191–195.

Stefanelli A (1951) The Mauthnerian apparatus in Ichthyopsida: Its nature and function and correlated problem of neurohistogenesis. Q Rev Biol 26:17–34.

Strutz J, Schmidt CL, Strürmer C (1980) Origin of efferent fibers of the vestibular apparatus in goldfish. A horseradish peroxidase study. Neurosci Lett 18:5–9.

Triller A, Korn H (1982) Transmission at a central inhibitory synapse. III. Ultrastructure of physiologically identified and stained terminals. J Neurophysiol 48:708–736.

Tuttle R, Masuko S, Nakajima Y (1987) Small vesicle bouton synapses on the distal half of the lateral dendrite of the goldfish Mauthner cell: Freeze fracture and thin section study. J Comp Neurol 265:254–274.

Will U (1986) Mauthner neurons survive metamorphosis in anurans: A comparative HRP study on the cytoarchitecture of Mauthner neurons in amphibians. J Comp Neurol 244:111–120.

Zottoli SJ (1977) Correlation of the startle reflex and Mauthner cell auditory responses in unrestrained goldfish. J Exp Biol 66:243–254.

Zottoli SJ (1978) Comparative morphology of the Mauthner cell in fish and amphibians. In Faber DS, Korn H (eds) Neurobiology of the Mauthner cell. New York: Raven Press, pp. 13–45.

Zottoli SJ, Faber DS (1979) Properties and distribution of anterior VIIIth nerve excitatory inputs to the goldfish Mauthner cell. Brain Res 174:319–323.

Zottoli SJ, Faber DS (1980) An identifiable class of statoacoustic interneurons with bilateral projections in the goldfish medulla. Neuroscience 5:1287–1302.

Zottoli SJ, Van Horne C (1983) Posterior lateral line afferent and efferent pathways within the central nervous system of the goldfish with special reference to the Mauthner cell. J Comp Neurol 219:100–111.

Zottoli SJ, Hordes AR, Faber DS (1987a) Localization of optic tectal input to the ventral dendrite of the goldfish Mauthner cell. Brain Res 401:113–121.

Zottoli SJ, Rhodes KJ, Mufson EJ (1987b) Comparison of acetylcholinesterase and choline acetyltransferase staining patterns in the optic tectum of the goldfish *Carassius auratus*: A histochemical and immunocytochemical analysis. Brain Behav Evol 30:143–159.

Part V Behavioral Analysis of Sensory Capabilities

24
Function of the Free Neuromasts of Marine Teleost Larvae

JOHN H.S. BLAXTER and LEE A. FUIMAN

24.1 Early Development of Free Neuromasts and Other Mechanoreceptors

Most species of marine teleosts have very small transparent larvae at hatching, with total lengths from about 1.5 to 8 mm. The eyes may or may not be pigmented at this time, but they always become functional when the larvae commence feeding a few hours to a few days after hatching. Feeding seems to be mainly a visual process, and the larvae of only a few species, such as Dover sole, *Solea solea*, are known to feed in darkness (Blaxter 1969). All species examined have free neuromast organs distributed over the head and body as superficial hillocks. These hillocks, which are very large relative to the body of the larva, have gelatinous cupulae projecting into the surrounding water. Larval neuromasts are well described by Iwai (1980) in several species including the goldfish (*Carassius auratus*), sea bass (*Lateolabrax japonicus*), black porgy (*Acanthopagrus schlegeli*), and right-eye flounder (*Kareius bicoloratus*). Disler (1971) followed the changes in number and distribution of the free neuromasts during the development of the sturgeon (*Acipenser stellatus*), chum salmon (*Oncorhynchus keta*), and several freshwater percids and cyprinids. Later work on gadoids (Fridgeirsson 1978), northern anchovy (*Engraulis mordax*) (O'Connell 1981), Atlantic herring (*Clupea harengus*) (Blaxter et al. 1983a), Atlantic halibut (*Hippoglossus hippoglossus*) (Blaxter et al. 1983b), and plaice (*Pleuronectes platessa*) and turbot (*Scophthalmus maximus*) (Neave 1986) confirmed the earlier findings, showing the increase in numbers with age and also the modifications that occur during the metamorphosis of flatfish.

The ontogeny of the free neuromasts and lateral line has been reviewed by Blaxter (1987). The primary neuromasts develop from placodes on the head region. They give rise to both canal neuromasts and free neuromasts, but the free organs may also arise as secondary neuromasts independently of the primordial placodes. At hatching, there are (in the few species studied) several free neuromasts distributed over the head and body, although in a flounder, *Paralichthys olivaceus*, there is only one on either side of the head behind the eye (Kawamura and Ishida 1985). During subsequent development the free neuromasts proliferate, and in the later larval stages there may be 100 or more distributed over the head in a lateral row on each myotome of the trunk and even on the caudal fin. In some species, like the chum salmon, which hatches at a relatively advanced state of development, 70 to 80 neuromasts are present at hatching, but they may not be functional until the larva (alevin) leaves the spawning redd (Disler 1971).

The organization of the neuromasts into lateral line canal systems occurs later in development: for example, at a body length of 18 to 20 mm in northern anchovy (O'Connell 1981); 17 mm in menhaden, *Brevoortia tyrannus* (Hoss and Blaxter 1982); 24 to 26 mm in Atlantic herring (Blaxter et al. 1983a); 19 mm (stage 4) in plaice; and 8 mm (stage 3) in turbot (Neave 1986). In flatfish, which metamorphose to lie with one side on the substratum, the canal system becomes more concentrated on the upper (eyed) side. In the plaice, the supraorbital canal degenerates on the blind side, but in the turbot it runs across the head to run adjacent to the supraorbital canal on the eyed side (Neave 1986).

The free neuromasts grow during development. The diameter of the hillocks may be as little as 20 to 30 μm in newly hatched larvae (Iwai 1967), although they are 40 to 60 μm in herring (Blaxter et al. 1983a) and 40 to 60 μm in plaice and turbot (Neave 1986). Few counts have been made of the number of hair cells. Blaxter et al. (1983a) found a mean of 8.5 ± 2.2 hair cells per neuromast on the trunk of 14-mm-long herring larvae, increasing to 28 ± 10 at 38 mm and 50 to 100 at 87 mm in the closely related sprat, *Sprattus sprattus*. In herring, most of the trunk neuromasts had anteroposterior polarity, but four out of four on the head had dorsoventral polarity. Iwai (1983) also found anteroposterior polarity in the trunk neuromasts of the anchovy, *Engraulis japonica*.

The cupulae at hatching are typically 50 to 60 μm long, although lengths of up to 141 μm long have been reported in the goby (*Tridentiger trigonocephalus*) by Iwai (1963), and 200 μm in the silverside (*Menidia beryllina*) by Cahn and Shaw (1962). In two species, herring and cod (Blaxter 1984a,b), the cupulae grow continuously and require about 24 h to regenerate fully if dislodged.

Very little is known about the early development of the other major mechanoreceptor, the ear. Otoliths are present at or before hatching; it therefore seems likely that gravistatic mechanisms are operating early in development. Dale (1984) investigated the development of the otoliths in the cod (*Gadus morhua*) and found a close association between the deposition of electron-dense granules in the otolith and the tips of the kinocilia, suggesting a secretory role. Much more is known about the inner ear of herring later in larval development, both in anatomical and behavioral terms (Blaxter et al. 1981; Blaxter and Batty 1985). Here, the elaboration of responses to sounds is related to the appearance of a gas-filled otic bulla.

24.2 Free Neuromasts and Behavior of Larvae

Although information on the role of the lateral line system in the behavior of adult teleosts is increasing (Bleckmann et al., 1986, Chapter 25; Hassan Chapter 10; Coombs and Janssen Chapter 15; Enger et al. Chapter 29; Montgomery Chapter 28), the importance of the free neuromasts in larvae, prior to lateral line canal formation, is not understood at all. There is much anecdotal information of larvae responding to water jets or escaping from nets in the dark. Yin and Blaxter (1987) measured the escape speeds of herring, cod, flounder, and plaice larvae when stimulated by probes and found that they were efficient in making escape movements away from the source of stimulation, even in the dark. The importance of escape responses to teleost larvae is obvious, and the role of startle or C responses mediated via Mauthner or other reticulospinal cells is well documented by Eaton and DiDomenico (1986). What is not at all clear is whether such escape behavior is linked to the presence and function of free neuromasts and what role might be played by the inner ear, either in terms of sound or gravistatic stimuli.

Cahn et al. (1968) related the onset of schooling behavior in the Atlantic silverside (*Menidia menidia*) to the elaboration of the free neuromasts and lateral line canal, reporting a surprising lack of innervation in the free neuromasts of the youngest stages. Such a lack of innervation has not been described in other species, and nerves can clearly be seen passing to the free neuromasts in newly hatched herring and cod, for example. Neave (1986) reported a dorsal light reaction in older larvae of plaice and turbot, indicating that the posture-maintaining mechanisms of the adult are present before metamorphosis.

Thus, our current knowledge of the free neuromasts in teleost larvae is based on isolated and disparate reports. The presence and size of the neuromasts, even at hatching, indicate that they must play an important role in the sensory awareness of larvae, especially in those species without functional eyes at hatching. It seems unlikely that they play a major role in feeding, since in most species feeding does not occur in the dark. Their most likely role is in the perception of physical stimuli from predators or conspecifics. Two recent reports (Bourgos 1986; Økland et al. 1988) suggest that confined rearing conditions in hatcheries cause abrasion of the cupulae in halibut, resulting in high mortality. An investigation of free neuromast function therefore has some immediate practical value in aquaculture.

24.3 Behavioral Study of Free Neuromast Function

The present experiments were initiated to test the role of the free neuromasts in the generation of escape responses in the larvae of herring (*Clupea harengus*), cod (*Gadus morhua*), plaice (*Pleuronectes platessa*), flounder (*Platichthys flesus*), and halibut (*Hipploglossus hipploglossus*). The larvae, at a number of different ages but all *before* lateral line canal development, were stimulated in the dark by a probe, and their responses were recorded by a video system using infrared light, which is invisible to the fish. Experiments were conducted on control larvae, larvae treated with turbulence (as a means of dislodging cupulae and so simulating the effect of abrasion), and larvae exposed to an ototoxic chemical, streptomycin sulfate, which is one of the antibacterial agents known to ablate mechanoreceptors (Wersäll and Flock 1964; Sanders and Sanders 1979; Shiozawa et al. 1981; Blaxter and Hoss 1981; Kaus and Fleischer 1982). Subsequent analysis of the video recordings allowed the responses to be quantified. Of particular interest was whether the larvae made fast startle responses or "C" starts. These are very rapid flexions of the body, probably involving Mauthner or similar reticulospinal cells with a large soma and large diameter axon that help to initiate very fast escape behavior (see review by Eaton and Hackett 1984 and Chapter 23 by Zottoli and Danielson).

24.3.1 Provision of Larvae

Herring eggs were obtained by stripping ripe fish on the spawning grounds. Cod eggs were provided from brood stock held by the Marine Laboratory of the Department of Agriculture and Fisheries for Scotland in Aberdeen; halibut larvae from brood stock held by the Seafish Industries Authority Marine Farming Unit at Ardtoe, Argyll; and plaice and flounder eggs from brook stock held in Oban. Eggs and larvae were reared in stock tanks at 7 to 9°C in the laboratory aquarium. After yolk resorption, cod and flounder larvae were fed on rotifers (*Brachionus plicatilis*), and plaice and herring on *Artemia* nauplii. The halibut larvae were only used in the early yolk sac stage.

24.3.2 Position of the Neuromasts

The position of the free neuromasts in the different species and stages used, as determined by Janus Green staining (Blaxter et al. 1983a), is shown in Fig. 24.1. The number at hatching varies from species to species, but, in general, the neuromasts are sparsely distributed on the head and trunk with a subsequent proliferation. In the later stages there may be a neuromast on each myotome and even several on the caudal fin.

24.3.3 Experimental Observations and Scoring of Responses

Larvae were transferred from the stock tanks by a wide-mouthed pipette (to minimize damage to cupulae) into transparent circular Perspex dishes 12 cm in diameter and 4 cm deep holding 350 ml seawater. Usually 50 larvae of a single species were used for each experiment and not reused. These dishes were kept for 20 to 24 h at 9°C in a constant-temperature room before testing, in order to allow regeneration of any dislodged cupulae.

The responses of the larvae to a probe were then observed and recorded in the dark using an infrared sensitive video camera placed above the dish, which was illuminated from below by a strobed, infrared, light-emitting diode (LED), giving light of about 910 nm wavelength that was invisible to the larvae (see Fig. 24.2 and Batty 1983 for details). The dish and camera were positioned under a hood of black curtain material. During a trial the observer sat outside the hood and held the probe through a slit in the hood. Individual larvae were then approached with the probe (an "encounter"), using a TV monitor outside the hood to guide the approach. The larvae were approached in a direction about normal to the

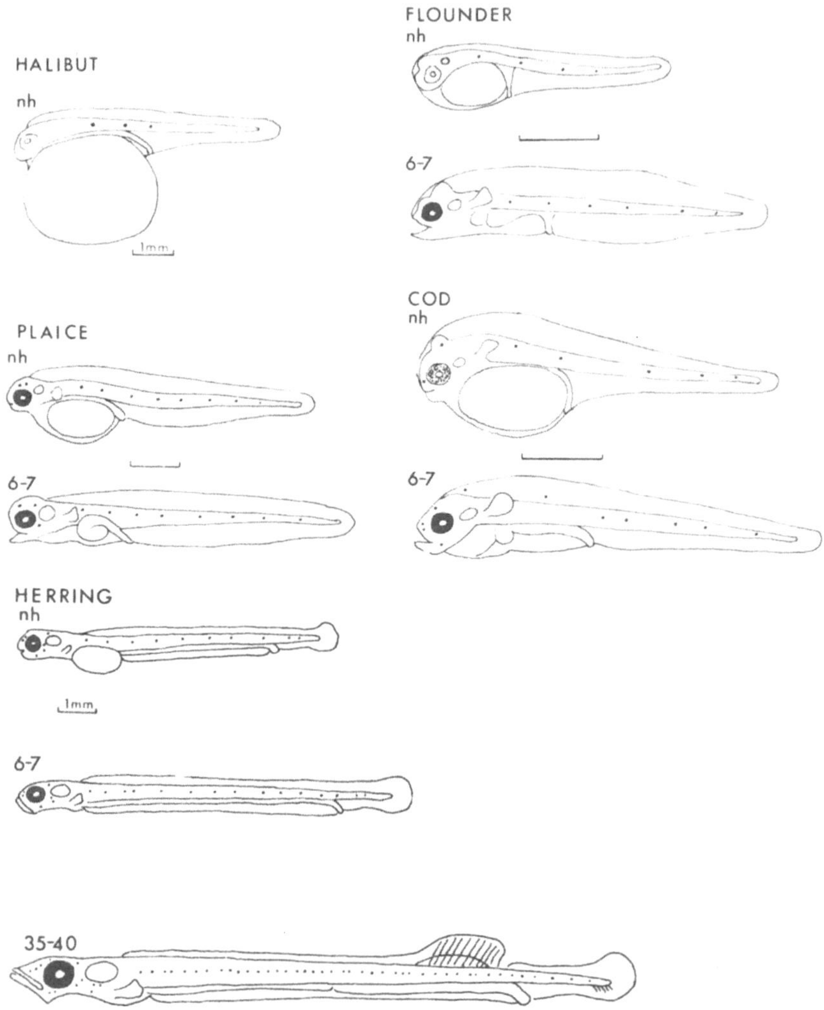

FIGURE 24.1. Drawings of species and stages used showing position of free neuromasts as black dots. All scale bars are 1 mm. Age given in days posthatching; nh, newly hatched.

body axis at as near a consistent speed as possible. Apart from the visible light emitted from the monitor, the room was in darkness. Light measurements under the hood were less than the minimum sensitivity of the light meter (10^{-4} meter candles (mc)), and the probe, a modified glass Pasteur pipette with a rounded tip about 1 mm in diameter, was invisible to the larvae.

Each trial with a group of larvae consisted of 15 encounters, followed by a wait of 5 min for the larvae to redistribute themselves in the dish, and then another series of 15 encounters. Care was taken to use different larvae, and, with 50 larvae in the dish, it is unlikely that many larvae were tested more than once, thus reducing the danger of habituation.

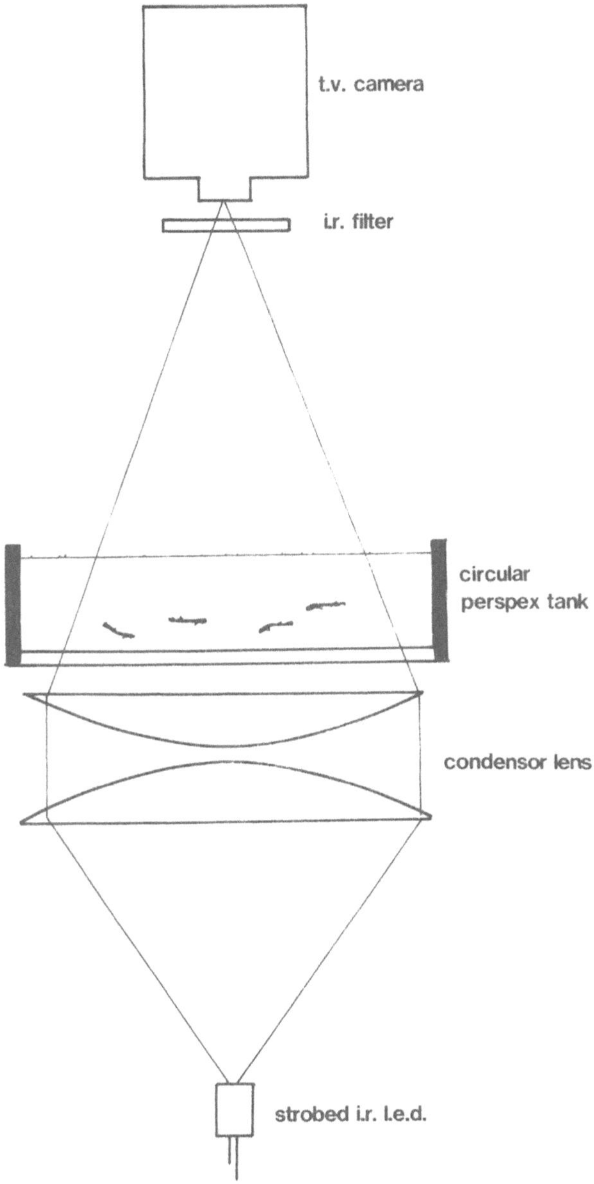

FIGURE 24.2. Diagram of apparatus used: i.r., infrared; l.e.d., light-emitting diode.

Larval responses were analyzed by replay of the videotapes, which had a frame rate of 50/s. The strobed LED illumination gave fairly sharp images with 20 ms between frames. During an encounter some larvae responded before they were touched by the probe. Other larvae responded only when touched, with varying latencies. Some larvae made C-starts with a latency of 20 to 40 ms; others made slower, less extreme responses. These responses were scored according to the following system:

Score

Score	
0	No response within 200 ms when touched
1	Turn or accelerate within 60–200 ms when touched
2	Turn or accelerate within 60 ms when touched
3	C-start when touched
4	C-start, turn, or accelerate *before* being touched

The mean score for the 30 encounters characterized each trial. When the larvae responded before being touched, the reaction distance was measured on the monitor. This measurement was the *minimum* distance of response in that the monitor screen could not show whether the probe was above or below the larvae. In some experiments a record was made of the "errors" made during responses. Thus a C-start *toward* the point of stimulation and escape swimming *toward* the direction of stimulation were considered errors.

24.3.4 Pretreatment of Larvae

Whereas ontogenetic and interspecific differences in responses were measured on control, untreated larvae, some experiments were done after treating the larvae with turbulence or different concentrations of streptomycin to ablate the neuromasts physically or chemically.

Turbulent conditions were obtained by keeping the larvae for 5 min in a 100-ml beaker containing an air stone (2.5 × 2.5 × 2.5 cm) giving off a vigorous output of bubbles. The effectiveness of this procedure in dislodging cupulae is shown in Table 24.1. There were considerable differences between individuals and species even in the control larvae. It is difficult to explain these, since it is necessary to both catch and stain the control larvae so that the state of the cupulae in the free-swimming untreated condition cannot be ascertained. It seems likely that the

TABLE 24.1. Effect of turbulence on the presence of cupulae in herring and flounder larvae.[a]

Species	Stage (days posthatch)	Percentage intact cupulae: Control $(n)^b$	Percentage intact cupulae: Treated by turbulence $(n)^b$	Significance
Herring	Newly hatched	27 (183)	13 (163)	c
	3–4	51 (174)	37 (170)	c
	6–7	60 (181)	25 (178)	d
Flounder	Newly hatched	85 (48)	23 (13)	d
	3–4	74 (78)	53 (88)	d

[a] At least 20 larvae were used for each species, stage, and treatment. The number of observations depends to a great extent on the success of the staining technique.
[b] Number of neuromasts observed.
[c] Statistically significant ($p < .01$).
[d] Statistically significant ($p < .001$).

TABLE 24.2. Mean reponse scores in 30 encounters of herring, place, and cod larvae in different molarities of streptomycin sulphate compared with a control.[a]

Species (days posthatch)	Time (hours)	Control	Streptomycin molarity (mM) 0.1	1.0	10.0
Herring (7)	0	1.60	1.17	1.60	0.17[b]
	24	1.90	1.67	1.60	2.03
Plaice (7)	0	1.97	1.97	1.57	0.43[b]
	24	1.43	2.07	1.37	1.60
Cod (4)	0	1.97	2.10	1.93	0.30[b]
	22	2.57	2.17	2.20	1.90

[a] Experiment was done at zero time followed by washing and recovery, with a further test at 22–24 h.
[b] Result was significantly lower than for the contemporaneous control (one-tailed test, $p \leq .05$).

cupulae are growing and being dislodged at intervals even under natural conditions (Blaxter 1984a,b). Although the number of intact cupulae was reduced by turbulence, they were never all dislodged. The larvae were always held for 30 min after this treatment before testing in order to reduce the effect of fatigue.

Streptomycin sulfate was used at concentrations of 0.1, 1.0, and 10 mM. The larvae were tested either in these solutions or after immersion for 1 to 2 min followed by four washes, estimated to reduce the original streptomycin level by 10^4 times. In most instances a second control experiment, termed recovery control, was done 20 to 24 h after treated larvae had been removed from the streptomycin solutions and washed as above or allowed to recover from turbulence.

In statistical comparisons between the various treatments (turbulence and streptomycin) and the appropriate untreated or recovery controls, treatments were expected to cause deterioration, not enhancement, of the response. Therefore, one-tailed tests were used on the assumption that the values for the treated larvae would be less than controls. For binomial variables (response score, percent C-starts, number of distance responses), chi-square statistics were used to judge the presence of a treatment effect. These results were always supported by the results of an analysis of variance (ANOVA). When a significant effect existed,

TABLE 24.3. C-starts as percent of positive responses in 30 encounters.

Species (days posthatch)	Time (hours)	Control	Streptomycin molarity (mM) 0.1	1.0	10.0
Herring (7)	0	27	20	43	0[a]
	24	27	27	43	50
Plaice (7)	0	33	43	43	10[a]
	24	30	43	23	30
Cod (4)	0	21	31	27	0[a]
	22	30	27	23	23

[a] Statistically significant ($p < .05$).

TABLE 24.4. Number of distance responses in 30 encounters.

Species (days posthatch)	Time (hours)	Control	Streptomycin molarity (mM) 0.1	1.0	10.0
Herring (7)	0	2	1	3	0[a]
	24	5	2	3	5
Plaice (7)	0	6	7	7	0[a]
	24	1	7	3	4
Cod (4)	0	8	5	2[a]	0[a]
	22	11	7	4[a]	4[a]

[a] Statistically significant ($p < .05$).

means were compared with Duncan's multiple range test. Normally distributed data (reaction distance) were analyzed only with ANOVA and Duncan's tests. Statistical procedures are described by Snedecor and Cochran (1967). Curves fitted to the data (Fig. 24.3, Table 24.8) were computed by least squares, nonlinear regression weighting by the number of observations in each point when they varied. Final selection of curves was based on minimal residual mean square.

24.3.5 Response to a Probe, Control Larvae

The responses of untreated (control) 7-day-old herring, 7-day-old plaice, and 4-day-old cod to the probe are shown in Tables 24.2 to 24.5. The mean response score (Table 24.2) lay between 1.5 and 2.6 (the maximum possible being 4), with cod being very responsive after 24 h. Of the responses 20% to 33% were C-starts (Table 24.3). The number of responses to the probe before it touched the larvae ("distance responses") varied from one to 11 out of 30, with the cod larvae much more prone to distance responses than the other species (Table 24.4). Considering all species, over 60% of distance response were C-starts. The mean reaction distance (Table 24.5) was, however, small, usually 1 to 3 mm (less than one body length), but in plaice after 24 h (recovery control) it reached 5 mm (about one

TABLE 24.5. Reaction distances (mm) of distance responses shown in Table 24.4. Points represent means of various numbers of observations (also shown in Table 24.4).

Species (days posthatch)	Time (hours)	Control	Streptomycin molarity (mM) 0.1	1.0	10.0
Herring (7)	0	1.0	1.0	3.3	–
	24	2.4	7.5	1.0	2.0
Plaice (7)	0	2.0	1.7	4.0	–
	24	5.0	2.7	2.3	1.8
Cod (4)	0	2.9	2.8	1.5	–
	22	2.7	2.2	2.4	1.3

TABLE 24.6. Mean response score in 30 encounters of various species and ages in 10 mM streptomycin sulfate compared with controls.

Species (days posthatch	Time (hours)	Control	10.0 mM
Herring (2)	0	2.47	0.5 [a]
	24	–	1.43
Herring (26)	0	1.70	0.03[a]
	22	2.33	1.33[a]
Plaice (4)	0	2.83	0.10[a]
	24	–	2.43
Flounder (4)	0	2.23	0.20[a]
	24	–	0.70
Flounder (30)	0	1.40	0.20[a]
	23	–	1.27
Halibut (4–8)	0	0.43	0.00[a]
	23	–	0.93

[a] Statistically significant ($p < .05$).

body length) for the single response observed. Only one parameter, reaction distance, was significantly improved over the untreated control level after 24 h of recovery, and only in 7-day-old herring. Thus, one may conclude that the handling of larvae prior to the treatment did not affect their responses.

24.3.6 Effect of Streptomycin

Larvae of the same three species and ages were tested in different concentrations of streptomycin (Tables 24.2 to 24.5). Only 10 mM streptomycin consistently reduced the levels of response in all aspects. The number of distance responses in cod was also significantly depressed by 1 mM streptomycin. Response parameters returned to control (recovery) levels 24 h after removal from streptomycin, except for the number of distance responses in cod larvae. However, these increased sufficiently in 24 h to be not significantly different from untreated (recovery) controls.

The deleterious effect of 10 mM streptomycin on response score (as compared with untreated controls) was further confirmed on herring and plaice of different ages, and on flounder and halibut (Table 24.6). In 2- and 26-day-old herring and 4-day-old flounder, the score was still below control levels 24 h after removal from streptomycin.

24.3.7 Effect of Turbulence

The larvae treated by turbulence showed a significant fall in response score in four out of seven experiments (herring, flounder, and cod; Table 24.7). All showed recovery to control levels in the subsequent 20- to 24-h period.

TABLE 24.7. Mean response scores in 30 encounters of various species and ages after treatment by turbulence to dislodge the cupulae compared with controls.[a]

Species (days posthatch)	Time (hours)	Control	Turbulence
Herring (2)	0	2.47	1.13[b]
	24	–	2.13
Herring (13)	0	1.30	0.57[b]
	20	1.67	1.70
Plaice (4)	0	2.83	1.60
	24	–	2.47
Plaice (19)	0	1.40	1.43
	24	1.40	1.36
Flounder (4)	0	2.23	0.70[b]
	24	–	1.87
Cod (4)	0	2.17	0.74[b]
	23	–	1.90
Halibut (4–8)	0	0.93	0.30
	41	–	1.20

[a] Experiment was done at zero time and then repeated after 20–24 h of recovery to allow the cupulae to regenerate.
[b] Result was significantly lower than for the untreated control (one-tailed test, $p < .05$).

24.3.8 Effect of Age

A series of experiments over a 41-day period was done on herring to see if there were changes of the response parameters of untreated larvae during development. There was a significant decline in responsiveness as judged by response score, percent C-starts, and response distance (Fig. 24.3). The percentage of distance responses increased to a maximum at 17 days posthatching, then declined rapidly. Two other parameters, percentage errors made in both the direction of the initial C-start and the subsequent escape path, were measured. The errors, in general, were low, below 20%, with a significant decline in errors of escape during development (Fig. 24.3, Table 24.8). The general decline in responsiveness with increasing age and size may have been caused by the decrease in the relative size of the stimulus compared with the larvae, the same probe being used throughout these experiments.

24.4 Discussion

24.4.1 Escape Behavior

These experiments allow some conclusions to be drawn about the role of the free neuromast system, but they raise some fundamental neurobiological questions. It is clear that teleost larvae of a wide range of species and ages can respond to mechanical stimuli from a probe in the dark. Sometimes the response occurs *before* the larvae are touched but nevertheless when the source of stimulation is rather close, only a few millimeters away. The larvae have the ability to respond

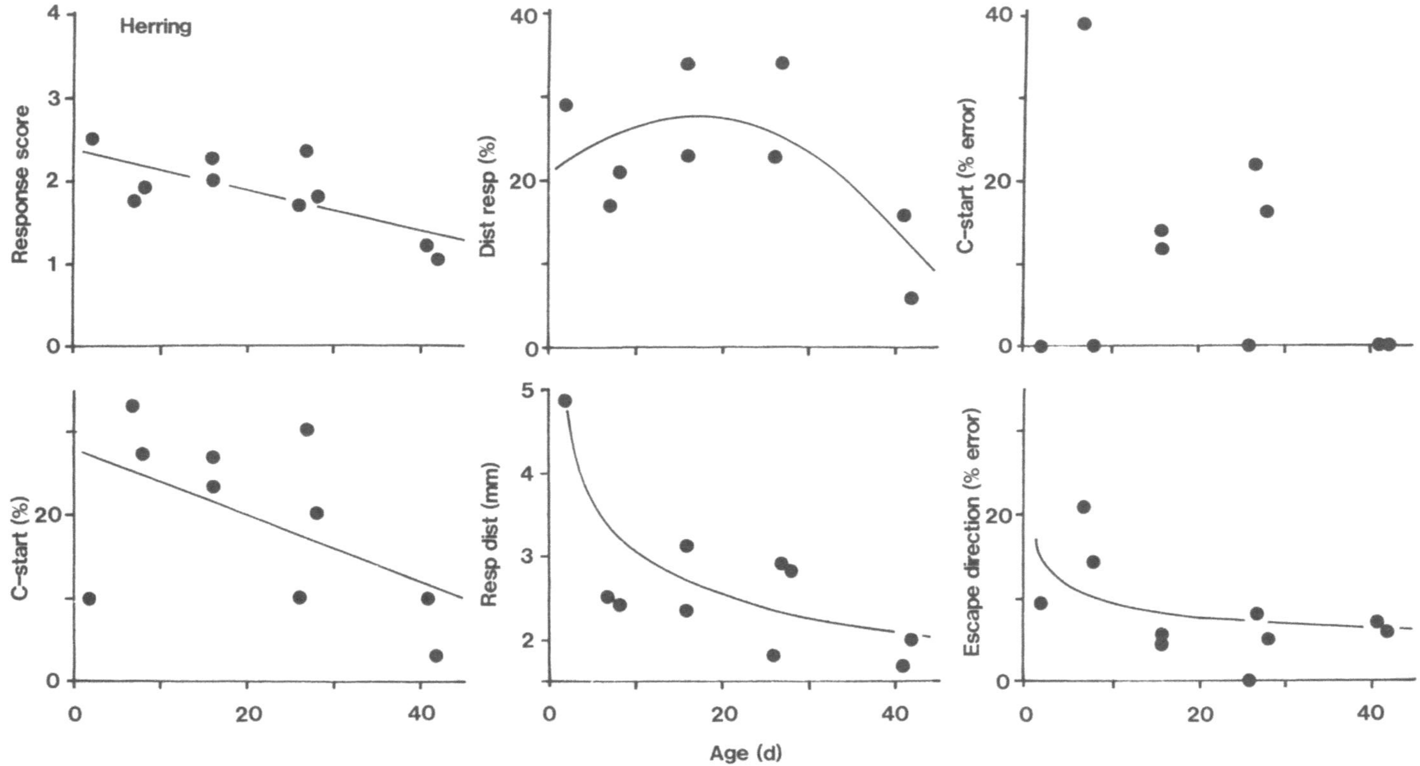

FIGURE 24.3. Effect of age in days posthatch on various response parameters of herring larvae. C-starts and distance responses are given as percent of positive responses. Response distance is the mean for all distance responses in that trial. C-start error is given as percent of those larvae making C-starts that make the initial turn toward the probe. Escape direction error is the percent of the escaping larvae that swam toward the probe. Each point plotted for response rate and C-starts represents 30 encounters. Points on other graphs represent varying numbers of observations. Equations for curves are given in Table 24.8. The growth in total length from 0 to 42 days posthatch is 8 to 15 mm.

TABLE 24.8. Equations for relationships shown in Fig. 24.3

Parameter (Y)	Equation (X = days posthatch)
Response rate	$Y = 2.355 - 0.024\ X$
Percent C-starts	$Y = 27.618 - 0.389\ X$
Percent distance responses	$Y = 28.145 - 0.026\ (X - 17.083)^2$
Response distance	$Y = 5.508\ X^{-0.261}$
C-start error	–
Escape direction error	$Y = 16.846\ X^{-0.271}$

with a C-start implicating Mauthner (M) or M-like cells in the brainstem and spinal cord. Over 60% of the distance responses are C-starts. Generally the ability to escape away from the stimulus is good, with errors being 20% or less. The responses are reduced or lost when the larvae are treated with streptomycin, which is known to ablate chemically the neuromasts, or held in turbulent conditions that dislodge some of the neuromast cupulae but may have other damaging effects.

The most likely explanation for these findings is that the approach of a probe (perhaps simulating a predator or other organism) close to a larva stimulates the free neuromasts, often leading to C-start behavior. It is also likely that C-start behavior is mediated via the Mauthner cells, indicating a functional connection of free neuromast input onto the Mauthner cells.

Some doubts need to be resolved in order to sustain this conclusion:

1. How certain is it that the probe is simulating the neuromasts and not other sensory systems in the skin or inner ear? It seems likely that the larvae may have Rohan-Beard cells under the skin, at least in the early posthatching stages, which respond to touch (Weis 1968; Roberts and Hayes 1977). Furthermore, although little is known of the structure of the inner ear of larvae, they possess otoliths even before hatching, and it is likely that the gravistatic sense is operating even if they are not sensitive to sounds.

It may well be that, in some instances, the larvae were responding when touched via touch receptors in the skin, but this is not possible for those larvae making responses at a distance from the probe. This leaves the possibility that the inner ear was involved in the distance responses. Could the treatment with 10 mM streptomycin, known to inhibit neuromasts, also have caused a loss of responses mediated via the labyrinth?

This was checked in the following way. Yolk sac herring larvae were transferred in two batches, one to natural seawater and the other to seawater containing 10 mM streptomycin. Swimming behavior in the light was then recorded by the video system in order to see whether there were signs of postural loss as a result of streptomycin entering the inner ear. No abnormalities of posture were observed, confirming that either streptomycin was not entering the labyrinth, or if it was, it was not affecting the sensory system.

There are some aspects of the streptomycin treatment that need further examination. The concentration required to diminish neuromast function was high.

According to Sand (1975), Shiozawa et al. (1981), and Kroese (personal communication), such high levels may be required because high levels of calcium ion (as in seawater) tend to antagonize the effects of suppressants such as La^{3+}, Co^{2+}, neomycin, and streptomycin. Furthermore, such high levels of streptomycin may also block the neuromuscular junctions (Sanders and Sanders 1979).

These points were checked experimentally. Yolk sac herring larvae were transferred to artificial seawater made up without calcium before attempting to check if lower levels of streptomycin were effective in ablating responsiveness. There was high mortality almost immediately in calcium-free artificial seawater, which did not occur in the controls, making it impossible to check the potency of streptomycin in calcium-free conditions.

Neuromuscular function was checked by recording the swimming behavior of two batches of yolk sac herring larvae, one in natural seawater, the other in seawater containing 10 mM streptomycin. Swimming was normal in both batches. An analysis of the video record showed that the larvae were significantly *less* active in the control than in the streptomycin solution, although there was no significant difference in routine swimming speeds (Table 24.9). There was thus no evidence for neuromuscular blockade, which would have interfered with swimming or would have reduced activity.

2. Are the free neuromasts exciting the M cells? To date, there is no published evidence that the lateral line can activate the M cells in any teleost (Eaton and Hackett 1984), either larval or adult. It is generally accepted that M cells appear very early in development (Eaton and DiDomenico 1986), even before hatching in zebra fish, *Danio rerio* (*Brachydanio rerio*), where they may be involved in violent movements during hatching (Eaton and Nissanov 1985). Meredith (personal communication) reported M cells in herring larvae, and further examination of her histological material shows the presence of axons passing down the spinal cord for several hundred micrometers. Cod, flounder, halibut, and plaice larvae have not been examined histologically.

It might seem that the existence of C-starts implies the presence of M cells, but work by Kimmel et al. (1980) and Eaton et al. (1984) on zebra fish larvae shows that C-starts can occur in their absence, although the latency is longer and the frequency much lower. There are, in fact, other cells in the reticulospinal system of zebra fish larvae (Kimmel et al. 1982) that appear to have a similar function, and these are also present in the herring.

There is good evidence that lateral line afferent fibers have inputs to the M cell in adult goldfish (Faber and Korn 1978; Zottoli and Van Horne 1983; see also Zottoli and Danielson Chapter 23) and larval zebrafish (Kimmel et al. 1982), but inputs in the goldfish appear to be to the ipsilateral side only (Zottoli and Van Horne 1983). Since the axons of the M and similar reticulospinal cells pass across to the contralateral side (Kimmel et al. 1982), a contraction of the body would be expected on the side away from the point of stimulation. This is the general finding in both the published and present work.

Although inputs from the free neuromasts to the M cell have not been established anatomically in the species used here, the circumstantial evidence (the

TABLE 24.9. Effect of 10 mM streptomycin on yolk sac herring larvae in the light.[a]

	Streptomycin-treated	Control	Significance
Percent of larvae active	32.4	19.8	$p < .001$ ($n = 559$)
Routine swimming speed (cm/s)	1.42	1.38	$p = .57$ ($n = 80$)

[a] Percentage of larvae active is based on counts in 20 frames over a 3-min period in both treated and control larvae. Routine swimming speed is based on measurements on 40 treated larvae and 40 control larvae. Significance pertains to a two-tailed test for the means for treated and control larvae.

presence of appropriate neurons, the existence of C-starts in response to stimulation of the free neuromasts) suggests that the free neuromasts are firing M or other recticulospinal cells to induce C-starts escape behavior, and this is generally away from the source of stimulation.

24.4.2 Neuromasts and the Boundary Layer

Most of the avoidance experiments described in this chapter were done on stationary larvae. In general it was more difficult to get moving larvae to respond. A discussion of boundary layers and their interaction with neuromast function is therefore relevant. The function of free neuromasts, being surface organs with a mechanical mechanism, may depend on the hydrodynamic environment. Characteristics of the flow of water around a fish change considerably during the lifetime of most species. This is because of the substantial increase in size that takes place and of the dependence of flow characteristics on size. Changes in the character of the boundary layer, the region of flow near a fish's body where there is a gradient in relative velocity, may be particularly important to the function of free neuromasts. With the exception of Daniel (1981), there have been no measurements made in the boundary layer of fishes, larval or adult. Consequently, predictions about changes in the boundary layer during growth must come from hydrodynamic theory, which is based on the flow around a stationary, rigid, flat plate. The nearest analogy for a fish would be during gliding.

The boundary layer is usually defined as the region in which flow around a body is reduced by at least 1%, but this is an arbitrary limit that gives a wrong impression that the layer is a discrete zone. At a given speed, the boundary layer extends farther from the body (is thicker) at posterior (downstream) locations than it does anteriorly. This means that a neuromast on the head experiences a steeper velocity gradient than one of the same height more posteriad. The thickness of the boundary layer also varies inversely with the speed of the fish relative to the water. As a fish grows, its swimming speed generally increases. This results in a thinner boundary layer at any given absolute distance from the snout, resulting in an increased velocity gradient at any point.

The application of these concepts to neuromast function is highly speculative, because we do not know how boundary layers are established around the oscillating body of a swimming fish or how much time is required for the boundary layer to become established in a dynamic situation. Furthermore the stationary larvae in the avoidance experiments would not have a boundary layer. Bearing these points in mind we can ask two questions: (1) Do neuromasts lie within the boundary layer of *swimming* fishes, and (2) Can individual free neuromasts function during swimming? Both questions can only be addressed at present by simple calculations, using equations based on the flat-plate analogy given by Vogel (1981). An average neuromast, including cupula, projects 100 μm from the body. The boundary layer (using the conservative definition, 10% flow reduction) close to the snout (0.5 mm) where it is narrowest is 2.5 mm at a swimming speed of 1 mm/s. It drops to 0.78 mm at 10 mm/s and 0.25 mm at 100 mm/s. Clearly, neuromasts do lie within the boundary layer. In larvae, such as 3- to 7-mm zebra fish, which cruise at 6 to 30 mm/s (Fuiman and Webb 1988), the neuromasts are within the inner half of the layer, experiencing little of the total velocity shear.

The second question asks whether the shear differential along the length of the cupula is so great that the motion caused by swimming stimulates the neuromasts. Assuming that the base of the cupula lies 20 μm from the body, the difference in velocities at the tip and the base of a cupula close to the snout is 36 μm/s at a swimming speed of 1 mm/s, and 1.1 mm/s and 36 mm/s at swimming speeds of 10 and 100 mm/s, respectively. These values drop to 11 μm/s, 0.4 mm/s, and 11 mm/s at the same speeds for neuromasts located 5 mm from the snout. The values for swimming speeds of larvae are small. Whether they are sufficient to trigger the hair cells depends on the drag of the cupulae. If swimming does trigger the hair cells and this response is sustained, the neuromasts could act as velocity meters. It is interesting in this respect that the herring larvae treated with streptomycin spent more time in activity (Table 24.9), although their velocity did not change.

If swimming movements are not sufficient to stimulate the free neuromasts in young larvae, then they may function as distance receptors until swimming speeds become great enough to thin the boundary layer and cause greater shear. At these faster speeds, the "noise" level created by the fish's own movements may preclude the function of free neuromasts as distance receptors. Yet, they could still operate as such when the fish is at rest. Perhaps the lateral line canal system, providing protection for the canal neuromasts, takes over the distance-receptor function of the free neuromasts when the shear in the boundary layer makes free neuromast function impractical. It seems significant that in all species examined, the lateral line canal system appears late in larval development when larvae are larger and swimming faster.

24.5 Conclusions and Summary

At hatching, the larvae of most teleost fishes have a few free neuromasts distributed over the head and trunk, but the numbers increase during development. The lateral line system, with the neuromasts sunk into pits or canals, appears sometime after hatching. Thus the younger larvae may survive for days or weeks with a very limited number, distribution, and type of mechanoreceptors at the body surface. During this time the larvae are capable of making avoidance responses when stimulated by probes in the absence of visual input. On the approach of a probe, as observed by a video system, utilizing infrared light invisible to the fish, larvae may respond on being touched by the probe, but they may also respond when the probe is a few millimeters away (distance responses). Evidence is presented that these distance responses are mediated via the free neuromast system. Both distance responses and responses to touch may involve C-start avoidance behavior. It is possible, therefore, that this is the first demonstration of C-start behavior being mediated via neuromasts.

Acknowledgments. We are grateful to the following, who made many suggestions to improve the chapter in manuscript (without necessarily agreeing with the conclusions): Sheryl Coombs, Eric Denton, Robert Eaton, Walt Metcalfe, and John Montgomery. L.A. Fuiman was supported by the National Science Foundation, under grant BSR-8503058 awarded in 1985.

References

Batty RS (1983) Observations of fish larvae in the dark with television and infra-red illumination. Mar Biol 76:105–107.

Blaxter JHS (1969) Visual thresholds and spectral sensitivity of flatfish larvae. J Exp Biol 51:221–230.

Blaxter JHS (1984a) Cupular growth in herring neuromasts. J Mar Biol Assoc UK 64:935–938.

Blaxter JHS (1984b) Neuromasts and cupular growth of cod larvae. Flodevigen Rapportserie 1:183–188.

Blaxter JHS (1987) Structure and development of the lateral line. Biol Rev 62:471–514.

Blaxter JHS, Batty RS (1985) The development of startle responses in herring larvae. J Mar Biol Assoc UK 65:737–750.

Blaxter JHS, Hoss DE (1981) Startle response in herring: The effect of sound stimulus frequency, size of fish, and selective interference with the acoustico-lateralis system. J Mar Biol Assoc UK 61:871–879.

Blaxter JHS, Denton EJ, Gray JAB (1981) The auditory bullae-swimbladder system in late stage herring larvae. J Mar Biol Assoc UK 61:315–326.

Blaxter JHS, Gray JAB, Best ACG (1983a) Structure and development of the free neuromasts and lateral line system of the herring. J Mar Biol Assoc UK 63:247–260.

Blaxter JHS, Danielssen D, Moksness E, Øiestad V (1983b) Description of the early development of the halibut *Hippoglossus hippoglossus* and attempts to rear the larvae past first feeding. Mar Biol 73:99–107.

Bleckmann H (1986) Role of the lateral line in fish behaviour. In: Pitcher TJ (ed) The Behaviour of Teleost Fish. London: Croom Helm, pp. 177–202.

Bourgos GD (1986) The development and operation of a new tank system suitable for rearing marine flatfish larvae on microencapsulated diets. Thesis submitted to Council for National Academic Awards, Plymouth Polytechnic, U.K.

Cahn PH, Shaw E (1962) First demonstration of lateral line cupulae in the Mugiliformes. Copeia 1:109–114.

Cahn PH, Shaw E, Atz EH (1968) Lateral line histology as related to the development of schooling in the atherinid fish *Menidia*. Bull Mar Sci 18:660–670.

Dale T (1984)Embryogenesis and growth of otoliths in the cod. Flodevigen Rapportserie 1:231–250.

Daniel TL (1981) Fish mucus: In situ measurements of polymer drag reduction. Biol Bull (Woods Hole) 160:376–382.

Disler NN (1971) Lateral Line Sense Organs and Their Importance in Fish Behavior. Jerusalem: Israel Program for Scientific Translations, 328 pp.

Eaton RC, DiDomenico R (1986) Role of the teleost escape response during development. Trans Am Fish Soc 115:128–142.

Eaton RC, Hackett JT (1984) The role of the Mauthner cell in fast starts involving escape in teleost fish. In: Eaton RC (ed) Neural Mechanisms of Startle Behavior. New York: Plenum, pp. 213–266.

Eaton RC, Nissanov J (1985) A review of Mauthner-initiated escape behavior and its possible role in hatching in the immature zebrafish *Brachydanio rerio*. Environ Biol Fishes 12:265–279.

Eaton RC, Nissanov J, Wieland CM (1984) Differential activation of Mauthner and non-Mauthner startle circuits in the zebrafish: Implications for functional substitution. J Comp Physiol A 155:813–820.

Faber DS, Korn H (eds) Neurobiology of the Mauthner Cell. New York: Raven Press, 290 pp.

Fridgeirsson E (1978) Embryonic development of five species of gadoid fishes in Icelandic waters. Rit Fiskideildar 5(6):68.

Fuiman LA, Webb PW (1988) Ontogeny of routine swimming activity and performance in zebra danios (Teleostei: Cyprinidae). Anim Behav 36:250–261.

Hoss DE, Blaxter JHS (1982) Development and function of the swimbladder–inner ear–lateral line system in the Atlantic menhaden *Brevoortia tyrannus* (Latrobe). J Fish Biol 20:131–142.

Iwai T (1963) Development of lateral line cupulae in the goboid fish *Tridentiger trigonocephalus*. Bull Misaki Biol Inst Kyoto Univ 4:1–20.

Iwai T (1967) Structure and development of lateral line cupulae in teleost larvae. In: Cahn PH (ed) Lateral Line Detectors. Bloomington: Indiana University Press, pp. 27–44.

Iwai T (1980) Sensory anatomy and feeding of fish larvae. In: Bardach JE, Magnuson JJ, May RC, Reinhart JM (eds) Fish Behaviour and Its Use in the Capture and Cultivation of Fishes. Manila, Philippines: ICLARM, pp. 124–145.

Iwai T (1983) Surface morphology of the naked neuromasts in anchovy larvae. Bull Jpn Soc Sci Fish 49:1935.

Kaus S, Fleischer K (1982) Ototoxische Wirkungen an der Haarzellen der Seitenlinieorgane von Fischen. Arch Ohren-Nasen Kehlkopfheilkunde 235:586.

Kawamura G, Ishida K (1985) Changes in the sense organ morphology and behaviour with growth in the flounder *Paralichthys olivaceus*. Bull Jpn Soc Sci Fish 51:155–165.

Kimmel CB, Eaton RC, Powell SL (1980) Decreased fast-startle performance of zebrafish larvae lacking Mauthner neurons. J Comp Physiol 140:343–350.

Kimmel CB, Powell SL, Metcalf WK (1982) Brain neurons which project to the spinal cord in young larvae of the zebrafish. J Comp Neurol 205:112–127.

Neave D (1986) The development of the lateral line system in plaice (*Pleuronectes platessa*) and turbot (*Scophthalmus maximus*). J Mar Biol Assoc UK 66:683–693.

O'Connell CP (1981) Development of organ systems in the northern anchovy *Engraulis mordax* and other teleosts. Am Zool 21:429–446.

Økland S, Øiestad V, Berg L (1988) Development of free neuromasts and their cupulae in halibut larvae (*Hippoglossus hippoglossus* L.) and destruction of the cupulae in different types of rearing systems. (In press.)

Roberts A, Hayes BP (1977) The anatomy and function of "free" nerve endings in an amphibian skin sensory system. Proc R Soc Lond B 196:415–429.

Sand O (1975) Effects of different ionic environments on the mechanosensitivity of lateral line organs in the mudpuppy. J Comp Physiol 102:27–42.

Sanders WE Jr, Sanders CC (1979) Toxicity of antibacterial agents: Mechanism of action on mammalian cells. Annu Rev Pharmacol Toxicol 19:53–83.

Shiozawa K, Asanuma A, Yanagisawa K (1981) Effects of calcium and neomycin on the mechano-sensitivity of the lateral line organ. J Physiol Soc Jpn 43:367.

Snedecor GW, Cochran WG (1967) Statistical Methods. Ames: Iowa State University Press.

Vogel S (1981) Life in Moving Fluids, the Physical Biology of Flow. Princeton, NJ: Princeton University Press, 352 pp.

Weis JS (1968) Analysis of the development of the nervous system of the zebrafish *Brachydanio rerio*. J Embryol Exp Morphol 19:109–119.

Wersäll J, Flock Å (1964) Suppression and restoration of the microphonic output from the lateral line organ after local application of streptomycin. Life Sci 3:1151–1155.

Yin MC, Blaxter JHS (1987) Escape speeds of marine fish larvae during early development. Mar Biol 96:459–468.

Zottoli S, Horne C Van (1983) Posterior lateral line afferent and efferent pathways within the central nervous system of the goldfish with special reference to the Mauthner cell. J Comp Neurol 219:100–111.

25 The Lateral Line System of Surface-Feeding Fish: Anatomy, Physiology, and Behavior

HORST BLECKMANN, GOTTFRIED TITTEL, and ELKE BLÜBAUM-GRONAU

25.1 Introduction

Fishes live in almost every type of aquatic habitat. Not surprisingly, many fishes show striking morphological, physiological, and behavioral adaptations. Some species, belonging to the families Cyprinodontidae, Hemirhamphidae, Gasteropelecidae, and Pantodonitidae, are specialized in foraging at the water surface. Schwartz (1965, 1971) has shown that surface-feeding fish detect and localize part of their prey, terrestrial insects fallen into the water, by means of capillary surface waves elicited by the prey's struggling. The receptive structures involved in prey detection are lateral line organs located on the fish's head and back (Schwartz 1970). Behavioral studies indicate that surface-feeding fish distinguish between different wave types. In addition, these fish determine the direction and the distance to a wave source under open loop conditions—i.e., even if only a single, short-lasting wave train is presented. In this chapter we review the literature on surface-feeding fish. In addition, we include recent unpublished anatomical, electrophysiological, and behavioral results. For more general reviews on water surface wave reception, the reader is referred to Bleckmann (1985a, 1988) and Wilcox (1988).

25.2 Physical Properties of the Water Surface

The particular properties of the water surface are created by strong cohesion between water molecules and the asymmetry of intermolecular forces between water and air. Water surface waves are boundary waves caused by moving a solid through the water, by shock, wind, and all similar forces that generate and maintain waves. Water waves transport energy but not mass.

Some of the results shown in this chapter are parts of the doctoral thesis of G. Tittel and E. Blübaum-Gronau.

25.2.1 Depth Impact

The maximal horizontal and vertical displacement components A_x and A_z caused by a surface wave can be calculated according to Lighthill (1980)*:

$$A_x = A_0/(\sinh(k \cdot h)) \cdot \cosh(k(h-z)) \text{ and}$$
$$A_z = A_0/(\sinh(k \cdot h)) \cdot \sinh(k(h-z)) \quad (1)$$

where $k = 2\pi/\lambda$ (λ = wavelength), A_0 = maximal displacement amplitude of the particles at the water surface above the observation point, z = vertical distance below the water surface, and h = water depth. For $h > \lambda$ and $z < h$, i.e., in deep water not too close to the bottom, A_x equals A_z; hence the particles describe a circular path (Fig. 25.1A). For $h < \lambda$ or $z \approx h$, i.e., in shallow water or in deep water very close to the bottom, particles move in an elliptical path (Fig. 25.1B). In general, depth impact of surface waves is very small and decreases with decreasing λ. For instance, at a depth of 1 λ (e.g., 23.6 mm at 10 Hz and 2.9 mm at 140 Hz, respectively), the vertical movement of water particles is less than 1/500 of that observed at the surface.

25.2.2 Dispersion Characteristics

Each short, nonsinusoidal local disturbance of the water surface generates a wave train (click) that has many frequencies. During horizontal propagation, such a click not only decreases in amplitude but also changes in frequency composition, duration, and time course (Fig. 25.1C). There are two reasons for this: (1) gravity waves ($\ll$ 13 Hz) have normal dispersion characteristics, i.e., the longer waves travel faster than the shorter waves; and (2) capillary waves ($\gg$ 13 Hz) are abnormally dispersive, i.e., the shorter waves travel faster than the longer waves (Sommerfeld 1970). For $h \gg \lambda$, the following equation describes the dispersion relationship of surface waves (Lighthill 1980)

$$C_{ph}^2 \equiv \frac{\omega^2}{k^2} = \left[g + \frac{Tk^2}{\rho}\right]k^{-1} \quad (2)$$

where $\omega = 2\pi f$ (f = frequency in Hz), g = gravitational acceleration, ρ = density of water, T = coefficient of surface tension, and C_{ph} = phase velocity in cm/s. For typical parameters (g = 981 cm/s² and T/ρ = 72 cm³/s²), the minimal phase velocity of water surface waves is 23.1 cm/s at a wavelength of about 1.7 cm (corresponding to a frequency of 13 Hz) (Fig. 25.1D).

After traveling a certain distance, a dispersive wave group consists of a band of different wavelength and frequencies. Locally λ can be defined by the crest-to-crest distance, but owing to dispersion (Eq. 2) the distance of successive crests

*Strictly speaking, Eqs. 1 hold valid only for gravity surface waves. Calculation of particle movements in capillary surface waves requires a more difficult, nonlinear approach (Lighthill 1980).

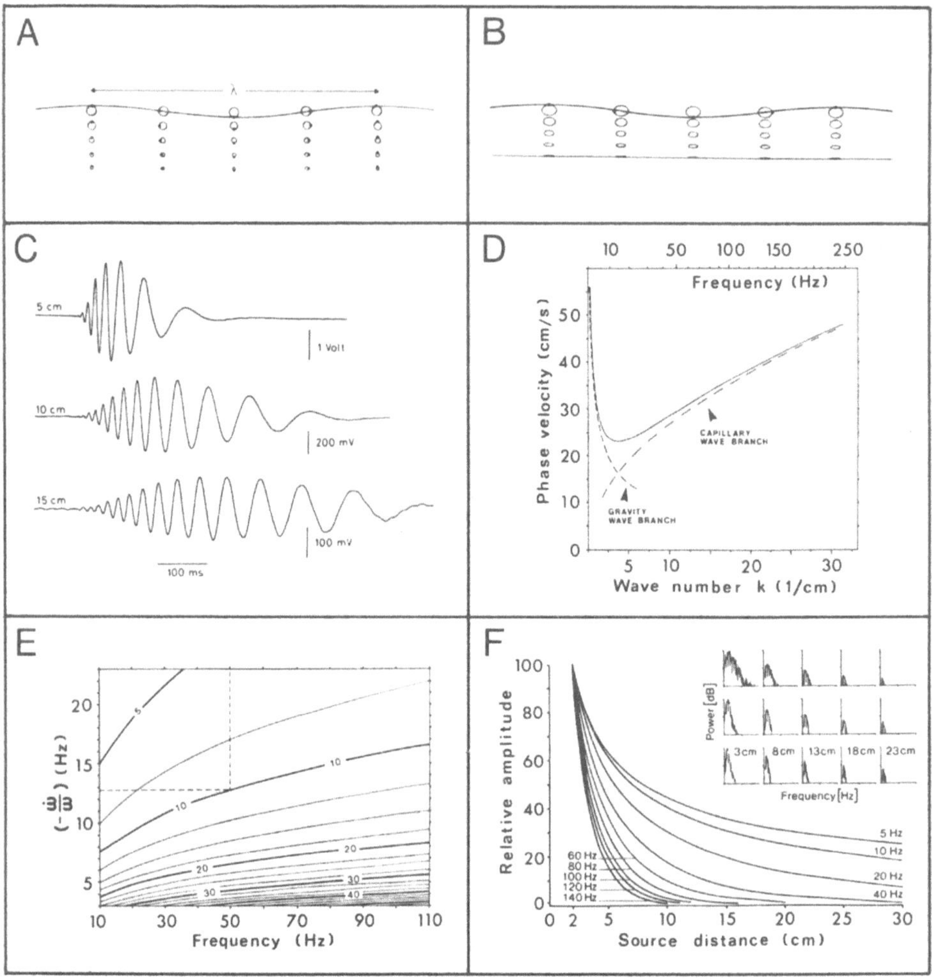

FIGURE 25.1. Motion of fluid particles in a sinusoidal wave of length λ traveling from left to right on deep water (A) and flat water (B). In (A) the maximum surface elevation is 0.02 λ, and particles on the surface describe circles of this radius. In each case the particle's instantaneous position on its circular path is shown. In (B) the maximum surface elevation again is 0.02 λ. Water depth is 0.16 λ. A particles instantaneous position on its elliptical path is shown only for those in the top row, but the motion of every particle in the same vertical line again is in phase. In (A) and (B) the movement of particles is to scale in relation to wave amplitude and depth of observation point (from Lighthill 1980). (C) The nature of a click stimulus as it passes points 5, 10, and 15 cm. (D) The wave speed C_{ph} for ripples on deep water (solid line). Asymptotes for pure capillary waves (dashed-dotted line) and gravity waves (dashed line) are also shown. Transition occurs around a frequency of 13 Hz, where phase velocity is minimal. (E) Range contours as function of relative frequency change and local frequency. For example, a stimulus with a local frequency of 50 Hz and a relative frequency change of 12.8 Hz has traveled a distance of 10 cm as indicated by the intersection of dashed lines. (F) Calculated attenuation of water surface waves (Eq. 4) as a function of stimulus frequency and distance from vibration source. Inset: change of spectral composition of water surface waves during stimulus propagation. (From Bleckmann 1985a; Käse and Bleckmann 1987).

differs by a small amount (Fig. 25.1C). These differences reflect both the distance-dependent frequency modulation of the wave group and the speed with which the wave packet spreads in space. Independent of frequency content, absolute frequency modulation, and wave amplitude, the source distance D can be calculated unequivocally if only two parameters, namely the local frequency ω and the frequency modulation $\dot{\omega}$ around this local frequency, are known. The exact relation between D phase velocity C_{ph}, ω, and $\dot{\omega}$ is given by (Käse and Bleckmann 1987):

$$D = -4.5\, C_{ph}\, (\omega/\dot{\omega}) \tag{3}$$

The dotted variable denotes the derivative with respect to time. Figure 25.1E displays the range contours calculated according to Eq. 3.

25.2.3 Attenuation

Concentrical surface waves are strongly attenuated during propagation. The slope of attenuation depends on both stimulus frequency and the distance the stimulus has traveled. For $D \geq 2$ cm, the damping of water waves can be approximated by (Bleckmann 1988):

$$A = A_i \sqrt{\frac{R_i}{R}}\, e^{(16\, \nu\pi^2/\lambda^2 C_{ph})(R_i\, -\, R)} \tag{4}$$

where A_i = displacement amplitude of the wave stimulus at distance R_i, A = actual wave amplitude at distance R, and ν = kinematic viscosity of water (=0.01 $cm^2\, s^{-1}$ at 20°C). Because of geometrical spreading, attenuation occurs mostly in the vicinity of a wave source. In addition, the water surface behaves like a low-pass filter; i.e., attenuation strongly increases with frequency (decreasing λ) (Fig. 25.1F).

25.2.4 Signals and Noise

The biologist who wants to study the mechanosensory lateral line faces the problem that little is known about natural lateral line stimuli. This especially holds true for midwater and bottom-dwelling fishes. The water surface is an exception in that surface waves can easily be observed, measured, and quantified. A common source of waterborne vibrations are aquatic, semiaquatic, and terrestrial insects that have fallen into the water (Lang 1980) and aquatic vertebrates such as small fish, frogs, and tadpoles (Bleckmann and Lotz 1987). "Noise" waves are caused by wind or water currents and by falling leaves, seeds, and twigs. According to present knowledge, the frequency content of prey (insect) stimuli and background noise (i.e., all other wave types) that potentially interferes with the relevant messages clearly differs close to the source. The spectrum of background noise is typically of narrow band width below 20 to 50 Hz, whereas insect stimuli are broad-banded and extend to frequencies above 50 Hz (Fig. 25.2).

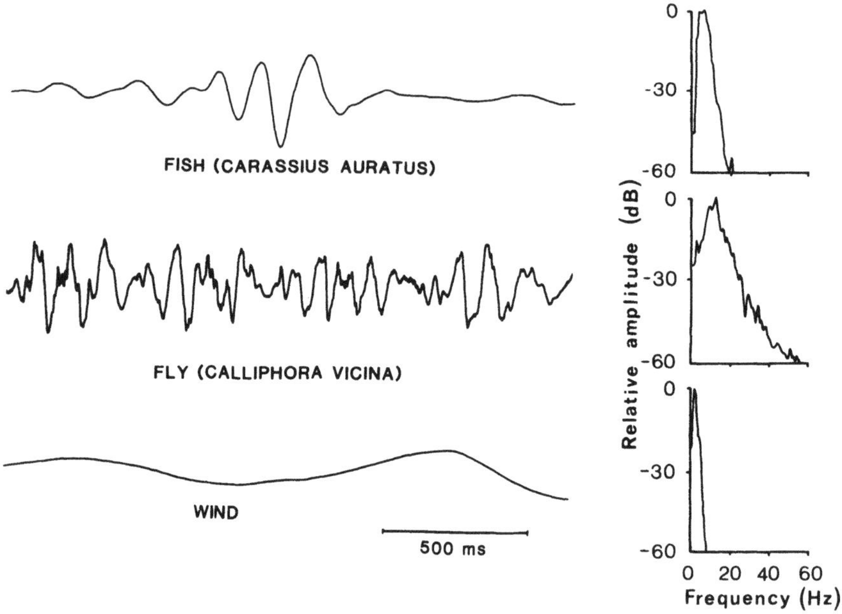

FIGURE 25.2. Examples of water surface waves caused by different sources. Amplitudes are not to scale. Right: Relative spectra of corresponding wave.

Water surface disturbances may differ not only in frequency but also in time structure and duration (Fig. 25.2). Insect stimuli are more irregular in the time course than all other wave types; their duration often exceeds 1.5 s, with values of 60 or more being common. Wind waves can last for hours or even days. However, the duration of all other wave types usually does not exceed 1.5 s (Bleckmann 1985b). Surface waves relevant in the present context are in the micrometer (insect waves) to centimeter range (wind waves, wind speed 6 to 12 km/h) (Lang 1980; Bleckmann and Rovner 1984; Bleckmann 1985b). In contrast to frequency content, time structure, and duration, wave amplitude alone is not sufficient for distinguishing between different stimuli.

25.3 Morphology

25.3.1 Receptors

Lateral line neuromasts are sensitive vibration receptors. In *Aplocheilus lineatus* (Cyprinodontidae) and *Pantodon buchholzi* (Pantodontidae), the neuromasts used for surface wave detection belong to the infraorbital line, the supraorbital line, and the supratemporal line. In addition, neuromasts innervated by a dorsal branch of the trunk lateral line contribute to surface wave perception (Müller and Schwartz 1982; Tittel et al. 1984). The lateral lines of surface-feeding fish may

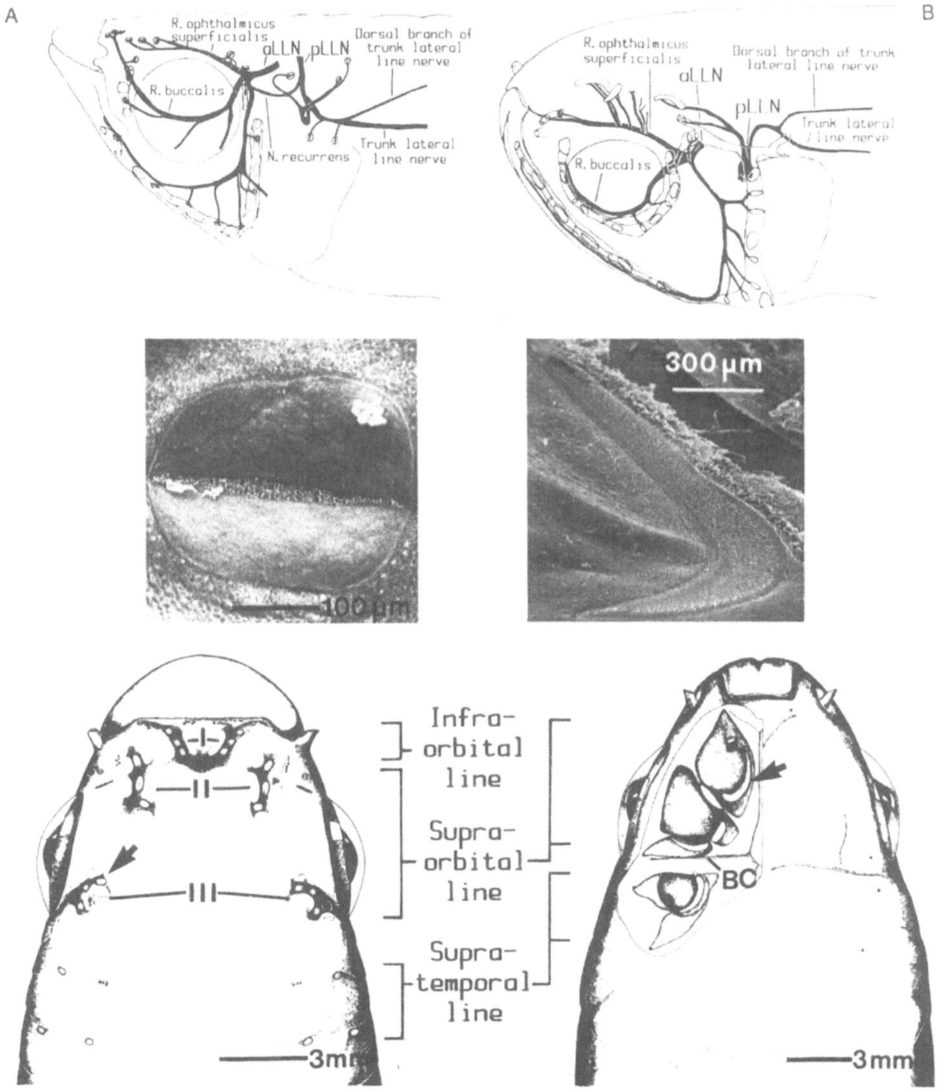

FIGURE 25.3. Schemes of organization of the cephalic and supratemporal lateral lines of *A. lineatus* (A) and *P. buchholzi* (B). Top: Innervation pattern of head lateral line neuromasts. Bottom: Dorsal view of the heads of *Aplocheilus* and *Pantodon*. In *Aplocheilus*, dorsal head neuromasts are arranged in three groups (labeled I, II, and III from rostral to caudal), each of which consists of three single organs (named in the text as 1, 2, and 3 from rostral to caudal). In *Pantodon*, the eight large neuromasts of the cephalic line and the two large neuromasts of the supratemporal line are covered by a thin epidermal membrane which is supported by a bony crest (BC). Insets: Scanning micrographs of the neuromasts marked by arrowheads. In *Aplocheilus*, the sensory epithelium of the neuromast covers only a small, medial stripe. In *Pantodon*, the hair cells are evenly distributed over the width of the entire banana-shaped neuromast.

differ with respect to number and size of neuromasts, neuromast arrangement, and neuromast fine structure (Schwartz 1970; Müller 1984). Based on afferent innervation, the supraorbital line includes 12 neuromasts in *Aplocheilus* and eight large and four small neuromasts in *Pantodon* (Fig. 25.3A,B). The head of *Pantodon* in addition contains six rows of up to 70 very small (about 50 × 30 μm) neuromasts which are innervated by branches of the ramus ophthalmicus superficialis. In *Aplocheilus*, the six most dorsorostrally located neuromasts of the infraorbital line are also involved in surface wave perception. These six infraorbital neuromasts (in *Aplocheilus*) as well as all supraorbital neuromasts (in *Aplocheilus* and *Pantodon*) are defined as cephalic lateral line in this chapter (Fig. 25.3A,B). In surface-feeding fish, the fine structure of neuromasts is similar to that observed in midwater and bottom-dwelling fishes (e.g., Jannsen et al. 1987). However, specific differences with respect to the cephalic lateral lines and neuromast fine structure do occur between different species of surface-feeding fish (Fig. 25.3A,B; Table 25.1).

25.3.2 Medullary Organization

Detailed information regarding the central organization of the lateral line are given by McCormick (Chapter 17) and Roberts and Meredith (Chapter 22). However, because surface-feeding fish have not been investigated with respect to central lateral line organization, some recent unpublished anatomical findings will be included in this chapter. It is well known that in teleosts the primary target of the mechanoreceptive lateral line is the medial octavolateralis nucleus (MON). In *Pantodon*, horseradish peroxidase–labeled lateral line afferents innervating supraorbital neuromasts course in the dorsal parts, those innervating mandibular neuromasts in the ventral parts of the dorsal and ventral root of the anterior lateral line nerve (aLLN). Three zones can be distinguished in the MON of *Pantodon* (Blübaum-Gronau and Münz 1987): a dorsolateral zone, which is the primary target of fibers innervating trunk neuromasts; a medial zone, which is the primary target of mandibular neuromasts; and a ventro-medial zone, which is the primary target of supraorbital neuromasts (Fig. 25.4).

A similar separation of trunk and supraorbital lateral line termination sites exists in *Aplocheilus* (Blübaum-Gronau et al. 1988). In *Pantodon*, different parts of the lateral line differ with respect to their collateral branching patterns. This is different from *Aplocheilus*, where different parts of the lateral line show a similar branching (Söhl, personal communication). Entering trunk afferents bifurcate to form ascending and descending fiber tracts which span the entire length of the MON. Both tracts give rise to collaterals that course medially (Fig. 25.4A). Some fibers of the trunk and mandibular lateral line continue into the nucleus caudalis, a nucleus clearly recognized only in *Amia calva* (McCormick 1981). Only collaterals of the mandibular afferents ramify extensively and reach the medial aspect of the contralateral MON (Fig. 25.4B). Afferents from the supraorbital line give rise to collaterals that ascend dorsorostrad in a common bundle up to the rostral boundary on the MON. The descending bundle continues without any further branching to its ventromedial termination site (Fig. 25.4C).

TABLE 25.1. Morphology and innervation of supraorbital, supratemporal, and trunk lateral line neuromasts.

	Aplocheilus lineatus		*Pantodon buchholzi*	
	Supraorbital line	Trunk lateral line	Supraorbital line	Supratemporal line
Nm length (mm)	0.3–0.6	0.015–0.15	0.81–2.08	2.2[a]
Nm depth (µm)	30–40	30–40		60–70[a]
Hc distribution	Narrow groove in Nm longitudinal axis	Circular in the Nm center	Nearly over the whole Nm epithelium	
HC orientation	Antagonistic		Antagonistic	
	Parallel to Nm longit. axis		Perpendicular to Nm longit. axis	
	Irregular		Paired	
	Peripheral and within the sensory epithelium		Peripheral	
Location of smaller HC				
HC number within a Nm	80–490	8–60	1,200–2,100	3,200[a]
HC length (µm)	15–25	About 15		–40[a]
HC diameter maximum	5–7	4–6		
Number of stereovilli	10–60	10–20		10–50[a]
Stereocilium length min/max (µm)	0.5–2	0.5–1		
Kinocilium length min/max (µm)	6–10	6–10		–18[a]
Relation of afferent/efferent terminals	About 1:4			About 1:4[a]
Number of myelinated nerve fibers supplying a Nm	45–130 ($N=5$; $n=19$)	7–14		About 800[a]
Innervation of Nm groups by axon coll.	No or few collaterals ($N=3$; $n=13$)	Mainly collaterals		
Mean convergence HC/nerve fiber	About 1:4	at least 1:6		About 1:4[a]

Data on trunk neuromasts are available only for *Aplocheilus*.
In each case at least 20 hair cells were investigated.
[a] Data from Müller (1984).

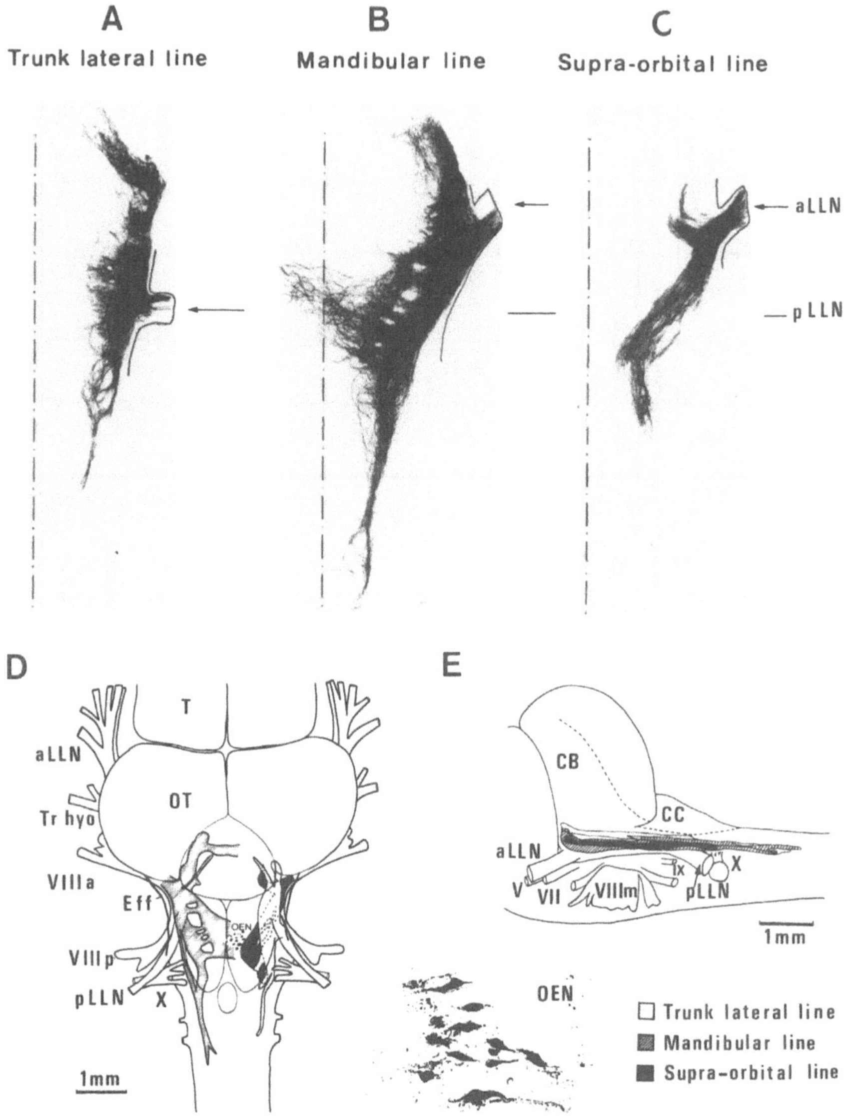

FIGURE 25.4. (A–C) Dorsal view of the branching pattern of *Pantodon* trunk lateral line, mandibular lateral line, and supraorbital lateral line afferents. In each case the dashed line symbolizes the midline of the medulla. Bottom: Dorsal (D) and lateral view (E) of the brainstem of *Pantodon*. Abbreviations: a, anterior; CB, cerebellum; CC, cerebellar crest; EFF, efferent fibers; LLN, lateral line nerve; m, medialis; OEN, octavolateralis efferent nucleus; OT, optic tectum; p, posterior; T, telencephalon; Tr hyo, truncus hyomandibularis. The termination sites of the anterior and posterior lateral line afferents (including their collaterals) in the MON are shown. Cranial nerves are labeled by Roman numerals. Inset D: Photomicrograph of HRP-labeled efferent cell bodies in the OEN.

In *Pantodon*, afferents that innervate mandibular or trunk neuromasts continue into the cerebellar eminentia granularis. Here they terminate in different layers: trunk lateral line fibers end in the dorsolateral, mandibular afferents in the ventrolateral eminentia granularis. Some fibers pass the ipsilateral eminentia granularis and terminate either in the granular layer of the corpus cerebellum or in the contralateral eminentia. In contrast, fibers of the supraorbital line do not course beyond the rostral border of the ipsilateral MON (Fig. 25.4C,D). In *Aplocheilus*, primary lateral line afferents also enter the valvula cerebelli (Blübaum-Gronau et al. 1988). In *Pantodon*, a small contingent of primary lateral line afferents travels rostrally and reaches the boundary zone between corpus and valvula cerebelli.

In surface-feeding fish, retrogradely labeled efferent cells can be found in the ipsilateral medulla either in a single media column (*Pantodon*) (Fig. 25.4D and inset) or in a rostral and a caudal nucleus (*Aplocheilus*) (Blübaum-Gronau et al. 1988). In *Pantodon* some contralateral efferent cells do occur. The efferent axons course rostrally, where they continue dorsolateral to the border of the medial longitudinal fasciculus. At the level where the facial nerve enters the medulla, they form a small fascicle which exits dorsal to the facial nerve and ventral to the aLLN. In *Pantodon* this fascicle bifurcates after leaving the brain: a small bundle travels rostrally to the aLLN, whereas the second bundle passes the octaval nerve to enter the posterior lateral line nerve.

25.4 Electrophysiology

25.4.1 Vibration Sensitivity of the Cephalic Lateral Line

If a surface-feeding fish is positioned close below the water surface, neural responses can be generated at least within the frequency range 10 to 250 Hz – i.e., within the frequency range relevant for prey (insect) detection (see above). If sinusoidal single frequency (sf) stimuli are presented, the cephalic neuromasts of *Aplocheilus* (only neuromast II.2 was tested) have minimal displacement thresholds in the range 75 to 150 Hz. Within this range, a peak-to-peak (p-p) water displacement of 0.2 to 0.4 μm is sufficient to generate a neural response (Fig. 25.5). As expected from the weak depth impact of water surface waves (Fig. 25.1A,B), neural responses decrease rapidly if the distance between the water surface and the neuromast recorded from is enlarged (Fig. 25.5, inset).

25.4.2 Input-Output Curves

For a given frequency there is an increase of neural response with increasing stimulus amplitude. In *Aplocheilus* and *Pantodon*, this increase is more pro-

►

FIGURE 25.6. Neural response of *Aplocheilus* (A) and *Pantodon* (B) to sinusoidal sf wave stimuli which varied in amplitude and frequency. In (A) the number of spikes per stimulus cycle, in (B) the amplitude of microphonic potentials was taken as a measure. Recordings were made from neuromast II.2 in (A) and a supratemporal neuromast in (B). (Modified from Topp 1983; Müller 1984.)

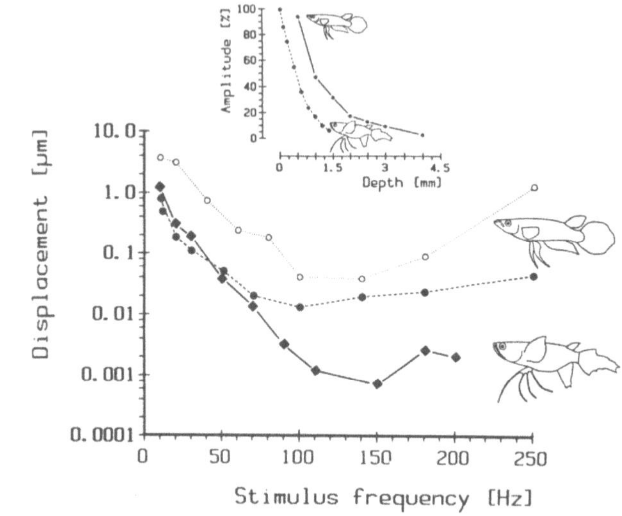

FIGURE 25.5. Displacement threshold of *A. lineatus* (○) and *P. buchholzi* (♦). Values were obtained in behavioral (solid symbols) and electrophysiological (open symbols) experiments. *Aplocheilus* recordings were made from a branch of the truncus supraorbitalis innervating neuromast II.2 (modified from Bleckmann 1980; Bleckmann and Topp 1981; Müller 1984). Inset: Displacement thresholds of cephalic neuromasts as function of distance between neuromasts and water surface. For both *Aplocheilus* (●—●) and *Pantodon* (●----●), the amplitude of microphonic potentials was taken as a measure. Ordinate: The largest response, defined as p-p amplitude of microphonic potential divided by p-p stimulus displacement, was set equal to 100%. Stimuli were 25 Hz, 20 µm p-p displacement for *Aplocheilus*, and 70 Hz, 6 µm p-p displacement for *Pantodon*. Abscissa: Distance between water surface and the neuromast recorded from. (From Unbehauen 1980; Müller 1984.)

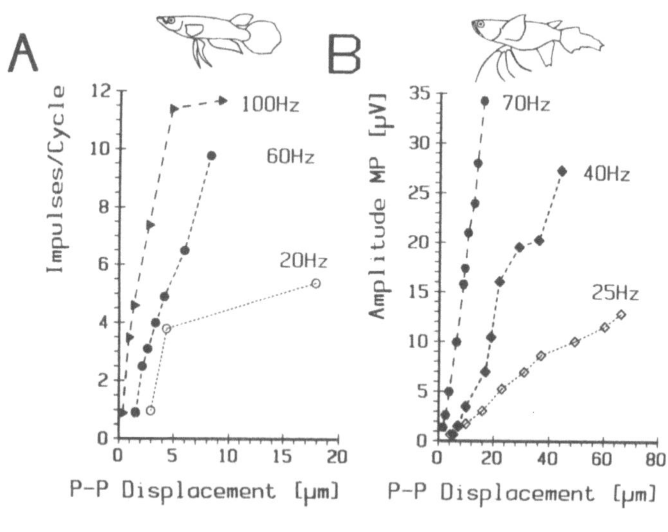

FIGURE 25.6.

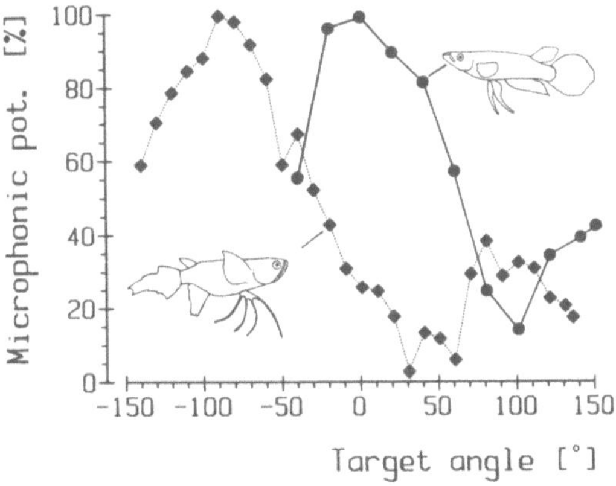

FIGURE 25.7. Directional sensitivity of neuromast II.2 (*Aplocheilus*) and a large supratemporal neuromast (*Pantodon*). For each stimulus angle the normalized amplitude of the microphonic potential is shown; 0° represents long axis of neuromasts. For technical reasons measurements could not be done for all angles. (From Unbehauen 1980; Müller 1984.)

nounced if low-amplitude high-frequency wave stimuli are presented (Fig. 25.6). Thus again, the cephalic lateral lines of surface-feeding fish are well suited to detect high-frequency, prey-caused wave components against low-frequency noise, irrespective of the small-displacement amplitudes prey caused stimuli may have (see spectra in Fig. 25.2).

25.4.3 Directional Sensitivity

Lateral line organs are directionally sensitive (Flock 1965). In *Aplocheilus*, maximum sensitivity of cephalic neuromasts (only neuromast II.2 was investigated) corresponds to the longitudinal axes of these neuromasts (Fig. 25.7). As expected from hair cell orientation (Table 25.1), maximum sensitivity of the supratemporal neuromasts of *Pantodon* is perpendicular to the neuromast's long axis (Fig. 25.7). As in *Xenopus* (see Görner, Mohr Chapter 27), there are no behavioral indications that surface-feeding fish make use of this directionality (see below).

25.5 Behavior

Under natural circumstances, waves generated by terrestrial insects trapped at the water surface serve as releasing and directing stimuli. If rewarded with food, surface-dwelling fish respond to artificial wave stimuli as well. Click stimuli, like the one shown in Fig. 25.1C, can be produced by a 10-ms punctate air blast onto the water surface. Up to 15 to 20 cm, the responses of surface-feeding fish to such a click are highly accurate with respect to both target angle and distance determination (Fig. 25.8).

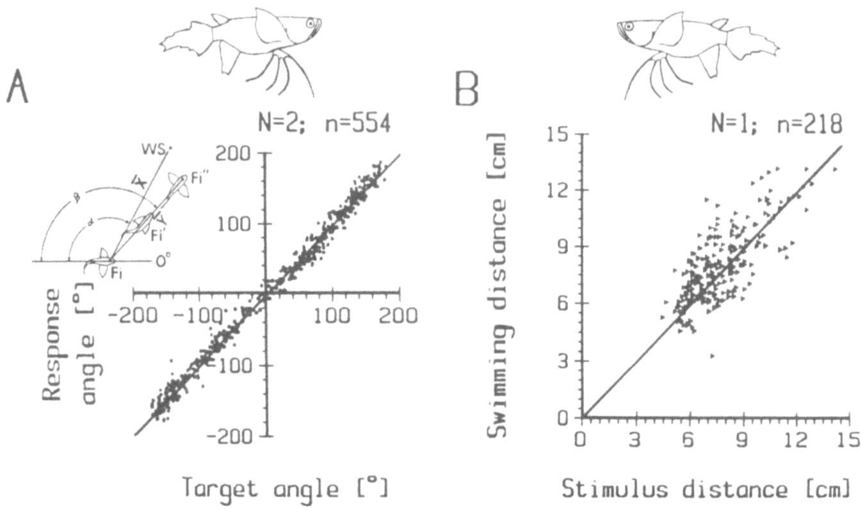

FIGURE 25.8. Orientation behavior of mature visually deprived *P. buchholzi*. The accuracy of target angle determination (A) and distance determination (B) is shown. Each dot represents one response. Stimulation was done under open loop conditions; i.e., in each case a short lasting click (see Fig. 25.2C) was generated by dipping a small rod once into the water. In this and all other figures, *N* refers to the number of fish and *n* to the number of experiments. Inset in (A): Definition of stimulus parameters and reaction values. Fi: animal at stimulus onset; Fi': animal during the phase of swimming; Fi'': animal after stopping (identified by the spreading of pectoral fins); WS: wave source, X: source distance, Y: swimming distance, α: target angle, β: turning angle.

25.5.1 Ontogeny of Localization Behavior

The larvae of *A. lineatus* hatch 15.1 ± 6.8 days (*N** = 81) after spawning (water temperature 24°C). One day after hatching, the cephalic neuromasts are already present. First spontaneous responses to surface waves occur in animals that are at least 9 or 10 days old. For the amplitudes tested, these animals respond only up to a distance of 4 cm. Compared with mature fish (i.e., fish 7 to 9 cm in length), the responses of larvae have long latencies (up to 3,000 ms) and are inaccurate with respect to both target angle and distance determination. However, even without rewarding, both error angle and response latency decrease significantly within 5 to 10 days (fig. 25.9A), and the response range is extended to about 9 cm. Only 18.5 ± 4.8 days (*N* = 57) after hatching, response latencies and error angles of juvenile *Aplocheili* no longer differ from those of mature animals, provided the fish were allowed to have contact with the water surface (Waldner 1981). If the larvae are rewarded, distance determination improves more rapidly than in fish that were not rewarded (Waldner, unpublished

*N refers to number of fish; *n*, to number of experiments.

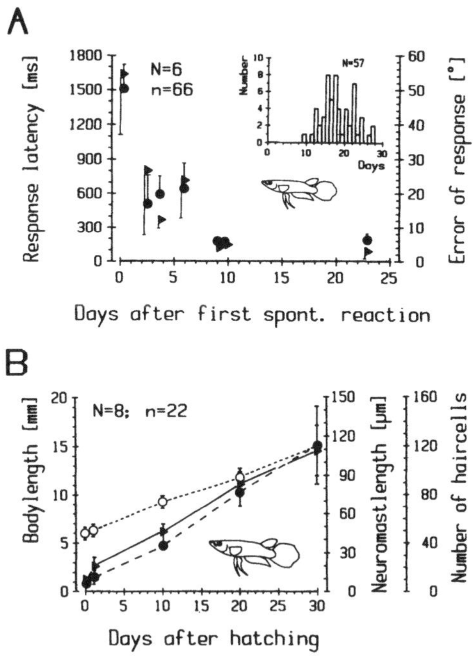

FIGURE 25.9. (A) Response latency (●) and error angle (►) as function of time after the first spontaneous response to surface waves occurred. Vertical bars represent the standard error of the means. Inset: Age distribution of immature *A. lineatus* whose localization accuracy was similar to that of mature animals. (B) Body length (O), neuromast length (►), and number of hair cells (●) as function of time after hatching (from Tittel 1988).

data). This may indicate that learning is involved in wave source (distance) localization. Parallel to behavioral alterations, morphological changes occur within the cephalic lateral line: both size of neuromasts and number of hair cells within a neuromast increase significantly ($N = 8$; $n = 22$) during the first 30 days after hatching (Fig. 25.9B). Besides quantitative changes, qualitative changes can be observed. In *Aplocheili* that were younger than 15 to 20 days and that did not respond to surface waves, only afferent synapses could be found. In contrast, in older fish and/or in younger fish already responding to surface waves, both afferent and efferent synapses were present (20 hair cells in two fish were examined for each age class) (Tittel 1987).

25.5.2 Behavioral Thresholds

Schwartz (1965, 1971) was the first to give behavioral thresholds for surface-feeding fish. He reports that amplitudes of 4 to 7.1 µm p-p water displacement are sufficient to trigger a behavioral response. However, later measurements, done with more sophisticated devices, have shown that the vibration sensitivity of surface-feeding fish is even higher. Minimal thresholds of carefully conditioned animals are between 0.0007 and 0.01 μm p-p displacement at 75 to 150 Hz (Fig. 25.5).

25.5.3 Determination of Target Angle

In *Aplocheilus* and *Pantodon*, determination of wave direction is highly accurate if clicks are presented (Figs. 25.8A and 25.10A). Thus the question arises how this performance is achieved. According to the strong damping and low propagation velocity of water surface waves, as well as the directional sensitivity of cephalic neuromasts (Fig. 25.7), wave direction in theory can be determined by (1) measuring the intensity gradient of the stimulus (Eq. 4), (2) comparing organ output differences due to neuromast directionality (Fig. 25.7), and (3) measuring arrival time differences between different neuromasts.

25.5.3.1 Localization Via Intensity Gradients

According to the strong increase of damping with wave frequency (Fig. 25.1F) and the fact that electrophysiologically determined stimulus-response curves are steeper at higher frequencies (Fig. 25.6), high-frequency surface waves especially should be suited to determine stimulus amplitude gradients (and thus wave direction). Compared with clicks, target angle determination is not significantly impaired in *Pantodon* or *Aplocheilus* if single-frequency (sf) stimuli, ranging from 30 to 112 Hz are presented (Hoin-Radkovski and Tittel, unpublished data). Thus it is unlikely that intensity gradients due to wave damping are important for target angle determination.

25.5.3.2 Localization Via Organ Output and Arrival Time Differences

Up to about $\pm 130°$ (0° is directly in front of the fish), target angle determination in an *Aplocheilus* with only the cephalic lateral line left intact is not different from that of untreated fish (Fig. 25.10C,D). In contrast, *Aplocheili* with only the supratemporal neuromasts and the neuromasts innervated by the dorsal branch of the trunk lateral line left intact show a fairly accurate target angle determination only in the range of 100° to 160° (Fig. 25.10E,F). Irrespective of target angle, unilaterally ablated fish always turn to the intact side (Schwartz 1965; Müller and Schwartz 1982).

An *Aplocheilus* with only one cephalic neuromast left intact remains sensitive to all wave directions but can no longer determine the target angle. Instead, the turning angle now correlates with the anterior-posterior position of the particular neuromast in that a more caudal neuromast induces a larger response angle than a more rostral neuromast ($r = .969, p < .001$) (Fig. 25.11A). There may be an additional correlation between neuromast alignment (with respect to the fish's long axis) and turning angle. If all neuromasts tested are taken into account, response angles do not correlate with neuromast alignment ($r = .502, p > .05$). However, if the cephalic neuromast III.3 is excluded (cf. Figs. 25.3, 25.11),

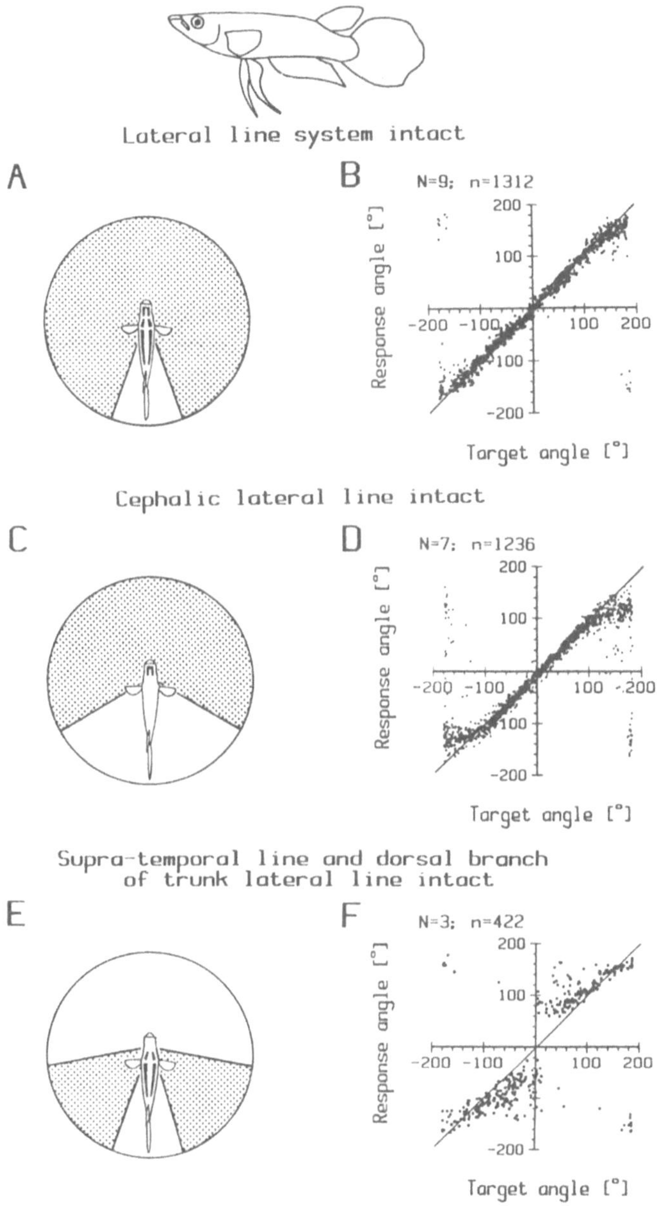

FIGURE 25.10. Right: Directional responses of visually deprived *A. lineatus*. Each dot represents one response. Left: symbolized *Aplocheili* (dorsal view) and the range where target angle determination was not impaired (stippled area). Neuromast groups left intact are indicated by heavy black lines (from Tittel 1988).

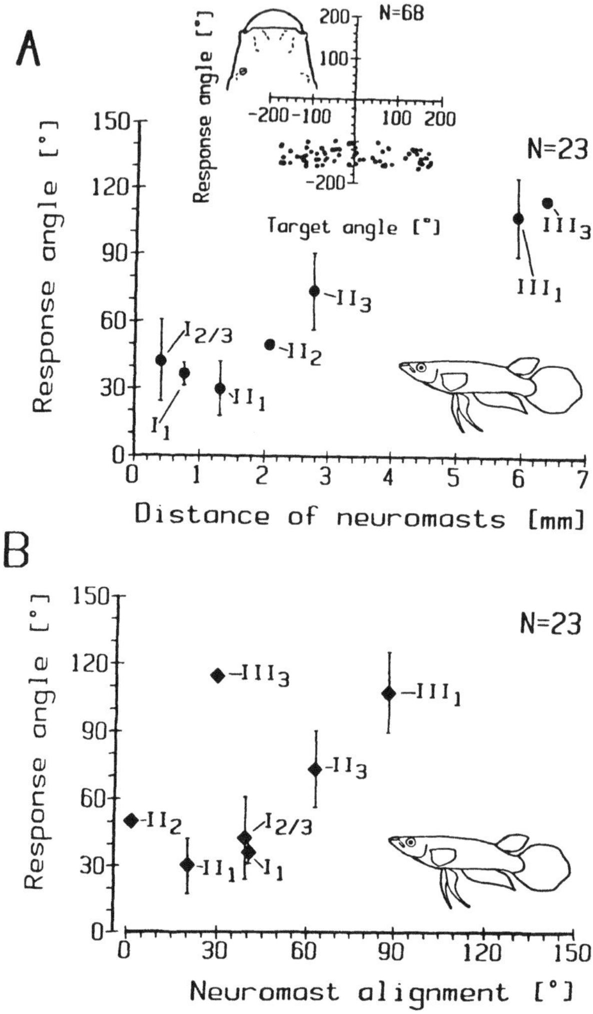

FIGURE 25.11. Response angle of mature *Aplocheilus* as function of neuromast distance from a reference point on the fish's snout (A) and of neuromast alignment with respect to the fish's long axis (B). Neuromast position alignment were determined for each single organ prior to the behavioral experiments. Numbers refer to the respective organs (cf. Fig. 25.3A) (from Müller and Schwartz 1982; Tittel 1988). Points represent means and their standard deviation. Inset A: Example of target angle localization in a fish whose neuromast have been destroyed except organ III.1.

response angle significantly increases with increasing neuromast alignment ($r = .81$, $p = .06$). Unfortunately only one animal was tested. Hence experiments with this ablation type should be repeated to clarify whether neuromast III.3 indeed is exceptional or the experimental animal used behaved in an abnormal fashion.

A *Pantodon* with only the two cranial temporal neuromasts left intact bilaterally (all branches of lateral line nerves innervating other neuromasts were cut) can no longer localize a wave source. However, a *Pantodon* in which one supratemporal neuromast in addition is left innervated can still determine the wave direction (Blübaum and Schwartz 1985), although the mean error angle now significantly (t-test, $p > .01$) increases from 8.5° ± 6.8 (intact fish) to 29.6° ± 28.4 (ablated fish). If one assumes that in *Pantodon* cupula movement somehow changes with target angle, this could be due to the striking radial arrangement of hair cells within the banana-shaped sensory epithelium of the supratemporal neuromast (cf. Table 25.1 and Fig. 25.3B, inset).

In a certain dynamic range (for definition see Tittel et al. 1984), *Aplocheili* with only two neuromasts left intact retain their ability to determine the target angle. To test whether intensity differences due to neuromast directionality and/or arrival time differences between different neuromast are important for target angle determination, ablation combinations were chosen to create slight differences in neuromast position, but large differences in neuromast alignment (and thus large differences in neuromast output) and vice versa (cf. insets Fig. 25.12A,B). Despite large differences in neuromast alignment, no differences with respect to size and location of dynamic ranges occurred between these two (and similar) neuromast combinations. This indicates that arrival time and/or phase differences between different organs—but not organ output differences—are the cues used for target angle determination. There is a second line of evidence that intensity differences due to organ directionality are less important: if sf waves (only one ipsilateral neuromast combination was tested) are presented, a 30-Hz stimulus leads to more accurate responses than an 80-Hz wave (Tittel 1985), despite the fact that input-output curves are steeper at higher frequencies (Fig. 25.6). It should be stressed, however, that the correlation between arrival time differences at different organs and the target angle is unequivocal only if the frequency-dependent wave propagation speed (Eq. 2) is taken into account. Obviously, surface-feeding fish do correct for this error.

25.5.4 Distance Determination

If one knows the frequency-dependent damping and phase velocity of surface waves, the distance to a punctate wave source can be determined by measuring (1) the curvature of the wave front, (2) the relative amplitude decrease per unit distance (Eq. 4), (3) the local frequency modulation of the initial part of the wave stimulus (Eq. 3), and (4) the amplitude spectrum (Eq. 4). In waves lacking higher frequencies, the amplitude spectrum can give some information about D only if compared with a commonly experienced standard.

25.5.4.1 Integration Time

The basic frequency of a wave train, its local frequency modulation, and amplitude spectrum can be obtained only if the stimulus is integrated over at least a

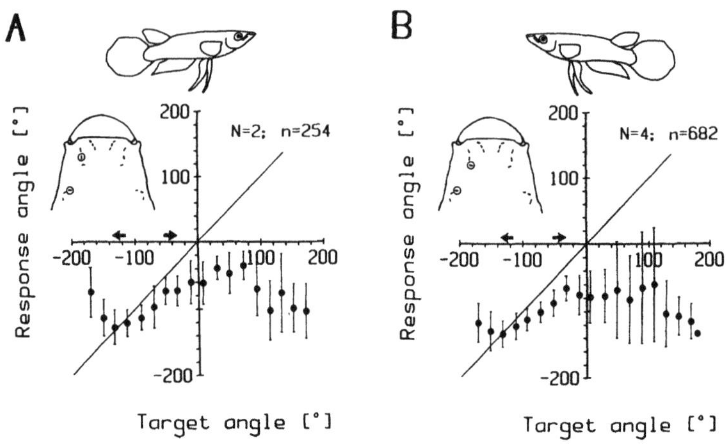

FIGURE 25.12. Directional responses of fish whose neuromasts have been destroyed except those of the preopercular line (which is not involved in surface wave perception) and the respective supraorbital neuromasts encircled. Points refer to mean values and their standard deviations. Directional and spatial differences between the two neuromasts left intact are respectively 83° and 4.1 mm in (A) and 19° and 2.8 mm in (B). Arrowheads indicate the upper and lower limits of the dynamic range (for definition see text).

few wave cycles. In *Aplocheilus*, first stimulus-evoked muscle potentials can be recorded after about 8.5 wave cycles have passed the fish (Bleckmann and Schwartz 1981). This value was found to be independent of stimulus amplitude, so in *Aplocheilus* the reaction time to water surface waves is mainly a function of stimulus frequency. Despite the fact that the cephalic neuromast of *Aplocheilus* have an efferent (inhibitory?) innervation (see above), neither the threshold to sf stimuli nor the reaction time to clicks or the ability to localize a wave source is impaired as long as the swimming speed of the fish does not exceed 10 cm/s (Bleckmann 1982).

25.5.4.2 Responses to Clicks

Because of the strong frequency-dependent damping of surface waves (Fig. 25.1F), a rough estimation of source distance is possible by evaluating the amplitude and frequency content of a wave train. If so, the determination of *D* could be based on the simple assumption that high-amplitude broad-band width stimuli on an average have traveled a shorter *D* than low-amplitude, low-frequency stimuli (see inset, Fig. 25.1F). However, if clicks of different amplitudes (and thus of different upper frequency limits) are presented, surface-feeding fish do not make this simple assumption. In both *A. lineatus* and *P. buchholzi*, the relative localization errors (for definition see Hoin-Radkovski et al. 1984) are independent of click amplitude, click frequency content, and source distance (Fig. 25.13A–C). Thus the determination of source distance cannot be based solely on the evaluation of the wave spectrum.

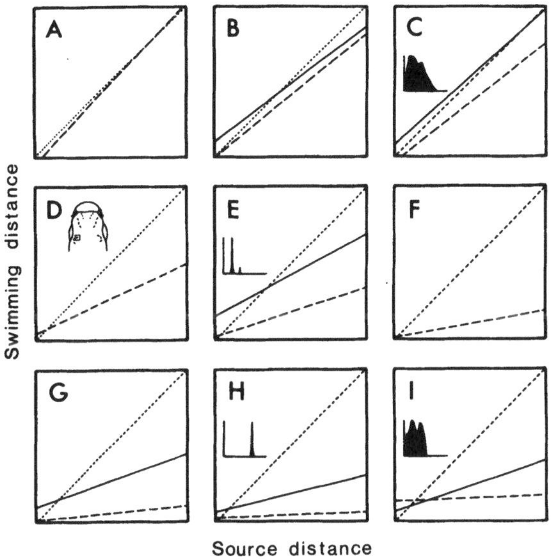

FIGURE 25.13. Mean swimming distance of visually deprived *A. lineatus* (----) and *P. buchholzi* (—) to click stimuli of 0.5-, 1.5-, and 4-V intensity (A–C); to sf stimuli of 30 Hz (*Aplocheilus*) and 35 Hz (*Pantodon*) (E); 50, 70, and 100 Hz (F–H), and to upward frequency-modulated wave stimulus (I). In (D) responses of *A. lineatus*, in which all except one head lateral line organ were destroyed (see inset), are shown. Dotted line, line of perfect localization (slope 1). The X axis (source distance) and the Y axis (swimming distance) are, in all cases, 3 to 18 cm. The number of fish for both species and all experiments is ≥ 3. Insets: Relative spectra obtained at source distances of 5 cm (C, E, and H) and 7 cm (I). In all cases, the highest power value was set equal to 0 dB. X axis, 0 to 150 Hz; Y axis, 0 to −60 dB. (From Bleckmann 1980; Bleckmann and Schwartz 1982; Müller and Schwartz 1982; Hoin-Radkovski et al. 1984.)

25.5.4.3 Responses to Single-Frequency Stimuli

Decoding of local frequency and frequency modulation of a wave stimulus is one way to determine *D* (Fig. 25.1E). Consequently the presentation of sf stimuli leads to an impaired distance determination in both *Aplocheilus* and *Pantodon*. The fish now tend to underestimate *D* if it exceeds 6 to 8 cm (Fig. 25.13E–H). However, there is still a weak but significant correlation between *D* and swimming distance (for *A. lineatus* this holds true only up to frequencies of 50 Hz (Bleckmann 1980)). In general the relative localization error at a given *D* rises with *f*, and at a given *f* it rises with *D* (Bleckmann 1988). Assuming that the curvature of the wave front is determined through arrival time differences, this is exactly what is to be expected (see Hoin-Radkovski et al. 1984). At a given *D*, surface-feeding fish swim progressively shorter distances in response to sf stimuli at higher frequencies. This may indicate that the amplitude spectrum is also evaluated; i.e., if no other cues are available, a high-frequency stimulus is "expected" to have traveled a shorter *D* than a low-frequency one. That surface-feeding fish do follow this strategy is also supported in experiments with artificial frequency-upward modulated stimuli containing high-amplitude, high-frequency components. These stimuli also cause an underestimation of *D* (Fig. 25.13I). Considering the filter properties of the water surface (Eq. 4), this probably reflects a good localization strategy.

25.5.4.4 Responses to Frequency-Modulated Stimuli

When surface-feeding fish are confronted with a computer-controlled wave stimulus that is generated at 7 cm but simulates the frequency modulation of a click at 15 cm, they swim on average 4 to 6 cm beyond the wave source (Fig. 25.14). This demonstrates again that the frequency modulation is an important cue used for distance determination. The use of frequency modulation for distance determination is further supported in experiments with *A. lineatus*, in which all but one cephalic neuromasts have been destroyed. Although these animals have no way to determine the curvature of the wave front or the amplitude decrease per unit of distance, they still show an increase of swimming distance with source distance if clicks are presented (Müller and Schwartz 1982). Distance determination in ablated fish, however, is somewhat impaired (Fig. 25.13D), thus indicating again that parameters other than frequency modulation are also used for distance determination.

25.5.4.5 Sound Waves

A splash made by the entry of any object into water radiates not only surface but also sound waves in both air and water (Franz 1959). In the African clawed toad *Xenopus*, some auditory neurons respond to water surface waves with a very short latency (H. Münz, personal communication). Assuming that similar neurons do exist in surface-feeding fish, these fish could calculate the wave source distance by comparing arrival time differences between pressure (acoustic) and surface waves. However, up to now the use of acoustic cues for wave source localization has not been demonstrated in surface-feeding fish (for a more detailed discussion see Bleckmann 1988).

25.5.5 Stimulus Discrimination

Tests to demonstrate that surface-feeding fish can distinguish complex wave stimuli have not been done. Behavioral studies show, however, that *A. lineatus* can discriminate between sf stimuli differing in only one parameter—frequency, amplitude, or slew rate. For instance, in the range of 10 to 150 Hz, *A. lineatus* can discriminate water wave frequencies to a resolution of 10% (Bleckmann et al. 1981). As expected from electrophysiological results, slew rate and amplitude discrimination are best when high-frequency, low-amplitude wave stimuli are presented. To give numerical examples: *A. lineatus* can discriminate between wave trains (p-p displacement 17 μm, duration 900 ms) whose slew rate differs

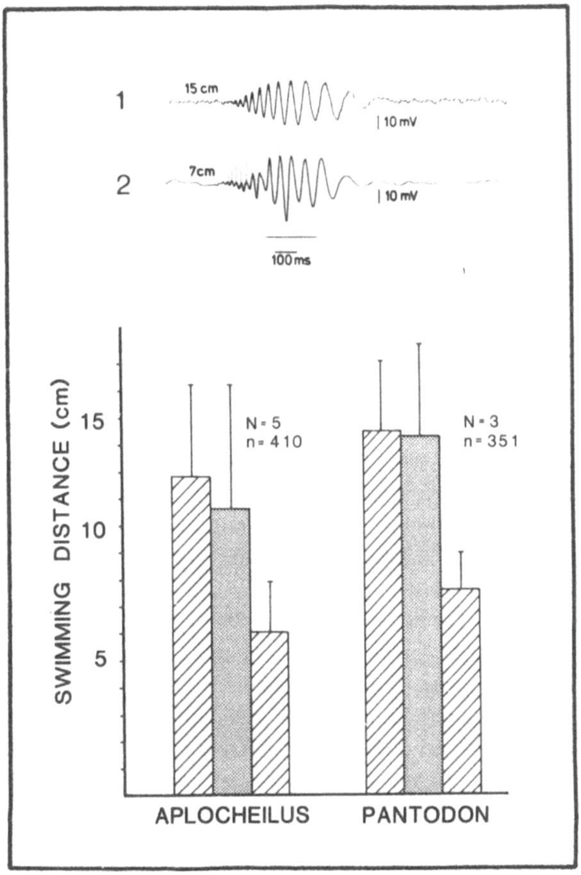

FIGURE 25.14. Mean swimming distance of visually deprived surface-feeding fish toward the wave source after presenting a surface wave stimulus. For each species from left to right: responses to a click at 13.5 to 16.5 cm source distance, to a wave stimulus at 7 ± 0.5 cm source distance simulating a click at 15 cm, and to a click at 7 ± 0.5 cm source distance. In all cases the middle of the fish's head (top view) served as reference. At top of the figure: (1) wave stimulus caused by a click at 15 cm source distance, and (2) simulation of this stimulus at 7 cm source distance. (From Bleckmann and Schwartz 1982; Hoin-Radkovski et al. 1984.)

by 2.5 μm/s (carrier frequency 70 Hz) and 12 μm/s (20 Hz). The corresponding values for amplitude discrimination (carrier frequency 40 Hz, stimulus duration again 900 ms) are 4 μm (amplitude of reference stimulus 21 μm) and 1 μm (0.15 μm) (Waldner 1981).

25.6 Summary and Conclusions

Behavioral experiments done over the past 20 years indicate that the cephalic lateral line of surface-feeding fishes is a highly sophisticated sensory system that is well suited to detect and decode prey (insect) waves. In addition, surface-feeding fish can determine the direction and the distance to a wave source. Irrespective of large differences with respect to the peripheral and central organization of the cephalic lateral lines of *Aplocheilus* and *Pantodon* (Figs. 25.3, 25.4), the mechanisms used for wave source localization appear to be fairly similar in these two species. Although for target angle determination, time and/or phase differences are the most important cues, distance determination is done by evaluating the curvature, the amplitude spectrum, and the local frequency and frequency modulation of a wave train.

It is still a matter of debate (see Kalmijn 1988, Chapter 9) whether the high displacement sensitivity of midwater teleosts and cartilaginous fishes (see Münz Chapter 14; Bleckmann and Bullock Chapter 19) for frequencies around 100 Hz has any biological meaning. However, in surface-feeding fish the impressive displacement sensitivity for the frequencies 50 to 150 Hz is probably advantageous. Those are the frequencies in which prey (insect) waves can be distinguished from all other wave types (see above).

In midwater and bottom-dwelling fishes, there may be a great emphasis on the analysis of transients and much less on the analysis of frequency (Denton and Gray 1988). However, in surface-feeding predators, the analysis of wave frequency enables both stimulus discrimination and wave source localization. Thus, it is not surprising that surface-feeding fish can discriminate between different sf wave stimuli (Bleckmann et al. 1981). The behavioral discrimination tests done so far should be extended so that more complex stimuli are offered in a choice situation. For instance, we do not know whether surface-feeding fish indeed use high-frequency wave components to identify prey (insect) stimuli as such. If so it would be worthwhile to test (1) the precision to which this ability can be achieved, and (2) whether these fish—like fishing spiders (Bleckmann 1988)—weigh the frequency content of a stimulus according to the distance the stimulus has traveled. Another intriguing question is whether surface-feeding fish can evaluate wave stimuli with respect to regularity, duration, and interstimulus time intervals. All these parameters are important for the distinction between prey and nonprey waves (see above). It should be added that up to now, all discrimination and localization experiments were made in an ideal, low-noise environment. Experiments that clarify the degree to which prey identification and localization are impaired under background noise conditions are clearly needed.

Irrespective of the intensive behavioral work that has been done with surface-feeding fish, nothing is known about the central physiology of the lateral line in these fish. Recordings from the medulla and higher brain centers are clearly needed. Those recordings should bridge the gap between peripheral and central anatomy, peripheral physiology, and orientation behavior. Fortunately, in surface-feeding fish neural recordings can easily be combined with quasinatural lateral line stimuli. This may not only facilitate our efforts to uncover the biological functions of higher lateral line centers (see Bleckmann and Bullock Chapter 20) but also ease all future attempts to learn whether and how the central lateral line physiology of surface-feeding fish differs from that of midwater species and aquatic amphibians.

Acknowledgments. We thank Drs. I. Waldner and U. Müller for permission to quote unpublished material. Dr. A. Reissland deserves thanks for deriving Eqs. 1. G.T. and E.B.-G. appreciate the constant advice of Dr. E. Schwartz during the course of their doctoral thesis. Last but not least we thank Drs. H. Münz, B. Fritzsch, and S. Coombs for commenting on the manuscript. Supported by grants of the DFG to H.B. (Bl 242/2-1).

References

Bleckmann H (1980) Reaction time and stimulus frequency in prey localization in the surface-feeding fish *Aplocheilus lineatus*. J Comp Physiol A 140:163–172.

Bleckmann H (1982) Reaction time, threshold values, and localization of prey in stationary and swimming surface feeding fish *Aplocheilus lineatus*. Zool Jahrb Abt Allg Zool Physiol Tiere 86:71–81.

Bleckmann H (1985a) Perception of water surface waves: How surface waves are used for prey identification, prey localization, and intraspecific communication. In: Ottoson D, Autrum MJ (eds) Sensory Physiology, Vol. 5. New York: Springer, pp. 147–166.

Bleckmann H (1985b) Discrimination between prey and non-prey wave signals in the fishing spider *Dolomedes triton* (Pisauridae). In: Kalmring K, Elsner N (eds) Acoustic and Vibrational Communication in Insects. Berlin: Paul Parey, pp. 215–222.

Bleckmann H (1988) Prey identification and prey localization in surface feeding fish and fishing spiders. In: Atema J, Fay RR, Popper AN, Tavolga W (eds) Sensory Biology of Aquatic Animals. New York: Springer, pp. 619–641.

Bleckmann H, Lotz T (1987) The vertebrate catching behaviour of the fishing spider *Dolomedes triton* (Arancae, Pisauridae). Anim Behav 35:641–651.

Bleckmann H, Rovner J (1984) Sensory ecology of the semiaquatic spider (*Dolomedes triton*). I. Roles of vegetation and wind-generated waves in site selection. Behav Ecol Sociobiol 14:297–301.

Bleckmann H, Schwartz E (1981) Reaction time of the topminnow *Aplocheilus lineatus* to surface waves determined by video- and electromyogram recordings. Experientia 37:362–363.

Bleckmann H, Schwartz E (1982) The functional significance of frequency modulation within a wave train for prey localization in the surface-feeding fish *Aplocheilus lineatus* (Cyprinodontidae). J Comp Physiol A 145:331–339.

Bleckmann H, Topp G (1981) Surface wave sensitivity of the lateral line system of the topminnow *Aplocheilus lineatus*. Naturwissenschaften 68:624–625.

Bleckmann H, Waldner I, Schwartz E (1981) Frequency discrimination in the surface-feeding fish *Aplocheilus lineatus*—a prerequisite for prey localization? J Comp Physiol A 143:485–490.

Blübaum E, Schwartz E (1985) Richtungsempfindlichkeit eines Einzelneuromasten vom Schmetterlingsfisch *Pantodon buchholzi*. Verh Dtsch Zool Ges 78:280.

Blübaum-Gronau E, Münz H (1987) Topologische Repräsentation primärer Afferenzen einzelner Seitenlinienabschnitte beim Schmetterlingsfisch *Pantodon buchholzi*. Verh Dtsch Zool Ges 80:268–269.

Blübaum-Gronau E, Seiss M, Söhl CH (1988) Untersuchung der Seitenlinenprojektion bei Teleostei—eine vergleichende HRP-Studie bei Osteoglossomorpha und Euteleostei. Verh Dtsch Zool Ges (in press).

Denton EJ, Gray ABJ (1988) Mechanical factors in the excitation of lateral lines of fishes. In: Atema J, Fay RR, Popper AN, Tavolga W (eds) Sensory Biology of Aquatic Animals. New York: Springer, pp. 595–617.

Flock Å (1965) The ultrastructure of the lateral line canal organ. Acta Otolaryngol (Suppl) 199:7–90.

Franz GJ (1959) Splashes as sources of sound in liquids. J Acoust Soc Am 31:1080–1096.

Hoin-Radkovsky I, Bleckmann H, Schwartz E (1984) Determination of source distance in the surface-feeding fish *Pantodon buchholzi* (Pantodontidae). Anim Behav 32:840–851.

Jannsen J, Coombs S, Hoekstra D, Platt C (1987) Anatomy and differential growth of the lateral line system of the mottled sculpin, *Cottus bairdi* (Scorpaeniformes: Cottidae). Brain Behav Evol 30:210–229.

Kalmijn A (1988) Hydrodynamic and acoustic field detection. In: Atema J, Fay RR, Popper AN, Tavolga W (eds) Sensory Biology of Aquatic Animals. New York: Springer, pp. 83–130.

Käse R, Bleckmann H (1987) Prey localization by surface wave-ray tracing—fish track bugs like oceanographers track storms. Experientia 43:290–292.

Lang HH (1980) Surface wave discrimination between prey and nonprey by the backswimmer *Notonecta glauca* L. (Hemiptera, Heteroptera). Behav Ecol Sociobiol 6:233–246.

Lighthill (1980) Waves in Fluids. London: Cambridge University Press.

McCormick CA (1981) Central projections of lateral line and eight nerves in the bowfin, *Amia calva*. J Comp Neurol 197:1–15.

Müller U (1984) Die morphologische und physiologische Anpassung des Seitenliniensystems von *Pantodon buchholzi* an den Lebensraum Wasseroberfläche. PhD Thesis, University of Giessen, West Germany.

Müller U, Schwartz E (1982) Influence of single neuromasts on prey localizing behavior of surface-feeding fish, *Aplocheilus lineatus*. J Comp Physiol A 149:399–408.

Schwartz E (1965) Bau und Funktion der Seitenlinie des Streifenhechtlings (*Aplocheilus lineatus* Cuv. u. Val.). Z Vergl Physiol A 50:55–87.

Schwartz E (1970) Ferntastsinnesorgane von Oberflächenfischen. Z Morphol Tiere 67: 40–57.

Schwartz E (1971) Die Ortung von Wasserwellen durch Oberflächenfische. Z Vergl Physiol A 74:64–80.

Sommerfeld A (1970) Vorlesungen über theoretische Physik, Vol. 2: Mechanik der deformierbaren Medien. Leipzig: Akademische Verlagsgesellschaft.

Tittel G (1985) Determination of stimulus direction by the topminnow, *Aplocheilus lineatus*. A model of two-dimensional orientation with the lateral line system. Verh Dtsch Zool Ges 78:242.

Tittel G (1987) Growth and development of the lateral-line organs of *Aplocheilus lineatus*. Verh Dtsch Zool Ges 80:281–282.

Tittel G (1989) Untersuchungen zum Beutefangverhalten des Streifenhechtlings *Aplocheilus lineatus* (Cu. U. Val.). Ein Modell zur Richtungsdetermination unter Berücksichtigung von ontogenetischen Prozessen des peripheren Seitenliniensystems. PhD Thesis, University of Giessen, West Germany.

Tittel G, Müller U, Schwartz E (1984) Determination of stimulus direction by the topminnow *Aplocheilus lineatus*. In: Varju D, Schnitzler HU (eds) Localization and Orientation in Biology and Engineering. New York: Springer, pp. 69–72.

Topp G (1983) Primary lateral line response to water surface waves in the topminnow *Aplocheilus lineatus* (Pisces, Cyprinodontidae). Pflügers Arch 397:62–67.

Unbehauen H (1980) Morphologische und elektrophysiologische Untersuchungen zur Wirkung von Wasserwellen auf das Seitenlinienorgan des Streifenhechtlings (*Aplocheilus lineatus*). PhD Thesis, University of Giessen, West Germany.

Waldner I (1981) Habituation von *Aplocheilus lineatus* auf Oberflächenwellen des Wassers, PhD Thesis, University of Giessen, West Germany.

Wilcox RS (1988) Surface wave reception in invertebrates and vertebrates. In: Atema J, Fay RR, Popper AN, Tavolga W (eds) Sensory Biology of Aquatic Animals. New York: Springer, pp. 643–663.

26 Wave Analysis by Amphibians

ANDREAS ELEPFANDT

26.1 Introduction

Perception of water movement on the body by means of lateral line organs is lacking in humans. Therefore, to understand stimulus analysis with the lateral line system, behavioral analyses that reveal the system's sensory capacities are particularly important. The results of psychophysical testing provide insight into what kind of information about the aquatic environment can be obtained with the system. They are further prerequisites for adequate electrophysiological and theoretical analyses of how the system's properties are accomplished. In this chapter, the results of behavioral analyses of the mechanoreceptive lateral line system (subsequently referred to as lateral line system) in amphibians are reviewed, and some parallels to stimulus analysis in the electroreceptive lateral line and auditory systems are pointed out.

Until about 10 years ago, little was known about the function of the amphibian lateral line system. Detection of wave direction with the system has been shown in the adult clawed frog, *Xenopus laevis* (Kramer 1933; Dijkgraaf 1947, 1967; Görner 1973, 1976). This frog, which remains aquatic after metamorphosis, retains its lateral line system throughout life. When hovering at or close to the water surface, it will often respond to impinging waves – such as may be produced by an insect floundering on the water – by turning toward the wave origin and, in response to frontal waves, will approach the wave source. This response, which can be easily elicited in the laboratory by lightly dipping a rod onto the water, persists after blinding of the animal, but it vanishes when the lateral line organs are destroyed as well (Kramer 1933). Perception of water jets with the lateral line system had been demonstrated in the tadpoles of *Xenopus laevis* (Shelton 1971) and of a salamander, *Ambystoma punctatum* (Scharrer 1932).

In recent years, progress in the understanding of the function and organization of the lateral line system has been made by studying the responses of the adult clawed frog, *Xenopus laevis*, under a variety of experimental conditions. Wave localization and its peripheral and central nervous organization have been investigated, wave frequency discrimination has been demonstrated, and it has been found that *Xenopus laevis* can analyze complex wave inputs. These results will

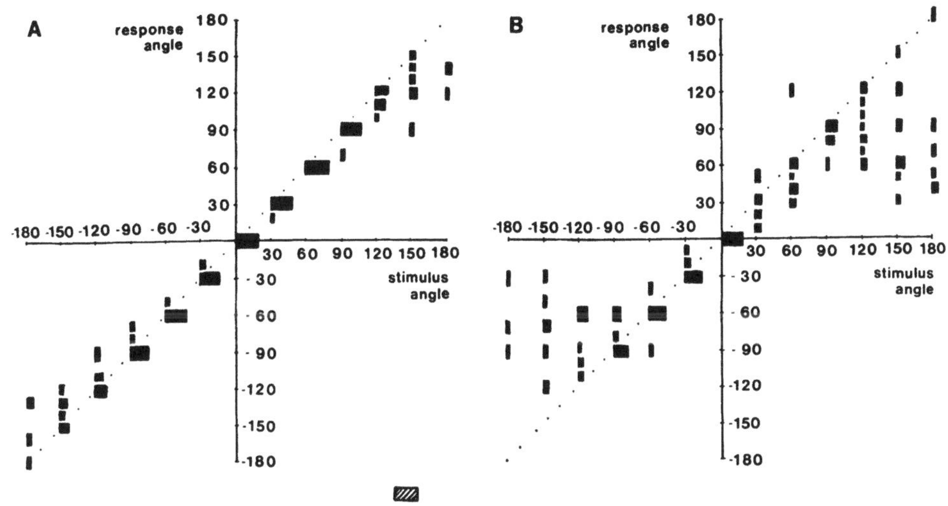

FIGURE 26.1. Histograms of the turn responses of a *Xenopus laevis* to surface waves at 2 successive days (sample specimen, large female). The animal was responsive to waves on the first day (A) but relatively unresponsive on the next day (B). A decrease in responsiveness typically correlates with reduced accuracy and undershoot of the turns, particularly toward posterior waves (B). At high responsiveness, accurate turns can be obtained for all wave directions (A). In the histograms the distributions of the percentage of responses made at different angles are shown for each stimulus angle. The hatched bar below the graph represents a scale bar of 100% for histogram bin height. An angle of 0° is directly in front of the animal, and positive and negative angles indicate right and left side, respectively. The dots along the major diagonal indicate where response angle is equal to stimulus angle. (From Elepfandt 1982.)

be described under respective headings. A final section will summarize the data on lateral line function obtained in other anurans.

26.2 Wave Localization in *Xenopus laevis* with Intact Lateral Line System

With the lateral line system intact, *Xenopus laevis* detects the direction of impinging water surface waves from all directions (Kramer 1933). The responsiveness as well as the accuracy of the response turns toward the wave source are variable, and turns to stimuli from posterior directions frequently undershoot (Fig. 26.1). Apparently, response accuracy is influenced by factors such as motivation, previous experience with type of food, size of specimen, and water depth (Elepfandt 1982). Occasionally, however, localization accuracy of at least $\pm 5°$ is seen for all directions (Elepfandt 1982). *Xenopus laevis* localizes surface waves not only when hovering at the surface but also when it is sitting on the bottom as much as 8 cm below the water surface (Dijkgraaf 1967; Elepfandt 1982).

Although at such depth only the low-frequency part of a surface wave is strong enough to stimulate the frog's lateral line organs (Elepfandt and Wiedemer 1987), localization accuracy appears undiminished (Elepfandt, unpublished observation). Variation of stimulus distance from 4 cm to 20 cm does not affect localization accuracy (Buschmann 1984). Since the high-frequency components of a surface wave decrease much more rapidly with distance than the other frequencies (Bleckmann 1985), these data, too, suggest that the wave's low- and midfrequency components suffice for wave localization with full accuracy.

In addition to horizontal surface wave localization, horizontal localization of subsurface waves has been shown (Kramer 1933), and indirect evidence for vertical wave localization has been presented (Elepfandt 1984a). There is also evidence of distance perception: the distance through which *Xenopus laevis* moves toward a wave source increases slightly with the distance of the source, the slope of the regression line being about 0.25 (Buschmann 1984).

Thus, wave source localization in *Xenopus laevis* comprises more than the mere detection of surface wave directions—the aspect primarily studied—and turning toward the wave source is not a simple reflexive action. The loose relation between the sensory abilities and response behavior of *Xenopus laevis* makes the exact determination of its sensory acuity in locating a wave source somewhat difficult. As compared to surface feeding fish (see Bleckmann et al. Chapter 25), wave source localization in *Xenopus laevis* appears less accurate, but on the other hand, it is not limited to the surface and uppermost 1 to 2 cm of the water.

26.3 Peripheral Organization of Wave Localization

Xenopus laevis possesses approximately 180 stitches, or lateral line organs, distributed over its body (Fig. 26.2A). The information from each organ is conveyed to the central nervous system by two afferent neurons (in some cases there are three or four), each of which contacts half of the organ's sensory hair cells (Görner 1963). Recordings from the afferent fibers have shown that each organ can be stimulated effectively by waves from many directions (Görner 1963; Elepfandt and Wiedemer 1987). This means that, as in hearing, the individual organ does not provide the animal with unambiguous directional information, so determination of stimulus direction requires the comparison of inputs from several organs. The organization of this comparison has been studied with behavioral tests of surface wave localization after destructions of various groups of organs (Elepfandt 1982, 1984a; Görner et al. 1984). It was found that—in contrast to the results with surface-feeding fish (Bleckmann et al. Chapter 25)—no particular group of organs is essential for wave localization. Instead, various groups, which may be small and even unilateral, suffice for localization of waves from any direction with the same accuracy as with an intact lateral line system (Fig. 26.2). The number of organs necessary for accurate wave localization varies considerably with location: accurate localization has been observed in animals with only four organs of the occipital group intact, whereas localization with only ventral organs intact requires more than half of them (Elepfandt 1982, 1984a; Görner

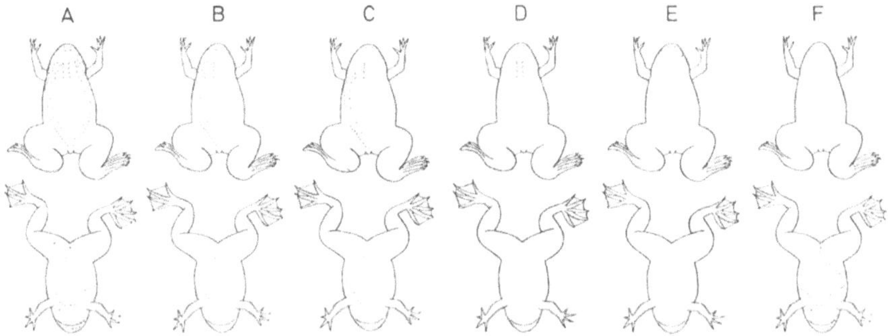

FIGURE 26.2. Drawings of sample destructions of lateral line organs after which localization accuracy in *Xenopus laevis* is unimpaired. Upper figure—dorsal side; lower figure—ventral side. The short lines on the body indicate the intact lateral line organs. (A) Intact lateral line system (reference). (B) Complete unilateral destruction of the organs. (C) destruction of all organs on the head and on one side of the body. (D) Destruction of all organs except the occipital group. (E) Destruction of all organs except the organs around the eyes. (F) Destruction of all dorsal organs. (Data from Elepfandt 1982, 1984a.)

et al. 1984). It seems that comparisons are not made within groups with fixed boundaries, but rather each organ's input is compared to that of its neighbors over some (as yet undetermined) distance. Accurate localization is also possible with organs that do not form a coherent group but are dispersed separately over the animal's body (Görner and Mohr Chapter 27).

This multiplicity of ways to detect wave direction seriously complicates the analysis of the mechanisms of input comparison. No cue for comparison—intensity, temporal differences, or differing directional sensitivity of the organs—has been found essential for all localizations. Rather, the cues seem to vary among the organs, and it may be that several cues are used simultaneously by particular organs (Elepfandt 1982, 1984a; Görner et al. 1984; Görner and Mohr Chapter 27). Temporal cues are used in horizontal wave localization by animals with only the occipital rows of lateral line organs left intact (Elepfandt 1982; Görner and Mohr Chapter 27). Comparison of input intensities is necessary between dorsal and ventral afferents for vertical localization. This is shown by the fact that destruction of all dorsal lateral line organs significantly increases the rate of diving responses (i.e., vertical deviations) to surface waves, whereas complete unilateral organ destructions do not impair localization accuracy (Elepfandt 1982, 1984a). So far, no evidence has been found for the use of the organs' directional sensitivity in wave localization (Elepfandt 1982; Görner and Mohr Chapter 27). Since with water waves (i.e., oscillatory motions) this cue has an 180° ambiguity, it may be used only in connection with other cues anyway.

The variability in the use of cues for comparison may be related to the fact that the physically available cues for wave localization vary with the distance of both the frog and the wave source from the water surface. Wave propagation is different for surface and subsurface waves, and the animal's wave shadow, which is a crucial factor for temporal and intensity differences of the stimulus at different organs, changes strongly with the animal's proximity to the water surface. Obviously, stimulus localization under such variable conditions requires highly developed and flexible central processing.

An additional complication in determining localization mechanisms results from the fact that *Xenopus laevis* retains residual localization abilities even when all its lateral line organs have been destroyed. Typically, *Xenopus laevis* no longer responds to waves after such a lesion, but some animals occasionally do, and their response angles are correlated with stimulus direction, though less than in intact animals (Görner 1973, 1976; Elepfandt 1982; Görner and Mohr Chapter 27). Although the sensory organ responsible for this localization is unknown, one candidate is the labyrinth (Görner 1976). It is unclear, furthermore, whether its input is used for wave localization when the lateral line system is intact. At best, however, it seems to play a secondary role in the orienting response, because, after certain lesions of lateral line organs, *Xenopus laevis* displays localization errors that are not seen after destruction of all lateral line organs (Elepfandt 1982, 1984a). This finding indicates that even erroneous input from a highly impaired lateral line system overrides the correct directional information from the other system. This and the great difference in responsiveness of animals with or without lateral line organs suggest that the intact *Xenopus laevis* relies on its lateral line system for wave localization and that the other system only substitutes when lateral line input is unavailable (Görner 1973, 1976; Elepfandt 1982, 1984a).

So far, all analyses of the system's organization have been made using *waves* as stimuli. Görner and Mohr (Chapter 27), however, present new evidence that the system's functional organization may be different for different kinds of stimulation: specimens in which the lateral line organs have been destroyed unilaterally are capable of localizing all wave directions correctly, but in response to water *jets* they always turn toward the intact side, irrespective of jet direction. This result might have considerable impact on further analysis of the system.

26.4 Central Nervous Organization of Wave Localization

The afferent fibers from the lateral line organs project into the lateral line nucleus in the ipsilateral medulla. From there, second-order neurons run in part to the contralateral lateral line nucleus, but mostly to the contralateral and ipsilateral magnocellular nucleus in the midbrain's torus semicircularis (see Will Chapter 18). Behavioral tests after brain lesions (Elepfandt 1984b, 1986c, 1988a,b) show that the midbrain is essential for wave localization: its removal extinguishes wave localization, but accurate localization is possible when merely diencephalon and telencephalon are cut off. They also demonstrate a topological organization within the midbrain with regard to wave direction: unilateral ablation of the sensory midbrain—i.e., tectum and torus semicircularis—abolishes localization of contralateral waves without impairing ipsilateral localization, and smaller unilateral lesions can extinguish wave localization throughout contralateral sectors of wave direction (Fig. 26.3).

This central topology indicates a marked contrast to the peripheral organization of the lateral line system. In the periphery, small and unilateral groups of lateral line organs are sufficient for localization of all wave directions, ipsilateral *and* contralateral ones, and no group has been found that would suffice for localization of some wave directions but fail completely at others. Localization failure after midbrain lesions, however, has been found throughout distinct angular sectors within the hemifield contralateral to the lesion. Thus, the midbrain topology does not reflect the topology of the lateral line organs on the body but is organized according to the angle of incidence of the *stimulus*—that is, to external coordinates. A parallel to this functional topology is found in sound localization. Unilateral lesions in the cat's midbrain only affect localization of contralateral sound sources, although sound localization requires comparison of the inputs from the left and right ears (Jenkins and Masterton 1982).

Only limited knowledge exists about the details of wave localization within the midbrain. Several types of effects have been found with lesion studies. One effect results in a widespread distribution of turn angles in response to a given stimulus direction, often found in combination with a tendency to turn through the right angle but toward the wrong side (Fig. 26.3) (Elepfandt 1984b, 1988a,b). It has been suggested that this effect results from lesions of localization circuitry in the lateral magnocellular nucleus of the torus semicircularis (Elepfandt 1988b). A second effect is seen when frogs swim straight ahead in response to lateral stimulation (Elepfandt 1986c, 1988b; see also Claas et al. Chapter 20). This error is not due to a motor deficit, because the animal is capable of turning through the appropriate angle. This effect is seen after lesions of the ventrolateral tectum (i.e., that part of the tectum that stretches around the lateral edges of the third ventricle). Similar lesion effects on the visual orienting response of frogs (Grobstein et al. 1983; Ingle 1983) make it likely that this effect results from affecting the sensorimotor transfer from the sensory directional information to the corresponding motor turn command (Elepfandt 1986c, 1988b). Third, the tectum proper also seems to be involved: electrophysiological recordings in the tectum revealed a maplike representation of wave directions (Zittlau et al. 1986) which, in contrast to the behavioral data, encompasses both contralateral *and* ipsilateral wave directions. Although large tectal lesions that spare the ventrolateral tectum do not abolish wave localization (Elepfandt 1988b), small lesions in the area of the tectal wave map lead to small systematic deviations from the correct response angle immediately after the lesion has been set (Claas et al. Chapter 20). As a fourth lesion effect, one often finds an increased percentage of overshoot turns (i.e., turns through a somewhat larger angle than would be necessary for accurate orientation, see Fig. 26.3A). In contrast to the first three effects, which are only contralateral, overshoots appear ipsilateral to the lesion. It is not yet clear whether overshoots occur in response to all ipsilateral stimuli or are confined to particular directions. In summary, a topological organization

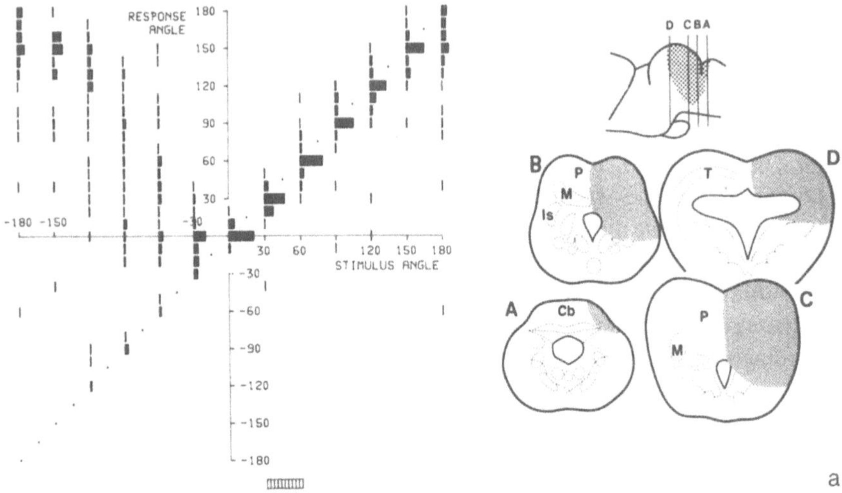

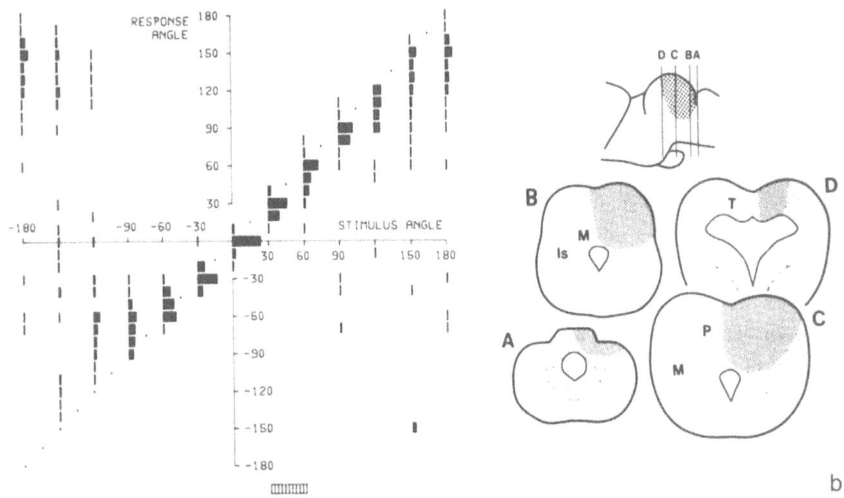

FIGURE 26.3. Representative samples of the effect of midbrain lesions on wave localization in the clawed frog, *Xenopus laevis*. (A) After unilateral removal of the mesencephalic tectum and torus semicircularis, all ipsilateral but no contralateral waves are localized. (B) After small unilateral midbrain lesions, wave localization is abolished within a sector of the contralateral hemifeld. In the lesion diagrams, the stippled areas indicate which parts of the brain had been eliminated. CB, cerebellum; Is, nucleus isthmi; M, magnocellular nucleus of the torus semicircularis; P, principal nucleus of the torus semicircularis; T, tectum. For explanation of the response histograms, see Fig. 26.1. (From Elepfandt 1988a,b.) With permission from S. Karger AG, Basel, Switzerland.

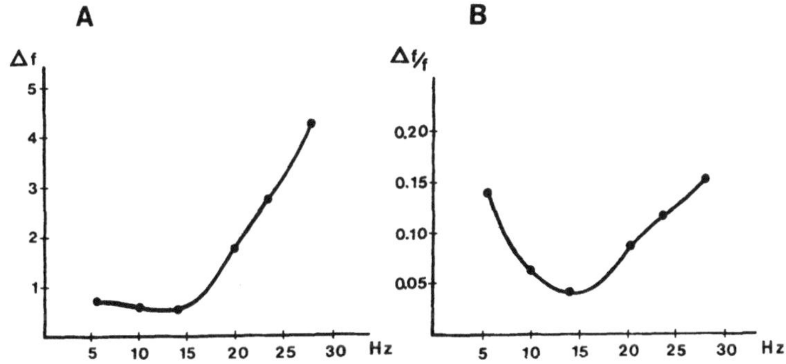

FIGURE 26.4. Absolute (A) and relative (B) frequency discrimination limens for monofrequent surface waves by the clawed frog, *Xenopus laevis*. Relative discrimination is optimal at 14 Hz and decreases toward lower and higher frequencies. Tests were made with monofrequent wave stimuli 3 s long; interstimulus intervals were 1 to 3 min. (From Elepfandt et al. 1985.)

with regard to wave *direction* is found in the midbrain, but several areas are involved in localization. The specific roles and exact midbrain locations of these areas remain to be elucidated.

26.5 Wave Pattern Recognition and Frequency Discrimination

For an adequate and adaptive behavior to occur, not only determination of wave direction but also the analysis of the wave's temporal and frequency characteristics is important. Because the lateral line organs in *Xenopus laevis* have identical frequency tuning (Kroese et al. 1978), it has been suggested that this frog might be incapable of wave pattern recognition. *Xenopus laevis* can, however, be trained to discriminate between waves of different frequency (Elepfandt et al. 1985). Discrimination data have been obtained for frequencies ranging from 5 to 30 Hz (Fig. 26.4). At about 15 Hz, the relative discrimination limen (just discriminable frequency difference, divided by test frequency) is 0.04 – i.e., as good as in wave frequency discrimination for surface-feeding fish (Bleckmann et al. Chapter 25) or in auditory frequency discrimination for nonmammalian vertebrates (Platt et al. Chapter 32). In hearing, it is an open question whether discrimination of low frequencies is based on differential frequency tuning of the receptors or on phase locking of the afferent discharges – i.e., on temporal cues (Pickles 1982). Wave frequency discrimination in *Xenopus laevis* must, however, be based on temporal cues because of the uniform frequency tuning of the lateral line organs (Kroese et al. 1978).

Wave frequency perception in *Xenopus laevis* is of absolute pitch quality (Elepfandt 1986a). In the hearing of humans, recognition of a previously presented tone usually deteriorates with time, but persons with absolute pitch recognize tones with the same accuracy after a week or more as after a minute (Bachem 1953). Similarly, *Xenopus laevis* recognizes a frequency with the same accuracy some weeks after training as during training.

The importance of wave pattern recognition for *Xenopus laevis* is stressed by the fact that the discrimination conditioning succeeded at all. Generally, efforts to condition amphibians have had very limited success (Macphail 1982). With regard to wave pattern discrimination, however, conditioning of *Xenopus laevis* is easy, and several forms of complex learning and long-term memory have been found (Elepfandt 1985).

26.6 Analysis of Complex Wave Forms

The test conditions usually used – i.e., individual circular waves spreading from a small stationary wave source over a smooth undisturbed water surface – are highly idealized. In natural environments, waves may be produced by moving and extended sources that also produce turbulences, and the waves may be distorted by background noise, refractions, or interference with other waves. Thus, simple test conditions may be insufficient to demonstrate the full analytical abilities of the lateral line system. In this section, therefore, processing of complex wave inputs in *Xenopus laevis* is considered, and evidence for common principles of stimulus processing in mechanoreceptive lateral line, electroreceptive lateral line, and auditory systems is pointed out.

One indication that a simple circular wave may not challenge the system's abilities is given by the apparent peripheral "redundancy" of the system with regard to localizing such a wave. Why should *Xenopus laevis* have some 180 lateral line organs distributed over its body if, as shown above, various small groups of them suffice for wave localization with undiminished accuracy? These groups provide the animal with information about water flow at their locations on the frog's body, and correct turns on the basis of such input indicate that the information from local groups of organs can be analyzed centrally in a meaningful way. Thus, analysis of wave forms more complicated than simple circular waves would be possible with simultaneous and parallel analyses of the inputs from several groups located in different areas of the body, with each group providing information about wave direction at its location.

Analysis of that kind – i.e., local stimulus determination by comparison within local groups of receptors and parallel analyses in several groups – is known in the electroreceptive lateral line system: the jamming avoidance response in the electric fish *Eigenmannia* can be driven by subjecting local groups of the animal's electroreceptors to an interference pattern composed of its own electric organ discharge and an electric stimulus of similar frequency. The stimulus analysis is based on the comparison of inputs from neighboring electroreceptive organs (Heiligenberg 1980). If different jamming patterns are presented to separate receptor groups, then the jamming avoidance response results from additive evaluation of "the effects of local stimulus regimes, applied to remote areas of the body surface" (Heiligenberg 1980).

Evidence for differential analysis of the inputs from local groups of receptors also exists in hearing. In the cochlea, hair cell receptors are aligned on the basilar membrane according to the best frequency of their tuning curves. Local groups, therefore, are receptors with similar tuning. From auditory psychophysics, however, it is known that the perception of complex sounds differs according to whether their frequencies are wider or narrower than a certain frequency range, the critical band (Moore 1982). The critical-band phenomenon, thus, is evidence for a specific analysis of local stimulus regimes on the basilar membrane. It also means that the auditory information is probably not encoded in the discharges of individual afferents but rather in the form of an ensemble code of inputs from neighboring afferents, as in the lateral line systems. The capacity for parallel analyses in separate groups of auditory receptors is shown in the ability to localize two tones of different frequency presented simultaneously.

Determination of stimulus quality by input comparison within local receptor groups is quite general in sensory systems with large receptor arrays. The comparisons are often made between receptors of different response characteristics as, for example, between cones in color vision. A common characteristic of the auditory and lateral line systems, however, is that the local comparisons are made between receptors with equal (or very similar) response properties, so that they are only concerned with the spatiotemporal difference of stimulus parameters at neighboring receptors. In view of these parallels in the analysis of local stimulus regimes, it is noteworthy that the encoding of stimulus parameters in the lateral line afferents of *Xenopus laevis* is very similar to that in low-frequency auditory afferents (Elepfandt and Wiedemer 1987).

The capacity for analyzing complex wave stimuli has been examined in *Xenopus laevis* by simultaneous presentation of two waves coming from different directions. In such wave interference, the frog's characteristic response is to turn toward one of the two wave sources if the interstimulus angle is more than 30° (Fig. 26.5) (Elepfandt 1986b). Even with smaller interstimulus angles, *Xenopus laevis* can orient directly toward one of the sources. Such oriented turns do not require differing wave patterns, but they are seen also when the two waves have the same frequency. If, however, a frog has been trained with single waves to respond to one frequency but not to the other, it turns toward the rewarded frequency when the two are presented simultaneously. That is, *Xenopus laevis* detects the direction and frequency of a component wave in the interference pattern of two waves.

Whereas at large interstimulus angles, wave identification might conceivably be done by analyzing the input from local organ groups that are stimulated by one wave but not by the other, wave detection at small interstimulus angles

FIGURE 26.5. Response of *Xenopus laevis* to two waves presented simultaneously. The frog turns toward that wave for which it had been trained previously, which indicates that it can detect the direction and frequency of the component waves in the interference. In the demonstrated case, 20 Hz – to which the frog has been trained – is presented from the left vibrator and 12 Hz is presented from the right vibrator.

obviously requires an additional mechanism. At interstimulus angles of 30° or less, the animal's wave shadow is insufficient to lead to stimulation of some parts of the body by only one of the waves. Thus, the animal has to determine frequency and direction of the component waves by comparing the patterns of wave superposition at individual organs, taking into account the positions of the organs—i.e., by spatiotemporal analysis of the interference. The basis for such an analysis is that the time course of the superposition at an organ is replicated in the discharge pattern of the organ's afferent fibers (Strelioff and Sokolich 1981).

Spatiotemporal analysis of interference patterns is known in electroreception: the jamming avoidance response is based on comparing, at neighboring electroreceptive organs, the fluctuations of the electric field that result from the superposition of the animal's own electric organ discharge and that of a neighbor (Heiligenberg 1980). Analysis of interference patterns also occurs in the perception of complex tones. The representation of a tone's frequency at a certain location on the basilar membrane of the inner ear is not so strict that each afferent neuron is stimulated by only one frequency. Instead, the net response of the neuron is determined by all stimulus frequencies in the range of its tuning curve, and their temporal patterns of superposition are reflected in the discharge pattern of the afferent fiber (Brugge et al. 1969; Rose et al. 1969, 1971). The identification of similar frequency components in a tone therefore requires central nervous comparison of the superposition patterns at neighboring places on the basilar membrane, taking into account the alignment of the receptors along the basilar membrane. This differential analysis corresponds to what has been described above for the analysis of wave interferences with lateral lines.

Thus, stimulus analysis seems to rely on similar principles in the auditory, mechanoreceptive lateral line, and electroreceptive lateral line systems. Local stimulus characteristics are determined by comparing the inputs from neighboring organs (or receptors) with equal response characteristics. The analysis implies comparison of the time course of wave superpositions at the organs. Finally, the encoding of stimulus parameters in the afferent discharges—the basis for the central nervous analysis— is very similar in lateral line afferents of *Xenopus laevis* and in low-frequency auditory afferents. These findings indicate an intriguing similarity in the organization of stimulus analysis in these systems, which is reminiscent of the acousticolateralis hypothesis that lateral line and auditory systems have a common evolutionary origin. Alternative explanations of this similarity are conceivable as well, however, and further data are necessary before conclusions can be drawn.

In summary, the data indicate that the mechanoreceptive lateral line system in anurans should no longer be considered simple. The previously supposed simplicity seems to reflect the simplicity of experimental approaches rather than a characteristic of the system. Comparison with stimulus processing in the electroreceptive lateral line system and the auditory system appears to be a valuable guideline for further investigation of the analytical capacities of the mechanoreceptive lateral line system.

26.7 Comparative Studies of Wave Analysis

Xenopus laevis is one of about 14 species of the genus *Xenopus* (with new species being continuously detected) that belongs to the family Pipidae. The pipids are presumably primitive frogs of South Africa and South America, all of which are strictly aquatic as adults and retain their lateral line system after metamorphosis (Escher 1925; Duellmann and Trueb 1986). Besides *Xenopus laevis*, wave localization with an intact lateral line system or after lesions of lateral line organs has been studied in another species of the genus, *Xenopus muelleri* (Elepfandt 1982; Görner et al. 1984). No difference from *Xenopus laevis* has been found. Furthermore, it has been suggested that the lateral line system plays a role in shoaling of the tadpoles of *Xenopus laevis*, but experimental evidence is lacking (Katz et al. 1981).

Recent anatomical investigations have shown lateral line organs not only in adult pipids but also in other aquatic anuran adults (see Fritzsch Chapter 5). The species in which the lateral line system is retained after metamorphosis do not form a systematic group but are scattered over various families. This discovery raises the question of the function of their lateral line system and whether the mechanisms of wave analysis in *Xenopus laevis* can be generalized to other anurans.

To answer that question, wave localization was studied in three adult anurans from different suborders that retain their lateral line system after metamorphosis: *Pipa carvalhoi* (a pipid frog like *Xenopus*), the yellow-bellied toad *Bombina variegata* (a discoglossid frog), and *Rana cyanophlyctis cyanophlyctis* (an aquatic ranid frog). The methods employed were the same as used with *Xenopus laevis* (Elepfandt 1982). All frogs were capable of detecting wave directions at an accuracy approximately the same as or, in *Bombina variegata*, somewhat below that of *Xenopus laevis* (Elepfandt and Simm 1985; Elepfandt 1988c). In other tests, it was found that *Rana cyanophlyctis cyanophlyctis* can discriminate between waves of different frequencies (Elepfandt 1988c). These data indicate that the basic capacities of wave analysis found in *Xenopus* also exist in other anurans with a lateral line system. On the other hand, it became evident that the species respond differently to the same stimuli and that analogous kinds of wave analysis may serve different behavioral functions (Elepfandt and Simm 1985; Walkowiak and Münz 1985; Elepfandt 1988c).

Acknowledgments. The author's experiments were supported by research grants from DFG (El 75/1 and 75/2). I am grateful to G. Ehret, G. Meredith, H. Bleckmann, and S. Coombs for helpful comments on an earlier version of this manuscript and to M.E. Bitterman for revision of the English text.

References

Bachem A (1953) Time factors in relative and absolute pitch discrimination. J Acoust Soc Am 26:751–753.

Bleckmann H (1985) Perception of water surface waves: How surface waves are used for prey identification, prey localization, and intraspecific communication. Prog Sensory Physiol 5:147–166.
Brugge JF, Anderson DJ, Hind JE, Rose JE (1969) Time structure of discharges in single auditory nerve fibers of the squirrel monkey in response to complex periodic sounds. J Neurophysiol 32:386–401.
Buschmann HP (1984) Untersuchungen zur Entfernungslokalisation beim Krallenfrosch *Xenopus laevis* D. Staatsexamensarbeit, Universität Bielefeld, Bielefeld, FRG.
Dijkgraaf S (1947) Über die Reizung des Ferntastsinnes bei Fischen und Amphibien. Experientia 3:206–208.
Dijkgraaf S (1967) Biological significance of the lateral line organs. In: Cahn P (ed) Lateral Line Detectors. Bloomington: Indiana University Press, pp. 83–95.
Duellmann WE, Trueb L (1986) Biology of Amphibians. New York: McGraw-Hill, 670 pp.
Elepfandt A (1982) Accuracy of taxis response to water waves in the clawed toad (*Xenopus laevis* Daudin) with intact or with lesioned lateral line system. J Comp Physiol 148:535–545.
Elepfandt A (1984a) The role of ventral lateral line organs in water wave localization in the clawed toad (*Xenopus laevis*). J Comp Physiol A 154:773–780.
Elepfandt A (1984b) Topological representation of water wave directions in the lateral line midbrain projection of the clawed toad, *Xenopus laevis* Daudin. Neurosci Lett (Suppl) 18:S241.
Elepfandt A (1985) Naturalistic conditioning reveals good learning in a frog (*Xenopus laevis*). Naturwissenschaften 72:492–493.
Elepfandt A (1986a) Wave frequency recognition and absolute pitch for water waves in the clawed frog, *Xenopus laevis*. J Comp Physiol A 158:235–238.
Elepfandt A (1986b) Detection of individual waves in an interference pattern by the clawed frog, *Xenopus laevis* Daudin. Neurosci Lett (Suppl) 26:S380.
Elepfandt A (1986c) Multiple topological representation of wave directions in the midbrain of the clawed frog, *Xenopus laevis* Daudin. Neurosci Lett (Suppl) 26:S381.
Elepfandt A (1988a) Central organization of wave localization in the clawed frog, *Xenopus*. I. Involvement and bilateral organization of the midbrain. Brain Behav Evol 31:349–357.
Elepfandt A (1988b) Central organization of wave localization in the clawed frog, *Xenopus*. II. Midbrain topology for wave localization. Brain Behav Evol 31:358–368.
Elepfandt A (1988c) Water wave localization in anurans other than *Xenopus*. In: Elsner N, Barth FG (eds) Sense Organs, Interfaces between Environment and Behaviour (Proc 16th Göttingen Neurobiol Conf Stuttgart: Thieme, p. 162.
Elepfandt A, Simm A (1985) Water wave taxis in the yellow-bellied toad, *Bombina variegata* L.: Evidence for lateral line function. Biol Behav 10:333–341.
Elepfandt A, Wiedemer L (1987) Lateral-line responses to water surface waves in the clawed frog, *Xenopus laevis*. J Comp Physiol A 160:667–682.
Elepfandt A, Seiler B, Aicher B (1985) Water wave frequency discrimination in the clawed frog, *Xenopus laevis*. J Comp Physiol A 157:255–261.
Escher K (1925) Das Verhalten der Seitenlinienorgane der Wirbeltiere und ihrer Nerven beim übergang zum Landleben. Acta Zool 6:307–414.
Görner P (1963) Untersuchungen zur Morphologie und Elektrophysiologie des Seitenlinienorgans vom Krallenfrosch (*Xenopus laevis* Daudin). Z Vergl Physiol 47:316–338.
Görner P (1973) The importance of the lateral line system for the perception of surface waves in the claw toad, *Xenopus laevis* Daudin. Experientia 29:295–296.

Görner P (1976) Source localization with labyrinth and lateral-line in the clawed toad (*Xenopus laevis*). In: Schuijf A, Hawkins AD (eds) Sound Reception in Fish. Amsterdam: Elsevier, pp. 171–184.

Görner P, Moller P, Weber W (1984) Lateral-line input and stimulus localization in the African clawed toad, *Xenopus* sp. J Exp Biol 108:315–328.

Grobstein P, Comer C, Kostyk S (1983) Frog prey capture behavior: Between sensory maps and directed motor output. In: Ewert JP, Capranica RR, Ingle DJ (eds) Advances in Vertebrate Neuroethology. New York: Plenum, pp. 331–347.

Heiligenberg W (1980) The jamming avoidance response in the weakly electric fish *Eigenmannia*. Naturwissenschaften 67:499–507.

Ingle DJ (1983) Brain mechanisms of visual localization by frogs and toads. In: Ewert JP, Capranica RR, Ingle DJ (eds) Advances in Vertebrate Neuroethology. New York: Plenum, pp. 177–226.

Jenkins WM, Masterton RB (1982) Sound localization: Effect of unilateral lesions in central auditory system. J Neurophysiol 47:987–1016.

Katz LC, Potel MJ, Wassersug RG (1981) Structure and mechanisms of schooling in tadpoles of the clawed frog, *Xenopus laevis*. Anim Behav 29:20–33.

Kramer G (1933) Untersuchungen über die Sinnesleistungen und das Orientierungsverhalten von *Xenopus laevis* Daud. Zool Jb Physiol 52:629–676.

Kroese ABA, van der Zalm JM, van den Bercken J (1978) Frequency response of the lateral-line organ of *Xenopus laevis*. Pflügers Arch 375:167–175.

Macphail EM (1982) Brain and Intelligence in Vertebrates. Oxford, U.K.: Clarendon Press, 423 pp.

Moore BCJ (1982) An Introduction to the Psychology of Hearing. London: Academic Press, 293 pp.

Pickles JO (1982) An Introduction to the Physiology of Hearing. London: Academic Press, 341 pp.

Rose JE, Brugge JF, Anderson DJ, Hind JE (1969) Some possible neural correlates of combination tones. J Neurophysiol 32:404–423.

Rose JE, Hind JE, Anderson DJ, Brugge JF (1971) Some effects of stimulus intensity on response of auditory nerve fibers in the squirrel monkey. J Neurophysiol 34:685–699.

Scharrer E (1932) Experiment on the function of the lateral-line organs in the larvae of *Ambystoma punctatum*. J Exp Zool 61:109–114.

Shelton PMJ (1971) The structure and function of the lateral line system in larval *Xenopus laevis*. J Exp Zool 178:211–232.

Strelioff D, Sokolich WG (1981) Stimulation of lateral-line sensory cells. In: Tavolga WN, Popper AN, Fay RR (eds) Hearing and Sound Communication in Fishes. New York: Springer-Verlag, pp. 481–505.

Walkowiak W, Münz H (1985) The significance of water surface waves in the communication of fire-bellied toads. Naturwissenschaften 72:49.

Zittlau KE, Claas B, Münz H (1986) Directional sensitivity of lateral line units in the clawed toad *Xenopus laevis* Daudin. J Comp Physiol 158:469–477.

27
Stimulus Localization in *Xenopus*: Role of Directional Sensitivity of Lateral Line Stitches

PETER GÖRNER and CLAUDIA MOHR

27.1 Introduction

Since the experiments of Kramer (1933) and Dijkgraaf (1947), it has been known that the clawed toad (*Xenopus* sp.) can localize moving objects in water or on the water surface. Kramer demonstrated that the lateral line system was responsible for the toad's ability to turn toward the source of disturbance. Ablation of all lateral line stitches on one side of the animal abolished its response to a hand-held vibrating sphere and to a weak water current delivered by a pipette on the side. Responses to stimuli delivered on the intact side remained unaffected. Dijkgraaf has investigated the reactions of the toad to surface waves created by dipping a small rod into the water. The toad responded by turning toward the wave center even when it rested on the substrate, 10 cm below the water surface. As it turned out, surface waves are well suited for a detailed study of lateral line function in *Xenopus*, since they elicit unconditioned orienting responses which do not habituate over the course of more than 140 successive stimulations (Görner 1976).

The lateral line system of *Xenopus* consists of several rows of stitches which can be found on both head and trunk (Fig. 27.1). Each stitch comprises a varying number of more or less linearly arranged neuromasts (two to 12 in *X. laevis*). There is anatomical and electrophysiological evidence that two afferent fibers innervate most stitches of the trunk and that up to 10 fibers innervate each stitch of the head (Will et al. 1985).

Several parameters may be used by an aquatic animal to localize the direction and/or the distance of the center of a surface wave. These include (1) the curvature and (2) the frequency spectrum of the surface waves that depend on the distance of the stimulus source, (3) the intensity and (4) time pattern "seen" by successively stimulated sense organs, and (5) the directional sensitivity of the sense organs. There is experimental evidence that all stimulus parameters, with the exception of 3 and 5, are important to surface-feeding fish (Bleckmann 1987; Bleckmann et al. Chapter 25) and perhaps to *Xenopus* as well. Nothing is known, however, about the behavioral significance of the directional sensitivity of the superficial neuromasts.

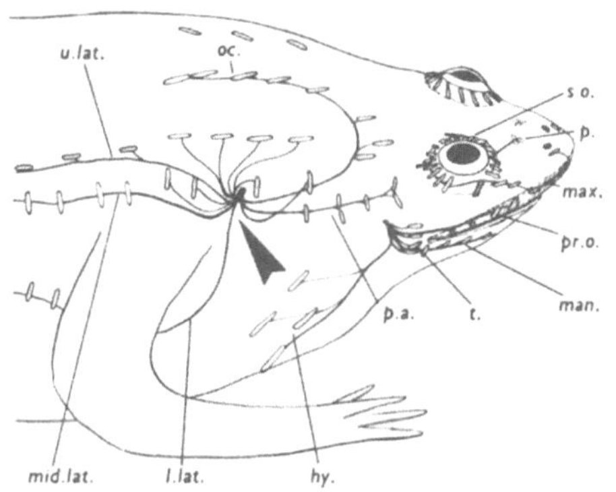

FIGURE 27.1. Location of lateral line stitches in *X. laevis*. Abbreviations: hy., hyomandibular lateral line; l. lat., lower trunk lateral line; man., mandibular lateral line; mid. lat., middle trunk lateral line; oc., occipital lateral line; max., maxillary lateral line; p., parietal lateral line; p.a., posterior auditory lateral line; pr.o., preorbital lateral line; s.o., supraorbital lateral line; t., tentacular lateral line, u. lat., upper trunk lateral line. Arrowhead indicates location of recording electrode in experiment 3. (Modified after Shelton 1970; Used with permission of the Company of Biologists Limited.)

This is most surprising, since it is known that hair cells are directionally sensitive (see Kroese and van Netten Chapter 13). In *Xenopus*, hair cells in neuromasts of the same stitch are all oriented in one of two opposite directions so that best axis of sensitivity is perpendicular to the long axis of the stitch (Görner 1963). Since one afferent fiber innervates only hair cells oriented in the same direction, this means that each of the two fibers innervating a single stitch will have two opposing responses. That is, if one fiber shows maximal excitation, the activity of the other is maximally suppressed (Fig. 27.2). Moreover, on the upper and middle trunk lateral line, stitches are perpendicular to each other. If the CNS is able to compare the afferent activities of these two rows, it should receive unequivocal information about the stimulus direction. Recent experiments by Elepfandt (1982) seem to provide evidence in favor of such a mechanism. In contrast to unilaterally lesioned toads, which still respond to surface waves, toads for which all but the upper trunk lateral line are ablated ceased to respond for a period of several months. This seems to indicate that central comparison between two orthogonally arranged stitches is essential for stimulus localization (Elepfandt 1982). However, our own (unpublished) experiments showed that toads with only the (equally aligned) stitches of one occipital line left intact were still able to turn correctly to the stimulus source.

Thus, to date, the biological significance of the directional sensitivity of the lateral line stitches has not yet been established. The experiments described below were designed to address the significance of this sensitivity to the ability of *Xenopus* to localize stimuli.

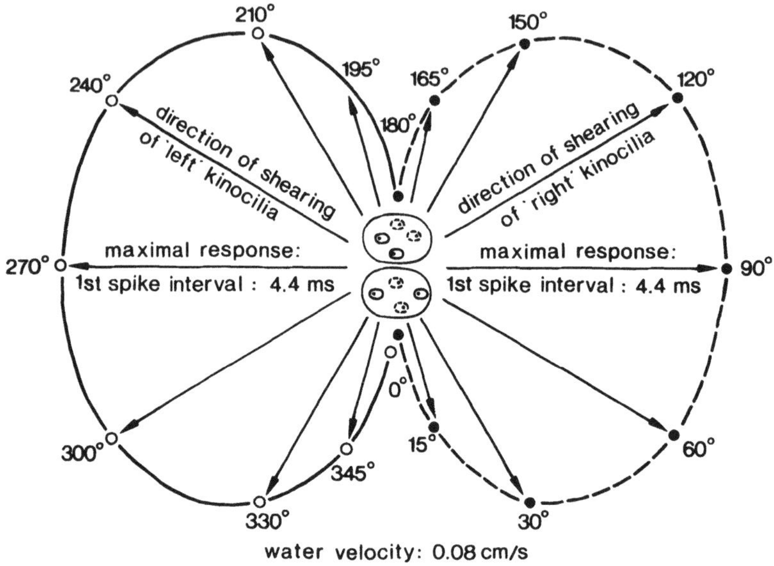

FIGURE 27.2. Directional sensitivity of two afferent nerve fibers of a stitch of *X. laevis*, stimulated with a laminar water current. Two neuromasts (large ovals) with the apical surface of four sensory cells each (small ovals) are shown schematically. The black dots within the sensory cells indicate the insertion of the kinocilium. The hair cells, drawn with solid lines, are depolarized by a water current from the right side which shears the kinocilium over the left rim of the hair cell. The arrows indicate the direction of shearing of the kinocilium. The length of the arrows and the outer solid line demonstrate the degree of excitation of the corresponding afferent nerve fiber (expressed as the reciprocal value of the first spike interval after onset of stimulation). The dashed line indicates the activation of the other afferent nerve fiber which innervates hair cells of opposite polarity, also drawn with dashed lines. These hair cells are depolarized by a water current from the left side. (Redrawn from Görner 1963, Fig. 8.)

Surface waves are not well suited to ask questions about the relationship between the directional sensitivity of a stitch and the toad's ability to turn toward the wave center. Not only do surface waves result in a bidirectional response of the activated stitch, but they also elicit responses from toads with a totally ablated lateral line system (Görner 1973, 1976; Elepfandt 1982; Görner et al. 1984). Most likely this is due to the inner ear (Görner 1976). Therefore, to get an answer to the question about the possible behavioral significance of the directional sensitivity of the lateral line system, a local stimulus such as a discrete water jet was used. This stimulus has a pronounced unidirectional component and can be adjusted in a way that it exclusively affects the lateral line stitches.

Three sets of behavioral experiments were performed. The first involved stimulating the animal with a jet of water from a hand-held pipette in order to determine the different modes of response and to find out whether this sort of stimulation is suited for a quantitative analysis of the behavioral responses. In the second set of experiments, the jet of water was generated by a device that allowed stimulation under defined conditions. At the beginning of this series, two control experiments, one behavioral and one electrophysiological, were performed. The behavioral experiment controlled for the possibility that the jet of water was detected by tactile receptors, as suggested by Kramer (1933). In the electrophysiological experiment, recordings were made from single afferent fibers innervating different stitches to confirm the directional response properties elicited with a jet of water. In a third series of experiments, the toads were stimulated with surface waves.

27.2 Materials and Methods

27.2.1 Animals

Experiments were conducted with enucleated male and female clawed toads, *Xenopus laevis* and *X. mülleri*. The toads, obtained from a dealer or from our own breeding stock, were kept individually in plastic containers and fed with cow heart. Stitches were ablated by means of a thermocauter in anesthesized (.1% tricainmethylsulfonate) animals. Unilateral ablations were made on either the left or right side, but results are presented as if only the left side had been ablated.

27.2.2 Behavioral Experiments

27.2.2.1 Stimulation with a Jet of Water from a Hand-Held Pipette

In a tank 50 × 35 cm with a water depths of 9 cm, local subsurface water currents were directed toward the resting animal from a hand-held Pasteur pipette at a distance of 1 to 2 cm. (At this distance the toad did not react to the small, unavoidable vibrations of the hand.)

27.2.2.2 Stimulation with a Controlled Jet of Water

A jet of water was generated by a device that allowed stimulation under defined conditions. A pipette was mounted on a holder that could be moved and adjusted in all three dimensions. From a distance of 1 to 3.5 cm, a jet of water was directed toward the lateral line stitches. The flow of water was electronically controlled by increasing the pressure in an air chamber that was connected to a water-filled pipette. After 50 ms of building pressure, a controlled amount of water was released from the opening of the pipette (1.1 to 1.25 mm in diameter). To measure water movement at the tip of the pipette, it was filled with a solution of India ink. As the dye passed from the pipette into the surrounding water, it appeared to cluster into a descrete "packet" (around 4 mm in diameter). By tracking this dye packet with single-frame analysis, it was possible to measure the duration and the velocity of water movement. Twenty-four registrations were analyzed. The duration of water flow was 50 ms, and the velocity at a distance of 1 cm from the tip of the pipette was 20 cm/s and decreased to 10 cm/s at a distance of 2.6 cm, thus being in the range of actively swimming zooplankton (see Montgomery Chapter 28).

27.2.2.3 Stimulation with Surface Waves

These experiments were conducted in a circular pool that had a diameter of 50 cm and a water depth of 3.5 cm. Surface waves were generated by dipping a plastic rod (3 mm in diameter) into the water about 10 cm from the resting toad (for further details see Görner et al. 1984).

27.2.2.4 Recording of the Response and Data Processing

The toad's behavioral responses were recorded on videotape. For data analysis the tapes were played back on a monitor, and the toad's position before and after the stimulus was drawn on a transparent plastic foil. From there all relevant coordinates were fed into a computer by means of an x–y tablet and treated by linear and circular statistics.

27.2.3 Electrophysiological Experiments

To record from stitches of the middle or upper trunk lateral line, an incision was made in the skin of the anesthesized toad on the dorsal side at the origin of the N. lateralis posterior (arrowhead in Fig. 27.1). The animals were immobilized by injection of three μg/kg body weight Alloferin. The nerve activity was recorded from single fibers with glass capillary microelectrodes. The stitches were stimulated with the same equipment as used in the behavioral experiments. In the electrophysiological experiments, the water depth was 6 cm, and Ringer solution, rather than tap water, was used. All toads recovered from immobilization, and the incision healed within 4 days. Spikes and trigger signals were stored on magnetic tape and analyzed by conventional computer programs.

27.3 Results and Discussion

27.3.1 Series 1: Stimulation with a Water Current from a Hand-Held Pipette

27.3.1.1 Experiment 1: Behavioral Response Pattern to Stimuli Directed Toward Different Skin Areas

The most striking result from experiments in which water jets were used was that the toad responded in a different manner from how it did to surface waves (Fig. 27.3). The response consisted of several components. In most cases, the toad

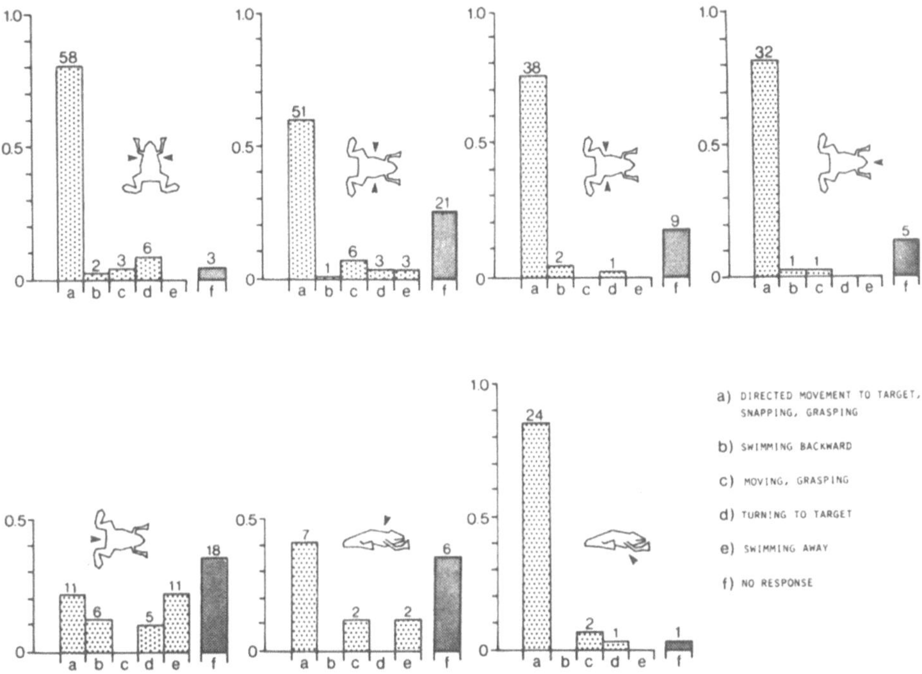

FIGURE 27.3. Responses of *X. laevis* after direct stimulation of the lateral line stitches with a jet of water from a Pasteur pipette. In all cases the toad rested on the bottom of the tank, except in the last diagram, where it "hung" under the water surface. Arrowheads indicate direction of water current.

grasped and tried to snap at the target. This was normally accompanied by a displacement and a turning to the target. Occasionally only one of these response components (e.g., moving, grasping) occurred; alternatively, the toad moved away, which was interpreted as no response to the stimulus. The toad reacted to a stimulus delivered at the left or right side near the forelegs, at the middle of the trunk, or near the hind legs. When stimulated directly from behind, it exhibited a feeding response only rarely. *Xenopus* also responded with directed movements to stimulation from above and from below. (Stimulations at the ventral side were made when the toad positioned himself at the water surface (see Fig. 20.2 in Claas et al. Chapter 20). Thus, *Xenopus* is able to localize a stimulus in all three dimensions (see also Kramer 1933; Elepfandt 1984a).

27.3.2 Series 2: Stimulation with a Defined Water Current

27.3.2.1 EXPERIMENT 2: BEHAVIORAL CONTROL FOR STIMULATION OF TACTILE RECEPTORS

In the behavioral control, six toads were used, three whose lateral line system was ablated with the exception of a few single stitches and three whose lateral line

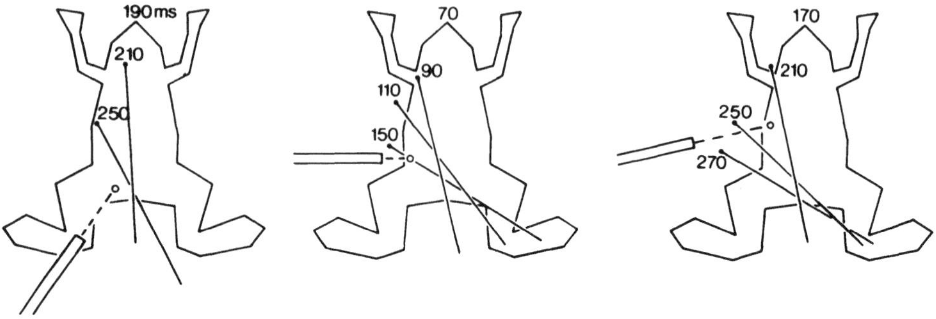

FIGURE 27.4. Responses of *X. laevis* to stimulation with a defined jet of water. The long axes of the toads and the corresponding times after onset of water current are shown. Numbers in front of the resting toad: time delay (in milliseconds) before response.

system was ablated on one side. When a jet of water was directed toward the stitches of the intact side, unilaterally lesioned toads responded 40% of the time ($n = 367$). The response pattern was the same as in intact animals. Following a delay of about 70 ms (median value of three intact toads, $n = 303$), toads turned and moved backward, thus bringing their mouth closer to the opening of the pipette (Fig. 27.4). When the operated animals were stimulated from the lesioned side, no response occurred ($n = 146$). Toads with only a few intact stitches reacted in 41% of cases ($n = 263$) when the jet of water was directed on the stitches but showed no response when areas of the skin were stimulated where the stitches were lesioned or absent ($n = 57$). Thus, the response to a jet of water in the experiments under controlled stimulus conditions was brought about by activating the lateral line neuromasts and not by the tactile epidermal receptors. Whether the stitches under these stimulus conditions show a clear directional response was tested in the following (electrophysiological) experiment.

27.3.2.2 EXPERIMENT 3: ELECTROPHYSIOLOGICAL CONTROL FOR DIRECTIONAL-DEPENDENT RESPONSES OF SINGLE AFFERENT NERVE FIBERS

Figure 27.5a–d shows four recordings from two different stitches on the trunk (see right inset figures in Fig. 27.5a and c) which were stimulated perpendicularly to their long axis – i.e., from a direction of maximum excitation or inhibition. For the fiber shown in Fig. 27.5a and b, stimulation from one side causes an increase whereas stimulation from the opposite side causes a suppression in the resting activity. Thus, the CNS receives clear information about the direction from which side the stitch was stimulated. Figure 27.5c shows a strong response in a different stitch when the water jet was delivered on the stitch. Figure 27.5d shows a weak response when the pipette was displaced a few millimeters parallel to the stitch's long axis so that the organ was not fully exposed to the water current. A similar effect was obtained when the distance between pipette opening and stitch was increased. Although the response of other afferent fibers could

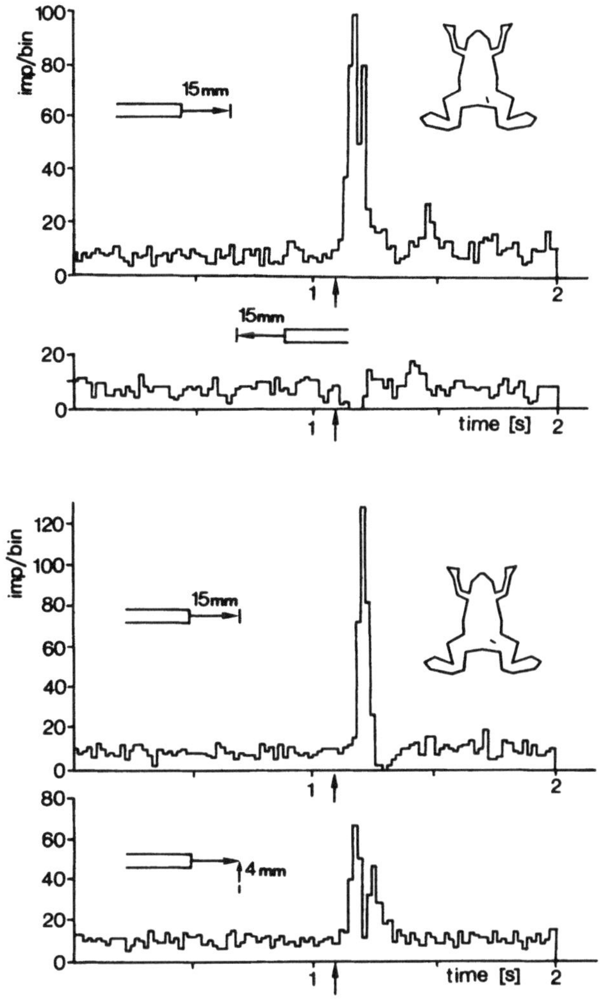

FIGURE 27.5. Activity of a single afferent fiber before and after stimulation with a jet of water. (a,b) Stimulation of a stitch from opposite directions. (c,d) Stimulation of a different stitch from the same direction. In (c) the water current was directed on the stitch; in (d) the pipette was moved 4 mm to the side. Arrow on x-axis: start of water current. Resting activity in (a) and (b): 11.8 imp/s; in (c) and (d): 15.6 imp/s. Sampling of 30 stimuli in each record. Bin width 20 ms. In the right inset figures (a and c), the location of the stitch of the upper trunk lateral line from which the nerve activity was recorded is indicated.

differ from the recordings presented here (e.g., an excitation followed an inhibition, and vice versa; sometimes two maxima (compare Fig. 27.5d) or minima occurred), the essential effect, namely an excitation or inhibition, was the main characteristic in all 98 fibers recorded from 10 toads. Thus, the stitches were activated by the water jet used in the behavioral experiments, and the degree of excitation depended on the stimulus direction.

27.3.2.3 Experiment 4: Behavioral Responses of Toads with an Intact Lateral Line System

If the directional sensitivity of the stitches is represented somewhere in the CNS, the toad should be able to recognize the direction of a water current, and, as a consequence, stimulation from different directions should cause it to turn in different angles. This is certainly true for the toad's responses to surface waves. However, when *Xenopus* is stimulated with a water current from different directions, it exhibits a different behavior. Depending on the stimulus angle, two different responses seem to occur. At stimulus angles from 0 to $\pm 29°$, the turning angle increases significantly with the stimulus angle ($r = .61$*, $n = 58$, $p < .0001$; Fig. 27.6a).

At greater stimulus angles (range: 30° to 180° and −30° to −180°), no such correlation is evident ($r = .06$, $n = 368$, $p > .1$). One problem with this analysis, however, is that different sets of stitches are stimulated at different stimulus angles. At stimulus angles of 0°, for example, stitches on the top of the head were stimulated, whereas for increasing stimulus angles, the more laterally located stitches were stimulated. For angles above ~ 30°, those on the side of the head and trunk were stimulated (Fig. 27.6c). Therefore, to answer the question whether the stimulus angle determines the turning angle, care must be taken that the same stitches are stimulated. Figure 27.6b shows results under these conditions where the jet of water was directed on stitches of a small region (12 mm of the body length) in the middle of the trunk (compare Fig. 27.6c, right arrows). In this case, the turning angle is not significantly related to the stimulus angle ($r = .18$, $n = 96$, $p = .05$). Thus, although the stitches on the trunk should provide the CNS with unequivocal information on the stimulus direction, this information is not reflected in the turning angle.

The conclusion that *Xenopus* is unable to localize the stimulus source, however, is not justified, because the animal repositions his body in such a way as to get its mouth closer to the pipette opening. As Fig. 27.7a shows, this repositioning results in a backward (and lateral) movement (dm), which increases as a function of distance from stitch to the tip of the snout (ds) ($p < .0001$). In addition to the backward displacement, the toad turns, whereby the turning angle also increases with increasing distance of the stimulated stitches from the tip of the snout (Fig. 27.8; $p < .0001$).

---

*When the correlation coefficient was calculated for experiments in which the toad turned either to the left (negative values) or to the right (positive values), all negative values were counted positive, as if the toad had always turned to the right.

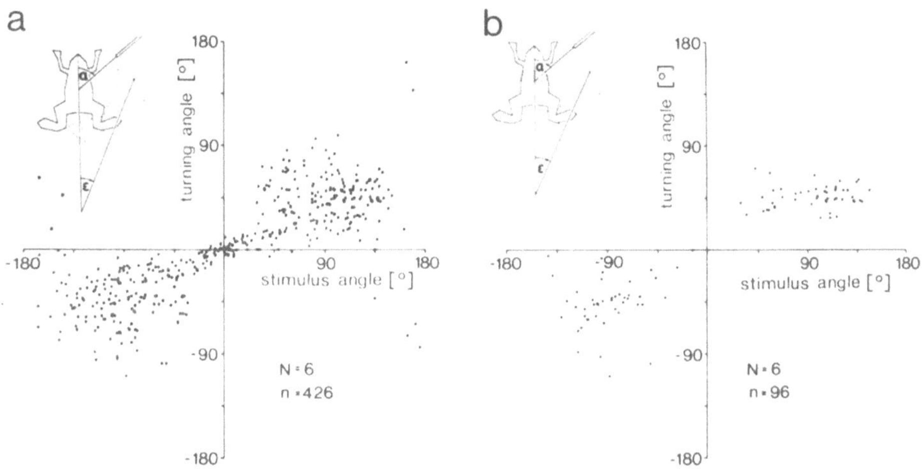

FIGURE 27.6. (a,b) Turning angle (ε) as a function of stimulus angle (α). Stimulation with a jet of water. In (b) only those reactions are plotted where the water current stimulated the same stitches in a small area (indicated by hatching in inset figure). (c) Simplified scheme to demonstrate that different stimulation angles between −30° and 30° in the head region mean that different stitches were stimulated, whereas angles greater than 30° may mean that the same stitches on the trunk were stimulated, as shown in (b).

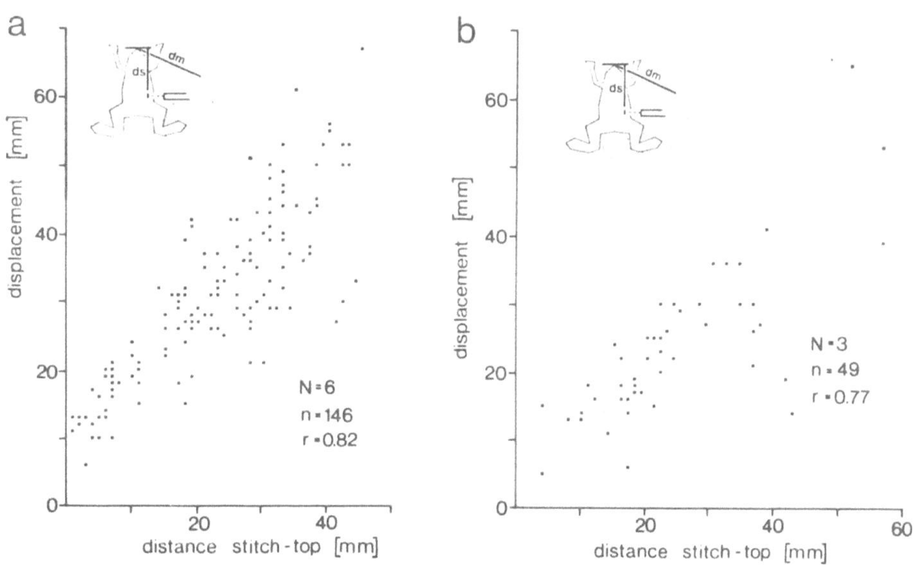

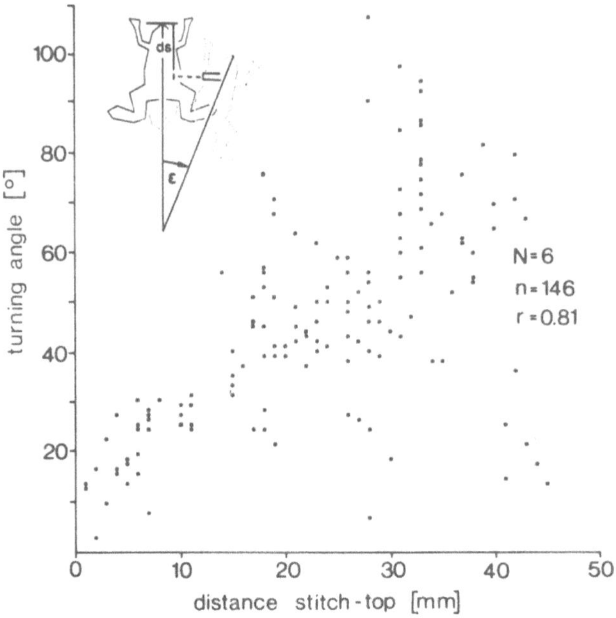

FIGURE 27.8. Correlation between the distance of the stitch from the top of the head and the turning angle in the same (intact) toads as in Fig. 27.6. As in Fig. 27.7a, only stimulation angles from the right and left side are plotted.

This does *not* mean that *Xenopus* is able to determine the distance to the pipette opening correctly. If it could do so, the poststimulus distance should be independent of the prestimulus distance. However, the distance between the pipette opening and the month after a response increased with the distance of the pipette to the stitch before stimulation ($p < .001$; Fig. 27.9). Therefore, *Xenopus* is not able to localize the pipette correctly at different distances.

The most plausible interpretation of the results is that *Xenopus* "knows" which of its stitches is stimulated and that the stimulus source is somewhere in the near vicinity of the activated stitches. If this conclusion is correct, then the result should be the same when only a single stitch is stimulated, as was done in the following experiment.

◀

FIGURE 27.7. Correlation between the distance (ds) of the stimulated stitch from the top of the head and the displacement (dm) of the toad (inset). Only stimulation angles from the right side (70° to 110°) and from the left side (−110° to −70°) are plotted. (a) Toads with an intact LLS. (b) Toads with only a few stitches on the head and trunk left intact.

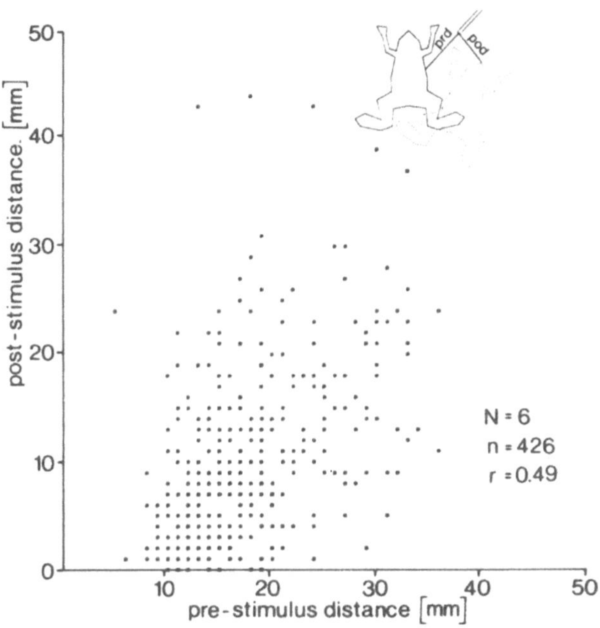

FIGURE 27.9. Correlation between the distance of the pipette from the stitch before stimulation and the distance of the pipette from the top of the head after stimulation. The same toads as in Figs. 27.6 and 27.7a. All stimulation angles from 0° to 360° are plotted. The significant correlation means that there is no distance localization.

27.3.2.4 Experiment No. 5: Behavioral Response of Toads with Only a Few Stitches Left Intact

In three toads, the lateral line system was ablated with the exception of a few single stitches. Care was taken that only one single stitch was stimulated by the water current. Results from this experiment (Fig. 27.7b) were very similar ($p < .0001$) to those from toads with intact systems (compare Fig. 27.7a). The results are the same if the responses after stimulation of neuromasts on the trunk (upper and middle lateral line, occipital line) are only taken into account ($n = 36$, $r = .77$, $p < .0001$).

This is important since one may argue that information about time of arrival or intensity at individual neuromasts is kept separate. However, this can only be valid for stitches of the head region, most of which receive more than two afferent fibers (Will et al. 1985), while neuromasts within a stitch of the trunk (except for those of the posterior auditory line) are innervated by the same two afferent fibers, and a comparison between the activities of single neuromasts is not possible. Thus, information about the location of the stimulus source must be derived from a topological representation of the activated stitch somewhere in the CNS.

So far, the experiments have shown that the directional sensitivity of the stitches does not determine the turning angle. The question remains whether the directional sensitivity enables the toad to determine whether a stimulus source at a fixed position is producing water flow in one of two opposing directions. This was tested in the following experiment.

27.3.2.5 Experiment 6: Behavioral Responses of Unilaterally Lesioned Toads

In this experiment, three toads for which the lateral line system was ablated on the left side were used. When the water jet was directed on the right (intact) side, the toads (with three exceptions) turned to the intact side, independent of the direction of water flow (Fig. 27.10a).

This result is quite surprising, since the directional sensitivity of the stitches should provide the CNS with unequivocal information about the side of the stimulus source. Moreover, there should be additional information in the time of arrival and/or intensity of the water flow as it reaches different stitches. However, since there are only a few stitches within reach of the small water current produced by the pipette, there may not be enough information for the CNS to compute the stimulus direction by means of time or intensity differences. To test this possibility, a surface wave that stimulates a much wider range of stitches was used.

27.3.3. Series 3: Stimulation with Surface Waves

27.3.3.1 Experiment 7: Behavioral Responses of Unilaterally Lesioned Toads

When the unilaterally lesioned toads of experiment 6 were stimulated with surface waves, all toads in most cases turned correctly, regardless of whether the wave was generated at the intact or at the lesioned side (Fig. 27.10b).

The most plausible explanation for the correct determination of the target angle is that the time or intensity pattern of the great number of successively stimulated stitches provided the essential information. It is also possible, however, that the labyrinth is involved in the toad's correct orientation. Therefore, a final experiment was carried out, in which the sensitivity of a few, equally aligned stitches was reversibly blocked so that the differences between the responses of toads with an intact and with a blocked lateral line system could be measured.

27.3.3.2 Experiment 8: Behavioral Responses of Toads with Only a Few Stitches Left Intact

For this experiment, all but five or six stitches were lesioned in three toads. The remaining stitches were located in different spots on the dorsal side of the animal, and their axes of best sensitivity were either perpendicular or parallel to the longitudinal axis of the toad (Fig. 27.11a). If the directional sensitivity of the stitches was the only cue in determining the stimulus direction, then two toads

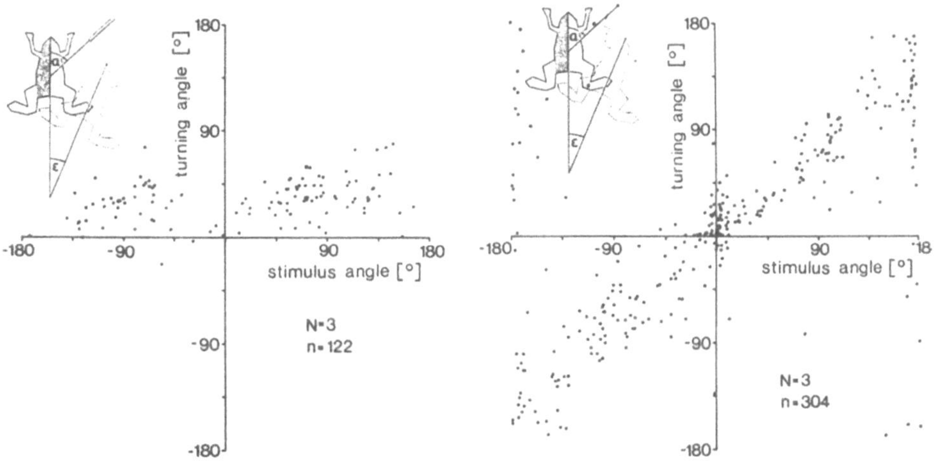

FIGURE 27.10. Correlation between stimulus angle and turning angle in toads with an intact lateral line system only on the right side. All stitches on the left side were lesioned (hatched in inset figures). (a) Stimulation with a water current. (b) Stimulation with surface waves. Positive values: stimulation from and turning to the right side. Negative values: stimulation from and turning to the left side.

with stitches parallel to the longitudinal axis of the animal (L07 and M11) should respond by turning either to the left or right side and the toad with stitches orthogonal to the longitudinal axis (L26) should respond by not turning or by turning 180°. However, all three toads were able to localize the direction of a wave center (Fig. 27.11b). By submerging the toads in a solution of 1% $CoCl_2$ for 1 h, the sensitivity of the lateral line system could be completely abolished without affecting other skin tactile receptors. The proper working of the tactile receptors was controlled by gently touching the toad's skin with a hair, which mediated a turning response to the target. (Details will be presented by Münz et al., in preparation. See also Karlsen and Sand 1987; Enger et al. Chapter 29.) One hour to 3 days after this procedure, two of the toads (L07 and L26; M11, the third, died) responded to surface waves, but there was no correlation between stimulus angle and turning angle (Fig. 27.11c). Five to 10 days later both toads recovered their ability to orient toward the stimulus.

Thus, toads are able to respond to surface wave stimuli with oriented responses when only a few stitches are stimulated, but they are unable to do so when information from the stitches is excluded. In this case, the animals are only able to turn to the correct side – a response most likely mediated by the inner ear. Therefore, the results from experiment 7 most probably reflect the involvement of the lateral line stitches and not the inner ear. Since the directional sensitivity of the stitches oriented in the same direction cannot have supplied this information, the directional component must have been obtained from the time or intensity pattern of the stimulated stitches.

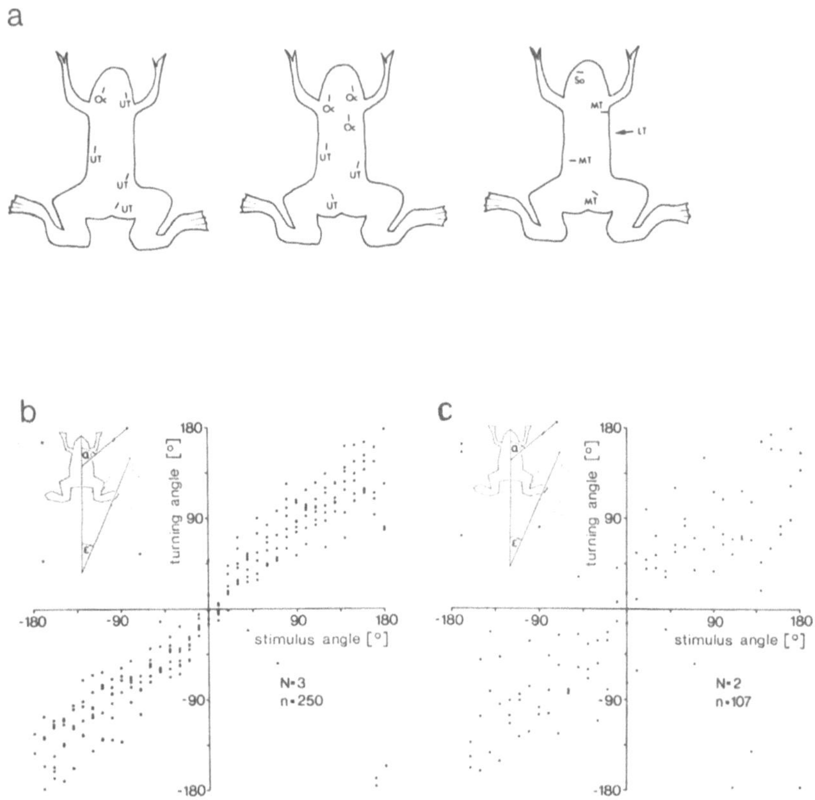

FIGURE 27.11. (a) Left: *X. laevis* L07. The lateral line system was lesioned with the exception of four stitches of the upper trunk (UT) lines and one stitch of the occipital (Oc) line. Middle: *X. mülleri* M11. Three stitches were left intact in the UT lines, and three in the Oc lines. Right: *X. laevis* L26. Three stitches of the middle trunk (MT) lines were left intact, one stitch of the supraorbital (So) line, and one of the lower trunk (LT) line (arrow). (b) Responses to surface waves of the toads before treatment. (c) One to 32 h after reversible blocking the lateral line activity with $CoCl_2$.

27.4 General Discussion

The experiments described here provide arguments against the hypothesis that the directional sensitivity of the lateral line stitches of *Xenopus* is of significance for the toad's ability to localize a stimulus source. Even in cases where the directional sensitivity could be responsible for *Xenopus*' stimulus localization, other explanations are possible and even more likely.

There are at least two arguments against the possibility that the intensity pattern is relevant. One is that the intensity with which the stitches are stimulated should always be maximal for the more dorsally located stitches, irrespective of the direction of the wave center. The dorsally located stitches are closer to the water surface than the lateral and ventral ones (compare Fig. 27.1) and should therefore be stimulated with a greater amplitude than the latter. Elepfandt (1982) came to a similar conclusion based on his results from unilaterally lesioned toads (see also Elepfandt Chapter 26). Another, similar argument is related to the way in which toads often "hang" from the water surface in deeper water with the posterior part of the body more deeply submerged. Stimulated with surface waves, the location of the maximally activated stitches should be drastically different from toads that, as in the experiments described here, rest more horizontally on the substrate in shallow water.

In obliquely positioned toads in deeper water, the stitches on the top of the head should be maximally activated. In toads on the substrate, the dorsal stitches on the head and trunk should be maximally activated. However, toads in both positions react to surface waves in practically the same way (Elepfandt 1982). The time pattern, however, is for both positions qualitatively the same irrespective of the different intensity differences in the dorsoventral and caudorostral axis. Recent experiments on the lateral line system of surface-feeding fishes also indicate that time and/or phase differences are important for target angle determination rather than intensity differences or the directional sensitivity of single neuromasts (see Bleckmann et al. Chapter 25).

Is the directional sensitivity of the stitch of behavioral significance at all? To answer this question, one has to realize that for a surface wave stimulus, not only is the direction of wave propagation important, but also the successive locations of the stitches on the body that are being stimulated. In lateral stitches (on the flank) that have a dorsoventral or frontocaudal orientation, the cupula will be moved up and down by a surface wave, independent of the wave's direction. For the many stitches on the toad's trunk or head region (where the axis of sensitivity is more or less inclined with respect to the horizontal plane of the water surface), the directional sensitivity to surface waves is quite poor (Zittlau et al. 1986). Furthermore, Elepfandt and Wiedemer (1987) have shown that even in more dorsally located stitches, the directional sensitivity of an afferent fiber is never as pronounced as with a laminar water current (compare Fig. 27.2). Thus, the directional sensitivity of many of the stitches is of little value for stimulus localization.

The morphological structures that lead to a directional sensitivity (eccentric position of kinocilia on the apex of the hair cells; two sets of hair cells with opposite maximum sensitivity; the flat, ribbonlike cupula) may have evolved to maximize sensitivity along a single axis. To maintain equal sensitivity in all directions, another, orthogonal axis has to be introduced, and the outputs of the two systems have to be compared centrally. In the upper and middle trunk lateral line, where stitches are oriented perpendicularly, such a system may have been realized. Similar anatomical patterns exist in the arrangement of the stitches on the head (compare Fig. 27.1). Thus, stimuli from all directions in the horizontal plane will be detected with practically the same sensitivity. Whether or not this hypothesized mechanism proves to be true will depend on future behavioral and neurophysiological studies in which the signal-to-noise processing capabilities of the system are measured as a function of stitch orientation.

27.5 Summary

The directional response of clawed toads (*Xenopus laevis* and *X. mülleri*) to surface waves and to water jets are quite different. To surface waves, they show a well-known, previously described behavior—i.e., they turn toward the wave center, whereby the turning angle increases linearly with the stimulus angle. To a jet of water, no such correlation is evident, but rather toads turn toward the simulus source and swim backward. Turning angle and backward displacement increase with stimulation of the more caudally located stitches.

When toads whose lateral line system is unilaterally ablated are stimulated with a jet of water, they turn predominantly to the intact side, irrespective of the direction of water flow. To surface waves, they turn correctly toward the wave center. Toads with only a few stitches left intact, the axes of which are either parallel or orthogonally oriented with respect to the longitudinal axis of the animal, react to surface waves nearly as well as toads with an intact lateral line system.

We conclude that in *Xenopus* the directional sensitivity of the stitches is not essential for stimulus localization. Most probably the time pattern of the successively stimulated stitches provides the relevant information. It is unlikely that the intensity pattern plays any role.

The directional sensitivity of the stitches, based on the orientation of ciliary hair cell bundles and other features of the neuromasts, may have evolved to maximize sensitivity. By adding information from two stitches of orthogonal orientation, the toad will be able to perceive stimuli from different directions with the same sensitivity.

Acknowledgments. We thank our friends and colleagues for many discussions and valuable suggestions for improving the manuscript. We want to mention particularly Dr. S. Coombs, Dr. A. Elepfandt, Dr. A.J. Kalmijn, Dr. H. Bleckmann and, for correcting an early version of the manuscript, Dr. U. Görner and Dr. P. Moller. Our thanks are also due Mrs. U. Markert for reliably conducting most of the behavioral experiments, U. Witte for writing the statistical and electrophysiological computer programs, Mrs. M. Koppers for drawing the figures, and Mrs. I. E. Appel for typing the manuscript. Supported by a grant of Deutsche Forschüngsgemeinschaft (GO 82-10).

References

Bleckmann H (1980) Reaction time and stimulus frequency in prey localization in the surface-feeding fish *Aplocheilus lineatus*. J Comp Physiol 140:163–172.

Dijkgraaf S (1947) Über die Reizung des Ferntastsinnes bei Fischen und Amphibien. Experientia 3:206–208.

Dijkgraaf S (1967) Biological significance of the lateral-line organ. In: PH Cahn (ed) Lateral Line Detectors. Bloomington: Indiana University Press, pp. 83–95.

Elepfandt A (1982) Accuracy of taxis response to water waves in the clawed toad (*Xenopus laevis* Daudin) with intact or with lesioned lateral line system. J Comp Physiol 148:535–545.
Elepfandt A (1984a) The role of ventral lateral line organs in water wave localization in the clawed toad (*Xenopus laevis*). J Comp Physiol 154:773–780.
Elepfandt A (1984b) Localization of water surface waves with the lateral line system in the clawed toad (*Xenopus laevis* Daudin). In: Varju D, Schnitzler H-U (eds) Localization and Orientation in Biology and Engineering. New York: Springer-Verlag, pp. 63–65.
Elepfandt A, Wiedemer L (1987) Lateral-line responses to water surface waves in the clawed frog, *Xenopus laevis*. J Comp Physiol 160:667–682.
Görner P (1963) Untersuchungen zur Morphologie und Elektrophysiologie des Seitenlinienorgans vom Krallenfrosch (*Xenopus laevis*). Z Vergl Physiol 47:316–338.
Görner P (1973) The importance of the lateral line system for the perception of surface waves in the claw toad *Xenopus laevis* Daudin. Experientia 29:295–296.
Görner P (1976) Source localization with labyrinth and lateral line in the clawed toad (*Xenopus laevis*). In: Schuijf A, Hawkins AD (eds) Sound Reception in Fish. New York: Elsevier, pp. 171–183.
Görner P, Moller P, Weber W (1984) Lateral-line input and stimulus localization in the African clawed toad *Xenopus* sp. J Exp Biol 108:315–328.
Karlsen HE, Sand O (1987) Selective and reversible blocking of the lateral line in freshwater fish. J Exp Biol 133:249–262.
Kramer G (1933) Untersuchungen über die Sinnesleistungen und das Orientierungsverhalten von *Xenopus laevis* Daud. Zool Jb Physiol Tiere 52:629–676.
Schwartz E (1971) Die Ortung von Wasserwellen durch Oberflächenfische. Z Vergl Physiol 74:64–80.
Shelton PMJ (1970) The lateral line system at metamorphosis in *Xenopus laevis* (Daudin). J Embryol Exp Morphol 24:511–524.
Will U, Luhede G, Görner P (1985) The octavo-lateralis area in *Xenopus laevis* I. The primary afferent projections. Cell Tissue Res 239:147–161.
Zittlau KE, Claas B, Münz H (1986) Directional sensitivity of lateral line units in the clawed toad *Xenopus laevis* Daudin. J Comp Physiol 158:469–477.

28
Lateral Line Detection of Planktonic Prey

JOHN C. MONTGOMERY

28.1 Introduction

This chapter is in three main sections. The first reviews the available evidence that some fish and aquatic amphibians are capable of using the lateral line to feed on zooplankton. This ability enables them to feed at night (and hence avoid diurnal predators or utilize a food resource enriched by demersal zooplankton) or to feed through the long periods of darkness at high latitude. Evidence is also presented that lateral line information is likely to be of importance in complementing vision in other species of plankton-feeding fish.

The second section deals with signals produced by swimming zooplankton. Available information is reviewed, and additional results presented on the signals produced by fast swimming zooplankton. In the third section, these signals are compared with the properties of mechanosensory lateral line sense organs, and the requirements on this system for prey detection and interception are discussed. These requirements show interesting similarities and differences with other tasks performed by lateral line systems, such as surface wave analysis (Bleckmann et al. Chapter 25).

28.2 Behavioral Evidence

Behavioral evidence for lateral line detection of planktonic prey has been obtained in the following species: mottled sculpin *Cottus bairdi* (Hoekstra and Janssen 1985, 1986), piper *Hyporhamphus ihi* (Montgomery and Saunders 1985; Saunders and Montgomery 1985), amblyopsid blind cave fish (Poulson 1963), and aquatic salamanders *Ambystoma maculatum* (Lannoo 1987). It is also implicated in the feeding behavior of the antarctic fish *Pagothenia borchgrevinki* (Montgomery and Macdonald 1987).

The most direct evidence for the use of the lateral line in plankton feeding comes from field and laboratory studies of the mottled sculpin (Hoekstra and Janssen 1985). Blinded fish will feed on live prey and react to other moving objects but ignore dead prey. Inactivation of the lateral line by external applica-

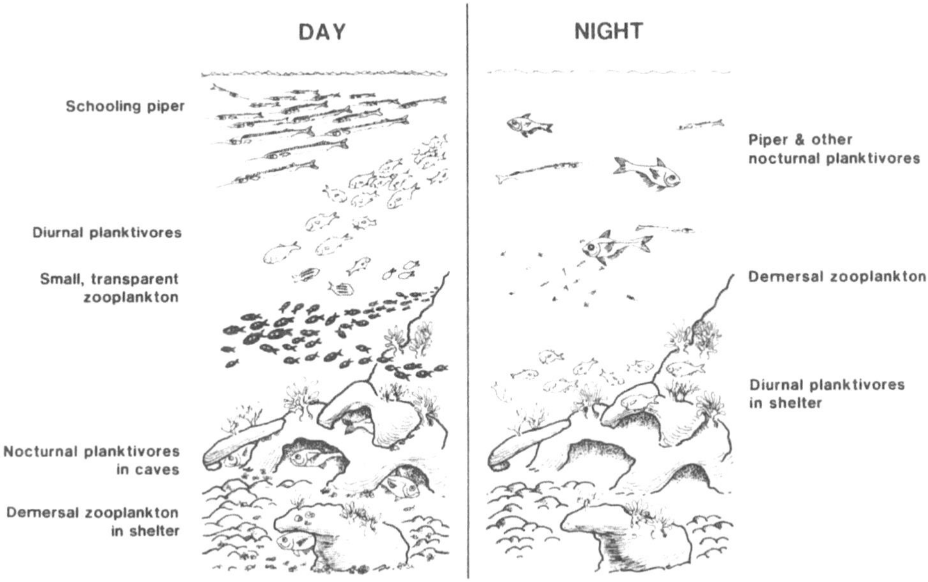

FIGURE 28.1. Diurnal and nocturnal plankton feeding fishes. During the day, piper school just below the water surface, and schools of visually feeding planktivorous fish swarm over the reef. Nocturnal plankton feeders that rely on vision, and the demersal zooplankton, are found close to the bottom in shelter. At night, the nocturnal plankton feeders are found in the water column feeding on the demersal zooplankton, and the diurnal plankton feeders seek shelter close to the reef.

tion of streptomycin, or mechanical blocking of canal pores, eliminates their feeding response. Night-feeding salamanders have also been shown to feed when deprived of vision, provided the prey are active (Lannoo 1987). However, given the electrosensory capabilities of salamanders (Himstedt et al. 1982), the relative contributions of electrosensory and mechanosensory lateral line systems to this behavior are uncertain.

Field studies of piper show that a major component of their diet is the demersal zooplankton, which only enter the water column after dark (Saunders and Montgomery 1985). Most zooplanktivorous fish are diurnal visual feeders, and it is thought that in response to this predation pressure, many of the larger, more visible zooplankton shelter during the day and only emerge at night (Fig. 28.1). Fish with specialized visual systems are able to utilize this food resource. For example, the "bigeye" (*Pempheris adspersa*) has the obvious visual specializations of a wide pupil, aphakic space, and conspicuous eyeshine, all of which bring gains in retinal illumination (Nicol and Zyznar 1973; Lythgoe 1979; Wang et al. 1980). Members of this family (Pricanthidae) are all active visual predators (Fishelson et al. 1971; Gladfelter 1979). They generally shelter in well-shaded holes and caves during the day. After sunset, they move into the water column and swarm

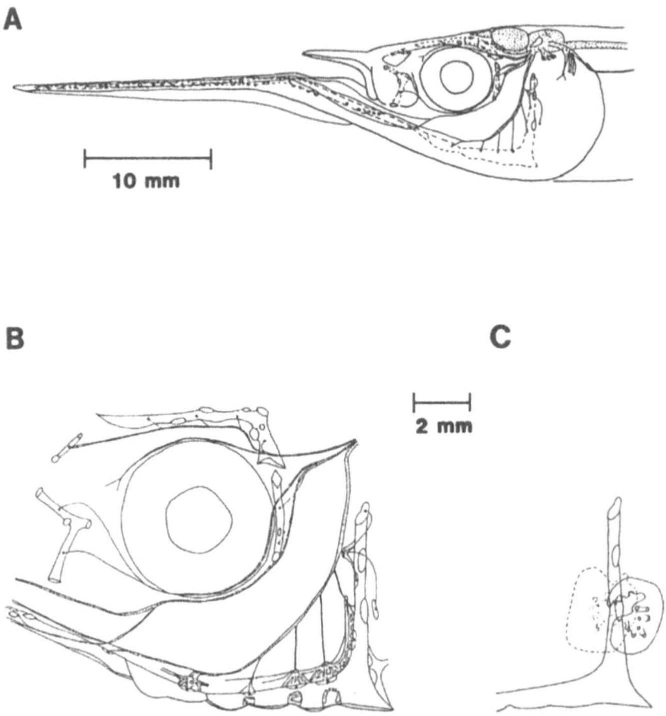

FIGURE 28.2. Anatomy of the anterior lateral line in the piper (*Hyporhamphus ihi*). (A) Lateral view of the head showing the lateral line canals (dotted) and the innervation of their neuromasts. (B) Closer view of the preorbital, nasal, supraorbital, postorbital, and the preopercular/mandicular canals and the innervation of their neuromasts. (C) A section of the preopercular/mandibular canal is shown along with the modified scales which overlie one of the pores in the canal. (After Montgomery and Saunders 1985.)

above the reef platform, where they feed on demersal zooplankton, including euphausids, amphipods, and isopods. A reverse migration occurs in the half-hour before sunrise when the whole school returns to the daytime refuges. In contrast, the piper, which is also a nocturnal planktivore, remains in the top 2 m of the water column throughout the day and night, the daytime schools giving way to dispersed slow-moving individuals at night.

Piper lack the obvious visual specializations of other nocturnal planktivores, and the elaborate development of the anterior lateral line appears to be the principal sensory organ involved in prey detection (Montgomery and Saunders 1985) (Fig. 28.2A). The most distinctive feature of the head canal system is the elongated preopercular/mandibular canal which runs out onto the extended lower jaw, or half-beak, of these fish. In addition, the system shows other specializations (Fig. 28.2B,C). The canals have a relatively wide bore (up to 0.5 mm) which opens to the exterior via pores. The pores (75–200 μm maximum diameter) are

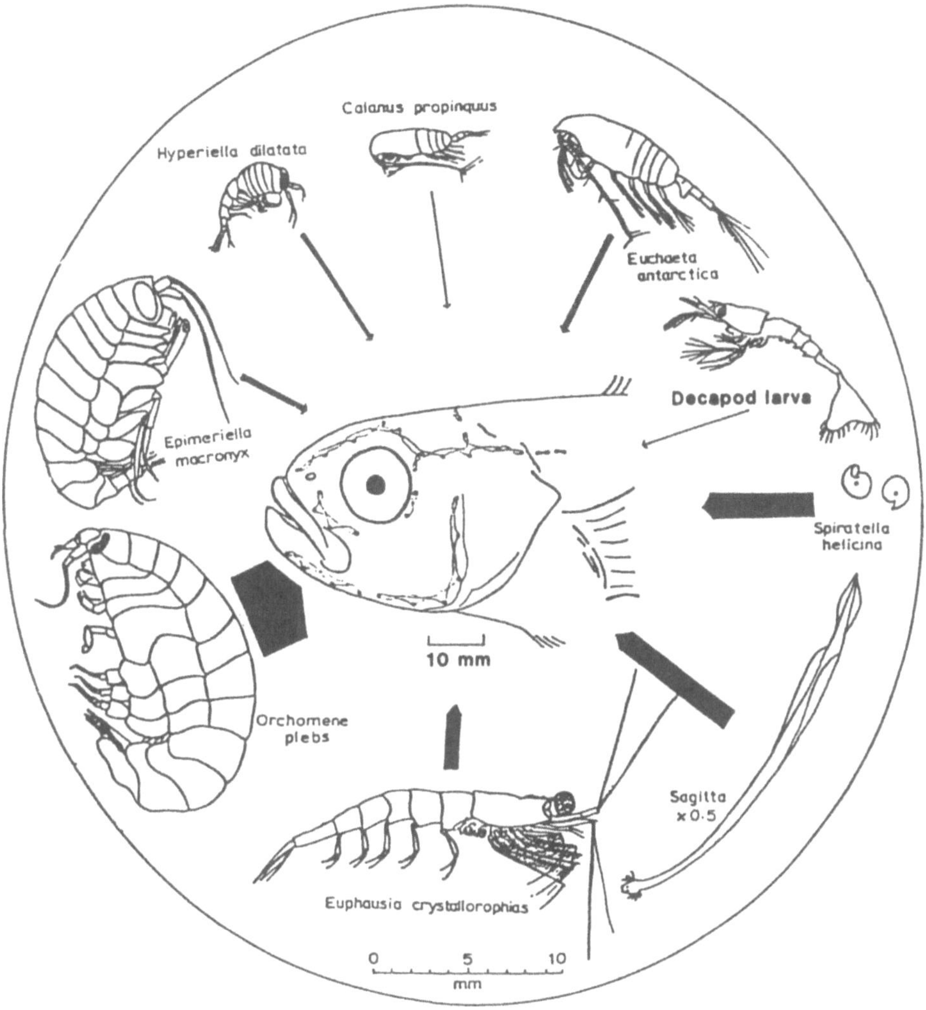

FIGURE 28.3. Anterior lateral line anatomy of the antarctic fish (*Pagothenia borchgrevinki*) and the prey species taken by this fish. Note the difference in scale between the illustrations of the fish and its prey. The relative contributions of the prey to the diet are shown by the relative thickness of the arrows. (After Foster et al. 1987.)

at the ends of branching tubules formed in the scales which overlie the canal openings (Fig. 28.2C). Between the canal openings are located discrete neuromasts, which in some cases (particularly along the horizontal arm of the preopercular canal) occur opposite constrictions in the canal (Fig. 28.2B). The mechanical consequences of this arrangement may amplify particle motions in the canal and are also likely to influence its frequency response characteristics (Denton and Gray 1988).

Piper apparently feed on demersal zooplankton at night, and there is a substantial body of structural and behavioral evidence that they utilize the lateral line to do so. This view is strengthened by laboratory experiments that demonstrate the piper's ability to preferentially take live prey under conditions of total darkness, and more recently by infrared video observation of prey capture in the dark (see Section 28.4).

The antarctic fish *Pagothenia borchgrevinki* feeds on several different species of macrozooplankton (Fig. 28.3) (Foster et al. 1987). Observations of feeding behavior indicate that vision is important (Foster et al. 1987), but given the low light levels that occur beneath snow and ice cover and the long polar winters, it is likely that nonvisual mechanisms of prey detection would also be significant. The suitability of the lateral line for the detection of crustacean prey has been demonstrated (Montgomery and Macdonald 1987), and will be dealt with later in the chapter. Other fish that are thought to be predominantly visual feeders, such as juvenile common bream (*Abramis brama*), have been shown to be quite capable of taking planktonic prey in total darkness (Townsend and Risebrow 1982). In the dark, they are able to feed on *Daphnia* at about half their normal rate, presumably using mechanosensory cues to take prey positioned in front of them within a distance of approximately 5 mm. This ability does not, however, extend to all plankton-feeding fish. For example, when light levels fall below the threshold for visual feeding (10^{-2} lux), herring no longer select individual plankters, but if they continue to feed, do so by filtration (Batty et al. 1986).

28.3 Signals Produced by Swimming Zookplankton

Water disturbances produced by swimming zooplankton have been determined with a variety of methods: variable speed stroboscope (Newbury 1972), impedance pneumograph (Gill and Crisp 1985), hot-wire anemometer (Kirk 1985), strain gauge (Montgomery and Macdonald 1987), and optoelectronic movement detector (described shortly). Different methods have their strengths and weaknesses. Those utilizing restrained animals are suitable for determining the output frequencies produced by the swimming animals, asymmetries in the produced fields, and field attenuation. But those employing a stationary detector and free-swimming animals provide better insight particularly into the temporal characteristics of the natural stimulus.

For smaller zooplankton (body lengths in the range 50 to 500 μm), the base frequency of swimming appendages is probably the only output of significance. Different copepod species show a range of limb beat frequencies from about 10 Hz up to about 45 Hz (Newbury 1972). Within one species the frequency of limb beats is inversely related to size and directly related to temperature (Gill and Crisp 1985). For the larger zooplankton (body lengths > 500 μm), vibration frequencies produced that are greater than the limb beat frequency are also likely to be of significance. For example, the macrozooplankton taken by the antarctic fish have relatively low limb beat frequencies (in the order of 5 to 10 Hz). However,

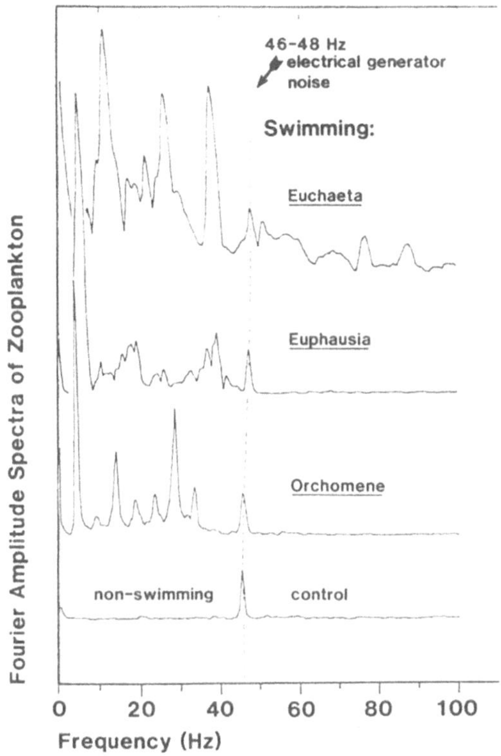

FIGURE 28.4. Amplitude spectra of swimming zooplanktonic crustaceans. Swimming vibration was recorded from a swimming plankter attached to a strain gauge, and processed by a spectrum analyzer. Vertical scale is linear, and each spectrum was derived from four to 32 leg beat cycles. Nearly all of the vibration produced by all three species is between 5 and 50 Hz. The small peak seen just below 50 Hz on all records is due to electrical noise from the laboratory power supply, which, when loaded, dropped below its nominal rating of 50 Hz.

they do produce significant frequency outputs in the 10- to 50-Hz range (Fig. 28.4). Similarly, copepods (body length 1.7 to 2.1 mm) also produce higher-frequency components of about 50 Hz, caused by flapping mouthparts (Kirk 1985) (Fig. 28.5C). The strength of these higher frequencies is quite low. In the above example, the 50-Hz peak-to-peak particle displacement at the detector would be about 1 μm.

Asymmetries in the flow field produced by swimming, but tethered, *Daphnia* are illustrated in Fig. 28.5A (from Kirk 1985). Since the power strokes of the antennae normally drive the animal forward, it is likely that these asymmetries would not be as pronounced in a free-swimming animal. Nevertheless, it is still to be expected that the water disturbance will be greater behind the animal than in front. During the measurement of attenuation (see Fig. 28.5B), the flow probe was moved outward along a radius in the lateral direction (Kirk 1985). In all measured cases, flow speed attenuates at a rate faster than d^{-14} (where d = the radial distance from the source), indicating that attenuation of flow speed around a complex flow source may be greater than the d^{-3} that occurs around a simple source such as a vibrating sphere.

Many of the demersal zooplankters are active swimmers, routinely maintaining speeds of 10 to 15 cm/s. Given the small signal size and its rapid attenuation,

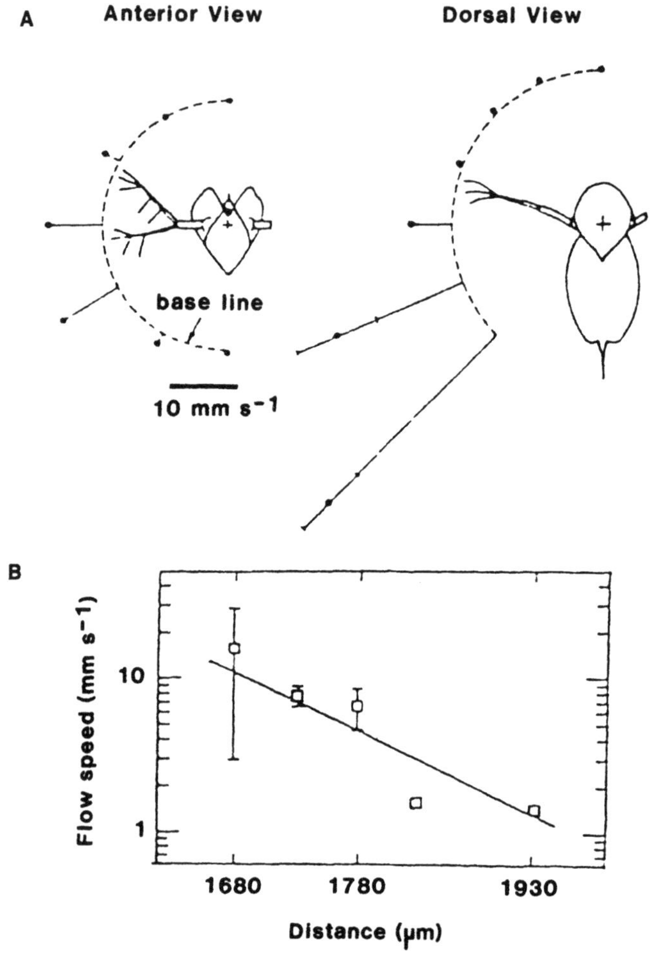

FIGURE 28.5. Signals produced by swimming *Daphnia* and *Diaphtomus* measured with a hot-wire anemometer (after Kirk 1985). (A) Asymmetries in the flow field produced by a tethered *Daphnia* as seen from an anterior and a dorsal view. The lengths of solid lines to the closed circle indicate mean maximum flow speed (± SE) at a fixed radial distance (dashed semicircle) from the *Daphnia*. (B) Attenuation of flow speed with distance as the probe was moved outward along a radius in the lateral direction. (C) Upper trace shows the flows produced by a tethered *Daphnia*. The transducer was positioned just beyond the tips of the antennal setae. Pulses occur in pairs, with the first pulse resulting from the antennal downstroke and the second from the recovery stroke. Lower trace shows the flows produced by *Diaptomus* swimming past a stationary transducer. The mouthparts of the copepod passed close to the transducer, producing the oscillatory component to the flow (about 50 Hz). *(Continued)*

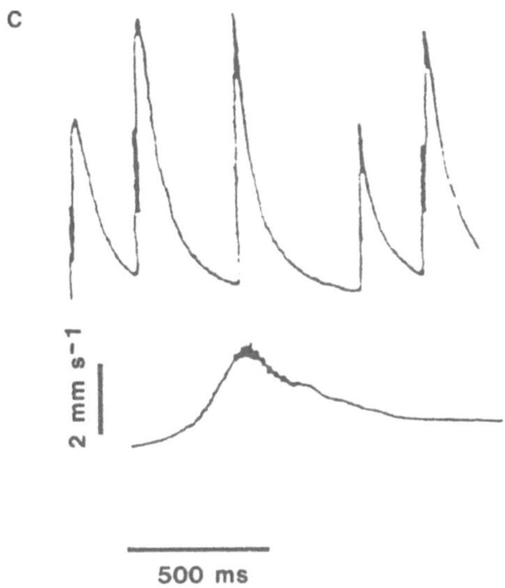

FIGURE 28.5. *Continued.*

a free-swimming zooplankter will produce only a brief stimulation at a stationary receptor. To demonstrate this point, a lateral line analog was set up to measure the water disturbance produced by swimming zooplankton. The transducer itself consisted of a Schottky-Barrier photodiode mounted in the film plane of a single lens reflex camera. The transducer generates an X and Y voltage out for movements of a bright image focused on it. Further details of the transducer are given in Montgomery et al. (1983). The camera containing the transducer was mounted on an inverted microscope which was focused on a capillary tube (internal diameter 1 mm) lying on the base of a Petri dish.

An ion exchange resin bead was located within the tube, between two windows cut in the wall of the capillary glass. The bead was illuminated from one side via a fiber-optic cable, and its movements could then be monitored with a resolution down to about 5 μm. At this sensitivity, the bead had to be placed in the capillary tube to prevent the larger-amplitude, low-frequency water movements from displacing the bead outside the dynamic range of the transducer. The signal recorded will be modified by the mechanics of the transducer, so this arrangement was used principally to illustrate the transient nature of the signals produced by swimming zooplankton at a stationary transducer.

An example is shown in the record of an ostracod (*Tetrdeium crassum*) swimming past the stationary particle movement transducer (Fig. 28.6). The overall pulse lasts only a little longer then 200 ms, and the higher-frequency components (corresponding to about 34 Hz) are clearly seen on the wave form. Ostracods are relatively modest swimmers in comparison to other members of the demersal zooplankton, which would produce even shorter stimulus pulses at a stationary receptor. For example, the signal produced by an amphipod (*Paradexamine*

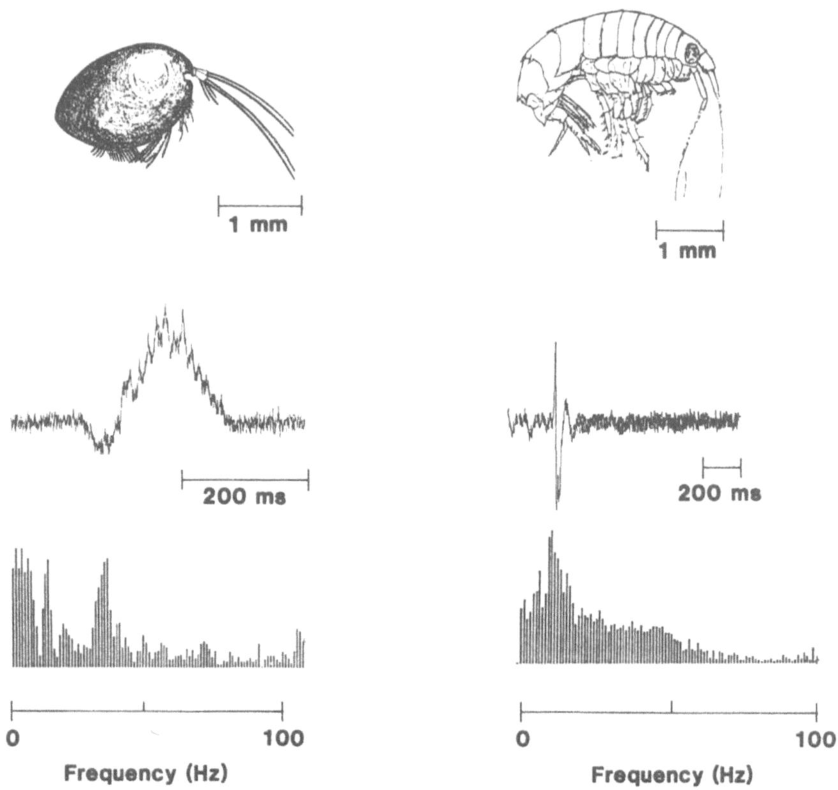

FIGURE 28.6. Signals produced by free-swimming demersal zooplankton measured with an optoelectronic transducer. (A) Middle trace shows the signal produced at the transducer by an ostracod swimming past. A higher-frequency component (37 Hz) is evident superimposed on the lower-frequency displacement of the transducer. The spectral composition of the displacement signal is shown in the lower trace. (B) Middle trace shows the signal produced at the transducer by an amphipod swimming past. The abrupt nature of the signal is reflected in the higher (20 to 50 Hz) components of the frequency spectrum shown in the lower trace.

houtete) swimming past the transducer is shown in Fig. 28.6B. In this case the abrupt wave form itself contributes to the higher-frequency components seen in the amplitude spectrum (Fig. 28.6B).

28.4 Prey Detection and Capture

The suitability of the lateral line for the detection of planktonic prey has been demonstrated in antarctic fish (Montgomery and Macdonald 1987). Single lateral line afferent fibers respond vigorously to the vibrations produced by a swimming amphipod (Fig. 28.7A). The average frequency response curve of

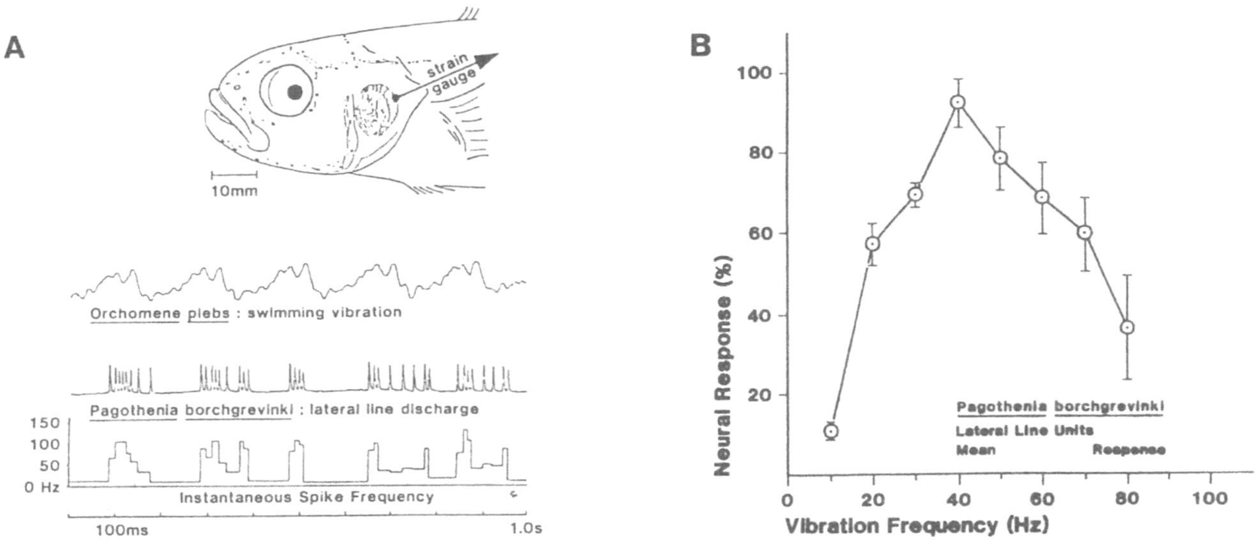

FIGURE 28.7. Lateral line responses to swimming prey and to a vibrating probe (after Montgomery and Macdonald 1987 © 1987, AAAS). (A) Simultaneous records of the swimming vibration of an amphipod, the spike discharge of a single lateral line afferent axon, and its associated instantaneous spike frequency: (B) frequency response characteristics of lateral line afferent fibers. Mean relative responses (based on phase-locking criteria) have been calculated for each vibration frequency for 13 fibers. Error bars indicate standard deviation.

primary afferent fibers stimulated by a probe vibrating at a constant amplitude (Fig. 28.7B) shows a best response around 40 Hz, corresponding to the higher-frequency components of the swimming movements of the crustacean prey. It is argued elsewhere in this volume (Kalmijn Chapter 9) that the effective stimulus to the lateral line system is probably particle acceleration. Plotting the relative neural response against stimulus acceleration would elevate the response curve at lower frequencies. In either case, it is clear that there is an effective overlap of the response characteristics of lateral line afferents and the signals produced by their planktonic prey.

Mottled sculpin and juvenile bream appear to detect and localize relatively slow-moving prey in the area in front of them, then strike at the object or reposition to bring the object into the strike zone (Hoekstra and Janssen 1986; Townsend and Risebrow 1982). Given a particular prey object within the receptive field of the fish, we still do not know precisely what the pattern of activity within the whole field of lateral line afferents will be or how this information is interpreted by the fish to localize the prey and engineer a successful strike. However, prey capture of this sort has certain "open-loop" features, where information is obtained and assessed and then the fish strikes, or repositions and then reassesses. This strategy is suitable for slow-moving prey, but for fast-swimming prey it is likely that a different strategy is employed.

For fast-swimming zooplankton the trajectory of swimming must be determined for successful interception. Individual receptors will receive only a transitory stimulus, and the trajectory is likely to be determined by the successive stimulation of adjacent receptors. An array of receptors extending symmetrically in front of, and behind, the mouth of piper (Fig. 28.2) seems particularly suited to the evaluation of the trajectory of objects moving parallel to its long axis. Infrared video recordings of successful prey capture by piper confirm that the trajectories of the plankton prior to capture most frequently lie along the long axis of the fish, either approaching from in front or behind (Fig. 28.8). A strike consists of a sharp lateral head movement to intercept the prey trajectory. Figure 28.8 plots the relative motion of the piper and the prey. In most cases piper cruise slowly in the dark, which will have the effect of improving the alignment of the relative motion of the prey with the long axis of the fish.

28.5 Summary and Conclusions

It is evident that the use of the lateral line in plankton feeding will grade from fish that depend exclusively on nonvisual cues for feeding (blind cave fish), through those species that customarily use the lateral line (mottled sculpin, piper, antarctic fishes), to those that can utilize lateral line cues when required (common bream), and finally to species that when bereft of vision cease to select individual zooplankters (herring). As with all sensory information, lateral line input will not be processed in isolation. Although the lateral line alone can in some species provide sufficient information for plankton feeding, whenever sensory capabilties

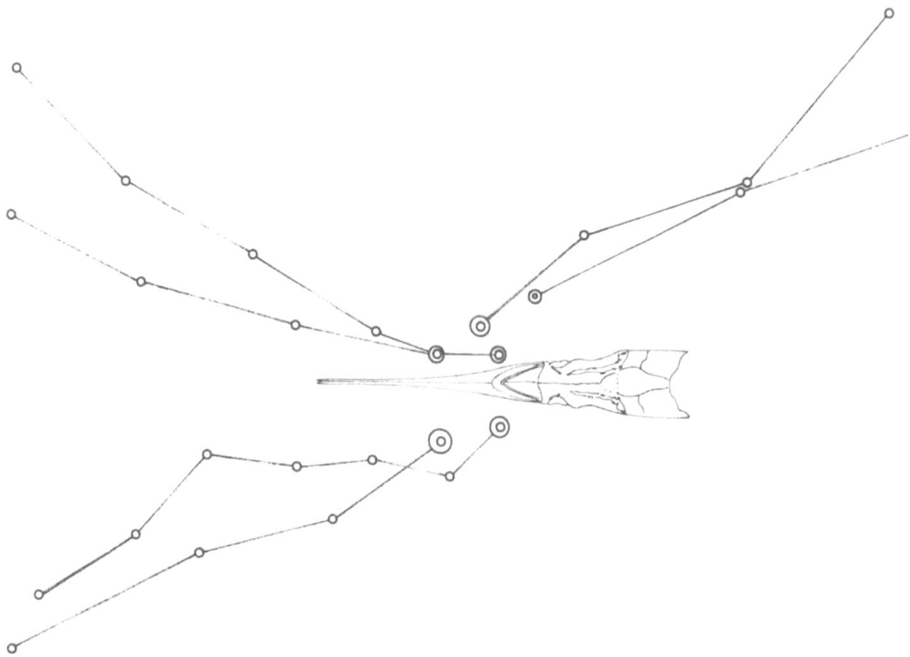

FIGURE 28.8. Trajectories of plankton movements prior to successful prey capture by the piper *Hyporhamphus ihi*. Piper were placed in a shallow (100 mm) test arena in a dark room along with freshly captured dermersal zooplankton. The water was covered with a Perspex sheet to prevent surface waves, and the area was illuminated from beneath by a low-intensity infrared source. An infrared video camera (National WV-1500E), positioned above the arena, recorded the trajectory of plankton prior to successful prey capture. Positions of the plankters (relative to the head of the fish) are marked at 40-ms intervals. The plankter was intercepted at the location marked by the circle. The species of individual zooplankters could not be identified from the video record, but the sample consisted mainly of amphipods and ostracods.

and conditions allow, lateral line information will probably be used synergistically with vision. It is a difficult task to partition the contributions of different sensory modalities to normal feeding behavior, but given the importance of plankton feeding in aquatic ecosystems, the ubiquitous presence of lateral line sensory systems in fish, and the advantages of feeding in conditions that are suboptimal for vision (twilight, night, murky water), it is likely that there is still much to learn of the manner in which the lateral line contributes to the feeding behavior of planktivorous fish.

The significant characteristics of the signals produced by swimming zooplankton can be summarized as (1) low-amplitude, high-frequency (20 to 50 Hz) components, and relatively high-amplitude, low-frequency components produced by larger zookplankton; (2) the localized and asymmetrical nature of the disturbance field around a free-swimming zooplankton; and 3) the transitory stimulus produced at a stationary receptor by faster-swimming prey. Lateral line sense organs have suitable properties to respond to these signals, and from the behavior of sculpins and piper it can be inferred that the central processing task is to extract the location or trajectory of the prey from the spatial and temporal characteristics of lateral line input. Thus consideration of the role of the lateral line system in natural behaviors also provides clues as to the important features of natural signals, how these are encoded by the sensory system, and the central processing required to extract the necessary information for the behavior.

Acknowledgments. I am grateful to Dr. Brian Foster for supplying identification and drawings of some of the crustacea, and to him and other colleagues (Dr. John Macdonald, Dr. Gary Housley, Dr. Larry Dill, Mr. Heiko Weix, Mr. Andrew Saunders) who have participated in these studies. I would also like to thank Drs. John Janssen, John Blaxter, and Sheryl Coombs for their comments on the manuscript. Aspects of this work were supported by grants from the New Zealand University Grants Committee and the Auckland University Research Committee. Logistic support was provided by the staff of the Leigh Marine Laboratory and Scott Base Antarctica.

References

Batty RS, Blaxter JHS, Liddy DA (1986) Herring filter feeding in the dark. Mar Biol 91:371–376.

Denton EJ, Gray JAB (1988) Mechanical factors in the excitation of the lateral lines of fishes. In: Atema J, Fay RR, Popper AN, Tavolga WN (eds) Sensory Biology of Aquatic Animals. New York: Springer-Verlag, pp. 595–617.

Fishelson L, Popper D, Gunermann N (1971) Diurnal cyclic behaviour of *Pempheis oualenois* Cuv & Val (Teleostei, Pempheridae). J Nat Hist 5:503–506.

Foster BA, Cargill JM, Montgomery JC (1987) Planktivory in *Pagothenia borchgrevinki* (Pisces: Nototheniidae) in McMurdo Sound Antarctica. Polar Biol 8:49–54.

Gill CW, Crisp DJ (1985) The effect of size and temperature on the frequency of limb beat of *Tempora longicornis* Muller (Crustacea: copepoda). J Exp Mar Biol Ecol 86:185–196.

Gladfelter WB (1979) Twilight migrations and foraging activities of the copper sweeper (*Pempheris schombougki*) (Teleostei: Pempheridae). Mar Biol 50:109–119.

Himstedt W, Kopp J, Schmidt W (1982) Electroreception guides feeding behaviour in amphibians. Naturwissenschaften 69:552.

Hoekstra D, Janssen J (1985) Non-visual feeding behaviour of the mottled sculpin, *Cottus bairdi*, in Lake Michigan. Environ Biol Fish 12:111–117.

Hoekstra D, Janssen J (1986) Lateral line receptivity in the mottled sculpin (*Cottus bairdi*). Copeia 1986:91–96.

Kirk KL (1985) Water flows produced by *Daphnia* and *Diaptomus*: Implications for prey selection by mechanosensory predators. Limnol Oceanogra 30:679–686.

Lannoo MJ (1986) Vision is not necessary for size-selective zooplanktivory in aquatic salamanders. Can J Zool 64:1071–1075.

Lythgoe JN (1979) Vision in fishes: Ecological adaptations. In Ali AN (ed) Environmental Physiology of Fishes. New York: Plenum, pp. 431–446.
Montgomery JC, Macdonald JA (1987) Sensory tuning of lateral line receptors in antarctic fish to the movements of planktonic prey. Science 235:195–196.
Montgomery JC, Saunders AJ (1985) Functional morphology of the piper *Hyporhamphus ihi* with reference to the role of the lateral line in feeding. Proc R Soc Lond B 224:197–208.
Montgomery JC, McVean AR, McCarthy D (1983) The effects of lowered temperature on spontaneous eye movements in a teleost fish. Comp Biochem Physiol 75A:363–368.
Newbury TK (1972) Vibration perception by chaetognaths. Nature 236:459–460.
Nicol JAC, Zyznar ES (1973) The tapetum lucidium in the eye of the bigeye *Priancanthus arenatus*, Cuvier. J Fish Biol 5:519–522.
Poulson TL (1963) Cave adaptations in amblyopsid fishes. Am Midl Nat 70:257–290.
Saunders AJ, Montgomery JC (1985) Field and laboratory studies of the feeding behaviour of the piper *Hyporhamphus ihi* with reference to the role of the lateral line in feeding. Proc R Soc Lond B 224:209–221.
Townsend CR, Risebrow AJ (1982) The influence of light level on the functional response of a zooplanktonivorous fish. Oecologia 53:293–295.
Wang TR, Nicol JAC, Thurston EL, McCants M (1980) Studies on the eyes of bigeyes (Teleostei: Pricanthidae) with special reference to the tapetum lucidum. Proc R Soc Lond B 210:499–512.

29 Behavioral Investigations on the Functions of the Lateral Line and Inner Ear in Predation

PER S. ENGER, AD. J. KALMIJN, and OLAV SAND

29.1 Introduction

The lateral line, comprising the canal organs and free neuromasts, is generally regarded as a sensory system for the detection of local water currents (Hofer 1908; Dijkgraaf 1934) and surface waves (Schwartz 1965; Bleckmann et al., Chapter 25). However, whether the lateral line responds to low-frequency sound as well has long remained an issue of debate. Dijkgraaf (1963, Chapter 2) strongly argued against an acoustic function of the lateral line, citing a lack of compelling behavioral evidence. Sand (1981, 1984) explained that the operation of the lateral line in free-moving fish is physically restricted to the immediate vicinity of the source. Kalmijn (1988a, Chapter 9) subsequently focused attention on the function of the inner ear in detecting the local flow fields of moving objects at distances beyond the limited range of the lateral line. The low-frequency nature of the two sensory systems is consistent with the results of earlier physiological studies (Suckling and Suckling 1950; Harris and van Bergeijk 1962; Enger 1966; Kalmijn 1988a).

Although it is commonly believed that the lateral line of midwater fish plays an important role in the detection of moving prey, conclusive behavioral evidence is scarce. Moreover, only sparse data are available on the distance of response and on the significance of the various frequency components in the prey fields. The present experiments were inspired by Wunder's (1927) earlier studies on the predatory behavior of the pike, *Esox lucius*. Whereas blinded pikes struck at live prey fish from distances of 5 to 10 cm, the animals failed to attack after permanent extirpation of the lateral line. Recently, Hoekstra and Janssen (1985, 1986) and Montgomery and MacDonald (1987) conducted similar experiments on bottom-dwelling and midwater plankton-feeding fish, respectively (see also Coombs and Janssen Chapter 15, and Montgomery Chapter 28).

This report adduces evidence for the functions of the lateral line and inner ear in detecting moving prey and discusses the operating ranges of the two sensory systems in a midwater predatory fish, the bluegill *Lepomis macrochirus*. The feeding behavior of the bluegill was observed under covert infrared illumination in animals whose lateral lines were either intact or reversibly blocked by cobalt ions. Added to the aquarium water, cobalt ions fully suppress the lateral line function without affecting the inner ear (Karlsen and Sand 1987). In addition, the hydrodynamic fields of moving prey were recorded with a newly developed underwater motion transducer to determine the frequency spectra of the natural stimuli (Kalmijn and Enger in prep.). In the behavioral tests, both live prey and hydrodynamically simulated targets were utilized.

29.2 Materials and Methods

Eleven bluegills (*Lepomis macrochirus*), 12 to 15 cm in length, were kept separately in 60-L glass aquaria. Twice a week, they were fed small goldfish (*Carassius auratus*), which were introduced into the bluegills' living quarters during the dark so as to accustom the animals to nighttime feeding. The evening before the trials, following a period of at least 2 days without food, two of the bluegills were transferred to the two experimental tanks.

The bluegills were tested for their responses to four different targets: (1) live goldfish, 2.5 to 3.5 cm in length; (2) artificial prey consisting of a slender elastic stopper, 3.0 cm in length and tapering from 0.6 to 0.3 cm in diameter, attached to a thin wire and manually moved about in the manner of a quietly swimming prey; (3) a fiberglas sphere, 1.9 cm in diameter, executing periodic motions in the form of slow advances and rapid retreats, synthesized of 10 discrete, equal-amplitude sinusoidal accelerations of frequencies ranging from 1 to 10 Hz; and (4) the same fiberglas sphere executing slow advances and rapid retreats, synthesized of 10 discrete frequencies ranging from 10 to 100 Hz. The acceleration amplitudes of the individual components were on the order of 10 cm/s^2.

The four targets were presented to bluegills with either intact or blocked lateral lines. To keep the lateral line intact, the animals were placed in an experimental tank filled with a freshwater solution containing 0.1 mM $CaCl_2$. To block the lateral line, they were placed in a tank with a calcium-free freshwater solution to which 0.1 mM $CoCl_2$ was added as described by Karlsen and Sand (1987). The blocking effect was reversed by keeping the fish for 1 day in a recovery tank with 0.25 mM $CaCl_2$ water before returning it to its living quarters. Atomic absorption measurements indicated that the originally calcium-free cobalt solution slowly, but insignificantly, increased in Ca^+ and decreased in Co^+ during the 2 days of experiments. Cobalt-treated bluegills were not tested again for at least 1 week. The blocking effect of the cobalt and the recovery in a calcium-rich medium were confirmed by monitoring the electrophysiological responses of the trunk lateral line nerve.

At the beginning of each trial, the bluegills were offered the sight of a goldfish in a small container outside the experimental tank. Without visual motivation, a long time could pass before the bluegill started its food-searching behavior. As soon as the bluegill charged at the goldfish, the room light was switched off, and the real or simulated prey was introduced into the experimental tank. The bluegills' behavior toward the target was further observed under covert infrared illu-

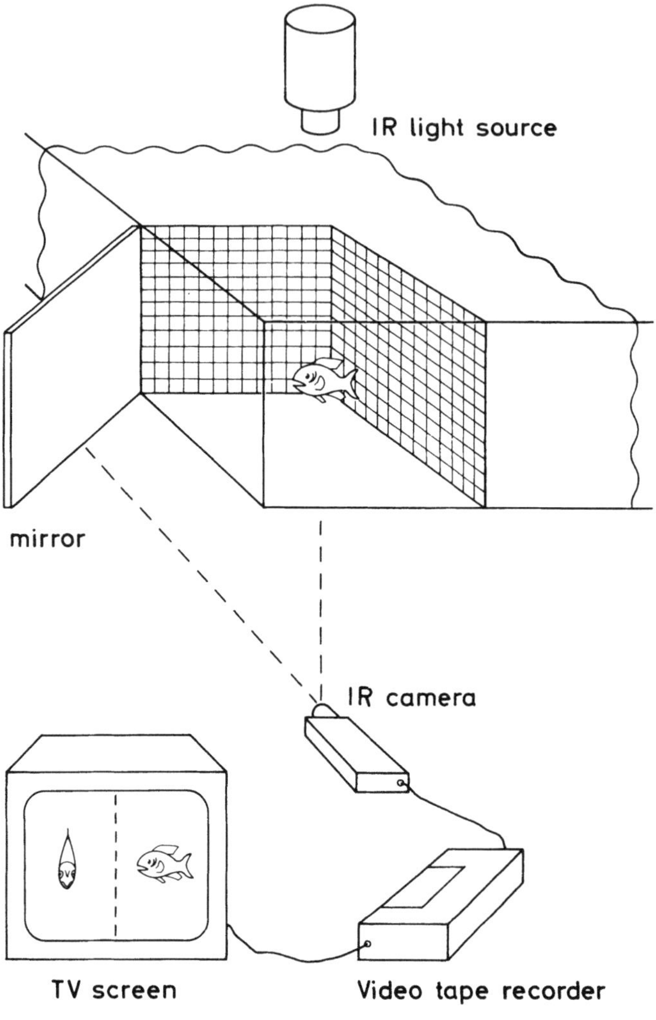

FIGURE 29.1. Experimental and setup for observing the behavioral responses of bluegills to live goldfish and artificial prey presented in the dark. The bluegill was confined to a screened-off 25 × 25 cm corner area of a large glass aquarium. A covert infrared light source was mounted above the fish's compartment. The animal was observed simultaneously through the front pane and through one of the side panes of the aquarium.

mination by use of an infrared video camera connected to a video monitor outside the darkroom. Both the video pictures and the verbal comments of the experimenters were stored on a video casette recorder for further analysis.

Because of the limited penetration depth of the infrared light, the bluegills were tested in rather small, screened-off compartments of larger aquaria. In series 1 and 2, the bluegills were confined to an area of 25 × 25 cm at a depth of 30 cm.

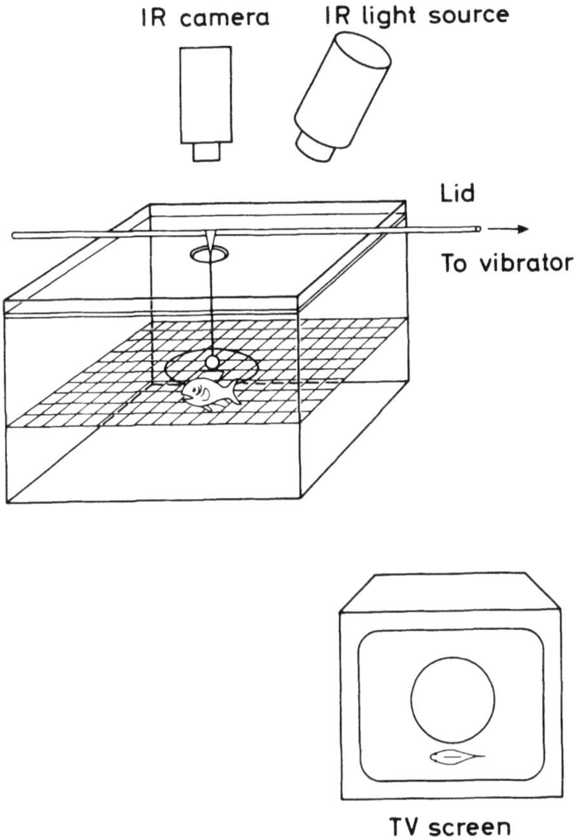

FIGURE 29.2. Experimental setup used to study the responses of the bluegills to a vibrating sphere. The tank was supplied with a heavy glass lid and completely filled with water to eliminate the air-water interface. A 43 × 43 cm false bottom made of plastic screening restricted the movements of the fish to the upper half of the tank. The vibrating sphere was positioned in the center of the tank, just above the false bottom, on which a circle of 10-cm radius was marked off.

The responses to real and artificial prey were observed three-dimensionally by use of a mirror (Fig. 29.1). In series 3 and 4, the movements of the bluegills were restricted by a false bottom of wide-mesh plastic screening to the upper half of a 20-cm-deep, 43 × 43 cm tank. The animal's behavior with regard to the vibrating sphere, mounted over the center of the false bottom, was viewed from above (Fig. 29.2). The false-bottom screen was marked with a circle of 10-cm radius with its center just below the vibrating sphere. In series 1 and 2, the bluegills nearly always attacked the target well below the free water surface. In series 3 and 4, the free water surface was eliminated by a tightly fitting glass plate to approximate midwater conditions. A 3-cm-diameter hole in the center of the plate allowed the target to be introduced into the water. Two similar setups were operated side by side.

29.3 Results

29.3.1 Daytime Predatory Behavior of the Bluegill

In daylight, bluegills rely mainly on vision in detecting midwater prey. When hungry, they often initiate their pursuits from distances of several body lengths and take the prey within seconds. They are also easily induced to dash at prey in neighboring aquaria and may even jump clear out of the water to catch objects in midair.

29.3.2 Experiments with Live and Artificial Prey (Series 1 and 2)

Bluegills are perfectly capable of catching prey in the dark. However, their night-time behavior appeared quite different from that during the day. In the dark, the bluegills typically glided smoothly through the water, driven only by an occasional tail flip. Often, they calmly approached a live goldfish to strike from a short distance with great vigor—or to veer off slowly if the target had stopped moving. The eventual strike was so fast that on the video screen merely a sudden forward jerk and a sucking motion of the bluegill were seen, usually followed by only a glitter of scattered goldfish scales. The bluegills attacked the artificial prey in a similar fashion, hitting the manually operated object with great force.

A small goldfish is a big meal for a bluegill. Hence, to avoid satiation, the artificial prey was presented in the first part of each test session. A positive outcome was recorded if the target was struck within 10 min. After an attack on the artificial prey or the passage of 10 min, a live goldfish was substituted for the artificial prey, and the observations were continued. Again, a positive outcome was recorded if the live prey was attacked within 10 min. If the goldfish was not taken in the allotted time, the room light was switched on, allowing the bluegill to catch the prey visually.

Table 29.1 lists two categories of responses—attacks from a distance and attacks after touch. An attack from a distance was recorded when the predator and prey were definitely separated by more than 0.2 cm at the onset of the final strike. An attack after touch occurred when a slowly gliding bluegill accidentally ran into the prey, or when the goldfish came closer than 0.2 cm or actually came into contact with the predator before the final attack. In case of doubt, due to either the short distance of response or the limited resolution of the video screen, an attack after touch was recorded. In the category *no attacks after 10 min*, the bluegill failed to strike, even though its behavior often seemed to indicate that it had sensed the target. During the presentation of the hand-held artificial prey, the number of attacks after touch was relatively low, for the person manipulating the target was directed to move it away from the bluegill if the distance between the two became critically close.

TABLE 29.1. Summary of results from four bluegills, listing number of successful attacks on live and artificial prey in regular water and in water containing cobalt.

	Regular water	Water containing Co^{2+}
Live goldfish		
Attacks from distance	37	0
Attacks after touch	28	28
No attacks after 10 min	11	30
Artificial prey		
Attacks from distance	32	0
Attacks after touch	15	10
No attacks after 10 min	14	53

In the dark, the difference in behavior between normal bluegills and bluegills without functional lateral lines was dramatic, as may be seen from Table 29.1. Animals with intact lateral lines successfully attacked the live goldfish and vigorously bit at the artificial prey from distances up to 2 cm, whereas those with inoperative lateral lines were unable to catch the goldfish and did not strike at the artificial prey unless they accidentally came into contact with the targets. Nevertheless, after failing to catch the goldfish within the 10-min period, the bluegills took the prey in a matter of seconds after the room light was turned on. Hence, the cobalt treatment did not noticeably affect the fish's appetite. In short, in the absence of visual and tactile cues, the bluegill clearly relies on its lateral line in performing its precise and successful attacks.

Interestingly, the bluegills often approached the live and artificial targets from distances of at least 5 cm in apparently deliberate moves, whether or not the lateral line function was blocked by cobalt ions. Whether the bluegills detected the targets from even larger distances was difficult to tell owing to the limited size of the test compartment. (See, however, Series 3 and 4 in Section 29.3.4.)

29.3.3 Measurement of Local-Flow Field Produced by Goldfish

The behavioral observations on the predatory behavior of the bluegill were originally conceived as a method of ascertaining the relative importance of the various frequency components in the local-flow fields of vertebrate prey. The ultimate goal was to resolve the old dilemma posed by the seemingly conflicting results of previous behavioral (Dijkgraaf 1963) and neurophysiological (Harris and van Bergeijk 1962) investigations. To obtain a first impression of the stimuli received by the bluegill and their frequency composition, recordings were made of the hydrodynamic flow fields produced by quietly swimming and hovering goldfish. The measurements were obtained with an underwater transducer developed to measure the local-flow field in the frequency range from zero to a few hundred hertz (Kalmijn and Enger in prep.).

Figure 29.3a–c shows the hydrodymamic flow field of an approaching and passing goldfish as detected by the stationary transducer. The resulting time

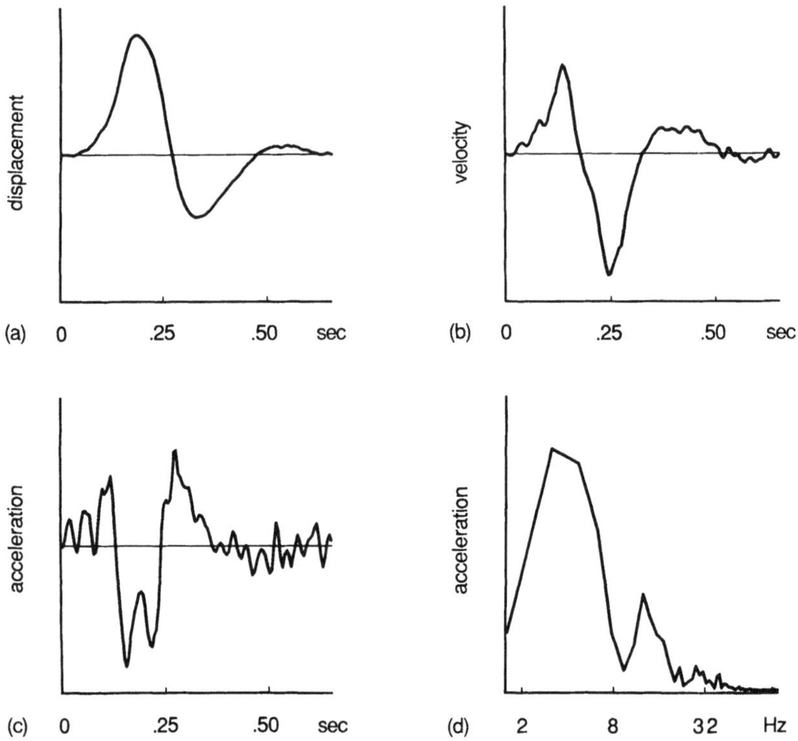

FIGURE 29.3. The hydrodynamic field of an approaching and passing goldfish. Recorded time series, 0.65 s in duration. (a) In units of displacement. (b) Same time series in units of velocity. (c) Same time series in units of acceleration. (d) Corresponding acceleration-versus-frequency amplitude spectrum. (Ordinates in arbitrary linear units.)

series of 0.65-s duration is expressed in units of displacement, velocity, and acceleration, respectively. The amplitude spectrum of the recorded field is plotted in Fig. 29.3d in units of acceleration, considered by Kalmijn (1988a, Chapter 9) to be the most relevant parameter for an evaluation of the lateral line's low-frequency response. The broad peak below 10 Hz is mainly the result of the fish's whole-body motions relative to the transducer. Obviously, the faster the prey swims and the closer it passes the receiver, the more this peak shift toward higher frequencies (see legends of Fig. 6.3 in Kalmijn 1988b). The frequencies above 10 Hz result not only from the fish's motion relative to the transducer but also from its periodic accelerations, fin movements, and tail beat.

29.3.4 Experiments with Advancing and Retreating Sphere (Series 3 and 4)

To compare the biological relevance of the frequencies below 10 Hz and the frequencies above 10 Hz received from quietly moving prey, a single bluegill was

TABLE 29.2. Results from one bluegill giving number of attacks on a vibrating sphere.

	Regular water	Water containing Co^{2+}
Vibrating sphere 1 to 10 Hz		
Attacks	7	0
Attacks after touch	0	0
Number of sessions	24	16
Vibrating sphere 10 to 100 Hz		
Attacks	0	0
Attacks after touch	0	0
Number of sessions	10	8

tested for its responses to a vibrating sphere producing calibrated dipole fields of known spectral composition. Since initial trials had shown that pure sine wave stimuli were relatively ineffective in evoking apparent feeding responses, a wave form of a more natural shape was chosen, namely, the periodic advance-and-retreat motions described above. Thus, stimuli composed of frequencies either from 1 to 10 Hz or from 10 to 100 Hz were applied.

The results of series 3 and 4 are summarized in Table 29.2. The observations must be regarded as preliminary, since thus far only a single bluegill has been tested. In about a third of the tests in cobalt-free water, the bluegill vigorously snapped at the sphere when presented with the advance-and-retreat motions of frequencies from 1 to 10 Hz. However, when presented with the motions of frequencies from 10 to 100 Hz, the bluegill performed only an occasional approach or avoidance, and never an attack. In water to which cobalt ions were added, the approaches toward the low-frequency moving sphere were still observed, but, instead of attacking, the bluegill slowly veered off when coming close to the target. Thus, it appears that the lower-frequency accelerations resulting from the prey's whole-body motions are biologically of utmost importance.

The directed approaches observed in bluegills of which the lateral lines were blocked by cobalt ions suggest that, in the dark, the fish are guided to the target from distances up to 10 cm by a sensory system other than the lateral line. These observations are consistent with the hypothesis that, beyond the limited range of the lateral line, a fish's responses to the hydrodynamic fields of moving objects are mediated by the inner ear (Kalmijn 1988a, Chapter 9). The ability of fish to detect accelerations well below 1 Hz has recently been shown for the Atlantic cod, *Gadus morhua*, by Sand and Karlsen (1986). Further studies on the role of the inner ear in the near field are in progress.

29.4 Discussion

During the dark of the night, the bluegill evidently relies predominantly on the lateral line for performing its short-range feeding attacks. Bluegills with inoperative lateral lines either did not strike from a short distance or, when they did, missed the prey. Visual cues were eliminated by depriving the animals of visible light. Smell and taste by themselves did not elicit any successful strikes from a distance in bluegills without functional lateral lines, even though the olfactory epithelium is not affected by cobalt ions (Yoshii and Kurihara 1983). The animal's tactile sense apparently also remained unaffected by the cobalt, since the attacks after touch occurred in fish both with and without functional lateral lines. The artifical prey and the dipole sphere were attacked from a distance only when actually moving.

Although the lateral line mediates the sudden attacks only within a range of 2 cm or less from the target in small bluegills, this does not necessarily mean that the sensory system cannot be used to detect prey from somewhat greater distances. Strategically, the bluegill may simply not initiate its final strikes until it is close enough to be sure that the prey has little chance to escape. Indeed, the detection range of the lateral line is theoretically expected to be more on the order of one body length of the recipient fish, depending on the strength of the target (Denton and Gray Chapter 10; Kalmijn Chapter 9).

Since the lateral line responds only to local water motions, it is virtually insensitive to noises from distant sources (beyond one body length), such as the low-frequency environmental noises in lakes and seas and the propeller noises from passing ships. Unless it is in contact with the bottom substrate, the nearly neutrally buoyant fish is passively accelerated along with the large-scale water motions. Clupeid fishes form a possible exception to this, as they have a direct connection between the lateral lines and the pressure-sensitive system of the inner ear (Denton and Gray 1983). The vortices and turbulence in the wake of a moving target may, however, be detected by the lateral line even after the sources that created them have long left the scene.

In the experiments with the slowly advancing and rapidly retreating sphere, the lower frequencies were not only of primary importance but also strictly necessary to elicit attacks in the dark from the single bluegill tested. This did not come as a surprise, since the whole-body motions of the approaching and passing goldfish gave rise to a predominance of frequencies below 10 Hz also, even when expressed in terms of acceleration. Nevertheless, in locating prey—or in escape, for that matter—higher-frequency accelerations, especially those resulting from the periodic fin and tail movements, may play an important role as well, for instance, in recognizing the nature of the source or to facilitate fast emergency responses. Montgomery (Chapter 28) also found high-amplitude, low-frequency components and low-amplitude, high-frequency components in the fields of swimming zooplankton. Thus, for catching prey—vertebrate or invertebrate—a sensitivity to low frequencies seems highly desirable.

The directed approaches of the bluegill with respect to both live goldfish and simulated prey were definitely not mediated by the lateral line, as they were also observed in cobalt-treated bluegills that did not attack the source, not even when coming close to it. Since the experiments were conducted in total darkness, not only with live prey, but also with chemically inert inanimate targets, it is, by elimination, tentatively assumed that the observed responses were mediated by

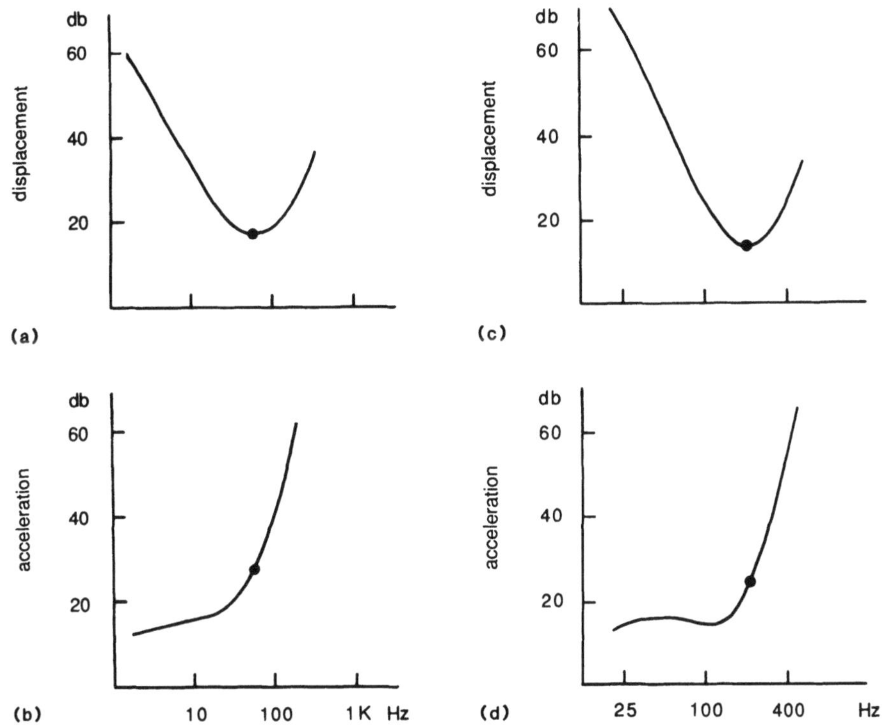

FIGURE 29.4. Frequency-response curves of lateral line canal and inner-ear otolith systems. (a) Lateral line displacement threshold versus frequency from Münz (1985). (b) Same data plotted as acceleration versus frequency (after Kalmijn 1988a). (c) Inner-ear displacement threshold versus frequency from Hawkins and Johnstone (1978). (d) Same data plotted as acceleration versus frequency (after Kalmijn 1988a). Note low-pass characteristics of acceleration plots. Dots in curves denote best frequency in displacement plots and approximate roll-off frequencies in acceleration plots.

the inner ear. On physical grounds, Kalmijn (1988a) has actually suggested that the acoustic function of the inner ear originated in the detection of the local-flow fields of moving objects. Moreover, Sand and Karlsen (1986) have experimentally shown that fish are capable of detecting imposed whole-body motions down to frequencies of less than 1 Hz, which led them to the hypothesis that the inner ear may play a role in the animals' orientation to distant sources of infrasound.

The historical lack of interest in the low-frequency response of the lateral line among neurophysiologists resulted largely from Harris and van Bergeijk's (1962) erroneous conclusion that the lateral-line canal system would respond to the displacement of the source, rather than its acceleration (Kalmijn Chapter 9). Replotted in terms of acceleration (Fig. 29.4), their data no longer show the familiar high-frequency tuning curves, but low-pass curves instead (Kalmijn 1988a). Thus, expressed in the proper stimulus parameter, the neurophysiological data fully support Dijkgraaf's (1963) conclusions concerning the low-frequency nature of the lateral line response. The recordings of approaching and passing goldfish reported here show that common sources of lateral line stimuli indeed give rise to sensory cues consisting predominantly but not exclusively of frequencies less than 25 Hz. These conclusions are in conformity with the original biophysical studies of Jielof et al. (1952), the work on the acceleration sensitivity of the canal organs in clupeid fishes (Denton and Gray 1983, Chapter 11) and in the trout, *Salmo gairdneri* (Kroese and Schellart 1987), the detailed physical analyses of van Netten and Kroese (1987, Chapter 12) on the cupular motions, and the recent neurophysiological and behavioral findings of Coombs and Janssen (Chapter 15).

29.5 Summary and Conclusions

To clarify the roles of the lateral line and inner ear in the detection of vertebrate prey, the feeding behavior of a midwater predatory fish, the bluegill *Lepomis macrochirus*, was observed under covert infrared illumination. In the dark, the animals quietly moved about, driven by only an occasional tail beat. When presented with live or simulated prey, bluegills with intact lateral lines performed sudden attacks when coming within a distance of 2 cm or less from the target. Prior to the final strikes, they often approach the real or artificial prey from distances of at least 5 cm in smooth, apparently deliberate moves. When the function of the lateral line was reversibly blocked by use of cobalt ions, the bluegills never made a successful attack at the goldfish, nor did they bite the simulated prey, unless they accidentally touched the target. The deliberate, directed approaches, however, were not affected by the cobalt treatment.

To assess the significance of the lower frequencies in the prey fields, one bluegill was tested for its responses to a sphere executing periodic advance and retreat motions synthesized of 10 sinusoidal accelerations from 1 to 10 Hz or from 10 to 100 Hz. The low-frequency source induced directed approaches from a distance of 10 cm and elicited final attacks from 2 cm or less when the fish's lateral line was intact. The high-frequency source, however, failed to evoke a noticeable response. When the lateral line was blocked by cobalt ions, the bluegill still approached the low-frequency source from a distance of 10 cm but did not attack the target, not even when coming close to it. These results are in conformity with the acceleration spectrum of a quietly swimming goldfish, showing its highest acceleration amplitudes at frequencies below 10 Hz, mainly resulting from the whole-body motions of the prey relative to the receiver.

In short, during the dark of the night, the bluegill relies on the lateral line in attacking moving prey. The final, decisive strikes takes place within a distance of 2 cm or less, which may, however, reflect the fish's prey-catching strategy rather than the ultimate performance of the lateral line. The deliberate approaches from distances up to 10 cm are, by exclusion of the lateral line, thought to be mediated by the inner ear. The lower frequencies resulting from the prey's whole-body motions relative to the recipient fish appear to be crucial for the bluegill's predatory behavior.

Acknowledgments. This study was supported by the National Science Foundation and the Office of Naval Research, A.J.K. Principal Investigator, La Jolla, CA, and by grants from the Norwegian Research Council for Science and Humanities providing for P.S.E. and O.S. to work in La Jolla and A.J.K. to work in Oslo, Norway. We extend our thanks to Vera Kalmijn for her invaluable participation in the endeavor.

References

Bleckmann H (1986) Role of the lateral line in fish behaviour. In: Pitcher TJ (ed) The Behaviour of Teleost Fishes. London: Croom Helm, pp. 177–202.

Denton EJ, Gray J (1983) Mechanical factors in the excitation of clupeid lateral lines. Proc R Soc Lond 218:1–26.

Dijkgraaf S (1934) Untersuchungen ueber die Funktion der Seitenorgane an Fischen. Z Vergl Physiol 20:162–214.

Dijkgraaf S (1963) The functioning and significance of the lateral-line organs. Biol Rev 38:51–105.

Enger PS (1966) Acoustic threshold in goldfish and its relation to the sound source distance. Comp Biochem Physiol 18:859–868.

Harris GG, Bergeijk WA van (1962) Evidence that the lateral-line organ responds to near-field displacements of sound sources in water. J Acoust Soc Am 34:1831–1841.

Hawkins AD, Johnstone ADF (1978) The hearing of the Atlantic salmon, *Salmo salar*. J Fish Biol 13:655–673.

Hoekstra D, Janssen J (1985) Non-visual feeding behavior of the mottled sculpin, *Cottus bairdi*, in Lake Michigan. Environ Biol Fish 12:111–117.

Hoekstra D, Janssen J (1986) Lateral line receptivity in the mottled sculpin, (*Cottus bairdi*). Copeia 1986(1):91–96.

Hofer B (1908) Studien ueber dei Hautsinnesorgane der Fische. I. Die Funktion der Seitenorgane bei den Fischen. Ber Kgl Bayer Biol Versuchsstation Muenchen 1:115–164.

Jielof R, Spoor A, Vries H de (1952) The microphonic activity of the lateral line. J Physiol 116:137–157.

Kalmijn AJ (1988a) Hydrodynamic and acoustic field detection. In: Atema J, Fay RR, Popper AN, Tavolga WM (eds) Sensory Biology of Aquatic Animals. New York: Springer-Verlag, pp. 83–130.

Kalmijn AJ (1988b) Detection of weak electric fields. In: Atema J, Fay RR, Popper AN, Tavolga WM (eds) Sensory Biology of Aquatic Animals. New York: Springer-Verlag, pp. 151–186.

Kalmijn AJ, Enger PS (in preparation) A linear electrolytic displacement transducer for the recording of hydrodynamic flow fields of biological origin. J Acoust Soc Am (in preparation).

Karlsen HE, Sand O (1987) Selective and reversible blocking of the lateral line in freshwater fish. J Exp Biol 133:249–262.

Kroese ABA, Schellart NAM (1987) Evidence for velocity-and acceleration-sensitive units in the trunk lateral line of the trout. J Physiol 394:13P.
Montgomery JC, MacDonald JA (1987) Sensory tuning of lateral line receptors in antarctic fish to the movements of planktonic prey. Science 235:195–196.
Münz H (1985) Single unit activity in the peripheral lateral line system of the cichlid fish *Sarotherodon niloticus* L. J Comp Physiol 157:555–568.
Netten SM van, Kroese ABA (1987) Laser interferometric measurements on the dynamic behaviour of the cupula in the fish lateral line. Hear Res 29:55–61.
Sand O (1981) The lateral line and sound reception. In: Tavolga WN, Popper AN, Fay RR (eds) Hearing and Sound Communication in Fishes. New York: Springer-Verlag, pp. 459–480.
Sand O (1984) Lateral-line systems. In: Bolis L, Keynes RD, Maddrell SHP (eds) Comparative Physiology of Sensory Systems. Cambridge, U.K.: Cambridge University Press, pp. 3–32.
Sand O, Karlsen HE (1986) Detection of infrasound by the Atlantic cod. J Exp Biol 125:197–204.
Schwartz E (1965) Bau und Funktion der Seitenlinie des Streifenhechtlings (*Apolocheilus lineatus Cuv. u. Val.*). Z Vergl Physiol 50:55–87.
Suckling EE, Suckling JA(1950) The electrical response of the lateral line system of fish to tone and other stimuli. J Gen Physiol 34:1–8.
Wunder W (1927) Sinnesphysiologische Untersuchungen ueber die Nahrungsaufnahme bei verschiedenen Knochenfischarten. Z Vergl Physiol 6:67–98.
Yoshii K, Kurihara K (1983) Role of cations in olfactory reception. Brain Res 274:239–248.

Part VI
Specialized Lateral Line and Related Sensory Systems

30
Specialized Lateral Line Receptor Systems in Elasmobranchs: The Spiracular Organs and Vesicles of Savi

MICHAEL A. BARRY and MICHAEL V.L. BENNETT

30.1 Introduction

The mechanoreceptive lateral line system shows considerable intra- and interspecific diversity in morphology (Coombs et al. 1988), but the functional ramifications of this diversity are poorly understood. The two receptor systems discussed in this chapter, the spiracular organs and the vesicles of Savi, are specialized lateral line organs, whose physiology has been studied in elasmobranchs.

The spiracular organs and vesicles of Savi were first observed and were recognized as parts of the lateral line sensory system in the mid- to late 19th century (Savi 1844; Boll 1875; Wright 1885). Norris and Hughes (1920) described the spiracular organs of *Squalus* and recognized that a number of organs described under different names were in fact spiracular organs of various species. Coggi (cited by Norris 1932) noted that in *Torpedo*, vesicles of Savi were specialized derivatives of the ordinary lateral line organs. The morphology and physiology of the vesicles of Savi have been characterized in a collection of papers by Szabo and colleagues (see Szabo 1968). The role of the spiracular organs of elasmobranchs has been examined in a recent series of studies (Barry and Boord 1984; Barry et al. 1988a,b).

30.2 Spiracular Organs

The spiracular organs occur in most gnathostome fish taxa including elasmobranchs, chondrosteans, holosteans, the Dipnoi, and possibly the Actinistia (Barry and Boord 1984; Northcutt 1986). They have not been observed in agnathans and were apparently lost in the teleosts (Patterson 1975) and again in the tetrapods (Northcutt 1986), although the paratympanic organ in the middle ear of birds resembles the spiracular organ (Jørgensen 1984, Chapter 6). A single spiracular organ is associated with the first (spiracular) gill cleft bilaterally. There is considerable interspecific variation in the morphology of the spiracular organ (Fig. 30.1). In all known cases, the spiracular organ is a pouch or tube whose lumen is lined with neuromasts containing typical vertebrate sensory hair

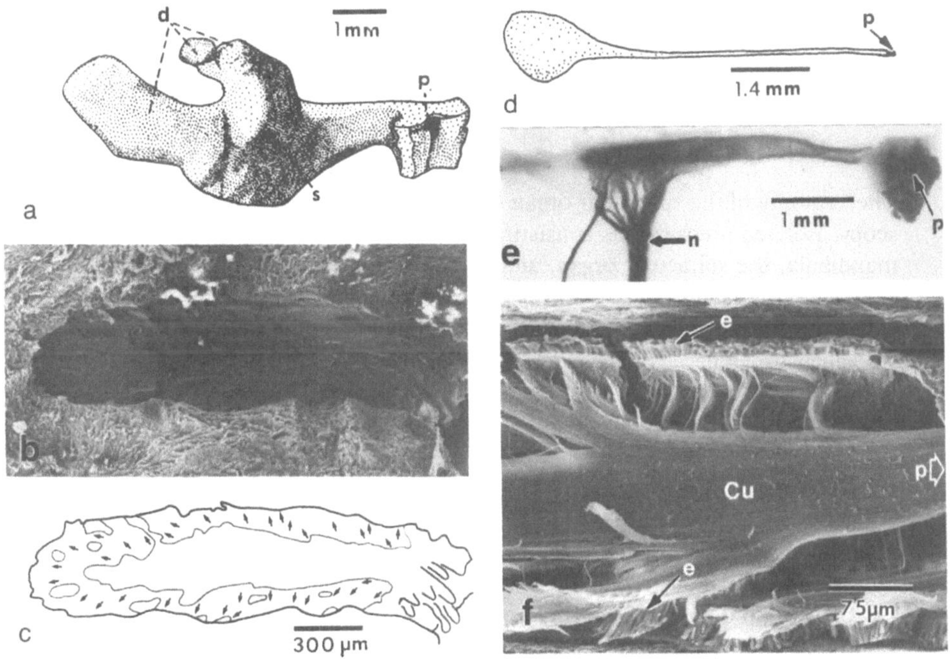

FIGURE 30.1. Spiracular organs from different elasmobranchs. The pore (p) is on the right in each figure. (a) Diagram of spiracular organ of spiny dogfish, *Squalus acanthias*. Norris and Hughes (1920) report that up to three diverticula (d) are present each with a neuromast, but Barry and Boord (1984) find a neuromast only in the largest diverticulum. (s) Saclike vestibule. (b) Scanning electron micrograph of spiracular organ of a dogfish, *Mustelus canis*, where the rostral wall and cupula have been removed. (c) Diagram of same micrograph with hair cell containing epithelia outlined. Arrows indicate the polarization of the hair cell ciliary bundles. Arrows point toward the kinocilia relative to the stereociliary bundles. (d) Diagram of spiracular organ from butterfly ray, *Gymnura*, which consists of a vesicle that contains a neuromast and is connected to the pore by a narrow tube. (e) Photograph of spiracular organ of little skate, *Raja erinacea*, stained with osmium. n, Nerve of spiracular organ. (f) Scanning electron micrograph of section of little skate spiracular organ with superficial (rostroventral) wall removed. The cupula (Cu) fills the spiracular organ lumen. e, Epithelial wall of spiracular organ. Panel (a) is from Norris and Hughes (1920); panels (b), (c), and (f) are from Barry et al. (1988a).

cells. A gelatinous cupula completely fills the lumen (Barry and Boord 1984; Barry et al. 1988a). In elasmobranchs, the lumen of the spiracular organ opens into the spiracular cleft or diverticulum of the cleft via a pore variable dimensions, but the pore is apparently of little functional importance (see section 30.2.2.4). In lungfishes, this opening is lacking (Northcutt 1986). Despite differences in morphology, in both the little skate, *Raja erinacea*, and the smooth dogfish, *Mustelus canis*, the spiracular organ appears to function as a proprioceptor sensitive to movements about the hyomandibula-cranial joint (Barry et al. 1988a,b).

30.2.1 Methods

The anatomy of the spiracular organ was studied with light and electron microscopy. Isolated preparations consisting of at least part of the cranium, the hyomandibula, the spiracular organ, and its nerve were utilized in all experiments (Barry et al. 1988a,b). The position of the hyomandibula was controlled with a string tied to its distal end, and multi- or single-unit records were made from the spiracular organ nerve during such manipulations.

30.2.2 Results

30.2.2.1 Morphology of Spiracular Organ in the Little Skate

The spiracular organ of the little skate is a 4- to 5-mm-long and 0.2- to 0.3-mm-wide tube that opens onto the posterior medial wall of the spiracle via a pore (< 0.1 mm in diameter) (Fig. 30.1e). The lumen of the organ is filled with a gelatinous cupula that is composed of strands (Fig. 30.1f). Patches of sensory hair cells line the sides and caudal wall of the proximal third of the lumen. Few hair cells occur on the rostroventral (superficial) wall. The hair cells form chemical synapses with afferent fibers, but no efferent synapses were identified. Each hair cell has a bundle of 10 to 20 stereocilia (up to 3 μm long) and a single kinocilium (about 14 μm). All the hair cells are oriented parallel to the long axis of the spiracular organ, with the kinocilium located on the side of the stereocilia bundles opposite the pore. The ciliary bundles are inclined toward the pore about 35° to 46° away from an axis perpendicular to the epithelium. The cupular strands are initially parallel to the cilia and then turn toward the pore to extend the length of the lumen.

30.2.2.2 Morphology of Spiracular Organ in the Smooth Dogfish

The spiracular organ of the smooth dogfish is a pouch 1.0 to 2.5 mm long and 0.5 to 1.5 mm wide with a single pore (0.3 to 0.6 mm wide) which opens ventrally into a medial and slightly anterior diverticulum of the spiracle. Hair cells are located in a U-shaped neuromast on the side and caudal walls but not on the thin rostral wall (Fig. 30.1b,c). The hair cells are polarized with their kinocilia on the side of the stereociliary bundles opposite the pore and middle of the caudal wall. The lumen is filled with a cupula, whose strands initially course largely in parallel with the cilia and then turn toward the pore. The spiracular organ morphology is similar to that in *Charcharinus plumbeus* (Barry and Boord 1984), except that in the latter species a small percentage of hair cells are oppositely oriented relative to the others.

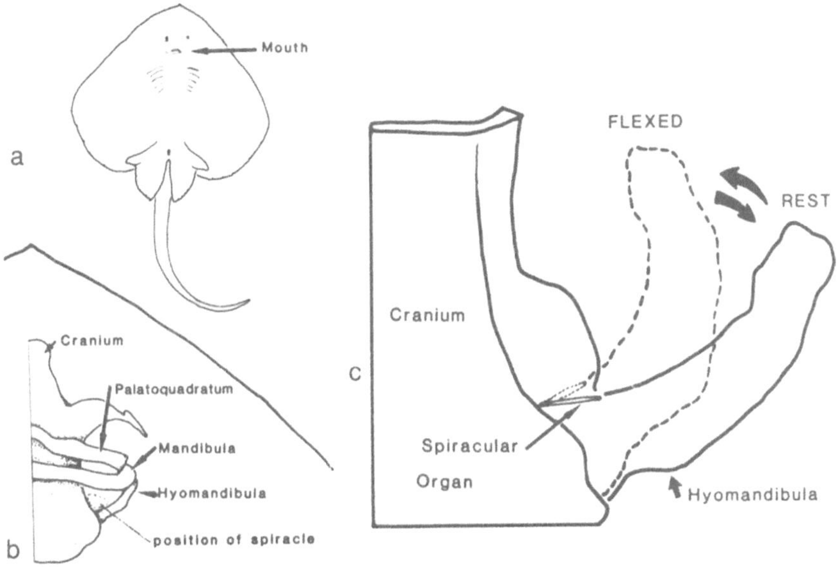

FIGURE 30.2. Diagrams to illustrate position of the spiracular organ in the little skate. (a) Ventral view of the body (female). (b) Ventral view of chondrocranium, in relation to the rostral border of the fish. The spiracle is indicated by a dotted line. (c) Ventral view of an isolated preparation consisting of part of the cranium, the hyomandibula, and the spiracular organ from the left side. The positions of the hyomandibula and spiracular organ are shown at rest and during flexion. From Barry et al. (1988a).

30.2.2.3 MECHANICS OF SPIRACULAR ORGANS IN RESPONSE TO STIMULATION

The hyomandibula, a suspensory element for the jaws, articulates proximally with the otic capsule and distally with the mandibula. Flexion of the hyomandibula at its cranial joint results in jaw protrusion, relative to the resting (almost fully extended) position. The skate spiracular organ is anchored at its pore end to the rostroventral aspect of the hyomandibula and at its proximal end by a ligamentous strand to the cranium (Fig. 30.2). Extension and flexion of the hyomandibula cause stretching and shortening, respectively, of the spiracular organ. Shortening is accompanied by a widening of the spiracular organ lumen. In the smooth dogfish, the spiracular organ is located between the ventral origin and lateral extension of the superior postspiracular ligament of Ridewood (1896) (Fig. 30.3). This ligament originates from the posterior wall of the orbit to insert on the distal end of the hyomandibula, such that when the hyomandibula is flexed, the ligament compresses the spiracular organ along its short axis (perpendicular to the long axis, which includes the pore).

The compression results in a change in cross-sectional shape, manifest as a narrowing from side to side and as a bulging out of the thin rostral wall of the spiracular organ. The cupula can be directly visualized following the removal of the

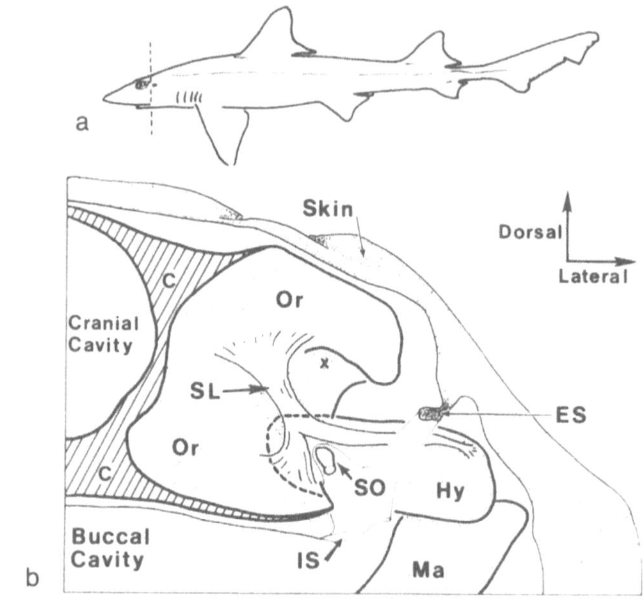

FIGURE 30.3. Diagrams to indicate position of the spiracular organ in a shark, the smooth dogfish. (a) Side view of a dogfish. Dashed line indicates position of transverse section shown in (b). (b) Diagram of upper left quadrant of transverse section at the level of the posterior wall of the orbit, viewed from the rostral aspect. The palatoquadratum (upper jaw) and palatoquadrate levator muscle (origin indicated by x) have been removed. Nerves and vessels are not illustrated. The hyomandibula (Hy; dashed line where hidden by more rostral part of cranium) articulates with the otic capsule caudal to the level of this section. The spiracular organ is a small pouch whose pore (dashed line) opens caudally onto the rostral wall of the diverticulum of the spiracle. The superior postspiracular ligament (SL) has a dorsal and a ventral origin from the posterior wall of the orbit (Or) and inserts on the distal end of the hyomandibula. c, Cut surface of chondrocranium; ES, external opening of spiracle; IS, internal (buccal) opening of spiracle; Ma, mandibula. From Barry et al. (1988a).

rostral wall without significantly effecting afferent discharges recorded in the spiracular organ nerve (see section 30.2.2.4). During hyomandibular flexion, the cupula is compressed and bulges rostrally (Fig. 30.4), but there is also some movement away from the pore. The observed cupular movement should cause a deflection of the ciliary bundles in the stereo- to kinociliary direction during flexion (Fig. 30.5).

In the little skate, we were unable to observe the movement of the cupula directly. When the spiracular organ lumen narrows during hyomandibular extension, the cupula will be compressed across the short axis and extended in the longitudinal axis of the spiracular organ. Because of the inclination of the cupular strands and the ciliary bundles, the cilia should be pushed down toward

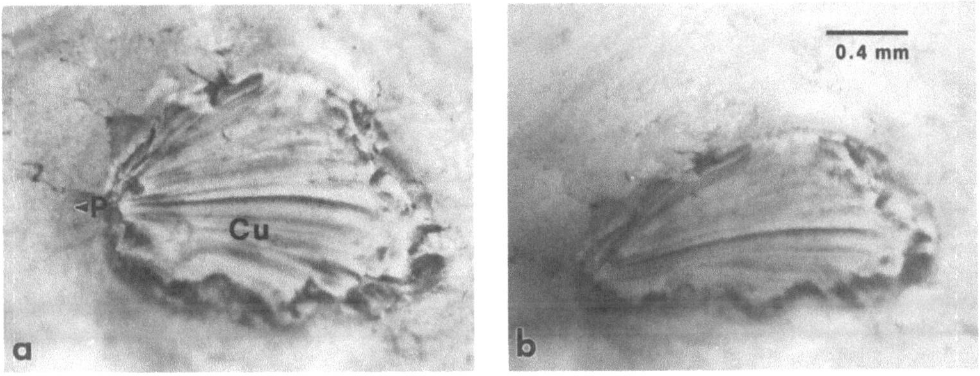

FIGURE 30.4. Photographs of a dogfish spiracular organ with the thin rostral wall removed, to show the change in cupula shape with the hyomandibula in the rest (a) and flexed (b) positions. Both photographs are shown at the same magnification. The surface of the cupula (Cu) was stained with a drop of methylene blue. P, Position of pore. From Barry et al. (1988a).

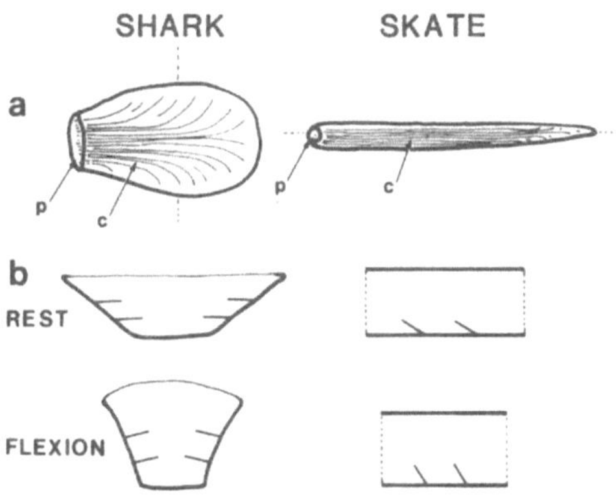

FIGURE 30.5. Diagrams of spiracular organs to illustrate deflection of cupula and ciliary bundles during jaw movements. (a) Shark and skate organs. Thin lines indicate the course of cupular strands. Dashed lines indicate locations of transverse section through shark organ and longitudinal section through skate organ shown in (b). (b) Sections through spiracular organs during rest and flexion of the hyomandibula. The thin rostral wall of the shark spiracular organ and rostroventral (superficial) wall of the skate spiracular organ are at the top of each diagram. The orientation of ciliary bundles is indicated by short, straight lines. Diagrams are not to scale. From Barry et al. (1988b).

the epithelium (deflection in the kino- to stereociliary direction) during hyomandibular extension (Fig. 30.5).

The deflection of the ciliary bundles of the little skate spiracular organ was measured utilizing transmission electron microscopy. In two skates, the spiracular organ from one side was fixed in situ in the extreme flexed position of the

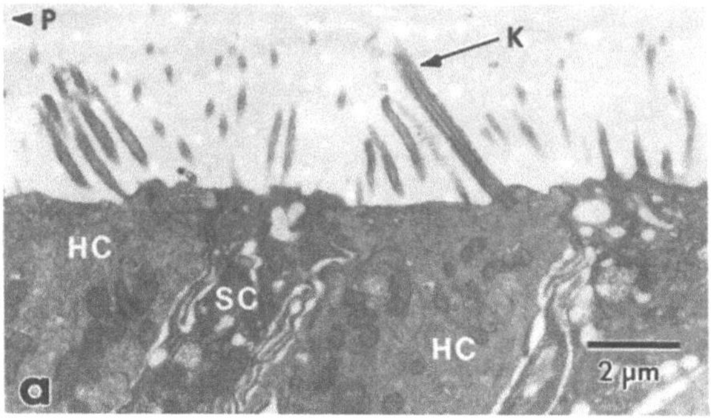

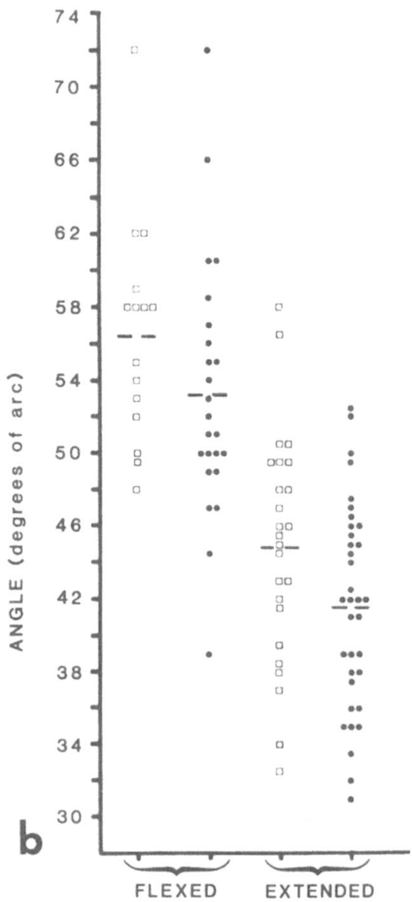

FIGURE 30.6. Movement of the hyomandibula in the little skate results in a measurable deflection of kinocilia. (a) Transmission electron micrograph of hair cells fixed with the hyomandibula in the extreme flexed position. HC, Hair cell; K, kinocilium; P, direction of pore; SC, supporting cell. (b) Plot of angle between base of kinocilium and apical surface of epithelium (measured on stereociliary side) for individual hair cells from spiracular organs fixed with the hyomandibula in the extreme flexed or extended positions. Solid circles and open squares represent the two animals. Horizontal dashed lines indicate the mean angle for each treatment and animal. From Barry et al. (1988a).

hyomandibula, and the other spiracular organ was fixed in the extended position. Extreme flexion and extension were defined by the limit the hyomandibula could be moved without straining the joint capsule and associated tissues. Flexion of the hyomandibula resulted in an increase in the angle between the kinocilia and

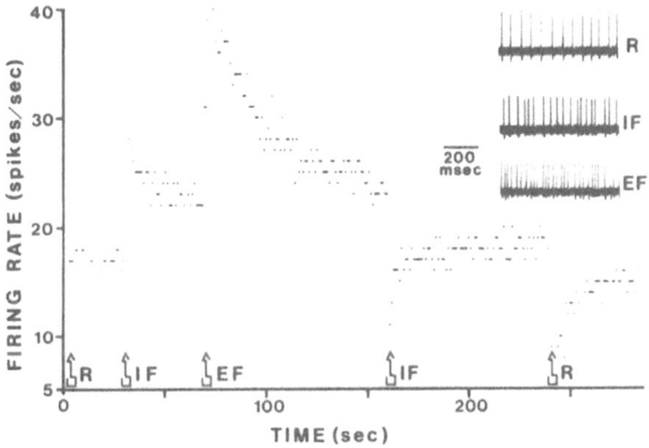

FIGURE 30.7. Response of a single afferent unit innervating the spiracular organ in the little skate. Firing rate is expressed as spikes per 1-s interval at the following positions of the hyomandibular joint: EF, extreme flexion; IF, intermediate flexion, R, rest position. The hyomandibula was moved to a different position (arrows) and held at that position until the next arrow. Inset shows original records for this unit. From Barry et al. (1988b).

apical surface on the stereociliary side of about 11.6° for both animals (Fig. 30.6). The mean angle during extension was slightly different in the two animals. The change in ciliary angle that was observed in the skate should result in depolarization of the hair cells and excitation of afferent fibers during hyomandibular flexion, based on the polarity of response of other octavolateralis hair cells (Flock 1965; Hudspeth and Corey 1977).

30.2.2.4 Physiological Response Recorded in Spiracular Organ Nerve

In both the little rate and the smooth dogfish, multiunit activity recorded in the spiracular organ nerve increased during hyomandibular flexion and decreased during extension. In the skate, multiunit and single-unit responses were essentially identical, and all single units were classified as tonic. Flexion of the hyomandibula resulted in an increase in single-unit activity from a spontaneous rate of 10 to 20 spikes per second up to 70 spikes per second at the extreme flexion position (Fig. 30.7). The unidirectional response pattern is consistent with the unidirectional polarization of the hair cells. The dynamic range of the neural response extended over the entire range of hyomandibular movement. The neural response was very slowly adapting. Following a rapid flexion (or extension from a flexed position), single-unit firing rate decayed exponentially over about 50 to 100 s to a rate that decayed over a much longer period (hundreds of seconds).

Multiunit recordings were made during various mechanical manipulations of the little skate spiracular organ to ascertain the role of the cupula. Tying a ligature around the organ distal to the neuromasts prevented volume changes of the spiracular organ lumen but did not change the direction of the neuronal response to hyomandibular flexion (Fig. 30.8a,b). Tying additional ligatures around the

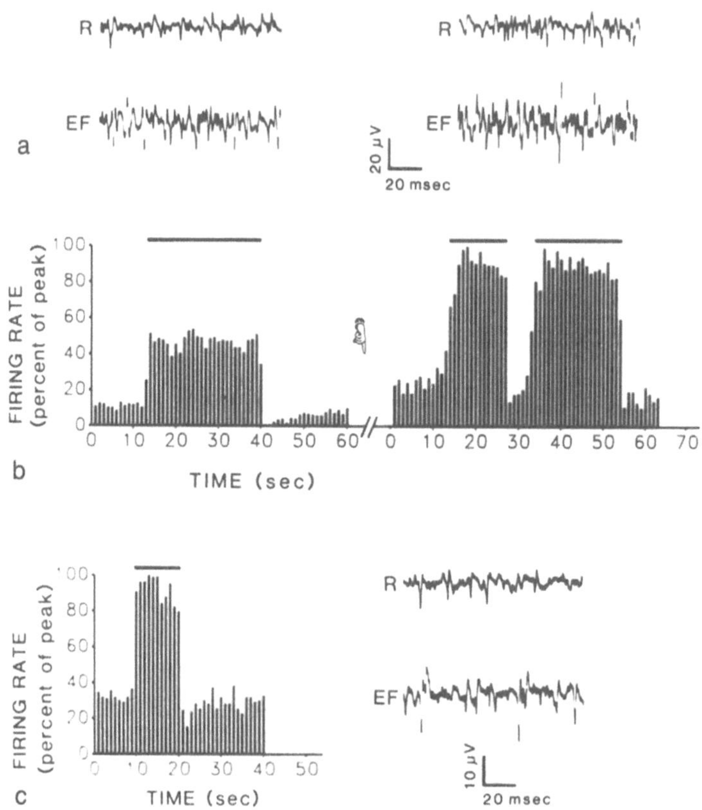

FIGURE 30.8. Neuronal response before and after ligation of the little skate spiracular organ to occlude its lumen. (a) Segments of a continuous multiunit record before (left) and after (right) ligation of the organ near the pore, for both rest (R) and extreme flexion (EF) positions of the hyomandibula. For calculating frequency histograms, a Schmitt trigger was set to give a moderate level of resting discharge. (b) Record of events per 1-s interval corresponding to data shown in (a). The record was interrupted during ligation (pointer). Firing rate is expressed as a percent of the peak rate during this record. Periods of extreme flexion are indicated by the solid black lines. After ligation, resting discharge and peak response were increased, presumably owing to displacement of the cupula toward the proximal end of the spiracular organ. (c) Multiunit records from the same preparation after two more ligatures were tied around the neural part of the spiracular organ, leaving only about the proximal tenth of the organ intact. Segments of original multiunit records are shown on the right. Total activity is reduced owing to damage to the neuromasts and nerve fibers, but the response sensitivity remains comparable. From Barry et al. (1988b).

proximal part of the spiracular organ reduced activity due to damage to the neuromasts, but the remaining receptors continued to respond normally (Fig. 30.8c). Thus a local deformation of the cupula is sufficient to excite the hair cells, which is consistent with the model of hair cell stimulation shown in Fig. 30.5.

The spiracular organs of the little skate and smooth dogfish are relatively insensitive to electrical stimulation. Anodal stimuli of 100 to 400 mV delivered at the skate spiracular organ pore, which should have made the lumen relatively positive, were excitatory, and cathodal stimuli were inhibitory. The opposite was true for stimuli delivered over the organ, which should have made the lumen relatively negative. The polarity of sensitivity is consistent with direct action of the stimuli on the presynaptic faces of the hair cells. Sensitivity to electrical stimuli was low and similar to the responses of other lateral line mechanoreceptors (Baumann and Roth 1986; Münz et al. 1984).

The spiracular organ afferents in the skate and dogfish were insensitive to a stream of water directed at the pore. The firing rate was also surprisingly insensitive to vibratory stimuli. There was no response to tapping the table forcefully or to loudspeaker-generated vibrations (10 to 1,000 Hz) of the entire preparation, up to an amplitude that prevented stable recordings due to movement of the preparation. However, spiracular organ afferents were sensitive to vibration of the hyomandibula in its axis of motion (1- or 2-mm movements of the distal end of the hyomandibula). Multiunit afferent discharges were synchronized with vibrations up to about 60 Hz, with a peak sensitivity (qualitatively assessed) of about 20 Hz.

30.3 Vesicles of Savi

Among elasmobranchs, and in particular among those that are dorsoventrally flattened, the mechanoreceptive lateral line system shows extensive morphological elaboration and interspecific variation on both the dorsal and ventral surfaces (Chu and Wen 1979). In members of the Torpediniformes, or electric rays, the canal system is lacking ventrally and has been replaced by the vesicles of Savi (Norris 1932). Similar lateral line receptors have been described in various species of the Myliobatiformes including *Urolophus* (Garman 1888; Norris 1932). Chu and Wen (1979) describe Savi-like vesicles in various dasyatid rays but not in *Urolophus*. Too little is known about these other vesicles to comment about their possible homology to vesicles of Savi in the electric rays. Lateral line canal organs with few or no pores have evolved numerous times among fishes (Coombs et al. 1988; Szabo et al. 1972).

30.3.1 Morphology of the Vesicles of Savi

Each vesicle of Savi is a closed off pouch approximately 2 to 3 mm long (Boll 1875). In all species the vesicles are located largely on the ventral surface, rostral to the mouth and along the rostral edge of the pectoral fin (Fig. 30.9) (Chu and Wen 1979). There are 100 to 120 vesicles (about 13% on dorsal surface) in *Torpedo* innervated by the superficial ophthalmic, buccal, and external mandibular divisions of the anterior lateral line nerve (Norris 1932). In *Torpedo*, each

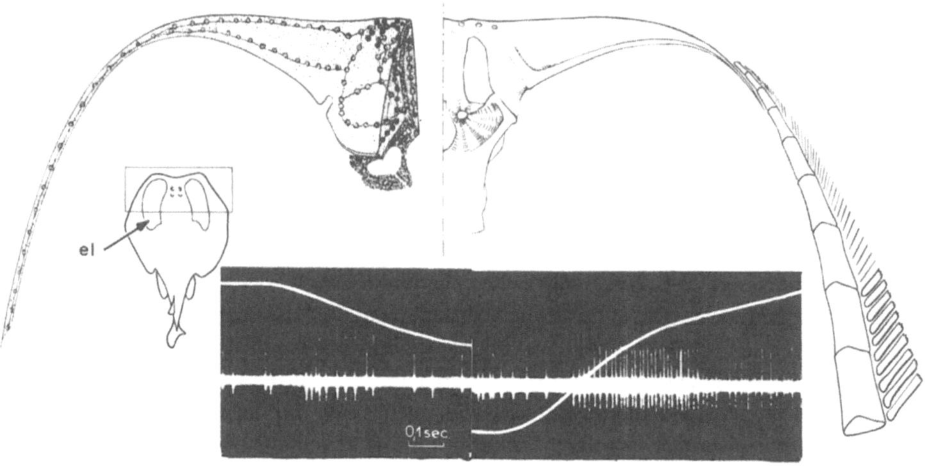

FIGURE 30.9. Vesicles of Savi of *Torpedo*, from a ventral aspect. The disposition of the vesicles with their associated connective tissue bands (within the gelatinous mass) is shown on the left, and the underlying cartilages are shown on the right. Vesicles located on the dorsal surface are indicated by black dots. Left inset shows a dorsal view of *Torpedo* indicating the region drawn above. el, electric organ. Right insets show responses of two phasic units, recorded in the nerve, to the deflection (solid line) of the cupula. Unpublished figures courtesy of T. Szabo.

vesicle contains three neuromasts about 0.5 mm in diameter (one central and two slightly smaller peripheral, whose centers fall on a straight line). Each neuromast has its own cupula. Mechanoreceptive hair cells are polarized bidirectionally in parallel with the line determined by the three neuromast centers (Szabo 1968). The hair cells have both afferent and efferent synapses (Derbin and Szabo 1966). Additional support for the presence of efferents is that acetylcholine blocks the neuronal response to stimulation of the vesicles of Savi (Roth and Heise 1969). We have further examined the morphology of the vesicles of Savi in *Narcine brasiliensis*.

As in *Torpedo*, the vesicles of Savi in *Narcine* are arranged in lines and are anchored at their bases (which are cartilaginous) to a connective tissue band, which is attached to the underlying cartilages. Superficial to this band, each line of vesicles is surrounded by a cylinder containing dense gelatinous material strengthened by septa of connective tissue which radiate from the walls of the vesicles of Savi to the edges of the cylinder. The skin overlies the gelatinous cylinders. In the snout, where there are many adjacent lines, the gelatinous compartments are continuous. Each vesicle contains a large central neuromast (0.4 mm in diameter) and two smaller (0.1-mm) peripheral neuromasts. The vesicles are about 2 mm in diameter and 1 mm in height at the center, but flatten out over the peripheral neuromasts. The height is maintained by a rigid arch formed by

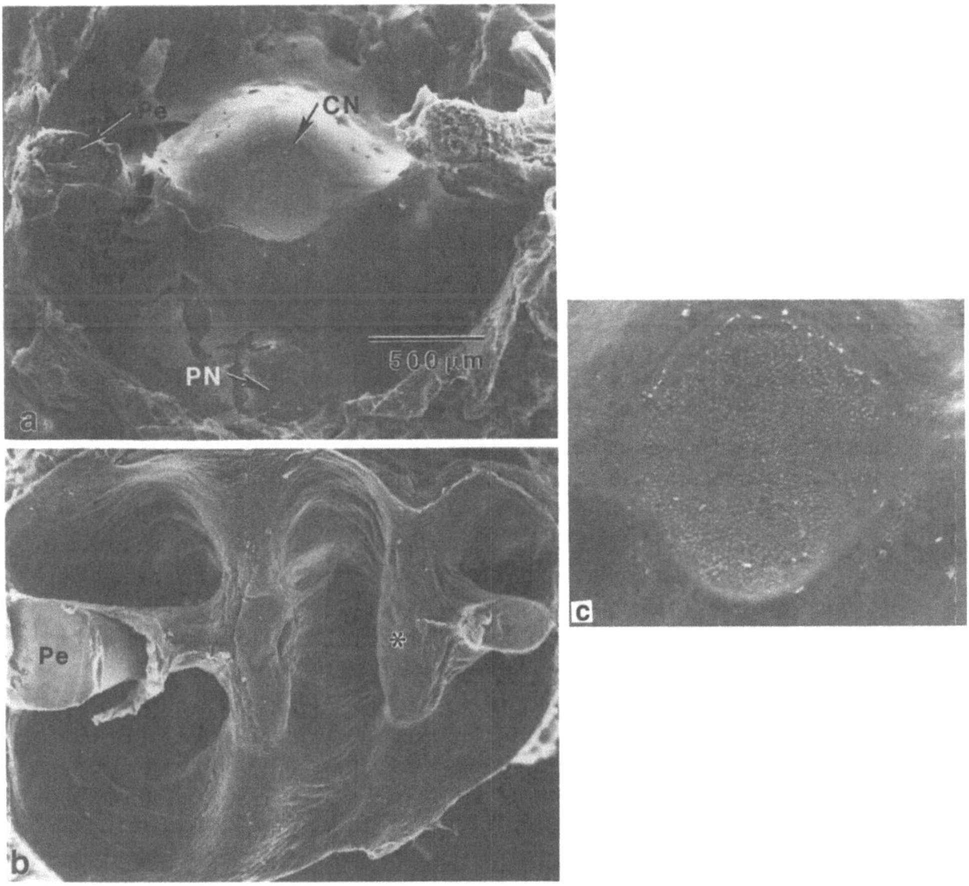

FIGURE 30.10. Scanning electron micrographs of vesicles of Savi from *Narcine brasiliensis*. (a) The basal wall of the lumen is exposed following the removal of the superficial wall of the vesicle. (b) The luminal face of the superficial wall of a vesicle of Savi (not from same vesicle as (a). Scale bar applies to both photographs. Note smooth epithelium (asterisk) at apex of pedicle. (c) Enlargement of the central neuromast showing detail. CN, Central neuromast; Pe, pedicle; PN, peripheral neuromast.

cartilaginous pedicles arising on either side of the central neuromast from the base of the vesicle (Fig. 30.10a). The cupula over the central neuromast fits partially into the space at the apex but also appears to contact the epithelium at the peaks of the pedicles (Fig. 30.10b). Hair cells are polarized similarly to those in *Torpedo*. The ciliary bundles consist of 40 to 60 stereocilia, up to 5 µm in height, with kinocilia only slightly longer in the central neuromast, but up to 10 µm long in the peripheral neuromasts.

30.3.2 Physiology of the Vesicles of Savi

Single units recorded in the nerves innervating each neuromast respond to displacements of the cupula (Szabo 1968). There are two types of phasic units, which differ in their sensitivity to cupular displacements. Phasic units respond linearly to the velocity of displacement with a firing rate up to 300 spikes per second (Fig. 30.9). Tonic fibers have a spontaneous discharge of five to 15 and a dynamic range of zero to 70 spikes per second in response to displacement of the cupula. Multiunit recordings reveal a sensitivity to vibrations in the range of 1 to 350 Hz, with a peak sensitivity at 150 to 200 Hz.

30.4 Summary and Conclusions

Based on its morphology and sensitivity to hyomandibular movements, the elasmobranch spiracular organ functions as a proprioceptor. Recordings from the intact animal should verify the proprioceptive function, although this procedure will be hindered by the inaccessibility of the nerve. However, the biological role of the spiracular organ in the animal's behavior is not clear. The hyomandibula may move slightly during resting respiration, but destroying the spiracular organs does not noticeably affect normal respiration (unpublished observation). The hyomandibula does move during intense respiration or during jaw protrusion. Jaw protrusion functions both to extend the mouth and create suction for pulling in food. The central projections of the primary spiracular afferents to the recticular formation (Fig. 30.11) (Barry and Boord 1984) are probably related to jaw movement or respiratory centers. Respiratory centers are found in similar regions of the reticular formation of teleost fishes (Ballintijn and Roberts 1976). However, fishes possess a variety of proprioceptors, including deflection-sensitive receptors in the branchial processes of the spiny dogfish (Satchell and Way 1962). A role as an exteroreceptor sensitive to vibrations transmitted via the jaws cannot be ruled out but is unlikely, owing to low vibration-sensitivity and because the coupling between the spiracular organ and the hyomandibula should be poor during jaw protrusion (less tension on the spiracular organ). Thus, the specific role of the spiracular organ requires further study.

The vesicles of Savi might be classified as proprioceptors, since movements of the underlying cartilages should excite the hair cells (Szabo 1958). However, their sensitivity to vibrations, as well as their rostral and largely ventral location, is consistent with a role for the vesicles of Savi as exteroreceptors particularly sensitive to substrate-borne vibrations and to deformations resulting from contact of the animal with objects in the substratum. Vibrations and deformations could be transmitted via the skin or the cartilages. The function of the arch in *Narcine* is not clear, but it appears to protect the central neuromast.

Both the spiracular organ and the vesicles of Savi are mechanoreceptors derived from the lateral line system. Their unique characteristics are due in part to

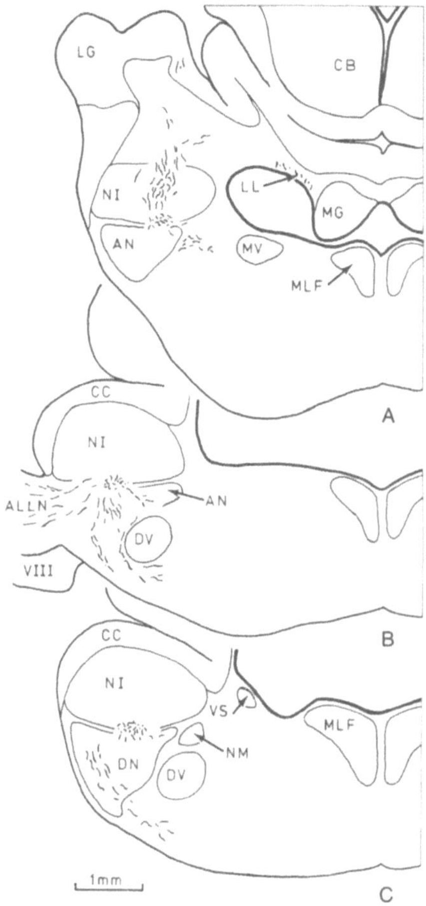

FIGURE 30.11. Diagram of a series of transverse sections through the rhombencephalon of the clearnose skate, *Raja eglanteria*, from rostral (A) to caudal (C) regions, showing the distribution of HRP-stained afferent fibers which innervate the spiracular organ. Afferents enter the medulla in the ventral branch of the anterior lateral line nerve (ALLN); course in the mechanoreceptive lateral line nucleus, nucleus intermedius (NI); and project to the vestibulocerebellum (LL and LG). Fibers also course through the anterior and descending octaval nuclei (AN and DN) to terminate in lateral parts of the reticular formation (not labeled). CC, cerebellar crest; CB, cerebellum; DV, descending nucleus of trigeminal nerve; LG, lateral granular layer; LL, lower lip; MG, medial granular layer of vestibulocerebellum; MLF, medial longitudinal fasciculus; MV, motor nucleus of trigeminal nerve; NM, nucleus magnocellularis; VIII, eighth nerve; VS, visceral sensory column. From Barry and Boord (1984), copyright 1984 by AAAS.

to the location and morphology of the organs, but also to the response characteristics of the hair cells and afferent fibers. Both the spiracular organ and the vesicles of Savi are effectively isolated from water movements and have specialized attachments to the skeleton. Other specializations may be largely related to differences in vibration sensitivity. The spiracular organ and vesicles of Savi represent extremes of lateral line mechanoreceptor vibration sensitivity. The spiracular organ is quite insensitive to vibrations, although it does respond to displacements of the hyomandibula occurring at up to 60 Hz, whereas the vesicles of Savi respond to vibrations up to 350 Hz. This is a higher frequency than that reported for most lateral line organs (see Münz 1985, Chapter 14) and in the range of that reported in a squirrelfish (Horch and Salmon 1973). It may be preferable for highly vibration-sensitive receptors to be isolated from gross deformations. This could be achieved in the vesicles of Savi by their distance from the skin, by the less efficient mechanical transduction of cupular movement due to short kinocilia, and/or by the apparent protection of the central neuromast by the arch in *Narcine*. The presence of phasic receptors may be another adaptation for high-frequency detection. The spiracular organ is organized for efficient transduction of gross deformations, because the cupula fills the lumen, and the receptors are well coupled to the cupula by long kinocilia. The peripheral neuromasts of the vesicles of Savi may function in a similar way.

The spiracular organs and vesicles of Savi exemplify the diversity of the lateral line system in terms of both its morphology and its function.

Acknowledgments. Thanks to Robert Johnson for supplying tissue from *Narcine*. This research was supported in part by grants NS-07512 and NS-07183.

References

Ballintijn CM, Roberts JL (1976) Neural control and proprioceptive load matching in reflex respiratory movements of fishes. Fed Proc 35:1983–1991.

Barry MA, Boord RL (1984) The spiracular organ of sharks and skates: Anatomical evidence indicating a mechanoreceptive role. Science 226:990–992.

Barry MA, Hall DH, Bennett MVL (1988a) The elasmobranch spiracular organ I. Morphological studies. J Comp Physiol A 163:85–92.

Barry MA, White R, Bennett MVL (1988b) The elasmobranch spiracular organ. II. Physiological studies. J Comp Physiol A 163:93–98.

Baumann M, Roth A (1986) The Ca^{++} permeability of the apical membrane of neuromast hair cells. J Comp Physiol A 158:681–688.

Boll F (1875) Die Savischen Blaschen von *Torpedo*. Arch Anat Physiol 1875:446–468.

Chu YT, Wen MC (1979) A study of the lateral-line canals system and that of Lorenzini ampullae and tubules of elasmobranchiate fishes of China. Monogr Fishes China 2:1–132.

Coombs S, Janssen J, Webb JF (1988) Diversity of lateral line systems: Evolutionary and functional considerations. In: Atema J, Fay RR, Popper AN, Tavolga WN (eds) Sensory Biology of Aquatic Animals. New York: Springer-Verlag, pp. 553–594.

Derbin C, Szabo T (1966) Ultrastructure de l'epithelium sensoriel de la vesicle de Savi. J Physiol Paris 5:508–509.

Flock Å (1965) Transducing mechanisms in the lateral line canal organ receptors. Cold Spring Harbor Symp Quant Biol 30:133–145.

Garman S (1888) On the lateral canal system of the Selachia and Holocephalia. Bull Mus Comp Zool (Harvard) 17:57–119.

Horch K, Salmon M (1973) Adaptations to the acoustic environment by the squirrelfishes *Mypristis violaceus* and *M. pralinius*. Mar Behav Physiol 2:121–139.

Hudspeth AJ, Corey DP (1977) Sensitivity, polarity, and conductance change in the response of vertebrate hair cells to controlled mechanical stimuli. Proc Natl Acad Sci USA 74:2407–2411.

Jørgensen JM (1984) Fine structure of the paratympanic organ in the avian middle ear. Acta Zool 65:89–94.

Münz H (1985) Single unit activity in the peripheral lateral line system of the cichlid fish *Sarotherodon niloticus* L. J Comp Physiol A 157:555–568.
Münz H, Claas B, Fritzsch B (1984) Electroreceptive and mechanoreceptive units in the lateral line system of the axolotl *Ambystoma mexicanum*. J Comp Physiol A 154: 33–44.
Norris HW (1932) The laterosensory system of *Torpedo marmorata*, innervation and morphology. J Comp Neurol 56:169–178.
Norris HW, Hughes SP (1920) The spiracular sense-organ in elasmobranchs, ganoids and dipnoans. Anat Rec 18:205–209.
Northcutt RG (1986) Lungfish neural characters and their bearing on sarcopterygian phylogeny. J Morphol (Suppl) 1:277–297.
Patterson C (1975) The braincase of pholidophorid and leptolepid fishes, with a review of the actinopterygian braincase. Phil Trans R Soc Lond B269:275–579.
Ridewood WG (1896) On the spiracle and associated structures in elasmobranch fishes. Anat Anz 11:425–433.
Roth A, Heise U (1969) Quelques caracteristiques du potentiel lent engendre au niveau d'un epithelium a cellules sensorielles ciliees. J Physiol Paris 61(S1):170–171.
Satchell GH, Way HK (1962) Pharyngeal proprioceptors in the dogfish *Squalus acanthias* L. J Exp Biol 39:243–250.
Savi P (1844) Etudes anatomiques sur le systeme nerveux et sur l'organe electrique de la *Torpille*. In: Matteucci (ed) Traite des Phenomenes Electrophysiologiques des Animaux. Paris: Chez L. Mechelsen, p. 272.
Szabo T (1958) Quelques precisions sur la morphologie de l'appareil sensoriel de Savi dans *Torpedo marmorata*. Z Zellforsch 48:536–537.
Szabo T (1968) Analyse morphologigicque et fonctionelle de l'epithelium sensoriel d'un mecanorecepteur. Actual Neurol 6:131–147.
Szabo T, Kalmijn AJ, Enger PS, Bullock TH (1972) Microampullary organs and a submandibular sense organ in the fresh water ray, *Potamotrygon*. J Comp Physiol 79:15–27.
Wright RR (1885) Some preliminary notes on the anatomy of fishes. Am Nat 19:187–190.

31 Hydrodynamic Receptor Systems in Invertebrates

BERND-ULRICH BUDELMANN

31.1 Introduction

Water currents are among the most basic events in the environment for almost all aquatic invertebrates. Since these animals are predominantly active at night or have no or only poorly developed eyes to localize a moving prey, predator, or conspecific, water currents—with or without chemical signals—are probably the most important source of information about events in their environment. Thus, in aquatic invertebrates one should expect a widespread occurrence of receptor systems that are sensitive to some kind of water movement.

Although behavioral reactions to water movements (such as feeding, attack, escape, vibrotaxis, rheotaxis) have commonly been reported in many invertebrate groups and many structures have been described as possible water movement receptors (for references, see Frings 1964; Horridge 1966; Frings and Frings 1967; Markl 1973, 1978; Creutzberg 1975; Laverack 1976; Ehlers and Ehlers 1977; Tautz 1979; Bush and Laverack 1982; Derby 1982), only a few physiological data are available that offer sufficient evidence to clearly characterize the mechanoreceptors described as water movement receptors. Moreover, basic difficulties exist in relating the function of an underwater mechanoreceptor to just one kind of mechanical energy (see Frings 1964; Markl 1978).

Thus, in the present review, a hydrodynamic receptor system will be understood in a very general way—i.e., as a flexible mechanoreceptor system that is sensitive to any kind of motion of a fluid medium surrounding it. This definition avoids all the difficulties of distinguishing between the various kinds of fluid motion and thus includes not only water flow receptors but also angular acceleration receptors, underwater vibration receptors, and underwater sound receptors (see Markl 1978 for classification; van Bergeijk 1964, Kalmijn 1988, for physical aspects; see also Chapter 9, Kalmijn).

This paper is dedicated to Professor Dietrich Burkhardt, Regensburg, on the occasion of his 60th birthday.

The ability to detect motion of a fluid medium is found throughout all major aquatic invertebrate groups, from protozoans to the lower chordates. Most hydrodynamic receptor systems occur on the animal's external body surface, but some are also enclosed in fluid-filled cavities or canals. They all are displacement-sensitive devices and, with the exception of the angular acceleration receptor systems, serve to detect and locate moving objects, such as prey, enemies, or social partners. Thus, invertebrate hydrodynamic receptor systems are comparable with the vertebrate (amphibian and fish) lateral line system.

Basically, any external or internal projection from an animal's body surface that can be displaced by fluid movements and is associated with a sensory structure may be part of, or serve as, a hydrodynamic receptor system. Most such flexible projections are cilia of sensory cells; others are cuticular sensory hairs. But there are also examples in which whole body appendages, such as antennae and spines, together with associated proprioceptors, serve as hydrodynamic receptor systems. The present review will follow this line of classification of the sensory structures involved in water movement detection. In a taxonomical order it will briefly outline those systems that in the various invertebrate groups serve, or may serve, as hydrodynamic receptor systems. Special emphasis will be placed on those systems that more closely resemble the vertebrate free and lateral line canal neuromasts.

31.2 Receptor Systems with Ciliated Sensory Cells

The flexible structures of the hydrodynamic receptor systems of almost all nonarthropod invertebrates are kinocilia of epidermal sensory cells. In general, these kinocilia are nonmotile, are often described as stiff, and have the typical internal $9 \times 2 + 2$ microtubuli arrangement. Some have modified internal or basal structures, and ciliary rootlets are often missing. The length and number of cilia per cell vary. When a single kinocilium occurs, it is generally closely surrounded by a collar of 6 to 10 stereovilli* of the same or of neighboring cells. In the cephalopod sensory cell, up to 200 kinocilia occur, but no stereovilli. The kinocilia are mechanically coupled with the surrounding fluid such that any lateral movement of that fluid will mechanically deflect the cilia and thus stimulate the sensory cells (see also Chapter 13, Kroese and van Netten). This coupling is either direct or indirect, via an accessory structure (cupula). The sense cells are generally primary sensory cells, but secondary sensory cells also occur (see also Chapter 6, Jørgensen).

---

*As suggested by Neugebauer and Thurm (1984), the term *stereovilli* (instead of *stereocilia*) will be used throughout the chapter for those villi that, in contrast to the microvilli, have specialized structures (i.e., are much thicker and contain microfilaments). The term *stereocilia* led to much confusion with *kinocilia* (containing the internal $9 \times 2 + 2$ microtubuli arrangement); furthermore, in the literature cited, the terms *stereocilia* and *microvilli* have been used interchangeably for the same structures.

31.2.1 Protozoa

Some protozoans are able to detect moving objects by water disturbances and respond to vibration (Schaeffer 1916). A positive vibrotaxis has been described for *Amoeba*, with a stimulus frequency of 50 Hz being most effective; this is the frequency of the ciliary beat of their prey *Glaucoma* (Kolle-Kralik and Ruff 1967). The sensory structures involved are not known, but in ciliates and flagellates it seems not impossible that their motile cilia might be sensitive to local medium currents (see Markl 1978).

31.2.2 Coelenterates

In coelenterates, several behavioral reactions strongly suggest the presence of hydrodynamic receptor systems associated with feeding; they may also be important in the symbiotic relationship between sea anemones and fish (Graefe 1964). The hydromedusa *Eutonia* is sensitive to low-frequency vibrations (Horridge 1966), and the hydroid polyp *Syncoryne* was observed to bend toward a vibrating object in its vicinity, but only at frequencies below 5 Hz (Josephson 1961). Also, in the sea anemone *Sagartia*, slow currents directed at the tentacles induce a bending toward the source; rapid currents, however, induce retraction (Frings and Frings 1967). In these and other studies on *Calliactis* and *Anemonia* (Passano and Pantin 1955), the sensory structures involved in the detection of fluid movements remained unknown.

However, in other coelenterate species, ciliary sensory cells have been described on the tentacles that most probably serve as hydrodynamic receptors (Fig. 31.1B) (Lyons 1973; Mariscal 1974; Peteya 1975). They bear a single kinocilium 10 to 25 μm in length that is surrounded by a collar of six to nine thick (0.3 to 0.5 μm in diameter) but somewhat shorter (2 to 9 μm) stereovilli of the same cell. These stereovilli are packed with microfilaments that continue as "rootlets" a few microns into the cytoplasm of the cell. The kinocilium/stereovilli "cone" complex is further surrounded by many thinner (0.05 to 0.2 μm in diameter) microvilli of neighboring cells such that their length and diameter decrease with distance from the center of the cone (Fig. 31.1B). This arrangement closely resembles that of the tentacle receptors of the hydroid polyp *Coryne* (Fig. 31.1A) (Tardent and Schmid 1972) in which a long (30 to 70 μm) modified kinocilium with a conspicuous basal lamellary body is surrounded by a collar of nine stiff rods that are electron-dense and much shorter than the kinocilium. Within the cell, each rod splits into two striated rootlets. The whole sensory cell is surrounded by a single accessory cell that forms a continuous collar around the cilium/rod complex. Both are suggested to represent a functional unit that provides information about direct contact, water movements, or vibrations.

31.2.3 Ctenophores

Clear behavioral reactions to small water movements have been shown in two ctenophore species. *Leucothea* is a large carnivorous comb jelly that bears finger-

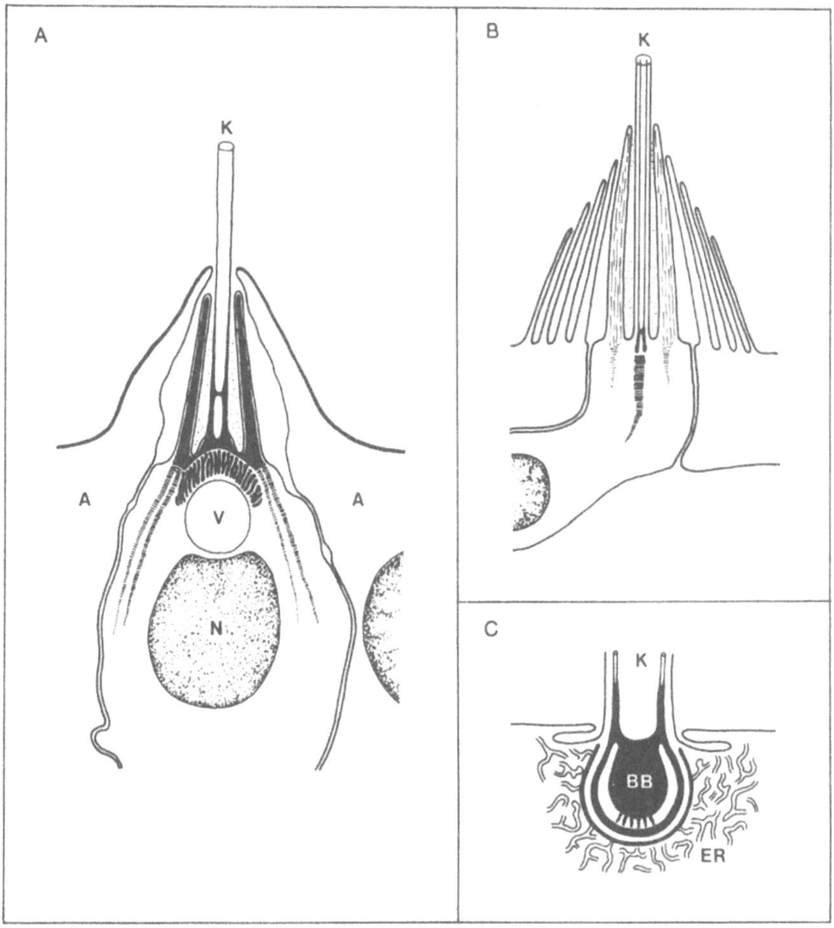

FIGURE 31.1. Presumed hydrodynamic receptors of coelenterates and ctenophores. (A) Tentacle receptor of the hydroid polyp *Coryne pintneri*. A, accessory cell; K, modified kinocilium; N, nucleus; V, vacuole. (Redrawn from Tardent and Schmid 1972.) (B) Distal part of a tentacle receptor ("ciliary cone") of anthozoan coelenterates. K, kinocilium. (Based on data from Lyons 1973; Peteya 1975.) (C) Receptor on the fingers of the comb jelly *Leucothea multicornis*; onionlike proximal part of the kinocilium. BB, central core of basal body; ER, endoplasmatic reticulum; K, kinocilium. (Redrawn from Horridge 1965.)

like projections on its surface. These fingers shoot outward toward a nearby source of low-frequency (10 Hz) vibrations, such as a copepod (Horridge 1965). The sea walnut *Pleurobrachia* is known to contract its tentacle when a vibrating needle is brought close to it (Horridge 1966). The sense organs that detect the small water displacements are epithelial cells that bear a very long (up to 100 μm) and stiff kinocilium with a specialized "onionlike" basal body and root (Fig.

31.1C). A round central core is surrounded by a spherical shell, and this fits concentrically inside another spherical shell. The outer shell is attached to the mass of tubules of the endoplasmatic reticulum that fills the distal part of the cell. A shearing between the two spherical shells is most probably involved in the transduction process of the cell (Horridge 1965; Hernandez-Nicaise 1974).

31.2.4 Flatworms

There is a large body of information about the structure of presumed mechanoreceptors in turbellarian, trematode, and cestode flatworms (for references, see Bedini et al. 1975; Ehlers and Ehlers 1977; Pariselle and Matricon-Gondran 1985); surprisingly, however, no physiological data exist. Therefore, on morphological evidence only, the "collar receptors" have been considered to be water movement detectors. Their somata lie subepidermally and bear a single kinocilium several microns in length. The proximal part of the cilium (2 to 3 μm) is embedded in the epithelium and again is surrounded by either a collar of eight separate stereovilli, or a collar with eight columnar ridges, which in either case is formed by the distal part of the cell. These stereovilli and columnar ridges are closely filled with microfilaments, more densely at their central side; ciliary rootlets are generally absent (Fig. 31.2A,B) (Bedini et al. 1975; Ehlers and Ehlers 1977).

Special types of collar receptors have been described in the cercariae of the trematode *Nicolla* (Fig. 31.2C–E) (Pariselle and Matricon-Gondran 1985). In one type (Fig. 31.2D), the kinocilium is surrounded by two concentric sets of long (4 to 5 μm) and thin (0.1 μm) stiff microvilli of tegumentary origin. In the other types (Fig. 31.2E), the collar forms a narrow sheath (1.5 to 1.8 μm in length) around the cilium and then ends in a crown of several long microvilli, almost perpendicular to the axis of the cilium. In *Nicolla*, there is behavioral evidence that its cercariae are sensitive to small water currents. They do not swim, but walk on their suckers like a leech. When standing on their tail and stimulated by small water currents, they stretch out in various directions in search for a mobile support, such as its vector host, the small crustacean *Gammarus* (Pariselle and Matricon-Gondran 1985).

31.2.5 Annelids

In leeches, several morphological, physiological, and behavioral data exist on water movement receptors. These animals exhibit a positive taxis when stimulated by water currents or surface water disturbances and orientation and swimming movements when stimulated by low-amplitude surface waves (see Young et al. 1981). The most likely receptor sites are the "segmental sensilla" (Friesen 1981). These are disklike sensory buds that occur all over the body surface (Fig. 31.3A–D). Their number, arrangement, and structure differ systematically from the anterior to posterior region of the leech (Blinn et al. 1986). Some are almost flush with the surface, but others clearly protrude (up to 50 μm). Each bud

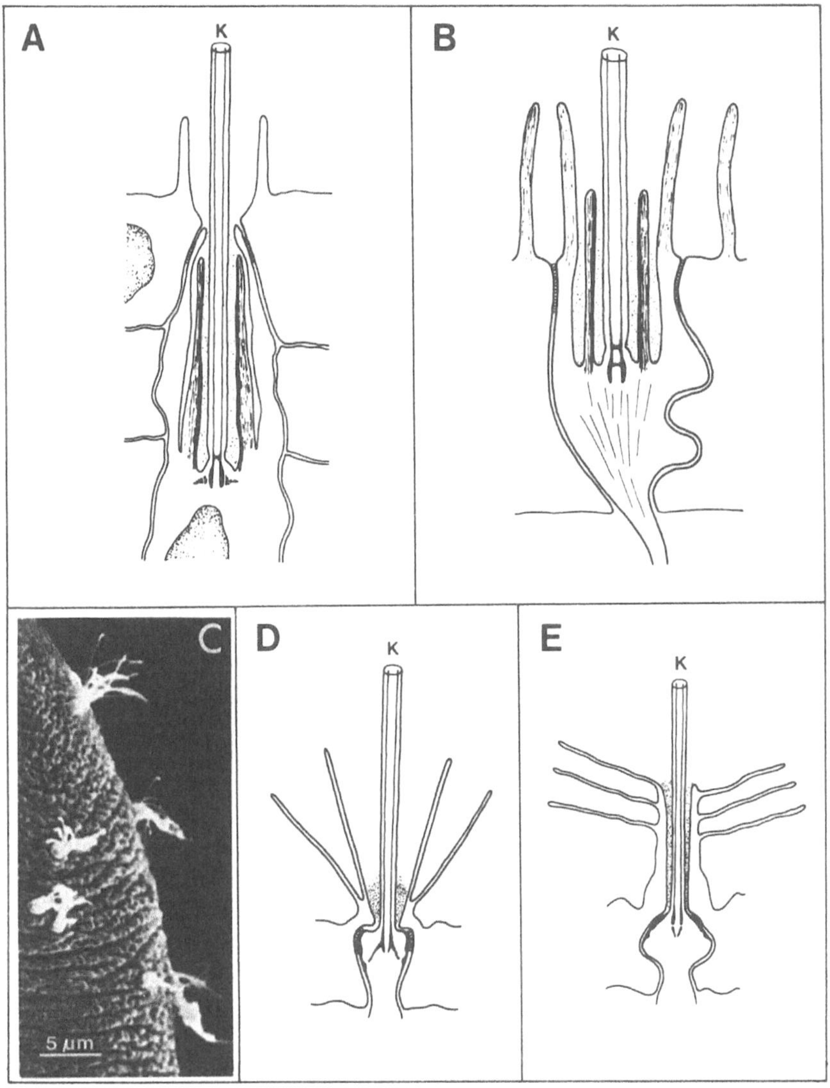

FIGURE 31.2. Presumed hydrodynamic receptors of flatworms. (A,B) Ciliary region of epidermal "collar receptors" of the turbellarians *Parotoplana capitata* (A) and *Schizochilus caecus* (B). (Redrawn from Ehlers and Ehlers 1977.) (C–E) Ciliary region of the "collar receptors" of the trematode *Nicolla gallica*. (C) Surface of head region with several ciliated receptors. (From Pariselle and Matricon-Gondran 1985.) (D) Ciliary region of receptor with long short collar and two concentric sets of microvilli. (E) Ciliary region of receptors with branched collar. (D and E, redrawn from Pariselle and Matricon-Gondran 1985.) K, kinocilium.

contains three types of ciliated sensory cells. Type I ("S hair," "S sensillum"; Fig. 31.3E) is a cell with a single kinocilium that projects at least 12 μm beyond the cuticle. At its base, the kinocilium is surrounded by a collar of nine or ten stereovilli that are filled with microfilaments, again more densely at their central side. There are no obvious ciliary rootlets. This type of receptor cell is suggested to be the most likely candidate for the water movement detection. Type II and type III ("G hairs," "G sensilla") are both multiciliated cells; one has two to four short (1 to 3 μm) kinocilia, and the other has four that remain below the cuticle. Both these types are presumed to be chemoreceptors (Derosa and Friesen 1981; Phillips and Friesen 1982; Blinn et al. 1986).

Besides leeches, sensitivity to water movements is known in other aquatic annelids as well. *Harmothoe* is able to locate a source of vibration (e.g., a prey) with receptors on its tentacular cirri and palps (Daly 1973). Tube worms withdraw rapidly into their tube when their crown of tentacles is stimulated with low-frequency vibrations, and the archiannelid *Trilobodrilus* shows a positive rheotaxis (Laverack 1968, for references). In the lugworm *Arenicola* (Jouin et al. 1985) and the ectoparasite *Branchiobdella* (Farnesi et al. 1982), sensory buds have been described that are similar to those of the leech. The lugworm's presumed water movement receptor (Fig. 31.3F–H), however, is different in that each sensory cell has three to seven kinocilia with ciliary rootlets and an apical sheathlike cylinder of fine filaments. This cylinder encloses the ciliary rootlets and microtubules, as well as many fine filaments that penetrate into the stout stereovilli (these do not form an obvious collar around the kinocilia). Still another presumed vibration receptor occurs on the tentacle tips of the tube worm *Sabella* (Santer and Laverack 1971). Its sense cells bear a conical bundle of about 40 stiff, but not immotile, kinocilia (30 μm in length); according to their 9 × 2 + 2 tubuli arrangement, they are all polarized in the same plane. Preliminary physiological experiments suggest that these ciliated cells are the receptors that sense vibration. There is no description of a cupula structure as being associated with any of the annelid sensory buds.

31.2.6 Mollusks, Except Cephalopods

There are suprisingly few morphological data on water movement receptors in chitons, gastropods, and bivalves, although these animals are commonly known to respond to water movements (Frings 1964; Frings and Frings 1967; Creutzberg 1975, for references). In trochid gastropods, the epipodial sense organ has cells that carry a single immotile kinocilium (80 μm in length). This is surrounded proximally by a collar of nine stereovilli which again are more electron-dense at their central side (Crisp 1981). A similar type of cell with a very long kinocilium (100 to 300 μm) has been described in the abdominal sense organ of the bivalve *Pteriomorpha* (Haszprunar 1985). Based on their morphological similarity to the sensory cells described in the other animal groups (e.g., in flatworms and annelids), both these ciliated cells might be water movement receptors.

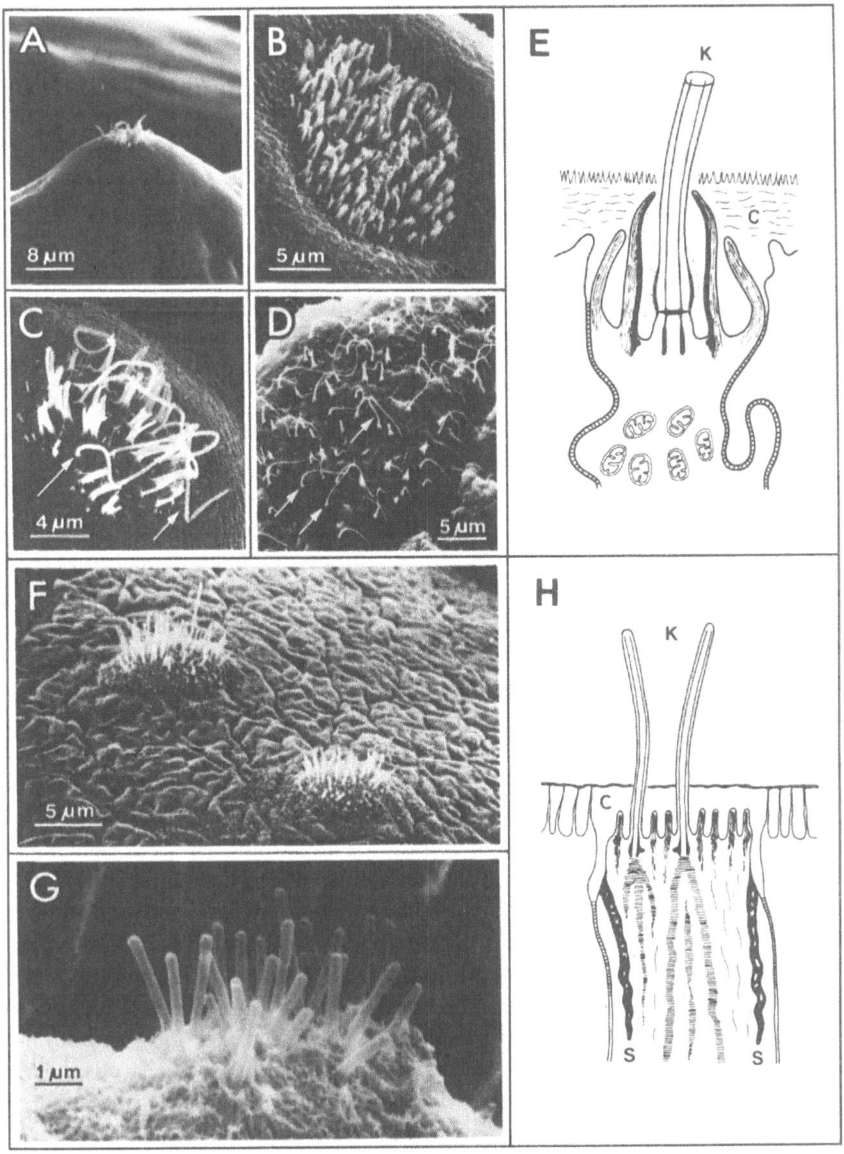

FIGURE 31.3. Presumed hydrodynamic receptor systems of annelids. (A–E) Sensory buds ("segmental sensilla") on the body surface of leeches. The single long kinocilia (arrows in C and D) belong to presumed hydrodynamic receptors, the groups of short kinocilia to presumed chemoreceptors. (A–C) Sensory buds of *Erpobdella montezuma*. (From Blinn et al. 1986; reprinted with the permission of the American Microscopical Society.) (D) Sensory bud of *Hirudo medicinalis*. (From Derosa and Friesen 1981; reprinted with the permission of Biological Bulletin, Woods Hole, MA.) (E) Ciliary region of the presumed hydrodynamic receptor of *Hirudo medicinalis*. (Redrawn from Phillips and Friesen 1982.)

31.2.7 Cephalopods

Among all the invertebrates, the cephalopod's water movement receptor systems are the most sophisticated of those with ciliated sensory cells. They are especially well known as angular acceleration receptor systems (Budelmann et al. 1987), but cephalopods also possess a system analog to the vertebrate lateral line (Budelmann and Bleckmann 1988) and are able to sense vibrations (see Hanlon and Budelmann 1987).

The cephalopod's angular acceleration receptor systems are part of their statocysts. These are highly sophisticated organs that show a remarkable degree of convergence with the vertebrate vestibular apparatus (Budelmann 1988). The two statocysts are fluid-filled cavities within the cranial cartilage and, with the exception of *Nautilus*, have receptor systems for both linear and angular accelerations. Although the organs differ in gross morphology in the various cephalopod genera (Young 1960, 1984; Stephens and Young 1976, 1982), the angular acceleration receptor systems are basically similar throughout all the groups (Budelmann 1977; Budelmann et al. 1987). Each system consists of ciliated sensory cells (hair cells) and an accessory structure (cupula) attached to them (Fig. 31.4).

Several rows of hair cells form a ridge (crista) that winds inside the cavity such that it covers three nearly orthogonal planes. This crista ridge is subdivided into nine sections in the octopod cephalopods and into four sections in the decapod cephalopods. To each crista section, a separate cupula is attached. The hair cells differ from all the ciliated cells described in that each cell carries up to 200 short kinocilia but no stereovilli (Fig. 31.4D,E). All kinocilia are of the same length (6 μm) and are arranged to form an elongated ciliary group. They are morphologically and physiologically polarized in only one direction—that is, at right angles to the long axis of the group, and thus at right angles to the course of the crista ridge (compare also octavolateralis hair cells; Chapter 13, Kroese and van Netten). Within each crista section, rows of oppositely polarized hair cells occur.

The neuronal and synaptic organization of the crista is highly complex in both the octopod and decapod species. This organization is especially well understood in *Octopus* (Fig. 31.4F). Each crista section is composed of primary sensory hair cells as well as two types (large and small) secondary sensory hair cells that are in afferent synaptic contact with two types (large and small) of first-order afferent neurons. All these elements receive many cholinergic and aminergic efferent endings, with most (77%) being inhibitory and few (16%) excitatory (Budelmann

---

(F–H) Presumed hydrodynamic receptors of the lugworm *Arenicola marina*. (F,G) Sensory buds on the body surface, with both presumed hydrodynamic receptors and presumed chemoreceptors. (From Jouin et al. 1985.) (H) Ciliary region of the presumed hydrodynamic receptor. (Redrawn from Jouin et al. 1985.) C, cuticle; K, kinocilium; S, apical cylinder.

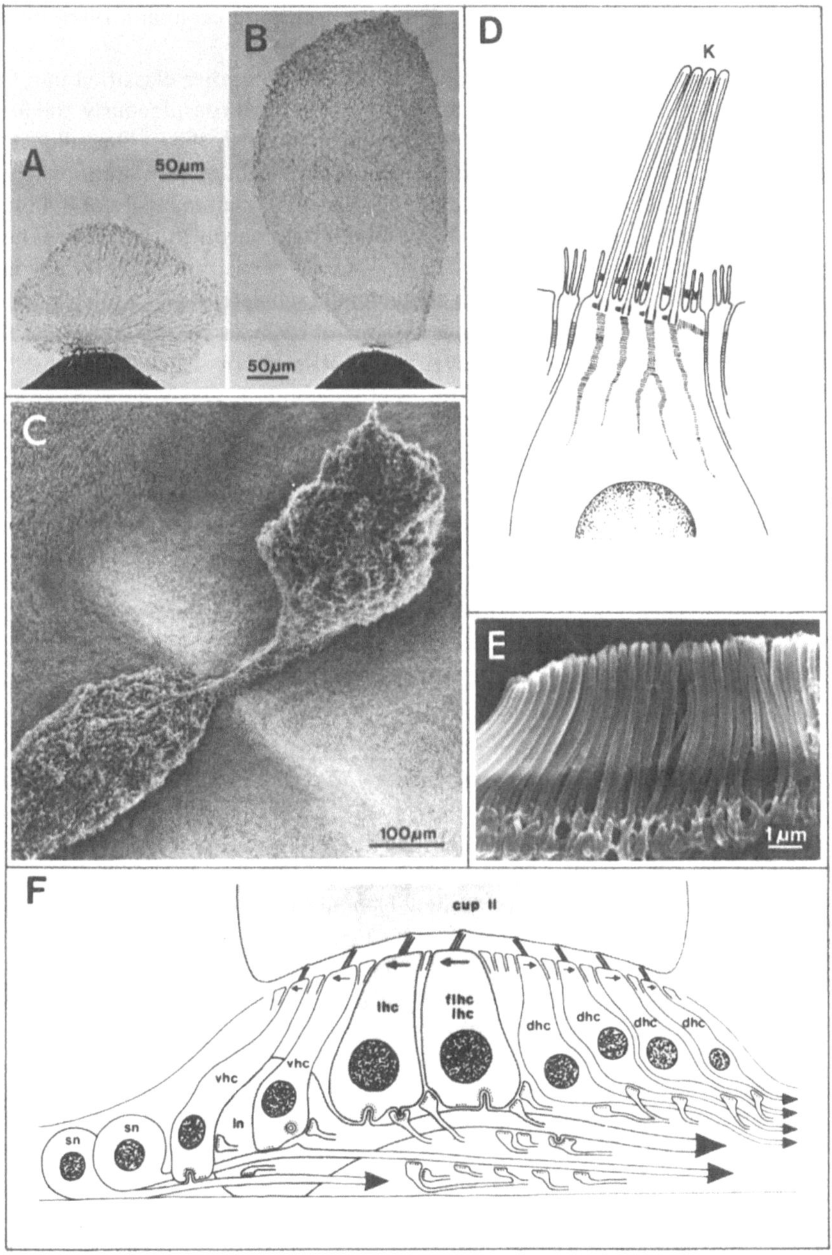
A
50µm
B
50µm
C
100µm
D
K
E
1 µm
F
cup II
sn
sn
vhc
ln
vhc
lhc
flhc
lhc
dhc
dhc
dhc
dhc

and Bonn 1982; Williamson 1985, 1986; Auerbach and Budelmann 1986; Budelmann et al. 1987).

In *Octopus*, the nine crista/cupula systems can be further classified into two subsystems on the basis of crista morphology and, most conspicuously, the form and size of their cupulae (Fig. 31.4A–C) (Budelmann et al. 1987). One subsystem has a small, wide-based cupula with a maximum height of 0.165 mm, whereas the other has a large, narrow-based cupula with a maximum height of 0.48 mm and a much smaller area of attachment (about one-fourth) to the crista. These cupulae protrude freely into the cyst cavity and can be easily moved, like a swinging door, by fluid movements. Both crista/cupula subsystems differ in their sensitivity ranges to angular accelerations by a factor of about 10, the crista with the large cupula being the more sensitive. This dual sensitivity can be understood as an adaptation to the octopod's two forms of locomotion, slowly crawling around and fast swimming by jet propulsion (Williamson and Budelmann 1985).

The cephalopod system that is analog to the vertebrate lateral line consists of 8 (cuttlefish) or 10 (squids) epidermal lines that run in longitudinal direction from the back of the head on to the arms. These lines have been described in *Sepia* and *Loligo* hatchlings (Fig. 31.5A,B) (Sundermann-Meister 1978; Arnold and Williams-Arnold 1980; Sundermann 1983). They are formed by primary sensory cells that each carry an elongated group of long (12 μm) kinocilia. All kinocilia of a single cell are aligned in parallel along the course of the epidermal lines. Although these lines were discovered 60 years ago (Naef 1928), their possible function remained unknown until recently. In *Sepia* hatchlings and *Lolliguncula* adults, stimulation of the lines on the head with local sinusoidal water movements, generated by a vibrating sphere, caused receptor potentials that show all the features that are known from microphonic potentials of the vertebrate lateral lines, with a minimum receptor amplitude threshold of about 0.2 μm water particle displacement at 75 to 100 Hz (Budelmann and Bleckmann 1988). Thus, the cephalopod epidermal lines are strikingly reminiscent of the vertebrate lateral lines and provide an interesting subject for future comparative research. Perhaps study of these lines will also help to elucidate how squids living under conditions of restricted vision (in deep or murky water) detect fast-moving predators or communicate in schooling behavior.

◄

FIGURE 31.4. Angular acceleration receptor system ("crista/cupula system") in the statocysts of the cephalopod *Octopus vulgaris*. (A–C) Small and large cupulae, respectively, attached to two of the nine crista sections. Cupula as seen (A,B) with the light microscope (cross sections) and (C) with the scanning electron microscope. (From Budelmann et al. 1987; reprinted with the permission of The Royal Society.) (D,E) Ciliary region of the receptor hair cell. Note in (D) the mechanical links between the bases of the kinocilia (K). (F) Diagram of a cross section of the crista. cup II, base of large cupula; dhc, primary sensory hair cell; flhc, lhc, secondary sensory hair cells; ln, sn, first-order afferent neurons. Arrows in distal ends of the hair cells indicate the direction of the cell's polarization. (From Budelmann et al. 1987; reprinted with the permission of The Royal Society.)

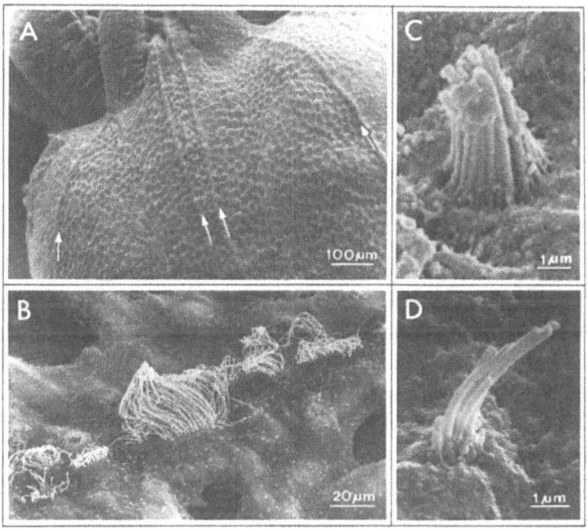

FIGURE 31.5. Receptor systems on the body surface of the cephalopods. (A,B) The lateral line analog: epidermal lines of ciliated cells (arrows) on a head of a young *Loligo forbesi*. (From Hanlon and Budelmann 1987; reprinted with the permission of The University of Chicago Press.) (C,D) Presumed hydrodynamic receptor systems: ciliary bundles on the mantle surface of *Loligo plei*. (C from Hanlon and Budelmann 1987; reprinted with the permission of The University of Chicago Press. D from Hulet 1982; reprinted with the permission of the Academy of Natural Sciences, Philadelphia, PA.)

The other systems that sense vibrations are only little understood, although morphological, physiological, and behavioral data exist (Hanlon and Budelmann 1987). The cuttlefish *Sepia* clearly responds with a color change to a vibration signal of 180 Hz (Dijkgraaf 1963), and the squid *Todarodes* is attracted by vibration signals of 600 Hz (Maniwa 1976); the latter fact is commercially used by fisherman in Japan to catch squid. Physiological evidence for a vibration sense comes from the fact that in *Octopus* the hair cells of the statocyst crista/cupula system are extremely sensitive to vibrations (Maturana and Sperling 1963); they respond in synchrony with vibration stimuli in the range of 20 to 100 Hz at 60 dB (Wolff and Budelmann, unpublished). In addition to the statocyst hair cells, there are others that occur scattered all over the animal's body surface; they bear a bundle of up to 30 short (about 6 µm) kinocilia. Those hair cells might also be involved in water movement detection (Fig. 31.5C,D) (Sundermann-Meister 1978; Arnold and Williams-Arnold 1980; Hulet 1982).

31.2.8 Chaetognaths

Arrowworms are predators of marine plankton, such as copepods and young fish. They wait motionless in the water until a prey or other source of vibra-

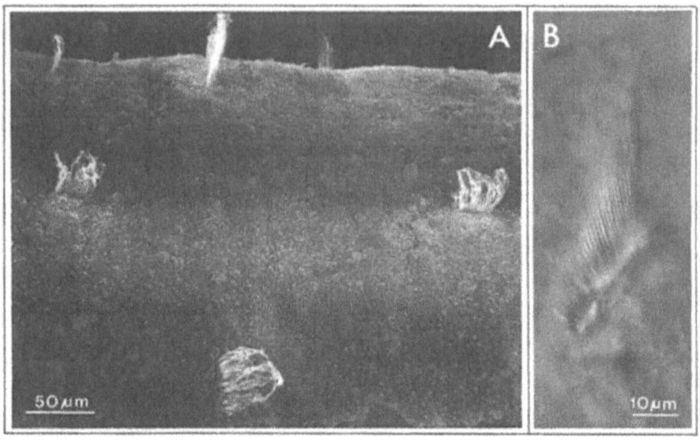

FIGURE 31.6. Hydrodynamic receptor cells of the chaetognath arrowworm *Spadella cephaloptera*. (A) Orientation of the "ciliary fences" on the body surface. (B) Oblique view of a single "ciliary fence" (Nomarski interference contrast). (From Bone and Pulsford 1978; reprinted with the permission of Cambridge University Press.)

tion comes near (3 mm or less). Then they turn in the appropriate direction and grasp for it with great accuracy. This attack occurs best at frequencies of 12 or 30 Hz (in *Spadella*, depending on the species) or 150 Hz (in *Sagitta*) (Horridge and Boulton 1967; Feigenbaum and Reeve 1977); arrowworms may even select specific copepods by the vibration rate the latter produce (Newbury 1972).

The underlying sensory structures on the animal's body surface have long been known, but there has been confusion about their terminology (Horridge and Boulton 1967; Reisinger 1969). From more recent papers (Bone and Pulsford 1978; Feigenbaum 1978) it is clear that the most numerous sensory structures are "fences" of 75 to 100 stiff kinocilia of up to 100 μm in length (Fig. 31.6). At right angles these "fences" look like the tapering "bristles" previously described. Ciliary rootlets extend far down into the cell. The plane of polarization of all the kinocilia of a single fence is at right angles to the long axis of the fence (the polarization direction of each fence is not known). The arrangement of the fences on the body, relative to the long axis of the animal, is either parallel to it, or across, the latter fences being about twice as abundant (Fig. 31.6A).

Each kinocilium of a fence belongs to a separate cell; it is still controversial whether these are primary or secondary in nature (Reisinger 1969; Bone and Pulsford 1978). It is almost certain, however, that these ciliary fence organs are able to detect the direction of water movements. In *Spadella*, a few other, less abundant epidermal structures occur, one of which, on the head and tail, might be a second type of vibration receptor. It is a cell with a modified cilium, about 20 μm in length and 0.5 μm thick, with a narrow joint and filled with microtubules (Bone and Pulsford 1978).

31.2.9 Urochordates

In these lower chordates, cupula sense organs exist that very closely resemble the vertebrate's free neuromast organ. In *Ciona*, they occur in abundance (75 to 100) in the exhalant siphon (Bone and Ryan 1978). Each forms a small hillock and bears an elongated cupula up to 250 μm in height (Fig. 31.7A); this cupula can be bent back and forth by water movements. The fibrous cupula material consists of carboxylated polysaccharides with a significant protein component. The hillock is formed by many ciliated sensory cells. Each cell bears a long kinocilium that passes at least 100 μm into the cupula; it arises from a deep invagination of the apical part of the cell (Fig. 31.7B). The internal $9 \times 2 + 2$ tubuli probably modify soon after the cilium passes into the cupula. Ciliary rootlets do not occur. The apical region of the cell is finely granular, is less electron-dense than the rest of the cell, and contains no mitochondria or other large cell organelles. At its apical surface, a collar of short microvilli surrounds the cilium. Like all sessile ascidian tunicates, *Ciona* is very sensitive to water movements, even after ablation of the brain; it responds by closing the siphons and retracting the body when stimulated by vibrations (25 to 400 Hz) 4 cm away from the exhalant siphon. The cupular organs are most certainly the structures that sense vibrations (e.g., from small predators); they may also be involved in monitoring and regulating the water current through the atrium cavity.

In the larvacean tunicate *Oikopleura*, another vibration-sensitive receptor ("Langerhans receptor") has been described on either side of the animal's trunk (Fig. 31.7C,D) (Bone and Ryan 1979). Its receptor cell forms a crater out of which arises a long (50 to 70 μm) and thick (2 to 3 μm) "bristle" that presumably is a modified cilium; it is filled with many parallel tubules. The bristle inserts by a narrow stalk to which run striated rootlets that recurve around the edge of the crater of a less electron-dense apical region of the cell. The receptor cell is a secondary sensory cell that is in electrical contact via a gap junction with a neuron of the caudal ganglion.

31.2.10 Cephalochordates

In the anterior region of the lancelet *Branchiostoma* (*Amphioxus*), there are two different types of ciliated secondary sensory cells that might well be involved in water current detection. One, on the buccal cirrhi, carries a normal kinocilium of up to 15 μm in length, and the other, on the velar tentacles, bears a short and straight modified cilium. The latter is about twice as thick as the former and contains a large number (up to 90 or more) of irregularly arranged microtubuli that are surrounded by nine tubuli doublets. In both types of cells, the base of the cilium is surrounded by numerous elongated microvilli (Bone and Best 1978; see also Chapter 6, Jørgensen). Experimental evidence suggests that these cells are mechanoreceptors (Bone 1961; Guthrie 1975), and it seems reasonable to suggest that they normally respond to water flow over the atrial surface or to inhalant water currents (Bone and Best 1978).

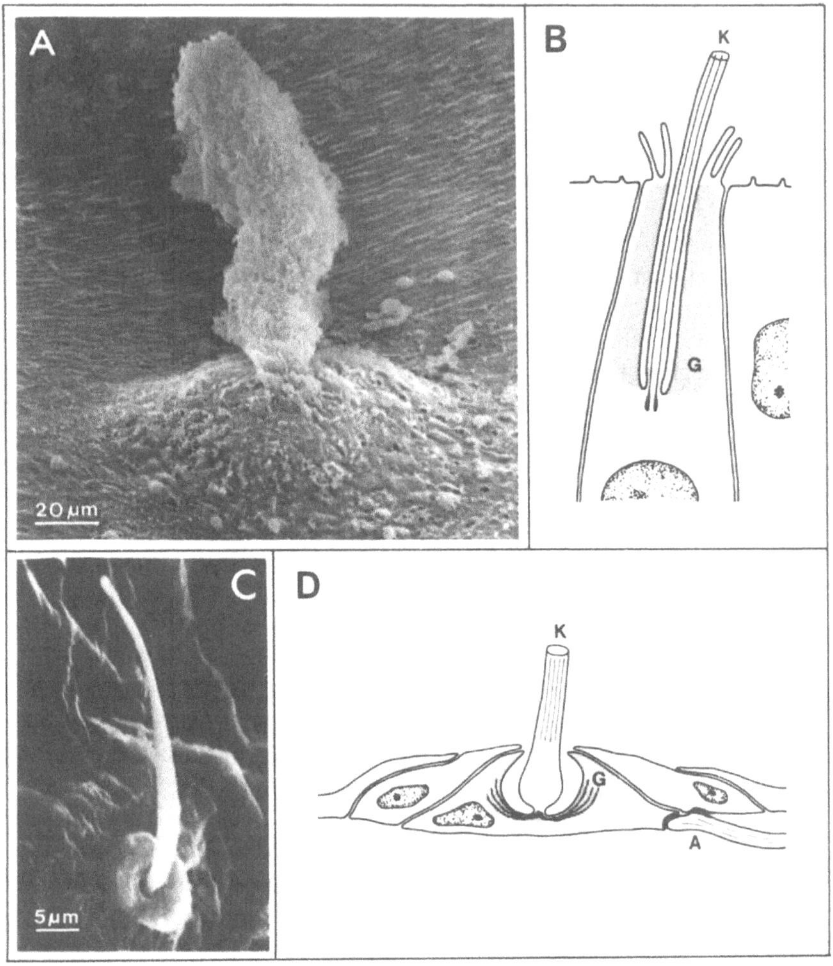

FIGURE 31.7. Hydrodynamic receptor systems of tunicates. (A) Cupula organ in the exhalant siphon of *Ciona intestinalis*. (From Bone and Ryan 1978; reprinted with the permission of The Zoological Society of London.) (B) Distal part of the ciliated receptor cell of the cupula organ of *Ciona intestinalis*. K, kinocilium; G, granular region. (Redrawn from Bone and Ryan 1978.) (C) "Langherhans receptor" of *Oikopleura labradoriensis*. (From Bone and Ryan 1979; reprinted with the permission of Cambridge University Press.) (D) Diagram of the "Langerhans receptor" and its two accessory cells. (Redrawn from Bone and Ryan 1979.) A, axon of central cell, in gap junction contact with receptor and accessory cell; K, modified kinocilium; G, granular region.

31.3 Receptor Systems with Cuticular Sensory Hairs

Hydrodynamic receptor systems are especially well known on the external body surface of crustacea (for references see Laverack 1968, 1976; Bush and Laverack 1982). As in all arthropods, their body surface is covered by a cuticular layer; the morphology of the flexible external part of the receptor system, therefore, is basically different from the cilia of all the nonarthropod species described. It is a cuticular hair that is mechanically coupled via a flexible joint, and often accessory cuticular elements, to a modified ciliary structure of one to four sensory cells (e.g., Schöne and Steinbrecht 1968; Ball and Cowan 1977; Vedel 1985; see also Bush and Laverack 1982). Of special importance for the process of stimulus transduction is the geometry of both the hair shaft and the hair joint; that of the latter often determines the range of movement of the hair (see Markl 1978). The hair shaft varies greatly in size (50 to 2,000 μm) and form; its surface area is often increased by many filamentous projections (Fig. 31.8) (Bush and Laverack 1982; Derby 1982; Vedel 1985). The cuticular hairs can easily be bent by fluid displacements whereby they mechanically affect and stimulate the sensory cells.

Most work on water movement-sensitive hairs has been done in decapod crustacea, particularly in lobsters and crayfish. The hairs occur all over their body surface (Laverack 1962a,b; Offutt 1970; Vedel and Clarac 1976; Derby 1982), antenna and antennules (Tazaki and Ohnishi 1974; Tazaki 1977; Vedel 1985; Phillips and Macmillan 1987), chelae (Laverack 1963; Solon and Cobb 1980; Tautz and Sandeman 1980), carapace (Laverack 1963; Mellon 1963), and telson (Wiese 1976). Only a few data are available for nondecapod crustacea (see Bush and Laverack 1982, for most references; Strickler and Bal 1973; Crouau 1985, 1986). The hairs are extremely sensitive to medium displacements, with a working range that is species-dependent (between 0.05 Hz and more than 300 Hz) and a minimum receptor amplitude threshold of about 0.2 μm water particle displacement (Tautz and Sandeman 1980). Behavioral experiments (Offutt 1970), however, indicate that receptors may exist with thresholds 3 orders of magnitude below (see Tautz and Sandeman 1980). Most hairs are innervated by two sensory cells, each responding to a different (opposite) deflection of the hair (see Laverack 1976; Bush and Laverack 1982).

In some decapod crustacea, there is a hydrodynamic receptor system resembling the vertebrate lateral line canals (Foxton 1969; Ball and Cowan 1977; Denton and Gray 1985). Certain penaeid shrimps have long second antennae, the proximal one-third of which is held roughly at right angles to the body, and the distal two-thirds of which, after a flexure, trails parallel to the body when the animal swims. On the distal trailing part of the antenna, paired curved, probably noninnervated, cuticular hairs form an almost closed tube (Fig. 31.9). At intervals along the tube, flexible cuticular hairs of gradually varying sizes project into the lumen of the tube (Fig. 31.9B). Each of these hairs is connected to four sensory cells and is very sensitive to water displacement. Thus, these animals have a hydrodynamic receptor system that, different from the vertebrate lateral lines

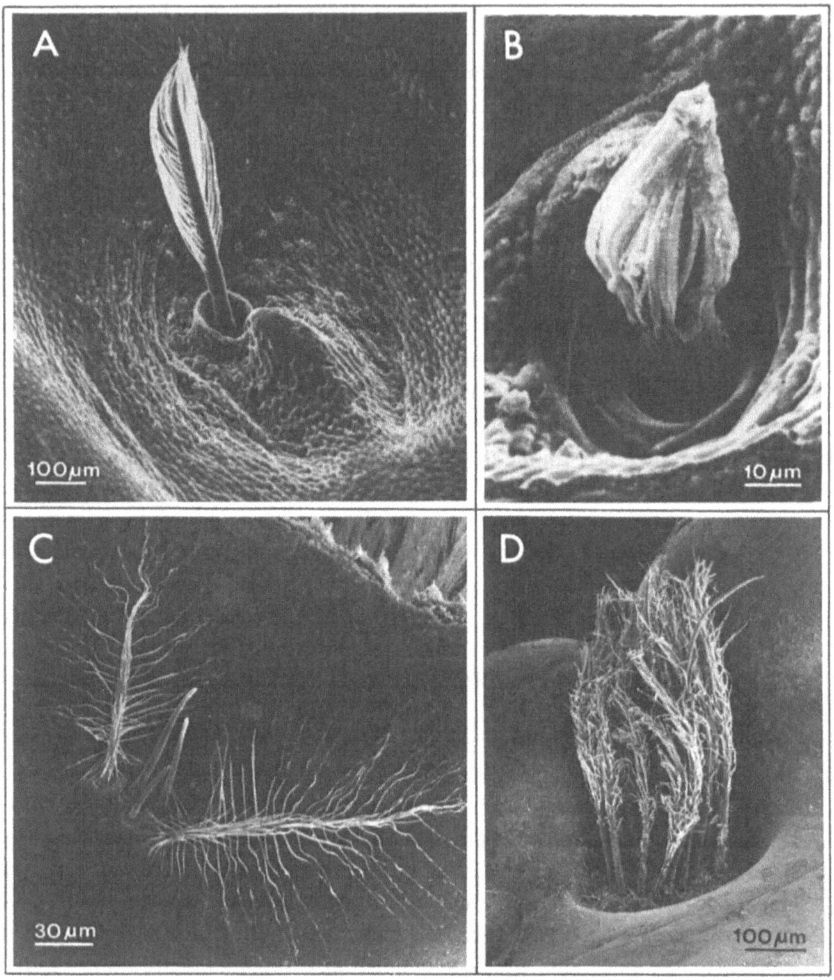

FIGURE 31.8. Hydrodynamic cuticular hair receptors of crustacea. (A) Feathered hair of the rock lobster *Palinurus vulgaris*. (From Vedel and Clarac 1976; © Gordon and Breach Science Publishers, S.A. Reprinted by permission.) (B) Hair-fan organ on the carapace of the lobster *Homarus gammarus*. (Courtesy of M.S. Laverack.) (C,D) Hair-pit organ of the young (C) and adult (D) crayfish *Cherax destructor*. (From Tautz and Sandeman 1980; reprinted with the permission of The Company of Biologists Ltd.)

and most of the corresponding invertebrate systems, is widely spaced from, and mechanically independent of, their body, making it more sensitive to changes in the position of an external source of vibration (Denton and Gray 1985).

In the statocysts of only decapod crustacea, special water movement receptors occur ("thread hairs" and "free-hook hairs"). They are basically similar in gross

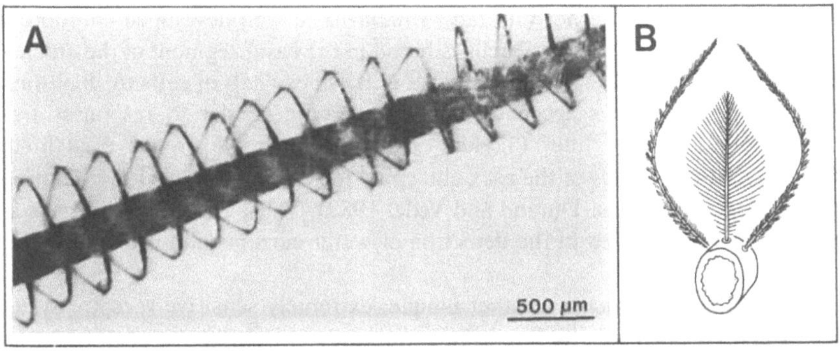

FIGURE 31.9. Lateral line-like receptor system on the antennae of certain of the penaeid crustacea. (A) Microphotograph of part of the trailing part of the antennal flagellum of *Funchalia* sp. (From Denton and Gray 1985; reprinted with the permission of The Royal Society.) (B) Diagram of a cross section of the trailing part of the antennal flagellum of *Funchalia villosa*, showing paired arched hairs forming a tubelike canal, and a sensory hair projecting into the lumen of this canal. (Redrawn from Foxton 1969.)

morphology to the superficial receptors. Because of their location in a bowl-shaped cavity (lobsters) or in canals (crabs), however, they serve as angular acceleration receptors (see Budelmann 1988 for references).

Besides crustacea, behavioral reactions of the mosquito larva *Chaoborus* indicate the presence of underwater vibration receptors in aquatic insects as well (Horridge 1966; Swift and Fedorenko 1975; Giguere and Dill 1979). Even in total darkness, these larvae feed on copepods and cladocerans and, within a distance of 5 mm, are clearly attracted by an object vibrating with 200 Hz and 5 µm amplitude. Although the relevant sense organs have not been identified, it might well be that some of the cuticular hairs that are abundant on the animal's surface (or chordotonal organs; see Section 31.4.1) serve as vibration receptors.

31.4 Receptor Systems That Use Body Appendages with Associated Sensory Cells

Flexible body appendages, together with associated sensory cells, are involved in water movement detection at least in crustacea and echinoderms. But this phenomenon might be more widespread among invertebrates; shell-carrying snails, for example, may sense shell displacements (brought about by water movements) via muscle proprioceptors (see Creutzberg 1975 for references).

31.4.1 Crustacea

Chordotonal organs are widespread among crustacea and are generally associated with joints of antenna, legs, or other body appendages (see Bush and

Laverack 1982). In the hermit crab *Petrochirus*, a well-developed chordotonal organ with two sets of sensory cells is found in the basal segment of the antennal flagellum (Taylor 1967a). The responses of these two sets of cells to small rapid flagellar displacements are antagonistic and quite similar to responses from the vertebrate lateral lines (Taylor 1967b). Comparable organs exist in the antenna and antennules of the rock and spiny lobsters (Laverack 1964; Hartman and Austin 1972; Rossi-Durand and Vedel 1982). Thus, chordotonal organs are most probably involved in the detection of water currents and near-field vibrations as well.

In the crayfish *Astacus*, a rather unique, extremely sensitive receptor system has been described for the intersegmental joints of the flagella of the first and second antenna (Tautz et al. 1981; Masters et al. 1982; Bender et al. 1984). A water motion first bends the flagellum. This bending, in turn, bends noninnervated procumbent "feathered hairs" that span the intersegmental joints of the flagellum. The force during bending the hairs is then most probably transmitted as stress and strain of the cuticle onto sensory cells of nearby smooth cuticular hairs.

31.4.2 Arthropods

In some arthropods, specialized organs are known to be involved in surface wave perception. These are the tarsal scolopidial organs of the back swimmer *Notonecta*, the Johnston organs in the antennae of whirligig beetles, and the metatarsal lyriform (slit sense) organs of some spiders (see also Bleckmann 1985 for references). Depending on broad definition, these organs may also be regarded as hydrodynamic receptors.

31.4.3 Echinoderms

Many behavioral data show that echinoderms are sensitive to various kinds of water movements (see Moore and Cobb 1986 for references). A recent neurophysiological and behavioral study on the brittle star *Ophiura* (Moore and Cobb 1986) clearly demonstrates that these animals are able to detect near-field vibrations (down to a few hertz) and far-field propagated pressure waves; they also can distinguish between different flow rates of water currents. This ability could be attributed to the large arm spines. Initially, cells with a long cilium were thought to be the responsible sensory structure (Cobb and Moore 1986). However, since all reactions persist after removal of these cells by stripping the epithelium, it is more likely that proprioceptors exist within the spines that monitor the spine's deflection (Moore and Cobb 1986).

31.5 Concluding Remarks

This review on hydrodynamic and presumed hydrodynamic receptor systems clearly shows that in invertebrates no characteristic morphological picture of such a receptor system exists. Two reasons for this are obvious: (1) invertebrates are not an evolutionary homogeneous monophyletic group, and (2) the definition of a hydrodynamic receptor system used in this review is very general and thus covers a broad spectrum of mechanoreceptors that sense motion of a fluid medium. However, as long as more precise physiological data are still missing for most of the invertebrate systems, it is premature to try any more specific differentiation of the hydrodynamic receptor systems. Basically, just two criteria are essential for such a receptor system: (1) the system must have a flexible sensory structure that can be moved by fluid motions, and (2) this flexible structure must project beyond the (frequency-dependent) stationary boundary layer to transmit the mechanical energy effectively from the fluid medium to the neuronal transducer (Tautz 1979; Tautz and Sandeman 1980; see also Chapter 9, Kalmijn). This surface boundary effect is especially critical for the systems with flexible ciliary structures, since these are often quite short. Surprisingly, data on the ciliary length are missing in many papers. Special effort should be made in any future research on these systems to check for possible accessory cupula structures.

In most of the systems with uniciliated sensory cells, either the proximal part of the kinocilium is surrounded by a collar of stereovilli that are densely packed with microfilaments (coelenterates, Figs. 31.1A,B; flatworms, Fig. 31.2; leeches, Fig. 31.3E) or it is deeply embedded into a modified distal part of the receptor cell (tunicates, Fig. 31.7). In all "collar-type" receptors, an amorphous, mucous or fibrillar material has been mentioned inside the collar, between the kinocilium and the stereovilli. It is of particular interest to check (with proper fixation methods) whether this "material" mechanically bridges the space between the kinocilium and the stereovilli, in a way similar to the mechanical links that have recently been described in the vertebrate vestibular and cochlear hair cells and that have been suggested to play an essential part in the transduction process of the hair cells (e.g., Hudspeth 1983; Comis et al. 1985; Neugebauer and Thurm 1985; Ernstson and Smith 1986). Signs already exist that such mechanical links also occur in the cephalopod hair cells between all the kinocilia of a single hair cell (Budelmann, unpublished; see Fig. 31.4D).

A basic problem in water movement detection is the discrimination between stimulation generated by an external fluid motion and that generated by a movement of the animal itself. So far, a peripheral efferent innervation of the afferent receptor units (compare Chapter 22, Roberts and Meredith) has been described only in the cephalopod angular acceleration receptor system (Williamson 1985, 1986; Budelmann et al. 1987). It will be interesting to see whether an efferent innervation also exists in any of the other invertebrate hydrodynamic receptor systems, or whether it is limited to only the more sophisticated systems of fast-moving invertebrates, such as cephalopods.

Acknowledgments. The author thanks Dr. Jürgen Tautz, Konstanz, for critically reading the manuscript.

References

Arnold JM, Williams-Arnold LD (1980) Development of the ciliature pattern on the embryo of the squid *Loligo pealei*: A scanning electron microscope study. Biol Bull 159:102–116.

Auerbach B, Budelmann BU (1986) Evidence for acetylcholine as a neurotransmitter in the statocyst of *Octopus vulgaris*. Cell Tissue Res 243:429–436.

Ball EE, Cowan AN (1977) Ultrastructure of the antennal sensilla of *Acetes* (Crustacea, Decapoda, Natantia, Sergestidae). Phil Trans R Soc Lond B Biol Sci 277:429–456.

Bedini C, Ferrero E, Lanfranchi A (1975) Fine structural observations on the ciliary receptors in the epidermis of three otoplanid species (Turbellaria, Proseriata). Tissue Cell 7:253–266.

Bender M, Gnatzy W, Tautz J (1984) The antennal feathered hairs in the crayfish: A non-innervated stimulus transmitting system. J Comp Physiol A 154:45–47.

Bergeijk WA van (1964) Directional and nondirectional hearing in fish. In: Tavolga WN (ed) Marine Bio-Acoustics. Oxford, U.K.: Pergamon Press, pp. 281–299.

Bleckmann H (1985) Perception of water surface waves: How surface waves are used for prey identification, prey localization, and intraspecific communication. In: Ottoson D (ed) Progress in Sensory Physiology, Vol. 5. New York: Springer-Verlag, pp. 147–166.

Blinn DW, Wagner VT, Grim JN (1986) Surface sensilla on the predaceous fresh-water leech *Erpobdella montezuma*: Possible importance in feeding. Trans Am Microsc Soc 105:21–30.

Bone Q (1961) The organization of the atrial nervous system of amphioxus (*Branchiostoma lanceolatum* (Pallas). Phil Trans R Soc Lond B Biol Sci 243:241–269.

Bone Q, Best ACG (1978) Ciliated sensory cells in amphioxus (*Branchiostoma*). J Mar Biol Assoc UK 58:479–486.

Bone Q, Pulsford A (1978) The arrangement of ciliated sensory cells in *Spadella* (Chaetognatha). J Mar Biol Assoc UK 58:565–570.

Bone Q, Ryan KP (1978) Cupular sense organs in *Ciona* (Tunicata: Ascidiacea). J Zool Lond 186:417–429.

Bone Q, Ryan KP (1979) The Langerhans receptor of *Oikopleura* (Tunicata, Larvacea). J Mar Biol Assoc UK 59:69–75.

Budelmann BU (1977) Structure and function of the angular acceleration receptor systems in the statocysts of cephalopods. Symp Zool Soc Lond 38:309–324.

Budelmann BU (1988) Morphological diversity of equilibrium receptor systems in aquatic invertebrates. In: Atema J, Fay RR, Popper AN, Tavolga WN (eds) Sensory Biology of Aquatic Animals. New York: Springer-Verlag, pp. 757–782.

Budelmann BU, Bleckmann H (1988) A lateral line analogue in cephalopods: Water waves generate microphonic potentials in the epidermal head lines of *Sepia* and *Lolliguncula*. J Comp Physiol A 164:1–5.

Budelmann BU, Bonn U (1982) Histochemical evidence for catecholamines as neurotransmitters in the statocyst of *Octopus vulgaris*. Cell Tissue Res 227:475–483.

Budelmann BU, Sachse M, Staudigl M (1987) The angular acceleration receptor system of the statocyst of *Octopus vulgaris*: Morphometry, ultrastructure, and neuronal and synaptic organization. Phil Trans R Soc Lond B Biol Sci 315:305–344.

Bush BMH, Laverack MS (1982) Mechanoreception. In: Atwood HL, Sandeman DC (eds) The Biology of Crustacea, Vol. 3. Neurobiology: Structure and Function. New York: Academic Press, pp. 399–468.

Cobb JLS, Moore A (1986) Comparative studies on receptor structure in the brittle star *Ophiura ophiura*. J Neurocytol 15:97–108.
Comis SD, Pickles JO, Osborne MP (1985) Osmium tetroxide postfixation in relation to the crosslinkage and spatial organization of stereocilia in the guinea-pig cochlea. J Neurocytol 14:113–130.
Creutzberg F (1975) Orientation in space: Animals. Invertebrates. In: Kinne O (ed) Marine Ecology, Vol. 2. Physiological Mechanisms, Part 2. London: John Wiley, pp. 555–656.
Crisp M (1981) Epithelial sensory structures of trochids. J Mar Biol Assoc UK 61:95–106.
Crouau Y (1985) Étude du comportement rhéotaxique d'un mysidacé cavernicole. Crustaceana 50:7–10.
Crouau Y (1986) Antennular mechanosensitivity in a cavernicolous mysid crustacean. J Crustacean Biol 6:158–165.
Daly JM (1973) The ability to locate a source of vibrations as a prey-capture mechanism in *Harmothoe imbricata* (Annelida, Polychaeta). Mar Behav Physiol 1:305–322.
Denton EJ, Gray J (1985) Lateral-line-like antennae of certain of the Penaeidea (Crustacea, Decapoda, Natantia). Proc R Soc Lond B 226:249–261.
Derby CD (1982) Structure and function of cuticular sensilla of the lobster *Homarus americanus*. J Crustacean Biol 2:1–21.
Derosa YS, Friesen WO (1981) Morphology of leech sensilla: Observations with the scanning electron microscope. Biol Bull 160:383–393.
Dijkgraaf S (1963) Versuche über Schallwahrnehmung bei Tintenfischen. Naturwissenschaften 50:50.
Ehlers U, Ehlers B (1977) Monociliary receptors in interstitial Proseriata and Neorhabdocoela (Turbellaria, Neoophora). Zoomorphologie 86:197–222.
Ernstson S, Smith CA (1986) Stereo-kinociliar bonds in mammalian vestibular organs. Acta Otolaryngol (Stockh) 101:395–402.
Farnesi RM, Marinelli M, Tei S, Vagnetti D (1982) Ultrastructural aspects of mechano- and chemoreceptors in *Branchiobdella pentodonta* (Annelida, Oligochaeta). J Morphol 173:237–245.
Feigenbaum DL (1978) Hair-fan patterns in the Chaetognatha. Can J Zool 56:536–546.
Feigenbaum D, Reeve MR (1977) Prey detection in the Chaetognatha: Response to a vibrating probe and experimental determination of attack distance in large aquaria. Limnol Oceanogr 22:1052–1058.
Foxton P (1969) The morphology of the antennal flagellum of certain of the Penaeidea (Decapoda, Natantia). Crustaceana 16:33–42.
Friesen WO (1981) Physiology of water motion detection in the medicinal leech. J Exp Biol 92:255–275.
Frings H (1964) Problems and prospects in research on marine invertebrate sound production and reception. In: Tavolga WN (ed) Marine Bio-Acoustics. Oxford, U.K.: Pergamon Press, pp. 155–173.
Frings H, Frings M (1967) Underwater sound fields and behavior of marine invertebrates. In: Tavolga WN (ed) Marine Bio-Acoustics, Vol. 2. Oxford, U.K.: Pergamon Press, pp. 261–282.
Giguere LA, Dill LM (1979) The predatory response of *Chaoborus* larvae to acoustic stimuli, and the acoustic characteristics of their prey. Z Tierpsychol 50:113–123.
Graefe G (1964) Zur Anemonen-Fisch Symbiose, nach Freilanduntersuchungen bei Eilat/Rotes Meer. Z Tierpsychol 21:468–485.

Guthrie DM (1975) The physiology and structure of the nervous system of amphioxus (the lancelet) *Branchiostoma lanceolatum* (Pallas). Symp Zool Soc Lond 36:43–80.

Hanlon RT, Budelmann BU (1987) Why cephalopods are probably not "deaf." Am Nat 129:312–317.

Hartman HB, Austin WD (1972) Proprioceptor organs in the antennae of decapod crustacea. J Comp Physiol 81:187–202.

Haszprunar G (1985) The fine structure of the abdominal organs of Pteriomorpha (Mollusca, Bivalvia). J Moll Stud 51:315–319.

Hernandez-Nicaise ML (1974) Ultrastructural evidence for a sensory-motor neuron in Ctenophora. Tissue Cell 6:43–47.

Horridge GA (1965) Non-motile sensory cilia and neuromuscular junctions in a ctenophore independent effector organ. Proc R Soc Lond B 162:333–350.

Horridge GA (1966) Some recently discovered underwater vibration receptors in invertebrates. In: Barnes H (ed) Some Contemporary Studies in Marine Science. London: Allen and Unwin, pp. 395–405.

Horridge GA, Boulton PS (1967) Prey detection by Chaetognatha via a vibration sense. Proc R Soc Lond B 168:413–419.

Hudspeth AJ (1983) Mechanoelectrical transduction by hair cells in the acousticolateralis sensory system. Annu Rev Neurosci 6:187–215.

Hulet WH (1982) Commentary on the international symposium on functional morphology of cephalopods. Malacologia 23:203–208.

Josephson RK (1961) The response of a hydroid to weak water-borne disturbances. J Exp Biol 38:17–27.

Jouin C, Tchernigovtzeff C, Baucher MF, Toulmond A (1985) Fine structure of probable mechano- and chemoreceptors in the caudal epidermis of the lugworm *Arenicola marina* (Annelida, Polychaeta). Zoomorphology 105:76–82.

Kalmijn AJ (1988) Hydrodynamic and acoustic field detection. In: Atema J, Fay RR, Popper AN, Tavolga WN (eds) Sensory Biology of Aquatic Animals. New York: Springer-Verlag, pp. 83–130.

Kolle-Kralik U, Ruff PW (1967) Vibrotaxis von *Amoeba proteus* (Pallas) im Vergleich mit der Zilienschlagfrequenz der Beutetiere. Protistologica 3:319–323.

Laverack MS (1962a) Responses of cuticular sense organs of the lobster, *Homarus vulgaris* (Crustacea). I. Hair-peg organs as water current receptors. Comp Biochem Physiol 5:319–325.

Laverack MS (1962b) Responses of cuticular sense organs of the lobster, *Homarus vulgaris* (Crustacea). II. Hair-fan organs as pressure receptors. Comp Biochem Physiol 6:137–145.

Laverack MS (1963) Responses of cuticular sense organs of the lobster, *Homarus vulgaris* (Crustacea). III. Activity invoked in sense organs of the carapace. Comp Biochem Physiol 10:261–272.

Laverack MS (1964) The antennular sense organs of *Panulirus argus*. Comp Biochem Physiol 13:301–321.

Laverack MS (1968) On the receptors of marine invertebrates. Oceanogr Mar Biol Annu Rev 6:249–324.

Laverack MS (1976) External proprioceptors. In: Mill PJ (ed) Structure and Function of Proprioceptors in the Invertebrates. London: Chapman and Hall, pp. 1–63.

Lyons KM (1973) Collar cells in planula and adult tentacle ectoderm of the solitary coral *Balanophyllia regia* (Anthozoa, Eupsammiidae). Z Zellforsch Mikrosk Anat 145: 57–74.

Maniwa Y (1976) Attraction of bony fish, squid and crab by sound. In: Schuijf A, Hawkins AD (eds) Sound Reception in Fish. Amsterdam: Elsevier, pp. 271–283.
Mariscal RN (1974) Scanning electron microscopy of the sensory surface of the tentacles of sea anemones and corals. Z Zellforsch Mikrosk Anat 147:149–156.
Markl H (1973) Leistungen des Vibrationssinnes bei wirbellosen Tieren. Fortschr Zool 21:100–120.
Markl H (1978) Adaptive radiation of mechanoreceptors. In: Ali MA (ed) Sensory Ecology. Review and Perspectives. New York: Plenum, pp. 319–344.
Masters WM, Aicher B, Tautz J, Markl H (1982) A new type of water vibration receptor on the crayfish antenna. J Comp Physiol A 149:409–422.
Maturana HR, Sperling S (1963) Unidirectional response to angular acceleration recorded from the middle cristal nerve in the statocyst of *Octopus vulgaris*. Nature 197:815–816.
Mellon D (1963) Electrical responses from dually innervated tactile receptors on the thorax of the crayfish. J Exp Biol 40:137–148.
Moore A, Cobb JLS (1986) Neurophysiological studies on the detection of mechanical stimuli in *Ophiura ophiura*. J Exp Mar Biol Ecol 104:125–141.
Naef A (1928) Die Cephalopoden (Embryologie). Fauna Flora Golf Neapel 35:1–357.
Neugebauer DC, Thurm U (1984) Chemical dissection of stereovilli from fish inner ear reveals differences from intestinal microvilli. J Neurocytol 13:797–808.
Neugebauer DC, Thurm U (1985) Interconnections between the stereovilli of the fish inner ear. Cell Tissue Res 240:449–453.
Newbury TK (1972) Vibration perception by Chaetognaths. Nature 236:459–460.
Offutt GC (1970) Acoustic stimulus perception by the American lobster, *Homarus americanus* (Decapoda). Experientia 26:1276–1278.
Pariselle A, Matricon-Gondran M (1985) A new type of ciliated receptor in the cercariae of *Nicolla gallica* (Trematoda). Z Parasitenkd 71:353–364.
Passano LM, Pantin CFA (1955) Mechanical stimulation in the sea-anemone *Calliactis parasitica*. Proc R Soc Lond B 143:226–238.
Peteya DJ (1975) The ciliary-cone sensory cell of anemones and cerianthids. Tissue Cell 7:243–252.
Phillips BF, Macmillan DL (1987) Antennal receptors in puerulus and postpuerulus stages of the rock lobster *Panulirus cygnus* (Decapoda: Palinuridae) and their potential role in puerulus navigation. J Crustacean Biol 7:122–135.
Phillips CE, Friesen WO (1982) Ultrastructure of the water-movement-sensitive sensilla in the medicinal leech. J Neurobiol 13:473–486.
Reisinger E (1969) Ultrastrukturforschung und Evolution. Ber Physik Med Ges Würzburg NF 77:5–47.
Rossi-Durand C, Vedel JP (1982) Antennal proprioception in the rock lobster *Palinurus vulgaris*: Anatomy and physiology of a bi-articular chordotonal organ. J Comp Physiol A 145:505–516.
Santer RM, Laverack MS (1971) Sensory innervation of the tentacles of the polychaete, *Sabella pavonina*. Z Zelforsch Mikrosk Anat 122:160–171.
Schaeffer AA (1916) On the feeding habits of ameoba. J Exp Zool 20:529–584.
Schöne H, Steinbrecht RA (1968) Fine structure of statocyst receptor of *Astacus fluviatilis*. Nature 220:184–186.
Solon MH, Cobb JS (1980) The external morphology and distribution of cuticular hair organs on the claws of the American lobster, *Homarus americanus* (Milne-Edwards). J Exp Mar Biol Ecol 48:205–215.

Stephens PR, Young JZ (1976) The statocyst of *Vampyroteuthis infernalis* (Mollusca: Cephalopoda). J Zool Lond 180:565–588.
Stephens PR, Young JZ (1982) The statocyst of the squid *Loligo*. J Zool Lond 197:241–266.
Strickler JR, Bal AK (1973) Setae of the first antennae of the copepod *Cyclops scutifer* (Sars): Their structure and importance. Proc Natl Acad Sci USA 70:2656–2659.
Sundermann G (1983) The fine structure of epidermal lines on arms and head of postembryonic *Sepia officinalis* and *Loligo vulgaris* (Mollusca, Cephalopoda). Cell Tissue Res 232:669–677.
Sundermann-Meister G (1978) Ein neuer Typ von Cilienzellen in der Haut von spätembryonalen und juvenilen *Loligo vulgaris* (Mollusca, Cephalopoda). Zool Jahrb Abt Anat Ontog Tiere 99:493–499.
Swift MC, Fedorenko AY (1975) Some aspects of prey capture by *Chaoborus* larvae. Limnol Oceanogr 20:418–425.
Tardent P, Schmid V (1972) Ultrastructure of mechanoreceptors of the polyp *Coryne pintneri* (Hydrozoa, Athecata). Exp Cell Res 72:265–275.
Tautz J (1979) Reception of particle oscillation in a medium—an unorthodox sensory capacity. Naturwissenschaften 66:452–461.
Tautz J, Sandeman DC (1980) The detection of waterborne vibration by sensory hairs on the chelae of the crayfish. J Exp Biol 88:351–356.
Tautz J, Masters WM, Aicher B, Markl H (1981) A new type of water vibration receptor on the crayfish antenna. I. Sensory Physiology. J Comp Physiol 144:533–541.
Taylor RC (1967a) The anatomy and adequate stimulation of a chordotonal organ in the antennae of a hermit crab. Comp Biochem Physiol 20:709–717.
Taylor RC (1967b) Functional properties of the chordotonal organ in the antennal flagellum of a hermit crab. Comp Biochem Physiol 20:719–729.
Tazaki K (1977) Nervous responses from mechanosensory hairs on the antennal flagellum in the lobster, *Homarus gammarus* (L.). Mar Behav Physiol 5:1–18.
Tazaki K, Ohnishi M (1974) Responses from tactile receptors in the antenna of the spiny lobster *Panulirus japonicus*. Comp Biochem Physiol 47A:1323–1327.
Vedel JP (1985) Cuticular mechanoreception in the antennal flagellum of the rock lobster *Palinurus vulgaris*. Comp Biochem Physiol 80A:151–158.
Vedel JP, Clarac F (1976) Hydrodynamic sensitivity by cuticular organs in the rock lobster *Palinurus vulgaris*. Morphological and physiological aspects. Mar Behav Physiol 3:235–251.
Wiese K (1976) Mechanoreceptors for near-field water displacements in crayfish. J Neurophysiol 39:816–833.
Williamson R (1985) Efferent influences on the afferent activity from the *Octopus* angular acceleration receptor system. J Exp Biol 119:251–264.
Williamson R (1986) Efferent activity in the *Octopus* statocyst nerves. J Comp Physiol A 158:125–132.
Williamson R, Budelmann BU (1985) The response of the *Octopus* angular acceleration receptor system to sinusoidal stimulation. J Comp Physiol A 156:403–412.
Young JZ (1960) The statocysts of *Octopus vulgaris*. Proc R Soc Lond B 152:3–29.
Young JZ (1984) The statocysts of cranchiid squids (Cephalopoda). J Zool Lond 203:1–21.
Young SR, Dedwylder RD, Friesen WO (1981) Responses of the medicinal leech to water waves. J Comp Physiol 144:111–116.

32 The Ear as Part of the Octavolateralis System

CHRISTOPHER PLATT, ARTHUR N. POPPER, and RICHARD R. FAY

32.1 Introduction

The inner ear and lateral line form the octavolateralis system in aquatic vertebrates. This system provides an array of electrosensory and mechanosensory inputs that are integrated with chemosensory and visual information to produce behavioral responses appropriate for the organism (see Blaxter 1988). As several chapters in this volume note (Enger et al. Chapter 29; Kalmijn Chapter 9), there are structural and functional parallels between the inner ear and the mechanosensory lateral line systems. Direct comparisons between these systems are useful with regard to the phylogeny, ontogeny, micromechanics, transduction processes, coding, central connections, and behavioral use of these organs. Just as research on the lateral line has advanced our understanding of inner ear function (see Flock 1974), insights gained from vestibular and auditory research might help to guide questions regarding lateral line function. This chapter addresses some parallels between the mechanosensory lateral line and the ear of fishes and considers how investigators of these two systems might learn from one another.

Recent reviews and books are available on the function of the inner ear in fishes, addressing vestibular (Platt 1983, 1988) and auditory mechanisms (Tavolga et al. 1981; Popper and Fay 1984; Popper et al. 1988). Fay (1988b) reviews the literature on behavioral studies of detection and discrimination mediated by the ears and the lateral line.

32.2 Similarities and Differences Between Ear and Mechanosensory Lateral Line

Differences in the major peripheral structures of the inner ear and lateral line of a fictitiously "representative" fish are schematically outlined in Fig. 32.1. The inner ear is fairly compact, containing six or seven end organs confined in a small space deep in the skull. The pouches and tubes of this membranous labyrinth contain a special fluid called endolymph and patches of sensory epithelia called cristae, maculae, or papillae, depending on their precise morphology. A

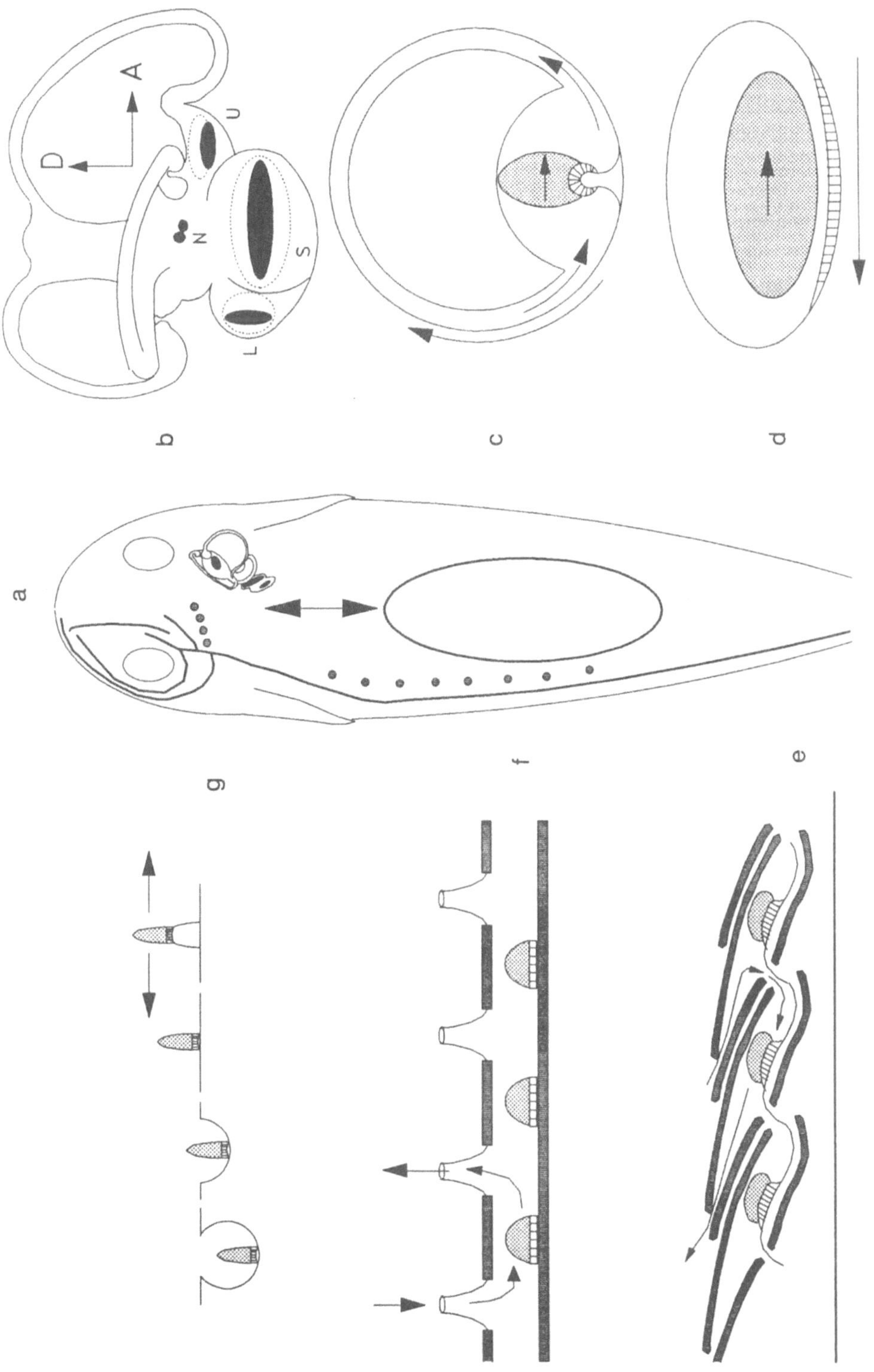
a
b
c
d
e
f
g
D
A
U
N
S
L

semicircular canal crista is a sensory ridge across a narrow tube. An otolith organ macula is a sensory patch or sheet in the wall of a pouch containing a dense calcification. Fishes and amphibians often have an additional accessory macula or papilla as a sensory patch in a tube or pouch. Fish having gas bladders (swimbladders) can also use their compressible volume as an accessory organ to the ear (see Sand and Enger 1973; Blaxter 1981).

In contrast, the lateral line system is extensive, with a large number of sensory end organs that are widely distributed over the body surface as either superficial or canal neuromasts (see Webb Chapter 4). Superficial neuromasts may be found singly or in groups on the skin surface, or in pits, or raised above the skin surface on pedestals. Canal neuromasts of the head lie in hollow tubes between pores of various dimensions that open to the surrounding water. Canal neuromasts of the body "line" often are located within scales that communicate with one another and with the outside water through pores.

There are similarities in the mechanoreceptors of both systems, their innervation by special cranial nerves, and the presence of efferent nerve supply to the periphery. The common mechanosensory receptors are hair cells that respond to deflection of some overlying structure such as a cupula or otolith (see Hudspeth 1983; Kroese and van Netten Chapter 13). The hair cells send information to distinct brain nuclei via separate peripheral branches of cranial nerves. The eighth (octaval) cranial nerve innervates the ear, and the anterior and posterior lateral line nerves innervate the head and trunk lateral line systems, respectively (Boord and Montgomery Chapter 16; McCormick Chapter 17; Will Chapter 18).

◄

FIGURE 32.1. Schematic representation of the organs of the inner ear and lateral line. (a) Layout of hair cell mechanoreceptor organs of a schematized teleost fish shown in dorsal view. Lateral line canals (thick lines) and free neuromasts (small shaded circles) are shown on the left side of the body. The inner ear with its semicircular canals and otolith organs are shown on the right in the head, and the gas bladder is indicated by the large ellipse in the abdominal area. (b) Enlarged lateral view of the right labyrinth shows the sensory organs. There are three semicircular canals with their ampullae and cristae and the sensory maculae of the utricle (U), saccule (S), lagena (L), and macula neglecta (N). Sensory surfaces are shown in black; dotted lines indicate the otolith outlines. D is dorsal, and A anterior. (c) Semicircular canal organ, schematic section. The fluid-filled narrow duct has an expansion, the ampulla, containing the crista with its sensory epithelium covered by the tall gelatinous cupula (shaded). Fluid in the canal acts inertially during head rotations to displace the cupula, which acts as an elastic diaphragm across the canal. (d) Otolith organ schematic section. The pouch contains the otolith (shaded) as a dense mass overlying the sensory macula. Relative movement of otolith is coupled to the epithelium by viscoelastic forces. (e) Typical arrangement of the lateral line trunk canal system showing neuromasts embedded within scales and arrows indicating flow paths. (f) Typical arrangement of the canal system of the head with a neuromast intervening between pores. Pressure difference at the two pores produces fluid flow in the canal between them. (g) Range of superficial neuromast types. Some are embedded within pits; others may project into the environment from the skin surface or from above the surface.

Efferent neurons have been demonstrated, suggesting some central control to modulate activity of the hair cells and afferent neurons in both systems (Claas et al. 1981; Roberts and Meredith Chapter 22; Zottoli and Danielson Chapter 23). An individual octavolateral efferent may innervate both ear and lateral line end organs (Fritzsch and Wahnschaffe 1987), but the functional significance of this arrangement is not clear.

The inner ear of fishes detects relative movement between the otoliths or fluid endolymph and the sensory structures, usually resulting from forces producing linear or angular acceleration of the head. The ear functions in controlling posture, locomotion, and eye stabilization and in the detection, analysis, and localization of underwater sounds in the "hearing specialists" among fishes (see Hawkins 1981; Platt and Popper 1981). The lateral line detects the spatial and temporal patterns of relative motion between the water environment and the fish's skin surface (see Kalmijn Chapter 9; Görner and Mohr Chapter 27). Its main functions are detection of moving neighbors, prey, predators, or other objects underwater or on the surface (see Sand 1981; Blaxter 1988; Blaxter and Fuiman Chapter 24; Coombs and Janssen Chapter 15; Münz Chapter 14). Both systems have directionally sensitive receptors in bilateral organs, allowing the use of possible directional, intensity, and timing comparisons from the bilateral organs for characterizing and localizing stimuli (see Claas et al. Chapter 20; Elepfandt Chapter 26).

These comparisons have historically led to questions about the degree of functional overlap and distinction between the two systems. Functional overlap is particularly great in some species where the ear and lateral line are physically contiguous, as in the clupeids (Blaxter et al. 1981; Denton and Gray Chapter 11). Several chapters in this volume address whether or not recent ontogenetic and phylogenetic evidence supports a common "octavolateralis" evolutionary origin for the inner ear and lateral line (Fritzsch Chapter 5; Jørgensen Chapter 6; Northcutt Chapter 3; McCormick Chapter 17; Will Chapter 18), as suggested in the earlier literature (see van Bergeijk 1967).

In this comparative approach, there are four structural and functional aspects of the inner ear and lateral line to which we would like to draw specific attention: (1) the multiple behavioral functions of the end organs; (2) the action of receptor organs as filters for particular stimulus features; (3) the role of nonneural elements in receptor function; and (4) the function of an organized pattern in a large array of receptors. These aspects of the ear and lateral line depend not only on properties of the receptor cells but on the receptor organ as a whole.

32.3 Functions of the Inner Ear

The inner ear mediates two major sensory functions. Vestibular responses are those to rotatory and linear head movements, and auditory responses are those to sound. The ear of fishes, lacking a cochlea, works chiefly as an inertial detector of acceleration, and the functions of the otolith organs in many fishes and amphibians are not fully understood, because they may respond not only to gravity (as in tetrapods) but also to vibration and to sound (see Lewis et al. 1985). Since a fish body is mostly water, a sound wave in water effectively accelerates the whole fish, and the denser otoliths have inertial lag (see Schuijf 1981; Kalmijn 1988). Separation of oscillating linear accelerations into a "vestibular," "vibratory," or "auditory" category is not always easy, and similarly it is not always easy to separate the ear of a fish neatly into "vestibular" and "auditory" regions.

All jawed vertebrates have three semicircular canal organs in the ear that respond to head rotations, two or three otolith organs, and often an additional macula or papilla. Of the otolith organs, the utricle is generally considered to be a vestibular organ, the saccule is specifically an auditory organ in many teleosts, and the lagena has been implicated in both vestibular and auditory function (see Lowenstein 1971, 1974; Platt and Popper 1981; Platt 1983).

Auditory function might exist in more than one otolith organ, since some teleosts have a utricle apparently specialized for audition, and some use primarily the saccule for balance (see Blaxter et al. 1981; Platt and Popper 1981; Hawkins and Horner 1981; Fay 1984). The accessory macula neglecta in elasmobranchs is believed to be an auditory organ (Corwin 1981), though its role in bony fishes is unknown. Some additional papillae of amphibians, the basilar papilla of reptiles, and the cochlea of birds and mammals act specifically as auditory organs, allowing a clearer distinction between vestibular and auditory structures in those animals (see Lewis et al. 1985; Popper and Fay 1980).

It may be difficult to define precisely a single end organ in the octavolateralis system. In the inner ear, a single semicircular canal crista is a clearly distinct epithelium with a relatively distinct function. But the macula neglecta often consists of two separate patches of epithelium with unknown function. There is some evidence that such paired maculae are phylogenetically derived from a single macula (Corwin 1981), suggesting that the two patches should be considered a single end organ. Additionally, the utriculus of many fishes contains a region called the cotillus, covered by otolith and otolithic membrane, and a region called the lacinia, which often is a narrow lateral strip covered only by otolithic membrane (Werner 1929).

In some elasmobranchs the lacinia is sensitive to vibration but not tilt, unlike the rest of the utricle, and roughly two-thirds of the saccule is sensitive to vibration, and the caudal one-third to tilt (Lowenstein and Roberts 1951). Records from the elasmobranch brainstem showed that some neurons respond to both tilts and vibration but that one class of such vibratory or acoustic responses is strongly inhibited by simultaneous static tilt or tilting movements, suggesting that the elasmobranch saccule and lacinia primarily provide postural rather than auditory information to the central nervous system (Plassmann 1983). The question arises whether an apparently polyfunctional sensory macula should be considered a single end organ.

Diverse morphology and spatial distributions of neuromasts can be observed both within a single lateral line system and between systems in different species, presenting even more complex questions. Can a single neuromast be considered a functional unit or end organ? Or is a functional unit better defined as a series of neuromasts in the same canal, or as a distinct cluster of superficial neuromasts? Do the superficial neuromasts, as a group, have functions distinct from those of the canal neuromasts in the head or trunk? Satisfactory answers to these questions are not simply semantic, since they involve consideration of specific function (in terms of both biomechanical response properties and behavioral significance), innervation, development, and evolutionary history. These questions lead toward the larger question of to what degree the inner ear systems and lateral line systems can be considered subsets of an octavolateralis system.

Nonneural elements that differ in different organs, and often in different regions of the same organ, include gelatinous overlying structures or calcified otoconial structures in a gelatinous matrix. The composition of the fluids and jellies affects chemical properties such as ionic gradients across the receptor cell membranes and physical properties such as viscosity, elasticity, and density. The membraneous pouches and ducts of the inner ear and the canals of the lateral line, along with the gelatinous or otoconial masses within the chambers and canals, constrain the fluid flows. The dimensions and compliances of the pouches, ducts, and canals likely determine the hydrodynamic properties of the system in combination with the properties of the overlying cupular or otolithic structures (see Lewis et al. 1985; Denton and Gray Chapter 11).

Morphological differentiation of the end organs in the ear is functionally important. Each semicircular canal crista is covered by a gelatinous cupula acting as a diaphragm across a circular tube of fluid; this diaphragm is deformed slightly by forces of fluid pressure against it, as the inertial fluid is rotationally accelerated by head motion. Each otolith organ macula is covered by a dense, calcified inertial mass, giving sensitivity to linear accelerations arising from movements of the head or from pressure-induced motions of the gas bladder. The papillae of the macula neglecta are covered by a gelatinous membrane much like the lateral line neuromasts. The elaborate variation in morphology found in the saccule and lagena of the ear (both within an individual and between species) is often associated with the processing of mechanical input arising from the gas bladder and so suggests adaptations for coding sound pressure signals (Popper et al. 1982). The less extensive variation in the utricle and canals may reflect more similar needs across all vertebrates for postural and locomotor guidance and eye stabilization (Ramprashad et al. 1986; Platt 1988).

Although each of the lateral line receptor organs has a gelatinous cupula overlying the neuromast, there may be substantial variation in the size, shape, and orientation of the cupulae and of their underlying neuromasts in canal structures from different body regions (Coombs et al. 1988; Webb Chapter 4). Just as differences in the morphology of the ear result in different functional capabilities, the structural diversity among receptors of the mechanosensory lateral line in a single animal may have important functional consequences. For example, there is now evidence that the form and locations of the superficial neuromasts allow these organs to respond best to water velocity and/or acceleration, and the canal neuromasts to respond best to acceleration (Coombs and Janssen

Chapter 15; Gray 1984; Kalmijn Chapter 9; Münz Chapter 14; van Netten and Kroese Chapter 12).

32.4 Filtering

The inner ear contains receptor epithelia with overlying accessory structures, such as the otoliths and otolithic membranes. Functions of these structures depend on the same kind of micromechanical and hydrodynamic considerations as those for the lateral line organs. The sensitivity of a given organ is determined by the complex biomechanics that transfer and filter a signal in the environment to a mechanical displacement of the ciliary bundle of the mechanosensory hair cell (see Allen et al. 1986).

The sensitivity of a particular patch of sensory epithelium to a relative acceleration, velocity, displacement, or pressure change depends on system parameters such as inertia, viscosity, compliance, elasticity, and compressibility. The semicircular canals are specialized for responding to angular accelerations at frequencies in the range of 0.01 to perhaps 100 Hz (Goldberg and Fernández 1975; Precht 1979). The otolith organs of fishes respond to static tilt and to vibrations and sound from 0.01 to more than 5,000 Hz in some fish species (see Popper and Fay 1984).

Some of the specializations that produce these frequency response characteristics also may exist in lateral line organs. The semicircular canal crista, for example, is a sensory ridge with an overlying gelatinous cupula that stands in a swelling of the fluid-filled semicircular duct called the ampulla. The crista has some similarity to a neuromast in the lateral line canal. The frequency response characteristics of the semicircular canals and lateral line canals are determined largely by the internal radius of the canal, the viscosity and density of the canal fluid, and the stiffness of the cupula (Denton and Gray Chapter 11; van Netten and Kroese Chapter 12).

There are some clear differences between these two kinds of "canal" organs. The semicircular canals are completely enclosed and contain a special fluid endolymph, but the lateral line canals in most species have pores along the length of the canal that are open to the outside water. The cupula of the semicircular canal forms a transverse partition like a diaphragm completely across the bulbous ampulla, the widest part of the canal. The lateral line canal cupulae are highly variable (see Denton and Gray Chapter 11). Some cupulae are thin, platelike structures elongated along the longitudinal axis of the canal; some are more rounded; and some are more transversely oriented. The lateral line canal neuromast, unlike the semicircular canal crista, typically lies in a region of the canal between two surface pores, where the canal diameter is reduced.

These structural differences between the lateral line and semicircular canals provide alternative mechanisms for driving the cupula, which may result in a different range of signals to which the two systems will respond. When the fluid in the semicircular canal is displaced, the partition is displaced directly by the fluid pressure over a fairly large area against one side of the cupula (Hillman and McLaren 1979). But when the fluid in a lateral line canal is displaced, the cupula is displaced by viscous drag (van Netten and Kroese Chapter 12). This drag depends on flow velocity within the canal, which in turn depends on the acceleration of the water environment outside the canal (Denton and Gray Chapter 11).

Acoustic sensitivity is determined to some degree by the mechanical properties of the system peripheral to the receptors. A volume of gas is compressible, unlike water, and will oscillate in volume in a sound field. Fishes having a compressible gas bladder in the abdominal cavity can have enhanced sensitivity (low thresholds) if the compressible bubble is somehow coupled to the fluid system of the ear. This increased sensitivity will be largely independent of frequency for frequencies below the resonant frequency of the bladder (Sand and Enger 1973), and some bladders response show virtually no resonance over many octaves (Blaxter et al. 1981).

Fishes with gas bladders also often have a higher-frequency range of hearing than fishes without bladders. For example, the otophysans (or ostariophysines: the goldfish, catfish, and relatives) have a direct mechanical bony linkage from gas bladder to inner ear and have high-frequency hearing better than most other teleosts tested, with some species able to detect sounds to over 4,000 Hz. If the gas bladder is deflated, sensitivity and frequency range drop drastically.

The gain from the presence of a gas bubble alone was demonstrated nicely by Chapman and Sand (1974). Flatfishes have no gas bladder, and their hearing normally does not extend much above several hundred hertz. Placing a small balloon close to the head of a flatfish caused an increase in sensitivity, and the behavioral thresholds linearly decreased with frequency. Presumably the sound-driven oscillations of the balloon volume caused radiation of water particle displacements that can affect the ear or lateral line if they are near enough. This indirect action as an amplifier of the initial waterborne sound depends on the ear's being in the "near field" of the bubble, where the reradiated sound wave has a large displacement component; without the bubble the ear is in the "far field" of the distant actual sound source, and the displacement component is negligible. For a given fish the gain in sensitivity can come from the gas bladder's presence, and the gain in bandwidth can come from both the general frequency independence of the gas bladder's response to sound and from special adaptations of the otolith organs for responding to higher frequencies.

Other simple mechanical differences also may enhance system sensitivity. The semicircular canals are not normally responsive to sound, but when the bony canal in a mammal was surgically opened, the semicircular canal organs became acoustically sensitive (Wit and Bleeker 1982). In the lateral line, Denton and Gray (1988) demonstrated that the structure of the lateral line canal may affect the frequency response of the neuromasts. The presence of canals of different diameter within a single animal might allow some degree of frequency tuning as a result of mechanical factors.

The dimensions and mechanical properties of the ciliary bundle on each hair cell might also confer some frequency filtering to the ear and/or lateral line. Data from the cochleae from a range of reptiles, birds, and mammals suggest that longer ciliary bundles are found in regions with best sensitivity to relatively lower frequencies and that regions with shorter ciliary bundles have a best sensitivity to higher frequencies (Saunders and Dear 1983). A range of ciliary bundle sizes is found in the ears of several teleost and elasmobranch fishes (Platt 1983; Platt and Popper 1981) and in amphibians, where bundle size has also been correlated with frequency sensitivity (Lewis et al. 1982). Similar ranges of ciliary dimensions are likely in the neuromasts of the lateral line, and it would not be surprising to discover that afferents innervating these different hair cells have corresponding frequency response functions.

However, there is no current evidence for hair cells in fishes having the kind of narrow electrical tuning found in the ears of reptiles (Art 1984; Kroese and van Netten Chapter 13). Some of these different bundle forms might reflect developmental history rather than, or perhaps in addition to, sensory functions (Corwin 1985).

32.5 Receptor Arrays

The sensory epithelium is an organized array of receptor cells and nerve terminals that might provide complex sensory processing. For example, the horizontal semicircular canal crista of an elasmobranch (the shovelnose guitarfish *Rhinobatos productus*) is innervated by a nerve branch that is divided into five bundles. The bundle innervating the center of the crista ridge has single fibers that have greater sensitivity to angular accelerations and shorter latencies to peak firing rates than fibers found in bundles innervating the ends of the crista (O'Leary et al. 1976). This functional separation demonstrates the sensory principle of range fractionation, in which different parts of a sensory system are best detectors of different aspects of a stimulus. Lateral line organs show great variation with regard to location and size of head and trunk canals and numbers of free neuromasts and canal neuromasts (Coombs et al. 1988). These differences could allow a similar type of range fractionation. The stimulus features being differentiated might be the stimulus motion component (displacement, velocity, or acceleration), source direction, signal amplitude, or the shapes of pressure gradients over the head and body.

The inner ear represents a very localized receptor array compared with the lateral line. The inner ear is used by aquatic animals for detecting sound waves that are centimeters to meters in wavelength, and from distances out to meters or more (Kalmijn 1988, Chapter 9). Yet the auditory maculae are typically only a few millimeters across, perhaps up to 1 or 2 cm in very large fish, and only a few millimeters or centimeters apart across the midline. From the auditory standpoint, the ears are acting more as a single-point detector than as a two-point comparator in a large field of sources.

Spatial localization of distant sound sources must require more than simple timing or amplitude differences between two closely spaced organs. The lateral line is apparently most useful for stimuli at fairly close ranges of a body length or less (see Sand 1981), and the lateral line itself extends over a large fraction of the body length and across most of the head surface in many fishes. The great longitudinal dimension of the lateral line organs could be useful for discriminating timing or amplitude cues along the body (see Görner and Mohr Chapter 27; Elepfandt Chapter 26). For surface waves, which are much slower and of shorter wavelength than underwater sound waves, source localization may apparently involve different local timing or amplitude cues from the lateral line (Görner et al. 1984; Elepfandt 1982; Müller and Schwartz 1982). Deeper underwater, successful escape or prey capture also depends on localization of the source, which at close range may involve both the ear and lateral line (Blaxter and Fuiman Chapter 24; Enger et al. Chapter 29).

The ability to determine the source of stimuli is a fundamental requirement of all sensory systems, including the octavolateralis system. Terrestrial animals use the interaural differences in intensity and time to "compute" sound source direction. But the octavolateralis system makes use of the morphological polarization that gives hair cells their directional sensitivity (see Lowenstein 1974; Hudspeth 1983; Kroese and van Netten Chapter 13). Each single hair cell has an asymmetrically placed kinocilium that gives the cell a direction of maximal sensitivity to shear. An array of such cells may have a complex pattern of directional sensitivity over the whole end organ (see Lowenstein 1974).

In the ear, the simplest case is that of a semicircular canal crista, which has all its hair cells polarized in the same direction relative to the axis of the canal. The presence of bilateral symmetry, though, adds the possibility of a "push-pull" differentiation across the midline. Each horizontal canal crista is excited by a turn of the head toward that side and inhibited by a turn toward the contralateral side, so head rotation to the right excites the right horizontal canal crista but inhibits the left horizontal crista. In addition, the right posterior vertical canal lies nearly coplanar with the left anterior vertical canal, and a rolling tilt in that plane excites or inhibits both those vertical canals together in a form of bilateral synergy instead of opposition. This system provides compensatory reflexes to the similarly reciprocal pairs of eye muscles to stabilize the visual gaze of the animal, a crucial function for acute vision while moving (see Lowenstein 1974; Goldberg and Fernández 1975; Precht 1979).

More complex patterns of directional polarization occur in other hair cell epithelia. All otolith organ maculae contain areas composed of cells having the same orientation, and there are always at least two differently oriented areas in each macula. Adjacent areas usually have cell populations oriented oppositely (at 180° to one another) or orthogonally (at 90°) (Lowenstein 1974). In all lateral line neuromasts that have been examined, hair cells of opposite orientations ("antiparallel") are intermingled, and orthogonal orientations have not been observed within a neuromast. Neuromasts in narrow bony canals have these oppositely polarized hair cells aligned parallel to the long axis of the canal (Janssen et al. 1987), providing an adaptation to maximize sensitivity to the fluid movement along the canal. In contrast, orthogonal axes do occur in pairs of neighboring superficial neuromasts, possibly allowing a resolution of water particle velocity vectors.

Since lateral line canals have different orientations on different parts of the head and trunk, vector resolution could be done by comparing inputs from neuromasts located in these different canals. (In crustaceans, analogous hydrodynamics are detected by superficial mechanosensory hairs, with directional analysis based on patterns of excitatory and inhibitory input to central neurons (Wiese 1988).) However, topological analysis based on such extensive input cannot explain the remarkable abilities of some surface-feeding fishes and amphibians to accurately locate wave sources even after ablation of large numbers of neuromasts (Bleckmann et al. Chapter 25; Görner and Mohr Chapter 27).

Opposing populations of hair cells in the utricle or lagena may allow a single macula to send appropriate opposing excitatory and inhibitory vestibular inputs to brain regions that integrate directional information. In some teleosts, the saccular macula may have four or more distinct areas of cells with opposing and orthogonal orientations, allowing the possible resolution of an acceleration vector by comparing inputs from cells within a single organ (Dijkgraaf 1962). Mapping of directional sensitivity of microphonics in codfish (Sand 1974) and of individual neurons from the otolith organs of cod (Hawkins and Horner 1981) and goldfish (Fay 1984, 1988) suggests that this encoding is responsible for the ability of some species to localize sound sources in three-dimensional space (e.g., Hawkins and Sand 1977). Proper processing of inputs from orthogonal and opposing hair cell populations has been invoked to explain the determination of signal frequency, intensity, direction, and other sound qualities (Popper et al. 1982; Rogers et al. 1988).

Little is known about the detailed patterns of afferent innervation of the organs of the inner ear or the lateral line (Jørgensen Chapter 6). The extent of branching of the single afferents in the eighth nerve in fish ears is quite variable; a single afferent neuron often innervates several hair cells, even those with different orientations (Sento and Furukawa 1979). Branching may affect several properties of coding in the afferent, such as the presence and statistical regularity of spontaneous discharge, phase-locking to different features of a stimulus waveform, and averaging mechanisms for extracting small signals from a noisy environment (Goldberg and Fernández 1975).

32.6 Sensitivity and Frequency Analysis of the Auditory and Lateral Line Systems

It is likely that the organs of the ear and lateral line canals respond best to acceleration forces, which lead to displacement at the hair cells. The ear is stimulated when the head moves, either from a postural or locomotory change, or when an acceleration field is uniform and large relative to ear size. The lateral line is stimulated when an acceleration field varies over the body surface so that a

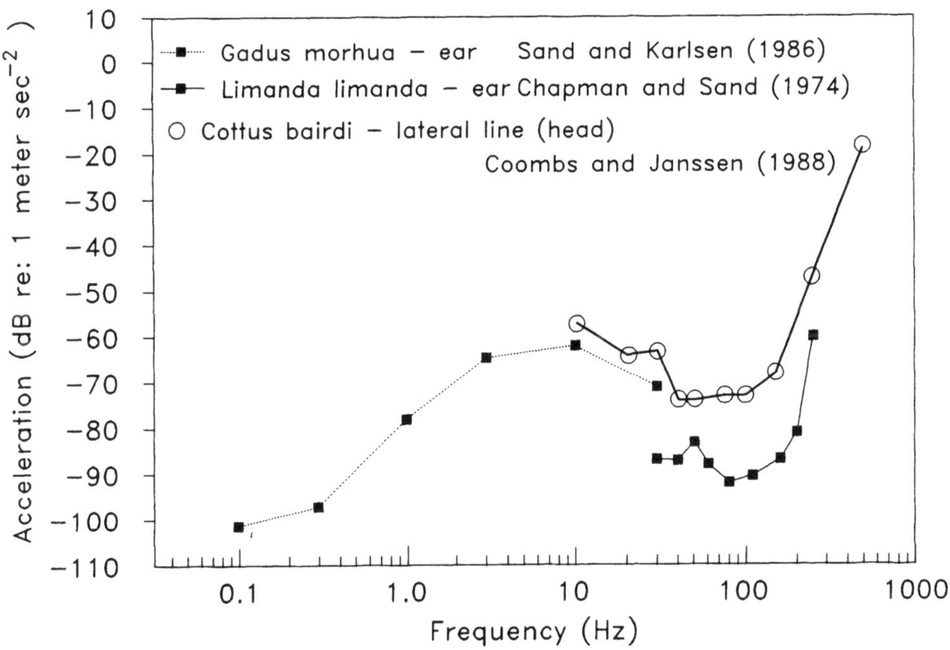

FIGURE 32.2. Behavioral measures of absolute acceleration sensitivity of ear of the cod (*Gadus morhua*) at "infrasonic" frequencies (Sand and Karlsen 1986), the ear of the dab (*Limanda limanda*) (Chapman and Sand 1974), and the head canal system of the mottled sculpin (*Cottus bairdi*) (Coombs and Janssen Chapter 15).

relative motion occurs between the surrounding water and neuromasts. Behavioral measures of absolute acceleration sensitivity are one of the very few dimensions studied that allow a direct comparison between the ear and lateral line system.

As shown in Fig. 32.2, behaviorally measured acceleration thresholds for the ear of the dab (*Limanda limanda*—a flatfish species without a gas bladder) (Chapman and Sand 1974) and the lateral line system of the mottled sculpin (*Cottus bairdi*) (Coombs and Janssen Chapter 15) are similar both in absolute sensitivity and in frequency range. Above about 30 Hz, the dab ear and sculpin lateral line detect particle acceleration best at about 100 Hz, with a sensitivity of about −75 to −90 dB re: 1 m s^{-2}. At lower frequencies, the otolith organs of the ear of the cod (*Gadus morhua*) are more sensitive (Sand and Karlsen 1986), suggesting that there may be a wide region of biologically significant acceleration signals for the otolith organs well below the acoustic frequency range normally tested in field or lab experiments. It is possible that the same may be true for the lateral line system.

Electrophysiological studies of single peripheral nerve fibers of the octavolateralis system also show a similarity between an otolith organ and the lateral

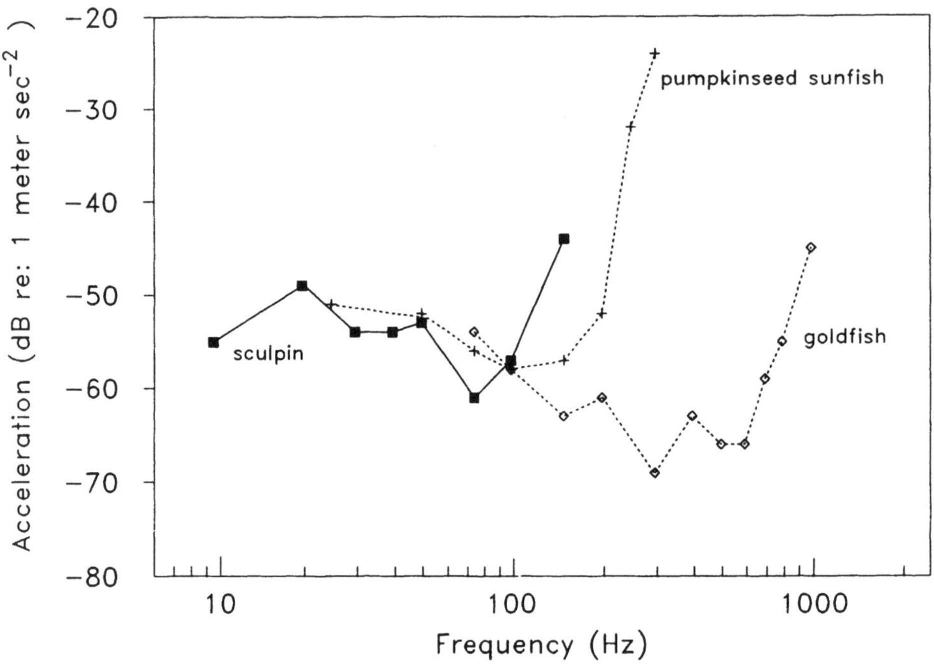

FIGURE 32.3. Tuning curves from afferent fibers innervating different octavolateralis organs in fish. Crosses – a saccular fiber from the pumpkinseed sunfish (*Lepomis gibbosus*) (Coombs and Fay unpublished observations). Solid squares – a lateral line fiber from the mottled sculpin (*Cottus bairdi*) (Coombs and Janssen Chapter 15). Diamonds – a high-frequency saccular fiber from the goldfish (*Carassius auratus*) (Fay 1981).

line in sensitivity and frequency response. Figure 32.3 shows tuning curves for a saccular fiber from the pumpkinseed sunfish (*Lepomis gibbosus*) (Coombs and Fay unpublished), another from the goldfish (*Carassius auratus*) (Fay 1981), and a lateral line afferent from the mottled sculpin (Coombs and Janssen Chapter 15). Clearly, the physiological responses of these two different organs show considerable overlap. This low-pass, acceleration-sensitive range is likely a common feature of hearing and lateral line organs of the octavolateralis system (see also Kalmijn Chapter 9). There are additional adaptations of the saccule of the goldfish for response to considerably higher frequencies of head acceleration.

These responses to head acceleration are independent of the gas bladder's response to sound pressure and thus provide evidence that the otolith organs themselves may be adapted for responding in quite different frequency ranges. In the goldfish saccule this extended frequency sensitivity is a property of some, but not all, fibers, and it may be an adaptation of hair cells for coding the high-frequency inputs that normally arrive from the gas bladder's response to sound pressure. It is not yet clear precisely what structures of the saccule are involved

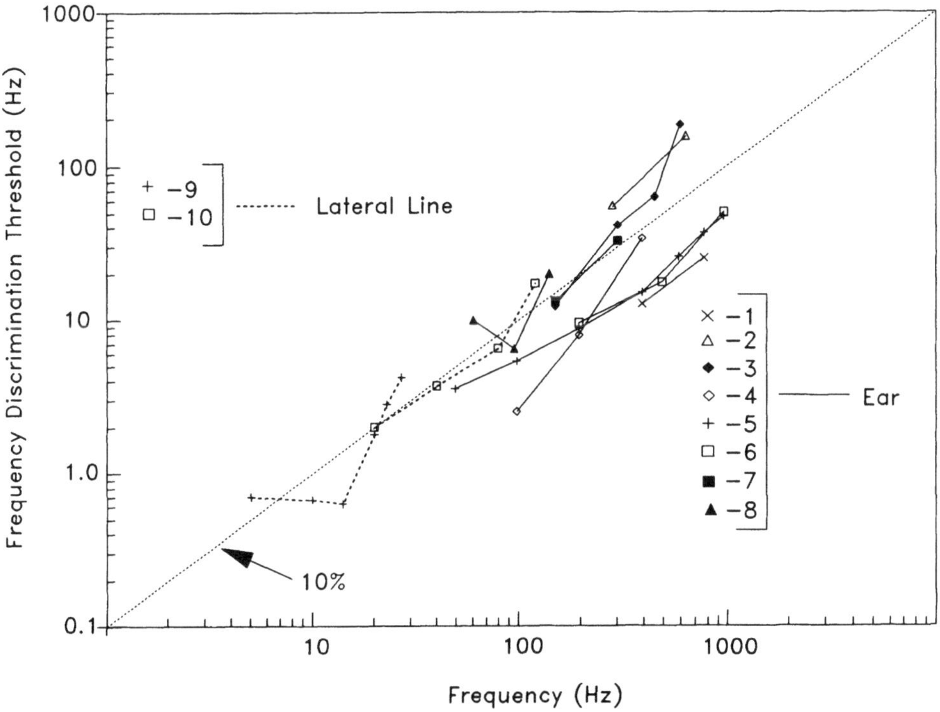

FIGURE 32.4. Behavioral pure-tone frequency discrimination thresholds from the auditory (1–8; solid lines) and lateral line (9, 10; dotted lines) systems of several species. 1. *Phoxinus laevis* (Dijkgraaf and Verheijen 1950). 2. *Phoxinus laevis* (Stetter 1929). 3. *Sargus annularis* (Dijkgraaf 1952). 4. *Arius felis* (Tavolga 1982). 5. *Carassius auratus* (Fay 1970). 6. *Carassius auratus* (Jacobs and Tavolga 1968). 7. *Gobius niger* (Dijkgraaf 1952). 8. *Cottus scorpius* (Pettersen 1980). 9. *Xenopus laevis* (Elepfandt et al. 1985). 10. *Aplocheilus lineatus* (Bleckmann et al. 1981).

in this response to high frequencies, and it is possible that lateral line end organs may also differ significantly in frequency response range.

A notable feature of the "tuning curves" of Fig. 32.3 is that these single neurons essentially act as acceleration low-pass filters of broad bandwidth comparable to the bandwidth of the otolith organ or lateral line system as a whole. So the frequency analysis of the stimulus spectrum seems far less acute in the peripheral parts of the octavolateralis system of fishes than in the peripheral auditory systems of amphibians and terrestrial vertebrates, where sharply tuned neurons are common (see Lewis et al. 1985). Some frequency or quality analysis performed by the octavolateralis system is likely to be done more centrally, in the brain, and be based on the processing of temporal neural codes. It is interesting that the acuity of the auditory and lateral line systems in discriminating sinusoidal frequency is quite similar, with relative discrimination thresholds remaining constant (in the region of 10%) in the frequency range between 5 and 1,000 Hz (Fig. 32.4). This constant percentage for discrimination acuity is consistent with both systems processing interspike times from phase-locked afferents (Fay 1978).

32.7 Summary

The mechanosensory lateral line and the inner ear of fishes have in common (1) similar hair cell mechanoreceptors; (2) similar nonneural accessory structures such as surrounding fluids, gelatinous matrices, and tubes; (3) similar epithelial organization, including bidirectional orientations of the hair cells; and (4) similar behavioral capacities for detecting and discriminating acceleration signals. These structural aspects of a whole end organ are important for its function, including the filtering of particular properties of its environmental stimulus features. The end organs in both the lateral line and inner ear appear to be differentiated into groups that mediate particular behavioral responses (see Dijkgraaf Chapter 2) but still have similar capacities for absolute sensitivity and frequency discrimination. Together, the inner ear and lateral line are complementary senses for disturbances in the water produced by self-motion and other sources over a range of distances and frequencies. We emphasize that functions of the inner ear and lateral line depend on factors beyond those of cellular features alone, since each end organ exploits particular relations between the mechanosensory hair cells and nonneural elements and on specific patterns of organization of the receptor epithelia and their innervation.

All these comparisons between these two major divisions of the "octavolateralis" system do not solve the problem of what to call the signal detected by the lateral line. The nose smells, the tongue tastes, the eye sees, and the ear hears. A comparable term for the lateral line was coined by the pioneer of lateral line work, Dr. Sven Dijkgraaf, who described this system for "Ferntastsinn" or "distant-touch sense" (Dijkgraaf 1947). This sense is not really hearing, and not really touch, and now it is also known to be not very distant. So to honor Dr. Dijkgraaf, as physical scientists often are honored by having their names given to physical parameters, we here propose the particular verb "sven" for this activity of lateral line sensing, as in the following:

Dear frog and fish, or newt and shark,
You needn't worry when it's dark;
You'll escape or dine just fine,
Svenning with your lateral line.

Acknowledgments. The writing of this chapter was supported in part by grant from the NINCDS, NIH to the Parmly Hearing Institute, NINCDS grant NS-15090 to A.N.P., and ONR contract N00014-87-K-0684 to A.N.P. The authors thank Sheryl Coombs, Olav Sand, and other conference participants whose comments improved the manuscript, and helped focus our collective thoughts.

References

Allen JB, Hall JL, Hubbard A, Neely ST, Tubis A (eds) (1986) Peripheral Auditory Mechanisms. Lecture Notes in Biomathematics 64. Berlin: Springer-Verlag.

Art JJ (1984) Auditory function at the receptor level in reptiles. In: Bolis L, Keynes RD, Maddrell SP (eds) Comparative Physiology: Sensory Organs. Cambridge, England: Cambridge University Press, pp. 47–66.

Bergeijk WA van (1967) The evolution of vertebrate hearing. In: Neff WD (ed) Contributions to Sensory Physiology, Vol. 2. New York: Academic Press, pp. 1–49.

Blaxter JHS (1981) The swimbladder and hearing. In: Tavolga WN, Popper AN, Fay RR (eds) Hearing and Sound Communication in Fishes. New York: Springer-Verlag, pp. 61–71.

Blaxter JHS (1988) Sensory performance, behavior, and ecology of fish. In: Atema J, Fay RR, Popper AN (eds) Sensory Biology of Aquatic Animals. New York: Springer-Verlag, pp. 203–232.

Blaxter JHS, Denton EJ, Gray JAB (1981) Acousticolateralis system in clupeid fishes. In: Tavolga WN, Popper AN, Fay RR (eds) Hearing and Sound Communication in Fishes. New York: Springer-Verlag, pp. 39–71.

Bleckmann H, Waldner I, Schwartz E (1981) Frequency discrimination of the surface-feeding fish *Aplocheilus lineatus*—a prerequisite for prey localization? J Comp Physiol 143:485–490.

Chapman CJ, Sand O (1974) Field studies of hearing in two species of flatfish, *Pleuronectes platessa* (L.) and *Limanda limanda* (L.) (family Pleuronectidae). Comp Biochem Physiol 47A:371–386.

Claas B, Fritzsch B, Münz H (1981) Common efferents to lateral line labyrinthine hair cells in aquatic vertebrates. Neurosci Lett 27:231–345.

Coombs SL, Janssen J, Webb J (1988) Diversity of lateral line systems: Evolutionary and functional considerations. In: Atema J, Fay RR, Popper AN, Tavolga WN (eds) Sensory Biology of Aquatic Animals. New York: Springer-Verlag, pp. 553–594.

Corwin JT (1981) Audition in elasmobranchs. In: Tavolga WN, Popper AN, Fay RR (eds) Hearing and Sound Communication in Fishes. New York: Springer-Verlag, pp. 81–105.

Corwin JT (1985) Perpetual production of hair cells and maturational changes in hair cell ultrastructure accompany postembryonic growth in an amphibian ear. Proc Natl Acad Sci USA 82:3911–3915.

Denton EJ, Gray JAB (1988) Mechanical factors in the excitation of the lateral line of fishes. In: Atema J, Fay RR, Popper AN, Tavolga WN (eds) Sensory Biology of Aquatic Animals. New York: Springer-Verlag, pp. 595–617.

Dijkgraaf S (1947) Über die Reizung des Ferntastsinnes bei Fischen und Amphibien. Experientia 3:206–208.

Dijkgraaf S (1952) Über die Schallwahrnehmung bei Meeresfischen. Z Vergl Physiol 34:104–122.

Dijkgraaf S (1962) The functioning and significance of the lateral line organs. Biol Rev 38:51–105.

Dijkgraaf S, Verheijen F (1950) Neue Versuche über das Tonunterscheidungsvermögen der Elritze. Z Vergl Physiol 32:248–256.

Elepfandt A (1982) Accuracy of taxis response to water waves in the clawed toad (*Xenopus laevis* Daudin) with intact or with lesioned lateral line system. J Comp Physiol 148:535–545.

Elepfandt A, Seiler B, Aicher B (1986) Water wave frequency discrimination in the clawed frog *Xenopus laevis*. J Comp Physiol 157:255–267.

Enger PS (1981) Frequency discrimination in teleosts – central or peripheral? In: Tavolga WN, Popper AN, Fay RR (eds) Hearing and Sound Communication in Fishes. New York: Springer-Verlag, pp. 243–255.

Fay RR (1970) Auditory frequency discrimination in the goldfish (*Carassius auratus*). J Comp Physiol Psychol 73:175–180.

Fay RR (1978) Phase-locking in goldfish saccular nerve fibers accounts for frequency discrimination capacities. Nature 275:320–322.

Fay RR (1981) Coding of acoustic information in the eighth nerve. In: Tavolga WN, Popper AN, Fay RR (eds) Hearing and Sound Communication in Fishes. Berlin: Springer-Verlag, pp. 189–222.

Fay RR (1984) The goldfish ear codes the axis of acoustic particle motion in three dimensions. Science 225:951–984.

Fay RR (1988a) Hearing in Vertebrates: A Psychophysics Databook. Winnetka, IL: Hill-Fay Associates.

Fay RR (1988b) Peripheral adaptations for spatial hearing in fish. In: Atema J, Fay RR, Popper AN, Tavolga WN (eds) Sensory Biology of Aquatic Animals. New York: Springer-Verlag, pp. 711–732.

Flock Å (1971) The lateral line organ mechanoreceptors. In: Hoar DJ, Randall RR (eds) Fish Physiology, Vol. V: Sensory Systems. New York: Academic Press, pp. 241–263.

Fritzsch B, Wahnschaffe U (1987) Electron microscopical evidence for common inner ear and lateral line efferents in urodeles. Neurosci Lett 81:48–52.

Goldberg JM, Fernández C (1975) Vestibular mechanisms. Annu Rev Physiol 37:129–162.

Görner P, Moller P, Weber W (1984) Lateral-line input and stimulus localization in the African clawed toad, *Xenopus laevis*. J Exp Biol 108:315–328.

Gray J (1984) Interaction of sound pressure and particle acceleration in the excitation of the lateral-line neuromasts of sprats. Proc R Soc Lond B 220:299–325.

Hawkins A (1981) The hearing capabilities of fish. In: Tavolga WN, Popper AN, Fay RR (eds) Hearing and Sound Communication in Fishes. New York: Springer-Verlag, pp. 109–137.

Hawkins A, Horner K (1981) Directional characteristics of primary auditory neurons from the codfish ear. In: Tavolga WN, Popper AN, Fay RR (eds) Hearing and Sound Communication in Fishes. New York: Springer-Verlag, pp. 311–328.

Hawkins A, Sand O (1977) Directional hearing in the median vertical plane by the cod. J Comp Physiol 122A:1–8.

Hillman D, McLaren JW (1979) Displacement configuration of semicircular canal cupulae. Neuroscience 4:1989–2000.

Hudspeth AJ (1983) Mechanoelectrical transduction by hair cells in the acousticolateralis system. Annu Rev Neurosci 6:187–215.

Jacobs DW, Tavolga WN (1968) Acoustic frequency discrimination in the goldfish. Anim Behav 16:67–71.

Janssen J, Coombs S, Hoekstra D, Platt C (1987) Anatomy and differential growth of the lateral line system of the mottled sculpin *Cottus bairdi* (Scorpaeniformes: Cottidae). Brain Behav Evol 30:210–219.

Kalmijn AJ (1988) Hydrodynamic and acoustic field detection. In: Atema J, Fay RR, Popper AN, Tavolga WN (eds) Sensory Biology of Aquatic Animals. New York: Springer-Verlag, pp. 711–732.

Lewis ER, Baird RA, Leverenz E, Koyama H (1982) Inner ear dye injection reveals peripheral origins of specific sensitivities. Science 215:1641–1643.

Lewis ER, Leverenz E, Bialek WA (1985) The Vertebrate Inner Ear. Boca Raton, FL: CRC Press.

Lowenstein O (1971) The labyrinth. In: Hoar WS, Randall DJ (eds) Fish Physiology, Vol. V: Sensory Organs. New York: Academic Press, pp. 207–240.

Lowenstein O (1974) Comparative morphology and physiology. In: Kornhuber HH (ed) Vestibular System: Basic Mechanisms, Handbook of Sensory Physiology, Vol. VI, Pt. 1. New York: Springer-Verlag, pp. 75–120.

Lowenstein O, Roberts TDM (1951) The localization and analysis of the responses to vibration from the isolated elasmobranch labyrinth. A contribution to the problem of the evolution of hearing in vertebrates. J Physiol (Lond) 114:471–489.

Müller U, Schwartz E (1982) Influence of single neuromasts on prey localizing behavior of the surface-feeding fish, *Aplocheilus lineatus*. J Comp Physiol 149:399–408.

O'Leary DP, Dunn RF, Honrubia V (1976) Analysis of afferent responses from isolated semicircular canal of the guitarfish using white-noise inputs. I. Correlation of response dynamics with receptor innervation. J Neurophysiol 39:631–644.

Pettersen L (1980) Frequency discrimination in the bullhead, *Cottus scorpius*, a fish without a swimbladder. Masters Thesis, University of Oslo, Oslo, Norway. Cited by Enger (1981).

Plassmann W (1983) Sensory modality interdependence in the octaval system of an elasmobranch. Exp Brain Res 50:283–292.

Platt C (1983) The peripheral vestibular system of fishes. In: Northcutt RG, Davis RE (eds) Fish Neurobiology, Vol. 1: Brainstem and Sense Organs. Ann Arbor: University of Michigan Press, pp. 89–123.

Platt C (1988) Equilibrium underwater: Signals, senses, and steering in the vertebrates. In: Atema J, Fay RR, Popper AN (eds) Sensory Biology of Aquatic Animals. New York: Springer-Verlag, pp. 783–809.

Platt C, Popper AN (1981) Fine structure and function of the ear. In: Tavolga WN, Popper AN, Fay RR (eds) Hearing and Sound Communication in Fishes. New York: Springer-Verlag, pp. 2–38.

Popper AN, Fay RR (1980) Comparative Studies of Hearing in Vertebrates. New York: Springer-Verlag.

Popper AN, Fay RR (1984) Sound detection and processing by teleost fish: A selective review. In: Bolis L, Keynes RD, Maddrell SHP (eds) Comparative Physiology of Sensory Systems. Cambridge, England: Cambridge University Press, pp. 67–101.

Popper AN, Platt C, Saidel WM (1982) Acoustic functions in the fish ear. Trends Neurosci 5:276–280.

Popper AN, Rogers PH, Saidel WM, Cox M (1988) The role of the fish ear in sound processing. In: Atema J, Fay RR, Popper AN (eds) Sensory Biology of Aquatic Animals. New York: Springer-Verlag, pp. 687–710.

Precht W (1979) Vestibular mechanisms. Annu Rev Neurosci 2:265–289.

Ramprashad F, Landolt JP, Money KE, Laufer J (1986) Comparative morphometric study of the vestibular system of the Vertebrata: Reptilia, Aves, Amphibia, and Pisces. Acta Otolaryngol (Suppl) 427:1–42.

Rogers P, Popper A, Hastings M, Saidel W (1988) Processing of acoustic signals in the auditory system of bony fish. J Acoust Soc Am 83:338–349.

Sand O (1974) Directional sensitivity of microphonic potentials from the perch ear. J Exp Biol 60:881–889.

Sand O (1981) The lateral line and sound reception. In: Tavolga WN, Popper AN, Fay RR (eds) Hearing and Sound Communication in Fishes. New York: Springer-Verlag, pp. 459–480.

Sand O, Enger PS (1973) Function of the swimbladder in fish hearing. In: Moller A (ed) Basic Mechanisms in Hearing. New York: Academic Press, pp. 893–910.

Sand O, Karlsen HE (1986) Detection of infrasound by the Atlantic cod. J Exp Biol 125:197–204.

Saunders JC, Dear SP (1983) Comparative morphology of stereocilia. In: Fay R, Gourevitch G (eds) Hearing and Other Senses: Presentations in Honor of E.G. Wever. Groton, CT: Amphora Press, pp. 175–198.

Schuijf A (1981) Models of acoustic localization. In: Tavolga WN, Popper AN, Fay RR (eds) Hearing and Sound Communication in Fishes. New York: Springer-Verlag, pp. 267–310.

Sento S, Furukawa T (1979) Intra-axonal labelling of saccular afferents in the goldfish, *Carassius auratus*: Correlations between morphological and physiological characteristics. J Comp Neurol 258:352–367.

Stetter H (1929) Untersuchungen über den Gehörsinn der Fische, besonders von *Phoxinus laevis* L. und *Amiurus nebulosus* Raf. Z Vergl Physiol 9:339–477.

Tavolga WN (1982) Auditory acuity in the sea catfish (*Arius felis*). J Exp Biol 96:367–376.

Tavolga WN, Popper AN, Fay RR (eds) (1981) Hearing and Sound Communication in Fishes. Berlin: Springer-Verlag.

Werner CF (1929) Experimente über die Funktion der Otolithen bei Knochenfischen. Z Vergl Physiol 10:26–35.

Wiese K (1988) The representation of hydrodynamic parameters in the CNS of the crayfish *Procambarus*. In: Atema J, Fay RR, Popper AN, Tavolga WN (eds) Sensory Biology of Aquatic Animals. New York: Springer-Verlag, pp. 665–683.

Wit HP, Bleeker JD (1982) Sensitivity of the vestibular system to acoustic stimuli. Arch Otorhinolaryngol 234:205–211.

33

Comparisons Between Electrosensory and Mechanosensory Lateral Line Systems

DAVID BODZNICK

33.1 Introduction

A volume on the lateral line would be incomplete without a consideration of its closely allied sense, electroreception. Electroreception is usually considered a specialized lateral line sense (Dijkgraaf 1962; Lissmann 1967; Szabo 1974). The marked similarities between mechanosensory lateral line (hereafter referred to as lateral line) and electrosensory systems, including embryology and morphology of receptors, receptor innervation and distribution, and CNS anatomy, are generally appreciated and held by most to reflect a close evolutionary relationship between these so-called lateralis senses. Since the discovery of electroreception 30 years ago in weakly electric teleost fishes (Lissmann 1958; Bennett and Grundfest 1959; Bullock et al. 1961; Fessard and Szabo 1961), the favored hypothesis has been that electroreceptors evolved as specialized derivatives of preexisting lateral line neuromasts (Lissmann 1958, 1967; Mullinger 1964; Bennett 1971). Major support for this hypothesis was the phylogenetic distribution of the two senses as it was then known. In the last volume on the lateral line, published 20 years ago, Lissmann (1967) stated the case simply: "The ordinary neuromast is the rule, the ampullary organ [electroreceptor] the exception."

Over the past 20 years, however, much has been learned of the anatomy and physiology of electrosensory systems, and in the past 10 years a different picture of the phylogenetic distribution of electroreception has emerged. In the light of this information it is appropriate to reexamine the relationship between the lateral line and electroreception. Although common evolutionary histories must account for many similarities between these sensory systems, similarities could also derive from ontogenetic constraints or from requirements for processing similar stimulus attributes. This chapter first considers the nature of the relationship between the lateral line and electroreception and briefly compares the senses with regard to phylogenetic distribution, ontogeny, and natural stimulus features. The second part of the chapter considers possible explanations for one fundamental distinction between the senses, the absence in all electroreceptors of efferent innervation, which is found in nearly all lateral line neuromasts.

TABLE 33.1. Similarities between lateral line and electrosensory systems.

1.	Dermal receptor organs distributed over the head (and body) innervated by lateral line nerves.
2.	Receptor cells (*hair cells*) in a support cell epithelium make occluding junctions with support cells and have presynaptic dense bodies, and apical microvilli and/or a kinocilium.
3.	Receptors and their innervation arise ontogenetically from dorsolateral epidermal placodes.
4.	Primary afferents project to alar medullary nuclei with cerebellarlike organization.
5.	Medullary nuclei have commissural connections and lemniscal projections to midbrain.

Although much of what is said in the chapter applies equally to the phasic, high-frequency electroreceptor systems (also termed tuberous) of weakly electric teleosts and the tonic, low-frequency electroreceptor systems (usually termed ampullary) found in all electroreceptive anamniotes, major emphasis in the chapter is given to the tonic, low-frequency systems that represent the more generalized or primitive condition. The term *tonic* electroreceptors, based on physiological properties (Bennett 1971), rather than the more common *ampullary* electroreceptors, based on structure, is used here when referring to all low-frequency electroreceptors, because the low-frequency electroreceptors of petromyzontids are not ampullary in structure (Ronan and Bodznick 1986). For comprehensive reviews of electroreception see Bullock (1982) and Bullock and Heiligenberg (1986).

33.2 Morphological Similarities Between Lateral Line and Electrosensory Systems

Although the exact relationship between the lateral line and electrosensory systems is uncertain, numerous shared features (Table 33.1), some unique to these systems, are evidence of a close evolutionary link. Foremost is the fact that electroreceptors and neuromasts are innervated by primary afferent fibers in the same lateral line nerves, the somata of which are found together in the same lateral line ganglia (McCready and Boord 1976; Northcutt 1985). The presence of their somata in the same ganglia may also reflect the embryonic origins of electroreceptors and neuromasts from the same ectodermal placodes, considered further below. Electroreceptors and neuromasts have similar wide distributions on the body surface, and individual lateral line nerve rami innervate both receptor types in a given area (Figs. 33.1, 33.2). The receptor cells of electroreceptors

▶

FIGURE 33.1. Distribution and innervation of lateral line and electrosensory organs in skates. Left: Lateral line canal system and neuromasts (black dots within canals innervated by lateral line nerves) are shown on the dorsal surface of *Raja batis* (from Ewart and Mitchell 1892). Right: Electroreceptor organs, ampullae of Lorenzini, on the dorsal surface of a similar skate, *Raja erinacea*. Note that ampullary organs occur in three clusters—hyoid (H), buccal (B), and superficial ophthalmic (SO)—each innervated by a separate anterior lateral line nerve ramus which also innervates lateral line neuromasts on the same part of the body.

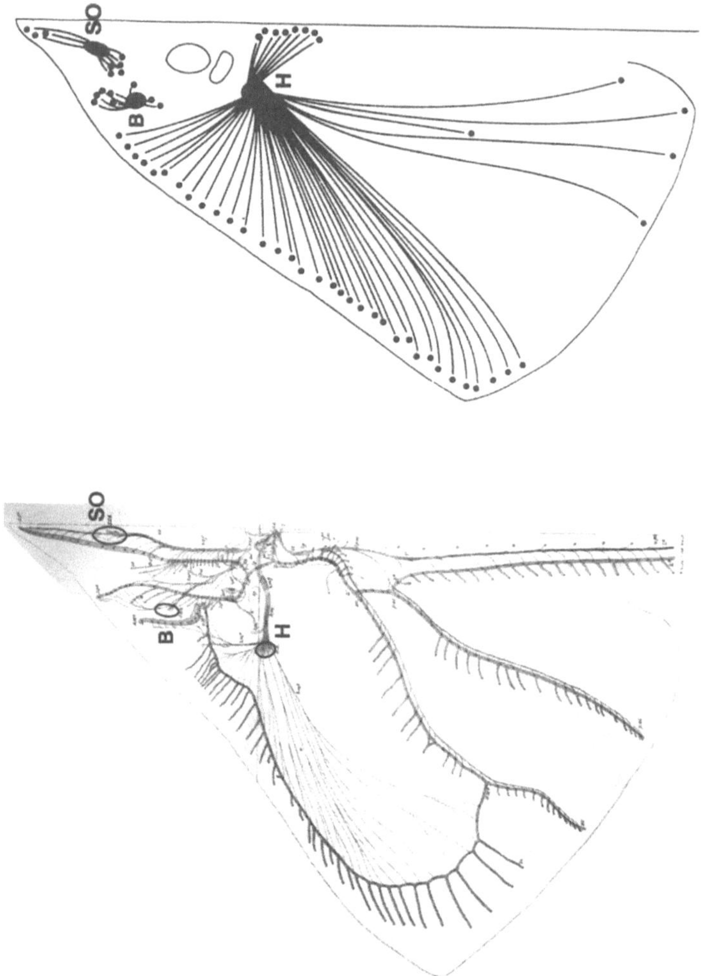

FIGURE 33.1. Distribution and innervation of lateral line and electrosensory organs in skates. Left: Lateral line canal system and neuromasts (black dots within canals innervated by lateral line nerves) are shown on the dorsal surface of *Raja batis* (from Ewart and Mitchell 1892). Right: Electroreceptor organs, ampullae of Lorenzini, on the dorsal surface of a similar skate, *Raja erinacea*. Note that ampullary organs occur in three clusters—hyoid (H), buccal (B), and superficial ophthalmic (SO)—each innervated by a separate anterior lateral line nerve ramus which also innervates lateral line neuromasts on the same part of the body.

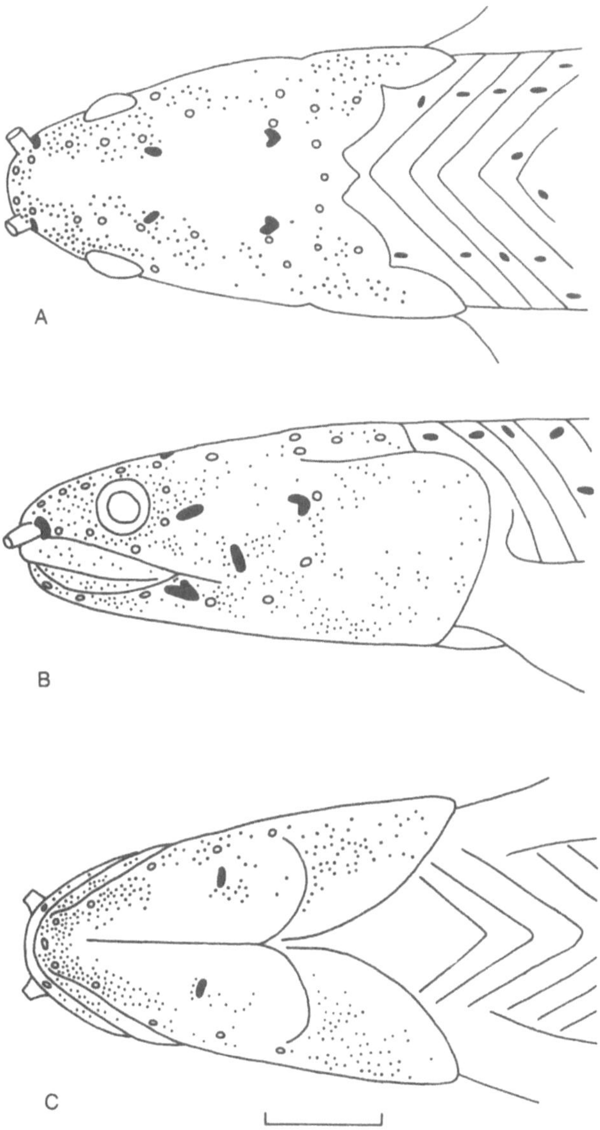

FIGURE 33.2. Distribution of ampullary electroreceptor organs (small dots), free neuromasts (solid black ovals), and lateral line canal pores (open circles) on the head of *Polypterus*, a primitive freshwater actinopterygian fish. Bar = 1 cm. (From Northcutt 1986b.)

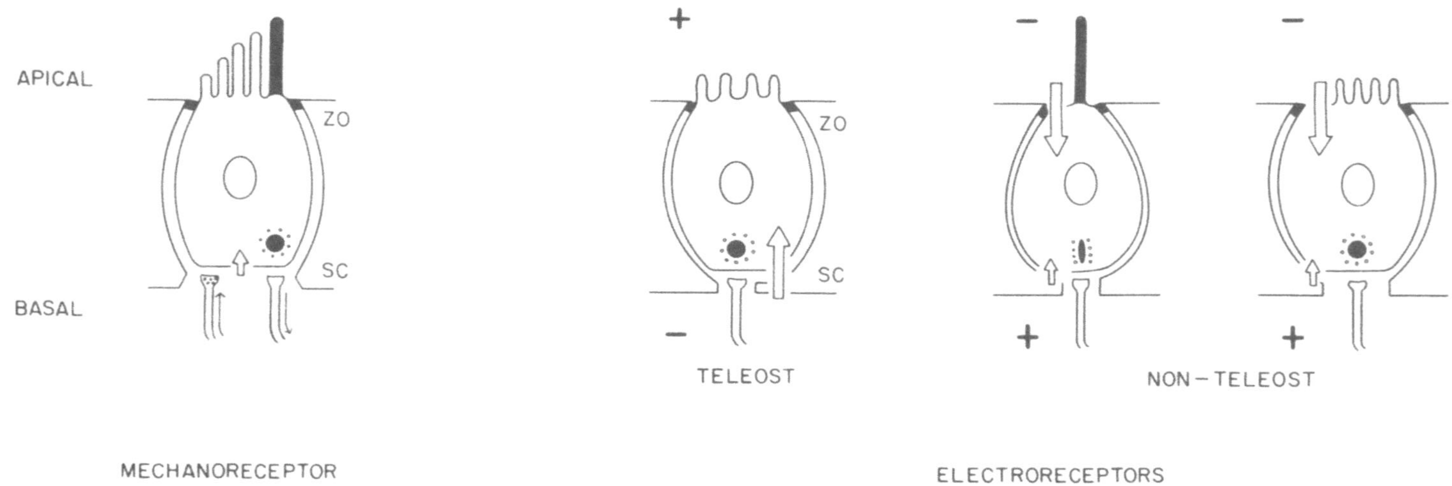

FIGURE 33.3. Diagrammatic representation of neuromast mechanoreceptor hair cell and typical teleost and nonteleost electroreceptor cells. Stimulus polarity for excitatory response of electroreceptors is indicated. SC, support cell; ZO, zonula occludens. Main inward Ca currents (demonstrated or presumed) responsible for voltage sensitivity are indicated by large arrows. Although teleost receptors typically have only apical microvilli and nonteleosts a kinocilium or microvilli, a kinocilium and microvilli occur on electroreceptors of some teleosts and nonteleosts. Only mechanoreceptor hair cell has efferent innervation. (Extensively modified after Obara and Sugawara 1984.)

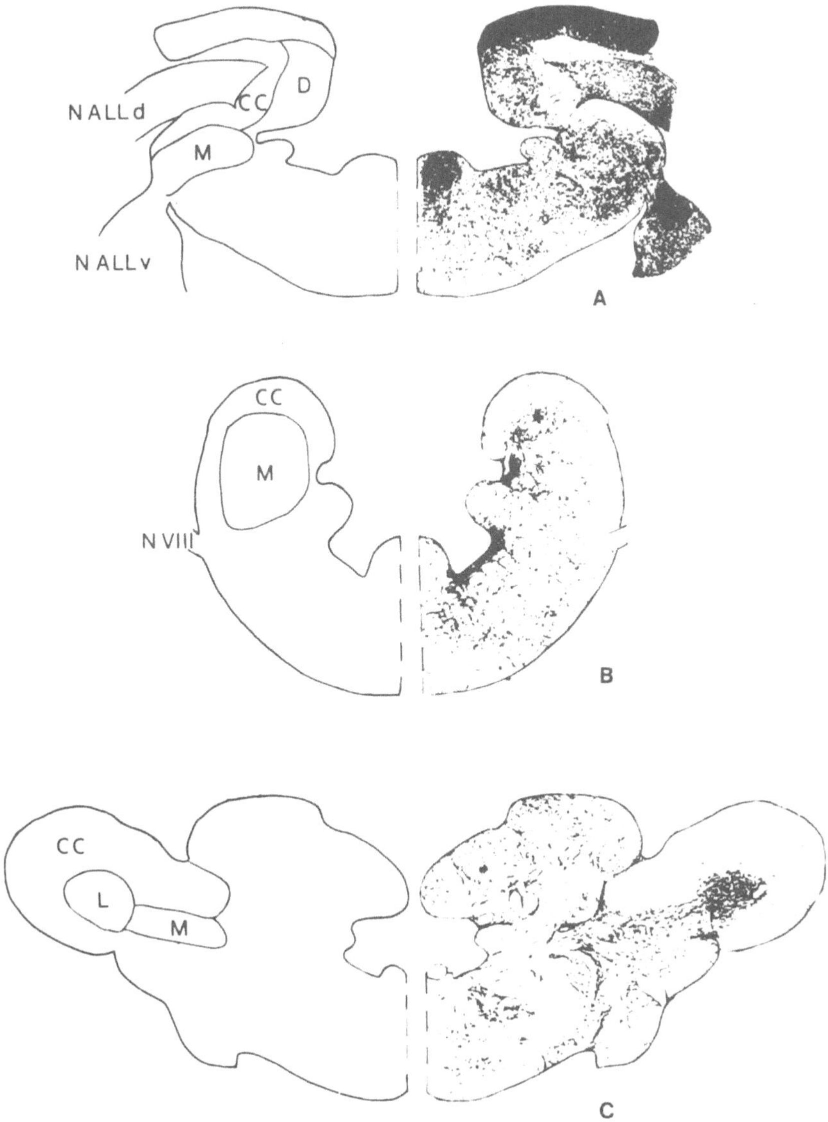

FIGURE 33.4. Transverse brain sections illustrating three patterns of medullary organization found in fishes and amphibians. (A) From the elasmobranch, *Squalus acanthias*, pattern representative of all *electroreceptive* nonteleost fishes and amphibians includes electrosensory dorsal nucleus (D) and lateralis mechanosensory medial nucleus (M). (B) From holostean fish, *Amia calva*, pattern representative of all holosteans, and nonelectroreceptive teleosts and amphibians includes only medial nucleus. Major exception is adult anurans that lack lateral line system and medial nucleus. (C) From teleost, *Ictalurus punctatus*, pattern found in electroreceptive teleosts includes medial nucleus and electroreceptive lateral line lobe (L). CC, cerebellar crest; N ALLd, anterior lateral line nerve dorsal ramus; N ALLv, anterior lateral line nerve ventral ramus; N VIII, eighth nerve. (From McCormick 1982.) A fourth pattern including dorsal nucleus but no medial nucleus is found in apodan amphibians that have electroreceptors but no lateral line system (McCormick and Braford 1988).

and neuromasts are also structurally similar (Fig. 33.3). In common with neuromast hair cells (Flock 1965), electroreceptor cells are secondary sense cells that make tight junctions with surrounding support cells and have apical microvilli and/or a kinocilium and presynaptic dense bodies (Szabo 1974; Zakon 1986). They have thus been considered modified hair cells, although, as noted previously (Northcutt 1986c), each of these features is found in other sensory cells besides hair cells.

The only feature clearly unique to hair cells—the polarized apical bundle of stereocilia and kinocilium (Hudspeth 1983; Flock 1965)—is missing in electroreceptor cells; perhaps this is not surprising given the structure's specialized role in mechanosensory transduction. The primary afferent fibers of neuromasts and electroreceptors project to separate but adjacent medullary nuclei which have similar anatomical organizations (Fig. 33.4) and connections with other CNS nuclei (for reviews see Northcutt 1981; McCormick and Braford 1988). Unique to these nuclei is a broad molecular layer (sometimes termed cerebellar crest) comprising parallel fibers supplied by the vestibulolateral cerebellum. The molecular layers of the electroreceptive and mechanoreceptive nuclei are continuous with each other, although separate cerebellar granule populations supply the parallel fibers to the two nuclei at least in many fishes (Boord 1977; Bass 1982; Schmidt and Bodznick 1987).

If one accepts the hypothesis of a common evolutionary lineage, there are still many different scenarios by which electroreceptors and neuromasts may have evolved. In addition, while the statements made thus far apply to all electrosensory systems and electroreception has been discussed as if it were a monophyletic sense, the phylogenetic distribution of electroreception and both anatomical and physiological evidence indicate that electrosensory systems have evolved several different times independently among anamniotes (Bullock et al. 1982, 1983). The same evolutionary scenario was apparently not followed in each case.

33.3 Phylogenetic Distribution and Evolution of the Lateralis Senses

Lateral line systems and neuromasts are found in virtually all aquatic anamniotes with the apparent exception of hagfishes (Fig. 33.5) (Fernholm 1985). This distribution and numerous characters shared by all of the systems indicate that they are homologous and that all have been passed down from the same early vertebrate ancestor (see Northcutt Chapter 3 for a comprehensive review). The heterostracans, which are the earliest vertebrates to appear in the fossil record, already had well-developed lateral line systems (Moy-Thomas and Miles 1971). In contrast, the distribution of electroreception indicates a polyphyletic origin for the sense, once in nonteleosts and two or more times in teleosts (Fig. 33.5) (Bullock et al. 1982, 1983). Among teleosts, where electrosensory systems were first

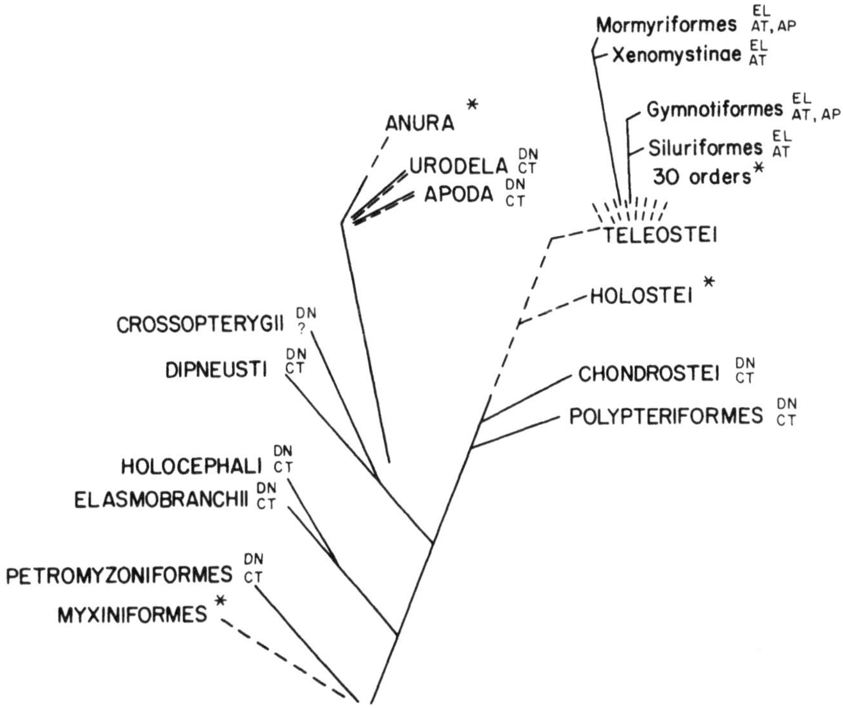

FIGURE 33.5. Distribution of electrosensory systems among the major taxa of anamniotes. Members of all taxa represented except hagfishes, Myxiniformes, possess mechanoreceptive neuromasts. Electroreception occurs in all taxa except those indicated by asterisk; loss of electroreception in a radiation (e.g., that leading to Holostei and Teleostei) is indicated by dashed line. All nonteleost electrosensory systems are apparently homologous and characterized by medullary dorsal octavolateralis nucleus (DN) and tonic electroreceptors preferring cathodal stimulus polarity (CT). Teleost electrosensory systems have independently evolved two to four times with medullary electroreceptive lateral line lobe (EL) and receptors preferring anodal polarity. AT, anodal tonic electroreceptors; AP, anodal phasic electroreceptors. (Modified from Bullock et al. 1982).

discovered, electroreception is the rare exception, as Lissmann (1967) stated. Fish in only four of more than 30 teleost orders are electroreceptive, and holostean fishes, the closest living relatives to teleost ancestors, also lack electroreception. The phylogenetic distribution indicates that electroreception evolved anew at least two different times in teleosts. Differences in receptor physiology and medullary anatomy also distinguish teleost from all nonteleost electrosensory systems and indicate that they are not homologous.

For teleosts the hypothesis that electroreceptors arose by modification of preexisting neuromasts still seems the most plausible. A class of free neuromasts, for example, could have been specialized for electrosensitivity (Northcutt 1986b). As mentioned previously, both neuromasts and electroreceptors are innervated by cells of the same lateral line ganglia. By virtue of their location in the skin, neuromasts are probably the receptors most sensitive to electric stimuli in nonelectroreceptive teleosts (Regnart 1931), and the switch from mechanoreceptor to electroreceptor may have involved only minor changes in receptor cells (Fig. 33.3), perhaps in conjunction with increased skin resistance (Bennett 1971). Changes in the receptors would include a loss of the apical specialization for mechanosensitivity and an increase in the voltage sensitivity of the basal face that already exists in mechanoreceptive hair cells for conversion of receptor potential to transmitter release (Bennett 1971). Teleost electroreceptor cells bear apical microvilli, but they do not have the geometrical staircase configuration of hair cells; a kinocilium is absent in all teleost electroreceptors except those in xenomystids (Jørgensen and Bullock 1987). An increase in voltage sensitivity of the basal membrane might have been accomplished by a simple increase in density of the voltage-sensitive calcium channels (Obara and Sugawara 1984). In all teleost electroreceptor cells, voltage sensitivity appears to be restricted to the basal membrane as in hair cells (Bennett and Obara 1986).

Also consistent with this origin for teleost electroreceptors are the facts that the receptor cells to afferent fiber synapses in teleost electroreceptors, as in neuromast hair cells, are characterized by a presynaptic density which is usually spherical or club shaped (Szabo 1974; Jørgensen and Bullock 1987) and that the medullary target of electrosensory afferents—the electroreceptive lateral line lobe—appears to be a derivative of the medial (octavolateralis) nucleus which is the medullary lateral line nucleus (Fig. 33.4) (McCormick 1982). Although electroreceptors evolved at least twice in teleosts, the similarities of the systems (Finger et al. 1986) suggest that similar evolutionary scenarios could have been followed in each case. In addition, as the more generalized condition, it is presumed that tonic (ampullary) electroreceptors arose first in teleosts and tuberous electroreceptors were secondarily derived in weakly electric teleosts from preexistent ampullary organs.

Although electroreception is rare in teleosts, all nonteleost fishes, except for the holostean fishes and hagfishes, and many amphibians outside of anurans are electroreceptive (Fig. 33.5) (Bullock et al. 1982, 1983; Fritzsch and Münz 1986). Surprisingly, it was an accurate interpretation of medullary organization that was the major breakthrough in this work. Boord and his colleagues (McCready and Boord 1976; Boord and Campbell 1977) correctly inferred from anatomical studies on elasmobranchs that the dorsal and medial (octavolateralis) nuclei were the primary medullary nuclei of electroreception and lateralis mechanoreception, respectively; this was later confirmed physiologically (Bodznick and Northcutt 1980). Based on this information McCormick (1982) hypothesized that (1) all animals with a dorsal nucleus including lampreys are electroreceptive; (2) holosteans and most teleosts, which have only a medial nucleus, lack electroreception; and (3) a few teleosts with no dorsal nucleus but rather an electrosensory lateral line lobe, which is topologically distinct from the dorsal nucleus, had reinvented electroreception (Fig. 33.4). This hypothesis has subsequently been well supported by physiological studies (Bodznick and Northcutt 1981; Bullock et al. 1982, 1983). Many amphibians of the orders Apoda and Urodela were later added to the list of those with a dorsal nucleus and electroreception (for review see Fritzsch and Münz 1986). A fourth pattern of medullary organization has recently been found in some apodans that have electroreceptors but no neuromasts (Braford, McCormick, and Saidel personal communication; Fritzsch and Münz 1986) and correspondingly have a medullary dorsal nucleus but no medial nucleus (McCormick and Braford 1988).

In addition to a dorsal nucleus, all nonteleost electroreceptive anamniotes have tonic electroreceptors that are activated by the cathodal pole of an electric field stimulus (teleost electroreceptors are activated by the anode) and that are innervated by only the anterior lateral line nerve. These characters, and the phylogenetic distribution, indicate that nonteleost electrosensory systems are homologous with each other but not with teleosts. The electrosensory systems in lampreys and primitive jawed fishes and amphibians apparently stemmed from the same early vertebrate that was ancestral to both petromyzontid agnathans and gnathostomes. Thus, nonteleost electroreceptors may be as old as or older than lateral line neuromasts, and it is equally likely that neuromasts evolved from preexisting electroreceptors, that electroreceptors evolved from preeixsting neuromasts, or that the two evolved separately from a third receptor such as the ciliated receptors of protochordates (Northcutt 1986c; Jørgensen 1982, Chapter 6). The phylogenetic distributions of electroreceptors and neuromasts do not favor any one of these possibilities over the others. Hagfishes, which may be considered a sister group to lampreys and gnathostomes, appear to lack electroreceptors (Bullock et al. 1982, 1983) as well as neuromasts (Fernholm 1985) and thus contribute little to clarifying this issue (Northcutt 1985, Chapter 3).

Finally, the physiological and morphological properties of the receptors also do not favor a particular scenario for the evolution of nonteleost electroreceptors and neuromasts (Fig. 33.3). Unlike neuromasts and teleost electroreceptors, the apical membrane of nonteleost electroreceptor cells is voltage-sensitive, and this accounts for their selectivity for the cathodal stimulus polarity (Bennett and Obara 1986). The apical structures of nonteleost electroreceptor cells are quite variable among taxa and include those with a single kinocilium only, microvilli only, or both kinocilium and microvilli (Szabo 1974; Jørgensen 1982, Chapter 6; Jørgensen and Bullock 1987). The presynaptic densities in nonteleost electroreceptor cells are most commonly in the form of synaptic ribbons or rods (Szabo 1974), but the electroreceptor cells of some nonteleost fishes and amphibians have spherical presynaptic densities like neuromasts (Jørgensen 1982; Fritzsch and Münz 1986).

33.4 Embryonic Origins of Electroreceptors and Neuromasts

A comparison of the ontogeny of neuromasts and electroreceptors is of interest because a common embryonic origin is evidence for the common evolutionary origin of the receptors. However, if neuromasts and electroreceptors arise by similar embryonic processes, whether or not they share a common evolutionary lineage, then ontogenetic constraints might also account for similarities between the two senses. Such ontogenetic constraints have been suggested as a possible reason for similarities between independently evolved electrosensory systems in teleosts (Finger et al. 1986).

Lateral line organs were a popular subject of early embryological studies on amphibians and fishes (Wright 1951 for review), and key findings from these studies have recently been confirmed (Winklbauer and Hausen 1983; Metcalfe 1985, Chapter 7). Experimental studies on amphibians demonstrate that both the sensory and supporting cells of lateralis neuromasts and the ganglion cells that innervate them arise entirely from a series of pre- and postotic epidermal placodes (Stone 1922; Winklbauer and Hausen 1983). There is no contribution to the organs from either neural crest or epidermal cells outside the placodal tissue. Despite early reports to the contrary (Landacre 1910; Bailey 1937), observations on normal embryos show that in fishes the neuromasts are also laid down by migratory pre- and postotic placodes (Metcalfe 1985, Chapter 7). Much less has been known about the embryonic origin of electroreceptors, as their existence was not appreciated by the early embryologists. However, circumstantial evidence (Northcutt 1986b) and recent observations on early embryonic stages of amphibians and teleost fishes indicate that electroreceptors and neuromasts develop from the same dorsolateral placodes. First is the fact that neurons innervating electroreceptors and neuromasts share the same ganglia. Second, in amphibians (Hetherington and Wake 1979; Fritzsch and Wahnschaffe 1983) and some embryonic (Vischer et al. 1987) and adult fishes (Fig. 33.2), the distribution of electroreceptors parallels that of the neuromasts as if they were laid down by the same migrating placodes. Finally, in recent observations of the skin of developing axolotls, *Ambystoma mexicanum*, stained to highlight the placodes and lateralis receptor primordia, the earliest recognizable electroreceptor primordia (three or four cells) were found within the same preotic placodes that give rise to the neuromasts of the head (Northcutt 1986a).

33.5 Comparisons of the Natural Stimuli for Lateral Line and Electrosense

As detailed in many chapters of this volume, neuromasts are mechanosensory organs whose hair cells respond to a wide range of natural stimuli that result in relative motion of water and fish. Early recordings from isolated ampullae of Lorenzini revealed a sufficient sensitivity to mechanical stimuli that they too were for a time considered mechanoreceptors (Murray 1960). However, recordings from the ampullary electroreceptors in freely swimming dogfish indicated little sensitivity to mechanical disturbances (Dijkgraaf and Kalmijn 1966), and mechanical stimuli have not been shown to be behaviorally relevant stimuli for any electroreceptors. Therefore, the natural stimuli for these two modalities appear to be distinct and nonoverlapping. Perhaps not coincidentally, the central projections of the lateral line and electrosensory afferents in the medulla are adjacent but also completely nonoverlapping (Maler et al. 1974; Bell and Russell 1978; Bodznick and Schmidt 1984). The same cannot be said about lateral line and eighth nerve fibers where several instances of overlap of central project ions are known (see Boord and Montgomery Chapter 16; McCormick Chapter 17; Will Chapter 18) that may reflect overlap in their stimulus ranges.

Despite the fact that the natural stimuli for the lateral line and electroreception are distinct, further comparison of the properties of each might permit insight into why the systems are organized as they are. It could be argued that some similarities that exist between the systems result from requirements for processing similar stimulus attributes such as the spatial distribution of stimulus intensities, orientation, or other features discussed below. However, perhaps most striking is one way in which the stimuli for the two modalities differ. Whereas the midwater local flow fields and surface waves that are stimuli for the lateral line sense spread away from their sources with finite velocities, important natural electric field stimuli are nonpropagating signals that appear in their entirety virtually instantaneously—i.e., they are electrostatic fields (Hopkins 1986). A major consequence of this distinction is that, unlike the lateral line, in the electrosense there are no differences in stimulus arrival time or phase at different receptors to serve as cues of source azimuth. These cues appear to be used by the lateral line system in localizing the source of surface waves in surface-feeding amphibians and fishes (Bleckmann et al. Chapter 25; Görner and Mohr Chapter 27). In addition, there is no distortion of electric field stimuli due to propagation-dependent phenomena. Although this permits preservation of important temporal features of electric signals such as the electric organ discharges of weakly electric teleosts (Hopkins 1986), it also precludes the use of propagation-dependent changes in the stimulus that might otherwise provide useful information. Such cues (e.g., changes in frequency composition or temporal dispersion of frequencies) are used by surface-feeding fishes to estimate the distance of surface wave sources (Bleckmann 1988; Bleckmann et al. Chapter 25).

Source localization in the electric sense must be based on the spatial distribution of electric field intensity and orientation. Perhaps this heavy dependence on spatial features accounts for the precise somatotopy that characterizes the medullary projections of primary afferents in all electrosensory systems. In contrast, the medullary projections of mechanosensory lateralis afferents in the medial nucleus appear to exhibit only a coarse somatotopy at best; although the terminal fields of the separate lateral line nerve rami are arranged in a somatotopic order in the medial nucleus, they overlap extensively (Bodznick and Schmidt 1984; Fritzsch et al. 1984; McCormick Chapter 17), and within those terminal fields the projections of the individual afferents are apparently not somatotopically arranged (Will et al. 1985; Will Chapter 18). Selective synaptic connections with second-order cells may result in a more precise medullary somatotopy than indicated by first-order projections, but it also seems conceivable that overlap in the terminal fields of afferents from distantly spaced neuromasts is required for the temporal comparisons useful for source localization.

Aside from the difference in availability of temporal cues, the natural stimuli for electroreception and the lateral line do have common attributes. In particular, for both senses important information is carried in the spatial distribution of the stimulus intensity, orientation, and polarity (direction). This accounts for the fact that both senses are characterized by widely distributed receptors (Figs. 33.1, 33.2) with directional sensitivity. Whereas the directional sensitivity of neuromasts is a product of aligned hair cells that are themselves directional receivers, the directional sensitivity of electrosensory organs is determined by the axis between the position of the receptor opening on the body and an internal reference point. For freshwater animals in which the body interior is nearly isopotential because of high skin resistivity, the effective reference point for all of the receptors is roughly the center of the body. For marine fishes where skin resistivity is low, the reference point is the location of the particular capsule in which the receptor's alveolus is located (Fig. 33.1).

The virtually uniform electric fields that result from electromagnetic induction (Kalmijn 1988a) represent one important category of electrosensory stimuli that in terms of stimulus attributes may have no close counterpart among lateral line stimuli. However, there appear to be several similarities between the dipole electric fields of other animals which are important stimuli for the electrosense and the dipole hydrodynamic flow fields which Kalmijn (1988b, Chapter 9) has argued are important and perhaps the most common stimuli for the lateral line. The movements of other animals are low-frequency sound sources that create such dipole flow fields. These local electric fields and hydrodynamic flows have in common their dipole spatial configuration (see also Hassan Chapter 10) and the fact that stimulus intensity attenuates exponentially with distance in both cases.

Within a couple of body lengths of local dipole electric fields, electroreceptive animals may potentially localize the field's source using spatial nonuniformities such as the curvature of field lines or by detecting the direction of the intensity gradient. Within a similar distance, the same features of the hydrodynamic flow field may be source location cues for the lateral line. However, more than a few body lengths from the source for both the dipole electric fields and hydrodynamic flow fields, these cues become extremely difficult to detect. Moreover, for such dipole sources the field direction, which is more easily determined at these distances, is not radial like the intensity gradient; that is, it does not point to the stimulus source except along the dipole axis.

Electric fishes must solve this problem to approach conspecifics from a distance of many body lengths using the dipole electric fields of the conspecific's electric organ discharges. Recent studies (Schluger and Hopkins 1987) indicate that they do so by simply aligning their bodies with the field lines and following them to their source. The fish's approach is thus indirect but always ends at the source (Fig. 33.6). A variation on this tactic, whereby the animal maintains a constant angle between its body and the field lines, is hypothesized to be used by elasmobranchs in approaching their prey (Kalmijn 1988a). A similar strategy is also suggested to guide fishes from distances of many body lengths into the vicinity of the sources of low-frequency dipole hydrodynamic flows (Kalmijn 1988b). However, Kalmijn (1988b, Chapter 9) suggests that, because at these greater distances the stimulus for the lateral line is extremely weak, the otolith organs of the ear are the detectors that determine the direction of the local hydrodynamic flow.

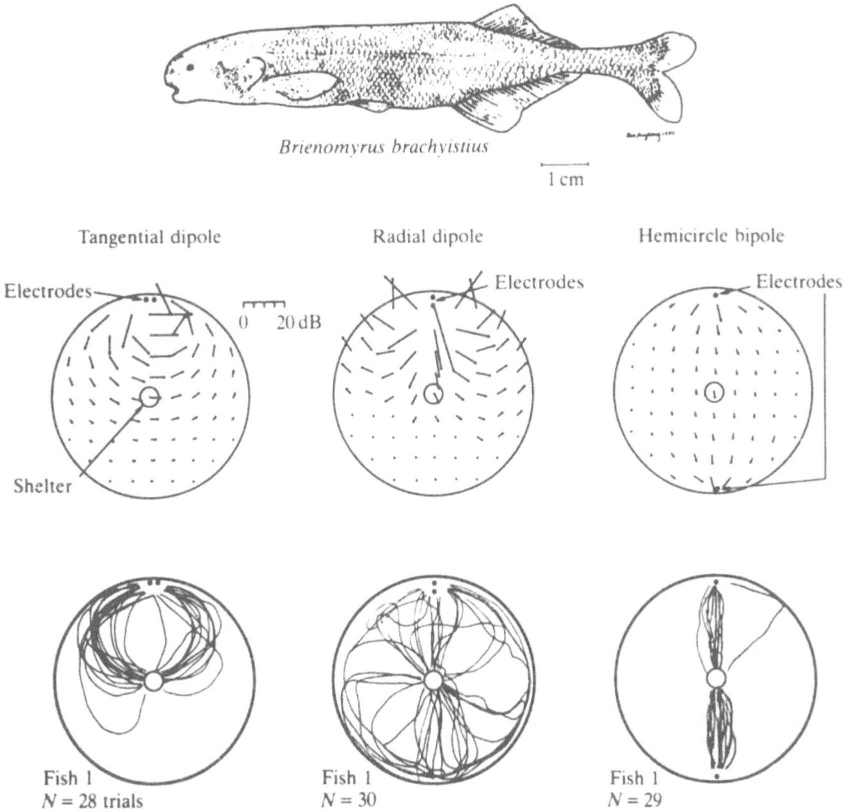

FIGURE 33.6. Approach paths of a resident mormyrid electric fish are shown to dipole electrodes producing a signal mimicking the electric organ discharge of a conspecific intruder. The dipole electrodes are positioned in three different configurations–tangential, radial, and hemicircle. These arrangements and the resultant dipole fields are shown as vector diagrams of field intensity and orientation in the top row. The paths taken by one fish over many trials for each configuration are charted on the bottom. Paths clearly indicate that the fish locates the source by following local current lines. (From Schluger and Hopkins 1987.)

33.6 Why Do Electroreceptors Lack Efferent Innervation?

The hair cells of nearly all lateralis neuromasts, except those of petromyzontids (Yamada 1973) and perhaps some free neuromasts in teleosts (Münz personal communication), have efferent innervation (Flock 1965; Roberts and Ryan 1971; Roberts and Meredith Chapter 22). Evidence from studies on both amphibians and fishes indicates that the efferent fibers make inhibitory synapses onto neuromast hair cells and suppress self-stimulation caused by the animal's own vigorous movements. Such self-generated sensory input is referred to as reafference (von Holst and Mittelstaedt 1950). Although electroreceptors are also subject to self-stimulation during an animal's activity, no electroreceptor organs have efferent innervation. What accounts for this fundamental distinction between neuromasts and electroreceptors? There are two kinds of explanations to be considered. First is the possibility that, although efferent innervation could play a useful function in electroreception, present-day electroreceptors lack such innervation owing to historical circumstances or pleiotropic effects of selection on other characters. The second possibility is that efferent innervation has been lost or has never evolved in electroreceptor organs because it has no functional importance for electroreception. Each of these will be considered in turn.

Assuming efferent innervation could serve a useful function in electroreception, what factors could explain its absence? As discussed earlier, the electroreceptors of nonteleosts may have antedated neuromasts. Even if nonteleost electroreceptors evolved from early neuromasts, this event may still have preceded the origin of lateralis efferent fibers, a possibility made plausible by the absence of efferents to lamprey neuromasts (Yamada 1973). Present-day electroreceptors might lack efferents because they have by chance not arisen in the course of electroreceptor evolution. This explanation is less satisfactory for the case of teleost electroreceptors, where it seems likely that electroreceptors evolved from preexisting neuromasts with efferent innervation. Even in this case, efferents to electroreceptors could have been lost owing to a stochastic process such as genetic drift or owing to a pleiotropic action of selection on other characters. However, such explanations are not very satisfying, given that electroreceptors have evolved several different times independently but in no case have efferent innervation. (It might be noted here that if some free neuromasts in teleosts indeed lack efferent innervation, then an origin of electroreceptors from these free neuromasts could also explain their lack of efferents in some teleosts.)

The alternative explanation is that efferent innervation has no functional importance for electroreception and so did not evolve in nonteleost electroreceptors or was lost in the evolution of teleost electroreceptors. This possibility can be assessed by further considering the role of efferents to neuromasts. Despite much recent interest in the function of efferent innervation in vestibular and auditory systems, lateral line efferents are still the most thoroughly explored and understood. Consistent findings from studies on amphibians (Görner 1967; Russell 1971), teleosts (Flock and Russell 1973a,b, 1976; Russell 1976), and elasmobranchs (Roberts and Russell 1972; Russell and Roberts 1972, 1974) indicate that the inhibitory action of lateral line efferents on the hair cells serves to protect these receptors and their synapses from overstimulation and adaptation or fatigue during particularly vigorous activity of the animal (also see Roberts and Meredith Chapter 22; Zottoli and Danielson Chapter 23). This preserves the system's sensitivity for the period immediately following the movement.

The evidence for this is that the intense firing of efferents (>20 to 40 imp/s) required to cause a significant reduction in primary afferent firing is only found immediately before and during vigorous self-generated movements or in response to stimuli that would elicit such movements in unparalyzed animals. In *Xenopus* efferent activation accompanies twitch but not slow muscle activity (Russell 1971), and in *Scyliorhinus* (Roberts and Russell 1972) efferent fibers are strongly

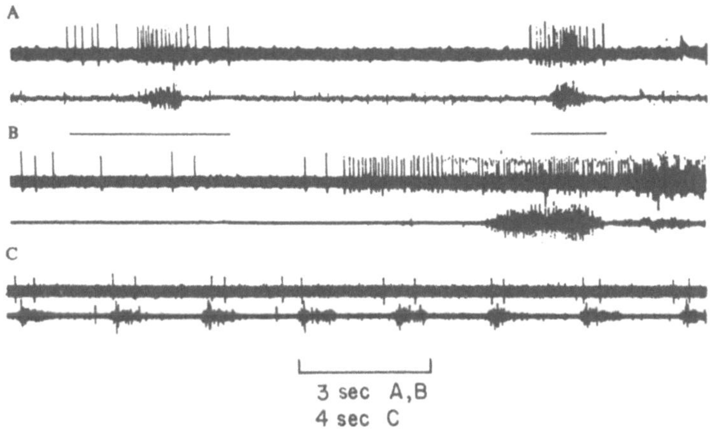

FIGURE 33.7. Activity of efferents in the posterior lateral line nerve of the dogfish, *Scyliorhinus canicula*, associated with body movement. In each case, top trace is a record of efferent impulses, bottom trace is EMG recorded from white muscle (A,B) or red muscle (C) of the trunk. (A) Efferent discharge directly precedes and accompanies white muscle activity in response to tactile stimulation of the trunk. Period of stimulation is indicated by lines beneath traces. (B) As in (A), but tactile stimulus is more vigorous and continuous. (C) Only weak efferent discharge accompanies steady swimming mediated by red muscle. (From Roberts and Russell 1972.)

activated only during vigorous escape-type movements mediated by white muscle (Fig. 33.7). Steady rhythmic swimming mediated by red muscle which results in modest lateral line reafference is accompanied by only weak efferent activation, which has little effect on afferent firing.

The function of the efferent system is apparently to protect the receptors from potentially deleterious overstimulation rather than to block all reafference due to movement (Roberts 1978). The lateral line efferent fibers cannot usually be activated by lateral line stimulation (Russell 1971; Roberts 1978). Thus they do not appear to mediate ordinary sensory feedback that might, for example, serve to enhance contrast between neuromasts on different parts of the body. Such functions might best be handled in the brain (as is true for electroreception, as described below). Efferents to the periphery may only be needed for protection of labile receptor or synaptic processes to preserve receptor sensitivity (Russell 1971).

The absence of efferent innervation of electroreceptors might then be understandable if stimuli that result from the animal's own activity are not so intense as to compromise electroreceptor sensitivity. This hypothesis is supported by our present knowledge of stimuli that generate electrosensory reafference. Since some electroreceptors including the ampullae of Lorenzini can be driven by at least strong mechanical stimuli (Murray 1960), movements might cause electrosensory reafference via either mechanical or electrical stimuli associated with them. However, the mechanosensitivity of electroreceptors does not approach that of lateralis neuromasts, and, as noted above, recordings from electroreceptor afferents in freely swimming dogfish, *Scyliorhinus* (Dijkgraaf and Kalmijn 1966), showed no responses to mechanical stimulation during swimming.

Several different types of electrical stimuli may also be generated by movements. The most potent of these for stimulating tonic electroreceptors are the low-frequency potentials generated by ventilatory activity. The actual source of these potentials is uncertain, although they are presumed to result from the modulation of dc potentials that exist across mucous and gill epithelia and skin by the opening and closing of mouth and gill openings (Kalmijn 1988a).

Ventilatory electrosensory reafference has been observed in many animals (Andrianov et al. 1974; Bodznick and Northcutt 1980; Bodznick and Preston 1983), and in some elasmobranchs primary afferents are modulated at rates that are 50% or more of the cells' dynamic range (Montgomery 1984; New and Bodznick 1987). However, even such vigorous self-stimulation does not appear to be strong enough to compromise the sensitivity of the electroreceptors. Montgomery (1984) found that electrosensory responses recorded in the brain of *Platyrhinoidis* to an extrinsic electric field stimulus were similar regardless of when the stimulus was presented in the ventilatory cycle. Other electric signals caused by movements appear to be less potent stimuli for electroreceptors. The electric fields from muscle impulse activity as recorded in the water adjacent to the fish are much weaker than ventilatory potentials (Kalmijn 1974), and they are also less effective stimuli for tonic electroreceptors because of their relatively high frequency composition. The phasic or tuberous electroreceptors of weakly electric teleosts are more sensitive to the high frequencies of muscle impulses, but their absolute sensitivity is lower than that of tonic receptors (Bullock 1982), and so they would not be expected to be strongly stimulated by muscle impulses.

Motional electric fields also accompany movements of an animal within the earth's magnetic field, but these are weak stimuli for even the most sensitive electroreceptors ($\sim$ 0.2 to 0.4 μV/cm at 1 m/s swimming speed) (Murray 1962; Kalmijn 1974). A last source of electric stimuli associated with movements is the streaming potentials that result from relative motion of water and fish (Kalmijn 1974, 1988a). The intensities of these fields are not well known but are likely to be quite weak. Consistent with this, electrosensory afferents in a skate (*Raja*) were not strongly stimulated by major changes in the flow of ventilatory water directed into the mouth and over the gills or by vigorous currents directed over the body surface (Bodznick unpublished observations).

The electric organ discharges of weakly electric teleosts can also be a source of unwanted reafference in both ampullary and one type of tuberous electroreceptors (Bell 1984). However, this reafference does not overstimulate the peripheral receptors but is effectively suppressed in the brain (see below). The intense electric organ discharge in strongly electric fish such as *Electrophorus* or *Torpedo* is one source of self-stimulation that would be expected to overstimulate tuberous and ampullary electroreceptors, although there have been no physiological studies of the electroreceptors in these animals. Based on the known sources of electrosensory reafference, it is possible to conclude that electroreceptor organs lack efferent fibers because, with rare exceptions, they are not subject to the same deleterious overstimulation due to the animal's activity as are neuromasts.

Recent attention has been given to a possible role of efferent innervation in controlling mechanical properties of outer hair cells of the cochlea (Brownell et al. 1985; Ashmore 1987). Although there is no direct evidence for such a role of lateral line efferents, Russell and Lowe (1983) have suggested that frequency-dependent changes in the microphonic potentials recorded from lateral line organs of perch, *Perca*, during efferent stimulation are consistent with changes in the mechanical properties of the lateral line hair cells. If an important function of lateralis efferents is to control mechanical properties of the hair cells, then the absence of mechanosensitivity and mechanoelectrical transduction in electroreceptors provides an alternative explanation for the absence of efferent innervation to these receptors (Bullock 1986).

33.7 Central Mechanisms of Reafferent Suppression

Although self-generated stimuli may not normally be deleterious to receptor processes in electroreception, electrosensory reafference could still mask or be confused with responses to extrinsic stimuli. Electrosensory systems do suppress unwanted reafference, but they do so by mechanisms within the brain rather than at the receptors. Since lateral line efferents only partially block vigorous self-stimulation of neuromasts and have little effect on reafference associated with more moderate movements, similar mechanisms could be used in the lateral line system to suppress unwanted mechanosensory reafference in the brain.

The Knollenorgan electroreceptors of mormyrid electric fishes are not used for active electrolocation but rather are specialized for the detection of the electric organ discharges of conspecifics. Responses of the Knollenorgans to the animal's own discharges are thus unwanted reafference. These responses, which consist of a single impulse at a short constant latency following each discharge, are blocked by a strong but very brief inhibitory corollary discharge signal in the medulla and midbrain that is accurately timed to coincide with the Knollenorgan discharge (Zipser and Bennett 1976; for review see Bell 1984). Evidence exists for a similar blockade of lateral line reafference in the medulla during escape responses mediated by Mauthner and other reticular motor neurons (Russell 1974; Zottoli and Danielson Chapter 23), although the inhibition is necessarily much longer in duration in this case.

The electric organ discharge of mormyrids also elicits unwanted reafference in ampullary electroreceptors which normally detect extrinsic low-frequency electric fields and are not involved in active electrolocation (for review see Bell 1984). However, unlike the Knollenorgan reafference, the ampullary reafference can be long-lasting and complex, including periods of both excitation and suppression. The particular response of individual receptors can differ and change with particular environmental conditions such as water conductivity or the presence of nonconductive barriers nearby. As simple blockade of receptor input throughout the reafference would interfere with the ability of the ampullary

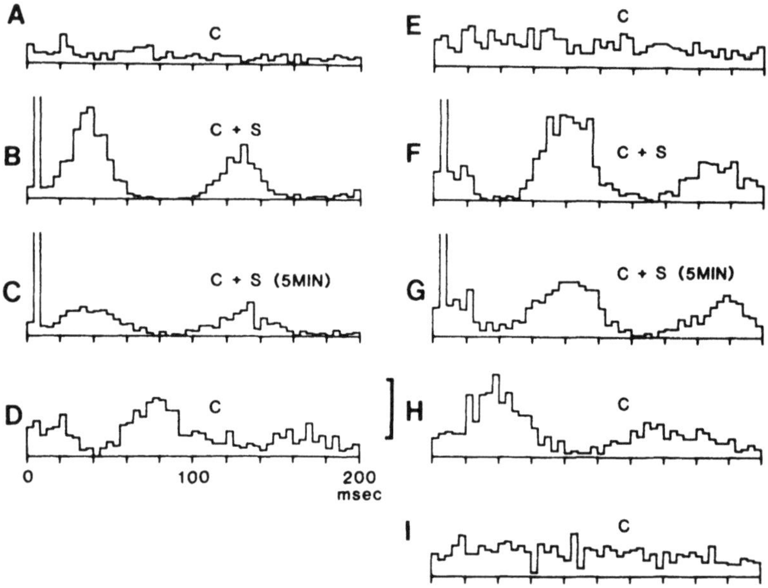

FIGURE 33.8. Demonstration of modifiable efference copy signal that normally cancels unwanted reafferent response in ampullary electrosensory cells of the medulla to electric organ discharge (EOD) in a mormyrid. In this test, fish's EOD is silenced by curare, but EOD command signal recorded near electric organ is used to trigger the presentation of extrinsic electric field stimuli and the histograms (number of impulses vs. time) of the medullary neuron responses. (A–D) Effect of pairing outside-negative, electric field stimulus (s) (5 μA magnitude, 5 ms duration, 5 ms delay) with EOD command. (A) Initial effect of command (c) alone. (B) Initial effect of command plus stimulus. (C) Command plus stimulus after pairing for 5 min. Note reduction due to addition of developing efference copy signal. (D) Effect of command alone immediately after 7 min of pairing stimulus and command; note negative-image efference copy signal. (E–I) Same cell tested as before but with outside-positive electric stimulus. (I) Effect of command alone 15 min after stimulus of F,G was turned off. (From Bell 1982.)

electrosensory system to sense meaningful extrinsic stimuli, the reafference is instead eliminated by the addition in medullary neurons of an efference copy signal that is the negative image of the unwanted reafference (Fig. 33.8). This efference copy signal is modifiable in the sense that it is continually updated to match changes that may take place in the reafferent response (Bell 1982). Such a central efference copy mechanism has clear advantages over direct inhibition of receptor cell activity for the suppression of long-lasting or complex reafferent responses. A central efference copy mechanism might also be useful for suppressing some kinds of lateral line reafference that are long-lasting and complex, so that simple blockade of incoming information throughout the reafference would interfere with normal sensing of extrinsic stimuli. The utility of such a true efference copy mechanism would require that either the particular activity and the lateral line reafference it generates be quite stereotyped so that a fixed efference copy signal

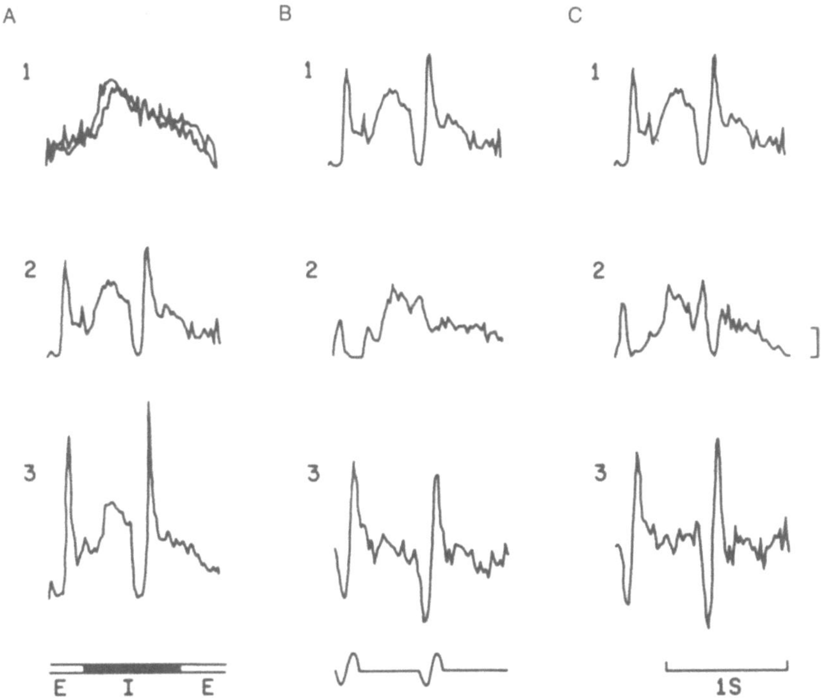

FIGURE 33.9. Histograms (number of impulses vs. time) of primary electrosensory afferent responses to ventilation and sinusoidal extrinsic electric fields in *Platyrhinoidis*, used to demonstrate efficacy of common mode rejection mechanism for canceling ventilatory reafference and enhancing differential responses to extrinsic fields. Animal's natural ventilation modulates activity of electroreceptors as in (A1). Ventilatory cycle (expiration, E; inspiration, I) depicted beneath (A3). (A2,3) Responses to ventilation and superimposed extrinsic fields [(A2), 25 μV/cm; (A3), 50 μV/cm] at times indicated below (B3). (B1) Receptor response to ventilation and extrinsic field as in (A2); receptor increases firing rate during head positive portion of stimulus cycle. (B2) As in (B1), but different receptor that is oppositely oriented on body and so increases firing during head negative portion of stimulus cycle. (B3) Computer subtraction of (B1)–(B2) demonstrates cancellation of common-mode ventilatory response and enhancement of extrinsic field response. (C1–3) Another example as in (B); (C2) is new receptor. (From Montgomery 1984.)

will always be a good negative image of the reafference, or else the activity must be repetitious over sufficient periods of time so that a modifiable efference copy signal can build up an accurate negative image of the reafference. Ventilatory movements or steady rhythmic swimming, as seen in many pelagic fishes, may be suitable activity for a central efference copy mechanism of reafference suppression.

A different central mechanism, not based on an efference copy signal (New and Bodznick 1987), used to block long-lasting and complex reafference in ampullary

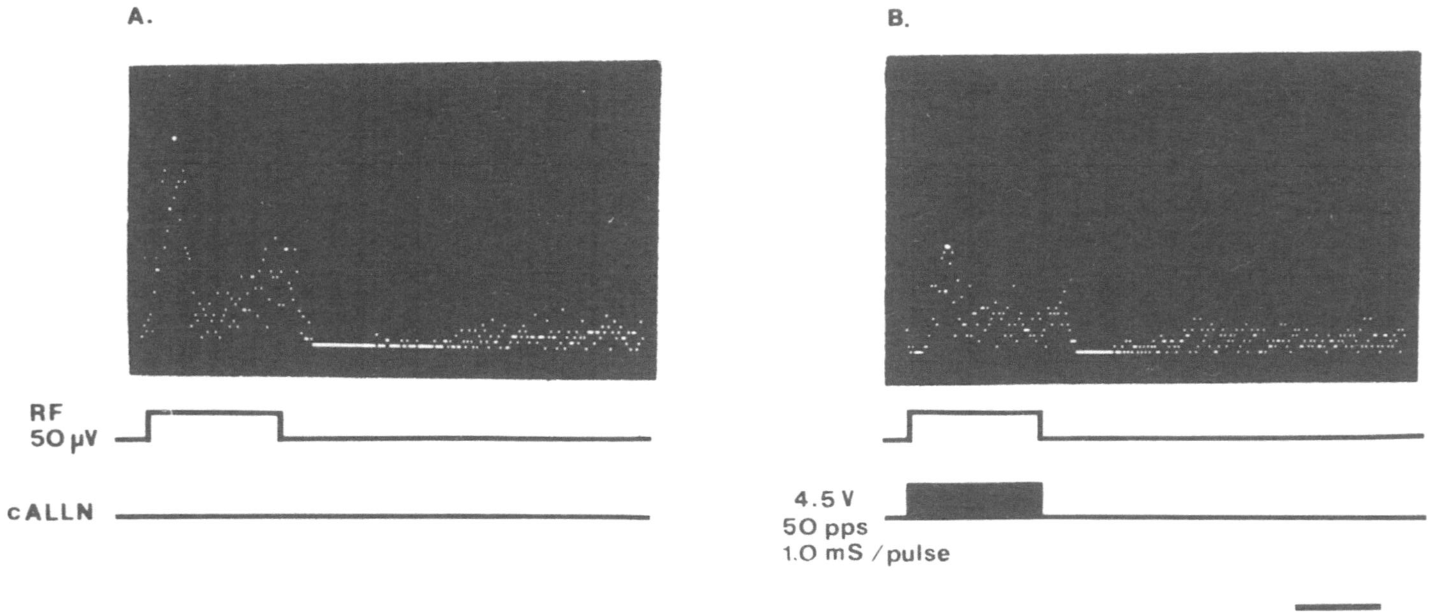

FIGURE 33.10. Histogrammed electrosensory responses (number of impulses vs. time) of an output neuron of the medullary dorsal nucleus in *Raja* demonstrate crossed inhibition between dorsal nuclei. (A) Peristimulus time histogram of the neuron's response to local electric field (RF, 50 μV) presented in receptive field center. (B) Histogram of response to same electric field stimulus and simultaneous direct stimulation (cALLN) of the contralateral anterior lateral line nerve (New unpublished data). As ventilatory movements result in electroreceptor reafference that is common mode (in phase) on the two sides of the animal, such crossed inhibition results in suppression of ventilatory reafference in medullary cells; transection of contralateral anterior lateral line nerve approximately doubles the ventilatory reafference in dorsal nucleus output cells. Bar = 0.5 s.

electroreceptors during ventilatory activity in elasmobranch fishes. In this case it appears that advantage is taken of the fact that ventilatory reafference is common mode among the receptors of a single ampullary cluster (Montgomery 1984) and even among receptors of different clusters including those on opposite sides of the body (New 1986; New and Bodznick 1987). Existing data indicate that medullary circuits subtract inputs from electroreceptors with openings on different parts of the body in order to eliminate common-mode signals. Such a mechanism was first inferred by Kalmijn (1974) based on the structural organization of ampullae of Lorenzini. By computer subtraction of primary afferent responses, Montgomery (1984) demonstrated that in addition to the cancellation of the common-mode ventilatory reafference, such a mechanism could also enhance sensitivity to extrinsic electric field stimuli such as uniform electric fields that affect receptors differentially (Fig. 33.9). Limited evidence exists for medullary circuits that subtract inputs from oppositely oriented electroreceptors of a single cluster in *Platyrhinoidis* (Montgomery 1984). In *Raja*, commissural projections between the bilateral electrosensory dorsal nuclei in the medulla mediate crossed inhibition (Fig. 33.10) that appears to be responsible for a major portion of the reduction of ventilatory reafference in medullary neurons (New 1986; New and Bodznick 1987). Unilateral transection of the anterior lateral line nerve in *Raja* results in a significant increase in ventilatory reafference in neurons of the contralateral dorsal nucleus. It is not known if any kinds of lateral line reafference are generally common mode and, therefore, suppressible by this mechanism. However, major commissural projections exist between medial (octavolateralis) nuclei in the medulla, and, in addition to a potential role in reafferent suppression, crossed inhibition could serve to enhance directional sensitivity in the lateral line sense. Crossed inhibitory projections may characterize all octavolateralis systems, as such crossed inhibition is known to enhance sensitivity to rotational stimuli in medullary neurons of the vestibular sense (Markham et al. 1977).

Acknowledgments. I thank Dr. M.C. Ronan for helpful comments on the manuscript and for stimulating discussion in the course of its preparation.

References

Adrianov GN, Brown HR, Ilyinsky OB (1974) Responses of central neurons to electrical and magnetic stimuli of the ampullae of Lorenzini in the Black Sea skate. J Comp Physiol 93:287–299.

Ashmore JF (1987) A fast motile response in guinea-pig outer hair cells: The cellular basis of the cochlear amplifier. J Physiol 388:323–347.

Bailey SW (1937) An experimental study of the origin of lateral-line structures in embryonic and adult teleosts. J Exp Zool 76:187–233.

Bass AH (1982) Evolution of the vestibulolateral lobe of the cerebellum in electroreceptive and non-electroreceptive teleosts. J Morphol 174:335–348.

Bell CC (1982) Properties of a modifiable efference copy in an electric fish. J Neurophysiol 47:1043–1056.
Bell CC (1984) Effects of motor commands on sensory inflow, with examples from electric fish. In: Bolis L, Keynes RD, Maddrell SHP (eds) Comparative Physiology of Sensory Systems. Cambridge, U.K.: Cambridge University Press, pp. 637–646.
Bell CC, Russell CJ (1987) Termination of electroreceptor and mechanical lateral line afferents in the mormyrid acoustico-lateral area. J Comp Neurol 182(3):367–383.
Bennett MVL (1971) Electroreception. In: Hoar WS, Randall DS (eds) Fish Physiology, Vol. 5. New York: Academic Press, pp. 493–574.
Bennett MVL, Grundfest H (1959) Electrophysiology of electric organ in *Gymnotus carapo*. J Gen Physiol 42:1067–1104.
Bennett MVL, Obara S (1986) Ionic mechanisms and pharmacology of electroreceptors. In: Bullock TH, Heiligenberg W (eds) Electroreception. New York: John Wiley, pp. 157–181.
Bleckmann H (1988) Prey identification and prey localization in surface-feeding fish and fishing spiders. In: Atema AA, Fay RR, Popper AN, Tavolga WN (eds) Sensory Biology of Aquatic Animals. New York: Springer-Verlag, pp. 619–663.
Bodznick D, Northcutt RG (1980) Segregation of electro-and mechanoreceptive inputs to the elasmobranch medulla. Brain Res 195:313–321.
Bodznick D, Northcutt RG (1981) Electroreception in lampreys: Evidence that the earliest vertebrates were electroreceptive. Science 212:465–467.
Bodznick D, Preston DG (1983) Physiological characterization of electroreceptors in the lampreys *Ichthyomyzon unicuspis* and *Petromyzon marinus*. J Comp Physiol 152: 209–217.
Bodznick D, Schmidt AW (1984) Somatotopy within the medullary electrosensory nucleus of the little skate, *Raja erinacea*. J Comp Neurol 225:581–590.
Boord RL (1977) Auricular projections in the clearnose skate, *Raja eglanteria*. Am Zool 17:887.
Boord RL, Campbell CBG (1977) Structural and functional organization of the lateral line system of sharks. Am Zool 17:431–443.
Brownell WE, Bader CR, Bertrand D, de Ribaupierre Y (1985) Evoked mechanical responses of isolated cochlear hair cells. Science 227:194–196.
Bullock TH (1982) Electroreception. Annu Rev Neurosci 5:121–170.
Bullock TH (1986) Significance of findings on electroreception for general neurobiology. In: Bullock TH, Heiligenberg W (eds) Electroreception. New York: John Wiley, pp. 651–674.
Bullock TH, Heiligenberg W (1986) Electroreception. New York: John Wiley.
Bullock TH, Hagiwara S, Kusano K, Negishi K (1961) Evidence for a category of electroreceptors in the lateral line of gymnotid fishes. Science 134:1426–1427.
Bullock TH, Northcutt RG, Bodznick D (1982) Evolution of electroreception. Trends Neurosci 5:50–53.
Bullock TH, Bodznick D, Northcutt RG (1983) The phylogenetic distribution of electroreception: Evidence for convergent evolution of a primitive vertebrate sense modality. Brain Res Rev 6:25–46.
Dijkgraaf S (1962) The functioning and significance of the lateral line organs. Biol Rev 38:51–105.
Dijkgraaf S, Kalmijn AJ (1966) Versuche zur biologischen Bedeutung der Lorenzinischen Ampullen bei den Elasmobranchiern. Z Vgl Physiol 53:187–194.

Ewart JC, Mitchell JC (1892) On the lateral sense organs of elasmobranchs. II. The sensory canals of the common skate (*Raja batis*). Trans R Soc (Edinb) 37:87–105.

Fernholm B (1985) The lateral line system of cyclostomes. In: Foreman RE, Gorbman A, Dodd JM, Olsson R (eds) Evolutionary Biology of Primitive Fishes. New York: Plenum, pp. 113–122.

Fessard A, Szabo T (1961) Mise en evidence d'un recepteur sensible a l'electricite dans la peau d'un mormyre. C R Acad Sci 253:1859–1860.

Finger TE, Bell CC, Carr CE (1986) Comparisons among electroreceptive teleosts: Why are electrosensory systems so similar? In: Bullock TH, Heiligenberg W (eds) Electroreception. New York: John Wiley, pp. 465–481.

Flock Å (1965) Electron microscopic and electrophysiological studies on the lateral line canal organ. Acta Otolaryngol (Suppl) 199:1–90.

Flock Å, Russell IJ (1973a) Efferent nerve fibres: Postsynaptic action on hair cells. Nature 243:89–91.

Flock Å, Russell IJ (1973b) The post-synaptic action of efferent fibres in the lateral line organ of the burbot *Lota lota*. J Physiol (Lond) 235:591–605.

Flock Å, Russell IJ (1976) Inhibition by efferent nerve fibres: Action on hair cells and afferent synaptic transmission in the lateral line canal organ in the Burbot, *Lota lota*. J Physiol (Lond) 257:45–62.

Fritzsch B, Münz H (1986) Electroreception in amphibians. In: Bullock TH, Heiligenberg W (eds) Electroreception. New York: John Wiley, pp. 483–496.

Fritzsch B, Wahnschaffe U (1983) The electroreceptive ampullary organs of urodeles. Cell Tissue Res 229:483–503.

Fritzsch B, Nikundiwe AM, Will U (1984) Projection patterns of lateral-line afferents in anurans: A comparative HRP study. J Comp Neurol 229:451–469.

Görner P (1967) Independence of afferent activity from efferent activity in the lateral line organ of *Xenopus laevis*. In Cahn PH (ed) Lateral-Line Detectors. Bloomington: Indiana University Press, pp. 199–216.

Hetherington TE, Wake MH (1979) The lateral line system in larval *Ichthyophis* (Amphibia: Gymnophiona). Zoomorphologie 93:209–225.

Holst E von, Mittelstaedt H (1950) Das Reafferenzprinzip. Naturwissenschaften 37: 464–476.

Hopkins CE (1986) Temporal structure of non-propagated electric communication signals. Brain Behav Evol 28:43–59.

Hudspeth AJ (1983) Mechanoelectrical transduction by hair cells in the acousticolateralis sensory system. Annu Rev Neurosci 6:187–215.

Jørgensen JM (1982) Fine structure of the ampullary organs of the bichir *Polypterus senegalus* Cuvier, 1829 (Pisces: Brachiopterygii) with some notes on the phylogenetic development of electroreceptors. Acta Zool 63(4):211–217.

Jørgensen JM, Bullock TH (1987) Organization of the ampullary organs of the African knife fish, *Xenomystus nigri* (Teleostei: Notopteridae). J Neurocytol 16:311–315.

Kalmijn AJ (1974) The detection of electric fields from inanimate and animate sources other than electric organs. In: Fessard A (ed) Handbook of Sensory Physiology, Vol. III/3. Berlin: Springer-Verlag, pp. 147–200.

Kalmijn AJ (1988a) Detection of weak electric fields. In: Atema J, Fay RR, Popper AN, Tavolga WN (eds) Sensory Biology of Aquatic Animals. New York: Springer-Verlag, pp. 151–186.

Kalmijn AJ (1988b) Hydrodynamic and acoustic field detection. In: Atema J, Fay RR, Popper AN, Tavolga WN (eds) Sensory Biology of Aquatic Animals. New York: Springer-Verlag, pp. 83–130.

Landacre FL (1910) The origin of the cranial ganglia in *Ameiurus*. J Comp Neurol 20(4): 309–411.

Lissmann HW (1958) On the function and evolution of electric organs in fish. J Exp Biol 35:156–191.

Lissmann H (1967) Some general comments: In: Cahn P (ed) Lateral Line Detectors. Bloomington: Indiana University Press, pp. 411–414.

Maler L, Finger T, Karten H (1974) Differential projections of ordinary lateral line and electroreceptors in gymnotid fish, *Apteronotus albifrons*. J Comp Neurol 158:363–382.

Markham CH, Yagi T, Curthoys IS (1977) The contribution of the contralateral labyrinth to second order vestibular neuronal activity in the cat. Brain Res 138:99–109.

McCormick CA (1982) The organization of the octavolateralis area in actinopterygian fishes: A new interpretation. J Morphol 171:159–181.

McCormick CA, Braford MR Jr (1988) Central connections of the octavolateralis system: Evolutionary considerations: In: Atema J, Fay RR, Popper AN, Tavolga WN (eds) Sensory Biology of Aquatic Animals. New York: Springer-Verlag, pp. 733–756.

McCready PJ, Boord RL (1976) The topography of the superficial roots and ganglia of the anterior lateral line nerve of the smooth dogfish, *Mustelus canis*. J Morphol 150(2): 527–538.

Metcalfe WK (1985) Sensory neuron growth cones comigrate with posterior lateral line primordial cells in zebrafish. J Comp Neurol 238:218–224.

Montgomery JC (1984) Noise cancellation in the electrosensory system of the thornback ray; common mode rejection of input produced by the animal's own ventilatory movement. J Comp Physiol 155A:102–111.

Moy-Thomas JA, Miles RS (1971) Palaeozoic Fishes. Philadelphia: W.B. Saunders, 259 pp.

Mullinger AM (1964) The fine structure of ampullary electric receptors in *Amiurus*. Proc R Soc Lond B Biol Sci 160:345–359.

Murray RW (1960) The response of the ampullae of Lorenzini of elasmobranchs to mechanical stimulation. J Exp Biol 37:417–424.

Murray RW (1962) The response of the ampullae of Lorenzini of elasmobranchs to electrical stimulation. J Exp Biol 39:119–128.

New JG (1986) Sensory processing in the medullary electrosensory nucleus of the little skate, *Raja erinacea*. PhD Dissertation, Wesleyan University, Middletown, CT.

New JG, Bodznick D (1987) Suppression of self-generated electrosensory interference via a commissural medullary pathway in the little skate. Neurosci Abstr 13(1):399.

Northcutt RG (1981) Audition and the central nervous systems of fishes. In: Tavolga WN, Popper AN, Fay RR (eds) Hearing and Sound Communication in Fishes. New York: Springer-Verlag, pp. 331–355.

Northcutt RG (1985) The brain and sense organs of the earliest vertebrates: Reconstruction of a morphotype. In: Forman RE, Gorbman A, Dodd JM, Olsson R (eds) Evolutionary Biology of Primitive Fishes. New York: Plenum Press, pp. 81–112.

Northcutt RG (1986a) Embryonic origin of amphibian electroreceptors. Soc Neurosci Abstr 12:103.

Northcutt RG (1986b) Electroreception in nonteleost bony fishes. In: Bullock TH, Heiligenberg W (eds) Electroreception. New York: John Wiley, pp. 257–285.

Northcutt RG (1986c) Evolution of the octavolateralis system: Evaluation and heuristic value of phylogenetic hypotheses. In: Ruben RW (ed) The Biology of Change in Otolaryngology. New York: Elsevier, pp. 3–14.

Obara S, Sugawara Y (1984) Electroreceptor mechanisms in teleost and non-teleost fishes. In: Bolis L, Keynes RD, Maddrell SHP (eds) Comparative Physiology of Sensory Systems. Cambridge, U.K.: Cambridge University Press, pp. 509–523.

Regnart HC (1931) The lower limits of perception of electrical currents by fish. J Mar Biol Assoc UK 17:415–420.
Roberts BL (1978) Mechanoreceptors and the behavior of elasmobranch fishes with special reference to the acoustico-lateralis system. In: Hodgson ES, Mathewson RF (eds) Sensory Biology of Sharks, Skates, and Rays. Arlington, VA: Office of Naval Research, Department of the Navy, pp. 331–390.
Roberts BL, Russell IJ (1972) The activity of lateral-line efferent neurons in stationary and swimming dogfish. J Exp Biol 57:435–448.
Roberts BL, Ryan KP (1971) The fine structure of the lateral-line sense organs of dogfish. Proc R Soc Lond B 179:157–169.
Ronan MC, Bodznick D (1986) End buds: Non-ampullary electroreceptors in adult lampreys. J Comp Physiol 158A:9–15.
Russell IJ (1971) The role of the lateral line efferent system in *Xenopus laevis*. J Exp Biol 54:621–641.
Russell IJ (1974) Central and peripheral inhibition of lateral line input during the startle response in goldfish. Brain Res 80:517–522.
Russell IJ (1976) Central inhibition of lateral line input in the medulla of the goldfish by neurones which control active body movements. J Comp Physiol 111A:335–358.
Russell IJ, Lowe DA (1983) The effect of efferent stimulation on the phase and amplitude of extracellular receptor potentials in the lateral line system of the perch (*Perca fluviatilis*). J Exp Biol 102:223–238.
Russell IJ, Roberts BL (1972) Inhibition of spontaneous lateral-line activity by efferent nerve stimulation. J Exp Biol 57:77–82.
Schluger JH, Hopkins CD (1987) Electric fish approach stationary signal sources by following electric current lines. J Exp Biol 130:359–367.
Schmidt AW, Bodznick D (1987) Afferent and efferent connections of the vestibulolateral cerebellum of the little skate, *Raja erinacea*. Brain Behav Evol 30:282–302.
Stone LS (1922) Experiments on the development of the cranial ganglia and the lateral line sense organs in *Amblystoma punctatum*. J Exp Zool 35:421–496.
Szabo T (1974) Anatomy of the specialized lateral line organs of electroreception. In: Fessard A (ed) Handbook of Sensory Physiology, Vol. III/3. New York: Springer-Verlag, pp. 13–58.
Vischer HA, Heiligenberg W, Northcutt RG (1987) Development of the mechano- and electrosensory systems in *Eigenmannia virescens*. Soc Neurosci Abstr 13:130.
Will U, Luhede G, Görner P (1985) The area octavo-lateralis in *Xenopus laevis*. I. The primary afferent projections. Cell Tissue Res 239:147–161.
Winklbauer R, Hausen P (1983) Development of the lateral line system in *Xenopus laevis*. I. Normal development and cell movement in the supraorbital system. J Embryol Exp Morphol 76:265–281.
Wright MR (1951) The lateral line system of sense organs. Q Rev Biol 26:264–280.
Yamada Y (1973) Fine structure of the ordinary lateral line organ. I. The neuromasts of lamprey, *Entosphenus japonicus*. J Ultrastruct Res 43:1–17.
Zakon HH (1986) The electroreceptive periphery. In: Bullock TH, Heiligenberg W (eds) Electroreception. New York: John Wiley, pp. 103;!156.
Zipser B, Bennett MVL (1976) Interaction of electrosensory and electromotor signals in lateral line lobe of a mormyrid fish. J Neurophysiol 39:713–721.

Part VII Prospectus

34
Lateral Line Research: Prospects and Opportunities

THEODORE H. BULLOCK

Lateral lines are difficult to understand, since we have no comparable system, and the presumed adequate stimuli overlap with those of other systems. Sensory biology in lower vertebrates is skewed in respect to well-developed and poorly developed areas. This volume shows substantial progress in some facets of lateral line research, such as anatomical and peripheral physiological aspects, at the same time that it reflects slower advance in others, such as behavioral and central physiological aspects. The following remarks can only be one man's personal view of the prospects and opportunities ahead, since it would take multiple standpoints to assure that some important challenge is not left out. I will confine these comments to three areas: evolutionary, general neurobiological, and neuroethological.

34.1 Evolutionary Problems

The skewness referred to includes an historic emphasis on the peripheral parts of the lateral line system and its anatomy. Less attention has been devoted to central representation, physiology, and the behavioral role, particularly in view of the remarkable specializations of the peripheral sense organs in different taxa. My first point, therefore, is to call for consideration of the whole system, from receptor to effector, through the CNS and its discrimination of normal stimuli to the selection or triggering of alternative responses. This consideration requires the evaluation of the relative roles of overlapping sense modalities, such as segmental somatosensory and acoustic systems. We cannot pretend to understand the evolution of diverse lateral line sense organs in the skin until we have some parallel appreciation of the central specializations and the behavioral uses of the whole pathway.

A prime evolutionary problem is recognition of which features of the lateralis system are primitive and which are derived, and of the latter, which are advanced and which are secondarily simplified. The chapters in this collection illustrate that we have not as yet any clear idea that distinguishes phylogenetic trends from adaptive radiation at the same level of derivedness. If an applicable dendrogram of lateral line characters were available, it would help to identify derived characters; but it might not suffice, since by itself, the dendrogram does not tell whether a particular trait is advanced or regressive. It may be necessary to remind ourselves that if a dendrogram is not available, it is not for us as students of the lateralis system to make one; that is the prerogative of the student of all the traits of all the related taxa. It is, however, our opportunity to mark a set of copies of the sufficiently inclusive dendrogram with the symbols for presence or absence of each trait of the peripheral and central parts of the lateral line system, remembering that each trait can evolve independently.

For advancing our knowledge of the lateral line system, determining homologies (e.g., of canals, free neuromast groups, nerves, or central nuclei) is not the only question. What came from what is one question; what is there in different taxa is another, and even more first-order question. Still another is whether one trait is more advanced (the opposite of simpler) or is merely derived (the opposite of primitive). Either dimension can be considered independent of any claim or knowledge of adaptedness. Merely derived traits, not necessarily advanced, may be most often those adapted to a specialized life-style, as in ultrasonic hearing or high-sensitivity chemoreception.

There are no accepted guidelines for this decision. I support the view that degree of complexity is in general the best guide to the more advanced condition, and that this can be objectively measured by estimating the number of different parts and processes and relations among them or, in principle, by estimating the bits of information needed to described them. This measure may not always work; one can at least imagine the case in human engineering or fashions in clothing where the more advanced, sophisticated design is more elegantly simple – but I am not aware that such cases are common in animal evolution.

34.2 Issues Relevant to General Neurobiology

Whereas the foregoing as well as the following sections deal mainly with differences among species, which is a deliberate emphasis, since nervous systems and behavior have been the favorite playing field for evolution, common mechanisms and general issues are at least equally important.

A host of contributions in the chapters of this volume are relevant to general neurobiological problems. Typically, however, in our eagerness to appeal to an audience of peers, we underplay the general relevance of our findings and fail to identify them with general problems. Lateral line systems are valuable as specially favorable cases, or for introducing new model species for general biological problems, for example, in the elucidation of developmental mechanisms, of mechanisms of transduction of impinging stimuli, of sensory coding in single channels, and across-channel coding for the differentiation of modalities and submodalities, of mechanisms and meaning of efferent control of sense organs, of parallel and distributed processing of afferent input in the successive levels of the brain. Lateral line systems may provide a good test of a general hypothesis, for example, about CNS sculpting by cell death (Wahnschaffe et al. 1987) or about continuous growth versus growth ceilings. They may turn up a novel mechanism for a common problem—for example, mapping stimulus parameters or dealing with reafference. Or they may simply underline the lesson that nature uses different solutions in different taxa—for example, centrifugal axons and corollary discharges here and not there. Whatever may be the relevance of our study to other systems or taxa, it should be a prime objective of each author to identify with the broadest general biological issue to which that study relates.

Selecting one example of a general issue, pointed up long ago by Katsuki et al. (1950, 1951a,b), lateral line afferent fibers raised the question of whether a common principle operates among many systems in which two populations of fibers, thick and thin, appear to provide duplicate channels with different threshold, gain, time constant, and maximum output frequency. Typically, but not invariably, the thicker fibers (Bullock 1953) have the higher threshold to adequate stimuli, more intensity discrimination, faster adaptation, and smaller receptive fields.

34.3 Neuroethological Problems

Shifting the style of formulation, I will state some of the prospects and opportunities for new research as needs to know, before we can claim a basic understanding of the lateral line system.

1. We need to know which behaviors depend on lateral line input and, of these, which are characteristic of species with principally or only free neuromasts; which are found in species dominated by canals with many pores; which in species with canals without pores; which in species with multiple, long parallel canals; which in species with short canals only on the head; and which in species with highly branched canals. Is the characteristic behavior primarily arousal, or is direction important? Is it identification or discrimination among similar stimulus sources? Is it food, enemies, partners, or mainly obstacles and landmarks? Is the sensory mode active, as in blind cave fish, or passive?

2. We need to know the adequate stimuli for eliciting behavior dependent on the lateral line—as Coombs and Janssen have done for one case in Chapter 15. Standard engineering methods used in systems analysis are useful, including signal manipulation and distortion, creating, in effect, illusions resembling normal stimuli. How important is pattern in time and space, for example, movement or combinations of frequencies? How is an effective signal different from noise? What are the differences in adequate stimuli for various taxa, from agnathans to anurans, from teleosts with mainly free neuromasts to those with highly branched canals? What differences between species can be found in the physiology of data processing and signal recognition? This is the kind of opportunity where we should separately compare taxa far apart in phylogeny and others within a given family that differ in anatomy or behavior or habit of life.

3. We need to know the structures and processes involved in the whole circuit, from sense organ to cerebral pallium and brainstem motor centers. We cannot pretend to understand the lateral line system from even an extensive knowledge only of receptors or of behavior or both, any more than we could pretend to understand the auditory system by knowing only the cochlea and the psychophysics of hearing. To ask the key questions, such as what is the mechanism of recognition of ethologically important signals, or what is the principle of the brain machine and its software, we need to know both the anatomy and the organization, including what I call the physiological personality of the components and their dependence on state, stage, and species. How separate are the pathways for different modalities, up to the pallium? What submodalities should be distinguished and what kinds of mappings – computed as well as topographic? What is the role of the torus semicircularis in lateral line mechanoreception? What are the roles of the tectum, the cerebellum, the thalamic nuclei, the striatum, the limbic system, the several pallial areas? Are there columns, modules, or templates, convergences or modulations from other inputs or states, such as circadian rhythms? How is reafference dealt with, and how are descending or centrifugal pathways used.?

4. We need to know what differences in behavior and adequate stimuli go with differences in anatomy and physiology, central or peripheral. This will require comparison of species in which both the behavior and the functional organization of the sensory system have been worked out to establish the relevance of the observed differences. It is the strong expectation of systems physiology that it can give clues to the specialization of the system for a particular habit of life , since the CNS is as adapted as any other part of the body.

These questions are based on the presumption, strongly suggested by the diversity of peripheral anatomy, that the mechanoreceptive lateral line system is not unifunctional, but instead operates in different species with different emphases among the candidate roles so far recognized, such as orientation in currents; detection of surface, bottom, and obstacles; and weather; schooling; courting; feeding; and predator avoidance.

34.4 Odds and Ends

This being a personal view, I indulge in a few items that amount to advice for future investigators.

1. Do not assume one species is representative even of a family, let alone an order. Many examples of central as well as peripheral anatomy and physiology of other octavolateral systems, especially electrosensory systems, have revealed sharp differences among genera of the same family and families of the same order.

2. Do not assume that adequate stimuli for normal ethological responses are unimodal; most stimuli that excite lateral line receptors probably excite also acoustic receptors of the eighth nerve, the degree of bimodality differing among taxa. Although the evidence in rays (Bleckmann et al. 1987) indicates that

central as well as peripheral units can be specific to one or the other class of stimulus, the behavior may sometimes depend on a central interaction of the two modalities.

3. When correlating behavior with sensory specialization, an elementary word of caution may be appropriate: Beware of correlations where the *n* is small; two or three species are quite likely to show a correlation purely by chance. The danger of this obvious pitfall arises in causal discussions and research planning when our guard is down, because plausibility is used as a sufficient test, though it is notoriously unreliable.

4. Do not forget the cerebellum when studying sensory systems. Although usually listed under motor systems in textbooks, the cerebellum is as much an analyzer of afferent input as a formulator of motor output. According to our present, surely incomplete information, lateral line nerve stimulation evokes activity in an elasmobranch (*Platyrhinoidis*) and a teleost (*Ictalurus*) in output (Purkinje) cells in a circumscribed part of the cerebellar cortex, a lateral line region, without convergence of other modalities. What transformation or operation is performed on the afferent information is not yet known, and it is this that I urge should receive attention.

If it sounds as though I am eclectic and advocate everything at once, it is so! This field is not exempt from the common affliction whose principal symptom is utterances like "THE question is . . ." or words implying that a certain approach is the best way to enter THE problem. Each author's data are the reality to him; other data are special cases to be ignored, if they are not outright suspect. The moral is eclectic humility—or, in the words of the late Blaine Ramsey, "enthusiastic tentativeness."

References

Bleckmann H, Bullock TH, Jørgensen JM (1987) The lateral line mechanoreceptive mesencephalic, diencephalic, and telencephalic regions in the thornback ray, *Platyrhinoidis triseriata* (Elasmobranchii). J Comp Physiol (A) 161:67–84.

Bullock TH (1953) Comparative aspects of some biological transducers. Fed Proc 12:666–672.

Katsuki Y, Yoshino S, Chen J (1950) Action currents of the single lateral-line nerve fiber of fish. I. On the spontaneous discharge. Jpn J Physiol 1:87–99.

Katsuki Y, Yoshino S, Chen J (1951a) Action current of the single lateral-line nerve fiber of fish. II. On the discharge due to stimulation. Jpn J Physiol 1:179–194.

Katsuki Y, Yoshino S, Chen J (1951b) Neural mechanism of the lateral-line organ of fish (fundamental neural mechanism of sensory organs in general). Jpn J Physiol 1:264–268.

Wahnschaffe U, Bartsch U, Fritzsch B (1987) Metamorphic changes within the lateral-line system of Anura. Anat Embryol 175:431–442.

Taxonomic Appendix

This appendix contains a partial systematic listing of extant genera mentioned in this volume. For specific references by page number, see the taxonomic index. See Chapter 3 for references to fossil forms. Fish systematic listings after Lauder and Liem (1983).

I. Invertebrate Genera (by Phylum)

Protozoa
- *Amoeba*
- *Glaucoma*

Coelenterata
- *Anemonia*
- *Balanophylla*
- *Calliactis*
- *Coryne*
- *Eutonia*
- *Gonionemus*
- *Hydra*
- *Pennaria*
- *Sagartia*
- *Syncoryne*

Ctenophora
- *Beroe*
- *Leucothea*
- *Pleurobrachia*

Platyhelminthes
- *Fasciola*
- *Neophasis*
- *Nicolla*

Annelida
- *Arenicola*
- *Branchiobdella*
- *Erpobdella*
- *Harmothoe*
- *Hirudo*
- *Sabella*
- *Trilobodrilus*

Rotifera
- *Brachionus*

Nematoda
- *Heterodera*

Nemertea
- *Lineus*

Mollusca
- *Loligo*
- *Lollinguncula*
- *Nautilus*
- *Octopus*
- *Pecten*
- *Pteriomorpha*
- *Sepia*
- *Todarodes*

Chaetognatha
- *Sagilla*
- *Spadella*

Crustacea
- *Artemia*
- *Astacus*
- *Cherax*
- *Daphnia*
- *Diaptomus*
- *Funchalia*
- *Gammarus*
- *Orchomene*
- *Palinurus*
- *Paradexamine*
- *Petrochirus*
- *Tetrdeium*

Uniramia
- *Chaoborus*
- *Notonecta*

Echinoderma
- *Amphipholis*
- *Asterias*
- *Echinus*
- *Ophiura*

Chordata
- Hemichordata
  - *Saccoglossus*
  - *Glossobalanus*
- Urochordata
  - *Ciona*
  - *Diplosoma*
  - *Oikopleura*
- Cephalochordata
  - *Branchiostoma*

II. Vertebrate Fish Genera (by class and various subclass taxa)

Agnatha
- Myxinoids (hagfishes)
  - *Entosphenus*
  - *Eptatretus*
  - *Myxine*
- Petromyzontoids (lampreys)
  - *Ichthyomyzon*
  - *Lampetra*
  - *Petromyzon*

Chondrichthyes (sharks, skates, rays, and chimeras)
- *Carcharhinus*
- *Cephaloscyllium*
- *Chlamydoselachus*
- *Gymnura*
- *Heterodontus*
- *Mustelus*
- *Narcine*
- *Platyrhinoidis*
- *Raja*
- *Rhinobatos*
- *Scyliorhinus*
- *Squalus*
- *Torpedo*
- *Urolophus*

Osteichthyes
- Sarcopterygii
  - Dipnoi (lungfishes)
    - *Lepidosiren*
    - *Neoceratodus*
    - *Protopterus*
  - Actinistia (coelacanth)
    - *Latimeria*
- Actinopterygii
  - Cladistia
    - *Calamoichthys*
    - *Polypterus*
  - Chondrostei
    - *Acipenser*
    - *Polyodon*
    - *Scaphyrhynchus*
  - Ginglymodi (gars)
    - *Lepisosteus*
  - Halecomorphi
    - *Amia*
  - Teleostei
    - Osteoglossomorpha
      - *Gymnarchus*
      - *Hiodon*
      - *Notopterus*

Pantodon
Xenomystus

Clupeomorpha
Brevoortia
Clupea
Engraulis
Sprattus

Elopomorpha
Anguilla
Astroconger

Euteleostei
Nonpercomorphs (including Ostariophysi, Protacanthopterygii, Stomiiformes, Aulopiformes, Myctophiformes, Paracanthopterygii, Atherinomorpha)
Aplocheilus
Astyanax (= *Anoptichthys*)
Brachydanio (= *Danio*)
Carassius
Clupea
Colisa
Cyprinus
Eigenmannia
Electrophorus
Fundulus
Gadus
Gasteropelecus
Hyporhamphus
Ictalurus (= *Ameiurus*)
Kryptopterus
Lebistes (= *Poecilia*)
Lota
Menidia
Oncohrynchus
Opsanus
Phoxinus
Poromitra
Salmo
Tinca
Percomorphs (inc. Lampridiformes, Gasterogleiformes, Pegasiformes, Dactylopteriformes, Beryciformes, Zeiformes, Perciformes, Scorpainiformes, tetraodontiformes, Pleuronectiformes, Channiformes, and Synbranchiformes)
Abramis
Acanthopagrus
Acerina (= *Gymnocephalus*)
Anoplogaster
Astronotus
Azurina
Bathybates
Cichlasoma
Cottus
Gobiocichla
Gobius
Hemibates
Hippoglossus
Kareius
Lateolabrax
Lepomis
Limanda
Macropodus
Oreochromis (= *Tilapia* = *Sarotherodon*)
Pagothenia
Paralichthys
Pempheris
Perca
Platichthys
Pleuronectes
Pseudorhombus
Sarotherodon
Scophthalmus
Sebasticus
Solea
Teleogramma
Trachurus
Tridentiger
Xiphister

III. Other Vertebrate Genera (by class)

Amphibia
Urodela
Ambystoma
Chioglossa
Cynops
Necturus

Proteus
Siren
Triturus

Anura
Ascaphus
Bombina
Bufo
Pipa
Rana
Xenopus

Gymnophiona
Ichthyophis
Typhlonectes

Reptilia
Gerrhonotus
Pseudomys

Aves
Melopsittacus

Mammalia
Ornithorhynchus

Author Index

Taxonomic Index

Subject Index

MIX
Papier aus verantwortungsvollen Quellen
Paper from responsible sources
FSC® C105338

If you have any concerns about our products,
you can contact us on
ProductSafety@springernature.com

In case Publisher is established outside the EU,
the EU authorized representative is:
Springer Nature Customer Service Center GmbH
Europaplatz 3, 69115 Heidelberg, Germany

Printed by Libri Plureos GmbH
in Hamburg, Germany